Note: The following credit line applies to all the articles in this issue:

This article is an invited paper selected from presentations at the 2008 International Thermal Spray Conference and has been expanded from the original presentation. It is simultaneously published in *Thermal Spray Crossing Borders, Proceedings of the 2008 International Thermal Spray Conference,* Maastricht, The Netherlands, June 2–4, 2008, Basil R. Marple, Margaret M. Hyland, Yuk-Chiu Lau, Chang-Jiu Li, Rogerio S. Lima, and Ghislain Montavon, Ed., ASM International, Materials Park, OH, 2008

Editorial Committee

International Board of Review

JOURNAL OF
Thermal Spray
TECHNOLOGY®

A publication of the Thermal Spray Society, an affiliate society of ASM International®

Editor
Christian Moreau

Associate Editors
Kendall Hollis
Seiji Kuroda
Armelle Vardelle

Editor Emeritus
Christopher C. Berndt

Published by
Springer Science+Business Media, LLC

ASM Staff
Manager, Technical Journals
Mary Anne Fleming
Manager, Production
Bonnie Sanders
Journal Coordinator
Ellyn T. Vander Kaay

This journal is covered in the following Institute for Scientific Information® products: Current Contents – Engineering, Computing & Technology, SciSearch®, Research Alert®, and the Materials Science Citation Index®.

This journal is indexed or abstracted by: Cambridge Scientific Abstracts (CSA), Compendex Engineering Index, Institute de L'Information Scientifique et Technique (INIST) and the American Institute of Aeronautics and Astronautics, Inc.

Change of Address: Allow six weeks for all changes to become effective. All communications should include both old and new addresses (with zip codes) and should be accompanied by a mailing label from a recent issue.

Back Volumes: Prices for back volumes are available on request.

Microform Editions: Available from: University Microfilms International, 300 N. Zeeb Road, Ann Arbor, MI 48106, USA.

Reprints: Articles are available as reprints from Sheridan Press, Reprint Service, 450 Fame Avenue, Hanover, PA 17331, USA, www.sheridan.com/springer/eoc or Phone 800-352-2110, Ext. 8134

JTTEE5 17:597
DOI: 10.1007/s11666-008-9275-0
1059-9630/$19.00 © ASM International

Editorial

Basil R. Marple | Margaret M. Hyland | Yuk-Chiu Lau | Chang-Jiu Li | Rogerio S. Lima | Ghislain Montavon

This special issue of the *Journal of Thermal Spray Technology* features a selection of papers presented at the International Thermal Spray Conference 2008 (ITSC 2008) held in Maastricht, The Netherlands, June 2-4, 2008. The theme of the conference, "Thermal Spray Crossing Borders," underlined a number of important elements: (i) to advance the field of thermal spray, information exchange, the sharing of ideas, and discussion must take place at an international level; (ii) thermal spray markets and business opportunities must cross borders and continents; and (iii) the use of coatings allows materials to cross into applications and operating conditions that would otherwise not be possible for the uncoated component.

This conference continued the tradition of international thermal spray events that have played an important role in advancing this field. The ITSC, held annually, is organized jointly by the ASM Thermal Spray Society (TSS), the German Welding Society (DVS), and the International Institute of Welding (IIW). The venue rotates on a 3-year basis among locations in Asia, Europe, and North America. It brings together key players, including researchers, developers, manufacturers, and end users, representing the full scope of activity in this area. The gathering provides an opportunity of hearing about cutting-edge research; identifying new applications and market opportunities; learning about current technical challenges and future needs; finding solutions for existing problems; generating ideas for orienting future research programs; and obtaining information on a wide array of the latest technology, materials, and equipment available in the field. A key aspect of the event is the networking opportunity, during which research collaborations are often initiated and business partnerships formed.

This volume contains over 50 papers drawn from the nearly 300 appearing in the Proceedings of ITSC 2008 published in CD-ROM format by DVS (Editor: E. Lugscheider). Authors of the selected papers were invited to submit manuscripts to the *JTST* where they were passed through the full peer review process required for publishing peer-reviewed papers in this journal. These papers represent a cross section of the topics covered at the conference and provide insight into the latest developments and current trends in the thermal spray field. We hope this expanded *JTST* issue serves as a valuable reference source.

Guest Editors

Basil R. Marple
National Research Council of Canada,
Gatineau, Quebec, Canada

Margaret M. Hyland
University of Auckland,
Auckland, New Zealand

Yuk-Chiu Lau
GE Energy,
New York, NY, USA

Chang-Jiu Li
Xi'an Jiaotong University,
Xi'an, Shaanxi, China

Rogerio S. Lima
National Research Council of Canada,
Gatineau, Quebec, Canada

Ghislain Montavon
University of Limoges,
Limoges, France

JTTEE5 17:598–602
DOI: 10.1007/s11666-008-9248-3
1059-9630/$19.00 © ASM International

Technical Note

Development of a Functional Sealing Layer for SOFC Applications

N. Caron, L. Bianchi, and S. Méthout

(Submitted June 26, 2008; in revised form September 30, 2008)

Solid oxide fuel cells (SOFCs) are widely considered as an alternative solution to the decrease in fossil energy consumption. However, to achieve high efficiency and long-term stability for a SOFC stack, it is essential to maintain a stable hermetic seal. To obtain efficient air tightness between two SOFC cells, a solid seal composed of a ceramic matrix charged with glass particles has been developed. Atmospheric plasma spraying was selected to produce the solid seal because it can be used on a wide range of substrates of various natures and shapes. The developed seal was found to be solid, undistortable, and adhesive to its support at ambient temperature. The sealing properties were acquired when the SOFC was put into service: the glassy phase migrated into the peculiar plasma-sprayed microstructure of the ceramic matrix toward the interface leading to the air tightness of the deposit. The performance of the seal was: the leak rate observed at 7 kPa was 0.43 Pa L/s and, as a comparison, the requirement of the US Department of Energy is 0.5 Pa L/s.

Keywords ceramic matrix, glass compound, sealing, SOFC

1. Introduction

Solid oxide fuel cells (SOFCs) are promising energy conversion devices that produce electricity by an electrochemical reaction of a fuel gas, such as hydrogen or methane, with oxygen (Ref 1). Besides a highly efficient process, advantages of SOFCs include low emission rates of air pollutants, the ability to use high-temperature exhaust for cogeneration or hybrid applications, and the ability of internal reforming (Ref 2). This family of fuel cells is mainly dedicated to high-power applications and mobile auxiliary power units.

Whether in the form of a stack or individual cells, it is essential that the fuel and oxidant be kept separate from one another. Otherwise, the efficiency in producing ion exchanges across the electrolyte would be lowered and there would be a strong risk of exothermal explosions. Generally, the required sealant characteristics include the following (Ref 3): coefficient of thermal expansion well adapted to the other components to minimize thermal stresses and to improve thermal cycling behavior, sufficient fluidity to seal gaps at the sealing/assembly temperature, tailored viscosity at the operating temperature, gastight/marginal leak rate, thermal as well as chemical stability, electrical insulation, low cost, and flexibility with regard to design.

This generic set of requirements is not an exhaustive list because the characteristics of the sealant are also linked to the operating conditions of the stack.

Both rigid and compressive seals are being developed to meet the challenge of the hermetic device, essential for the development of planar SOFC (Ref 4). A recent review by Fergus on this subject explained the differences between various seals and presented both advantages and drawbacks (Ref 5). A major advantage of compressive seals is the fact that they are not rigidly fixed to the other SOFC components. Thus, an exact match of the thermal expansion is not required. However, a constant pressure must be applied to operate the gas-tight function during operation. For an effective compression, the seal must deform in response to the applied stress. So the main directions for the technical development consist in using ductile metals (Ref 6), materials of deformable shapes, or mica-based materials (Ref 7). On the other hand, rigid seals, which are mainly constituted of glass (Ref 8) and glass-ceramic sealants or metallic brazes, do not require this applied pressure. They have more stringent requirements for adherence, cracking and thermal expansion matching. However, the main advantage of glass seals is the possibility to adapt the glass composition to get optimized physical properties (Ref 9, 10). To go beyond the stress caused by differences in thermal expansion coefficients, fiber-reinforced glass seals have been studied. Their advantages are a certain elasticity at the operating

This article is an invited paper selected from presentations at the 2008 International Thermal Spray Conference and has been expanded from the original presentation. It is simultaneously published in *Thermal Spray Crossing Borders, Proceedings of the 2008 International Thermal Spray Conference,* Maastricht, The Netherlands, June 2-4, 2008, Basil R. Marple, Margaret M. Hyland, Yuk-Chiu Lau, Chang-Jiu Li, Rogerio S. Lima, and Ghislain Montavon, Ed., ASM International, Materials Park, OH, 2008.

N. Caron, L. Bianchi, and **S. Méthout**, Atomic Energy Commission, F-37260 Monts, France. Contact e-mail: nadege.caron@cea.fr.

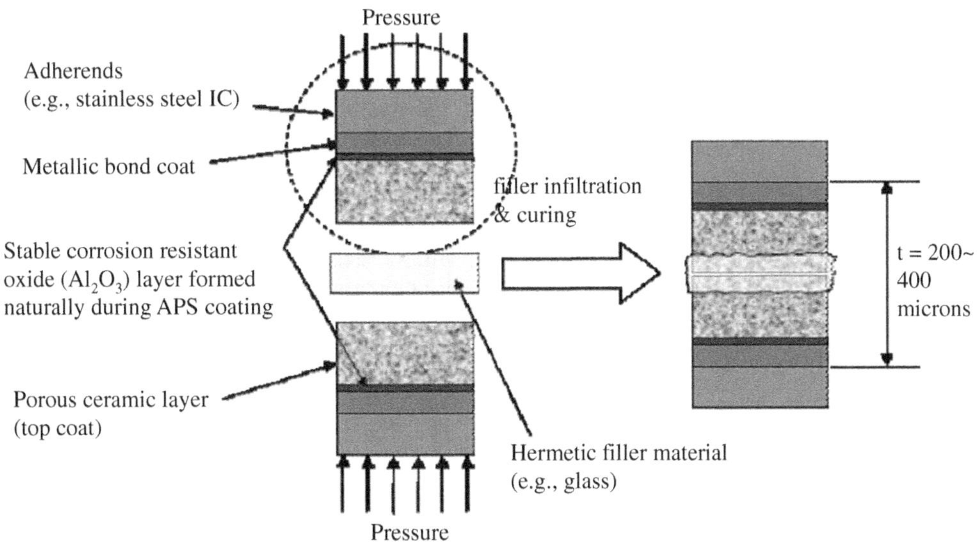

Fig. 1 Scheme of the conceptual structure of the integrated composite seal (Ref 13)

temperature and a better compatibility with the thermal cycles of the SOFCs (Ref 11, 12).

The main processes used for elaborating glass seals are screen printing or pneumatic spraying. In the case of fiber seals, a mechanical pre-compaction allows to elaborate auto-structured seals. In the literature, only a single report mentioning the use of plasma spraying could be found. Huang et al. (Ref 13) investigated a multilayered composite seal structure consisting of thin layers of oxidation-resistant metals, porous ceramics, and fillers/glass as shown in Fig. 1. The seal is an integrated part of the interconnected cells since it is directly coated onto the surfaces of matching parts owing to the use of atmospheric plasma spraying. During the stack assembly, the cells were joined through a heat/pressure-assisted curing process. The coating demonstrated an excellent mechanical robustness (thermal shock, tensile adhesion), high electrical resistivity, and low gas permeability.

The seal developed at the Laboratory of Thermal Spraying in CEA (Monts) differed rather significantly from those previously reported, and has led to a patent filing (Ref 14). The plasma spray technique (Ref 15) was used to elaborate a coating composed of glassy particles dispersed into a matrix that remained solid during high-temperature operations.

Details of elaboration as well as the mechanisms are presented in the next sections. Moreover, initial tests with regard to the leak rate and thermo-cycling resistance are reported.

2. Experimental Procedure

The materials were selected according to their coefficients of thermal expansion (CTE), which is one of the fundamental characteristics for an efficient seal. The ceramic matrix was ZrO_2-8wt.%Y_2O_3 (particle size range 5-25 μm, Medicoat France). Whereas the glassy compound was a borosilicate (particle size range 2-22 μm, DMC^2 France S.A). Its detailed composition (supplier data) was B_2O_3 36 wt.%, SiO_2 31 wt.%, Na_2O 16.7 wt.%, CaO 7.6 wt.%, MgO 3.6 wt.%, and Al_2O_3 0.87 wt.%.

The properties of this glass are: half-sphere temperature = 710 °C; softening point under pressure = 580 °C; and thermal expansion coefficient = 8.9×10^{-6} K^{-1}. The half-sphere temperature is the temperature at which the sample changes its shape so that its base radius equals its height, taking the shape of a half-sphere.

Boron oxide is an important additive to silicate glasses since it decreases the glass viscosity, as well as the softening point and glass transition temperature of the SOFC sealant (Ref 16). The presence of calcium and magnesium in the C105 is known to promote adherent and stable interfaces with yttria-stabilized zirconia (YSZ) (Ref 17).

The composite powder was manufactured by mechanically blending 20 wt.% glassy C105 compound with ceramic YSZ powder. The choice of the plasma spraying process was dictated by the need for cost reduction compared to other manufacturing routes. Atmospheric plasma spraying (APS) was carried out with an automatic plasma system equipped with a Sulzer-Metco F4VB plasma gun (Wolhen, Switzerland), a turntable, and a six-axis robot. The plasma spray parameters were optimized with respect to the Ar/H_2 working gases, the spray distance, and the powder injection.

The APS seal was characterized with scanning electron microscopy (SEM) in back-scattered electron (BSE) mode to differentiate the different components while the porosity and the glass percentage were measured by the combination of two techniques: Archimedes Porosimetry and Helium Picnometry. These volumetric methods involved the elaboration of free-standing samples.

3. Principles of the Plasma-Sprayed Seal

As shown in Fig. 2, the coating consisted of flattened droplets with interlamellar contacts, microcracks, and globular porosities. The total void content of this typical APS coating is around 13%. The percentage of glass compound after plasma spraying is more important (up to 40%) than the one contained in the blending feedstock because of a heterogeneous treatment in the plasma jet. Unfortunately, the weak contrast difference does not allow highlighting this phenomenon. However, some glass inclusions could be seen forming dispersed within the matrix of YSZ droplets.

In fact, during the SOFC operation, the glass compound softened whereas the matrix remained solid. Owing to a low viscosity and no reaction with the YSZ matrix, glass could migrate by capillarity through the three-dimensional network of interconnected cavities toward the free surface of the ceramic matrix involved in the glassy seal coating. At the same time, the glass filled most of the defects linked to the plasma spraying process. This gave rise to a continuous network and effectively blocked the continuous 3D leak path.

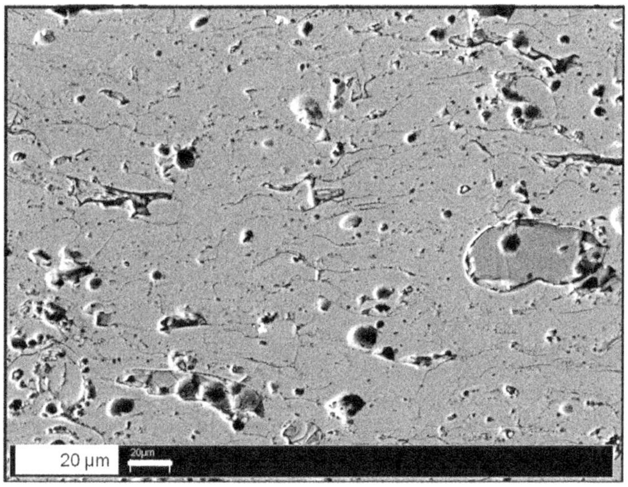

Fig. 2 SEM image of a cross section of the plasma-sprayed seal

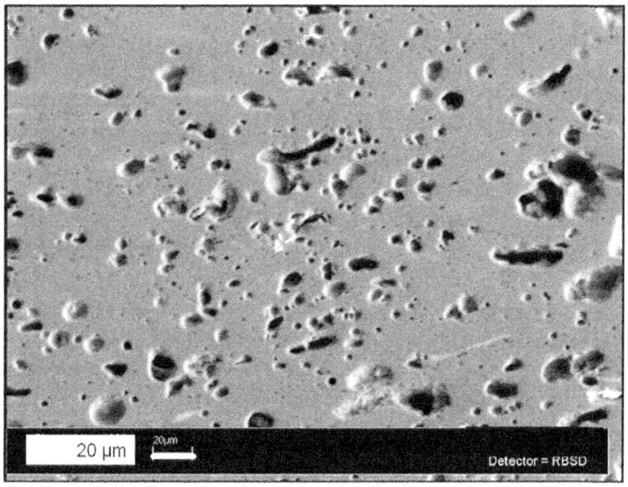

Fig. 3 Cross section of a plasma-sprayed seal after annealing at 900 °C during 30 min in air

Figure 3 displays a SEM image of a cross section of plasma-sprayed seal after annealing during 1/2 h at 900 °C in air. Due to the poor optical contrast between the glass compound and the ceramic matrix, the plasma-sprayed microcracks that are usually filled with glass were no longer visible. The only thing that remained was a closed porosity linked to cavities that initially contained the glass compound, but were emptied due to the glass migration. The overall porosity remained constant.

4. Characterization of the Plasma-Sprayed Seal

4.1 Leak Rate Measurements

To qualify leak rate of the plasma-sprayed seal, the "pressure loss" method was chosen. This method allows the measurements at the SOFC operating temperature. The setup for the leak rate test was composed of a gas feeder, an insulation valve, a data recording system, a buffer volume, and an oven that was able to reach a temperature of 1200 °C. The seal (thickness = 150 μm) was produced on a conical form corresponding to the tested design, on which a load of 19.6 kPa was applied (Fig. 4).

The seal was heated to 900 °C at 2.5 °C/min. After the furnace reached the desired temperature, the inner part of the conical form was flushed with nitrogen to perform the sealing tests for several pressure differentials as shown in Fig. 5.

Three domains were observed: [0-40 kPa]: leak rate ranged between 0 and 2.4 Pa L/s; [40-70 kPa]: leak rate was stabilized; [>50 kPa]: leak rate showed a linear increase from 3.15 to 5.94 Pa L/s.

According to the US Department of Energy, the requirement of a good sealing is 0.5 Pa L/s under a differential pressure of 7 kPa. The plasma-sprayed seals obtained in our work resulted in the leak rate of 0.43 Pa L/s and are therefore suitable for application in SOFC.

4.2 Thermal Cycling

Thermal cycle testing was conducted by heating the specimen in air at a rate of 2.5 °C/min to 900 °C, keeping constant this temperature for 30 min, and then cooling to ambient temperature, under a pressure differential of 20 kPa, before reheating under the same conditions. No degradation of the leak rate was observed after four cycles.

4.3 Rupture Strength Test

The second application of this particular plasma-sprayed seal in the possibility is to join two interconnectors in a SOFC stack as shown in Fig. 6. The seal was plasma sprayed onto the first sand-blasted and degreased

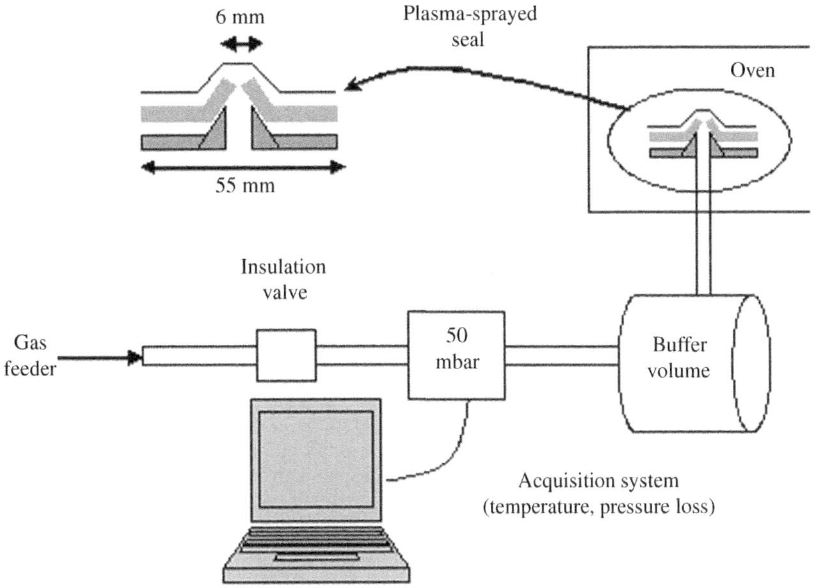

Fig. 4 Experimental setup for leak rate testing

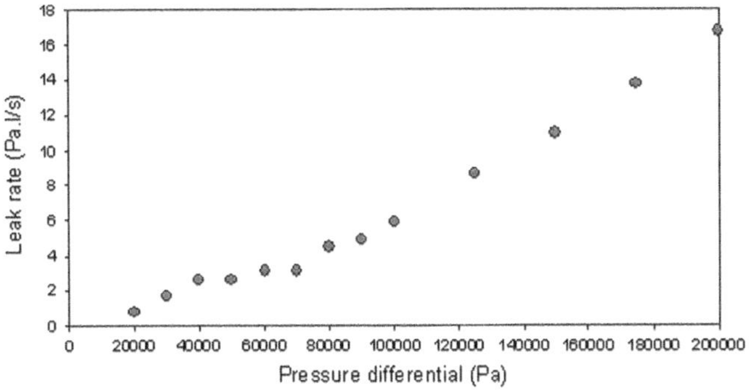

Fig. 5 Measured leak rate as a function of pressure difference

interconnector (CROFER APU 22-Thyssen Krupp), and the second interconnector was placed on the free surface of the plasma-sprayed seal. At the conditioning temperature (900 °C) after a heating rate of 2.5 °C/min, due to its viscous flow, the glass compound filled the free space between the two pieces giving rise to an air-tight layer as shown in the top part in Fig. 6.

To quantify the rupture strength of the joining, a test was developed. It consisted of clamping the bonded assembly into a fixture then increasing the air pressure until the interface, composed of the seal between the two interconnectors, breaks. The rupture pressure for a seal with a thickness of 220 μm was measured to be around 250 kPa.

5. Conclusions

One of the challenges for implementation of SOFCs relies in maintaining a hermetic sealing that remains resistant and stable over the lifetime of the stack. To efficiently and hermetically join two SOFC repetitive units or components, a plasma-sprayed solid seal constituted of a ceramic matrix charged with glass particles was developed.

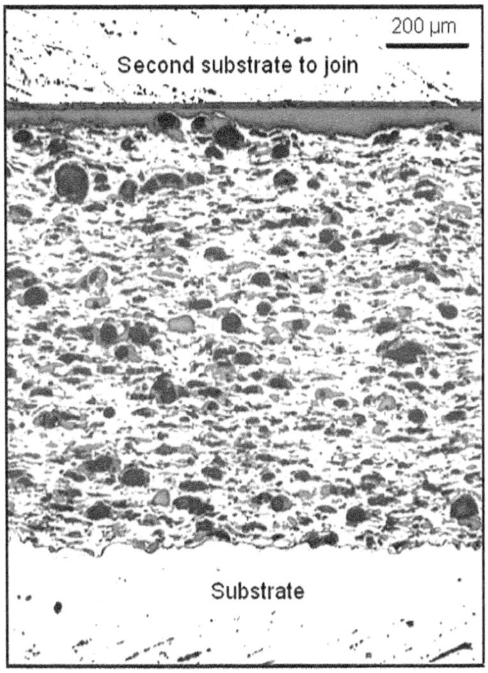

Fig. 6 Optical micrography of a cross section of two surfaces (CROFER 22 APU) joined by the migration of the glass compound from the plasma-sprayed seal

This seal was found to be solid, undistortable, and bonded to its support at ambient temperature. It could also be realized as a self-supported body with a thickness from 50 μm to several millimeters.

Even though the initial structure of the plasma-sprayed seal was microcracked, the principle of the glass filler migration leads to an increase in the gas tightness of the matrix and to the possibility of sealing two surfaces owing to the glassy coating at the free interface.

This seal was tested in a stack of five SOFC cells under normal operating conditions. The performance of the stack was nominal and no internal combustion between hydrogen and oxygen was observed.

Acknowledgments

The authors wish to thank Mr. Bories and Mr. Toulc'hoat for their contribution in the development of the seals and Dr G. Renouard-Vallet for her help in the testing of the plasma-sprayed seals.

References

1. N.Q. Minh, Ceramic Fuel Cells, *J. Am. Ceram. Soc.*, 1993, **76**(3), p 563-588
2. A. Boudghene Stambouli and E. Traversa, Solid Oxide Fuel Cells : A Review of an Environmentally Clean and Efficient Source of Energy, *Renew. Sustain. Energy Rev.*, 2002, **6**, p 433-455
3. K. Scott Weil, C.A. Coyle, J.S. Hardy, and J.Y. Kim, Alternative Planar SOFC Sealing Concepts, *Fuel Cells Bull.*, 2004, **2004**(5), p 11-16
4. A. Khandkar, S. Elangovan, and J. Hartvigsen, SOFC System Design Interactions with Stack Fabrication Technology, *Ceram. Trans.*, 1996, **65**, p 263-277
5. J.W. Fergus, Sealants for Solid Oxide Fuel Cells, *J. Power Sources*, 2005, **147**, p 46-57
6. V.A.C. Haanappel, V. Shemet, I.C. Vinke, M. Gross, Th. Koppitz, N.H. Menzler, M. Zahid, and W.J. Quadakkers, Evaluation of the Suitability of Various Glass Sealant-Alloy Combinations Under SOFC Stack Conditions, *J. Mater. Sci.*, 2005, **40**, p 1583-1592
7. M. Bram, S. Peckers, P. Drinovac, J. Monch, R.W. Steinbrech, H.P. Buckkremer, and D. Stover, Deformation Behavior and Leakage Tests of Alternative Sealing Materials for SOFC Stacks, *J. Power Sources*, 2004, **138**(1-2), p 111-119
8. S. Simner and J.W. Stevenson, Compressive Mica Seals for SOFC Applications, *J. Power Sources*, 2001, **102**(1-2), p 310-316
9. K.A. Nielsen, M. Solvang, S.B.L. Nielsen, A.R. Dinesen, D. Beeaff, and P.H. Larsen, Glass Composite Seals for SOFC Application, *J. Eur. Ceram. Soc.*, 2007, **27**(2-3), p 1817-1822
10. R. Wang, Z. Lü, C. Liu, R. Zhu, X. Huang, B. Wei, N. Ai, and W. Su, Characteristics of a SiO_2-B_2O_3-Al_2O_3-BaO-PbO_2-ZnO Glass-Ceramic Sealant for SOFCs, *J. Alloys Compd.*, 2007, **432**(1-2), p 189-193
11. Composite Sealant Materials for Solid Oxide Fuel Cell, Patent US 6,271,158 B2, 7 August 2001
12. S. Taniguchi, M. Kadowaki, T. Yasuo, Y. Akiyama, Y. Miyake, and K. Nishio, Improvement of Thermal Cycle Characteristics of a Planar-Type Solid Oxide Fuel Cell by Using Ceramic Fiber as Sealing Material, *J. Power Sources*, 2000, **90**, p 163-169
13. X. Huang, K. Ridgeway, S. Narasimhan, K. Reifsnider, and X. Ma, Application of Plasma Sprayed Coatings in a Novel Integrated Composite Seal for SOFCs, *Building on 100 years of Success*, B. Marple et al., Ed., May 15-18, 2006 (Seattle, WA), ASM International, 2006
14. Solid Seal Which is Obtained by Means of Thermal Spraying, Patent WO/2007/042505, 19 July 2007
15. P. Fauchais, A. Vardelle, and B. Dussoubs, Quo vadis Thermal Spraying?, *J Therm. Spray Technol.*, 2001, **10**(1), p 44-66
16. R. Zheng, S.R. Wang, H.W. Nie, and T.-L. Wen, SiO_2-CaO-B_2O_3-Al_2O_3 Ceramic Glaze as Sealant for Planar ITSOFC, *J. Power Sources*, 2004, **128**(2), p 165-172
17. R.E. Loehman, H.P. Dumm, and H. Hofer, Evaluation of Sealing Glasses for Solid Oxide Fuel Cell, *Ceram. Eng. Sci. Proc.*, 2002, **23**(3), p 699-710

JTTEE5 17:603–607
DOI: 10.1007/s11666-008-9226-9
1059-9630/$19.00 © ASM International

Technical Note

Microstructure of Plasma Sprayed La_2O_3-Modified YSZ Coatings

Y. Liu, Y.F. Gao, S.Y. Tao, X.M. Zhou, W.D. Li, H.J. Luo, and C.X. Ding

(*Submitted May 12, 2008; in revised form August 20, 2008*)

Yttria-stabilized zirconia (YSZ) has received great attention as a thermal barrier coating (TBC) material for its excellent thermal and mechanical properties. However, the grain growth of YSZ, particularly under temperatures higher than 1200 °C, limits its further applications to a great extent. In our present study, to develop better understanding of the aforementioned phenomenon and explore effective methods to conquer this challenge, TBCs using traditional and La_2O_3-modified YSZ powders were deposited by atmospheric plasma spraying and their microstructures were investigated. The results show that the La_2O_3 addition can effectively alleviate the grain growth of coatings under high temperatures.

Keywords grain growth, heat treatment, microstructure stability, plasma spray, thermal barrier coating

1. Introduction

State-of-the-art aircraft and industrial gas turbine engines should operate at as high temperature as possible to maximize their thrust and fuel efficiency. However, the limits of melting point of hot path alloy components in turbines make increasing the temperature difficult. This problem can be facilitated by three principal methods: improved alloy design, cooling by air flow through internal channels cast into the component, and usage of thermal barrier coatings (TBCs), which is generally considered to be the most economic and feasible method (Ref 1-3).

TBCs are widely applied on transition pieces, combustion liners, first-stage blades and vanes, combustion chambers, and other hot-path components of gas turbines to enable modern gas turbine engines to operate at gas temperatures well above the melting temperature of the super alloy, thereby improving engine efficiency and performance without increasing the surface temperature of the alloy or reducing the requirements for the cooling system. Along with internal cooling of the underlying super alloy component, a 250 μm thick TBC can reduce the average metal temperature by 111 to 167 °C (Ref 4). As a result, it helps to extend the life of alloy components in the hottest sections in an engine (Ref 5-7). Being exposed to extremely aggressive thermal and mechanical environment, TBCs should meet several performance constraints, such as relatively low thermal conductivity and a thermal expansion coefficient comparable to that of the substrate. The coatings should avoid phase transformation during cycling between room temperature and operation temperature. Another less stringent but nevertheless rather practical requirement is that the coating material is thermodynamically compatible with the oxide formed by oxidation of the bond coat. Also, some other factors need to be considered and explored such as high melting point, chemical inertness to the combustion gases, and low sintering rate of the porous microstructure (Ref 1). Among all the requirements for TBC materials, thermal expansion coefficient and low thermal conductivity seem to be the most important (Ref 2). The number of materials that can be used as TBCs is very limited. So far, only a few materials have been found to basically satisfy these requirements (Ref 6).

Yttria-stabilized zirconia (YSZ) possesses a suite of desirable properties that make it proven to be a highly durable TBC material. It is likely to remain the material of choice for turbines with current operating temperatures. First, of all ceramics, YSZ has one of the lowest thermal conductivities at elevated temperature and a relatively high thermal expansion coefficient. Second, it has a relatively low density, which is important for parasitic weight considerations in rotating engine components. It also is resistant to ambient and hot corrosion. Finally, YSZ has a high melting point (2700 °C), making it suitable for high-temperature applications (Ref 5). One of the striking findings from the numerous studies is that yttria-stabilized zirconia coatings containing 7-8 mol% $YO_{1.5}$ having the

This article is an invited paper selected from presentations at the 2008 International Thermal Spray Conference and has been expanded from the original presentation. It is simultaneously published in *Thermal Spray Crossing Borders, Proceedings of the 2008 International Thermal Spray Conference*, Maastricht, The Netherlands, June 2-4, 2008, Basil R. Marple, Margaret M. Hyland, Yuk-Chiu Lau, Chang-Jiu Li, Rogerio S. Lima, and Ghislain Montavon, Ed., ASM International, Materials Park, OH, 2008.

Y. Liu, Y.F. Gao, S.Y. Tao, X.M. Zhou, W.D. Li, H.J. Luo and **C.X. Ding**, Chinese Academy of Sciences, Shanghai Institute of Ceramics, Dingxi 1295, Shanghai 200050, People's Republic of China; and **Y. Liu**, Graduate University of the Chinese Academy of Sciences, Beijing 100049, People's Republic of China. Contact e-mail: yfgao@mail.sic.ac.cn.

tetragonal-prime structure exhibited the longest lifetime under thermal cycling (Ref 8).

However, a major disadvantage of YSZ is its limited operation temperature (<1200 °C) for long-term application. At higher temperatures, sintering occurs, giving rise to changes in coating morphology and performance. There are two principal concerns. One is that sintering inevitably increases the elastic modulus and thereby decreases the strain compliance of the coating. At the present stage in development, the reduction in strain compliance would generate very large stresses and lead to spontaneous failure on cooling. The other is that densification decreases the volume fraction of porosity, which, in turn, causes the thermal conductivity to increase (Ref 8). So suppressing sintering would be a big challenge in the development of YSZ coatings. Several efforts were made to stabilize the coating microstructural morphology under high temperature (Ref 2, 9). M. Matsumoto and his co-workers (Ref 9) reported that the addition of rare earth oxides could develop TBCs with high resistance to sintering and La_2O_3-containing YSZ coatings have a greatly reduced thermal conductivity as well as high structural stability. A better understanding of the dopant effects will help to develop advanced sintering-resistant TBC coatings in the future. In our present work, TBCs were plasma sprayed using lanthana precursor coated YSZ powders as feedstocks and the effects of lanthana on the sintering process and grain growth of YSZ were evaluated.

2. Experimental Procedures

Commercially available 7-8YSZ (7-8wt.% YSZ) powder and lanthanum nitrate (99% pure) were used as starting materials. The commercial powders have a particle size range of 20-40 μm. YSZ powders were coated with lanthana precursor through a wet chemical method. Commercial YSZ powders were dispersed in $La(NO_3)_3$ and urea mixed solution and then sonicated in an ultrasonic cleaner for 10 min to break up the agglomerates. The suspension was heated in a sealed beaker at 90 °C for 30 h with constant stirring to prevent the YSZ powders from settling down or agglomerating. After reactions, the products were then centrifuged and washed with deionized water and ethanol, in turn, followed by drying in an oven at 70 °C for 24 h. The dried powders were subsequently calcined in an alumina crucible at 800 °C in air for 2 h to obtain La_2O_3-coated YSZ powders. The additive of lanthana was fixed at 10 wt.% of YSZ powder.

Both the pure and lanthana-modified YSZ coatings were prepared by using a Metco A-2000 atmospheric plasma spraying equipment with F4-MB plasma gun (Sulzer Metco AG, Rigackerstrasse 16, CH-5610 Wohlen, Switzerland) with Ar and H_2 as plasma gases at a flow rate of 30 and 15 standard liters per minute (slpm), respectively. More details about the spraying parameters are listed in Table 1. The coatings were sprayed about 3 mm in thickness on nickel-based alloy substrates. After deposition, the coatings were mechanically removed from the substrates.

Table 1 Plasma spraying parameters

Parameters	Value
Arc current intensity, A	660
Primary gas (Ar) flow rate, slpm	30
Secondary gas (H_2) flow rate, slpm	15
Spray distance, mm	130
Powder carrier gas flow rate, slpm	4
Injection diameter, mm	1.8

To investigate the effect of heat treatment on the microstructure of coatings, the freestanding coatings were heated at a rate of 5 K/min, followed by a hold time typically of 50 h, and then cooled in the furnace. For morphology and grain growth studies, scanning electron microscopy (SEM) images of the feedstocks and coatings were collected on a JSM 6700F field emission scanning electron microscope (JEOL Ltd., Tokyo, Japan). The option of energy dispersive X-ray spectroscopy (EDS) was used to determine the chemical composition of the powder within the same equipment that was used for SEM. The element distribution of the coatings was determined with an EPMA-8705QH2 electron probe analyzer (Shimadzu Co., Tokyo, Japan). The phase structure was analyzed on a Japan Rigaku D/max 2550V (Rigaku Co., Tokyo, Japan) X-ray diffractometer using Cu Kα ($\lambda = 0.15406$ nm) radiation at 40 kV and 100 mA.

3. Results and Discussions

3.1 Morphology of Feedstock and As-sprayed Coatings

The morphologies of the commercially available YSZ powders and the as-coated powders are shown in Fig. 1, which reveals that the YSZ powders have a smooth and flat surface. Compared with the starting powders, the modified ones have some tiny grains with particles sizes ranging from 50 to 500 nm. The grains were identified as lanthanum precursor by EDS analysis (at the positioned point in Fig. 1b). According to our XRD results and Ref 10, the lanthanum precursor is affirmed to be lanthanum oxycarbonate, which can be decomposed into lanthana above 720 °C.

The surface and cross-sectional morphologies of the two kinds of coatings were extensively investigated in our experiments. As shown in Fig. 2, the La_2O_3-modified YSZ coatings show a typical morphology of as-sprayed APS coatings. It can be seen that there are many pores and microcracks distributed homogeneously in the coatings (Fig. 2a and b). The pores are isolated and most of them are less than 10 μm in diameter, which offers an effective barrier to heat transportation. Figure 2(c) and (d) presents the fractured cross section of the coatings, demonstrating that the coatings consist of splats aligned predominantly parallel to the substrate surface, which can enhance the thermal insulation property of coatings. The splat structures have a thickness of about 3-6 μm with columnar grains of a diameter of about 1 μm. It can be speculated

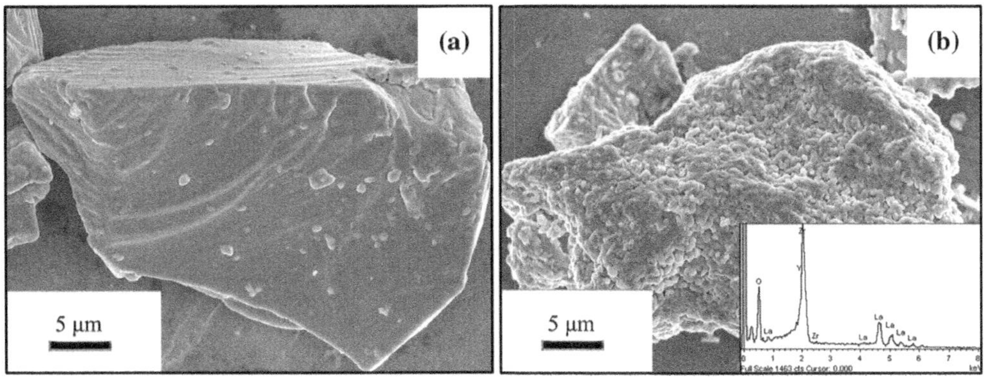

Fig. 1 Morphologies of pure YSZ powders (a) and the modified YSZ powders (b). The inset is EDS profile of A region

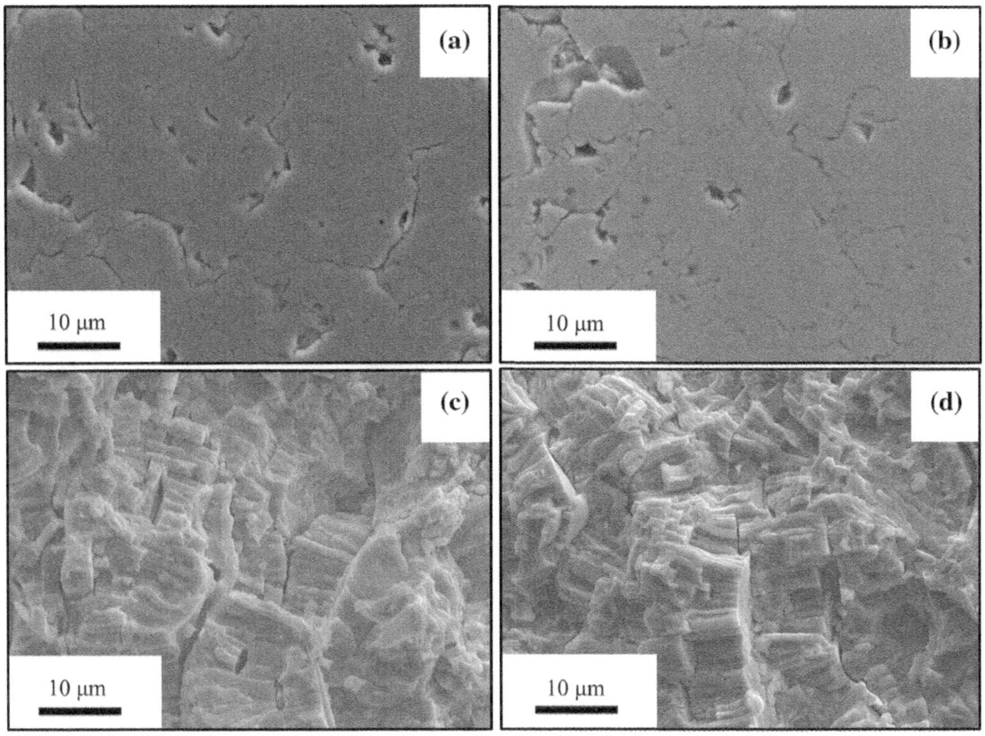

Fig. 2 SEM micrographs of cross section and fractured surface of YSZ coatings (a, c) and lanthana-modified YSZ coatings (b, d)

from the smooth surface perpendicular to the heat flux direction that large pores also exist at the splat boundaries.

3.2 The Effect of Heat Treatment on the Grain Growth of the Coatings

One of the concerns in developing reliable and robust TBCs is how the coating changes during use at high temperatures. Figure 3 and 4 give the SEM micrographs of traditional YSZ coatings and lanthana-modified YSZ coatings after annealing treatment at 1200 °C and 1400 °C for 50 h, respectively. The SEM image reveals that the mean grain size of traditional YSZ coatings aging at 1200 °C for 50 h is about 500 nm with a few abnormal larger ones, whereas a mean grain size under 200 nm was observed for the lanthana-modified YSZ coatings. After annealing at 1400 °C for 50 h, both the kinds of coatings have a grain growth to some degree, but compared with the traditional YSZ coatings, the grains in the lanthana-modified YSZ coatings still have a small grain size distribution and showed a rather slow grain growth rate.

The element of lanthanum probably comprises three kinds of existing state: lanthana, stabilizer dissolved in zirconia, and lanthanum zirconate, which would be formed through the reaction of lanthana and zirconia. The solubility of La_2O_3 in ZrO_2 at equilibrium state is below 1 mol% (Ref 11). There is no visible evidence of $La_2Zr_2O_7$ in the coatings by XRD. Element distribution analysis carried out by electron probe analyzer clearly confirms that a few lanthana particles segregated along the zirconia grain boundaries. As described in the literature

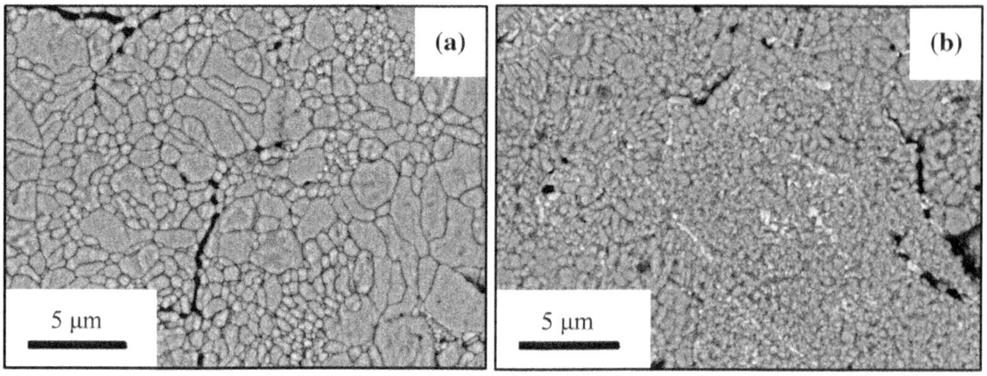

Fig. 3 Microstructures of traditional (a) and La_2O_3-modified YSZ coatings (b) annealed at 1200 °C for 50 h

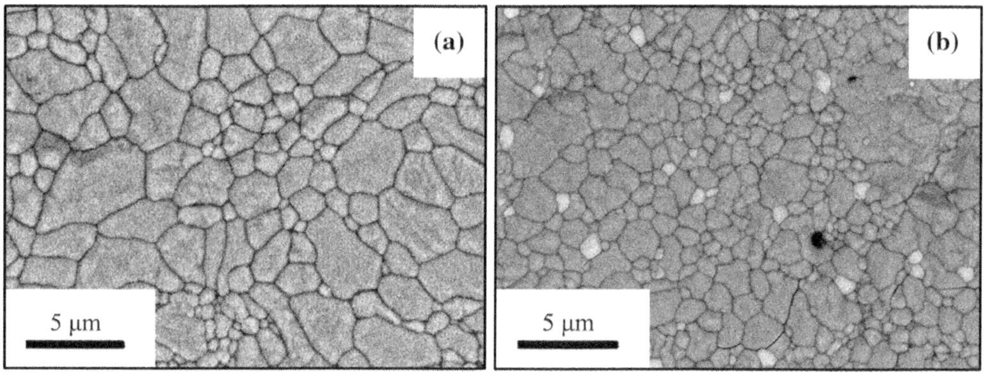

Fig. 4 Microstructures of traditional (a) and La_2O_3-modified YSZ coatings (b) annealed at 1400 °C for 50 h

(Ref 12), trivalent and large cation could segregate on the grain boundaries of zirconia and have a strong effect in suppressing the mobility of grain boundaries and consequently the grain growth. The phenomenon can be satisfactorily rationalized using the space charge concept and the model of impurity drag. This could explain the slow grain growth in the lanthana-modified YSZ coatings under heat treatment.

4. Conclusions

Morphology stability of La_2O_3-modified YSZ coatings under high temperature treatment was investigated. It was revealed that addition of La_2O_3 was effective to suppress the grain growth in TBC. The other thermophysical characterization of the coatings for determining the effects of heat treatment is to be performed and will be reported in our next work.

Acknowledgments

The authors would like to thank Ms. J. H. Gao and Mr. W. Wu for the SEM and EDS measurements. YFG thanks the Century Program (One-Hundred-Talent Program) of the Chinese Academy of Sciences for special funding support.

References

1. D.R. Clarke and S.R. Phillpot, Thermal Barrier Coating Materials, *Mater. Today*, 2005, **8**(6), p 22-29 (in English)
2. X.Q. Cao, Application of Rare Earths in Thermal Barrier Coating Materials, *J. Mater. Sci. Technol.*, 2007, **23**(1), p 15-35 (in English)
3. F. Cernuschi, L. Lorenzoni, S. Ahmaniemi, P. Vuoristo, and T. Mäntylä, Studies of the Sintering Kinetics of Thick Thermal Barrier Coatings by Thermal Diffusivity Measurements, *J. Euro. Ceram. Soc.*, 2005, **25**(4), p 393-400 (in English)
4. A. Maricocchi, A. Bartz, and D. Wortman, PVD TBC Experience on GE Aircraft Engines, *J. Therm. Spray Technol.*, 1997, **6**(2), p 193-198 (in English)
5. N.P. Padture, M. Gell, and E.H. Jordan, Thermal Barrier Coatings for Gas-Turbine Engine Applications, *Science*, 2002, **296**(4), p 280-284 (in English)
6. X.Q. Cao, R. Vassen, and D. Stoever, Ceramic Materials for Thermal Barrier Coatings, *J. Eur. Ceram. Soc.*, 2004, **24**(1), p 1-10 (in English)
7. S. Paul, I.O. Golosnoy, A. Cipitria, T.W. Clyne, L. Xie, and M.R. Dorfman, Effect of Heat Treatment on Pore Architecture and Associated Property Changes in Plasma Sprayed TBCs, *Proceedings of the International Thermal Spray Conference 2007*, 1st ed., B. Marple, M. Hyland, Y.C. Lau, C.J. Li, R. Lima, and G. Montavon, Eds., May 14-17, 2007 (Beijing, China), ASM International, 2007, p 411-416 (in English)

8. D.R. Clarke and C.G. Levi, Materials Design for the Next Generation Thermal Barrier Coatings, *Annu. Rev. Mater. Res.*, 2003, **33**, p 383-417 (in English)
9. M. Matsumoto, N. Yamaguchi, and H. Matsubara, Low Thermal Conductivity and High Temperature Stability of ZrO_2-Y_2O_3-La_2O_3 Coatings Produced by Electron Beam PVD, *Scr. Mater.*, 2004, **50**(6), p 867-871 (in English)
10. Y.F. Gao, Y. Masuda, and K. Koumoto, Micropatterning of Lanthanum-Based Oxide Thin Film on Self-Assembled Monolayers, *J. Colloid Interf. Sci.*, 2004, **274**(2), p 392-397 (in English)
11. B. Bastide, P. Odier, and J.P. Coutures, Phase Equilibrium and Martensitic Transformation in Lanthana-Doped Zirconia, *J. Am. Ceram. Soc.*, 1988, **71**(6), p 449-453 (in English)
12. S.L. Hwan and I.W. Chen, Grain Size Control of Tetragonal Zirconia Polycrystals Using the Space Charge Concept, *J. Am. Ceram. Soc.*, 1990, **73**(11), p 3269-3277 (in English)

JTTEE5 17:608–611
DOI: 10.1007/s11666-008-9265-2
1059-9630/$19.00 © ASM International

Plasma Spraying of Metal Coatings Using CO_2-Based Gas Mixtures

L. Pershin, L. Chen, and J. Mostaghimi

(Submitted May 16, 2008; in revised form September 29, 2008)

Compared to argon, plasmas generated by molecular gases and operated at the same electric current level contain much higher enthalpy. This is because molecular gases must dissociate before ionization, which requires larger energy input. The authors have developed a new DC plasma torch, which operates with a mixture of carbon dioxide (CO_2) and methane (CH_4). The torch was employed for plasma spray coating with input power in a range of 30 to 45 kW. This paper reports on the effect of CO_2/CH_4 mixtures on the particle conditions during spraying of nickel alloy powders. Results of gas composition analysis at various distances from the nozzle exit are presented. The particle in-flight conditions, coating microstructure, and deposition efficiency are analyzed.

Keywords coating oxidation, graphite cathode, molecular gases plasma, plasma composition

1. Introduction

Molecular gases are commonly used for high power plasma generation in waste or coal gasification and in steelmaking. In comparison with inert gases, the higher energy that is required for dissociation and ionization of the molecular gases translates into a higher plasma enthalpy. High plasma enthalpy is desirable as it allows higher material feed rate and higher materials processing rates. Carbon-containing molecular gases, in particular a mixture of carbon dioxide and hydrocarbons, have a number of advantages. In addition to high plasma enthalpy, such plasmas have high thermal conductivity resulting in efficient heat transfer to treated materials. For example, at one atmosphere pressure and a temperature of 7000 K, the thermal conductivity of mixture containing two parts CO_2 and one part CH_4, is almost 20 times greater than that of argon (Ref 1, 2).

Another feature of carbon-containing plasma gases, which was exploited in the design of the torch in this study, is as follows. Positive carbon ions are produced in the arc region, and because of the Coulomb attractive forces they flow toward the negatively charged cathode and deposit on its surface. The formation of a complex surface morphology of the deposit, which consisted of carbon nanotubes and nanoparticles, has already been reported in a previous publication (Ref 3). If the cathode is made of graphite, the induced ionic current could partially or totally compensate for the sublimation and loss of the cathode material, which occurs due to high cathode heat fluxes. This phenomenon may then greatly increase the cathode life (Ref 3, 4).

Based on the above principle, a new DC plasma torch with graphite cathode for thermal spraying was designed, constructed, and tested at the *Centre for Advanced Coating Technologies* (CACT). Because the arc voltage is relatively high (130-220 V), it was able to generate a high power plasma at much lower arc currents in comparison with conventional argon mixtures. Lower arc current is beneficial as it considerably reduces the thermal load on the electrodes, resulting in longer electrode life. The torch was operated in a power range of 20 to 70 kW at arc currents that were no more than 400 A. The thermal efficiency was measured and was found to be high (60-85%). This paper presents the effect of the volumetric ratio of CO_2 to CH_4 on the plasma plume composition and on the particle parameters during spraying of Ni alloy powder.

2. Experiment

The new torch used for the coating application is operated with CO_2/CH_4 gas mixtures. One important feature of the torch is a water-cooled, button type cathode made of a highly structured graphite. The torch has a cylindrical anode with 7 mm in diameter and a vortex gas distributor, and the powder is externally injected (Ref 3).

This article is an invited paper selected from presentations at the 2008 International Thermal Spray Conference and has been expanded from the original presentation. It is simultaneously published in *Thermal Spray Crossing Borders, Proceedings of the 2008 International Thermal Spray Conference*, Maastricht, The Netherlands, June 2-4, 2008, Basil R. Marple, Margaret M. Hyland, Yuk-Chiu Lau, Chang-Jiu Li, Rogerio S. Lima, and Ghislain Montavon, Ed., ASM International, Materials Park, OH, 2008.

L. Pershin, L. Chen, and **J. Mostaghimi**, Mechanical and Industrial Engineering Department, Centre for Advanced Coating Technologies, University of Toronto, 5 King's College Rd., Toronto, ON, Canada M5S 3G8. Contact e-mail: pershin@mie.utoronto.ca.

Table 1 Torch operating conditions and thermal efficiency (measured)

Gas flow	Cathode material	Current, A	Voltage, V	Power, kW	Thermal efficiency, %
$CO_2 + 17\%CH_4$ (30 lpm)	C	305	138	42.1	65
$CO_2 + 30\%CH_4$ (30 lpm)	C	310	132	40.9	60
Ar (55 lpm)	W	600	47	28.2	35

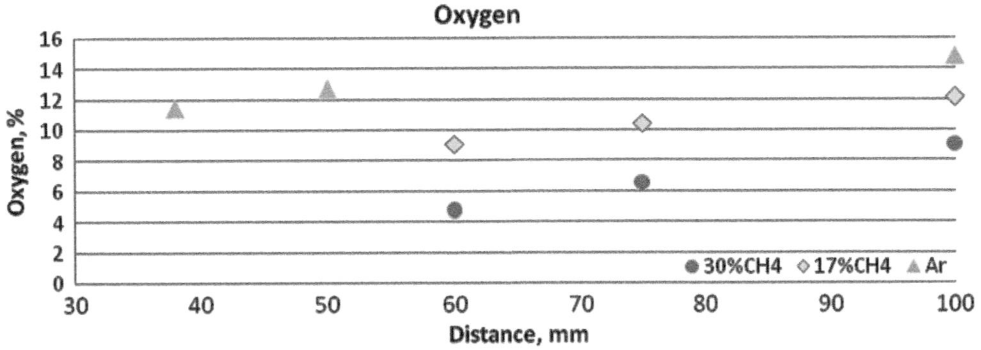

Fig. 1 Oxygen content variation along the torch axis for CO_2 with 30% and 17%CH_4 gas mixtures and Ar

Two gas compositions were considered with 17 and 30% volumetric CH_4 content; the total gas flow rate was, however, kept constant at 30 lpm. For comparison, the torch was also operated with argon. In this case the graphite cathode was changed and tungsten was used as cathode materials. Table 1 shows the torch operating conditions used in this study.

The gas composition in the plasma jet was measured by a mass-spectrometer (Microvision Plus from MKS Instrument, Crewe, UK) attached to a water cooled Pitot probe. The probe tip was placed along the torch centerline at distances of 60, 75, and 100 mm. To prevent overheating of the probe, the closest distance to the torch exit was limited to 60 mm. In contrast, when argon was used as plasma gas, the minimum distance was 40 mm.

The sprayed Ni-based alloy powder was Metco 461NS (Sulzer-Metco, Westbury, NY, US) with particle size distribution of $-150+22$ μm. Powder feed rate was 14 g/min; the relatively low feed rate was limited by the powder deposition inside the nozzle for SG-100 torch. Measurements of the in-flight particle conditions were made by a DPV-2000 monitoring system (Tecnar Automation, St-Bruno, QC, Canada). Cross sections through the sprayed coatings were polished and examined under a scanning electron microscope (SEM). Oxygen concentrations of the initial powder and coatings were measured using a LECO TC-136 (Leco, St-Joseph, MI, USA) oxygen analyzer.

3. Results and Discussion

3.1 Gas Composition

Plasma spray process in the ambient atmosphere is always accompanied by entrainment of air into the plasma plume, which intensifies in-flight metal particle oxidation. It is important to know oxygen content of the plasma specifically for the noninert carbon dioxide gas mixtures. Results of the measurements showed that the oxygen content was much lower for CO_2/CH_4 plasma at a distance of 60 mm in comparison with argon plasma at a distance of 50 mm (Fig. 1). As expected, the mixture with higher methane content, i.e., with 30%CH_4, had a lower concentration of oxygen, approximately one-third of that for Ar plume at 60 mm distance.

Farther from the torch exit, in spite of strong air entrainment, the oxygen level remains significantly lower for CO_2/CH_4 mixtures. For instance, at a distance of 100 mm the oxygen concentration for the mixture with 30%CH_4 and for pure argon plasma was 9 and 14.9%, respectively. It is important to note that in addition to lower oxygen content, CO_2/CH_4 plume contains a significant amount of CO. The presence of CO may be the reason for lower oxygen content in the CO_2/CH_4 plasma, since the entrained oxygen may react with CO and produce CO_2. The gas chemistry in the plume is thus very advantageous as it reduces the possibility of in-flight oxidation of the powder (Fig. 2). At 60 mm distance, for the methane-rich mixture, the measurements also showed the presence of hydrocarbons (1.6%) and hydrogen (0.7%).

3.2 Particles Conditions

The design of the CACT torch and its cooling capacity does not allow operating the torch at currents above 450 A. For comparison, the SG-100 (Praxair, Concord, NH, USA) torch for coating applications in pure argon was used. SG-100 was operated at a similar input power, which required an arc current of 700 A. The argon flow rate was 55 lpm. Earlier results have proved that the thermal properties of plasma generated by CO_2/CH_4 mixture are much better than argon plasma and that

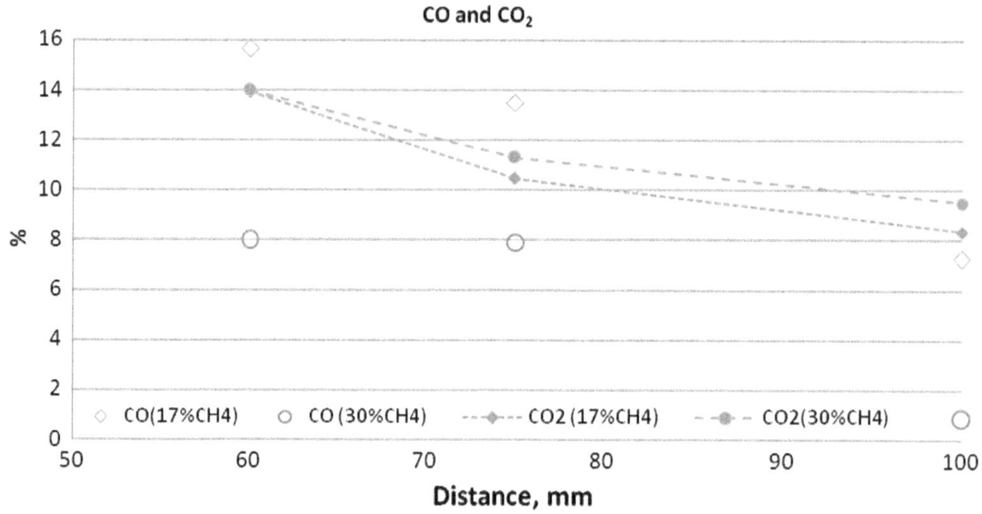

Fig. 2 CO and CO_2 content in plasma plume along the torch axis for CO_2 with 30% and 17%CH_4 gas mixtures

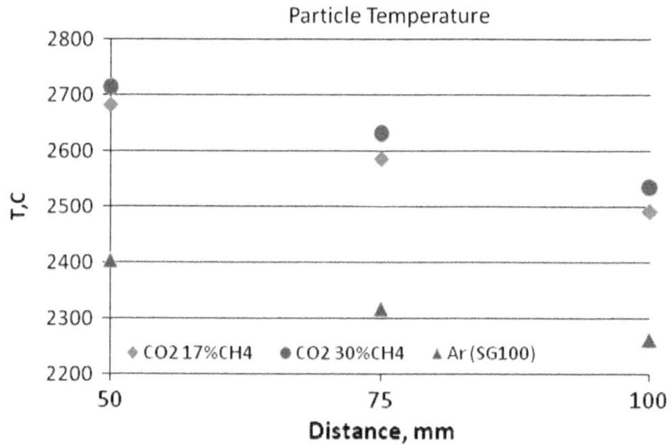

Fig. 3 Particle in flight temperature evolution

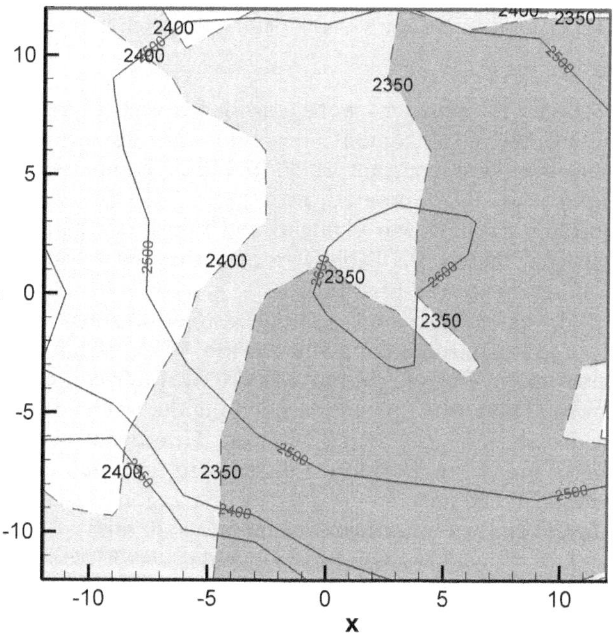

Fig. 4 Particle temperatures at 75 mm spray distance sprayed in mixture with 17%CH_4 (solid lines) and Ar (shaded area on background)

injected powders are heated and melted much more effectively (Ref 1, 3). The results of measurements (Fig. 3) demonstrate that the particle temperatures at all spray distances were about 300 °C above of the particles temperature sprayed in Ar despite internal powder injection in SG-100 torch as against the external injection of the powders in this study. It is worth noting that higher methane results in higher hydrogen concentration, which in turn increases the thermal conductivity of the plasma (Ref 2). Effect of these conditions could be seen in higher particle temperatures for the mixture with 30%CH_4.

Particle velocities along spray were very close for both torches. The velocities between 50 and 100 mm spray distances decreased from 175 to 96 m/s for the SG-100 torch and from 155 to 109 m/s for the new torch.

Deposition efficiency of the spray is an important parameter as it affects the operating cost. To improve deposition efficiency, it is important to achieve a high degree of melting of the injected powders. To measure the degree of melting, a scan of the plasma plume was performed by DPV-2000 at 9 × 9 points with 3 mm step after "*auto centering*" routine at all spray distances. This gives an opportunity to compare particle conditions not only along the centerline but also within the cross section of the jet. Figure 4 shows a map of particle temperature distribution at 75 mm distance for 83%CO_2 + 17%CH_4 and for argon plasmas.

The shaded area with dash lines on the background of Fig. 4 represent the particles temperature distribution sprayed in Ar, whereas solid lines show isotherms for particles sprayed with 83%CO_2 + 17%CH_4 gas mixture. These measurements demonstrate that, with the new torch, a significantly larger number of particles were heated to higher temperatures. In any thermal spray

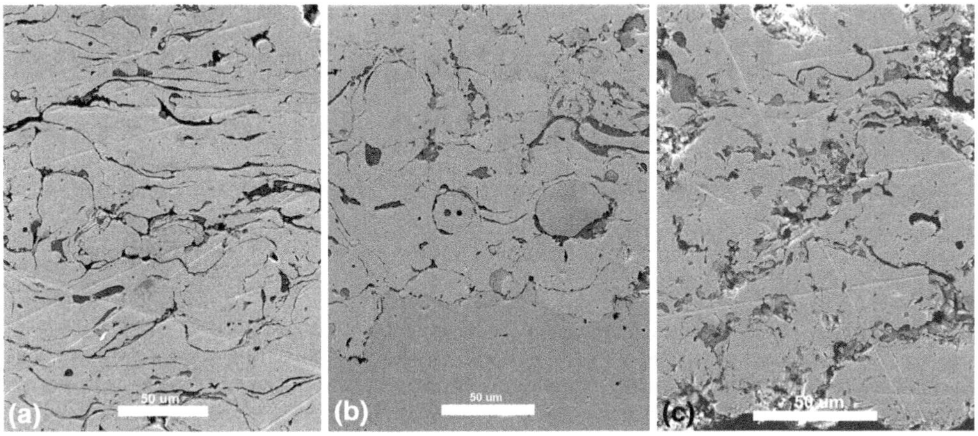

Fig.5 Micrographs of the coatings deposited at 75 mm distance in mixtures with 30% CH_4 (a), 17% CH_4 (b), and Ar (c)

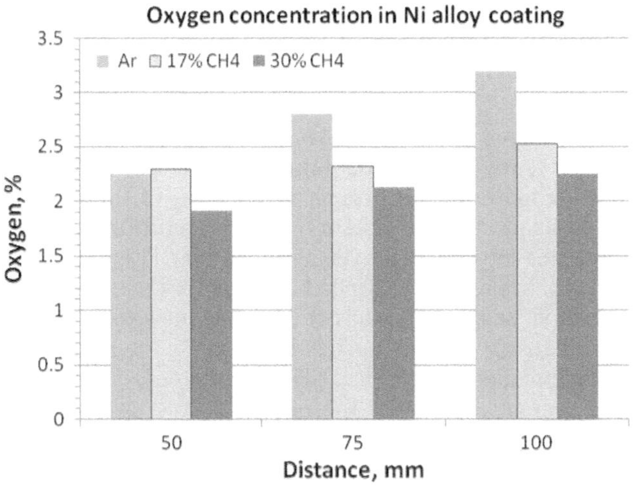

Fig. 6 Oxygen content measured in deposited coatings

process, it is important to attain the maximum deposition efficiency to lower the cost and time of the process. It should be noted that, to improve particle heating, SG-100 typically operates with the addition of an auxiliary gas such as H_2 or He.

A large number of molten particles at spray distance will result not only in higher deposition efficiency but also in producing dense, better quality coatings. The thickness of the coatings per pass of the plasma torch with a traverse velocity of 10 m/min at 50 mm standoff distance was measured. For the 70% CO_2 + 30% CH_4 composition, the deposit thickness per pass was approximately 40 μm; the thickness was 33 μm for the mixture with 17% CH_4 and 11 μm for Ar.

Metallographic observations of the deposited coatings also revealed positive effect of higher thermal properties of CO_2/CH_4 plasma on the coating microstructure (Fig. 5). The coatings have lamellar structure with porosity of 9.8% (Fig. 5a) and 10% (Fig. 5b). For porosity measurements, image analysis software Clemex Application Suite was used. It is worth noting that the large particles of initial powder (up to 150 μm) were molten or softened enough to be deposited and form flattened lamellae. The coatings had well-adhered interface.

The coating deposited by argon plasma (Fig. 5c) was formed by less-flattened particles with extensive inter lamellar porosity of 12.4%, which is indicative of insufficient melting. Adhesion of the coating was poor, which resulted in delamination during deposition.

Oxygen concentration of the deposited coatings accurately reflects differences in the plasma plume composition (Fig. 6). Higher oxygen content of Ar plasma resulted in a more in-flight oxidation of the powder. The particles oxidation deposited by the CACT torch is much lower despite higher particle temperatures during deposition. Oxygen content in the feedstock powder was measured at 2.43% mass.

4. Conclusions

The new torch operating with a CO_2/CH_4 mixture generates a plasma plume, which creates favorable conditions for spraying metals due to lower oxygen level and the presence of reducing gases.

Better thermal properties of the CO_2-based mixtures produced dense well-adhered coatings from powders with relatively large particle size and at about three times higher deposition rate than in argon plasma.

References

1. L. Chen, J. Mostaghimi, and L. Pershin, Numerical Simulations of Cascaded Plasma Torch Using Ar and Molecular Gases, *Global Coating Solutions,* B. Marple, et al., Eds. (Materials Park, OH) ASM International, 2007
2. L. Pershin, L. Chen, and J. Mostaghimi, Comparison of Molecular and Argon Gases for DC Plasma Generation, *Proceedings of the 18th ISPC*, K. Tachibana, O. Takai, K. Ono, T. Shirafuji, Eds. (Kyoto, Japan), 2007
3. L. Pershin, L. Chen, and J. Mostaghimi, Deposition of YSZ Coatings by a High Efficiency DC Plasma Torch, *Proceedings of the 21 Int. Conf. on Surface Modification Technologies*, T. Sudarshan and M. Jeandin, Eds. (Paris, France), Sept 24-26, 2007
4. R. Baird, Method for Increasing Electrode Life, US Pat. 3,307,011

JTTEE5 17:612–616
DOI: 10.1007/s11666-008-9236-7
1059-9630/$19.00 © ASM International

Technical Note

Tailor-Made Coatings for Turbine Applications Using the Triplex Pro 200

Katharina Richardt, Kirsten Bobzin, Dieter Sporer, Thomas Schläfer, and Petr Fiala

(*Submitted May 20, 2008; in revised form September 24, 2008*)

In modern jet engines, the efficiency of compressor stages is highly dependent on the clearance between blade tip and casing. To improve efficiency of gas turbines, the gap between the turbine blades and casing has to be minimized. Abradable coatings permit a minimization of the clearance and control of the over-tip leakage by allowing the blade tips to cut into the coating. Thermal sprayed abradables aim at a well-balanced profile of properties relevant for abradable seals. Amongst others these include: abradability, aging resistance, corrosion and oxidation resistance, and bond strength to substrate. Here, abradable coatings consisting of a multiphase material, comprising a metal matrix and a solid lubricant as well as a defined porosity, were developed using the Triplex Pro 200 to increase reproducibility and deposition efficiency. In addition, the influence of the process parameters on coating characteristics (porosity, hardness) and, as a result, erosion properties and abradability was investigated.

Keywords advantages of TS, APS coatings, coatings for gas turbine components, influence of spray parameters, TBC topcoats, TS coating process

1. Introduction

The plasma spraying process is a well-accepted process for applying functional coatings for different applications. Process optimization and innovative material applications gain more and more interest in regard to continuously increasing functional and structural demands on thermal sprayed coatings. This study describes the parameter development for an abradable composite coating material, relevant for the aerospace industry.

The reduction of the leakage by improved sealing can result in significantly improved efficiency as well as power output, time-on-wing, and compressor stall margin (Ref 1-4). Manufacturing tolerances and in-service behavior of the rotor that allow a reduction of the over-tip leakage have been a fundamental field of research in recent years. Abradable linings between housing and rotating blades are a promising solution, as they allow a self-regulation of the system by a purposeful cutting of the blade tips into the relatively soft coating. The blade wear during that process has to be kept to a minimum, while the abradable seal coating has to be resistant against erosive wear caused by the particle loaded jet, aging, and/or oxidation during service. To meet these demanding requirements, complex thermally sprayed material and coating systems have been introduced to the market and successfully used. In this work, the coating characteristics of an experimental abradable coating material Durable 2621 were investigated. The powder material is a multiphase material, comprising a metal matrix (Ni-based) in addition to a solid lubricant, mostly hexagonal boron nitride, with a well-defined level of porosity.

2. Experiments

The spraying experiments were carried out using the Sulzer Metco Triplex Pro 200 APS-coating system (nozzle diameter: 2.3 mm, powder injector diameter: 1.8 mm).

As plasma gases, combinations of argon and nitrogen were used. With the use of a combination of one and two atomic plasma gases, an optimal acceleration and heating of the powder particles can be reached. In a plasma gas mixture consisting of nitrogen and argon, nitrogen is mostly responsible for heating up the powder material; however, due to its high atomic weight, which favorably influences the pulse transmission, argon affects the particle velocity. Using only argon as a plasma gas, the particle velocity is comparably high, the melting time of the particles in the plasma jet low. This results in a too high polyester content in the coating, and reduces the

This article is an invited paper selected from presentations at the 2008 International Thermal Spray Conference and has been expanded from the original presentation. It is simultaneously published in *Thermal Spray Crossing Borders, Proceedings of the 2008 International Thermal Spray Conference*, Maastricht, The Netherlands, June 2-4, 2008, Basil R. Marple, Margaret M. Hyland, Yuk-Chiu Lau, Chang-Jiu Li, Rogerio S. Lima, and Ghislain Montavon, Ed., ASM International, Materials Park, OH, 2008.

Katharina Richardt, Kirsten Bobzin and **Thomas Schläfer**, Surface Engineering Institute (IOT), RWTH Aachen University, Aachen Germany; **Dieter Sporer**, Sulzer Metco, Schattwald, Austria; and **Petr Fiala**, Sulzer Metco, Fort Saskatchewan, Canada. Contact e-mail: richardt@iot.rwth-aachen.de.

hardness. In this work, the experimental powder material Durabrade 2621 (Sulzer Metco, Wohlen, Switzerland), consisting of NiCrAl/boron nitride and 10 wt.% polyester, was used. The abradable coatings with a required coating thickness between 2 and 2.2 mm were deposited on CoNiCrAlY-bondcoats (Amdry 995C). After spraying and prior to any further testing, the coatings were heat-treated in air at 435 °C for 5 h to evaporate the polyester filler deposited in the coating, and thereby creating additional coating porosity. During the heat treatment, the coating thickness is decreased by 200 ± 80 µm, and therefore, any effects of the substrate in hardness measurements can be excluded.

For analysis of the microstructure, the coatings were embedded by vacuum infiltration; a micro-structural evaluation was carried out using optical microscopy of cross sections of the coatings. The homogeneous distribution of the material components was secured by the comparison of optical image analyses of several micrographs. In-flight particle characteristics were measured using the optical diagnostic systems DPV2000 (Tecnar, St. Bruno, QC, Canada) as well as Accuraspray (Tecnar). Rockwell hardness of the coatings was measured at MTU (Munich, Germany), using a ball indenter, pressed into the surface of the coating. The indentation is based on a two-step application—a preliminary force P followed by an additional test force. The result of the test is read directly from the tester as:

$$\mathrm{HR} = E - e$$

where E = indenter constant-here 100 units and e = permanent indentation depth

Based on the indenter constant $E = 100$ used here, the hardness scale is not limited on positive values; values close to 0 do not mark the lowest hardness possible. Surface roughness (R_a) was measured using a laser profilometer.

After a general parameter development, in order to deposit homogeneous coatings within a required hardness range and a homogeneous particle temperature and velocity distribution within the spray plum (measured with DPV 2000), the influence of the process parameters on coating characteristics and particle properties was investigated. To do so a full factorial DoE was generated. The hardness as a coating property is one main influencing factor concerning the wear behavior of a coating system, and therefore significantly influencing the abradability and erosion resistance (Ref 6). Based on a first coating optimization and erosion, aging, and abradability tests, a hardness range between 25 and 45 HR 15Y is required to reach an adequate coating performance. Arc current and N_2 flow were found to have the most pronounced influence on coating hardness, and hence those two parameters were varied systematically. To evaluate an optimal powder feed rate, it was included in the DoE as well. Fixed parameters were argon flow, standoff distance, and robot velocity. The varied process parameters are shown in Table 1. The samples produced within the DOE were characterized concerning their hardness HR 15Y and microstructure.

Table 1 Varied parameter values for the DoE

Factor name	Lowest value	Highest value
A Current, A	280	320
B Nitrogen, SLPM	3	7
C Powder, g/min	60	120

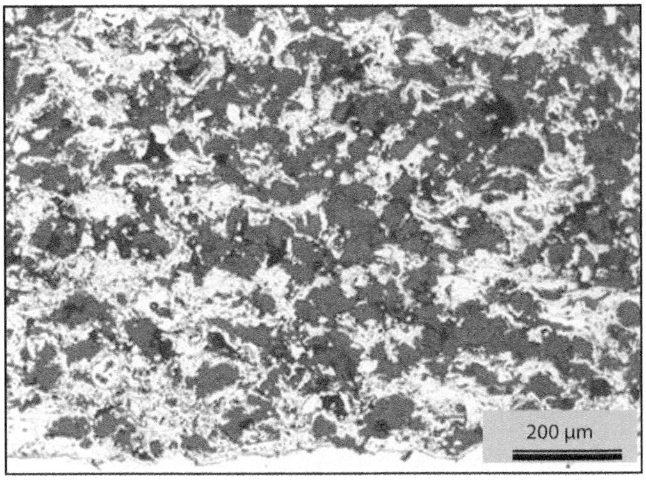

Fig. 1 Microscopic image (SM 2621, Ar: 45 SPLM, N2: 5 SLPM, current: 300 A, powder feed rate: 90 g/min, standoff distance: 180 mm)

3. Results

A microscopic image of the coatings microstructure, developed for further analysis of the correlations between coating characteristics and process parameters, is shown in Fig. 1.

A homogeneous distribution of the ceramic (hbN) and metallic (NiCrAl) components in the coating can be observed by the comparison of different coating sections by light microscopy.

The polyester content, deposited in the coating, was burnt out in a heat treatment, which resulted in a porous microstructure with a porosity, measured with optical methods, of 27% after heat treatment. Erosion tests showed a good erosion resistance and a low blade wear against titanium blades.

3.1 Deposition Efficiency

The highest deposition efficiency (up to 45%) can be observed using a powder feed rate of 30 g/min for each of the three powder feeders (90 g/min total). Using either 60 or 120 g/min total, the deposition efficiency decreases to values of 20-30%, independent of the spray parameters current and nitrogen.

3.2 Coating Properties Surface Roughness, Porosity, and Coating Hardness

The coating properties surface roughness and coating hardness as a response to the varied spray parameters are shown in Table 2. A statistical interpretation of the DoE shows that the strongest three parameters on the hardness of the coating in the order of importance are the nitrogen secondary plasma gas flow, the current, and an interaction of the nitrogen flow and the current (Fig. 2). Significant factors are those extending to the right of the red dotted line (significance limit).

In the DoE each spray parameter was repeated and the hardness results compared. In most cases, the hardness differences of the replicates are in an acceptable range of 0 to 3 HR 15Y; a good reproducibility of coating hardness for the same spray parameters could be observed. Only when using low nitrogen gas flows (3 SLPM) the hardness of the replicates shows differences up to 16 HR 15Y. The hardness of the coating can be controlled by changing the nitrogen flow and/or using higher current values. In general, the biggest changes in hardness can be achieved by using a low nitrogen flow and changing the current or using a low current and changing the nitrogen flow. For a smaller variation of hardness, a higher nitrogen flow and current should be used. To adjust the coating hardness, the current has been identified as a suitable parameter, as it does not affect the deposition efficiency.

Figure 3 shows the hardness development with changing current values. Increasing current values result in increasing coating hardness.

Regarding the surface roughness, no parameter shows a significant influence as shown in the pareto chart (Fig. 4). Surface roughness is primarily determined by the particle size of the thermal spray powder using the parameter window described here. Interaction plots show that the surface roughness cannot be controlled by selectively changing the main influencing spray parameters.

Table 2 Coating properties hardness and surface roughness in dependency on the spray parameters

Powder feed rate, g/min	N_2, SLPM	Current, A	Hardness, HR 15Y	R_a, µm
60	3	320	26	20.5
120	7	320	42	25.8
60	7	320	47	20.4
60	7	280	42	23.4
60	3	280	−31	22.1
90	5	300	34	24.0
120	3	280	−30	15.9
120	7	320	40	18.3
120	7	280	24	28.5
60	3	280	−15	18.9
120	3	320	10	25.7
60	7	280	40	22.8
120	3	320	8	17.4
60	7	320	45	16.9
90	5	300	35	20.6
120	7	280	31	16.5
120	3	280	−33	18.1
60	3	320	42	28.4

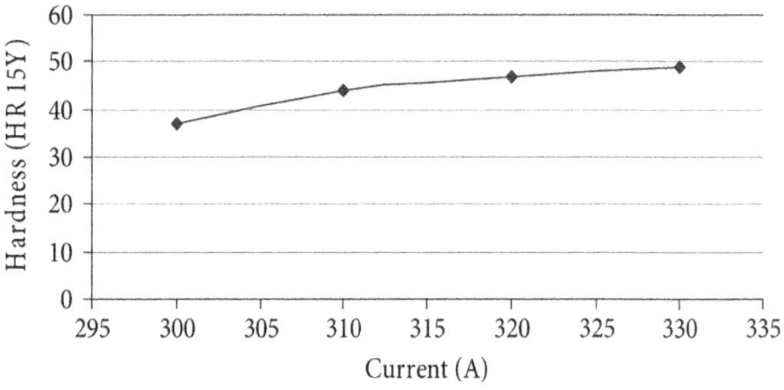

Fig. 3 Current influence on coating hardness

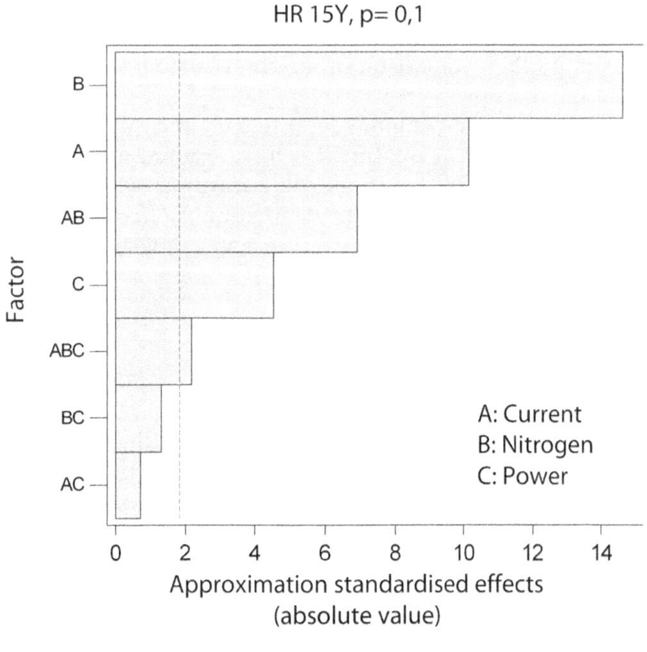

Fig. 2 Pareto chart of standardized effects in response to the hardness

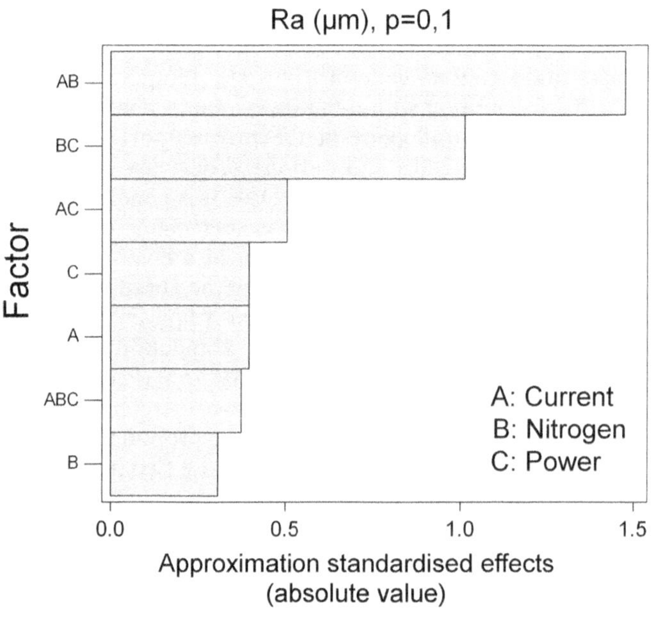

Fig. 4 Pareto chart of standardized effects in response to surface roughness

Before heat treatment, the measured coating porosity ranges at about 5 ± 2%. The measured porosity after heat treatment ranges between 19 and 35%, decreasing with increasing plasma power. This can be explained by a higher degree of heating of the spray powder material with increasing plasma power. This causes more polyester to burn off during the spray process, consequently causing less polyester to be deposited in the as-sprayed coating, which finally causes a lower level of coating porosity. The coating hardness can be correlated with the measured porosity. Higher porosity values lead to a lower coating hardness; samples in the required hardness range show a coating porosity between 23% (HV 15Y = 40) and 35% (HV 15Y = 25).

3.3 Particle Properties Temperature and Velocity

Particle property measurements with the DPV and Accuraspray only allow a measurement of hot (metallic/ceramic) particles, while cold particles (polyester compound) will not be detected. The measured particle temperatures and velocities (Accuraspray) are shown in Table 3.

The nitrogen flow followed by arc current can be observed as the process parameters having the largest influence on particle velocity and temperature. With a higher nitrogen gas flow, which is primarily responsible for heating up the powder particles, the particle temperature and the velocity increases, where the increased particle speed is mainly an effect of the increased total gas flow.

The average particle temperatures increase with increasing plasma power, either by using higher current values or higher nitrogen gas flows. The measured temperatures of the hot particles range from 1980 to 1500 °C. The statistical interpretation shows the significance of the parameters nitrogen followed by the current on the particle temperature. There is only a very weak interaction of the current and the nitrogen flow. Quite significant differences were also assessed for particle speeds. The measured particle speeds range from 76 to 111 m/s. Particle speeds increase with increasing gun power. Significant factors for particle velocity are the nitrogen flow and the current; their interactions do not show any effects on the particle velocity.

Table 3 Particle in-flight temperature and speed as a result of varied spray parameters

Powder feed rate, g/min	N_2, SLPM	Current, A	Particle temp., °C	Particle speed, m/s
60	3	320	1685	98
120	7	320	1870	107
60	7	320	1978	116
60	7	280	1820	108
60	3	280	1520	91
90	5	300	1794	108
120	3	280	1500	85
120	7	320	1980	111
120	7	280	1770	103
60	3	280	1580	93
120	3	320	1750	96
60	7	280	1850	103
120	3	320	1720	99
60	7	320	1940	110
90	5	300	1800	101
120	7	280	1790	98
120	3	280	1470	76
60	3	320	1800	102

With an increased particle velocity the particle temperature also increases and vice versa. This can be explained by the influence of a bigger current and a secondary plasma gas flow on higher plasma temperature and a plasma plume constriction that produces a faster, hotter plasma jet. As a result, particle velocity and temperature exhibit a linear correlation.

A strong correlation between the coating hardness and particle velocity and temperature and a reasonably good correlation between the particle velocity and temperature can be observed as shown in Fig. 5.

Monitoring particle temperature and velocity can be used for online process control, even for the multiphase material used in this work. Both increasing particle velocity and particle temperature cause an increase in coating hardness (Fig. 6 and 7).

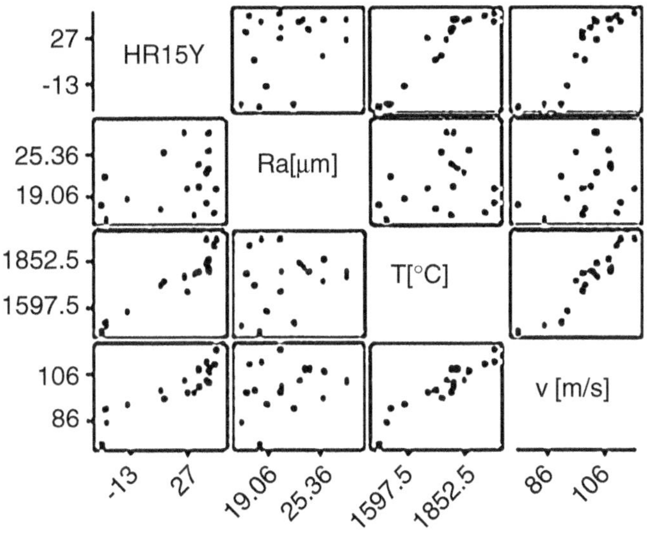

Fig. 5 Correlation of the responses within the DOE

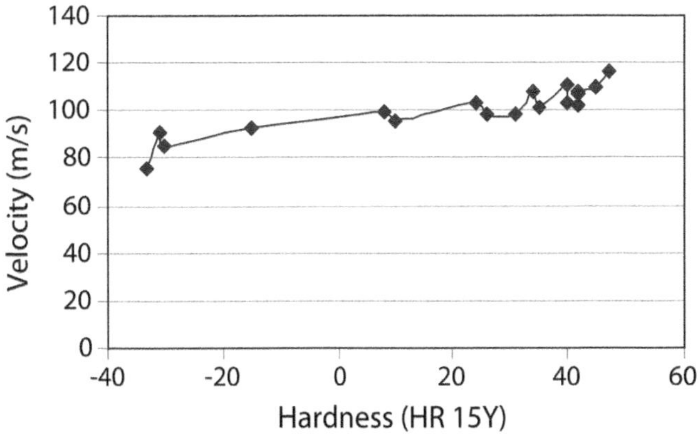

Fig. 6 Development of the hardness with particle in-flight velocity

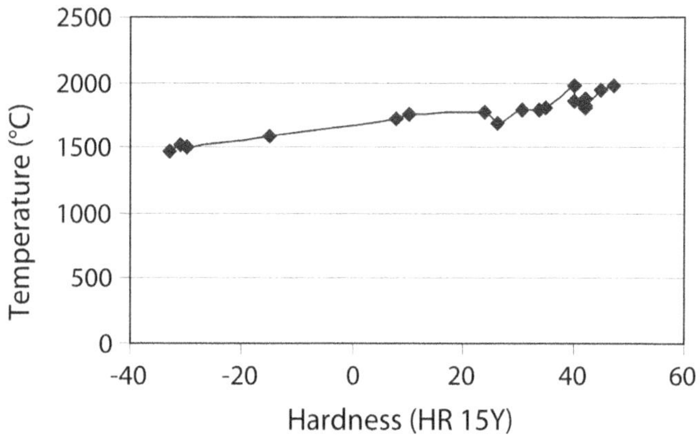

Fig. 7 Development of the hardness with particle in-flight temperature

4. Conclusions and Outlook

In this work, the relationship between coating hardness and spray parameters verify the statistical significance of the process parameters current and nitrogen gas flow. The observed influence of the nitrogen flow on the hardness is based on the influence of the secondary plasma gas flow on plasma temperature as well as on gas velocity. Increasing nitrogen flow leads to increased particle temperatures as nitrogen is mostly responsible for the heating of the powder material. Due to its high atomic weight, which favorably influences the pulse transmission, argon mainly affects the particle speeds. In addition, a strong influence of the current on particle properties, especially the particle temperature, can be observed. The influence of current is based on the higher power input and increasing temperatures of the plasma plume and gas jet, and corresponding effects on particle properties and coating hardness. In addition a higher temperature of the gas jet leads to higher hardness values as a result of enforced oxidation of the metal phase on one hand and a reduced content of polyester deposited in the coating on the other hand.

Concerning the powder feed rate, a optimal value of 90 g/min could be established. Lower or higher powder feed rates result in reduced deposition efficiencies.

The analysis of the correlation between diagnostic data and hardness reveals a significant influence of particle velocity and temperature on coating hardness for the parameter field studied. Nevertheless, a relationship between diagnostic data and hardness could be found that allows a prediction of resulting coating properties with a reasonable level of confidence.

References

1. K.J. Melcher and J. Kypuros, Toward a Fast-Response Active Turbine Tip Clearance Control, *International Symposium on Air Breathing Engines* (Cleveland, Ohio), 2003
2. A.M. Wallis, J.D. Denton, and A.A.J. Demargne, The Control of Shroud Leakage Flows to Reduce Aerodynamic Losses in a Low Aspect Ratio, Shrouded Axial Flow Turbine, *International Gas Turbine and Aeroengine Congress and Exhibition* (Munich, Germany), 2000
3. B.M. Steinetz, R.C. Hendricks, and J. Munson, Advanced Seal Technology Role in Meeting Next Generation Turbine Engine Goals, *Propulsion and Power Systems First Meeting on Design Principles and Methods for Aircraft Gas Turbine Engines* (Toulouse, France), 1998
4. S.B. Lattime and B.M. Steinetz, Turbine Engine Clearance Control Systems: Current Practices and Future Directions, *Joint Propulsion Conference and Exhibit* (Indianapolis, Indiana), 2002
5. D. Sporer, M. Dorfman, L. Xie, A. Refke, I. Giovannetti, and M. Giannozzi, Processing and Properties of Advanced Ceramic Abradable Coatings, *Proceedings of the International Thermal Spray Conference* (Beijing, China), 2007
6. K. Bobzin, E. Lugscheider, J. Zwick, F. Ernst, D. Sporer, N. Hopkins, M. Hertter, and J. Matejicek, Microstructure and Properties of New Abradable Seal Coatings for Compressor Applications, *Proceedings of the International Thermal Spray Conference* (Seattle, Washington, USA), 2006

JTTEE5 17:617–622
DOI: 10.1007/s11666-008-9267-0

LARGE—A Plasma Torch for Surface Chemistry Applications and CVD Processes—A Status Report

Stephan Zimmermann, Eckart Theophile, Klaus Landes, and Jochen Schein

(Submitted May 30, 2008; in revised form September 19, 2008)

The LARGE (*LONG ARG GE*NERATOR) is a new generation DC-plasma torch featuring an extended arc which is operated with a perpendicular gas flow to create a wide (up to 45 cm) plasma jet well suited for large area plasma processing. Using plasma diagnostic systems like high speed imaging, enthalpy probe, emission spectroscopy, and tomography, the LARGE produced plasma jet characteristics have been measured and sources of instability have been identified. With a simple model/simulation of the system LARGE III-150 and numerous experimental results, a new nozzle configuration and geometry (LARGE IV-150) has been designed, which produces a more homogenous plasma jet. These improvements enable the standard applications of the LARGE plasma torch (CVD coating process and surface activation process) to operate with higher efficiency.

Keywords advantages of TS, diagnostics and control, influence of process parameters, influence of spray parameters, plasma spray forming, stability of TS process, TS coating process

1. Introduction

Chemical vapor deposition (CVD) processes are promising paths toward more cost-effective technologies for wide area coatings on strips or sheets of steel, glass, polymeric web, etc.

In order for economic coating technologies to be compatible with industrial requirements, the following aspects need to be addressed (Ref 1, 2):

- Availability of scalable wide-area plasma sources
- Stability of plasma jets without fluctuations
- PE-CVD process for continuous air-to-air conditions
- Characterization of coatings/deposition rate

This article is an invited paper selected from presentations at the 2008 International Thermal Spray Conference and has been expanded from the original presentation. It is simultaneously published in *Thermal Spray Crossing Borders, Proceedings of the 2008 International Thermal Spray Conference*, Maastricht, The Netherlands, June 2-4, 2008, Basil R. Marple, Margaret M. Hyland, Yuk-Chiu Lau, Chang-Jiu Li, Rogerio S. Lima, and Ghislain Montavon, Ed., ASM International, Materials Park, OH, 2008.

Stephan Zimmermann, Eckart Theophile, Klaus Landes and **Jochen Schein**, Universität der Bundeswehr München IPM (LPT), 85577 Neubiberg, Germany; and **Eckart Theophile**, Reinhausen Plasma, Regensburg, Germany. Contact e-mail: Stephan.Zimmermann@UniBw.de.

The DC plasma source LARGE (Ref 3-6) was developed and evaluated for continuous plasma-enhanced (PE) CVD under atmospheric pressure. The LARGE is characterized by two rod-shaped electrodes arranged opposite to each other on a common axis with adjustable distance (up to 45 cm) (Fig. 1).

The name of the linear DC arc jet plasma source LARGE stands for *LONG ARG GE*NERATOR. Added to the name is usually a code where the first part indicates the development/application number and the second part the length of the plasma sheet in mm (e.g. LARGE III-150 = version 3, 150 mm length). Table 1 shows a detailed overview of the basic parameters used to operate LARGE III-150 as a standard module. Using multiple modules, a LARGE III-300 or LARGE III-450 can be constructed.

Different from conventional torch systems the plasma gas is injected laterally to the linear electrical arc stretching along the torch axis. Passing through the arc, the gas transferred into the plasma state leaves the torch as a flat homogenous plasma flow. Shrouding the electrodes with inert gases allows to use any desired plasma gas composition in order to set plasma specific parameters, as temperature, radical, or ion densities. Moreover, plasma chemical reactions are initiated. Conventional DC plasma generators are built by conical cathode and concentrically hollow anode. Because of the limited spot size of the plasma jet, the treatment of large surfaces or band materials requires a scanning process of the torch. The goal for the LARGE development was to have access to a flat, homogenous plasma sheet in order to enable plasma chemistry for large surface technologies and to minimize any non-uniformity of the treatment (Fig. 2). In this case, the arc is ignited between these two electrodes while plasma gas is fed through axially, resulting in an emanating plasma jet exhibiting a rectangular shape of distinct width, height, and length.

Fig. 1 Picture of LARGE III-150

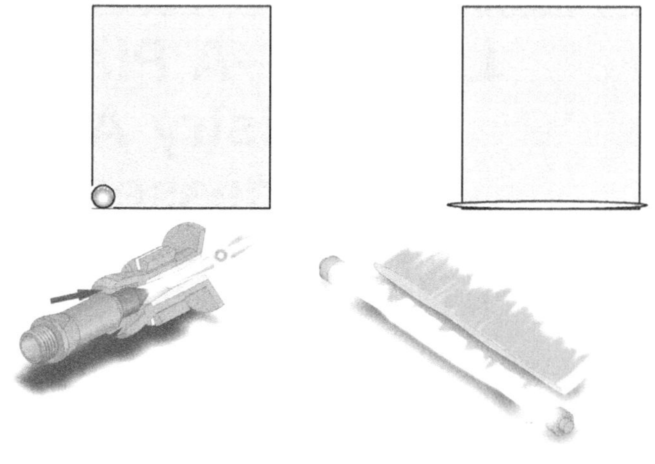

Fig. 2 Scanning treatment of planar surfaces by a conventional plasma torch and flat plasma geometry (LARGE)

Table 1 Basic parameters of LARGE III-150 and LARGE IV-150

Parameter	Range
Plasma gas	O_2, H_2, He, Ar, N_2, CH_2
Current	$I = 10\text{-}100$ A
Voltage	$U = 100\text{-}300$ V
Electrical power (linear power density)	1-30 kW (7-2 kW/cm)
Gas flow	10-100 slpm
Gas temperature at the exit of the torch	100-3500 °C

2. Principle and Operation of the Linear DC Arc Jet Plasma Source LARGE

The ignition of the arc over a long distance is crucial to the operation of the LARGE. The arc is initially ignited over a short distance and subsequently elongated step-by-step by opening of semiconductor switches (CIPASS—*C*ombined *I*gnition by *P*ilot *A*rcing and *S*uccessive *S*witching). The principle is shown in Fig. 3. All cascade plates can be connected through semiconductor switches to the anode potential of the power supply (Ref 5, 6). Having closed all switches the controller starts the ignition process of a high-voltage power supply in order to ignite a pilot arc between the cathode and the first cascade plate. Subsequent opening of the switches forces the anodic arc root to move step-by-step toward the anode. In this way, controlled by the CIPASS system, the arc is elongated until it reaches the anode, completing the ignition process. The DC power supply (TOPCON) has been developed for operation of the LARGE. The voltage and power range of up to several hundreds of kW, together with a fast response, makes this power supply a suitable tool for plasma applications.

The plasma gas is injected perpendicular to the torch axis. Passing through the arc, the gas is transferred to the plasma state and leaves the torch laterally through a flat slit. A main characteristic of the LARGE torch is the innovative form to stabilize the long plasma arc. In the case without a cross-flow of cold gas, the arc is

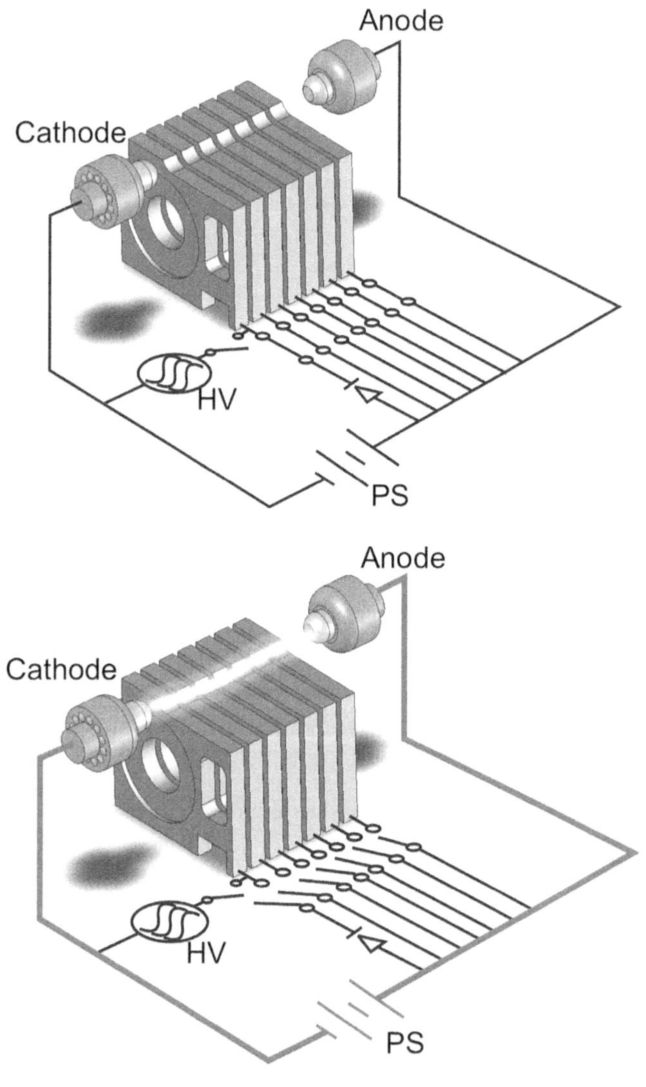

Fig. 3 Principle of the CIPASS method, before and after LARGE ignition (HV, high voltage power supply; PS, power supply)

wall-stabilized: according to the Steenbeck's principle, a transferred arc adjusts itself to a configuration of minimum power (Ref 7). For the LARGE torch, consisting of a straight arc connecting both electrodes, the minimum power configuration corresponds to a straight arc line. The additional cross flow of the cold plasma gas produces, due to the viscous force, a global bending of the arc downstream, curvature which has to be compensated. This compensation is achieved by the Lorentz force (Ref 8, 9) created by a the interaction of the electric current of the plasma arc to the strong magnetic field produced by two long permanent magnets located above and below the plasma arc (Fig. 4).

The torch system LARGE can be operated under low pressure as well as under atmospheric pressure conditions. The torch is water cooled and has a thermal efficiency of about 40-60% depending on the operating parameters. The torch behavior is controlled by two different gas flows: the plasma gas and additionally a gas flow surrounding the electrodes to ensure a fixed arc attachment point. As electrode gas typically inert gas flows of 2 × 1.5 slpm argon shroud the cathode and the anode. This shrouding of the electrodes permits the use of almost any desired gas mixture as plasma gas as any oxidation of the electrodes can be avoided. Actual plasma gas flows (Ar, H_2, N_2, O_2, CO_2,CH_4,...) are in the range from 0.5 to 5 slpm per centimeter plasma jet width. The utilization of chemical reactive gases, which are partly transferred to radicals, ions, and molecule fragments, allows the torch to be operated in CVD and surface treatment processes.

3. Experimental Diagnostics

Electrical, acoustical, magnetical, and optical measurements (Ref 8, 10) have been carried out on the system LARGE to elucidate the operating parameter range and to clarify the specific plasma jet properties.

As a preliminary diagnostic, the distribution of the magnetic field produced by the long magnets at the torch outlet was measured by means of a Tesla probe (Fig. 4). Characteristic for this distribution is the rapid decrease in the magnetic strength at locations near the cathode/anode.

To qualify the plasma source LARGE, information about the homogeneity of the broad plasma jet is necessary. Therefore, measurements of the integral light emission were taken as indicator. A CCD camera is mounted to look straight down to the plasma jet. The video data is transmitted to a personal computer to analyze the plasma jet contour. If the plasma jet is deviating from the desired contour, gas flow controlling valves can adjust the flow rate. A periodic variation of the integral light emission of the plasma jet is observed (Fig. 7, above) by the gas

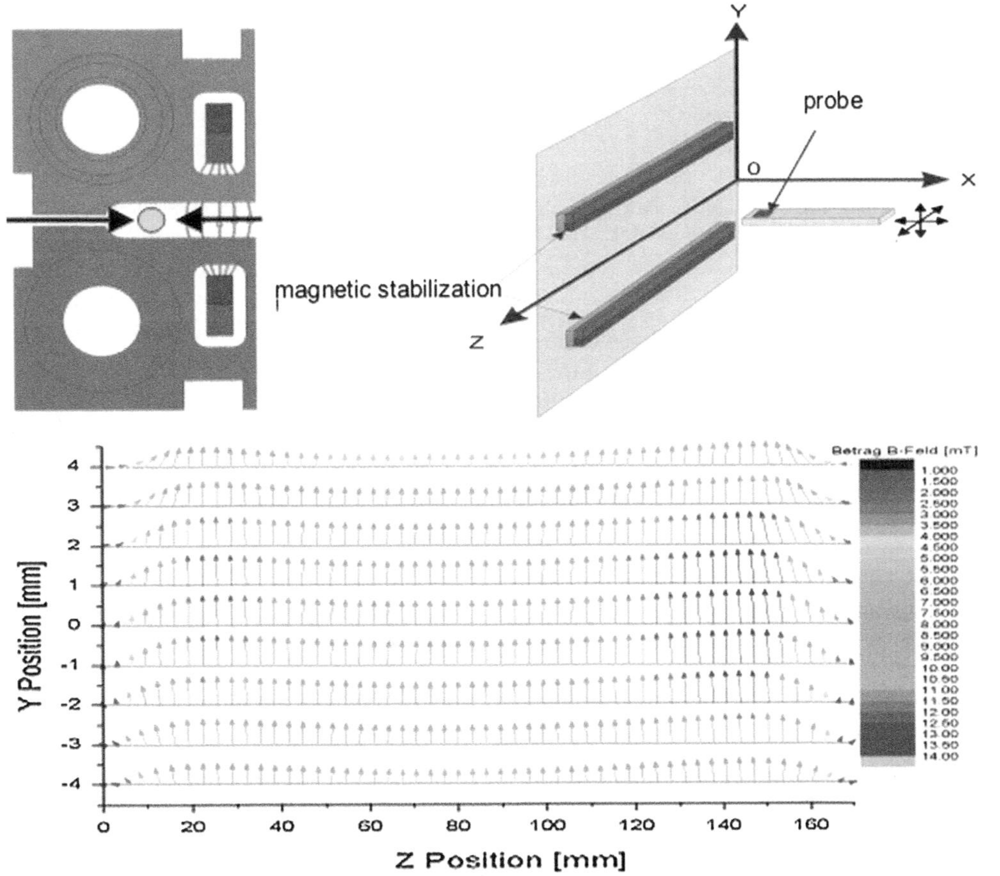

Fig. 4 Principle of arc stabilization and dependence on temperature of conductivity and tesla-probe measurement of the magnetic field B_y und B_z in a distance of 14 mm

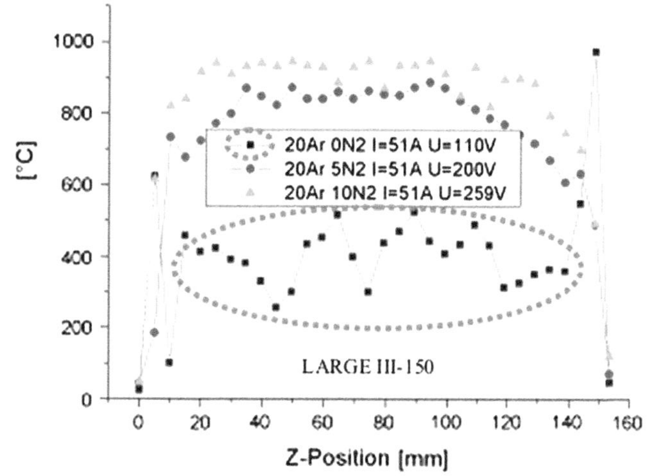

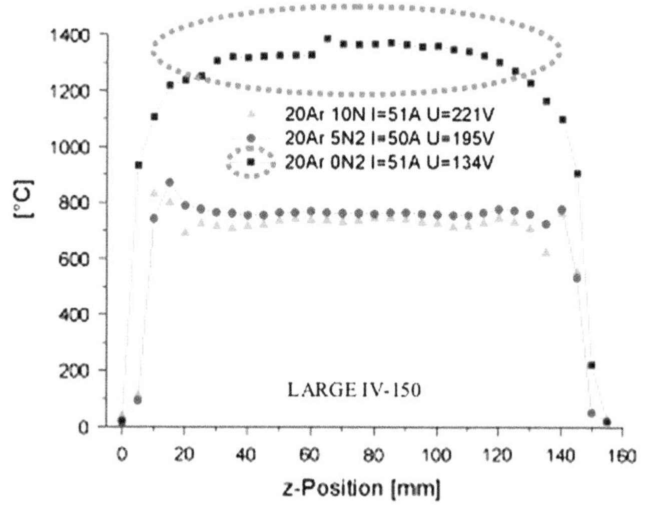

Fig. 5 Distribution of temperature measured with enthalpy probe at a location 4 mm downstream for LARGE III-150 (above) and LARGE IV-150 (below)

injection through discrete ports. An improved gas divider is already in production which is supposed to reduce these fluctuations. The new gas divider will inject the plasma gas as linear constant sheet overcoming the constraints of discrete gas injection. Additionally, plasma jet temperatures and plasma jet velocities were measured using an enthalpy probe (Tekna-ENP 04 in combination with probe type ENP-476). In comparison to the imaging technique mentioned above, which produces a 2D image, the enthalpy probe is a 3D spatially resolving method. As the plasma source LARGE generally makes use of gas mixtures, a quadruple mass spectrometer (Balzers QMG311) was used to determine the gas composition. However, the overall length of the extraction system impedes the determination of radical concentrations, e.g. of ozone in oxygen containing gas mixtures, due to rapid chemical reactions at atmospheric pressure. These concentrations have to be measured with spatially resolving optical techniques. The resulting temperature distribution can be seen in Fig. 5, and detailed results can be found in Ref 3. This topic will be the subject to a latter publication comparing enthalpy probe measurements to spectroscopic investigations. So far only qualitative information about the optical emission spectrum (OES) has been gathered (Ref 8, 11, 12). The optical emission spectroscopy signal was recorded by an optical cable to a SpectraPro-0.275 Meter Digital Scanning Mono-chromator (Acton Research Corporation). The OES spectrum of a pure Ar plasma is shown in Fig. 6.

4. Modeling and Simulation of the LARGE

By means of the software package ANSYS, a simple model was developed to calculate the flow without interaction to the magnetic field and assuming a laminar flow (Ref 8, 10, 13). The objective is to evaluate the fluctuations and instabilities of the produced plasma jet as influenced by the nozzle geometry. The goal was to design a new nozzle to produce a more homogenous plasma jet (Fig. 7).

The velocity distribution of the cold gas flow (LARGE III-150) was calculated by using software ANSYS. In Fig. 7 (middle), the momentum of the cool gas flow during its travel through the plasma canal is depicted for the LARGE III-150. The resulting velocity distribution displays a well pronounced vortex previous to the torch outlet. This effect may be responsible for the inhomogeneous plasma flow issuing from the torch, just as experimentally measured in the temperature distribution by means of the enthalpy probe (Fig. 5, above). In order to reduce this effect, a new modification of the cascade plates has been modeled. Herein, the arc is constricted into a small semicircular cavity in the upper cascade plates and the gas flow is pushed into this cavity through a slight elevation in the lower cascade plates (Fig. 7 and 8, below). This cascade plate modification corresponds to the improved model LARGE IV-150. This new design shows improved behavior for continuous wide-area PE CVD processing (Ref 1, 3) through better stability, even though a lower jet temperature has to be taken into account. In Fig. 8, examples of the gas flow profile, at different locations and with special nozzle geometries, are depicted. The LARGE IV-150 source featuring the new nozzle design limited the fluctuations significantly at the exit (measurement location 2).

5. Results and Conclusions

In the torch system LARGE, the arrangement of the anode and the cathode opposite to each other on a common axis leads to a substantially different plasma jet compared to conventional torch systems. Utilizing this flat, broad plasma jet brings advantages particularly if large-scale workpieces have to be treated. By shrouding the electrodes with inert gases, reactive plasma gases can be used for plasma chemical reactions in the central part of the torch. Electrical measurements show the stability of the discharge and constant power consumption. Noise

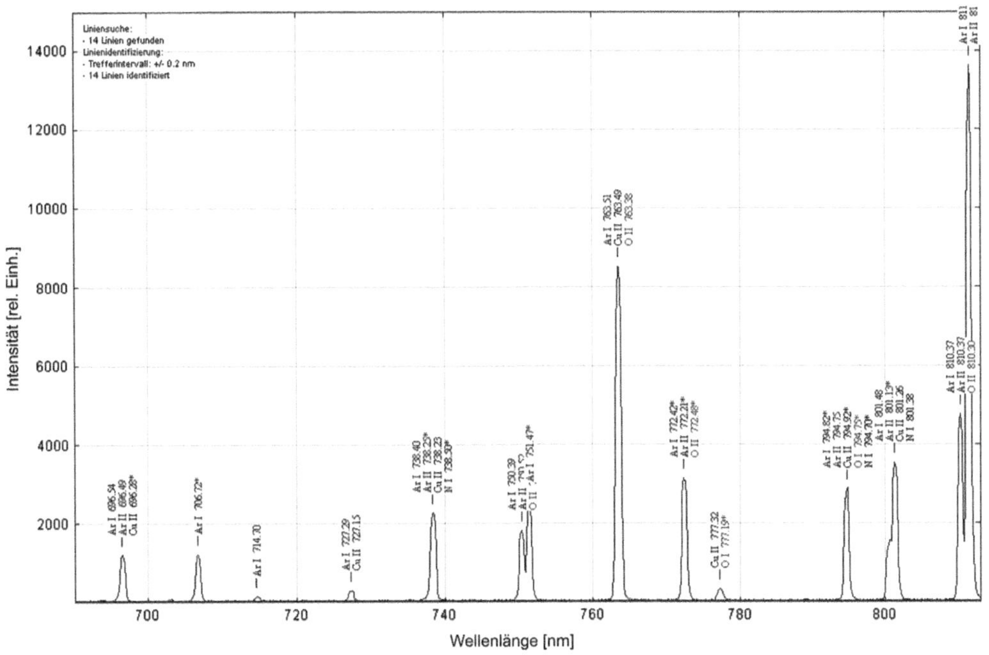

Fig. 6 Emission spectroscopy investigation of LARGE III-150, 20 slpm Ar, $I = 50$ A

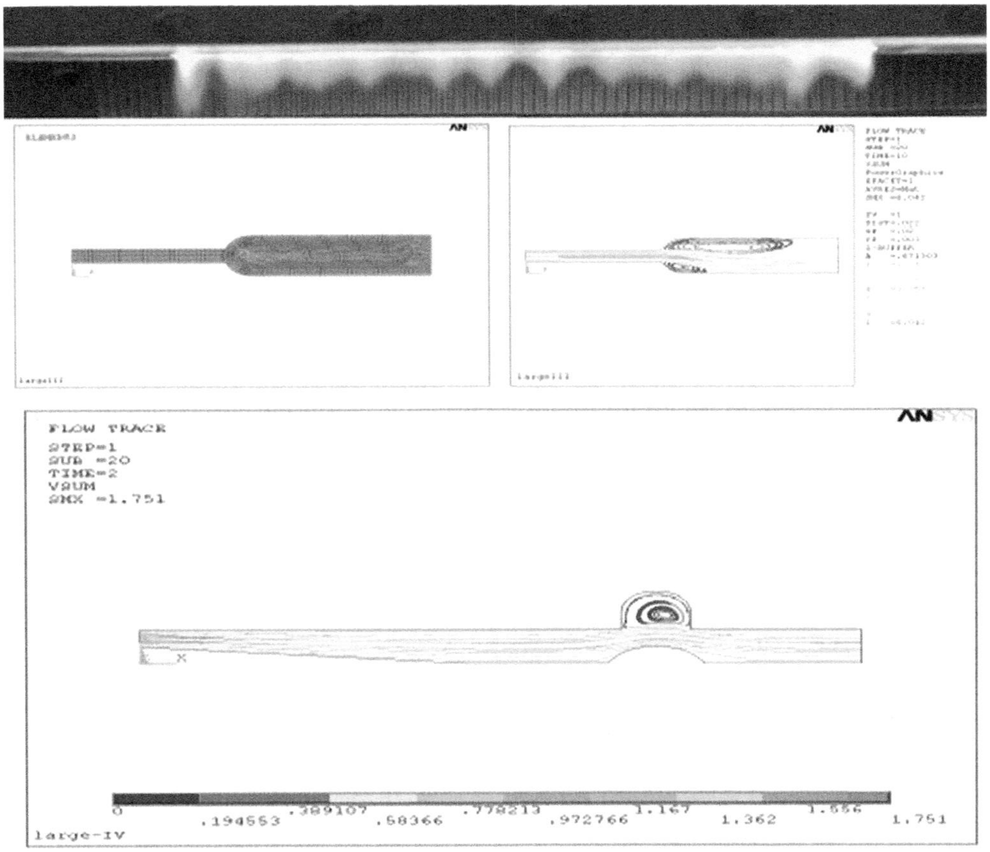

Fig. 7 Fluctuations and instabilities of plasma jet measured with CCD camera, LARGE III-150 (above). Simulated velocity distribution in LARGE III-150 (middle) and LARGE IV-150 (below)

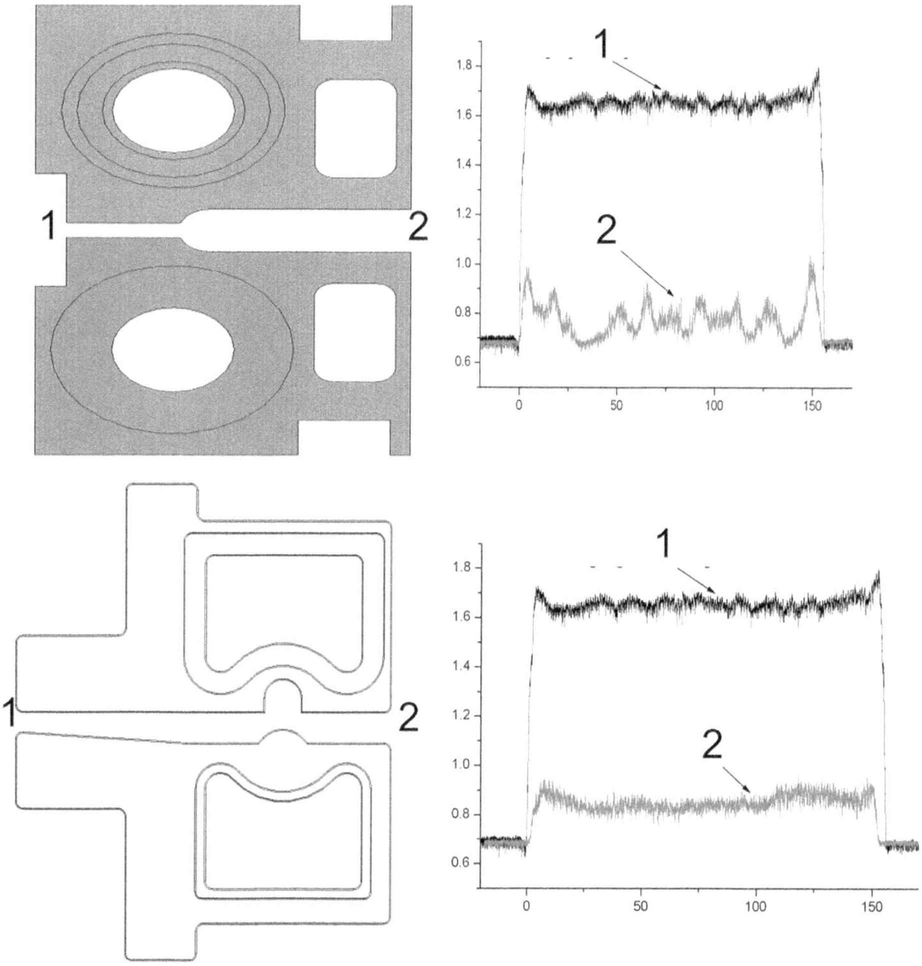

Fig. 8 LARGE III-150 geometry and velocity of gas flow on different measurement locations 1/2, and LARGE IV-150 geometry and velocity of gas flow on different measurement locations 1/2

levels of the torch are remarkably low (<80 dB(A)). Optical and enthalpy probe measurements show fluctuations of the plasma jet, which appear to be correlated to the plasma gas injection through discrete ports. An improved design of the gas divider with a homogeneous plasma gas injection system should provide better stability.

The potential application range of the newly developed Arc Jet-CVD technology was evaluated by screening studies on several substrates. Plasma pretreatment was found to be very efficient for improving layer adhesion (Ref 1, 2). More optical and enthalpy probe measurements as well as modeling will be used to find out the optimum nozzle geometry for present and future applications.

References

1. V. Hopfe, D. Rogler, G. Maeder, I. Dani, K. Landes, E. Theophile, M. Dzulko, C. Rohrer, and C. Reichhold, Linear Extended ArcJet CVD – A New PECVD Approach for Continuous Wide Area Coating Under Atmospheric Pressure, *Chem. Vapor Depos.*, 2005, **11**, p S.510-S.522
2. M. Koppers and E. Theophile, umweltgerechtes Verfahren zur Folienbehandlung und Folien-kaschierung im Atmosphärenplasma, BMBF – gefördertes Projekt, öffentlicher Endbericht, 2006
3. E. Theophile, "Entwicklung eines Plasmagenerators mit atmosphärischem DC-Langlichtbogen zur Modifikation von Werkstoff-oberflächen," Dissertation UniBw München, 2006
4. R. Hartmann, G. Kraut, and K. Landes, CIPASS—An Ignition Method for the Innovative Plasma Torch "LARGE", *Eur. Phys. J. Appl. Phys.*, 1999, **8**, p 253
5. R. Hartmann, J. Zierhut, and K. Landes, European Patent 97810823, 1997
6. TOPCON, http://www.regatron.ch/
7. Th. Peters, Über den Zusammenhang des Steenbeckschen Minimumprinzips mit dem thermodynamischen Prinzip der minimalen Entropieerzeugung, *Zeitschrift für Physik*, 1956, **144**, p 612-631
8. R. Huddlestone and S. Leonard, *Plasma Diagnostic Techniques*, Academic Press, New York, San Francisco, London, 1965
9. H. Rosenbauer, "Messung der durch ein Quermagnetfeld erzeugten Gasströmung in einem wandstabilisierten Niederstrombogen," Dissertation, Technische Universität München, 1970
10. M. Boulos, P. Fauchais, and E. Pfender, *Thermal Plasmas – Fundamentals and Applications*
11. R. Payling and P. Larkins, *Optical Emission Lines of the Elements*, Wiley Verlag, Chichester, 2000
12. A. Capetti and E. Pfender, Probe Measurements in Argon Plasma Jets Operated in Ambient Argon, *Plasma Chem. Plasma Process.*, 1989, **9**(2), p 329-341
13. T. Drabek, Numerische Simulation der turbulenten Strömung in ebenen Diffusoren, Diplomarbeit, TU Wien, 2001

JTTEE5 17:623–630
DOI: 10.1007/s11666-008-9221-1
1059-9630/$19.00 © ASM International

Peer Reviewed

In Situ Visualization of Impacting Phenomena of Plasma-Sprayed Zirconia: From Single Splat to Coating Formation

Kentaro Shinoda, Hideyuki Murakami, Seiji Kuroda, Kohsei Takehara, and Sachio Oki

(Submitted May 13, 2008; in revised form August 26, 2008)

The authors have developed an in situ monitoring system for particle impacts under atmospheric dc plasma spraying conditions. This system utilized a high-speed video camera coupled with a long-distance microscope, and was capable of capturing the particle-impinging phenomena at one million frames per second. To understand the coating formation mechanism, two approaches were attempted, i.e., observation of the single splat formation and the subsequent coating formation. In the former case, the deformation and cooling processes of yttria-stabilized zirconia (YSZ) droplets impinging on substrates were successfully captured. In the latter case, multiple-droplet-impacting phenomena were observed as an ensemble treatment. Representing the coating process, the tower formation (0-dimensional) and bead formation (1-dimensional) were observed under typical plasma spray conditions for thermal barrier coatings using a triggering system coupled with the motion of a robot. The obtained images clearly showed the coating formation resulting from the integration of single splats.

Keywords coatings for gas turbine components, diagnostics and control, influence of spray parameters, TBC topcoats

1. Introduction

Plasma spraying of yttria-stabilized zirconia (YSZ) has been widely used for thermal barrier coatings (TBC) on turbine blades. It can also be applied to solid oxide fuel cells (Ref 1, 2). It is important to understand the fundamental formation mechanism of plasma-sprayed coatings, which can be roughly classified into five regions: plasma jet formation, powder injection, particle heating and acceleration, single particle impact accompanying spreading and solidification, and coating formation by successive impact of particles. Based on the three-dimensional transient modeling work of arc root fluctuation (Ref 3) and in-flight particle diagnostics (Ref 4, 5), the three upstream processes are now well understood. On the contrary, the subsequent two phenomena, i.e., single splat and coating formations, have not been fully understood, especially due to the experimental difficulty. As a result, the process parameters could be related to the coating properties, e.g., adhesion strength, porosity, and crack density, only by the design of experiments such as the Taguchi method (Ref 6), leaving the impact processes in the state of black boxes. Although the Taguchi method itself is a strong tool for optimizing coating properties in industry, and although recent development in on-line control of spray processes enables the advanced evaluation utilizing in-flight particle characteristics, such as velocity and temperature, these parameters could be only optimized within the limitation of conventional plasma spray processes. In other words, if we can elucidate the splat formation and subsequent coating formation, we can reconstruct the plasma spray process from phenomenological aspects, which will then give us new insights into plasma spray technologies.

To observe these two processes, we need to estimate the characteristic time scale of the plasma spray process. As Vardelle and Fauchais summarized (Ref 7, 8), for a single splat formation, flattening and splashing occur within a few microseconds, followed by solidification within 10 μs; two successive impacts are on the order of 10-100 μs; one layer/pass, which is the unit of coating, is on the order of ms. Thus, the measurement system with the time resolution of less than 1 μs will be required to capture the single splat formation, while a system that can record the event for ms for the coating formation is needed.

This article is an invited paper selected from presentations at the 2008 International Thermal Spray Conference and has been expanded from the original presentation. It is simultaneously published in *Thermal Spray Crossing Borders, Proceedings of the 2008 International Thermal Spray Conference,* Maastricht, The Netherlands, June 2-4, 2008, Basil R. Marple, Margaret M. Hyland, Yuk-Chiu Lau, Chang-Jiu Li, Rogerio S. Lima, and Ghislain Montavon, Ed., ASM International, Materials Park, OH, 2008.

Kentaro Shinoda, Hideyuki Murakami, and **Seiji Kuroda**, Coating Materials Group, Composites and Coatings Center, National Institute for Materials Science (NIMS), 1-2-1 Sengen, Tsukuba, Ibaraki 305-0047, Japan; and **Kohsei Takehara** and **Sachio Oki**, School of Science and Engineering, Kinki University, 3-4-1 Kowakae, Higashi Osaka, Osaka 577-8502, Japan. Contact e-mail: shinoda.kentaro@nims.go.jp.

Meanwhile, Etoh et al. developed an image sensor for a high-speed video camera (HSV) based on the concept of "ISIS" (in situ storage image sensor) (Ref 9). It enables us to capture 100 consecutive images at 1 million frames per second (fps) with the exposure time of 250 ns. Recent progress in charge-coupled device ultrahigh-speed video cameras were reviewed by Thoroddsen et al. (Ref 10). This HSV may enable us to observe not only a single splat formation, but also the subsequent coating formation.

In this context, we have developed an in situ monitoring system utilizing the HSV of 1 million fps to capture the impacting phenomena during atmospheric dc plasma spraying (APS) (Ref 11). First, the impact phenomena of an YSZ droplet were observed by coupling the HSV with a long-distance microscope. The tower and bead formations were then observed using a triggering system coupled to the motion of a dc plasma torch to demonstrate the TBC formation.

Table 1 Experimental conditions at (a) single splat, (b) tower, and (c) bead formations; (d) reference conditions

Parameters	a, b, c	d(a)
Current, A	800	800-1200
Voltage, V	32	32
Primary gas, Ar, SLM(b)	50	50
Secondary gas, He, SLM	13	13
Powder feeding rate, g/min	3, 7, 14	20
Carrier gas, Ar, SLM	8.5	8.5
Spray distance, mm	100(c)	60-100
Powder size, μm	−53, −53, −75/45	−120/20
Preheating temperature, K	300	643-973
Traverse speed, mm/s	200, 0, 50	150-200
Grit blasting	No, Yes, Yes	Yes

(a) Guo et al. (Ref 15, 16); (b) SLM: standard liters per minute; and (c) a shield was inserted between the torch and the substrate

2. Experimental

2.1 Single Splat Formation

Detailed descriptions and a schematic diagram of the developed in situ monitoring system are shown elsewhere (Ref 11, 12). A conventional dc plasma torch with a 40 kW subsonic anode (SG-100 with type 2083-165 anode, TAFA, Inc., A Praxair Surface Technologies, Inc.) was used as the APS system. A single particle was collected using a V-shaped roof-type shield with a hole and a single particle collector (Ref 13, 14).

An HSV (HPV-1, Shimadzu Corp., Japan) coupled with a long-distance microscope (QM100, Questar Corp.) was utilized to monitor the droplet impact. The pixel count and gray levels of the HSV at 1 million fps are 81,120 (= 312 in width × 260 in height) and 10 bits, respectively. The working distance and the angle between the substrate and the microscope were 11.4 cm and 31.4 degrees, respectively. Other optics to collect the thermal radiation history during impact were utilized as a trigger source for the HSV.

The experimental conditions are shown in Table 1(a). Hollow-oven spherical powder (HOSP) of 8-wt.%YSZ (204NS, Sulzer Metco Inc.) sieved to 45-53 μm was sprayed onto a quartz glass plate with the thickness of 0.5 mm. The substrate surface temperature was kept at 300 K during the experiments.

2.2 Coating Formation

We have tried to observe the formation of a tower by fixing a plasma torch (0 dimension, 0D), and the formation of a bead by scanning the plasma torch on a line (1 dimension, 1D). Experimental conditions for the 0D and 1D observations are shown in Table 1(b) and (c), respectively. The HOSP of the 8-wt.%YSZ (204NS sieved to 45-53 μm for 0D and 204B-NS, Sulzer Metco, for 1D measurements, respectively) were sprayed onto grit-blasted steel substrates (50 × 100 × 5 mm) with the dc plasma torch. The experimental conditions were designed so that they were identical to those under which the segment TBC could be formed, as shown in Table 1(d) (Ref 15, 16) except for the substrate temperature.

Prior to the coating formation observations, the flow rate, velocity, temperature, and size distributions of the in-flight particles were measured by an in-flight particle monitoring system (DPV2000, Tecnar Automation Ltd., Canada). The spray conditions were the same as the 1D measurement shown in Table 1(c).

For the 0D experiment, capturing of the images was conducted at arbitrary times. For the 1D experiment, a robot was programmed to scan the dc torch 10 times back and forth in the straight line, in which one scan of 50 mm length took 7 s including the wait time. A simple triggering system was constructed to trigger the HSV linked to the motion of the robot. The trigger timing was adjusted so that the HSV could record the impinging phenomena when the flow rate of the particles impinging on the center of the field of view (FOV) of the HSV became a maximum value, based on the result of the in-flight particle diagnostics. The substrate temperature was monitored using a k-type thermocouple inserted in the middle of the substrate.

3. Results

3.1 Single Splat Formation

Figure 1 shows the deformation and cooling processes of a droplet impinging on a substrate taken by the HSV. Each frame interval t_I was 1 μs and the exposure time t_E was 500 ns. From the images, this droplet was evaluated to be 50 μm in diameter and impacted on the substrate at a speed of 170 m/s. First, a sprayed droplet entered the FOV of the HSV from right to left in frame 1, and impinged on the substrate in frame 2. As soon as the droplet impinged on the substrate, a liquid sheet jetted out from the droplet to form a splat. At this stage, the splat was disk-shaped, although a part of the periphery began to be notched as

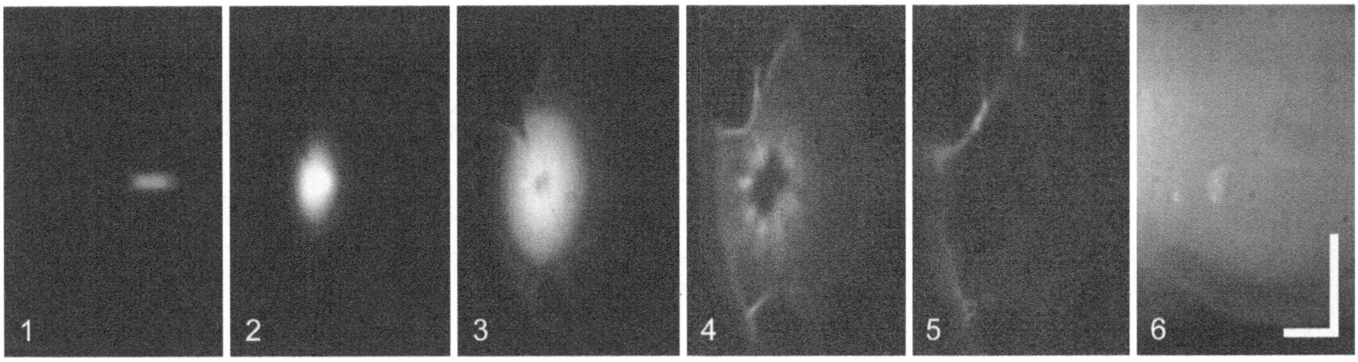

Fig. 1 Thermal radiation images of a plasma-sprayed YSZ droplet impacting on a smooth cold quartz glass substrate (frames 1 to 5). Interval between each frame is 1 μs and exposure time of each frame is 500 ns. The resultant splat is shown in frame 6. Contrasts of the images were adjusted. The white scale bar indicates 200 μm in each direction on a substrate. Frames 1 to 5 were adapted with permission from Shinoda et al. (Ref 11)

observed in frame 2. The spreading sheet then started to slightly warp while the notch at the periphery was developing in frame 3. The center region of the splat already became darker than the surroundings, suggesting that this region experienced a more rapid cooling. In frame 4, the sheet became detached from the substrate and was still growing without disintegration. Detachment of the sheet can be judged by the asymmetry of the deforming splat on the both sides of the vertical centerline of the splat. Since the substrate was inclined against the HSV, the radius of the left hand side of the splat looks smaller than that on the right when detachment of the periphery occurs. Moreover, the rim could not maintain the disk shape, which indicates that the rim was no longer stable. The center part of the splat was far below the sensitivity of the HSV, which meant that it had already cooled at this stage. High-temperature regions were nonuniformly distributed along off-center. The detached sheet then began to rupture in frame 5. After rupture of the splat, only the central region was attached to the substrate as shown in frame 6.

3.2 In-Flight Particle Diagnostics

Figure 2 shows the in-flight particle distributions obtained by the in-flight particle monitoring system. The regions of 11 × 35 points with a 5 mm interval were measured. The measurement time for each point was 4 s. The trace of the maximum particle flow rate R region was inclined to the x-axis; i.e., the maximum R was leaving from the center with increasing the axial position x, suggesting that the injected powder penetrated the plasma jet and deviated from the central torch axis as was expected. The deviation angle off the x-axis, which was derived from the linear approximation, was 10.7 degrees. In the same manner, deviation angles of the particle temperature T, velocity V, and diameter d were found to be 7.6, 6.7, and 11.9 degrees, respectively; i.e., the deviation became larger in the order of V, T, R, and d.

The radial position of the maximum flow rate zone, y, at $x = 100$ mm, was −15 mm. Therefore, when the torch was scanned in the direction of minus y in the 1D observation, the HSV, which was focused at $x = 100$ mm, was triggered when the center axis of the torch reached 15 mm before the center of the FOV.

3.3 Coating Formation

Figure 3 shows the appearance of the obtained samples. When fixing the torch (0D), as was expected, a tower was formed. This tower was sprayed at the powder feed rate of 7 g/min for 186 s. The deposition region was 36 mm in diameter and the height of the tower was 21 mm. For a scanning torch (1D), a bead was formed. At the spray distance of 100 mm, the width and height of the bead were 19 mm and 607 μm, respectively. The reproducibility of the bead formation was relatively good so that the averaged bead width and thickness were approximately 20 mm and 600 μm for six time trials, respectively. This corresponds to the fact that the average thickness of 30 μm was deposited during each single pass. The substrate temperature during the 1D experiment increased from room temperature to 550 K during the bead formation.

Figure 4 shows the formation of a tower when fixing the torch (0D) as an example of the in situ observations. The tower was formed by the impact and deposition of particles. During the initial stage of the tower generation (a), the impacted particles isotropically splashed, because the impacted surface was flat. At (b), anisotropic particle splashing occurred, because the impacted plane formed a gradient according to the formation of the tower. At stage (c), the tower further grew and the top of the tower was out of the FOV. At any stage, the degree of splashing was greater at the center part of the tower, while the frequency of splashing decreased at the periphery of the tower. At the periphery, secondary particles were ejected and slowly flew in the radial direction. This can be the rebound phenomena of solid particles rather than the splashing of the liquids.

Bead formation was observed while scanning the torch on a line (1D). Since the scanning speed of the plasma torch was 50 mm/s, the motion of the torch during the video capturing was considered to be negligible. There were no obvious visual differences between the initial and

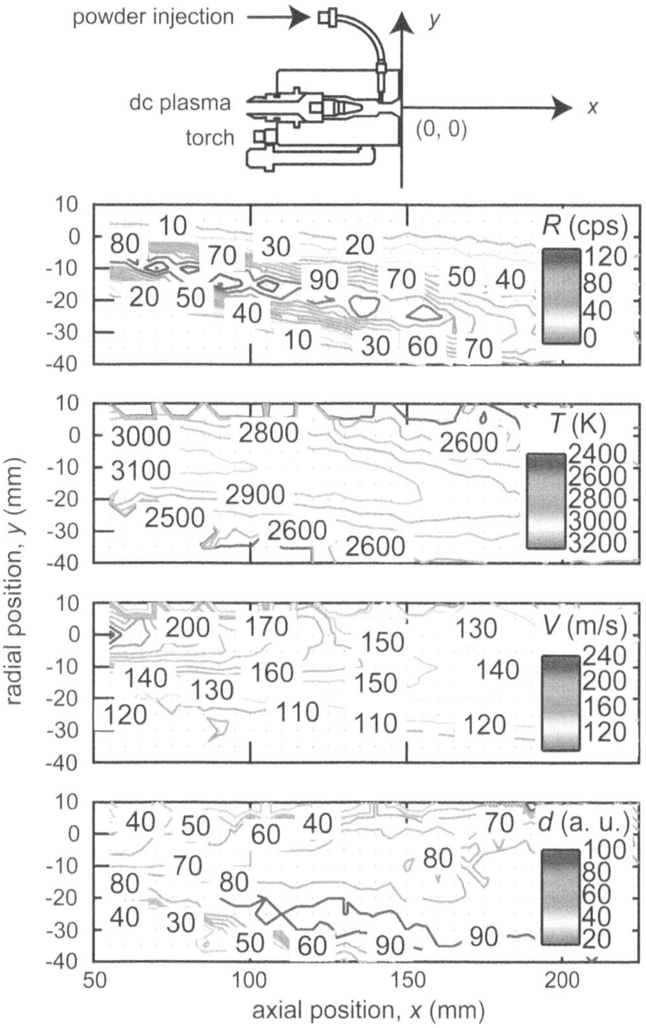

Fig. 2 (color online) In-flight particle distributions concerning flow rate R (counts/s (cps)), temperature T (K), velocity V (m/s), and diameter d (arbitrary unit (a.u.)) on the axis of a dc plasma torch—that of particle injection port plane. The axial position x (mm) is the distance from the head of the torch in the axial direction, and the radial position y (mm) is the distance from the center of the torch perpendicular to the torch axis. The powder was injected from the plus side of the torch. The unit of d is indicated not in meters but by a.u. due to the difficulty in calibration. The value of R was derived from the number of particles that passed the measurement volume of an in-flight particle monitoring system divided by the measurement time

latter stages of the deposition. Hence, a detailed analysis of captured video was conducted. In this study, we showed the analyzed results for the first pass as the first step. Every particle that impacted the center region along the y-axis of the FOV in focus was captured, and the trajectory of the particle was tracked. From the trajectory, the velocity and impact angle to a substrate were derived as a function of the impact position in the direction of y. Figure 5 shows the relationship between these parameters. In the direction of the powder injection (y-axis), the velocity widely varied; the velocity reached 300 m/s when close to the center of the torch while 150 m/s when further from the torch. This means that the velocity difference was more than twice along the direction of the powder injection. The impact angle varied from 2 to 8 degrees according to the impacted position from the center to the further side. The velocity just before impact and the velocity during flight were almost the same, suggesting that in the case of plasma spraying, the existence of a substrate does not result in a significant deceleration of the particle.

Figure 6 shows the distribution of the particle impact velocity as a function of time. The velocity was distributed between 150 and 300 m/s, but occasionally exhibited narrow distribution (for instance $t \sim 400$ μs) where $180 < V < 220$ m/s. These results may be related to the plasma fluctuation, but a further investigation will be needed to confirm it.

Finally, Fig. 7 shows the cooling time of the impinged particle on a substrate as a function of the impact velocity. The cooling time was defined here as the time from which a particle impacted to when the brightness of the center of the impacting particle decreased to one-tenth, although this definition is not based on any physical model. Pearson's correlation coefficient r between the cooling time and the impact velocity was -0.21, suggesting the tendency in which the cooling time decreased with the increase in the particle impact velocity.

4. Discussion

4.1 Coating Formation in Relation to Single Splat Formation

For single splat formation, interesting findings are that the center region of the splat had already cooled while the periphery was still spreading at a higher temperature. Assuming that the dynamic range of the HSV in temperature was 500 K, the cooling rates at the center and periphery were estimated to be 3-5 $\times 10^8$ and 5-10 $\times 10^7$ K/s, respectively. This means that the cooling rate at the center and that at the periphery would differ up to one order of magnitude. Thus, for the impact phenomena on a cold surface, the temperature distribution within a deforming droplet is not uniform, which cannot be disregarded. For the shape of a spreading droplet, McDonald et al. also captured similar images of YSZ droplets impacting on a cold smooth glass substrate: the droplets did not splash until they spread to the maximum spread extent (Ref 17). Our study was consistent with their images. However, their images lacked temperature information due to utilizing a short pulse laser as an illumination. The obtained thermal images contain complement temperature information to their shadow images.

We reported that the droplet spreading and splashing phenomena on a cold substrate in plasma spraying could be categorized as a splash between a prompt splash and a corona splash (Ref 11). According to previous studies (Ref 18-21), in the prompt splash, secondary droplets emerge and detach at the advancing liquid-substrate contact line. The corona splash is reminiscent of a crown. A symmetric corona is first formed, followed by secondary

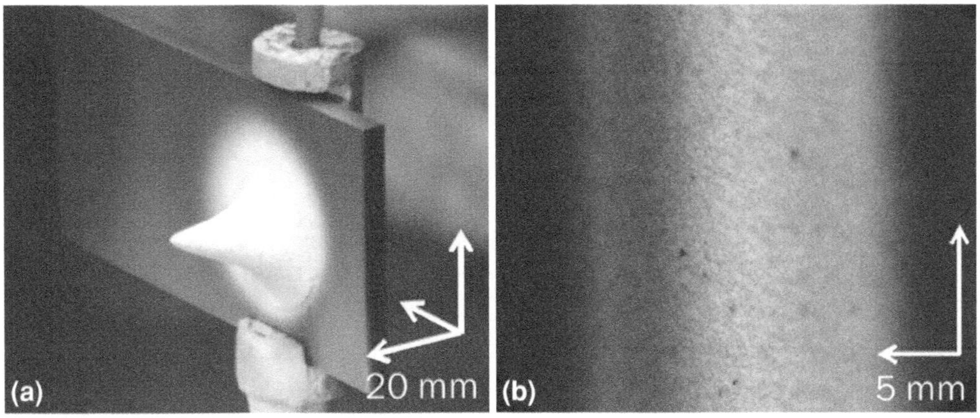

Fig. 3 (a) Obtained tower and (b) bead at 0D and 1D measurements, respectively

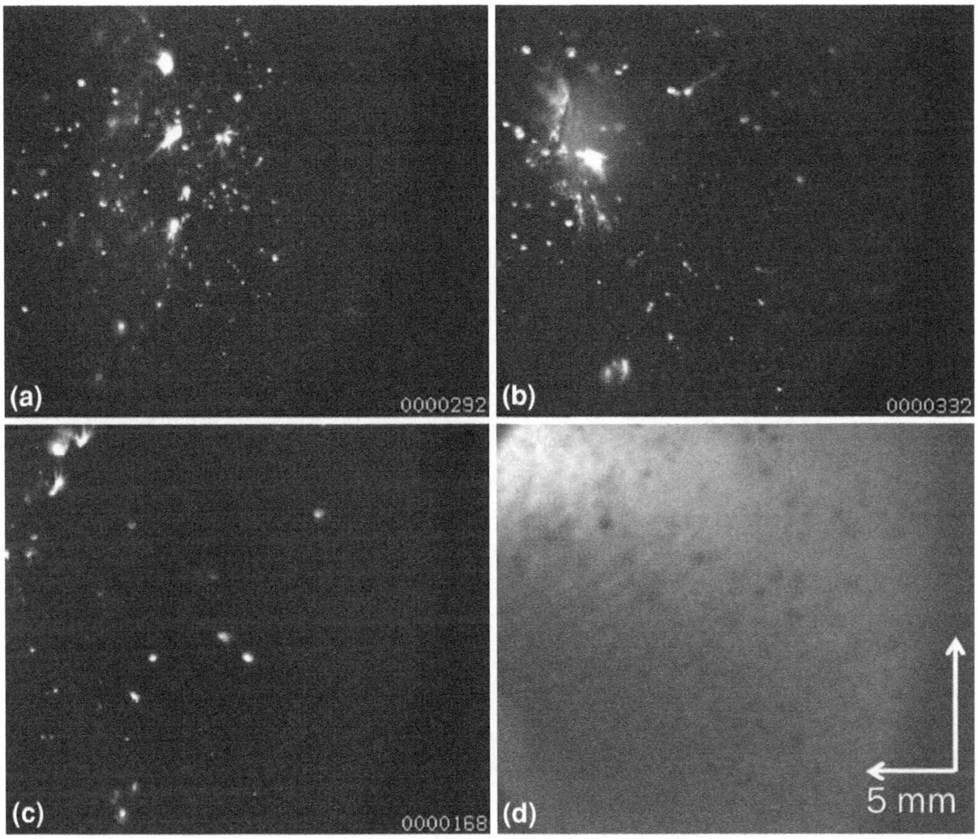

Fig. 4 In situ observation of tower formation at fixing torch (0D): (a) 12 s, (b) 43 s, and (c) 64 s after powder injection. (d) After deposition. Monitoring was conducted at 43° from a substrate. The scale is for a substrate plane

droplets ejected from the expanding corona. In parallel to our previous research (Ref 11), Xu et al. gave simple and clear explanations to this matter. The prompt splash and corona splash are caused by the surface roughness and by instabilities produced by the surrounding gas, respectively (Ref 20, 21). In our observations, during the first splat formation on a cold smooth glass substrate, the droplet spread to its maximum extent without breakage, while on the actual surface of the tower and bead formations, particles splashed instantly after impact. In the latter case, the surface was considered not as smooth from the previously deposited layer. Thus, the splash on a cold smooth substrate in plasma spraying seemed as a type of corona splash, while the splash observed in the tower and bead formations appeared as the prompt splash.

4.2 Parameters Measured and to be Measured

In this study, we have measured the velocity and trajectory of impacting particles. For the comparison, the

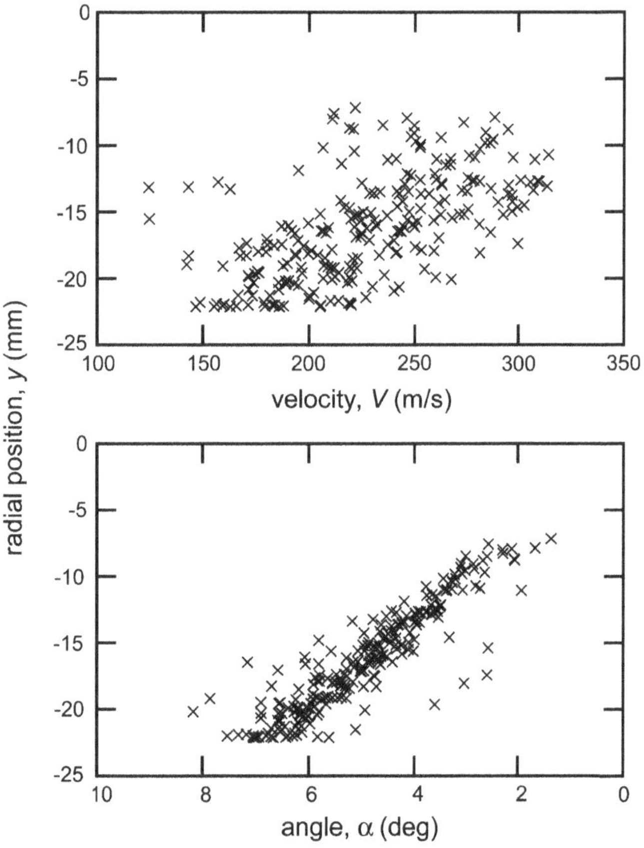

Fig. 5 Particle impact velocity V (m/s) and angle α (°) distributions along radial position y (mm). Position y (mm) is the distance from the center of the torch perpendicular to the torch axis. The powder was injected from the plus side of the torch

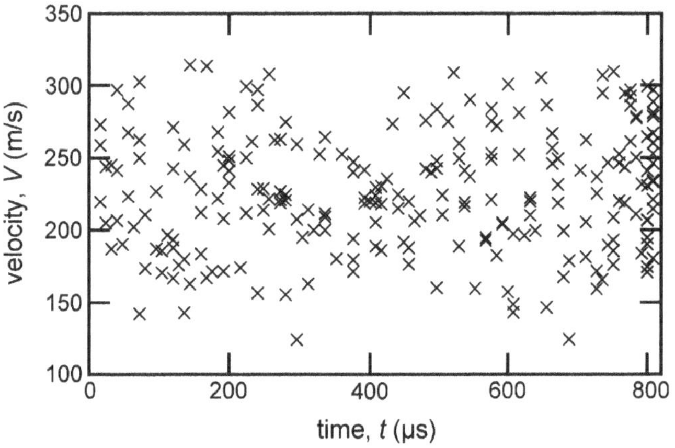

Fig. 6 Time-resolved particle impact velocity V (m/s). The time $t = 0$ corresponds to the start of recording of a high-speed video camera

velocity measured with the HSV was plotted with that measured with DPV2000 in Fig. 8. The velocity of the HSV tended to show the larger value than that of DPV2000; the velocity of the HSV was approximately 20% greater than that of DPV2000 at the maximum velocity at each position y; meanwhile, lower velocity particles could not be detected with the HSV. Although there were such discrepancies in the absolute values, the tendency was consistent. The method to derive the velocity and trajectory was similar to particle imaging velocimetry, thus the results were also consistent with the phase Doppler particle anemometry as reported by Cetegen and Yu (Ref 22).

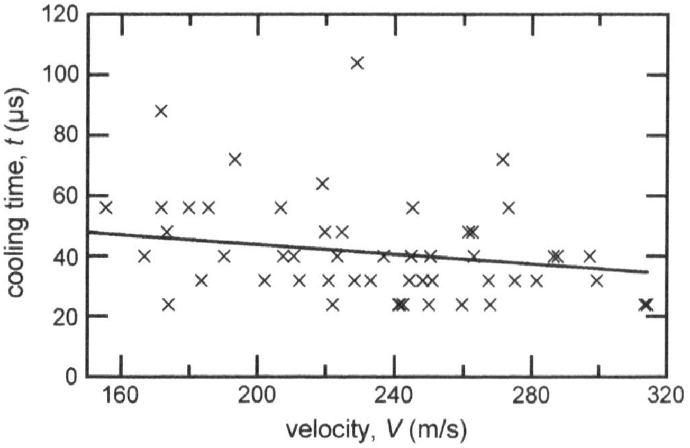

Fig. 7 Relationship between particle cooling time t (μs) and impact velocity V (m/s)

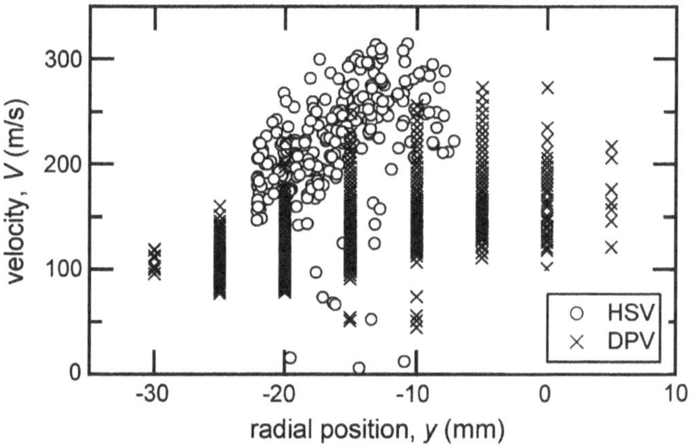

Fig. 8 Comparison of impact velocity V (m/s) derived from a high-speed video camera (HSV) with that measured with an in-flight particle monitoring system (DPV) as a function of radial position y (mm). The position y is the distance from the center of the torch perpendicular to the torch axis. The powder was injected from the plus side of the torch. The data of DPV was at the axial position of 100 mm

Since we did not measure the temperature in this study, we utilized the cooling time instead of the cooling rate for coating formation observations. The definition of the cooling time in this study was rather arbitrary and had no specific physical meanings. However, as Moreau et al. correlated the cooling time to the cooling rate, they will have an inverse relationship (Ref 23). Hence, our results in Fig. 7 suggested that increasing the impacting velocity increased the cooling rate of droplet. Since the impact velocity has a positive relation with the Reynolds number, the result was consistent with the previous study on a single splat by Vardelle et al. in which the increase in the Reynolds number increased the cooling rates (Ref 24).

We measured some parameters in situ, yet at least two other factors should be measured to understand the plasma spray process quantitatively. One is the molten degree of in-flight particles. From the previous study on the melting index (Ref 25), some portion of particles would not be fully molten when they reached to the substrate. Even with the precise surface temperatures, molten status inside droplets would not be known (Ref 26). The other is the degree of splashing. Most particles splashed during the coating formations, but even with recent advanced simulation model on the coating formation process the degree of splashing is left out (Ref 27). The experimental study to quantify the degree of splashing especially in volume fraction will be needed.

Last, we have showed the feasibility of direct observations on the coating formation process with the HSV including cases of the stationary plasma torch conditions (0D) and of the bead formation (1D). Since the bead is a unit of the coating, our approach is expected to form a basis to extend to the actual coating process in plane (2D) or in solid (3D).

5. Conclusion

In this study, we have tried to visualize two important phenomena during plasma spraying, that is, single splat formation and the subsequent coating formation. Utilizing the in situ monitoring system coupled with the HSV of 1 million fps enabled us to clearly capture the consecutive images of the deformation and cooling behavior of an YSZ droplet impinging on a cold smooth substrate: The liquid sheet jetting out sideways from the droplet became detached from the substrate and kept spreading without disintegration until its maximum extent; the center part of the flattened droplet rapidly cooled while the liquid sheet was still spreading. Representing a coating process, tower (0D) and bead (1D) formations were observed. The successive evaluation of the splat formations as a function of their impact position and time became possible. Particle impact velocities varied by two times along the direction of the powder injection. The cooling rate of the splats tended to become faster with increasing the impact speed.

Acknowledgments

We thank Prof. Takehar-Goji Etoh (Kinki University) for his useful discussions. We also thank Mr. Masayuki Komatsu and Mr. Nobukazu Kakeya (NIMS) for help with the experiments.

References

1. T. Yoshida, Toward a New Era of Plasma Spray Processing, *Pure Appl. Chem.*, 2006, **78**(6), p 1093-1107
2. F. Gitzhofer, M. Boulos, J. Heberlein, R. Henne, T. Ishigaki, and T. Yoshida, Integrated Fabrication Processes for Solid-Oxide Fuel Cells Using Thermal Plasma Spray Technology, *MRS Bull.*, 2000, **25**(7), p 38-42
3. E. Moreau, C. Chazelas, G. Mariaux, and A. Vardelle, Modeling the Restrike Mode Operation of a DC Plasma Spray Torch, *J. Therm. Spray Technol.*, 2006, **15**(4), p 524-530
4. J.F. Bisson, B. Gauthier, and C. Moreau, Effect of Plasma Fluctuations on In-Flight Particle Parameters, *J. Therm. Spray Technol.*, 2003, **12**(1), p 38-43
5. J.F. Bisson and C. Moreau, Effect of Direct-Current Plasma Fluctuations on In-Flight Particle Parameters: Part II, *J. Therm. Spray Technol.*, 2003, **12**(2), p 258-264
6. P. Saravanan, V. Selvarajan, M.P. Srivastava, D.S. Rao, S.V. Joshi, and G. Sundararajan, Study of Plasma- and Detonation Gun-Sprayed Alumina Coatings Using Taguchi Experimental Design, *J. Therm. Spray Technol.*, 2000, **9**(4), p 505-512
7. P. Fauchais, Understanding Plasma Spraying, *J. Phys. D: Appl. Phys.*, 2004, **37**, p R86-R108
8. A. Vardelle, C. Moreau, and P. Fauchais, The Dynamics of Deposit Formation in Thermal-Spray Processes, *MRS Bull.*, 2000, **25**(7), p 32-37
9. T.G. Etoh, D. Poggemann, G. Kreider, H. Mutoh, A.J.P. Theuwissen, A. Ruckelshausen, Y. Kondo, H. Maruno, K. Takubo, H. Soya, K. Takehara, T. Okinaka, and Y. Takano, An Image Sensor which Captures 100 Consecutive Frames at 1 000 000 Frames/s, *IEEE Trans. Electron Dev.*, 2003, **50**(1), p 144-151
10. S.T. Thoroddsen, T.G. Etoh, and K. Takehara, High-Speed Imaging of Drops and Bubbles, *Annu. Rev. Fluid Mech.*, 2008, **40**, p 257-285
11. K. Shinoda, H. Murakami, S. Kuroda, S. Oki, K. Takehara, and T.G. Etoh, High-Speed Thermal Imaging of Yttria-Stabilized Zirconia Droplet Impinging on Substrate in Plasma Spraying, *Appl. Phys. Lett.*, 2007, **90**(19), Art. No. 194103 (3 pp)
12. K. Shinoda, H. Murakami, S. Kuroda, S. Oki, K. Takehara, and T.G. Etoh, High-Speed Thermal Imaging of Yttria-Stabilized Zirconia Droplets Impinging on a Substrate in Plasma Spraying, *Proceedings of the 18th International Symposium on Plasma Chemistry*, K. Tachibana, Ed., August 26-31 2007 (Kyoto), International Plasma Chemistry Society, 2007, Art. No. 28C-a3 (4 pp)
13. K. Shinoda, Y. Kojima, and T. Yoshida, In Situ Measurement System for Deformation and Solidification Phenomena of Yttria-Stabilized Zirconia Droplets Impinging on Quartz Glass Substrate under Plasma Spraying Conditions, *J. Therm. Spray Technol.*, 2005, **14**(4), p 511-517
14. K. Shinoda, T. Koseki, and T. Yoshida, Influence of Impact Parameters of Zirconia Droplets on Splat Formation and Morphology in Plasma Spraying, *J. Appl. Phys.*, 2006, **100**(7), Art. No. 074903 (6 pp)
15. H.B. Guo, H. Murakami, and S. Kuroda, Effect of Hollow Spherical Powder Size Distribution on Porosity and Segmentation Cracks in Thermal Barrier Coatings, *J. Am. Ceram. Soc.*, 2006, **89**(12), p 3797-3804
16. H.B. Guo, S. Kuroda, and H. Murakami, Microstructures and Properties of Plasma-Sprayed Segmented Thermal Barrier Coatings, *J. Am. Ceram. Soc.*, 2006, **89**(4), p 1432-1439
17. A. McDonald, C. Moreau, and S. Chandra, Thermal Contact Resistance Between Plasma-Sprayed Particles and Flat Surfaces, *Int. J. Heat Mass Transfer*, 2007, **50**(9-10), p 1737-1749
18. R. Rioboo, C. Tropea, and M. Marengo, Outcomes from a Drop Impact on Solid Surfaces, *At. Sprays*, 2001, **11**, p 155-165
19. A.L. Yarin, Drop Impact Dynamics: Splashing, Spreading, Receding, Bouncing, *Annu. Rev. Fluid Mech.*, 2006, **38**, p 159-192
20. L. Xu, L. Barcos, and S.R. Nagel, Splashing of Liquids: Interplay of Surface Roughness with Surrounding Gas, *Phys. Rev. E*, 2007, **76**(6), Art. No. 066311 (5 pp)
21. L. Xu, Liquid Drop Splashing on Smooth, Rough, and Textured Surfaces, *Phys. Rev. E*, 2007, **75**, Art. No. 056316 (8 pp)
22. B.M. Cetegen and W. Yu, In-Situ Particle Temperature, Velocity, and Size Measurements in DC Arc Plasma Thermal Sprays, *J. Therm. Spray Technol.*, 1999, **8**(1), p 57-67
23. C. Moreau, M. Lamontagne, and P. Cielo, Influence of the Coating Thickness on the Cooling Rates of Plasma-Sprayed Particles on a Substrate, *Surf. Coat. Technol.*, 1992, **53**, p 107-114

24. M. Vardelle, A. Vardelle, A.C. Leger, P. Fauchais, and D. Gobin, Influence of Particle Parameters at Impact on Splat Formation and Solidification in Plasma Spraying Processes, *J. Therm. Spray Technol.*, 1995, **4**(1), p 50-58
25. L. Li, A. Vaidya, S. Sampath, H.B. Xiong, and L.L. Zheng, Particle Characterization and Splat Formation of Plasma Sprayed Zirconia, *J. Therm. Spray Technol.*, 2006, **15**(1), p 97-105
26. G. Mauer, R. Vassen, and D. Stover, Detection of Melting Temperatures and Sources of Errors Using Two-Color Pyrometry During In-Flight Measurements of Atmospheric Plasma-Sprayed Particles, *Int. J. Thermophys.*, 2008, **29**(2), p 764-786
27. R. Ghafouri-Azar, J. Mostaghimi, S. Chandra, and M. Charmchi, A Stochastic Model to Simulate the Formation of a Thermal Spray Coating, *J. Therm. Spray Technol.*, 2003, **12**(1), p 53-69

JTTEE5 17:631–636
DOI: 10.1007/s11666-008-9250-9

Peer Reviewed

The Effect of Surface Chemistry and Morphology on the Properties of HVAF PEEK Single Splats

B.P. Withy, M.M. Hyland, and B.J. James

(Submitted April 28, 2008; in revised form August 3, 2008)

Thermal spray of polymers has had limited investigation due to the narrow processing windows that are inherent to polymer powders, especially their low temperatures of thermal degradation. The polymer poly aryl ether ether ketone (PEEK) has a continuous use temperature of 260 °C, does not suffer significant thermal degradation below 500 °C (Lu et al., Polymer, 37(14):2999-3009, 1996), and has high resistance to alkaline and acidic attack. These properties led to PEEK being selected for investigation. To minimize thermal degradation of the particles, the high velocity air fuel technique was used. To investigate the effect of substrate pretreatment on single splat properties, single splats were collected on aluminum 5052 substrates with six different pretreatments. The single splats collected were imaged by scanning electron microscopy and image analysis was performed with ImageJ, an open source scientific graphics package. On substrates held at 323 °C, it was found that substrate pretreatment had a significant effect on the circularity and area of single splats, and also on the number of splats deposited on the substrates. Increases in splat circularity, area, and the number of splats deposited on the surface were linked to the decrease in chemisorbed water on the substrate surface and the decrease of surface roughness. This proved that surface chemistry and roughness are crucial to forming single splats with good properties, which will lead to coatings of good properties.

Keywords coating-substrate interaction, HVAF, influence of process parameters, polymer, single splats

1. Introduction

Poly aryl ether ether ketone (PEEK) is a polymer that can be used to form wear- and corrosion-resistant coatings. It is also a food safe polymer that makes it ideal for coatings in the food processing industry. PEEK is a highly stable polymer and suffers no chemical attack by water and steam. It also retains its mechanical strength and toughness at elevated temperatures. PEEK is able to be used in direct contact with food in the temperature range −196 to 300 °C (Ref 1). PEEK also shows high resistance to a wide range of solvents, acids, and alkalis, meaning that in the cleaning processes employed in food processing plants PEEK will not suffer degradation during the high-temperature alkali- or acid-cleaning regimes (Ref 2). PEEK has a low coefficient of friction compared to most metals and ceramics and good wear resistance compared to other polymers (Ref 3); in combination with its chemical resistance this makes PEEK suitable for use for components in pumps and other high-pressure, wear-sensitive situations. A summary of the physical properties of PEEK is shown in Table 1.

In high velocity air fuel (HVAF) thermal spraying, compressed air is used as the combustion agent. The presence of nitrogen in the combustion gases results in HVAF guns having a cooler, less-oxidizing flame than high velocity oxygen fuel (HVOF) guns, which should result in less thermal degradation of polymer particles (Ref 4). Typically an HVOF system has a flame temperature of approximately 2650 °C, while the Browning HVAF system has a flame temperature of approximately 1870 °C (Ref 5). Particle velocities of the HVAF system are also typically lower than those of HVOF systems, which at a combustion chamber pressure of 0.4 MPa are about 600 and 800 ms^{-1}, respectively (Ref 5).

The interaction of thermal spray particles with the substrate upon impact has predominantly focused on metallic and ceramic particles. Studies have shown that room temperature substrates result in fingered splats, while heating the substrate results in a higher proportion of disc splats (Ref 6-9). Surface roughness has also been

This article is an invited paper selected from presentations at the 2008 International Thermal Spray Conference and has been expanded from the original presentation. It is simultaneously published in *Thermal Spray Crossing Borders, Proceedings of the 2008 International Thermal Spray Conference*, Maastricht, The Netherlands, June 2-4, 2008, Basil R. Marple, Margaret M. Hyland, Yuk-Chiu Lau, Chang-Jiu Li, Rogerio S. Lima, and Ghislain Montavon, Ed., ASM International, Materials Park, OH, 2008.

B.P. Withy, Applied Vehicle Systems, Defence Technology Agency, Auckland, New Zealand; and **M.M. Hyland** and **B.J. James**, Chemical and Materials Engineering, University of Auckland, Auckland, New Zealand. Contact e-mail: b.withy@dta.mil.nz.

shown to have an effect, with rough surfaces resulting in increased fingering of splats (Ref 6, 8). If the interactions can be optimized to produce disc splats with intimate contact between splat and substrate, coatings with improved adhesion and porosity should result. This study has approached the interaction of particles with the substrate by controlling substrate conditions and observing the resulting splats. Common splat forms for polymer splats include disc, 'fried egg,' and splashed or fingered morphologies (Ref 10, 11). Disc splats, of which 'fried egg' splats are a subgroup, tend to result in coatings with good adhesion and cohesion with low coating porosity, while conversely, coatings formed from splashed splats have high porosity and poor adhesion and cohesion (Ref 10-13). Polymer particles are usually sprayed in a semimolten state where the center of the particle remains solid while the outer layer melts due to the low thermal conductivity of polymers (Ref 11). Substrate temperature is also critical when spraying polymers, and whether it is above or below the glass transition temperature (T_g) and above or below the melting temperature (T_m) will have an effect on the splat and coating properties (Ref 10). Splats deposited on a substrate above the T_m can receive additional thermal energy from the substrate allowing a high degree of conformity with the substrate, similar to that achieved with fully molten particles. Substrates above the T_g that cool slowly after splat deposition tend to lead to semicrystalline coatings, and the coating stresses associated with change in specific volume of the polymer must be considered. Finally, polymers that are deposited on substrates below T_g, or which are quenched, form amorphous coatings with properties analogous to those of the amorphous polymer (Ref 10, 11, 14-17).

In this study, PEEK splats were deposited on substrates subjected to 6 different pretreatments to change the surface chemistry and morphology of the substrates. In a wider study a range of substrate temperatures during spraying was investigated, but the discussion here is restricted to splat formation on substrates held at a temperature between T_g and T_m during spraying. Splats on each substrate were characterized by the number of splats deposited on a given area of substrate (splat area density), the area of individual splats, and their circularity. Characterization of splats was carried out by image analysis of backscatter scanning electron microscopy images.

Table 1 Physical properties of PEEK 150PF as supplied by Victrex

Glass transition temperature (T_g)	143 °C
Melting temperature (T_m)	343 °C
Typical crystallinity	35%
Density	
Amorphous	1260 kg m^{-3}
Crystalline	1320 kg m^{-3}
Water absorption (equilb. at 23 °C)	0.5%
Coefficient of thermal expansion	
Above T_g	4.7 × 10^{-5} °C^{-1}
Below T_g	10.8 × 10^{-5} °C^{-1}

2. Experimental Procedure

Single splats of PEEK were collected on 20 mm square AA5052 substrates that had been pretreated by one of the six pretreatments and were held at 323 °C (below the T_m and above the T_g of PEEK). Substrates were mounted on a heated copper block whose temperature was monitored by two K-type thermocouples.

Six pretreatments were used to prepare the aluminum substrates, polished (P), polished and thermally treated (PT), etched (E), etched and thermally treated (ET), boiled (B), and boiled and thermally treated (BT). Substrates were received in the polished condition and degreased with acetone. Etching involved a 10 min etch in AcidBrite, a Henkel (Auckland, New Zealand) etchant composed of 15% HF and 10% H_2SO_4. Boiling involved immersion in boiling deionized water for 30 min to form a layer of aluminum oxyhydroxide (AlOOH) on the substrate surface. Thermal treatment was a 350 °C soak for 90 min to drive off organic adsorbates and adsorbed water from the substrates. The etching process removed the native oxide layer and replaced it with a thin hydrated oxide layer, boiling built up a thick layer of aluminum oxyhydroxide on top of the native oxide layer, and polishing removed the native oxide layer and allowed it to reform under controlled laboratory conditions. Thermal treatment of these substrates resulted in the dehydration of the oxide layers to varying extent, but did not form additional oxide. These pretreatments were selected for their effect on the oxide layer composition and their effect on surface roughness as discussed in the Results and Discussion and summarized in Table 2 and Fig. 1. Pretreated substrates of different oxide layer composition and similar roughness and, conversely, substrates of different roughnesses but similar chemistry were produced by these pretreatments, and thus the influence of surface roughness on single splat morphology could be differentiated from the influence of oxide layer composition.

Particles were sprayed with an HVAF Browning Aerospray 150 (Browning Thermal Systems Inc., Enfield, New Hampshire) with a 100 mm nozzle and 200 mm spray distance. These operating parameters were selected based on a simple optimization that considered three nozzles (50, 100, and 200 mm) and three spray distances (100, 200, and 300 mm). The 100 mm nozzle with a 200 mm spray distance was found to result in the greatest deposition efficiency. The Aerospray burns kerosene and compressed air to generate a high velocity flame through a barrel nozzle. The system generated a combustion chamber pressure of 0.4 MPa, which corresponds to a flame velocity of ~600 ms^{-1}; particle velocity and temperature could not

Table 2 Surface roughness measurements of the different surface pretreatments

	B	BT	E	ET	P	PT
R_a, nm	34	18	253	254	12	12
Skewness	−0.41	0.22	0.53	−0.21	−0.14	−0.14

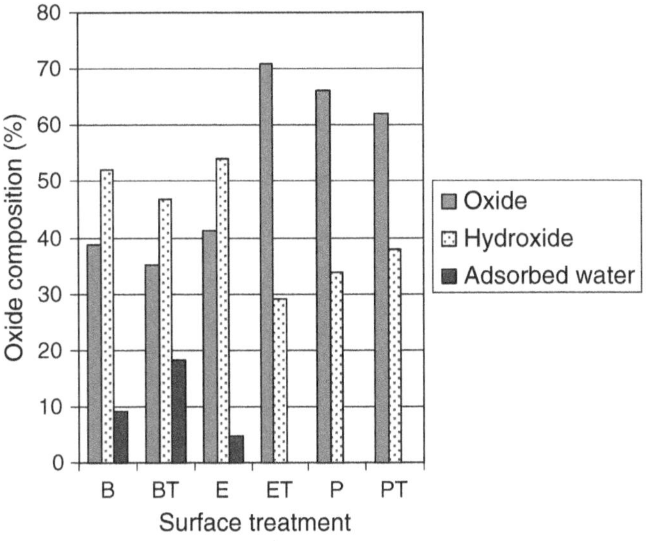

Fig. 1 Fraction of aluminum oxide, aluminum hydroxide, and chemisorbed water on substrate surfaces

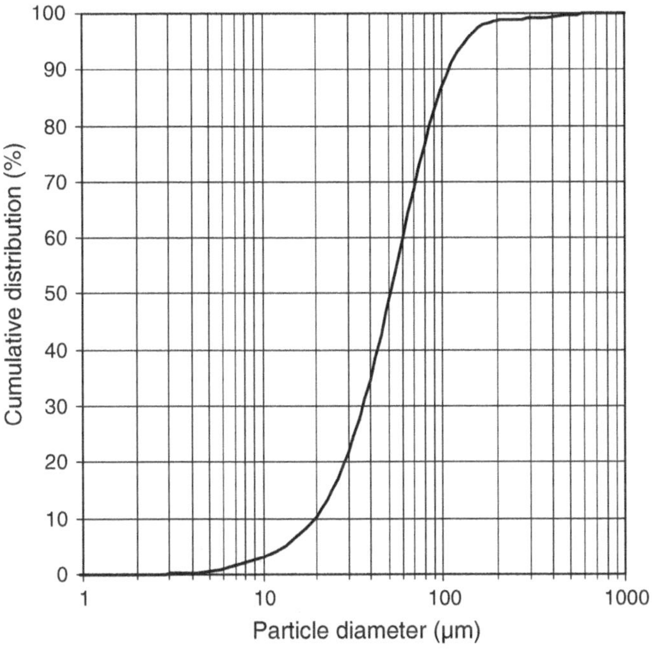

Fig. 2 Cumulative size distribution of PEEK feedstock powder

be measured with this system. Powder was fed at a rate of 7 g/min from a vibrating screw type powder feeder with nitrogen as the carrier gas. The feedstock had a nominal median diameter of 50 μm and a range of 0 to 100 μm, and a measured distribution as displayed in Fig. 2 . Single splats were achieved with a 0.5 ms^{-1} horizontal swipe across the substrates. Splats analyzed were from the center of the spray pattern.

The surface chemistry and roughness resulting from each pretreatment were determined by X-ray photoelectron spectroscopy (XPS) and atomic force microscopy (Digital Instruments NanoScope IIIa, Santa Barbara, CA), respectively. XPS (Kratos Ultra DLD, Manchester, UK) revealed the elemental composition of the substrate surface, the oxidation states of the aluminum oxide species, and the relative proportions of oxide, hydroxide, and adsorbed water. This technique is extremely surface-sensitive, analyzing only the outermost 5 nm of the surface.

Splats were imaged using a FEI Quanta 200 FEG environmental scanning electron microscope (Hillsboro, OR), utilizing backscatter imaging to provide improved definition between polymer splats and aluminum substrates. Image analysis was performed with the open source software ImageJ (Ref 18), which enabled splat counting, measurement of splat area, and circularity. Five images of each substrate (from the area in the center of the spray pattern) were analyzed, with each image containing at least 50 splats of area greater than 200 μm^2. Splats smaller than 200 μm^2 were not counted as they were determined to be the result of splashing, and not primary splats. The area of a splat was computed by summing the number of pixels in a splat and multiplying by the area of a single pixel as defined by the scale. Circularity is calculated by the formula $4\pi A/(\text{perimeter}^2)$, where A is the area of the splat. A circularity of 1 equates to a perfect circle, and the closer to zero the value, the more elongated or fingered the splat.

3. Results and Discussion

The six pretreatments resulted in different surface roughnesses as listed in Table 2. Surface roughness was measured over a 50 μm square area, the typical interaction area of a PEEK splat in this study. Boiling resulted in a slight increase in roughness over the polished surfaces, while the etching resulted in a very rough surface. The thermal treatment had little or no effect on the substrate surface roughness.

The oxide layer present on the surface of aluminum is a combination of various alumina phases, from aluminum trihydroxide ($Al(OH)_3$), to aluminum oxyhydroxide (ALOOH) to transition aluminas and corundum (Al_2O_3) (Ref 19, 20). This results in three resolvable oxygen components in the XPS analysis, one due to an Al-OH bond, one due to the Al=O bond in AlOOH, and one due to chemically adsorbed water (Ref 21). The six pretreatments used result in different ratios of oxide, hydroxide, and chemisorbed water on the aluminum substrate surfaces, as shown in Fig. 1. The surface of P and PT substrates are dominated by oxide, with some hydroxide from the reaction with atmospheric moisture, while E and ET surfaces have a higher proportion of hydroxide to oxide. These oxide layers are very thin, approximately 4 nm, compared to 225 nm for the B and BT oxide layers. While the boiled surface appears to have a similar surface chemistry to that of E, it is in fact quite a thick oxyhydroxide. This is evidenced in the thermal treatments when the very thin etched oxide rapidly loses its surface hydroxide, whereas the thicker boiled sample changes

Peer Reviewed

very little, because the oxide is so thick. The increase in chemisorbed water on the BT surface is due to a thin surface layer of aluminum oxyhydroxide dehydrating with the heat energy of thermal treatment. This then adsorbs atmospheric moisture increasing the chemisorbed water content of the surface relative to the B surface. Dehydration starts to occur at temperatures above 300 °C according to the reaction (Ref 20):

$$2AlOOH \rightarrow \gamma\text{-}Al_2O_3 + H_2O$$

The change in hydroxide concentration on the etched surfaces with thermal treatment, the different surface chemistries and morphologies of boiled and etched surfaces, the chemical differences between polished and boiled surfaces, and the similar chemistries yet differing morphologies of ET and PT surfaces will allow us to interpret the shape and spreading of PEEK splats with clearly defined knowledge of the properties of that surface and comparison surfaces. Particularly, the similarity in chemistry of the ET, P, and PT surfaces shown in Fig. 1 will be beneficial in separating the effects of surface morphology from those of surface chemistry on the properties of single PEEK splats.

Splats deposited on substrates held at 323 °C during spraying were affected by surface pretreatment. Increasing substrate temperature will not have a significant effect on the surface chemistry of the substrates as the dehydration reactions driven by thermal energy take a significant length of time to reach equilibrium (Ref 20), and the splats were deposited on substrates immediately after the desired substrate temperature was reached. Characteristic images of the PEEK splats on polished, boiled, and etched substrates are shown in Fig. 3. As shown in Fig. 4, fewer splats were deposited on an E surface than were deposited on a B surface, which had fewer splats than a P surface. This trend was repeated with the thermally treated surfaces with ET having fewer splats than BT, which had fewer splats than a PT surface. However, there was no difference between the thermally treated and nonthermally treated samples, with B and BT, E and ET, and P and PT surfaces having similar numbers of splats. This relationship appears to be governed by surface roughness primarily, although it is likely to have a surface chemistry aspect as well. The lower splat area density due to surface roughness is likely due to a mechanism associated with reduced contact area resulting in reduced bonding area and therefore adherence of fewer splats. Results following a similar trend are exhibited for average splat area, but due to greater variability of these results, as displayed in Fig. 5, they must be considered statistically the same. The variability of splat area on each substrate was primarily determined by the variability of the PEEK particles that are impacting and adhering to the aluminum surfaces, which show an even distribution over the size range of 0-100 μm, as shown in Fig. 2. The smaller diameter of splats on E and ET surfaces (although not statistically significant) is likely due to increased fingering and fragmentation of particles, which is consistent with the lower circularity of splats on these substrates, as shown in Fig. 6.

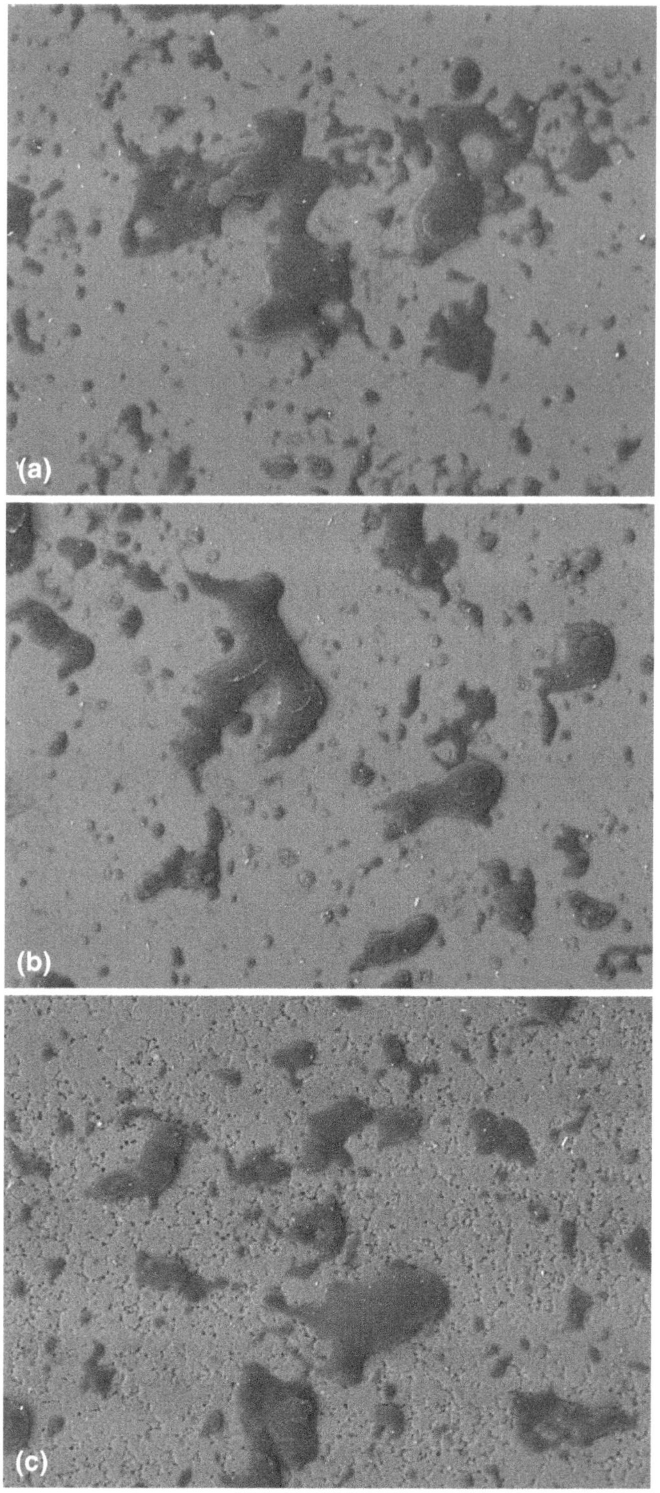

Fig. 3 Characteristic splat images on (a) polished, (b) boiled, and (c) etched substrates

The trend of surface morphology influencing splat morphology is continued to splat circularity, as displayed in Fig. 6. For substrates held at 323 °C, surface roughness appears to govern splat circularity, with E and ET surfaces

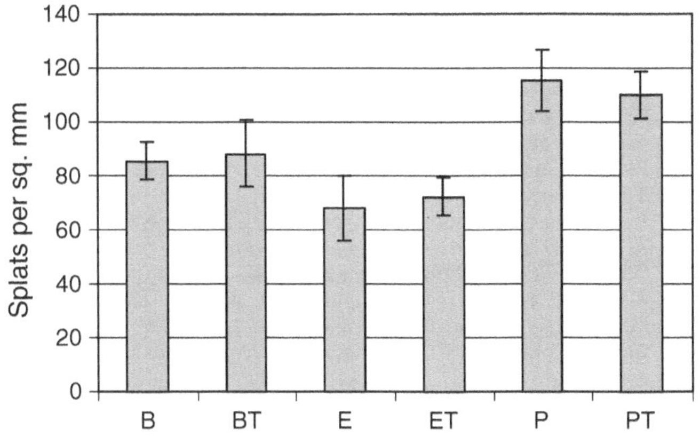

Fig. 4 Area density of splats deposited on each of the 6 pretreated substrates. Error bars represent the standard deviation associated to the average value

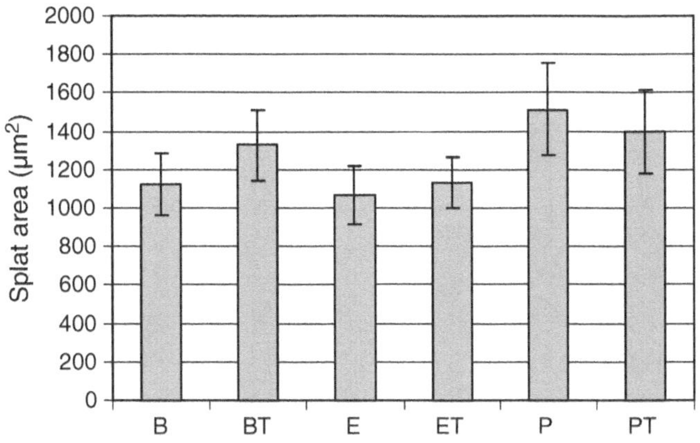

Fig. 5 Average area of individual splats deposited on each of the 6 pretreated substrates. Error bars represent the standard deviation associated to the average value

having splats with the lowest circularity, and B, BT, P, and PT surfaces having splats of similar circularity and similar surface roughness. This is an expected result as rougher surfaces will lead to more instability and variability of flow within the splats upon impact and, as such, result in a greater degree of fingering and less circular splats.

The majority of these results can be solely attributed to the different roughnesses of the 6 substrates, especially for the splat area and circularity results. Surface chemistry differences between the substrates can be discounted as ET and PT surfaces have essentially identical surface chemistries, yet the differences in splat morphology and surface roughness between these two pretreatments are marked.

Splat area density however appears to be affected by both surface chemistry and substrate morphology, and while substrate roughness definitely plays a contributing role to the splat area density, the difference in splat adhesion between B and P substrates and between BT and PT substrates cannot be solely ascribed to the small difference in roughness between these surfaces. It is thought that this difference in splat area density can be ascribed to

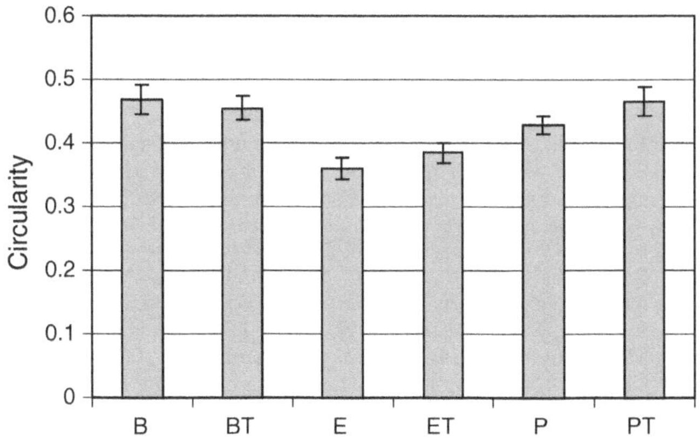

Fig. 6 Average circularity of the splats deposited on each of the 6 pretreated substrates. Error bars represent the standard deviation associated to the average value

the differing surface chemistries of the substrates in question. The boiled substrates, with an AlOOH surface layer, were significantly different from the 65% oxide, 35% hydroxide surfaces of P and PT substrates. It is thought that heating of the substrate by impacting PEEK particles releases water vapor due to desorption of chemisorbed water and dehydration of the AlOOH, by the reaction (Ref 20):

$$2\mathrm{AlOOH} \rightarrow \gamma\text{-}\mathrm{Al_2O_3} + \mathrm{H_2O}$$

The release of water vapor from B and BT substrates would therefore act as a vapor cushion between impacting PEEK particles and the substrate, resulting in particles of lower velocities not achieving sufficient contact area to adhere to the substrate. This low adhesion of PEEK to boiled surfaces has been seen in other, unpublished, areas of this study, and in the work of Trompetter, who found that NiCr particles exhibit poor deposition on boiled aluminum substrates (Ref 22).

4. Conclusions

Surface morphology was the main factor found to influence the splat area density, the area of single splats, and the circularity of single splats of HVAF-sprayed PEEK on substrates maintained at 323 °C. Substrate surface chemistry was however found to have an influence on the area density of deposited single splats, where surface morphology alone cannot explain the significant differences between the 6 different pretreated substrates.

References

1. Victrex plc, Food Processing Flyer, *Victrex Technical Information*, Victrex plc, 2006. http://www.victrex.com/adm/literature_files/Food_Processing.pdf
2. Victrex plc, Chemical Resistance Flyer, *Victrex Technical Information*, Victrex plc, 2006. http://www.victrex.com/adm/literature_files/Chemical_Resistance.pdf
3. Z.P. Lu and K. Friedrich, On Sliding Friction and Wear of PEEK and Its Composites, *Wear*, 1995, **181-183**(Part 2), p 624-631

Peer Reviewed

4. L. Jacobs, M.M. Hyland, and M. De Bonte, Comparative Study of WC-Cermet Coatings Sprayed Via the HVOF and the HVAF Process, *J. Therm. Spray Technol.*, 1998, **7**, p 213-218
5. J.A. Browning, Hypervelocity Impact Fusion – A Technical Note, *J. Therm. Spray Technol.*, 1992, **1**, p 289-292
6. J. Cedelle, M. Vardelle, P. Fauchais, and Y. Tanaka, Effect of Substrate Surface Topography and Temperature, on Millimeter and Micrometer Sized Splat Formation and on Thermal Contact Resistance, *ITSC 2006: Pushing the Envelope of Materials Performance*, ASM International, 2006
7. P. Fauchais, M. Fukumoto, A. Vardelle, and M. Vardelle, Knowledge Concerning Splat Formation: An Invited Review, *J. Therm. Spray Technol.*, 2004, **13**(3), p 337-360
8. M. Fukumoto, H. Nagai, T. Yasui, Influence of Surface Character Change of Substrate Due to Heating on Flattening Behaviour of Thermal Sprayed Particle, *ITSC 2006: Building on 100 Years of Success*, B. Marple, Ed., ASM International, 2006
9. A. McDonald, M. Raessi, S. Chandra, J. Mostaghimi, and C. Moreau, Fragmentation of Plasma-Sprayed Molybdenum Particles on Glass Surfaces, *ITSC 2006: Building on 100 Years of Success*, B. Marple, Ed., ASM International, 2006
10. E. Petrovicova and L.S. Schadler, Thermal Spraying of Polymers, *Int. Mater. Rev.*, 2002, **47**(4), p 169-190
11. M. Ivosevic, R.A. Cairncross, and R. Knight, 3D Predictions of Thermally Sprayed Polymer Splats: Modelling Particle Acceleration, Heating and Deformation on Impact with a Flat Substrate, *Int. J. Heat Mass Transf.*, 2006, **49**, p 3285-3297
12. M. Fukumoto, E. Nishioka, and T. Nishiyama, New Criterion for Splashing in Flattening of Thermal Sprayed Particles onto Flat Substrate Surface, *Surf. Coat. Technol.*, 2002, **161**, p 103-110
13. S. Sampath and X. Jiang, Splat Formation and Microstructure Development During Plasma Spraying: Deposition Temperature Effects, *Mater. Sci. Eng. A*, 2001, **304-306**, p 144-150
14. J.R.T. Branco and S.V. Campos, Wear Behaviour of Thermally Sprayed PET, *Surf. Coat. Technol.*, 1999, **120-121**, p 476-481
15. R.S. Lima, A.S. Takimi, M.D. Lima, and C.P. Bergmann, Production of Recycled Polyethylene Terephthalate Coatings by HVOF and Plasma Spray, *Thermal Spray: A United Forum for Scientific and Technological Advances*, C.C. Berndt, Ed., ASM International, 1997
16. T.E. Twardowski, X.H. Fang, and R. Knight, Structure and Properties of Thermally Sprayed Amorphous Polymer Coatings, *ITSC 2000: 1st International Thermal Spray Conference*, May 8-11, 2000 (Montreal, Quebec), Canada, 2000
17. G. Zhang, H. Liao, M. Cherigui, and C. Coddet, Effect of Crystalline Structure on the Hardness and Interfacial Adhesion of Flame Sprayed PEEK Coatings, *ITSC 2006: Building on 100 Years of Success*, B. Marple, M. Hyland, Y. Lau, R.S. Lima, and J. Voyer, Eds., ASM International, 2006
18. W.S. Rasband, *ImageJ*, U.S. National Institutes of Health, 1997-2006
19. R.S. Alwitt, The Aluminium-Water System, *Oxides and Oxide Films*, J.W. Diggle and A.K. Vijh, Eds., Marcel Dekker, 1976, p 169-254
20. K. Wefers and C. Misra, *Oxides and Hydroxides of Aluminium*, Aluminium Company of America, 1987
21. E. McCafferty and J.P. Wightman, Determination of the Concentration of Surface Hydroxyl Groups on Metal Oxide Films by a Quantitative XPS Method, *Surf. Interface Anal.*, 1998, **26**, p 549-564
22. W.J. Trompetter, "Splat-Substrate Interactions in High Velocity Thermal Spray Coatings: A Thesis Submitted in Partial Fulfilment of the Requirements for the Degree of Doctor of Philosophy in Engineering," Ph.D. Thesis, University of Auckland
23. S.X. Lu, P. Cebe, and M. Capel, Thermal Stability and Thermal Expansion Studies of PEEK and Related Polyimides, *Polymer*, 1996, **37**(14), p 2999-3009

JTTEE5 17:637–645
DOI: 10.1007/s11666-008-9237-6
1059-9630/$19.00 © ASM International

Peer Reviewed

Effects of Surface Chemistry on Splat Formation During Plasma Spraying

A.T.T. Tran, M.M. Hyland, T. Qiu, B. Withy, and B.J. James

(Submitted May 2, 2008; in revised form September 21, 2008)

Ni-Cr single splats were plasma-sprayed at room temperature onto aluminum and stainless steel substrates, which were modified by thermal and hydrothermal treatments to control the oxide surface chemistry. The proportions of the different splat types were found to vary as a function of substrate pretreatment, especially when the pretreatment involved heating. It was observed that surface roughness did not correlate with changes in splat morphology. Substrate surfaces were characterized by X-ray photoelectron spectroscopy using in situ heating in vacuum to determine the effect of thermal pretreatment on substrate surface chemistry. It was found that the surface layers were composed primarily of oxyhydroxides. When the substrates were heated to 350 °C, water vapor was released from the dehydration of oxyhydroxide. Preheating the substrate can remove the water prior to spraying: preheated substrates had improved the physical contact between the splat and substrate, which enhanced the formation of disk splats and increased the number of splats.

Keywords plasma spray, preheating substrate, roughness effect, surface chemistry

1. Introduction

The nature of the interface between splat and substrate in thermal spray coating is important in determining the mechanical, thermal, and electrical properties of the coatings. Bonding formation and splat morphology are profoundly influenced by the substrate temperature and the substrate surface oxide layer (Ref 1-11). Fukumoto (Ref 1-5) and others (Ref 8, 10, 11) have shown that splat morphology changes from an unfavorable splash splat shape to a favorable disk splat shape when the substrate is heated over a narrow range of temperatures. The underlying cause of the change in splat morphology is not clearly understood. It has been attributed to desorption of gases (Ref 7, 8, 11, 12), oxidation of the surface (Ref 3, 9), and changes in surface roughness (Ref 1, 3). Jiang (Ref 11) showed that the contact area between splat and substrate is increased and the adhesion is improved on substrates where the adsorbed gas has been removed or reduced. However, heating can also result in changes in the thickness and roughness of the surface layer, which may also affect the splat formation process. Fukumoto et al. (Ref 1) suggested that the thickness of the oxide layer and increased surface roughness on the nanoscale level result in a better wetting ability at the interface of substrate and splat when substrates were heated or preheated, consequently promoting disk splat formation. In the same study, no changes in chemical composition could be detected, supporting the greater importance of surface roughness rather than surface chemistry on splat morphology. Later studies (Ref 3, 4), however, suggested that the nanoscale roughness was not a dominant factor, since gold-coated preheated substrates, having increased surface roughness on the nanoscale, did not result in disk splat morphology. There were also no significant changes in splat morphology of Mo particles on nonheated and heated glass substrates (Ref 10). McDonald et al. (Ref 12) have recently shown that the average roughness on a nanometer scale had lesser influence on the splat morphology than attached adsorbates. Thus, surface roughness does not appear to be a critical character in splat morphology.

All metals are covered with oxide layers of variable composition and thickness, depending strongly on the local environment, pretreatments, and a variety of other parameters. The surface layers on aluminum and stainless steel formed under ambient conditions are mixtures of oxides and oxyhydroxides (Ref 13-18). Heat treatment causes changes in the relative amounts of these species and consequent release of gases from the surface (Ref 19-23) and can also induce segregation of elements to the surface due to differences in free energy of formation oxides (Ref 24, 25). Such changes can only be detected by high resolution surface spectroscopic analysis. A recent study of single splats of HVAF-sprayed PEEK on aluminum substrate has shown that surface chemistry can

This article is an invited paper selected from presentations at the 2008 International Thermal Spray Conference and has been expanded from the original presentation. It is simultaneously published in *Thermal Spray Crossing Borders, Proceedings of the 2008 International Thermal Spray Conference*, Maastricht, The Netherlands, June 2-4, 2008, Basil R. Marple, Margaret M. Hyland, Yuk-Chiu Lau, Chang-Jiu Li, Rogerio S. Lima, and Ghislain Montavon, Ed., ASM International, Materials Park, OH, 2008.

A.T.T. Tran, M.M. Hyland, T. Qiu, B. Withy, and **B.J. James**, Chemical & Materials Engineering Department, The University of Auckland, Private Bag 92019 Auckland, New Zealand. Contact e-mail: atra021@aucklanduni.ac.nz.

affect the polymer splat formation process (Ref 26). Hence, the aim of this work was to understand the role of both surface chemistry and surface roughness on splat formation and splat morphology. The methodology was to control the surface chemistry and roughness of aluminum and stainless steel substrates by subjecting them to particular surface treatments, to characterize the resultant surface chemistries using high resolution X-ray photoelectron spectroscopy, and to look at the correlation of surface chemistry and surface morphology with the resultant NiCr splat morphologies.

2. Experimental Details

In this work, aluminum 5052 (Al) and stainless steel 304 (SS) substrates were mechanically polished to a mirror-like surface finish. Four different pretreatments were used to control the chemistry of the surface oxide: polished (P), polished and thermally treated (PT), boiled (B), and boiled and thermally treated (BT). The hydrothermally treated samples (B, BT) were boiled in deionized water for 30 min. The subsequent thermal treatment of boiled or polished samples was carried out in air at 350 °C for 90 min. All samples were stored in a dessicator to preserve the surface chemistry until they were ready to be sprayed. The time between removing from the dessicator and spraying ranged from 10 to 60 min. One set of the substrate samples was used to evaluate the effect of surface roughness on the splat shape by atomic force microscope (AFM, NanoScope IIIa) using a scan size of 100 μm. The other set of 8 samples was sprayed with molten Ni80-Cr20 alloy particles (Sulzer Metco 43VF-NS, Switzerland, $-45+5$ μm) at room temperature using a plasma spray technique. All substrates were sprayed at the same time, to allow for direct comparison of the splat morphologies, since all samples would have been exposed to the same particle size distributions and splat temperature. Plasma spraying was carried out with a Sulzer Metco 7MB gun operating at a current of 550 A and a voltage of 74 V. The spray distance was 80 mm. The powder was injected at a feeding rate of 1 g/min. The plasma gas mixture was nitrogen and hydrogen, at a flow rate of 47.6 SLPM and 5.4 SLPM, respectively. Collected splats on (25 × 50 × 3 mm) substrates were characterized qualitatively and quantitatively by Scanning Electron Microscopy (SEM) and ImageJ imaging software (National Institute of Health, Washington DC). The SEM is a Philips FEGXL30.

To examine the effect of temperature on the substrate surface chemistry, in particular the release of water or adsorbates, the polished and hydrothermal-treated samples were analyzed with X-ray Photoelectron Spectroscopy (XPS), using a Krastos Ultra Axis DLD. The samples were heated in situ in the XPS to determine the variation of surface composition with temperature. Wide scans (0-1000 eV) and narrow scans of Al 2p, Fe 2p, O 1s and C 1s at different temperatures were collected; the measured binding energies were referenced to C1s at 285.0 eV. The narrow scans were fitted by CasaXPS software with a mixture of Gaussian (G) and Lorentzian (L) for the peak shape. The fitting method was based on the established literature for aluminum (Ref 19, 20, 22) and stainless steel (Ref 21, 23).

3. Results and Discussion

3.1 Splat Characterization

Collected splats were examined using SEM to assess the effect of substrate pretreatment on the morphology of the splat. Backscattered images were used to distinguish between splat and substrate. Approximately 40 backscattered electron images at low magnification (100× to 200×) and secondary electron images at high magnification were acquired at random locations for every sample. From these images, a minimum of five random areas with around 300 splats were analyzed using ImageJ to obtain quantitative and qualitative information on splat morphologies. The obtained information included feret diameter, area, circularity, and perimeter of the individual splats. The feret diameter is the longest distance between any two points on the boundary of the splat. Typical images of collected NiCr splats on aluminum and stainless steel substrates are shown in Fig. 1. A splat classification

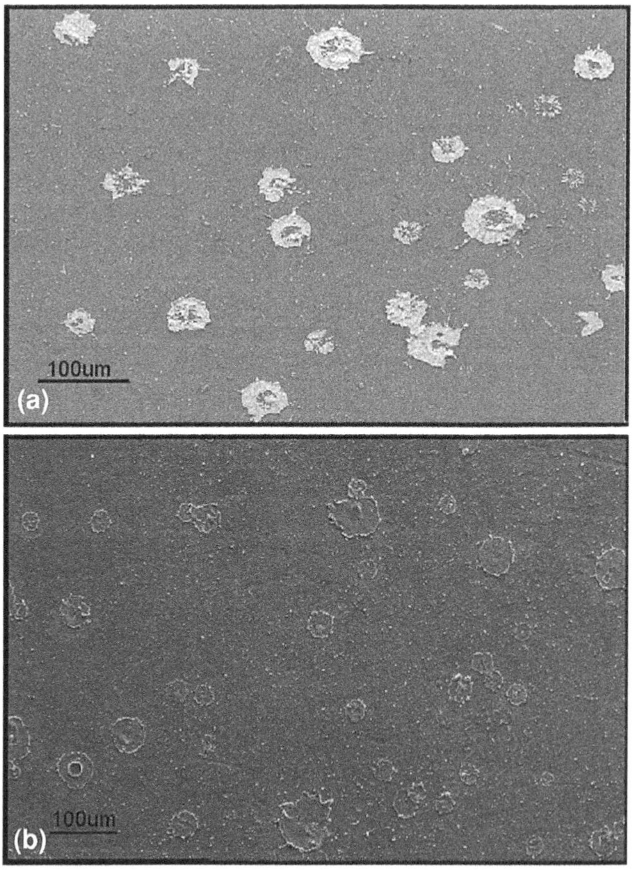

Fig. 1 Typical images of collected NiCr splats on (a) aluminum and (b) stainless steel substrates

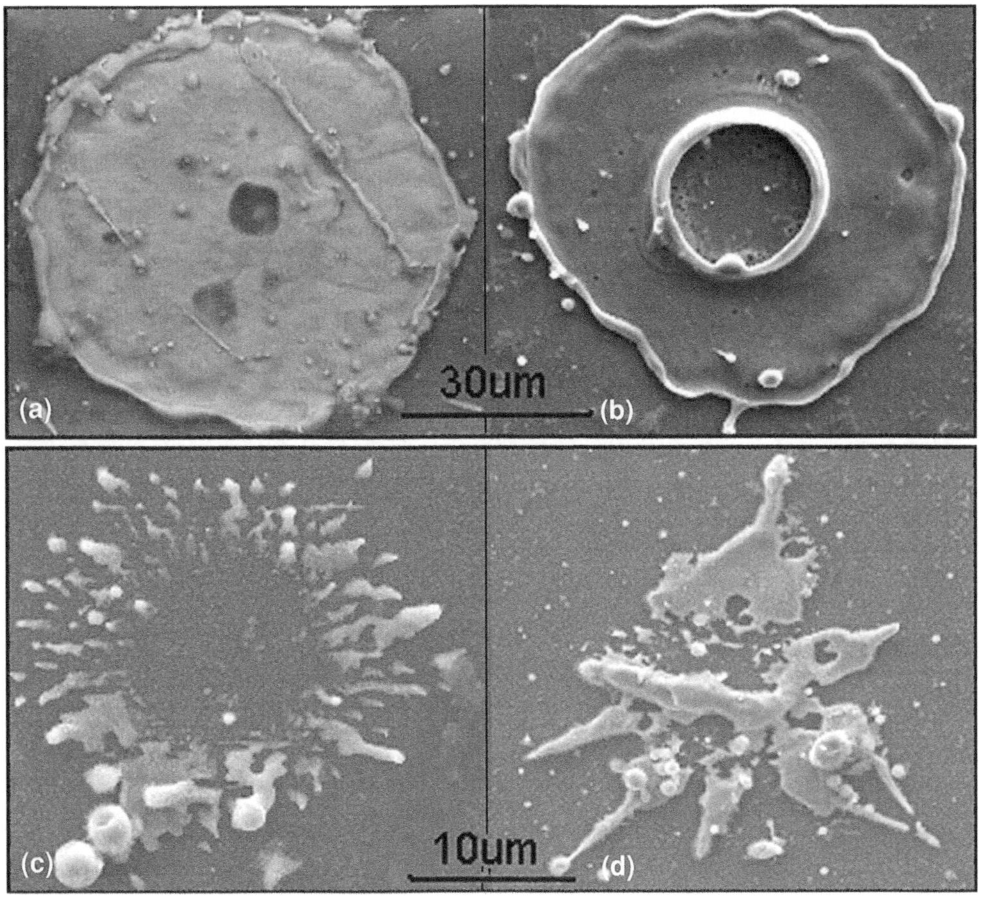

Fig. 2 Two subcategories of disk splat: (a) round disk splat and (b) doughnut disk splat; and of splash splat, (c) round splash splat and (d) irregular splash splat

scheme was developed based on two primary morphologies: splash and disk-type splats. In addition, these splat types were divided into two subcategories: (a) round and (b) doughnut disk splats, and (c) round and (d) irregular splash splats as shown in Fig. 2. The proportion of each of these splat types was quantified for all sprayed samples.

Collected disk splats had a reasonably round shape, a flat surface, and some small fingers indicating the deposition of completely molten NiCr droplets. Round disk splats are here defined as splats whose center pore diameter was smaller than 4 μm. Doughnut disk splats are defined as round disk splats with a center bubble-like hole having a diameter larger than 4 μm. Micropores of approximately 1 μm and curling up at the edges of splats were also observed. It was also observed that there was no splat formation on boiled aluminum substrates. Instead, impact marks were observed on the surface as in Fig. 3.

The results of image analysis using ImageJ tool are shown in Fig. 4. Because there was no splat formation on boiled aluminum substrates, no data were available for these samples. It was found that the splat distribution and splat morphology on aluminum and stainless steel had a similar trend with pretreatment. For both substrate types, the splat density, diameter, and percentage of splats as disk splats increased markedly on preheated substrates compared to non-preheated substrates. For example, as shown in Fig. 4, the splat density on the polished aluminum substrates increased by 40% as a result of preheating. Disk splats were found on both non-preheated and preheated aluminum substrates, but in very different proportions (Fig. 4b). It was observed that there were no round disk splats and only approximately 5% doughnut disk splats (with the average diameter of 42.5 μm) formed on Al_P surface. However, the disk splat proportion was sharply increased for preheated substrate with 17% of round disk splat and 26% of doughnut disk splat. In addition, the average diameter of disk splats increased by about 12% for preheated substrates compared to non-preheated substrates.

A general linear trend showing increasing splat density and disk splat proportion with preheating substrate was also observed on stainless steel substrate. The image analysis results show that for the stainless steel substrates, the boiled substrate surface clearly had the least splats adhering to it. The SS_P and SS_BT surfaces had a similar number of splats whereas SS_PT surface had the most splats. The increase of splat density on the preheated stainless steel substrates was 14% (PT) and 4% (BT) relative to their non-preheated counterparts. In addition, the number of micropores noticeably decreased upon heat treatment. More micropores were found on boiled stainless steel substrates than on polished surface; however, the

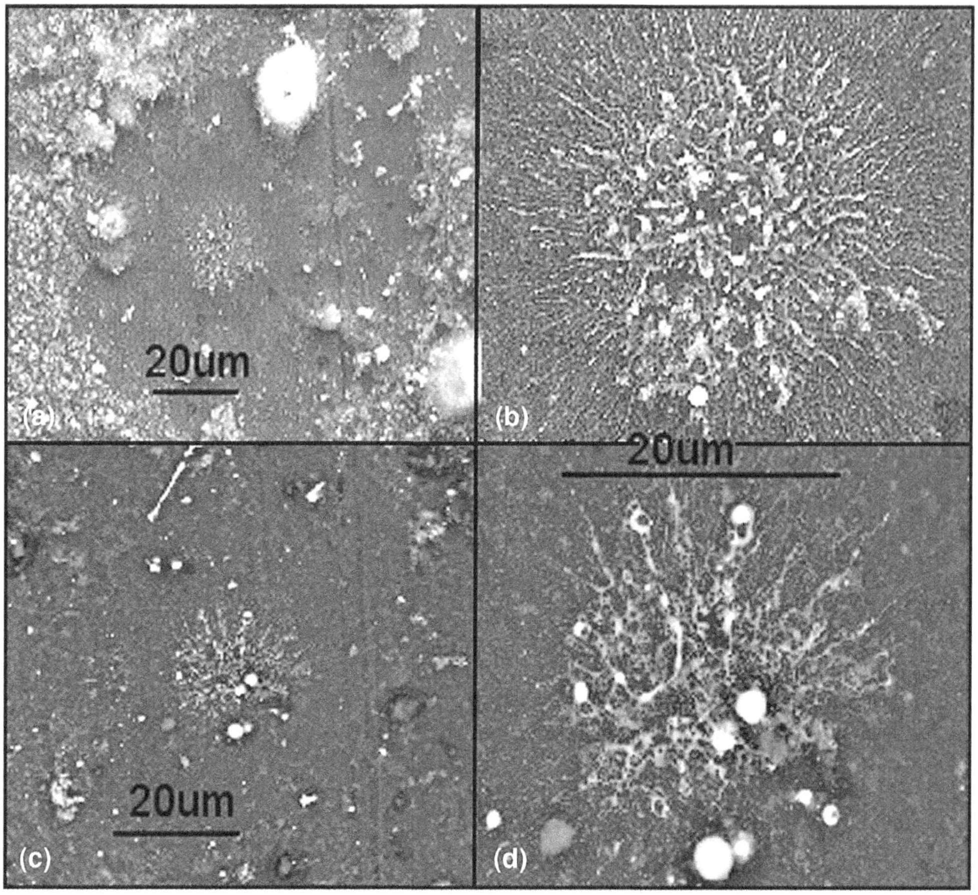

Fig. 3 Impact marks by NiCr droplets on (a) Al_B, (b) enlargement of (a), (c) Al_BT, and (d) enlargement of (c)

number of micropores was similar after heat treatment. Unexpectedly, micropores were mainly found in splats deposited on stainless steel substrate compared to aluminum substrates.

On non-preheated stainless steel substrates, the SS_P surface clearly had a smaller proportion of disk splats (45%), with a similar diameter to those on Al_P, while SS_B had higher proportion disk splats of (63%) with an average diameter of 50.2 μm. More important, preheating the substrates resulted in a significant decrease in the amount of irregular splash splats and an increase in the number of round disk splats. It was expected that some doughnut disk splats converted to round disk splats on preheated substrates. The average diameter of disk splats on stainless steel also increased by 12 to 14% for preheated substrates compared to non-preheated substrates.

It was also found that splat density and proportion of disk-type splats were significantly higher on stainless steel than on aluminum substrates, indicating that the adhesion of NiCr splats on stainless steel was much better than on aluminum. Unlike boiled aluminum substrates, there was splat formation on boiled stainless steel. It was expected that the wetting ability of impacting NiCr splats on boiled sample would decrease compared to that on polished sample. However, experiments showed the reverse trend with disk splat fraction of 45% on polished sample and 63% on the boiled one. When these substrates were preheated, they had similar proportions of disk splats.

3.2 Surface Roughness

Surface roughness was evaluated by two parameters: the average surface roughness (Ra) and skewness (Sk). Ra gives the average distance between the surface and mean centerline, but does not discriminate very well between peaks and valleys. Skewness is used to measure the height distribution. A negative skewness value indicates that the sample has more valleys than peaks, and the reverse for a positive skewness value (Ref 27).

The roughness results of aluminum substrate with different pretreatment types, evaluated by AFM on a scan size of 100 μm, are tabulated in Table 1. It was observed that there were no significant differences in both Ra and Sk for polished aluminum and polished and thermally treated aluminum. In these cases, the small negative value of Sk was obtained because peaks are more easily polished than valleys (Ref 27).

In contrast, surface roughness changed slightly on the hydrothermally treated samples. Boiling the aluminum substrate in deionized water for 30 min induced a change to a surface topography consisting of more valleys with slightly higher value of Ra compared to that of polished

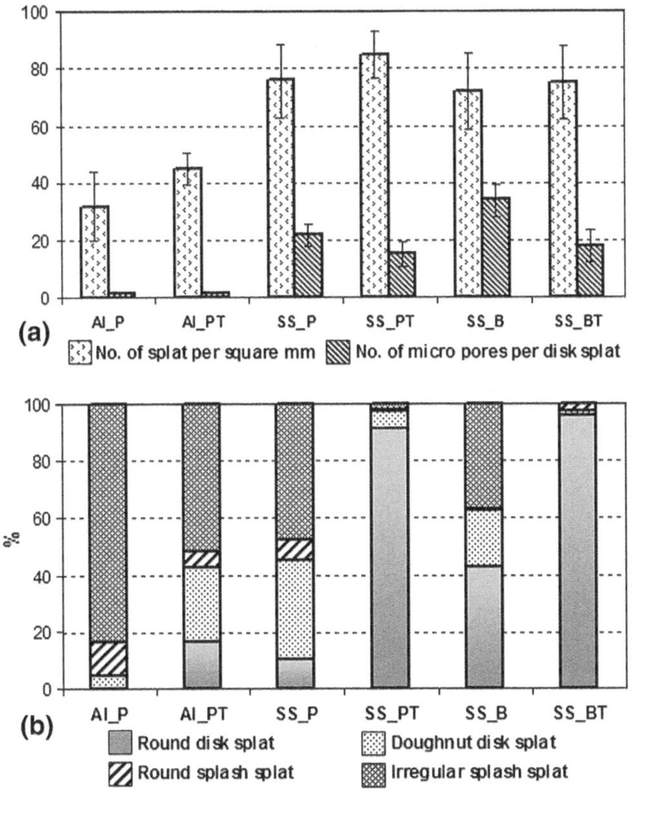

Fig. 4 Effect of surface pretreatment on the (a) splat density and number of micropores per disk-type splat with a standard deviation of 90% confidence and (b) splat distribution

Table 1 Surface roughness results

Substrates	Ra, nm	Sk
Al_P	11.6	−0.14
Al-PT	14.7	−0.14
Al-B	39.8	−0.41
Al-BT	15.1	0.22

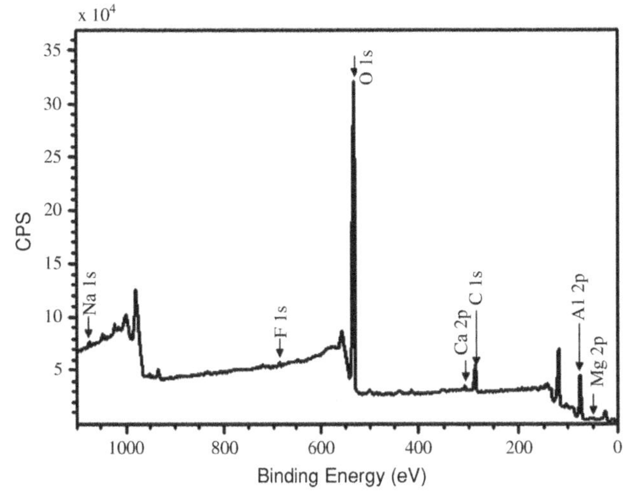

Fig. 5 Wide scan of Aluminum sample

samples. Preheating the boiled aluminum produced a smoother surface with a positive Sk value and an Ra value similar to that of polished samples.

Although there was no significant difference in the measured surface roughness between non-preheated and preheated polished aluminum substrates, the splat density and disk splat fraction greatly improved with preheat treatment. In addition, there was no splat formation on boiled aluminum substrates regardless of different surface roughness values. These results suggest that the differences in splat morphology between preheated and non-preheated substrates are not due to a surface roughness effect.

3.3 XPS Results

The change in surface chemistry due to preheating of aluminum and stainless steel was studied using XPS over a range of temperatures from ambient to 350 °C. A typical wide scan of polished aluminum sample at ambient temperature is shown in Fig. 5. The sample contained minor levels of impurities—Na, Ca, and F—due to contamination from tap water during washing and polishing.

The Al 2p envelopes were fitted by three different photopeaks. They are metal peak (~73.2-73.3 eV), oxide peak (~76.1-76.3 eV), and oxyhydroxide peak (~78.3-78.6 eV) (Fig. 6a) (Ref 20, 28). The surface oxide thickness (d, nm) was estimated from the Al 2p peak oxide/metal peak ratio using literature value for escape depth for the Al 2p photopeak (Ref 19).

The calculated oxide thickness of polished aluminum sample at room temperature and 350 °C was 5 and 6.4 nm, respectively. In contrast, the absence of metal peak in the Al 2p core level of boiled samples at ambient condition and 350 °C (Fig. 6b) indicated that the films formed by these processes were thicker than XPS analysis depth (8 nm). It was estimated from Ion Beam Analysis (IBA) that the oxide layer is approximately 220 nm thick.

The O 1s peak was resolved into three different photopeaks representing the oxide (~532.6-532.9 eV), oxyhydroxide (~533.9-534.5 eV), and chemisorbed water (~535.4 V) from lowest to highest binding energy, respectively (Ref 20, 28). The narrow scans of O 1s of polished and boiled aluminum samples at specified temperatures are illustrated in Fig. 7(a) and (b), respectively. It was observed that oxyhydroxide was present on all aluminum surfaces. In the polished sample, the surface sample consisted of a layer of chemisorbed water and oxyhydroxide. The presence of these peaks was attributed to the hydration reaction of the surface with moisture, which occurs rapidly under ambient conditions (Ref 13-16). However, the oxyhydroxide proportion was much smaller than the oxide proportion. In contrast, their proportion was equal in boiled sample, as expected, since the form of oxyhydroxide here, AlOOH, should have these two species in equal stoichiometries. In addition, the

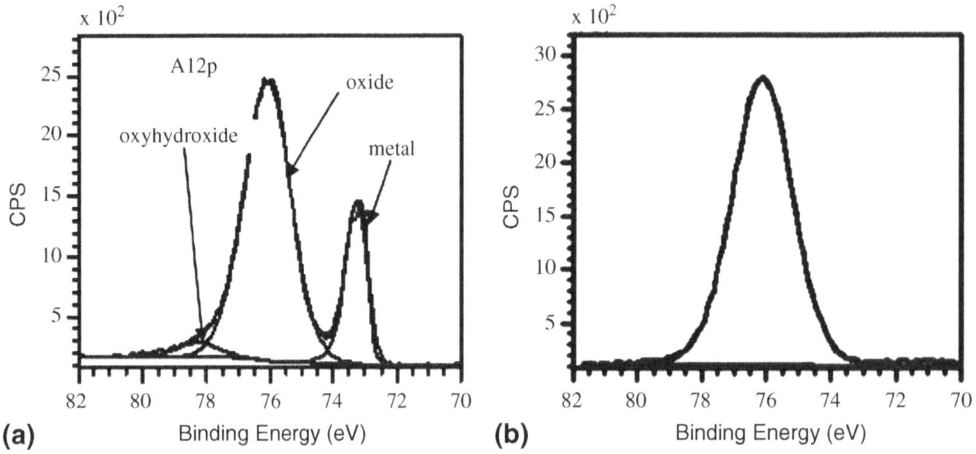

Fig. 6 Narrow scan of Al 2p (a) polished and (b) boiled/boiled and thermally treated sample

atomic ratio of O: Al in boiled sample was 2, which confirmed the formation of a surface layer of pseudo-boehmite, AlOOH (Ref 19, 20).

There was a significant change in the surface chemistry with heating, as evidenced by changes in the O 1s peak. When the temperature was raised to 100 °C with holding time of 30 min, the chemisorbed water was completely driven off. As the temperature was raised further, the oxyhydroxide gradually decreased in intensity and the oxide increased, on both polished and boiled aluminum. This is evidence that the oxyhydroxide is dehydrating, releasing water vapor as a product, and converting to oxide. This process, which also takes place in bulk aluminum oxides, begins at about 275 °C and is completed by 350 °C (Ref 16). For the thin surface oxyhydroxide/oxide layers on polished substrates, the entire surface layer was completely dehydrated in the 30-min treatment time. However, this cannot be confirmed for the much thicker layers on boiled substrates. Certainly the outermost 5-6 nm were dehydrated, but because the XPS can only probe the top 8 nm of the surface, the degree of dehydration of the underlying layers could not be established. The pretreatment also caused the surface oxide to increase slightly in thickness, from 5 to 6.4 nm. XPS was also carried out on stainless steel substrate. The results, not shown here, confirmed that, like the aluminum substrates, the surface of the stainless steel was composed of a thin layer of oxyhydroxide.

Some surface enrichment of minor elements, Mg in the case of aluminum and Cr in the case of stainless steel, was observed as a result of preheating (Table 2). However, there was no Mg segregation on pretreated boiled aluminum samples.

In short, the considerable change in surface chemistry at 350 °C suggests that under spraying conditions, when the hot droplet impacts the sample, the oxyhydroxide will dehydrate, releasing water as a gas that can affect the splat spreading process. In addition, the segregation of Mg and Cr to the surface may also affect the wettability (Ref 29) of the surfaces and thus also affect the splat formation process.

4. Discussion

The surface layer of all non-preheated substrates consists of a mixture of oxide, oxyhydroxide, and chemisorbed water, in varying proportions. The existence of this layer is due to the hydration reaction of the surface oxide with atmospheric moisture. Preheating results in dehydration: i.e., the release of water, first from the chemisorbed species at about 100 °C and then again between 300 and 350 °C, due to the conversion of the oxyhydroxide to oxide. A further change in the surface chemistry due to preheating was the segregation of Mg (for aluminum substrates) and Cr (for stainless steel substrates) at the surface. At the same time, the thickness of the oxide layer on polished substrates increases, but the roughness of this layer does not change significantly. Thus, the significant improvements of splat density, disk-type splat proportion, and the average diameter of disk splat with preheating are probably the result of the chemical changes at the substrate surface (prior removal of oxyhydroxide), rather than surface roughness. It is proposed that when the molten droplet impacts the substrate at high temperature, the dehydration of oxyhydroxide layer on the substrate surface to oxide occurs, and releases water vapor, which inhibits the adhesion of impinging splat. Preheating drives off water vapor resulting in better contact between splat and substrate, and hence enhancing disk splat and reducing splashing. For the relatively thin surface oxide/oxyhydroxide layers on the polished substrates, 5 nm, a 90-min pretreating time was sufficient to completely or nearly completely dehydrate the surface layer. Thus, preheating treatment was is an important parameter contributing to the formation of splat and splat shape on aluminum and stainless steel substrates because of the chemical changes, dehydration, induced at the surface.

In addition, the observed segregation of magnesium on the surface with heating is expected to affect the wetting properties of aluminum substrate. It has been reported that magnesium disrupts the oxide layer, reducing the

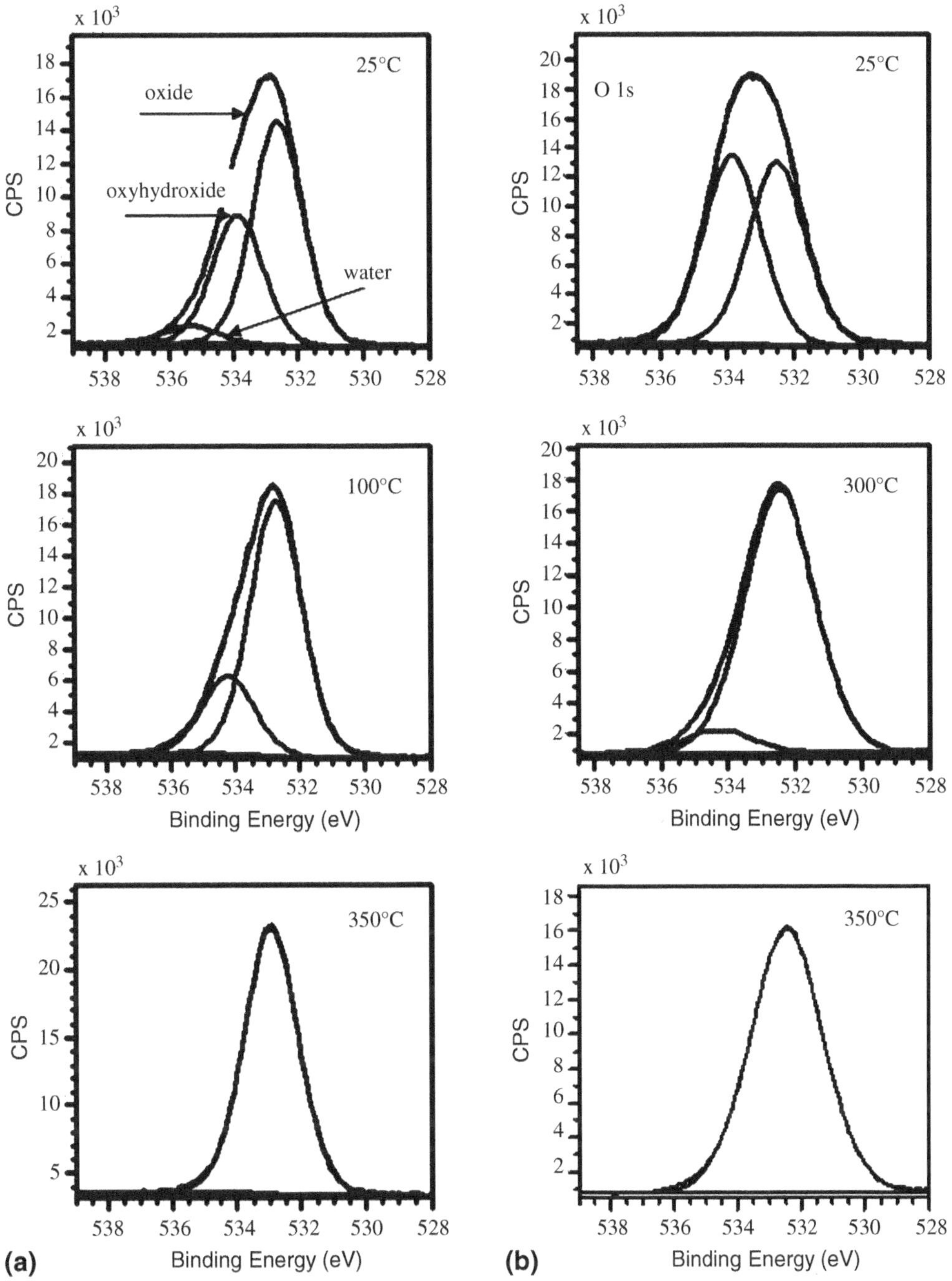

Fig. 7 Narrow scan of O1s (a) polished and (b) boiled aluminum with heating temperatures and holding time of 30 min

alumina proportion and producing clean and wettable surfaces (Ref 29). The greater the Mg proportion was on the surface, the more disk splats were formed. On stainless steel substrates, the existence of chromium on the surface improved wettability of NiCr splat on substrate thanks to its metal-like property. Thus, the difference of Cr proportion between polished and boiled stainless steel may explain the variations in disk splat fraction and also explain the difference in splat formation between boiled aluminum and boiled stainless steel substrates.

There was a complete absence of splats on boiled aluminum substrates. The boiled aluminum substrate represented the extreme example of water release from conversion of surface aluminum oxyhydroxide to oxide. These surfaces consisted of relatively thick layers (>200 nm) of an almost pure aluminum oxyhydroxide (pseudoboehmite, AlOOH). It appears that the desorption of large amount of water vapor at high temperature and thick layer of oxyhydroxide or oxide reduced the wettability of the substrate and impeded the spreading and adhesion of the splat. The reason for the lack of splat formation on the pretreated boiled aluminum is still unclear, but the greater similarity of the surface roughness with the polished substrates suggests that the answer

Table 2 Surface composition from various substrate treatments at different temperatures

Substrates	T, °C	O, %	Al, %	Mg, %
Al_P	20	48.2	35.2	1.2
	350	52.2	39.2	3.1
Al_B	20	60.8	30.8	0
	350	53.2	38.1	0
			Fe, %	Cr, %
St_P	20	48.9	10.9	1.4
	350	45.2	21.4	7.5
St_B	20	45.6	9.5	3.8
	350	40.3	17.5	7.2

probably lies in the surface chemistry, not surface topography. It may be that the oxyhydroxide was not completely dehydrated in the 90 min of pretreatment and subsequent water release during splat impact prevented splat formation or, alternatively, that the absence of Mg had a negative impact on the wettability.

The splat density and disk-type splat proportion were significantly higher on stainless steel substrates than on aluminum substrates, indicating that the adhesion of NiCr splats on stainless steel was much better than on aluminum. The average diameter of disk-type splat on stainless steel substrates was a little larger than that on aluminum substrates. It is speculated that the heat transfer between splat and substrate contributed to the splat density and final splat diameter. Because the thermal diffusivity of stainless steel is much smaller than that of aluminum, the solidification rate of NiCr splats on stainless steel is lower than that on aluminum. It has been found that fast solidification rate triggers the phenomena of freezing-induced splashing (Ref 30), explaining the high percentage of splash splat on aluminum substrates. In contrast, delaying solidification made the molten droplet to spread more to a round disk-type splat on stainless steel substrate before solidification occurs. As a results, the splat density, disk-type splat, and average diameter of splat were much larger on stainless steel substrates than on aluminum substrates.

5. Conclusion

The influence of preheating on splat morphologies suggests that surface chemistry, not surface roughness, is the dominating factor in the splat formation and splat shape. When the aluminum and stainless steel substrates were heated, Mg/Cr segregation and the dehydration of oxyhydroxide to oxide occurred. These changes correlated with measurable changes in splat number and morphology. The release of water due to the conversion of surface oxyhydroxide to oxide, which occurs as the hot splat impacts the surface, generates a gas layer that impedes the adhesion of the splat to the substrate. Prior conversion of the oxyhydroxide to oxide during thermal pretreatment, and thus prior removal of water, prevented this gas release, resulting in improved physical contact between the splat and substrate. In addition, the segregation of magnesium and chromium enhanced wettability of the substrate. As a result, the segregation of magnesium and chromium and/or the increase of the physical contact of splat and substrate enhanced the formation of disk splats, decreased the number of pores evident in the splats, and increased the number of splats and their diameter in preheated substrate. In addition, the complete absence of splats on boiled aluminum samples was due to the desorption of water vapor at high temperature from the thick layer of oxyhydroxide and the lack of segregation of Mg on the surface. All these factors reduced the wettability of the substrate and impeded the spreading and adhesion of the splat. However, the relative importance of each factor is not clearly identified. Therefore, more work is required to determine the separate effects of dehydration and segregation separately on the splat formation and splat shape.

Acknowledgments

The authors would like to acknowledge Tertiary Education Commission of New Zealand for Top Achiever Doctoral Scholarship. The authors thank S. Enright for surface roughness measurements.

References

1. M. Fukumoto, I. Ohgitani, M. Shiiba, and T. Yasui, Effect of Substrate Surface Change by Heating on Transition in Flattening Behavior of Thermal Sprayed Particles, *Mater. Trans.*, 2004, **45**(6), p 1869-1873
2. M. Fukumoto and Y. Huang, Flattening Mechanism in Thermal Sprayed Ni Particles Impinging on Flat Substrate Surface, *J. Thermal Spray Technol.*, 1999, **8**(3), p 427-432
3. M. Fukumoto, H. Nagai, and T. Yasui, Influence of Surface Character Change of Substrate due to Heating on Flattening Behavior of Thermal Spray Particle, *J. Thermal Spray Technol.*, 2006, **15**(4), p 759-764
4. Y. Tanaka and M. Fukumoto, Investigation of Dominating Factors on Flattening Behavior of Plasma Sprayed Ceramic Particles, *Surf. Coating Tech.*, 1999, **120-121**, p 124-130
5. M. Fukumoto, T. Yamaguchi, M. Yamada, and T. Yasui, Splash Splat to Disk Splat Transition Behavior in Plasma-Sprayed Metallic Materials, *J. Thermal Spray Technol.*, 2007, **16**(5-6), p 905-912
6. V. Pershin, M. Lufitha, S. Chandra, and J. Mostaghimi, Effect of Substrate Temperature on Adhesion Strength of Plasma-Sprayed Nickel Coatings, *J. Thermal Spray Technol.*, 2003, **12**, p 370-376
7. C. J. Li and J. L. Li, Evaporated-Gas-Induced Splashing Model for Splat Formation during Plasma Spraying, *Surf. Coating Tech.*, 2003, **184**(1), p 13-23
8. A. McDonald, M. Lamontagne, C. Moreau, and S. Chandra, Impact of Plasma-Sprayed Metal Particles on Hot and Cold Glass Surfaces, *Thin Solid Films*, 2006, **514**(4), p 212-222
9. A. A. Syed, A. Denoirjean, B. Hannoyer, P. Fauchais, P. Denoirjean, A.A. Khan, and J.C. Labbe, Influence of Substrate Surface Conditions on the Plasma Sprayed Ceramic and Metallic Particles Flattening, *Surf. Coating Tech.*, 2005, **200**(7), p 2317-2331
10. A.G. McDonald, M. Lamontagne, S. Chandra, and C. Moreau, Photographing Impact of Plasma Sprayed Particles on Metal Substrates, *J. Thermal Spray Technol*, 2006, **15**(4), p 708-716
11. X. Jiang, W. Yuepeng, H. Herbert, and S. Sanjay, Role of Condensates and Adsorbates on Substrate Surface on Fragmentation

of Impinging Molten Droplets during Thermal Spray, *Thin Solid Films*, 2001, **385**(1-2), p 132-141

12. A. McDonald, C. Moreau, and S. Chandra, Effect of Substrate Oxidation on Spreading of Plasma-Sprayed Nickel on Stainless Steel, *Surf. Coating Tech.*, 2007, **202**, p 23-33
13. W. Vedder and D. A. Vermilyea, Aluminum + Water Reaction, *Trans. Faraday Soc.*, 1969, **65**, p 561-584
14. R.S. Alwitt, The Growth of Hydrous Oxide Films on Aluminum, *J. Electrochem. Soc.*, 1974, **121**(10), p 1322-1328
15. W.J. Bernard and J.J. Randall, Jr., An Investigation of the Reaction between Aluminum and Water, *J. Electrochem. Soc.*, 1960, **107**(6), p 483-487
16. K. Wefers, and C. Misra, *Oxide and Hydroxides of Aluminium, No. 19*, Alcoa Laboratories, 1987, p 22-68
17. U.R. Evans, *The Corrosion and Oxidation of Metals. Scientific Principles and Practical Applications,* Edward Arnold, London, 1960
18. B. Withy, M. Hyland, and B. James, Pretreatment Effects on the Surface Chemistry and Morphology of Aluminium, *Int. J. Modern Phys. B*, 2006, **20**(25-27), p 3611-3616
19. M.R. Alexander, G.E. Thompson, and G. Beamson, Characterization of the Oxide/Hydroxide Surface of Aluminium Using X-ray Photoelectron Spectroscopy: A Procedure for Curve Fitting the O 1s Core Level, *Surf. Interface Anal.*, 2000, **29**(7), p 468-477
20. M.R. Alexander, G. Beamson, P. Bailey, T.C.Q. Noakes, P. Skeldon, and G.E. Thompson, The Distribution of Hydroxyl Ions at the Surface of Anodic Alumina, *Surf. Interface Anal*, 2003, **35**(8), p 649-657
21. A.P. Grosvenor, B.A. Kobe, and N.S McIntyre, Examination of the Oxidation of Iron by Oxygen using X-ray Photoelectron Spectroscopy and QUASES, *Surf. Sci.*, 2004, **565**(2-3), p 151-162
22. M.R. Alexander, G.E. Thompson, X. Zhou, G. Beamson, and N. Fairley, Quantification of Oxide Film Thickness at the Surface of Aluminium using XPS, *Surf. Interface Anal.*, 2002, **34**(1), p 485-489
23. A.P. Grosvenor, B.A. Kobe, and N.S. McIntyre, Studies of the Oxidation of Iron by Water Vapour using X-ray Photoelectron Spectroscopy and QUASES, *Surf. Sci.*, 2004, **572**(2-3), p 217-227
24. C.R. Werrett, D.R. Pyke, and A.K. Bhattacharya, XPS Study of Oxide Growth and Segregation in Aluminium-Silicon Alloys, *Surf. Interface Anal.*, 1997, **25**, p 809-816
25. N. Birks, G.H. Meier, and F.S. Pettit, *Introduction to the High-Temperature oxidations of Metals*, Cambridge University Press, 2006
26. B.P. Withy, M.M. Hyland, and B.J. James, The Effect of Surface Chemistry and Morphology on the Properties of HVAF PEEK Single Splats, *International Thermal Spray Conference: Thermal Spray Crossing Borders*, Maastricht, The Netherlands, DVS-Verlag GmbH, Düsseldorf, June 2-4, 2008, p 983-986
27. T.R. Thomas: *Rough Surfaces*, 2nd ed., Imperial College Press, 1999, Chap. 7, p 135-145
28. T. Do and N.S. McIntyre, Pressure Effects on Aluminium Oxidation Kinetics using X-ray Photoelectron Spectroscopy and Parallel Factor Analysis, *Surf. Sci.*, 1999, **440**, p 438-450
29. S. Lopez, J.R. Butruille, H.M. Dunlop, G. Tourillon, J.P. Petit, J.B. Metson, B.W. Yates, Hydration, Acido-Basicity, and Adhesion Properties of Passive Films on Aluminium: A Multitechnique Study, *Proc. Intl. Symp. Aluminum Surface Science & Technology*, Antwerp, Belgium, ATB Metallergie, May 12-15, 1997, p 143–150
30. R. Dhiman and S. Chandra, Freezing-Induced Splashing During Impact of Molten Metal Droplets with High Weber Numbers, *Int. J. Heat Mass Transfer*, 2005, **48**(25-26), p 5625-5638

JTTEE5 17:646–654
DOI: 10.1007/s11666-008-9238-5
1059-9630/$19.00 © ASM International

Peer Reviewed

The Effect of Undercooling on Solidification of YSZ Splats

Haibo Liu, Markus Bussmann, and Javad Mostaghimi

(Submitted May 21, 2008; in revised form September 2, 2008)

We consider the rapid solidification of a molten YSZ particle, by solving the so-called hyperbolic equations for heat and mass transfer. The hyperbolic model predicts the interface undercooling (due to thermal and solutal effects) and velocity as a function of time, as well as the yttria redistribution within the solid phase. Results are then compared to corresponding ones that we obtained from a parabolic model, to assess the extent to which YSZ solidification is influenced by nonequilibrium effects. Results indicate that these effects are limited to the early part of the solidification process when undercooling is most significant. At this stage, the interface velocity is unsteady, and solute redistribution is most evident. As solidification decelerates, the nonequilibrium effects wane and solidification can then be properly modeled as an equilibrium process.

Keywords hyperbolic, interface tracking, nonequilibrium, rapid solidification, relaxation time, undercooling, YSZ

1. Introduction

Thermal spray coating is an enabling technology, which has wide application in many industries including the gas turbine, aerospace, automotive, and biomedical (Ref 1). Zirconia-based ceramic coatings in particular are used as effective thermal barriers for metal structures in gas turbines (Ref 2, 3). These coatings can delay thermally induced failure, thus increasing the durability of machine components.

In the plasma spray process, yttria-stabilized zirconia (YSZ) powders are injected into a plasma jet, particles melt and accelerate until they collide with a substrate, spread, and solidify. The substrate material can vary, depending on use. Deposition of high quality coatings requires knowledge that involves materials and interface science, fluid mechanics, and heat transfer. This multidisciplinary character of the spray coating process implies that the development of new products and techniques requires a high investment. The simulation of spray coating formation can help optimize a design, and so save time and money.

This article is an invited paper selected from presentations at the 2008 International Thermal Spray Conference and has been expanded from the original presentation. It is simultaneously published in *Thermal Spray Crossing Borders, Proceedings of the 2008 International Thermal Spray Conference*, Maastricht, The Netherlands, June 2-4, 2008, Basil R. Marple, Margaret M. Hyland, Yuk-Chiu Lau, Chang-Jiu Li, Rogerio S. Lima, and Ghislain Montavon, Ed., ASM International, Materials Park, OH, 2008.

Haibo Liu, Markus Bussmann, and **Javad Mostaghimi**, Department of Mechanical and Industrial Engineering and Centre for Advanced Coating Technologies, University of Toronto, 5 King's College Road, Toronto, ON, Canada M5S 3G8. Contact e-mail: bussmann@mie.utoronto.ca.

List of Symbols

C	concentration (wt.%)
C_0	initial concentration (wt.%)
C_p	specific heat capacity (J kg^{-1} K^{-1})
b	splat thickness (m)
D	mass diffusivity (m^2 s^{-1})
h	heat transfer coefficient (W m^{-2} K^{-1})
J_C	concentration flux (wt.% m s^{-1})
k_f	nonequilibrium partition coefficient
k_e	equilibrium partition coefficient
m	slope of nonequilibrium liquidus (K wt.%$^{-1}$)
m_e	slope of equilibrium liquidus (K wt.%$^{-1}$)
Q	latent heat of solidification (J kg^{-1})
T	temperature (K)
T_m	equilibrium melting temperature (K)
T_0	initial temperature (K)
V_D	mass diffusion velocity (m s^{-1})
V_i	interface velocity (m s^{-1})

Greek Symbols

α	thermal diffusivity (m^2 s^{-1})
κ	thermal conductivity (W m^{-1} K^{-1})
μ	linear kinetics coefficient (m s^{-1} K^{-1})
ρ	density (kg m^{-3})
τ	relaxation time (s)

Subscripts and Superscripts

i	interface
j	solid or liquid phase
−	liquid interface
+	solid interface

Pure zirconia has a complex allotropic structure (Ref 4). A cubic crystal structure exists between 2400 and 2700 K; it then transforms into a tetragonal structure as it further cools. When the temperature reaches about 1000 K, the tetragonal crystals transform to monoclinic form. This phase transformation involves volume change that may cause compression or tension in the coating, which can induce surface cracking and poor mechanical flexibility. To circumvent these problems, a zirconia yttria (6-8 wt.%) mixture, which has no transformation to a monoclinic phase at low temperature is used. Further, research (Ref 4) has found that when rapidly solidified, the zirconia-rich corner of the phase diagram has a tetragonal-prime phase, which is meta-stable.

Few numerical studies of the rapid solidification of YSZ have been published (Ref 2). Pure metal models are only useful in conditions far different from those of ceramic solidification (e.g., Ref 5), and so here we solve the coupled thermal and concentration equations for a binary system. Also, Fourier's model is usually used to simulate heat conduction during solidification, but this model is only appropriate for relatively slow processes. To solve a rapid solidification problem, this so-called parabolic model should be replaced by a hyperbolic one, by adding a relaxation term to Fourier's model to take into account nonequilibrium effects related to the interface velocity-induced heat wave (Ref 6). Using a similar strategy based on the theory of so-called extended irreversible thermodynamics, Sobolev (Ref 7) added the relaxation term to the equilibrium equation for concentration diffusion, to model alloy solidification as a hyperbolic process. And to simplify the problem, Sobolev and Galenko assumed a constant interface velocity at the solidification front, and so were able to obtain analytic solutions for planar and dendritic interfaces (Ref 7, 8, 9).

Numerical models of spray coating consider a single droplet, or a spray of them, impacting the surface of a substrate. As the droplet contacts the surface, heat is released that fluxes into the substrate; grains begin to grow when nuclei are beyond the critical size of heterogeneous nucleation. The microstructure of a spray coating, including the grain shape, density, and orientation, is affected by the interface velocity (Ref 2), which is determined via an extra undercooling constraint (Ref 2, 5).

Nonequilibrium effects associated with the relaxation term occur if the solid/liquid interface velocity is comparable to the diffusive speed (Ref 7). Determining an accurate interface velocity profile can help to explain the microstructure of a solidified material. The interface velocity also affects the splat morphology, as it has been observed that splat structure can vary with the substrate temperature. Note that an important factor that affects the solidification speed is the change of thermal contact resistance with temperature. For metal substrates, thermal contact resistance increases with increasing substrate temperature, because the oxide layer formed at the surface becomes thicker. For nonmetal surfaces, the situation is reversed, due to the desorption of adsorbed gases on the surface. Figure 1 shows YSZ splats deposited on stainless steel at different temperatures, and illustrates how the change in substrate temperature and thus thermal contact resistance affects the spread time and the cooling rate of the splats, and thus the overall splat morphology. Similar experiments were conducted by Bianchi et al. (Ref 10) and Fukomoto et al. (Ref 11), who presented scanning electron micrographs of individual alumina and zirconia splats deposited by plasma spraying on a stainless steel

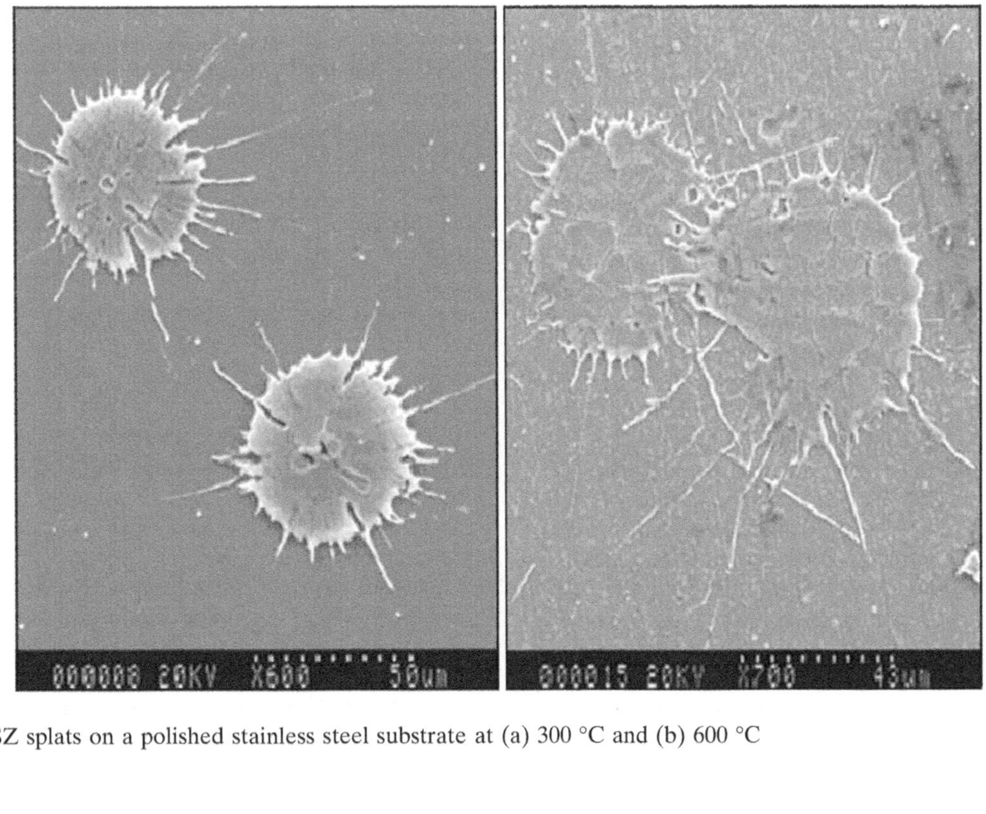

Fig. 1 Single YSZ splats on a polished stainless steel substrate at (a) 300 °C and (b) 600 °C

plate. They found that droplets landing on a cold substrate were more irregular in shape, with fingers radiating out from their centers, while droplets that landed on a hot substrate were usually shaped like disks. They explained that the solidified layer next to the substrate triggers radial jetting of the liquid and so leads to splashing (Ref 12). Therefore, the faster the interface velocity, the faster the solid layer forms, and the more likely the splat will splash. This explanation suggests that interface velocity can affect the morphology of a splat. The interface velocity also can affect the splat microstructure. According to the grain formation model of Kurz and Fisher (Ref 13), the morphology, tip radius, and cell and dendrite spacings are all closely related to interface velocity. Therefore, determining an accurate interface velocity enables us to analyze splat morphology. The interface tracking model (Ref 5, 14) we present here, for the rapid solidification of YSZ, yields an accurate interface velocity profile.

At the splat/substrate interface, it is well known that the interface velocity of the solidification front is strongly affected by the substrate temperature (Ref 12) as well as by the condition of the substrate (e.g., contamination (Ref 15), oxidation (Ref 16), roughness (Ref 17)). To simplify our simulation, these effects are represented by two parameters that we varied: the surface temperature and a heat transfer coefficient.

To calculate the local yttria concentration at the interface, the kinetic solute partition function k_f is a critical coefficient. k_f is the ratio of the solute concentration on the solid side of the dynamic solidification front to the concentration on the liquid side. The concentration difference at the solid/liquid interface is what drives solute redistribution. k_e is the ratio of the equilibrium solidus to liquidus concentrations based on the phase diagram. At high interface velocity, k_f deviates from k_e. Aziz (Ref 18, 19) derived an expression for k_f via a continuous growth model that assumes a "rough interface" that provides sites for crystallization events (Ref 19); Sobolev and Galenko (Ref 7, 9) extended this expression to obtain a nonequilibrium partition coefficient based on a modified Fick's law.

Another coefficient that affects the local yttria concentration at the interface is the slope of the liquidus m. m_e is the equilibrium liquidus slope based on the phase diagram. At higher interface velocities, m deviates from m_e. Aziz and Boettinger (Ref 19) obtained the slope of the nonequilibrium liquidus by expressing a chemical potential balance of a binary concentration system, from which the total Gibbs free energy could be expressed; as with k_f, Galenko (Ref 9) extended the expression for m to include hyperbolic effects.

Both physical and mathematical complications of a binary rapid solidification process lead to equations that are challenging to solve numerically and theoretically. Wang (Ref 14) simulated planar rapid solidification using an interface tracking method, but solved the parabolic rather than the hyperbolic equations. Sobolev and Galenko (Ref 7, 9) solved the hyperbolic equations analytically; they obtained a steady-state solution by assuming a constant interface velocity. For this work, we implemented the iterative interface tracking method of Wang (Ref 14), which is both accurate and efficient, but applied the methodology to solve the hyperbolic equations to predict solidification characteristics that include the solid/liquid interface velocity, temperature, and the partition coefficient. This information in turn can be used to interpret experimental results such as those illustrated in Fig. 1.

2. Physical Model and Methodology

To solve for the nonequilibrium phase change of YSZ rapidly solidifying on a substrate, we have extended the 1D model of Wang (Ref 14), based on the following assumptions and idealizations:

1. an impacting droplet rapidly forms a thin layer of molten YSZ, with uniform initial temperature; the droplet is assumed to spread rapidly so that solidification can be modeled as a 1D phenomena; the validity of this assumption depends on the particular parameters of the droplet impact, but given the complexity of the heat and mass transfer analysis presented here and the focus on solidification just above the substrate, an analysis that also incorporates flow would make the problem intractable;
2. the interface velocity is linearly related to the undercooling;
3. the droplet and substrate are in contact along a stable planar interface, and the heat transfer coefficient at the droplet/substrate interface is assumed to be constant, inferred from experimental measurements;
4. the heat transfer coefficient at the upper surface of the splat is much smaller than that at the droplet/substrate interface, and so we assume it to be zero.

Fick's First Law states that the flux of a component across a concentration field is proportional to the concentration gradient, and so assumes an infinite velocity of concentration propagation:

$$J_c + D_j \nabla C_j = 0 \qquad \text{(Eq 1)}$$

where C is the concentration of the solute, D is the solute diffusivity, J_c is the diffusion mass flux, and the subscript j represents the solid or liquid phase. Diffusion of a dissolved component in a solid or liquid phase can be expressed as a concentration balance:

$$\nabla \cdot J_c + \frac{\partial C_j}{\partial t} = 0 \qquad \text{(Eq 2)}$$

where t is time. It is important to note that the above equations assume microscopic reversibility, which means that the concentration flux is uniquely linked to diffusion, and that the inertial effects of the concentration flux acceleration are ignored. Namely, a signal change at one point anywhere in a concentration field will instantaneously be felt everywhere else. The above equations accurately predict most experimental results. However, when mass transport is relatively high, the relaxation time to accelerate the mass flux will be relatively large; in such a case, a relaxation term should be included in the transport equation.

Cattaneo (Ref 20) first proposed a damped version of Fourier's heat conduction law in 1948. In the case of mass transport, a modified Fick's law can be expressed in a similar way. It includes a mass flux relaxation term that involves a relaxation time τ_D (Ref 5). In a local non-equilibrium situation, τ_D is the time interval necessary for atoms to jump onto a neighboring lattice, and thus for the mass flux to reach a stable state.

Adding a relaxation term to Fick's First Law yields a general Fick's Law (Ref 9):

$$\tau_D \frac{\partial J_c}{\partial t} + J_c + D_j \nabla C_j = 0 \qquad \text{(Eq 3)}$$

where $\tau_D = D_j/V_D{}^2$ (Ref 21). According to previous work on pure metal solidification (Ref 5), in which we solved only for the thermal field, a value of τ_κ (the thermal equivalent of τ_D) on the order of 10^{-9} s yielded negligible differences between the parabolic and hyperbolic models. But as YSZ has a relatively low mass diffusivity and a low diffusive speed V_D ~1 m s^{-1} (Ref 22), τ_D for the concentration equation is also on the order of 10^{-9} s, yet one would expect a more dramatic effect of this term. Also, the solute partition coefficient and the liquidus slope are both related to V_D, which led us to investigate the effect of τ_D on the solidification of YSZ.

To analyze the extent of nonequilibrium behavior is to compare the magnitudes of the solid/liquid normal interface velocity V_i and the diffusive speed V_D of the melt. When V_i is comparable to but smaller than V_D, the diffusive velocity cannot be assumed to be infinite. Solute propagation is then a joint process of diffusion and wave propagation (Ref 7). As $V_i \to V_D$, the wave mechanism dominates. When $V_i > V_D$, diffusion in the liquid ceases, and solidification is then unrelated to the concentration distribution. Hence, the extent of nonequilibrium behavior is related to the magnitudes of V_i and V_D: if these velocities are comparable, then concentration-induced nonequilibrium effects will be more pronounced.

Combining the above equations, a hyperbolic equation for the concentration is obtained (Ref 7):

$$\frac{\partial C_j}{\partial t} + \tau_D \frac{\partial^2 C_j}{\partial t^2} = D_j \nabla^2 C_j \qquad \text{(Eq 4)}$$

and a hyperbolic heat conduction equation can be expressed similarly:

$$\frac{\partial T_j}{\partial t} + \tau_\kappa \frac{\partial^2 T_j}{\partial^2 t} = \alpha_j \nabla^2 T_j \qquad \text{(Eq 5)}$$

where α_j is the thermal diffusivity, T is the temperature, and τ_κ represents the time lag between the temperature gradient and the commencement of heat flow in a medium. Here we assume this thermal time lag is also associated with the time required for atoms to jump from the liquid side to the solid side, and so we assume τ_κ and τ_D to be the same. This model, then, describes heat wave propagation with finite velocity. But, as already noted, τ_κ in its proper range will have little effect on the thermal field, and so although we assume τ_κ and τ_D to be equal, we expect differences between the parabolic and hyperbolic models to result from the concentration equation.

At the solid/liquid interface we impose an extra constraint on the governing equations (Ref 2):

$$T_i = T_m + m \cdot C_i^- - V_i/\mu \qquad \text{(Eq 6)}$$

where μ is the kinetic coefficient of interface motion, T_m is the equilibrium melting temperature, and C_i^- is the interface concentration on the liquid side. Because we assume a planar interface, there is no curvature-induced undercooling.

The magnitude of nonequilibrium effects depends on V_i. As $V_i \to V_D$, the local concentration at the interface is affected, because the magnitude of the concentration flux at the interface is dominated by the wave mechanism instead of by species diffusion. As $V_i \to V_D$, partitionless phenomena occur, and a generalized partition function for solute portioning is required. Meanwhile the liquidus curve on the phase diagram is shifted because of solute trapping, which requires the use of a generalized form of the liquidus slope to consider such a shift. The generalized equations for the partition coefficient and for the liquidus slope are presented below.

Regarding the temperature field, local thermal equilibrium of an alloy is established much more quickly than the local equilibrium of the diffusion field, to the extent that Galenko (Ref 21) ignored the thermal nonequilibrium term in the governing equations. In our simulations, we keep the nonequilibrium term, but we ignore the nonequilibrium thermal effects (Ref 5). However, as the dynamic interface condition (Eq 6) is a constraint on the hyperbolic governing equations (Eq 4, 5), the temperature field is still affected by hyperbolic effects.

At the interface, the concentration undercooling is given by Eq 6. To calculate the concentration at the planar interface, the concentration segregation

$$C_i^+ = k_f C_i^- \qquad \text{(Eq 7)}$$

is required (C_i^+ is the interface concentration on the solid side), and k_f is the nonequilibrium partition coefficient, expressed by Aziz et al. (Ref 18, 19) as:

$$k_f = \frac{k_e + V_i/V_D}{1 + V_i/V_D} \qquad \text{(Eq 8)}$$

where k_e is the equilibrium partition coefficient.

By introducing an effective diffusion coefficient $D^* = \left(1 - \frac{V_i^2}{V_D}\right)$ for $V_i < V_D$ and following the derivation of Aziz (Ref 18, 19), Sobolev and Galenko (Ref 7, 9) obtained a more general partition coefficient:

$$k_f = \frac{k_e(1 - V_i^2/V_D^2) + V_i/V_D}{1 - V_i^2/V_D^2 + V_i/V_D}, \quad V_i < V_D \qquad \text{(Eq 9)}$$

and

$$k_f = 1, \quad V_i \geq V_D \qquad \text{(Eq 10)}$$

since solute is then completely trapped. Both expressions for k_f (Eq 8, 9, 10) have been used previously by researchers investigating rapid solidification.

Peer Reviewed

In Eq 6, the nonequilibrium liquidus slope m is also unknown. Based on arguments described in the Introduction, Aziz and Boettinger derived the following expression:

$$m = m_e[1 + (k_e - k_f(1 - \ln(k_f/k_e))/(1 - k_e)] \quad \text{(Eq 11)}$$

where m_e is the slope of the equilibrium liquidus. An alternative expression for m was derived by Galenko (Ref 9) based on the hyperbolic model; adding a diffusion relaxation term to the steady diffusion flux leads to:

$$m = \frac{m_e}{1 - k_e}\left[1 - k_f + \ln(k_f/k_e) + (1 - k_f)^2(V_i/V_D)\right], \quad V_i < V_D \quad \text{(Eq 12)}$$

$$m = \frac{m_e \ln k_e}{k_e - 1}, \quad V_i \geq V_D \quad \text{(Eq 13)}$$

As for k_f, both expressions for m (Eq 11-13) have been used to investigate rapid solidification.

All that remains is to specify the initial and boundary conditions. The initial conditions are:

$$T(x,0) = T_0, \quad C(x,0) = C_0, \quad J(x,0) = J_C(x,0) = 0 \quad \text{(Eq 14)}$$

At the upper surface of the splat ($x = b$):

$$\frac{\partial T(b,t)}{\partial x} = 0, \quad \frac{\partial C(b,t)}{\partial x} = 0 \quad \text{(Eq 15)}$$

At the base of the splat, we simplify the thermal boundary condition (Ref 5), and assume no mass transfer:

$$J^+ = h[T(x,0) - T_\infty], \quad \frac{\partial C(0,t)}{\partial x} = 0 \quad \text{(Eq 16)}$$

where T_∞ is a specified substrate temperature, δy is the thickness of the solidified layer, and h is the heat transfer coefficient. Since the mass diffusivity on the solid side is several orders of magnitude less than on the liquid side, we ignore concentration diffusion in the solid.

At the moving solid/liquid interface the following is an expression of energy conservation:

$$T_Q \cdot V_i = J^- - J^+ \quad \text{(Eq 17)}$$

where $T_Q = Q \cdot C_p^{-1}$ is the temperature of adiabatic solidification, Q is the latent heat, C_p is the specific heat, and J^- and J^+ represent the heat fluxes from the liquid and solid sides of the solidification front.

The corresponding conservation equation for concentration at the solid/liquid interface is the following flux balance:

$$\left(C_i^- - C_i^+\right) \cdot V_i = J_C^- - J_C^+ \quad \text{(Eq 18)}$$

where C_i^- and C_i^+ are the concentrations on the liquid and solid sides of the solidification front, and J_C^- and J_C^+ represent the corresponding fluxes.

Finally, at the solid/liquid interface, the expressions for k_f and m derived both by Aziz (Eq 8, 11) and by Galenko (Eq 9, 10, 12, 13) were applied to solve the parabolic and hyperbolic sets of equations, respectively.

By assuming that solidification begins when the interface temperature between the splat and substrate reaches the nucleation temperature (below the equilibrium melting temperature), the interface velocity is then linearly related to the local undercooling via a crystal growth kinetics relationship (Eq 6), which is an extra constraint on the governing equations (Eq 4, 5).

To solve the system of equations numerically, a common solution strategy was applied. Rather than solve concentration Eq 4, we solve Eq 2 and 3, which are two first-order partial differential equations, as well as Eq 5. We fix the moving solid/liquid interface by transforming the physical coordinate x to a computational one η (see Fig. 2):

$$\eta = \left(\frac{x}{\varepsilon}\right)^p \quad \text{(Eq 19)}$$

p adjusts the mesh density in the liquid region (Ref 5); this non-uniform transformation is necessary to solve the system efficiently. The governing equations in the computational domain ($0 \leq \eta \leq 1$) are then the following:

$$\frac{\partial C}{\partial t} - p\frac{d\varepsilon}{dt}\frac{\partial C}{\partial \eta} + p\eta^{\left(\frac{p-1}{p}\right)}\frac{1}{\varepsilon}\frac{\partial J_c}{\partial \eta} = 0 \quad \text{(Eq 20)}$$

$$\tau_D\frac{\partial J_c}{\partial t} - \tau_D p\frac{\eta}{\varepsilon}\frac{d\varepsilon}{dt}\frac{\partial J_c}{\partial \eta} + J_c + p\eta^{\left(\frac{p-1}{p}\right)}\frac{D}{\varepsilon}\frac{\partial C}{\partial \eta} = 0 \quad \text{(Eq 21)}$$

$$\rho c\frac{\partial T}{\partial t} - \rho c p\frac{d\varepsilon}{dt}\frac{\partial T}{\partial \eta} + p\eta^{\left(\frac{p-1}{p}\right)}\frac{1}{\varepsilon}\frac{\partial J}{\partial \eta} = 0 \quad \text{(Eq 22)}$$

$$\tau_D\frac{\partial J}{\partial t} - \tau_D p\frac{\eta}{\varepsilon}\frac{d\varepsilon}{dt}\frac{\partial J}{\partial \eta} + J + p\eta^{\left(\frac{p-1}{p}\right)}\frac{\kappa}{\varepsilon}\frac{\partial T}{\partial \eta} = 0 \quad \text{(Eq 23)}$$

MacCormack's predictor-corrector scheme was used to solve this set of equations: at each time step, we iteratively solved for a V_i that also satisfied the thermal and mass conservation conditions at the interface (Ref 5).

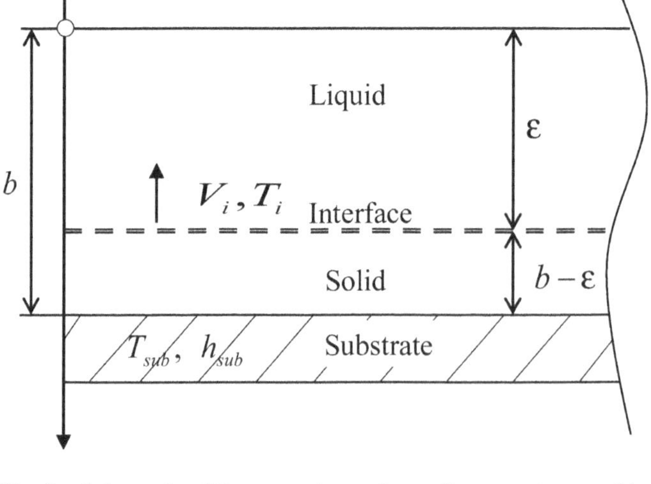

Fig. 2 Schematic of the geometry and coordinate system used in the numerical simulations (Ref 5)

3. Results and Discussion

The physical properties of YSZ (Ref 2, 4, 22) used in these simulations are specified in Table 1. YSZ has a relatively low thermal conductivity and thus a low diffusive speed $V_D \sim 1.0$ m s^{-1}, which is why we solve the hyperbolic heat and concentration equations, although we also solved the parabolic equations, and compared the results. At the melt/substrate interface, we specified a heat transfer coefficient $h = 1.0 \times 10^7$ or 5.0×10^7 W m^{-2} K^{-1} (Ref 2). The substrate temperature was maintained at a constant 500 K.

To solve the concentration equation we used 1000 nodes, concentrated near the solid/liquid interface to resolve the concentration gradient. We required only 100 nodes distributed uniformly to solve the energy equation over the same domain. The equilibrium melting temperature of YSZ was specified as 2950 K (Ref 2) and the initial undercooling as 150 K; thus when the interface temperature at the melt/substrate interface initially reached 2800 K, grain growth began. Finally, we set the kinetic coefficient $\mu = 0.0012$ m s^{-1} K^{-1} (Ref 2).

As indicated in Fig. 3, the parabolic and hyperbolic equations yield very similar curves of interface velocity V_i during the early stages of solidification; the hyperbolic results always predict a slightly higher velocity, and the difference between the models is more pronounced for the higher value of h. However, compared to results obtained for pure metal solidification (Ref 5), the difference between the parabolic and hyperbolic models (Fig. 3) is actually more significant. With $h = 1.0 \times 10^7$ W m^{-2} K^{-1}, V_i increases sharply from 0.1 m s^{-1} to 0.8 m s^{-1} at the very beginning of solidification, before decreasing slowly. With $h = 5.0 \times 10^7$ W m^{-2} K^{-1}, the maximum velocity is 1.5 m s^{-1}, which is larger than V_D. These velocity curves are very similar to those obtained by Wang (Ref 2).

Profiles of interface temperature T_i are presented in Fig. 4, and again, the difference between results obtained by the parabolic and hyperbolic models is relatively small, although more pronounced at the higher value of h. With $h = 1.0 \times 10^7$ W m^{-2} K^{-1}, T_i decreases from 2850 K to 2250 K abruptly at the initial stage of solidification, before T_i plateaus. The variation of T_i when $h = 5.0 \times 10^7$ W m^{-2} K^{-1} is even more dramatic, as the temperature falls to about 1600 K before recovering. In both cases the rapid initial temperature decrease is related to the initially low interface velocity, which in turn implies a weak latent heat release; as V_i increases, the latent heat release becomes intense, and compensates for the heat loss to the substrate, which causes the temperature to stabilize. As for why the hyperbolic model yields temperatures that are lower than those obtained from the parabolic model, consider that interface velocity is related to undercooling; because the hyperbolic model predicts a higher interface velocity, the undercooling is greater, and so T_i is lower.

Figure 5 shows k_f versus location of the solidification front, and unlike the previous results, the parabolic and hyperbolic models yield dramatically different results. With $h = 1.0 \times 10^7$ W m^{-2} K^{-1}, the parabolic model predicts that k_f decreases from 1.9 to 1.7; the hyperbolic model predicts that k_f decreases to 1.3. Much more

Table 1 Physical properties of YSZ used in the calculations (Ref 2, 4, 22)

Parameter	Units	YSZ
T_0	K	3250
T_m	K	2950
D_L	m^2 s^{-1}	1.0×10^{-9}
D_S	m^2 s^{-1}	$\sim 10^{-15}$
k_e		2.1
κ_L	W m^{-1} K^{-1}	3
κ_S	W m^{-1} K^{-1}	2
C_{PL}	J kg^{-1} K^{-1}	713
C_{PS}	J kg^{-1} K^{-1}	580
ρ_L	kg m^{-3}	5890
ρ_S	kg m^{-3}	5890
m_e	K wt.%$^{-1}$	2.9
C_0	wt.%	8
Q	J kg^{-1}	8.12×10^5
V_D	m s^{-1}	1~2
μ	m s^{-1}K^{-1}	0.0012

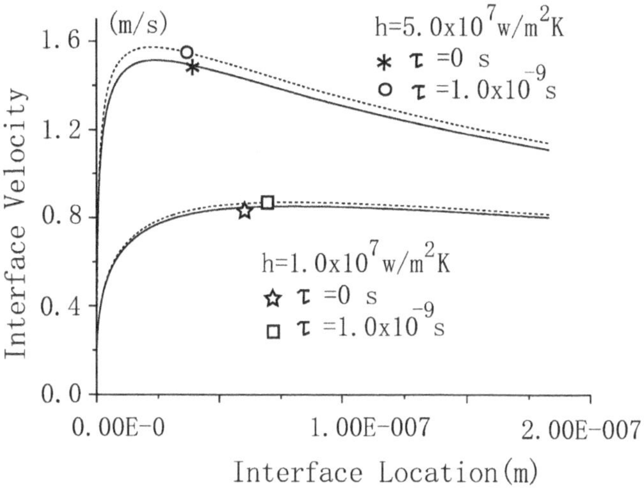

Fig. 3 Interface velocity versus interface location

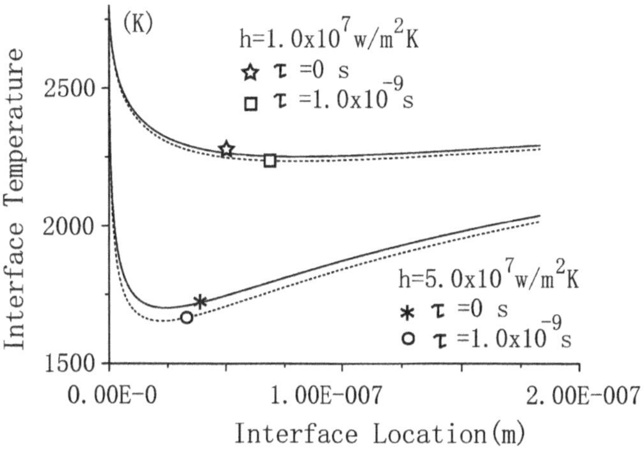

Fig. 4 Interface temperature versus interface location

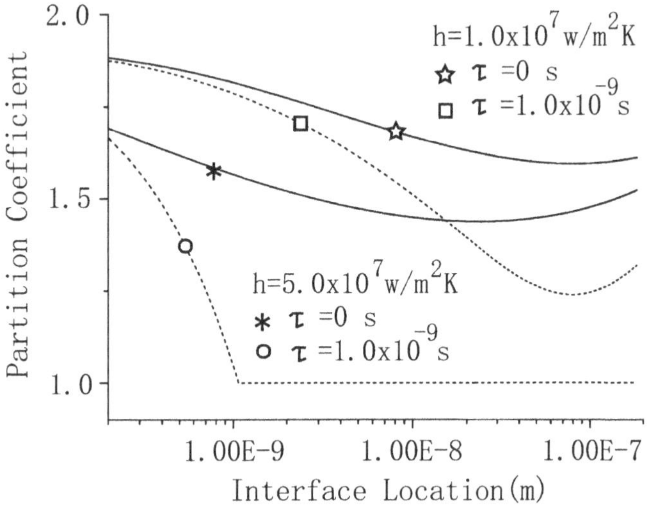

Fig. 5 Partition coefficient versus interface location

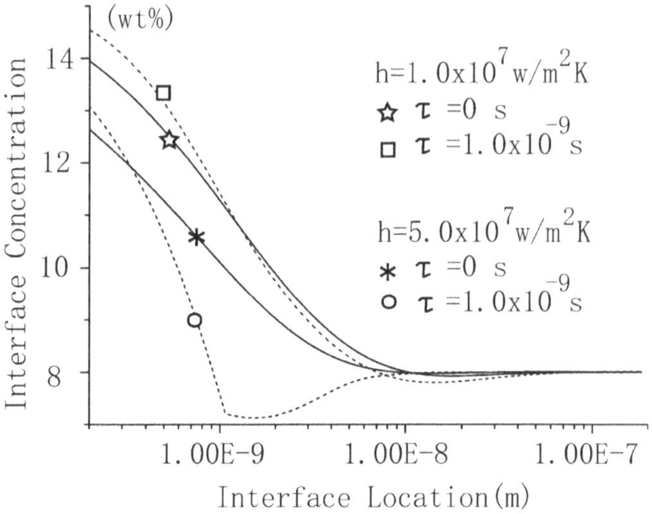

Fig. 6 Interface concentration at solid side versus interface location

interesting is when $h = 5.0 \times 10^7$ W m^{-2} K^{-1}, because the hyperbolic model predicts a discontinuous variation of k_f, which quickly falls to 1, after which there is no segregation. The reason, according to Galenko's model (Eq 10, 13), is that $V_i \rightarrow V_D$ is a critical condition, because $k_f \rightarrow 1$ and concentration trapping begins. The parabolic model based on Aziz's equations, on the other hand, has no critical condition and so the curves are smooth. Finally, note that because the hyperbolic model predicts a higher V_i, it then approaches V_D more rapidly than the parabolic model.

Figure 6 presents C^+ (the yttria concentration on the solid side of the interface) versus location of the solidification front. At the very beginning of solidification, because the liquidus slope of YSZ is positive (according to the phase diagram), the partition coefficient is greater than 1 and so $C^+ > C_0$ for all results. With $h = 5.0 \times 10^7$ W m^{-2} K^{-1}, C^+ (predicted by both the parabolic and hyperbolic models) is lower than when $h = 1.0 \times 10^7$ W m^{-2} K^{-1}, because $k_f > 1$, and according to Eq 9, k_f decreases as V_i increases. When V_i is higher, segregation is lower, and therefore C^+ is lower. With $h = 5.0 \times 10^7$ W m^{-2} K^{-1}, the hyperbolic model yields a concentration that actually drops to a value less than C_0 before reaching a minimum; after this point, C^+ approaches C_0 when concentration is totally trapped. Finally, note that all of the curves show that C^+ approaches C_0 (8 wt.%) at steady state, which is evidence of mass conservation.

Figure 7 presents C^- (the yttria concentration on the liquid side of the interface) versus location of the solidification front. Both the hyperbolic and parabolic models predict C^- to be less than 8.0 wt.%. When $h = 1.0 \times 10^7$ W m^{-2} K^{-1}, C^- drops rapidly from 8.0 wt.% to a minimum value, then quickly approaches an equilibrium that depends on whether the parabolic or hyperbolic equations are being solved. The physical explanation for this phenomenon is as follows: at the beginning of solidification, because interface velocity is relatively low, the partition coefficient is relatively high, and so C^- decreases

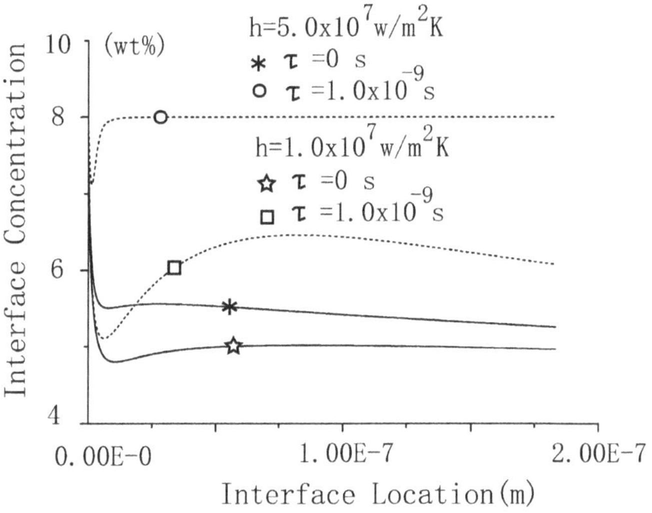

Fig. 7 Interface concentration at liquid side versus interface location

to maintain a higher concentration on the solid side. As V_i increases, the partition coefficient drops, and so C^- begins to rise. Also worth noting is that C^- rises faster with the hyperbolic model, because V_i is higher, and so C^- approaches steady state more quickly. When $h = 5.0 \times 10^7$ W m^{-2} K^{-1}, $V_i \rightarrow V_D$ and so solute is trapped, which explains why the hyperbolic curve shows that C^- rapidly equals C_0. Finally, note that the difference between the hyperbolic and parabolic curves of C^- is much more pronounced than the differences in the curves of V_i and T_i, and that the difference becomes larger as h increases.

The surface temperature beneath a splat will affect the strength of adhesion. Profiles of this temperature are presented in Fig. 8, and again, the difference between the parabolic and hyperbolic models is relatively small, although more pronounced at the higher value of h. With $h = 1.0 \times 10^7$ W m^{-2} K^{-1}, the temperature decreases

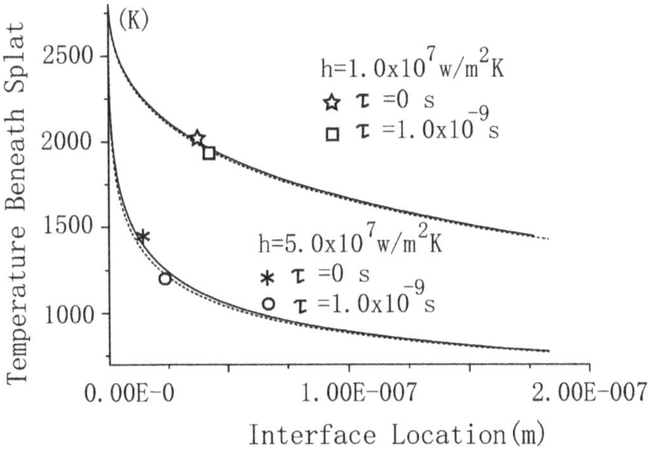

Fig. 8 Temperature beneath a splat versus interface location

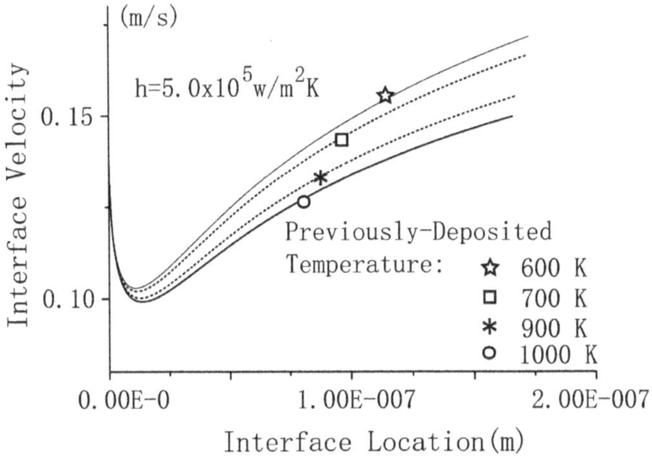

Fig. 9 Interface velocity versus interface location, for solidification onto a previously deposited splat

from 2850 K to 1500 K abruptly at the initial stage of solidification, and then plateaus. The variation of the temperature at $h = 5.0 \times 10^7$ W m^{-2} K^{-1} is even more dramatic, as the temperature immediately falls to about 700 K. Clearly the splat/substrate temperature is closely related to h. To obtain strong adhesion, the substrate surface is usually prepared before spraying, to get rid of surface contamination and oxidation. Those factors will strongly affect the heat transfer coefficient.

Finally, the results so far have considered a single splat solidifying on a stainless steel substrate. More generally, there are multiple layers of splats that fall upon each other, and the previously deposited splats can be either solid or liquid. To simulate the solidification of multiple splats, we assumed a previously deposited temperature in the range from 600 to 1000 K. The heat transfer coefficient was also assumed lower: $h = 5.0 \times 10^5$ W m^{-2} K^{-1}; this value was estimated based on the splat thickness and the heat conduction properties of YSZ (heat resistance into previously deposited YSZ will be far greater than that into a stainless steel substrate). The parabolic and hyperbolic equations yielded very similar curves of interface velocity V_i during the early stages of solidification, and so only the parabolic results are presented. Figure 9 illustrates that V_i decreases as the temperature of the previously deposited layer increases, and the differences between the V_i profiles are almost linearly related to the differences between temperatures of the previously deposited splats. Comparing this result to that of Fig. 3, V_i for multiple splat deposition is much smaller, and so the nonequilibrium behaviors are significantly less.

4. Conclusions

We have previously investigated nonequilibrium phase change during rapid solidification of a pure metal (Ref 5). In this work we considered a coupled set of hyperbolic equations for temperature and concentration. Using an interface tracking method, the model simulates 1D nonequilibrium phase change of a binary system effectively. YSZ was chosen for the simulation based on the relatively low magnitude of the diffusive speed. The results show that for a realistic relaxation time (10^{-9} s), the profiles of V_i and T_i versus interface location are significantly affected by h, while the effect of hyperbolic behavior is relatively minor. Only the interface concentration shows pronounced differences between the hyperbolic and parabolic models, and these become larger as h increases. Also, the hyperbolic model yields a turning point when $V_i \rightarrow V_D$, as the concentration on the liquid side of the solidification front is frozen or trapped. We also examined a droplet impacting onto a previously deposited splat; results show that V_i is lower because the heat transfer coefficient is presumed lower, and so nonequilibrium behavior is not obvious.

Peer Reviewed

References

1. Y.K. Chae, J. Mostaghimi, and T. Yoshida, Deformation and Solidification Process of a Super-cooled Droplet Impacting on the Substrate Under Plasma Spraying Conditions, *Sci. Technol. Adv. Mat.*, 2000, **1**, p 147-156
2. G.-X. Wang, R. Goswami, S. Sampath, and V. Prasad, Understanding the Heat Transfer and Solidification of Plasma-sprayed Yttria-partially Stabilized Zirconia Coatings, *Mater. Manuf. Process.*, 2004, **19**, p 259-272
3. T. Streibl, A. Vaidya, M. Friis, V. Srinivasan, and S. Sampath, A Critical Assessment of Particle Temperature Distributions During Plasma Spraying: Experimental Results for YSZ, *Plasma Chem. Plasma Process.*, 2006, **26**, p 53-72
4. D.R. Clarke and C.G. Levi, Materials Design for the Next Generation Thermal Barrier Coatings, *Annu. Rev. Mater. Res.*, 2003, **33**, p 383-417
5. H. Liu, M. Bussmann, and J. Mostaghimi, A Comparison of Hyperbolic and Parabolic Models of Phase Change of a Pure Metal, in press, *Int. J. Heat Mass Tran.*, 2008. doi:10.1016/j.ijheatmasstransfer.2008.08.030
6. A.M. Mullis, Rapid Solidification Within the Framework of a Hyperbolic Conduction Model, *Int. J. Heat Mass Tran.*, 1997, **40**, p 4085-4094
7. S.L. Sobolev, Rapid Solidification Under Local Nonequilibrium Conditions, *Phys. Rev.*, 1997, **55**, p 6845-6854
8. P.K. Galenko and D.A. Danilov, Linear Morphological Stability Analysis of the Solid-Liquid Interface in Rapid Solidification of a Binary System, *J. Phys. Rev. E*, 2004, **69**, p 051608, 1-13

9. P.K. Galenko and D.A. Danilov, Local Nonequilibrium Effect on Rapid Dendritic Growth in a Binary Alloy Melt, *Phys. Lett. A*, 1997, **235**, p 271-280
10. L. Bianchi, F. Blein, P. Lucchese, M. Vardelle, A. Vardelle, and P. Fuchais, Effect of Particle Velocity and Substrate Temperature on Alumina and Zirconia Splat Formation, *Thermal Spray Industrial Applications*, C. Berndt and S. Sampath, Ed., (Materials Park, OH), ASM International, 1994, p 569-574
11. M. Fukomoto, Y. Huang, and M. Ohwatari, Flattening Mechanism in Thermal Sprayed Particle Impinging on Flat Substrate, *Thermal Spray Meeting the Challenges of the 21st Century*, C. Coddet, Ed., (Materials Park, OH), ASM International, 1998, p 401-406
12. M. Pasandideh-Fard, V. Pershin, S. Chandra, and J. Mostaghimi, Splat Shapes in a Thermal Spray Coating Process: Simulations and Experiments, *J. Therm. Spray Technol.*, 2002, **11**, p 206-217
13. W. Kurz and D.J. Fisher, *Fundamentals of Solidification*, 4th ed., Trans Tech Publications, Aedermannsdorf, 1998, p 63-89
14. G.-X. Wang and E.F. Matthys, Modeling of Nonequilibrium Surface Melting and Resolidification for Pure Metals and Binary Alloys, *J. Heat Transf.*, 1996, **118**, p 944-951
15. C. Li, J. Li, and W. Wong, The Effect of Substrate Preheating and Surface Organic Covering on Splat Formation, *Thermal Spray Meeting the Challenges of the 21st Century*, C. Coddet, Ed., (Materials Park, OH), ASM International, 1998, p 473-480
16. J. Pech, B. Hannoyer, A. Denoirjean, and P. Fauchais, Influence of Substrate Preheating Monitoring on Alumina Splat Formation in DC Plasma Process, *Thermal Spray Surface Engineering via Applied Research*, C. Berndt, Ed., (Materials Park, OH), ASM International, 2000, p 759-765
17. H.R. Salimi Jazi, J. Mostaghimi, T. Coyle, S. Chandra, C.Y. Lau, L. Rosenzweig, and E. Moran, Effect of Droplet Characteristics and Substrate Surface Topography on the Final Morphology of Plasma Sprayed Zirconia Single Splat, *J. Therm. Spray Technol.*, 2007, **16**, p 291-299
18. M.J. Aziz and T. Kaplan, Continuous Growth Model for Interface Motion During Alloy Solidification, *Acta Metall.*, 1998, **36**, p 2335-2347
19. M.J. Aziz and W.J. Boettinger, On the Transition from Short-range Diffusion-limited to Collision-limited Growth in Alloy Solidification, *Acta Metall. Mater.*, 1994, **42**, p 527-537
20. C. Cattaneo, Sulla conduzione del calore, *Atti Sem. Mat. Fis. Univ. Modena*, 1948, **3**, p 83-101
21. P.K. Galenko and D.A. Danilov, Model for Free Dendritic Alloy Growth Under Interfacial and Bulk Phase Nonequilibrium Conditions, *J. Cryst. Growth*, 1999, **197**, p 992-1002
22. H.-B. Xiong, L.-L. Zheng, and T.A. Streibl, Critical Assessment of Particle Temperature Distributions During Plasma Spraying: Numerical Studies for YSZ, *Plasma Chem. Plasma Process.*, 2006, **26**, p 53-72

JTTEE5 17:655–661
DOI: 10.1007/s11666-008-9264-3
1059-9630/$19.00 © ASM International

Peer Reviewed

Introduction to High-Velocity Suspension Flame Spraying (HVSFS)

Rainer Gadow, Andreas Killinger, and Johannes Rauch

(*Submitted May 14, 2008; in revised form October 2, 2008*)

High-velocity suspension flame spraying (HVSFS) has been developed to thermally spray suspensions containing micron, submicron, and nanoparticles with hypersonic speed. For this purpose, the suspension is introduced directly into the combustion chamber of a modified HVOF torch. The aim in mind is to achieve dense coatings with a refined microstructure. Especially from nanostructured coatings superior physical properties are expected for many potential applications. Direct spraying of suspensions offers flexibility in combining and processing different materials. It is a cost-saving process and allows the allocation of entirely new application fields. The paper gives an overview of the HVSFS spray method and will present some actual results that have been achieved by spraying the nanooxide ceramic materials Al_2O_3, TiO_2, 3YSZ, and Cr_2O_3.

Keywords nanocrystalline composites, nanopowders, nanostructured materials

1. Introduction

Thermal spray technology has been applied for decades with great success and still bears a great potential for new applications in many industrial fields. One of the great benefits is the wide variety of spray materials that can be applied to all kind of substrate materials.

However, standard spray techniques face restrictions regarding their coating thickness and resulting microstructure. From the fact that the processed spray material (rods, wires, or standard spray powders) results in spray particles with a certain particle size that cannot be reduced, there are limitations to decreasing coating thickness significantly below a value in the range of 30 μm because individual splats are too coarse to form a homogeneous and sealed coating structure. For most applications, a coating thickness of 200 to 300 μm matches well with the technical requirements; however, there are new applications where thin coatings in the range of 5 to 50 μm would be of great interest, for example, in the field of sensor technology, in the development of solid electrolytes (e.g., for solid oxide fuel cell, or SOFC), and for a new generation of nanostructured tribocoatings.

A reduction of particle sizes below 5 μm means an extraordinary effort in powder feeding technique, as fluidization of the spray powder gets more and more challenging with a decreasing particle size. Therefore, suspensions seem to offer a promising solution of this problem. Moreover, using a suspension opens up a completely new field of specially designed spray materials, including nanopowders that are currently not available on the spray powder market. The processing of a nanopowder using the standard thermal spray procedure first of all requires the agglomeration of the nanopowders by a spray drying process to form spray particles with appropriate grain size that are suitable for a standard powder feeding device. These powder qualities are expensive and not available on the market for every combination of materials. Oxide-based materials are supplied by several companies, whereas cermet-based powder materials containing nanosized hard-phase particles such as carbides or borides are relatively rare and expensive. Spraying these powder types, however, leads to spray particles with particle diameters comparable to standard spray powders. The aim to form thin-film spray coatings with a refined internal coating structure cannot be solved by this approach, although (in the case of mixed-phase powders) a dispersion of submicron or even nanoscaled hard phases can be realized. A general problem is the fact that during the melting process in the flame, the agglomerated particles lose their nanostructure (in the case of a single-phase agglomerate), so it is not yet clear if there is any benefit compared to using standard powders.

Suspension spraying has been carried out using plasma spraying, flame spraying, and high-velocity oxyfuel (HVOF) (Ref 1-4). In the case of plasma spraying, the injection of the liquid generally is performed perpendicular

This article is an invited paper selected from presentations at the 2008 International Thermal Spray Conference and has been expanded from the original presentation. It is simultaneously published in *Thermal Spray Crossing Borders, Proceedings of the 2008 International Thermal Spray Conference*, Maastricht, The Netherlands, June 2-4, 2008, Basil R. Marple, Margaret M. Hyland, Yuk-Chiu Lau, Chang-Jiu Li, Rogerio S. Lima, and Ghislain Montavon, Ed., ASM International, Materials Park, OH, 2008.

Rainer Gadow, Andreas Killinger, and **Johannes Rauch**, Institute for Manufacturing Technologies of Ceramic Components and Composites (IFKB), University of Stuttgart, Allmandring 7b, D-70569 Stuttgart, Germany. Contact e-mail: rainer.gadow@ifkb.uni-stuttgart.de.

to the plasma flame. In the case of conventional flame and HVOF spraying, an axial injection into the flame or combustion chamber respectively is preferred as it circumvents the problem of strong flame disturbance causing instabilities in the spray process (Ref 1).

Introducing a liquid into the spray process raises some basic problems. Any liquid, being water or an organic solvent, starts to evaporate rapidly when introduced into the flame causing two important effects: (i) a significant cooling of the flame and (ii) a strong disturbance of the free expanding hot gas stream caused by the expanding vapor through the evaporation process, as is especially the case in plasma spraying. When the vapor is introduced into a combustion chamber, a significant rise of the combustion chamber pressure will occur as can be observed in the high-velocity suspension flame spraying (HVSFS) process.

Particle size strongly influences the particle thermal history. Small particles rapidly heat up, but they also rapidly cool down, resulting in a far different behavior concerning the splat formation on the surface. It also affects interlamellar adhesion of the splats and thereby influences the mechanical properties (e.g., Young's modulus, hardness) of the coating. When thermal spraying suspensions, the liquid is fragmented through an atomization process into small droplets not necessarily having the size of the primary particles (e.g., nanoparticles); this issue has been extensively discussed in literature (Ref 5). It has been reported for plasma spraying of suspensions (Ref 6), but can also be observed in the HVSFS process, as shown by the authors in Ref 4. The goal is to optimize the atomization in the combustion chamber to achieve the smallest suspension droplets possible. This leads to finer splats in the coating with the above-described improvements.

2. Experimental

The principle of HVSF spraying has been introduced and described in our recent publication (Ref 4); an improved configuration of the HVSFS equipment is shown in Fig. 1. It consists of a suspension container connected to a pump feeder system. A flexible suspension line feeds the suspension axially and concentrically into the combustion chamber of a modified HVOF gun. A second container holds the pure solvent to flush the suspension line and injection nozzle when shutting down the process. The HVSFS torch is based on a TopGun-G system (manufactured by GTV, GTV Verschleiss-Schutz GmbH Gewerbegebiet "Vor der Neuwiese", Luckenbach) and has been modified to meet suspension spray requirements. Major modifications concern a suspension injection nozzle that is capable of evenly injecting a suspension into the combustion chamber of the HVSFS torch. The injection nozzles have a smaller diameter than those used in conventional HVOF torches. Two different shapes, cylindrical and conical, were used. It was found that cylindrical injection nozzle geometries show a strong tendency to clog during the process, whereas a conical geometry allows a stable coating process.

Propane and ethene were preferentially chosen as combustion fuels as they allow stable operation of the HVSFS process. Basically, using acetylene can be of advantage when spraying oxide materials, as it delivers higher flame temperatures (Ref 7). However, with the insertion of liquid fuel, the combustion chamber pressure rises significantly, exceeding the maximum gas line pressure of 1.5 bar that is allowed for acetylene; therefore, expensive high-pressure lines would be necessary to run a stable HVSFS process.

For all suspensions, isopropanol was chosen as the organic solvent. Although stabilization of particles is more difficult than in aqueous media, this alcohol delivers fairly high enthalpy values during the combustion process. In case of the TiO_2, the suspension also contained water (10 wt.% of liquid phase) to optimize the viscosity by using the dispersing agent Dolapix CE64 (Zschimmer und Schwarz GmbH & Co KG, Chemische Fabriken, Lahnstein/Rhein).

3. Preparation of the Suspensions

Table 1 summarizes all powders used in this work. In a first step, the powders were attrition milled in isopropanol

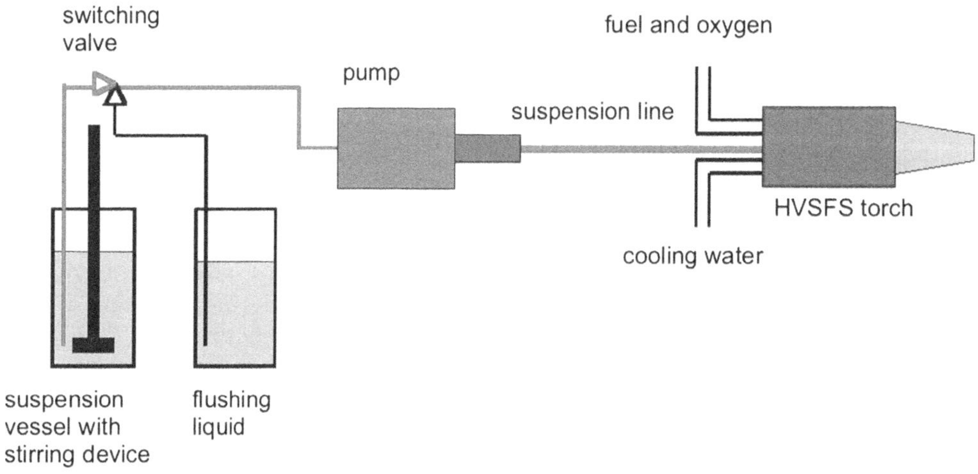

Fig. 1 Experimental setup for HVSFS (pump-based system)

solvent to force deagglomeration and particle breakup. This is necessary because most nanopowders show agglomerates with grain sizes that might clog the suspension injection nozzle (diameter 0.3-0.5 mm). To remove milling balls and possible agglomerates, the suspensions were sieved after milling process with a mesh of 50 μm. Suspensions were then applicable for spraying. The solid content of all suspensions in this work was in the range of 15 to 20 wt.%. A summary of the suspension compositions is given in Table 2.

Table 1 Summary of applied powder materials

Brand name (manufacturer)	Chemical composition	Primary grain size, nm	Phases analysis derived from XRD
TM-DAR (Taimei Chemicals)	n-Al_2O_3	150	α-Al_2O_3
Aeroxide P25 (Degussa)	n-TiO_2	15	Anatase, rutile
Nanofiller (DGTec)	n-Cr_2O_3	100	Eskolaite
VP PH (Degussa)	n-3YSZ	15	Tetragonal, monoclinic

Table 2 Composition of the different suspensions

Suspension	Solid content, wt.%	Isopropanol content, wt.%	Additives
n-Al_2O_3	20	80	HNO_3
n-TiO_2	10	90/10	Dolapix CE64
n-Cr_2O_3	15	85	HNO_3
n-3YSZ	20	80	HNO_3

Table 3 Parameters for HVSFS

Suspension	Fuel gas, slpm	O_2, slpm	Torch speed, mm/s	No. of cycles	Spray distance, mm	Feeding rate, g/min
n-Al_2O_3	P 65	350	600	10	150	10
n-TiO_2	P 50	220	800	10	100	8
n-Cr_2O_3	P 65	350	600	5	125	8
n-Cr_2O_3	E 90	300	600	5	125	8
n-3YSZ	E 90	300	600	10	120	10

P, propane; E, ethene

4. Coating Experiments

All suspensions were sprayed using propane or ethene as fuel gases; relevant spray parameters are summarized in Table 3. When choosing propane as a fuel gas, it was found that an oxygen/fuel ratio adjusted to approximately 5.4 achieved the best deposition rates. In case of ethene, the oxygen/fuel ratio was approximately 3.3. It should be kept in mind that flame temperature is strongly influenced by the presence of isopropanol in the flame. It also strongly changes the oxidation behavior of the flame.

Coatings were applied either on planar mild steel (St37) or aluminum alloy (AlMg3) samples (50 × 50 mm) and were mounted vertically on a sample holder. The HVSFS torch was operated on a six-axis robot system using a simple meander kinematic with a 1 mm offset. Sample cooling was performed using two air nozzles mounted on the torch. Substrate preparation prior to spraying included degreasing of the surface with acetone. For all substrates, a mild injection grit-blasting operation was carried out using white corundum particles (FEPA 120; ø = 100 to 125 μm) and 5 bar compressed air. Spray parameters are given in Table 3.

5. Characterization of the HVSFS Sprayed Coatings

From all samples, cross sections were prepared for image analysis using light microscopy and/or scanning electron microscopy (SEM). Surface topography was taken in SEM to reveal the structure of single splats. Microhardness was measured on coating cross section using a Fisherscope equipped with a Vickers indenter. All measurements are performed dynamically (universal hardness) according to DIN 50359-1. The given values are the average of 10 indentations. Coating porosity was determined from digital image analysis (Software AQUINTO, Analysis Software, Olympus Deutschland GmbH, Hamburg). Surface roughness was measured using a contact profilometer (Perthometer Concept, Mahr, Mahr GmbH Esslingen, Esslingen), R_z and R_a values were determined according to DIN EN ISO 4287. All results are summarized in Table 4.

5.1 Al_2O_3

Aluminum oxide suspension was HVSF sprayed on aluminum substrate using propane as fuel gas. A conical-shaped injection nozzle was used. The feeding rate was

Table 4 Mechanical and structural data of the HVSF sprayed coatings

Spray material	Vickers hardness	Coating porosity, %	R_z, μm	R_a, μm	Phase composition derived from XRD
n-Al_2O_3	620-880 $HV_{0.1}$, SD = 76	5	3.14	0.58	Mainly γ
n-TiO_2	~1000 $HV_{0.05}$	0-0.5	3.5	0.65	Mainly anatase
n-Cr_2O_3 propane	1400-1800 $HV_{0.01}$, SD = 145	4-5	2.75	0.47	Hexagonal
n-Cr_2O_3 ethene	1100-2000 $HV_{0.01}$, SD = 283	7-10	2.79	0.48	Hexagonal
n-3YSZ	715-816 $HV_{0.03}$, SD = 38	<1	8.68	1.75	Tetragonal

SD, standard deviation

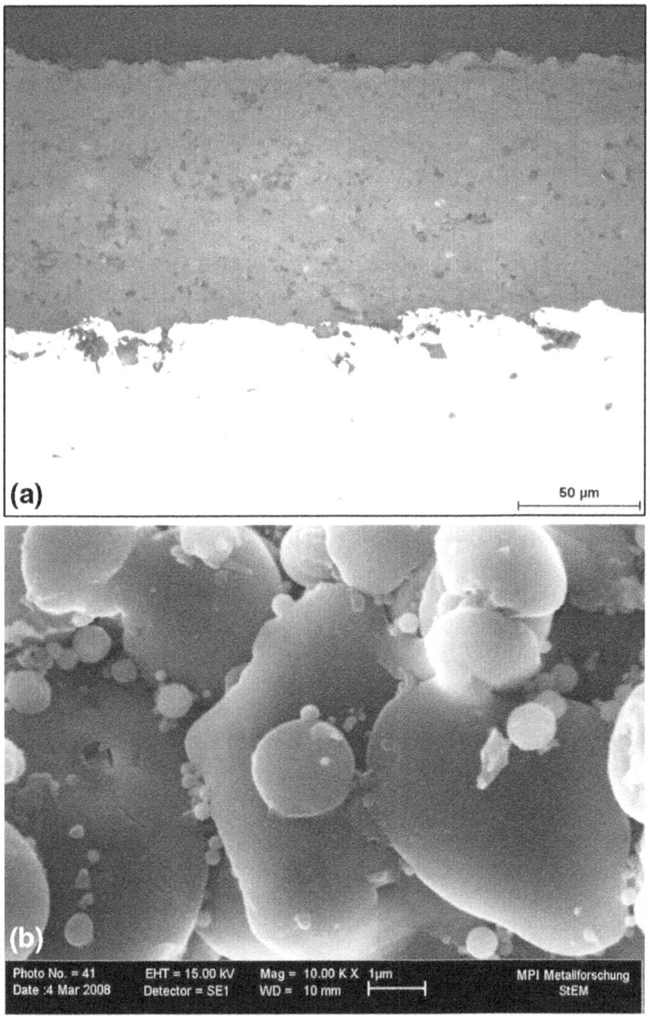

Fig. 2 (a) Light microscope image of HVSF sprayed Al_2O_3 coating. (b) SEM image of the coating surface

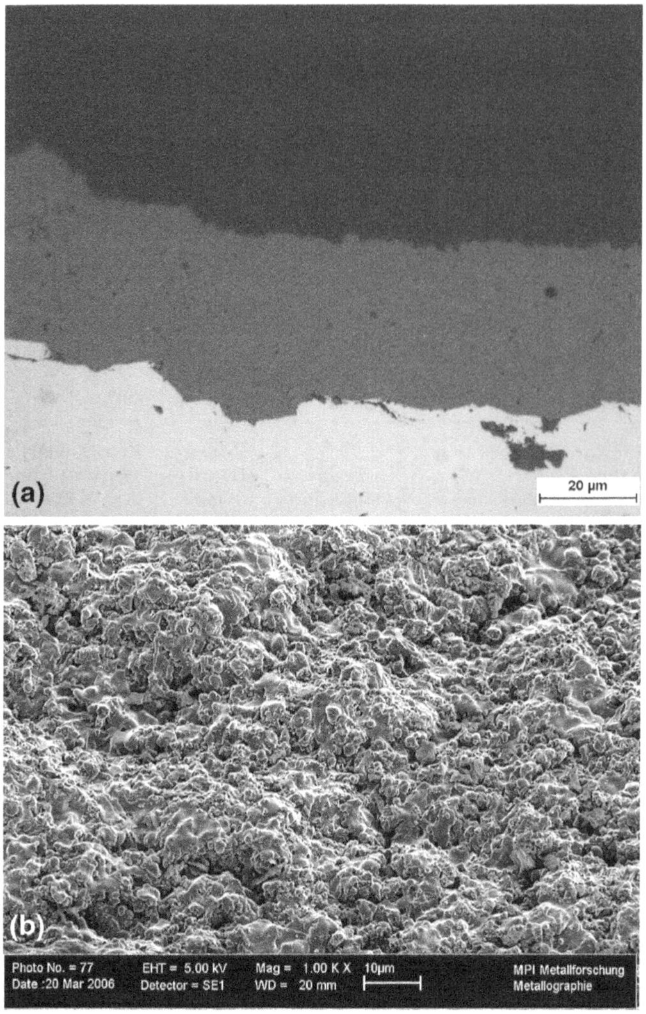

Fig. 3 (a) Light microscope image of HVSF sprayed TiO_2 coating. (b) SEM image of the coating surface

approximately 10 g/min, and the coating thickness was about 100 µm. Deposition rate was approximately 10 µm per transition. The alumina coating exhibits a good adhesion to the aluminum substrate (Fig. 2a). The porosity was approximately 5%; microhardness was in the range of 600 to 900 $HV_{0.1}$.

The coating surface shows fully molten droplets in the range of 1 to 5 µm, in the intersplat region small spherical particles with a size of 200 up to 500 nm are visible (Fig. 2b). Surface roughness is $R_z = 3.14$ µm and $R_a = 0.58$ µm.

5.2 TiO_2

The titanium oxide suspension was HVSF sprayed on aluminum substrate using propane as a fuel gas. The feeding rate was approximately 8 g/min; a conically shaped 0.3 mm nozzle was used. The applied coating thickness was approximately 30 µm; deposition rate was approximately 4 µm/cycle. In cross section, the TiO_2 coating appears with a very low porosity (Fig. 3a). The coating appears homogeneous; no internal structure is visible in cross section, and no unmolten particles are detected. The SEM surface image reveals fully molten splats ranging from a few 100 nm up to 10 µm with a good intersplat connection (Fig. 3b).

From the x-ray diffraction (XRD) data, it can be stated that the main fraction of the coating consists of anatase (approximately 75%); the rest of the material consists of rutile.

5.3 Cr_2O_3

Chromium oxide suspension was HVSF sprayed with propane and ethene to compare the influence of the fuel gases. Suspension feeding rate, torch setup, robot kinematics, and substrate were identically for both coatings, refer to Table 3. The n-Cr_2O_3 suspension was sprayed on steel substrates with a feeding rate of approximately 8 g/min using a cylindrically shaped 0.4 mm nozzle. Deposition rate was in the range of 2 µm/cycle for both fuel gases. The achieved coating has a thickness of 10 to 15 µm, exhibits good adhesion to the steel substrate, and shows a porosity of approximately 5% in the case of propane and approximately 10% in case of ethene; refer to Fig. 4(a) and 4(b); for porosity data see Table 4.

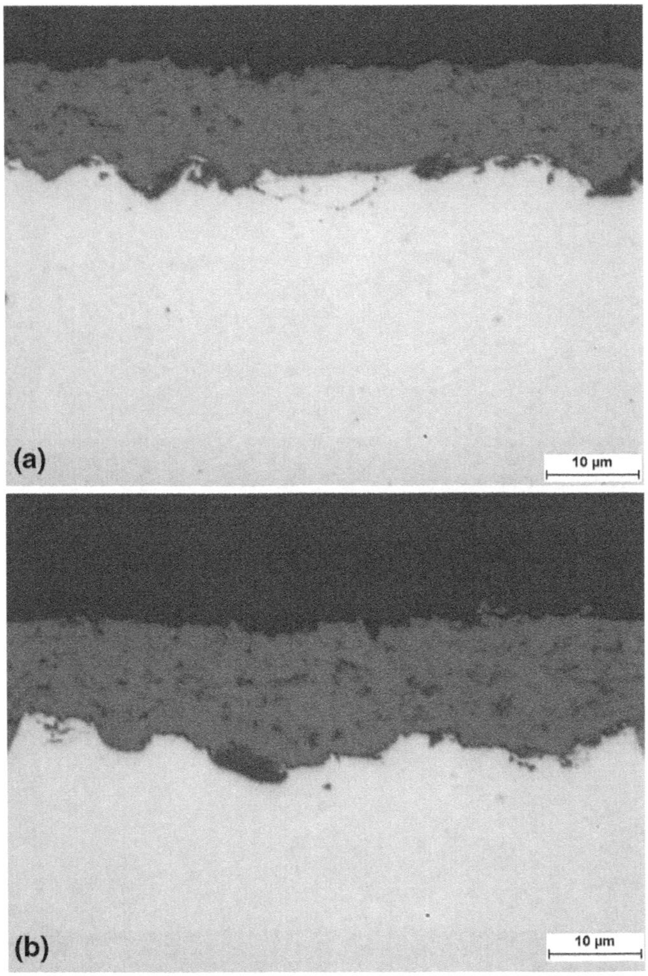

Fig. 4 Light microscope images of HVSF sprayed chromium oxide cross sections. (a) Sprayed with fuel gas: propane. (b) Sprayed with fuel gas: ethene

Fig. 5 SEM images of HVSF sprayed chromium oxide surfaces. (a) Fuel gas: propane. (b) Fuel gas: ethene

The microhardness of the propane sprayed coating was in the range 1400 to 1800 $HV_{0.01}$; hardness of the ethene sprayed coating was in the range of 1100 to 2000 $HV_{0.01}$. In case of the ethene sprayed samples, the lower microhardness corresponds to the higher porosity in the coating. There is a fairly high variance of the measured values due to the low indenter forces.

In SEM, coatings show different surface structures. The surface of the coating sprayed with propane appears coarser with fully molten droplets in the range of 1 to 3 µm (Fig. 5a). A refined splat structure is visible on the ethene sprayed coating surface (Fig. 5b). Small spherical droplets in the range of 100 nm are present together with larger droplets (1 to 3 µm). The surface roughness of both coatings is comparable; $R_z = 2.75$ µm, $R_a = 0.47$ µm for the propane sprayed, and $R_z = 2.79$ µm, $R_a = 0.48$ µm for the ethene sprayed.

5.4 3YSZ

The YSZ suspension was flame sprayed on steel substrate with a feeding rate of approximately 10 g/min using a 0.4 mm conical nozzle. In this case, ethene was chosen as the fuel gas to achieve a high flame temperature. The deposition rate was in the range of 3 µm/cycle. In cross sections, the YSZ coating appears dense and homogeneous. The thickness of the coating is about 30 µm (Fig. 6a). In SEM images stretched pores with a diameter of about 200 nm are visible (Fig. 6b). The microhardness of the coating was measured 766 $HV_{0.03}$. From XRD it can be stated that the coating consists of a pure tetragonal phase.

The surface of the 3YSZ coating appears as a fully molten splat structure with splats in the range of 1 to 3 µm and a well-established intersplat connection. Fine spherical particles with a diameter of 50 to 200 nm are distributed between the larger agglomerates (Fig. 7a and b). The surface roughness was measured to $R_z = 8.68$ µm and $R_a = 1.75$ µm.

6. Discussion

All HVSF sprayed coatings exhibit a refined splat structure that is at least one order of magnitude below standard spray coatings. This is also valid for the pore

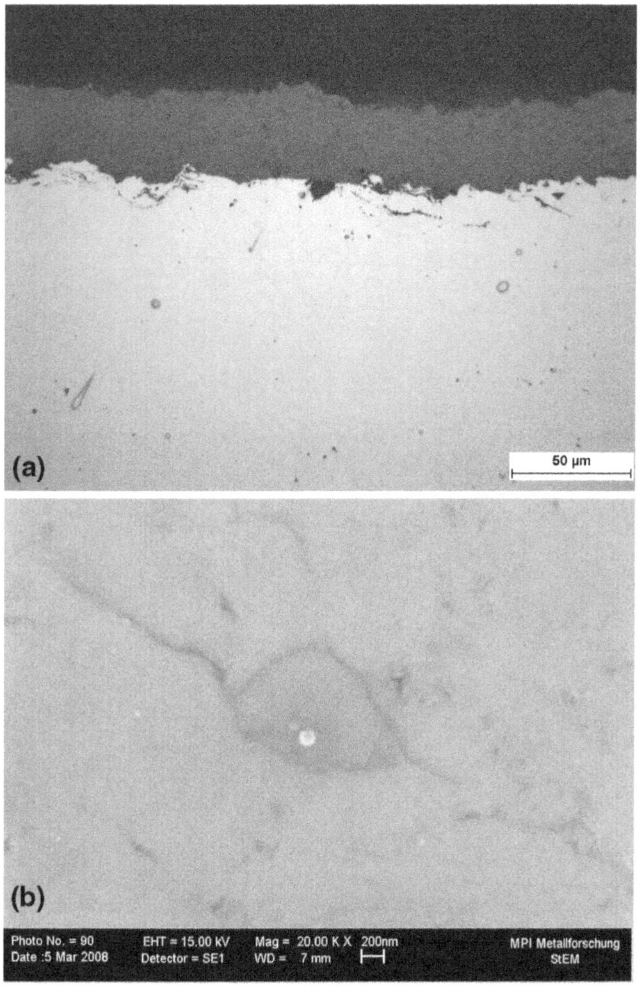

Fig. 6 (a) Light microscope image of a HVSF sprayed 3YSZ coating. (b) SEM image of HVSF sprayed 3YSZ, pore structures are in the range of 50 nm

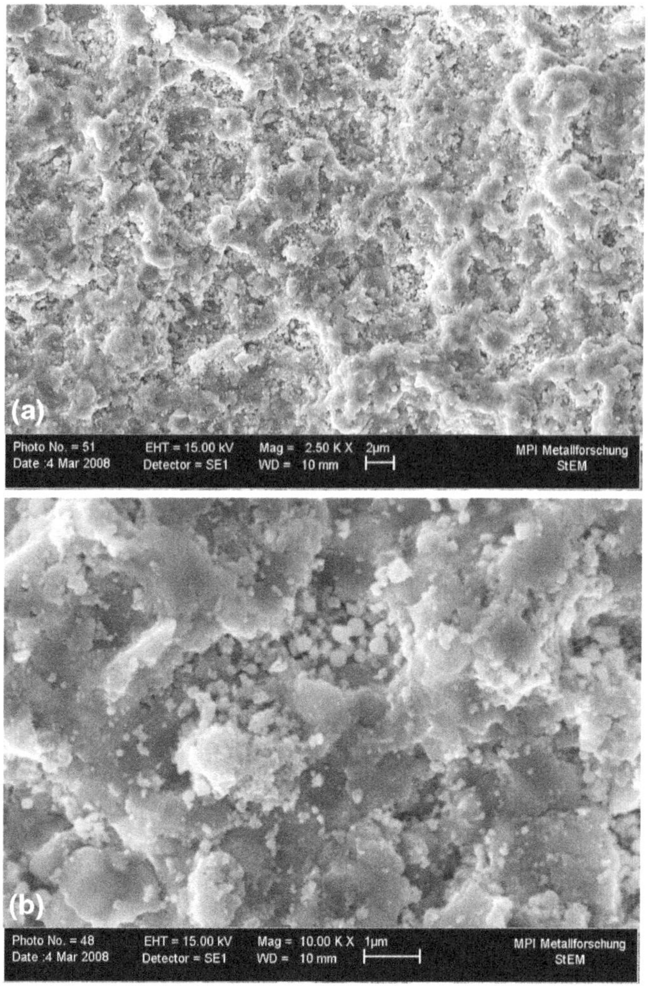

Fig. 7 SEM images of the surface of HVSF sprayed 3YSZ at different orders of magnitude

structure. Therefore, all HVSF sprayed coatings exhibit remarkably low values of surface roughness. From all investigated coatings, the titanium and zirconium oxide coatings exhibit lowest porosity values.

Generally, there is a fairly high variance of the measured hardness values because of the reduced coating thickness in chromium and zirconium oxide; further investigations will include nanoindenter analysis to achieve more precise values for coatings below 50 μm.

For all coatings, there is no visible evidence that particles were insufficiently molten. So far, there is no clear tendency that porosity would directly correspond to the melting point of the sprayed material, titanium oxide (melting point 1800 °C) and 3YSZ (melting point 2700 °C) show highest densities. Deposition efficiency is in the range of 3 to 4 μm/cycle for all investigated suspensions. So far, using ethene as a fuel gas providing a higher flame temperature did not show any advantage concerning the deposition efficiency and coating porosity achieved in the HVSFS process.

From XRD data, it is found that a crystallite structure in the coating significantly exceeds the size of the initial nanopowders. However, from our earlier publication it was found that spraying of mixed nanooxides can preserve a nanostructure in the coating (Ref 8).

7. Summary

The HVSFS method was evaluated to manufacture coatings from n-Al_2O_3 n-TiO_2, n-Cr_2O_3, n-3YSZ nanooxides. Appropriate suspensions were prepared in-house using isopropanol as a solvent. All materials were sprayed using propane or ethene as a fuel gas.

For all investigated materials, homogeneous oxide coatings featuring a significantly lower surface roughness and a highly refined microstructure compared to standard spray coatings could be manufactured. So far, measured microhardness values are comparable or slightly lower than those of standard coatings, but nanoindenter measurements are required to further characterize the coatings more precisely. All coatings clearly show a crystalline structure in XRD.

References

1. R. Gadow, A. Killinger, M. Kuhn, and D. Lopéz, Published patent application DE 10,2005,038,453 A1, 2007
2. C. Delbos, J. Fazilleau, J.F. Coudert, and P. Fauchais, Plasma Spray Elaboration of Finely Structures YSZ Thin Coating by Liquid Suspension Injection, *Thermal Spray 2003: Advancing the Science & Applying the Technology*, C. Moreau and B. Marple, Eds., (Materials Park, OH), ASM International, 2003, p 661-669
3. T. Poirer, A. Vardelle, M.F. Elchinger, M. Vardelle, A. Grimaud, and H. Vesteghem, Deposition of Nanoparticle Suspensions by Aerosol Flame Spraying: Model of the Spray and Impact Processes, *Thermal Spray 2003: Advancing the Science and Applying the Technology,* Vol 2, B.R. Marple and C. Moreau, Eds., May 5-8, 2003 (Orlando, FL), ASM International, 2003, 393 pages
4. A. Killinger, M. Kuhn, and R. Gadow, High-Velocity Suspension Flame Spraying (HVSFS), A New Approach for Spraying Nanoparticles with Hypersonic Speed, *Surf. Coat. Technol.*, 2006, **201**, p 1922-1929
5. E. Bouyer, F. Gitzhofer, and M. I. Boulos, Powder Processing by Suspension Plasma Spraying, *Thermal Spray: A United Forum for Scientific and Technological Advances,* C.C. Berndt, Ed., Sept 15-18, 1997 (Indianapolis, IN), ASM International, 1998, 353 pages
6. R. Rampon, G. Bertrand, F.-L Toma, and C. Coddet, Liquid Plasma Sprayed Coatings of Yttria Stabilized Zirconia for SOFC Electrolytes, *Proceedings of the 2006 International Thermal Spray Conference* (Seattle, WA), ASM International, 2006
7. H. Kreye, Vergleich der HVOF-Systeme—Werkstoffverhalten und Schichteigenschaften, 4. Kolloquium Hochgeschwindigkeitsflammspritzen, 1997, p 13-21
8. R. Gadow, F. Kern, and A. Killinger, Manufacturing Technologies for Nanocomposite Ceramic Structural Materials and Coatings, *Materials Science and Engineering B,* Vol 148 (No.1-3), P. Kumta, Ed., (Amsterdam), Elseveir B. V., 2008, p 58-64. ISSN 0921-5107

JTTEE5 17:662–670
DOI: 10.1007/s11666-008-9218-9

Peer Reviewed

Suspension Plasma-Sprayed Alumina Coating Structures: Operating Parameters Versus Coating Architecture

O. Tingaud, A. Grimaud, A. Denoirjean, G. Montavon, V. Rat, J.F. Coudert, P. Fauchais, and T. Chartier

(Submitted May 8, 2008; in revised form July 21, 2008)

Suspension plasma spraying (SPS) is able to process sub-micrometric-sized feedstock particles and permits the deposition of layers thinner (from 5 to 50 µm) than those resulting from conventional atmospheric plasma spraying (APS). SPS consists in mechanically injecting within the plasma flow a liquid suspension of particles of average diameter varying between 0.02 and 1 µm, average values. Upon penetration within the DC plasma jet, two phenomena occur sequentially: droplet fragmentation and evaporation. Particles are then processed by the plasma flow prior their impact, spreading and solidification upon the surface to be covered. Depending upon the selection of operating parameters, among which plasma power parameters (operating mode, enthalpy, spray distance, etc.), suspension properties (particle size distribution, powder mass percentage, viscosity, etc.), and substrate characteristics (topology, temperature, etc.), different coating architectures can be manufactured, from dense to porous layers. Nevertheless, the coupling between the parameters controlling the coating microstructure and properties are not yet fully identified. The aim of this study is to further understand the influence of parameters controlling the manufacturing mechanisms of SPS alumina coatings, particularly the spray beads influence.

Keywords alumina coating, microstructure, spray bead, suspension plasma spraying

1. Introduction

Suspension plasma spraying (SPS) is an emerging technology and an alternative to atmospheric plasma spraying (APS) to manufacture thinner layers (from 5 to 50 µm, average values) (Ref 1, 2). It consists in axially or radially injecting within a plasma flow a suspension of sub-micrometer-sized particles. In the present study, mechanical injection has been considered, that is to say that a continuous stream of suspension exiting from an injector of a few tens of micrometers is injected into the plasma flow where it experiences first fragmentation, the resulting droplets encountering then solvent evaporation prior simultaneous acceleration and melting of particles until their impact onto the substrate to be covered (Ref 3, 4).

The main differences between SPS and APS are (Table 1): (i) the carrier gas used to provide micrometric-sized particles with adequate momentum to be injected into the plasma jet in APS that is replaced by a carrier liquid which experiences fragmentation (in a few µs) and vaporization (in a few tens of µs) when interacting with the plasma jet; (ii) the stand-off distance which is shorter in SPS (~40 mm) compared to APS (~100-140 mm) due to the lower thermal and kinetic inertia of sub-micrometric particles; (iii) a higher heat flux transferred from the plasma to the substrate that can reach values as high as 25 MW m^{-2} in some cases (Ref 5).

In SPS, the liquid stream or drops penetrate the plasma jet if their densities of momentum are significantly higher than the one of the plasma (since upon penetration, the size of the drops decreases due to solvent vaporization). Upon penetration within the plasma flow, the liquid stream encounters two mechanisms, fragmentation and vaporization (Ref 6). In a first approximation and whatever the droplet characteristic dimension, the vaporization duration is two orders of magnitude longer than the fragmentation one. Indeed, the fragmentation occurs for a liquid flow characteristic dimensionless Weber number, *We*, higher than 12-14 (Ref 7). *We* represents the ratio of inertia forces to surface tension forces and is expressed as follows:

$$We = \frac{\rho_{\text{liquid}} \Delta V^2 d_{\text{liquid}}}{\sigma_{\text{liquid}}} \text{(dimensionless)} \quad \text{(Eq 1)}$$

where ρ_{liquid} is the liquid (suspension) specific mass (kg m^{-3}), d_{liquid} the liquid (suspension) stream

This article is an invited paper selected from presentations at the 2008 International Thermal Spray Conference and has been expanded from the original presentation. It is simultaneously published in *Thermal Spray Crossing Borders, Proceedings of the 2008 International Thermal Spray Conference*, Maastricht, The Netherlands, June 2-4, 2008, Basil R. Marple, Margaret M. Hyland, Yuk-Chiu Lau, Chang-Jiu Li, Rogerio S. Lima, and Ghislain Montavon, Ed., ASM International, Materials Park, OH, 2008.

O. Tingaud, A. Grimaud, A. Denoirjean, G. Montavon, V. Rat, J.F. Coudert, P. Fauchais, and **T. Chartier**, SPCTS-UMR CNRS 6638, Faculty of Sciences, University of Limoges, 123 Avenue Albert Thomas, 87060 Limoges Cedex, France. Contact e-mail: ghislain.montavon@unilim.fr.

Table 1 Major differences between APS and SPS processes

Characteristics	APS	SPS
Particle average size, μm	20-50	0.02-1.0
Particle relative mass, –	1	1.10^{-9}–1.10^{-3}
Feedstock carrier	Gas (Ar usually)	Liquid (H_2O or Et-OH) + additives
Particle mass flow rate, g min^{-1}	20-50	0.75-6
Stand-off distance, mm	100-140	30-50
Heat flux at stand-off distance, MW m^{-2}	2-0.5	25-10

characteristic dimension (m), σ_{liquid} the liquid (suspension) surface tension (considered at ambient temperature in contact with air at ambient pressure) (N m^{-1}), and ΔV is the difference in velocities between the liquid (suspension) stream and the plasma flow (m s^{-1}). The Weber number evolves along the penetration of the suspension within the plasma jet and can reach values as high as 350, leading to a catastrophic break-up (Ref 8).

One can envisage the following simplified sequences for the processing of a suspension made of micrometric-sized particles (Ref 9): (i) fragmentation of the liquid stream occurred before solvent evaporates (two orders of magnitude); (ii) solvent evaporation leads to the formation of single particles or aggregates constituted by a few grains; (iii) these particles melt and form liquid drops which impact, spread, and solidify to form flattened lamellae of equivalent diameters between a few hundred nanometers to a few micrometers.

Lamellae resulting from impact, spreading, and solidification onto the substrate of impinging particles are characterized, apart from their typical size, by (i) almost the absence of peripheral projections around the lamellae; this signifies that the dimensionless flattening Sommerfeld number, K_f, is lower than 6 (Ref 10); and (ii) the absence of intralamellar cracks within the lamellae, *a contrario* to lamellae collected under conventional plasma spray conditions, indicating that the residual quenching stress developing within the lamellae upon rapid solidification and cooling is lower than the intrinsic mechanical resistance of their constitutive material.

The layer results from the stacking of such lamellae and exhibits hence a typical granular structure. Indeed, SPS layers are made of lamellae (molten particles, W), angular particles (unmolten particles, U), and small spherical grains (molten particles resolidified prior their impact upon the substrate, R) (Fig. 1).

This very typical architecture results from several effects/mechanisms:

1. SPS is particularly sensitive to the arc voltage fluctuations induced by the arc instabilities. These voltage fluctuations (Fig. 2) depend upon the plasma torch operating mode (Ref 11, 12), namely the take over mode that develops when the plasma torch is operated with monoatomic plasma forming gases (Ar, He) and characterized by a relative voltage fluctuation, $\Delta V/V$, lower than 0.3 or the restrike mode that develops when the plasma torch is operated with diatomic plasma forming gases (N_2, H_2) and for which $\Delta V/V \geq 1$. The frequency at which the torch voltage fluctuates is about a few kHz, typically from 3 to 5. The available energy in the plasma flow fluctuates together with the voltage fluctuations. The plasma flow average velocity is fluctuating together with the available energy so that the different steps in suspension processing (liquid fragmentation, solvent vaporization, and solid particles melting and acceleration) are not achieved at the same rates all along the time of flight of the feedstock (Ref 13, 14). Indeed, the variations in velocity of the plasma flow, together with its variations in length and position, lead to different locations of the fragmentation of the liquid stream due to the variations in shearing stresses (consecutive to the variation in the plasma momentum density, ρv^2). Moreover, the plasma flow is characterized by strong thermal and kinetic radial gradients (Ref 15)

Fig. 1 Typical Al_2O_3 SPS layer architecture (feedstock d_{50} = 500 nm, solvent: Et-OH, particle mass load in suspension MP = 20%, mechanical suspension injection, Ar-H_2 45-15 slpm plasma forming gases, plasma arc current intensity I = 500 A, torch internal nozzle diameter Ø = 5 mm, plasma mass enthalpy h = 14 MJ kg^{-1})

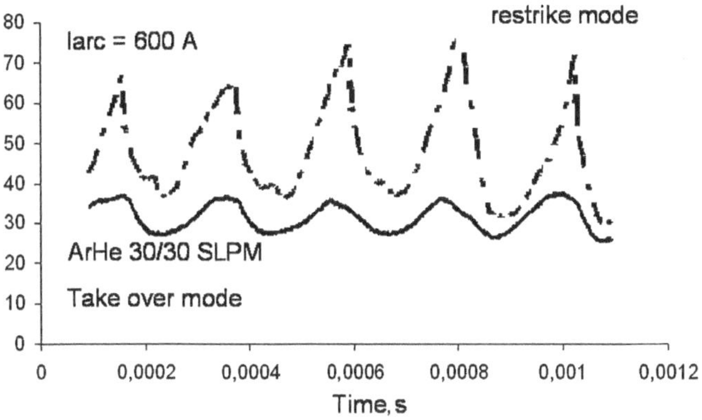

Fig. 2 Arc voltage temporal fluctuations of a DC plasma torch operated under the restrike (Ar-H_2, 45-15 slpm) and the take-over (Ar-He, 30-30 slpm) modes (arc current intensity I = 600 A, anode internal diameter at exit Ø = 5 mm)

Peer Reviewed

providing a heterogeneous energetic transfer from the plasma flow to the droplets and later on to the particles.

2. If the particle size distribution of the solid particles in the suspension is fairly large, this leads to very different characteristic times in the particles processing (Ref 16).
3. In interaction with the plasma jet, the suspension jet crumbles. The first droplets, containing solid particles, fragment from the suspension jet and remain in the plasma fringe, whereas the jet continues its trajectory toward the plasma core (Ref 17). Since the plasma jet is characterized by sharp thermal and kinetic radial gradients (Ref 18, 19), this leads to an inhomogeneous treatment of particles along their trajectories. The solid particles that penetrated into the plasma core will impact upon the substrate in a molten state and form splats, whereas those traveling in the plasma fringes of lower temperature will remain untreated and form angular embedded particles. Spherical particles originate from particles that have been entrained deep into the plasma jet core where they were fully molten before being resolidified either due to the lumped mass heat transfer showing an exponential decrease evolution as a function of the particle diameter (the smaller the particle, the higher the cooling rate) or due to the thermophoresis and related effects (Ref 20) when crossing high thermal gradient regions between the core and the developed jet. Those particles could be hence ejected again in the fringes where they re-solidified due to their low thermal inertia. Even at very short spray distance (30 mm), some spherical grains are embedded within the coating structure. One could consider nevertheless that at such a short spray distance, those particles remain in region of the flow warm enough to not induce solidification. This is why one can estimate that thermophoresis and related effects are not negligible. Nevertheless, at the 50-mm spray distance, the relative fraction of small spherical particles is increased compared to the 30 mm one. One can state hence that a synergy takes place in these conditions between the thermophoresis and the lumped mass heat transfer.

Hence, depending on their size and trajectories, the particles experience different thermal histories and trajectories, leading to different molten state and impact location onto the substrate. Depending upon the fraction of poorly treated particles (processed or re-solidified in the plasma jet core fringes) to the one of appropriately treated (processed in the plasma core and in a molten state when impacting), the coating architecture will evolve from fairly dense (low fraction of poorly treated) to fairly porous (high fraction of poorly treated) (Fig. 3). Densifying the layer architecture requires hence reducing the fraction of poorly treated particles.

Apart from this mechanism, layer architecture depends also upon the manner successive spray beads overlap (Ref 21, 22). Indeed, coating is build up by overlapping the successive spray beads to form the spray pattern. The geometry of the spray bead can be approximated by a Gaussian function that size and shape depend upon the deviation and dispersion of the particle stream at the spray distance (depending themselves upon the plasma power and feedstock injection parameters), the particle state upon impact and plasma torch scan velocity (Ref 21). For fixed spray bead size and shape, the overlapping depends upon the lateral plasma torch scanning step. This is why apart from the optimization of the injection of the suspension and its processing by the plasma flow, the study of spray pattern influences also directly the microstructure of the resulting layer and is hence important to understand the coating formation and to optimize the operating parameters (Ref 23).

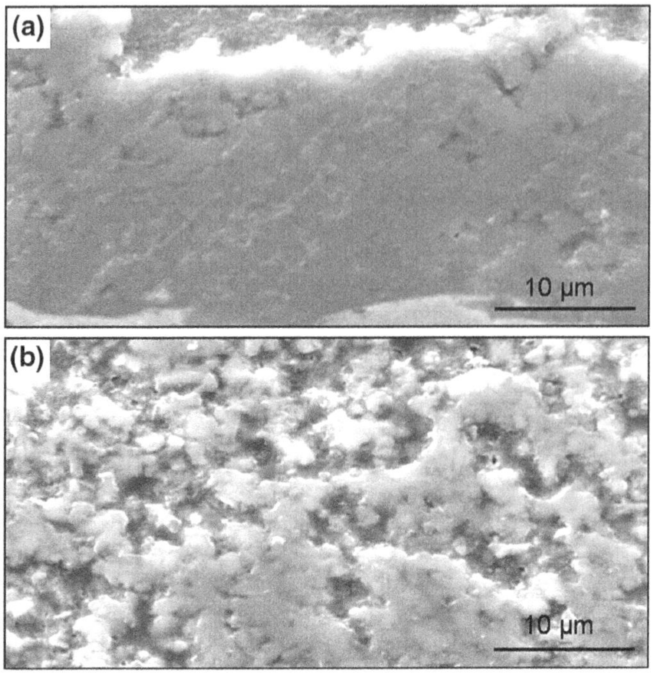

Fig. 3 Typical Al_2O_3 SPS layer architectures manufactured with (a) a low fraction of poorly treated particles (spray distance: 30 mm) and (b) a high one (spray distance: 50 mm) (feedstock $d_{50} = 500$ nm, solvent: Et-OH, MP = 20%, mechanical suspension injection, Ar-H_2 45-15 slpm, $I = 500$ A, Ø = 5 mm, $h = 14$ MJ kg^{-1})

The aim of this study is hence to assess in particular the effects of spray bead geometry and overlapping on coating structure in view of densifying to the maximum possible extend SPS Al_2O_3 layers. For reference power, suspension, and injection parameters, the spray distance, the number of passes, the spray velocity, and the lateral scanning step were varied to manufacture either spray patterns or layers.

2. Experimental Set-ups

2.1 Operating Parameters and Substrates

A stick-cathode DC plasma torch developed at SPCTS equipped with a 5-mm nozzle internal diameter was used to carry out experiments. It was operated with a binary

Ar-He (30-30 slpm) plasma forming gas mixture and with an arc current intensity of 600 A resulting in a plasma flow mass enthalpy of 12 MJ kg^{-1}. The injection system was composed of a diaphragm with a calibrated hole (150 μm, average diameter). The suspension to be sprayed was stored into a pressurized tank. Depending on the compressed air pressure applied in the tank, the velocity of the suspension at the injector exit varied between 18 and 25 m s^{-1} (momentum density of 0.40 and 0.56 MPa, respectively). The suspension of alumina was made of α-Al_2O_3 P152 SB (Alcan, Saint-Jean de Maurienne, France) of 0.5 μm average diameter (d_{50}), with different mass percentages, MP, of 5, 10, and 20%, respectively, into pure ethanol. To manufacture spray beads and layers, the injection has been optimized to have a penetration of the suspension deep into the plasma flow core. The plasma torch scan velocity (SV) was varied between 0.5 and 1.5 m s^{-1}. The spray distance (SD) was varied from 30 to 50 mm. The substrate temperatures during spraying were estimated using a monochromatic infrared pyrometer (5 μm wavelength). The emissivity was constant during the spray run (whereas it varies with the temperature) and fixed to 0.7. The integration area of the radiation emitted by the substrates was corresponding to surface of equivalent diameter of about 18 mm. Meanwhile such a control does not permit an accurate monitoring of the temperature during spraying; it permits nevertheless to estimate the substrate preheating average temperature to be around 250 °C and the coating average temperature during manufacturing to be about 500 °C.

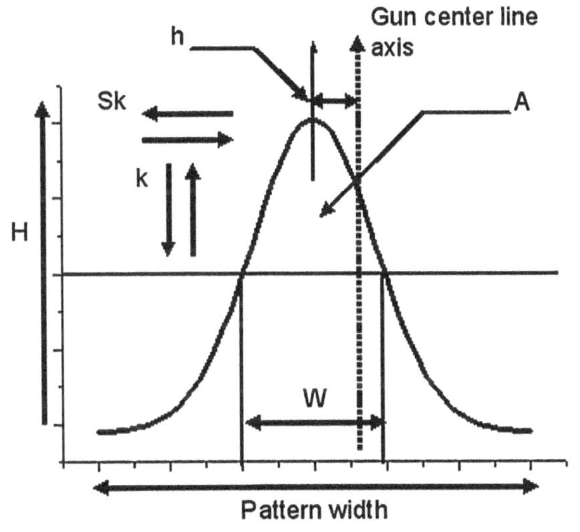

Fig. 4 Spray bead size and shape descriptors

Spray beads were sprayed onto 316L stainless steel plates (120 × 50 × 5 mm) that were previously degreased by immersion in acetone vapors and pre-polished using SiC papers and polished using diamond slurries to obtain different average roughnesses ranging from 0.06 to 0.20 μm. In this case, the plasma torch was moved back and forth in front of the substrates without scanning step.

Coatings were sprayed onto 316L stainless steel buttons (25 mm in diameter and 20 mm in thickness) exhibiting similar average roughnesses. The scanning step (SS) was varied in this configuration from 10 to 20 mm per pass, corresponding to velocities ranging from 0.025 to 0.05 m s^{-1}.

2.2 Measurement of Profiles of Spray Beads

The spray bead profile and its surface average roughness were assessed using a diamond stylus profilometer (Dektak IIA surface profilometer, Sloan Technology, Santa Barbara, USA). Ten measurements randomly located along the bead were carried out on each sample. After adjustment (by discarding the highest and the lowest values), data were averaged. Gaussian functions were selected to fit, with correlation factors ranging from 0.95 to 0.99, the bead profiles. Additional size and shape factors were defined (Fig. 4) such as (Ref 24): (i) the spray bead cross-sectional area A (mm^2) representative of the intrinsic deposition efficiency, the spray bead width L (mm), its maximal height H (μm), its width at half-height W (mm), the mean value m (dimensionless) of the Gaussian distribution, see Eq 2, and its standard deviation σ (dimensionless), see Eq 3; (ii) the spray bead offset h (mm) from the gun centerline axis and the scattering angle α (°) representing the dispersion cone angle of deposited materials ($\tan \alpha = SD/L$); (iii) the pattern skewness Sk (dimensionless), see Eq 4, representing the third moment of the height distribution profile and quantifying the asymmetry of the profile height distribution; and (iv) the spray pattern kurtosis k (dimensionless), see Eq 5, representing the fourth order moment of the height distribution profile and quantifying the flatness/sharpness of the profile height distribution.

Table 2 Profile deposit beads characteristics for different operating conditions

	Sample	H, μm	L, mm	W, mm	A, mm^2	α, °	m, −	Sk, −	k, −	R_a, μm	R_z, μm
NP (−) (for SD = 30 or 40 mm, MP = 10% and SV = 1 ms^{-1})	76(30 mm)	18.20	13.30	4.48	0.112	23.9	5.81	0.98	−0.54	0.80	5.16
	38(30 mm)	7.85	13.70	5.25	0.052	24.5	2.90	0.33	−1.22	0.42	3.36
	152(30 mm)	53.00	18.48	6.64	0.441	31.6	17.81	0.04	−1.13	2.14	8.25
	76(40 mm)	16.10	16.60	6.70	0.135	22.5	6.12	0.16	−1.27	1.17	6.53
	38(40 mm)	6.10	15.00	5.49	0.046	20.5	2.28	0.88	−0.88	0.51	4.11
	152(40 mm)	58.50	17.40	7.47	0.548	23.5	21.93	0.03	−1.37	2.90	10.56
SV (m s^{-1}) (for NP = 75, SD = 40 mm, MP = 10%)	1.5	10.20	12.80	4.80	0.065	17.7	3.86	0.96	−0.85	0.86	5.86
	0.5	17.40	21.50	9.07	0.187	28.3	6.95	0.39	−1.42	1.11	6.71
MP (%) (for NP = 75, SV = 1 ms^{-1}, SD = 40 mm)	5	28.80	21.50	8.08	0.292	28.2	10.08	0.72	−1.18	1.49	10.93
	20	24.50	20.50	8.44	0.260	27.1	9.53	0.56	−1.36	1.32	6.90

$$m = \frac{1}{L}\int_0^L z\,dx \quad \text{(Eq 2)}$$

$$\sigma = \sqrt{\frac{\int_0^L (z-m)^2\,dx}{L}} \quad \text{(Eq 3)}$$

$$Sk = \int_0^L \frac{(z-m)^3\,dx}{\sigma^3 L} \quad \text{(Eq 4)}$$

$$k = \int_0^L \frac{(z-m)^4\,dx}{\sigma^4 L} \quad \text{(Eq 5)}$$

where $z(x)$ is the pattern height function (μm) that represents the local probability density function of the particle impact distribution onto the substrate.

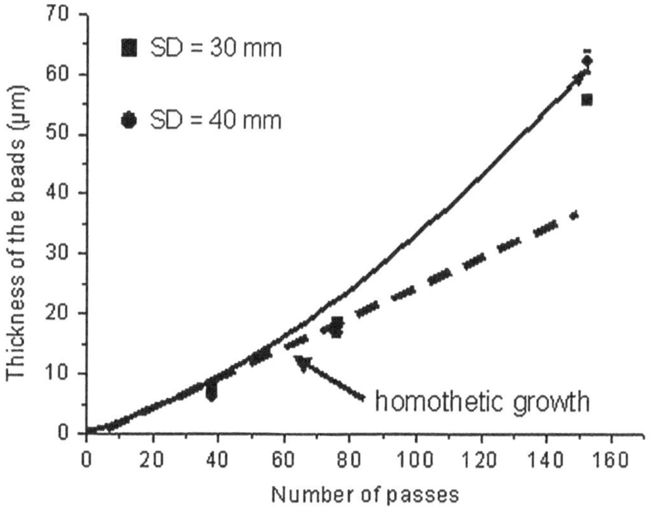

Fig. 5 Evolution of the bead thickness versus the number of passes

Moreover, two characteristics were evaluated in order to define the surface roughness. The average roughness, R_a (μm), that is the arithmetic mean of the absolute heights of roughness profile on the assessment length and the average maximum peak-to-valley height of the roughness profile within the sampling length, R_z (μm). The measurements were made parallel to the spray pattern (with a cut-off of 0.8 mm). All the profile characteristic measurements are listed in Table 2.

Cross sections and top of views of spray pattern and coating architectures were observed by secondary electron-scanning electron microscopy (SE-SEM).

3. Results and Discussion

3.1 Effect of the Number of Passes

At moderated numbers of passes (NP), the spray bead exhibits a homothetic evolution corresponding to a linearly growth of the deposited material (Fig. 5) that disappears then at higher numbers of passes. This can be explained as follows. When the number of passes increases, i.e., when the time spraying increases, the temperature of the sample increases too, up to 500 °C. The smaller particles (<1 μm) can stick at such temperature to the deposited beads (Ref 25) generating stacking defects between two successive passes. One could consider also that the increase in surface roughness of the bead could also form preferential impact locations for impinging particles.

This denotes that for a high number of passes the cohesion between the particles/lamellae decreases and so the surface irregularities and the layer porosity increase (Fig. 6). The average thickness deposited per pass can be hence estimated as varying from 0.20 to 0.25 μm per pass (for MP = 10% and SD = 30 mm).

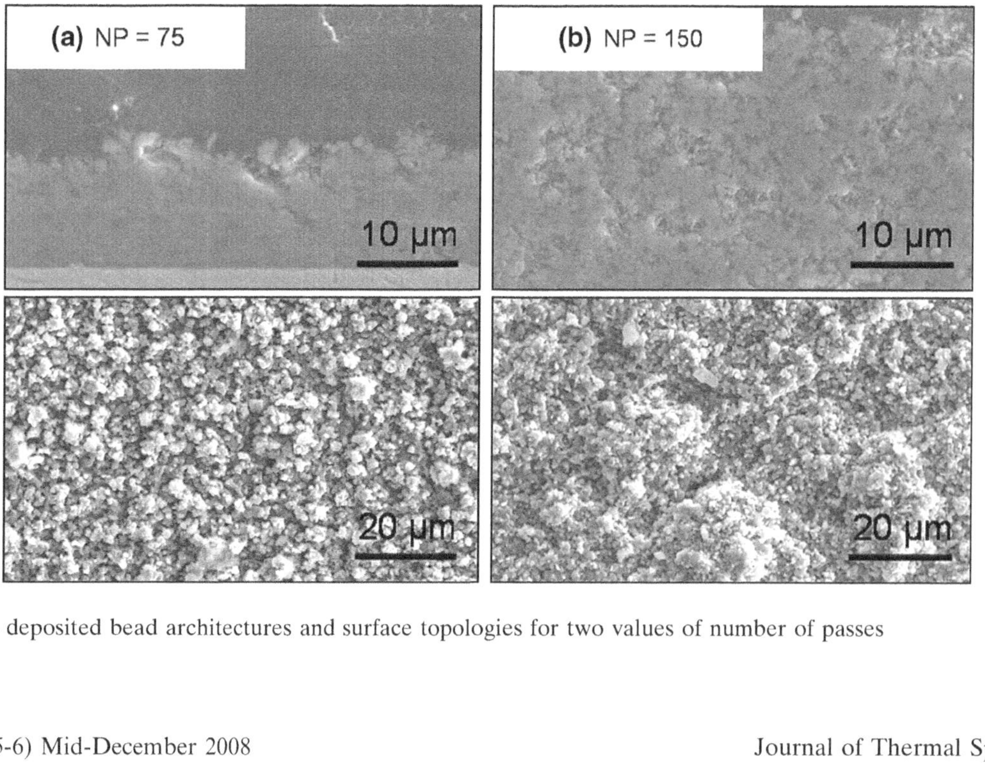

Fig. 6 Al_2O_3 SPS deposited bead architectures and surface topologies for two values of number of passes

3.2 Effect of the Spray Distance

The height and area of the spray beads present a quadratic evolution as a function of the spray distance (Fig. 7). This evolution is analogous to the one of the intrinsic deposition efficiency versus the spray distance. The highest values correspond to SD = 30 mm (average deposited thickness per pass higher than 0.20 μm), whereas they very significantly decrease to almost 0 for SD = 50 mm, demonstrating the criticity of the spray distance in the SPS process. Moreover, as the spray distance increases, the fraction of poorly treated particles is increased (mostly due to the low thermal inertia of particles) and the resulting coating architecture and surface topology are affected, being more porous and irregular (Fig. 8).

3.3 Effect of the Spray Velocity

In a first approximation, the spray bead size and shape criteria evolve linearly with an increase in the plasma torch scan velocity, SV, as shown in Fig. 9, which depicts the evolutions of the normalized bead area, A, width at half-height, H, and the scattering angle, α, versus the spray velocity. Those criteria decrease when SV increases. Indeed, the poorly treated particles generate stacking defects and hence promote porous and heterogeneous architectures (Fig. 10). At low velocity, those particles are better removed from the coating and the high heat flux transferred from the plasma to the substrate can keep locally the successive impinging particles in a molten state, or close to it, leading to denser coatings.

3.4 Effect of the Powder Mass Load

The powder mass percentage in suspension seems to influence directly the dispersion and deviation angles of the sprayed material. Indeed, the spray pattern characteristics do not present a continuous evolution but show an

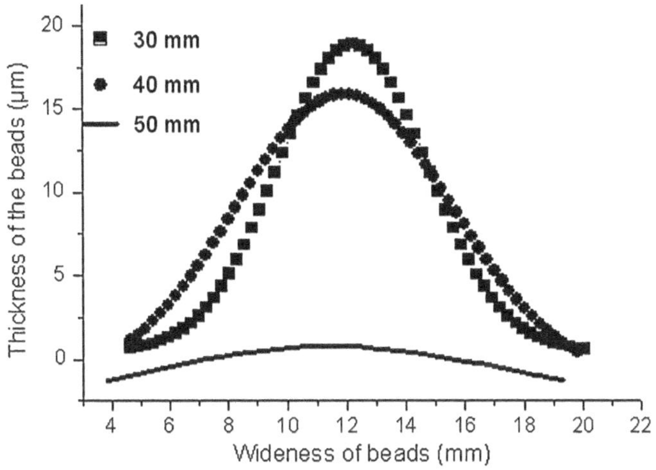

Fig. 7 Bead profiles manufactured at different spray distances

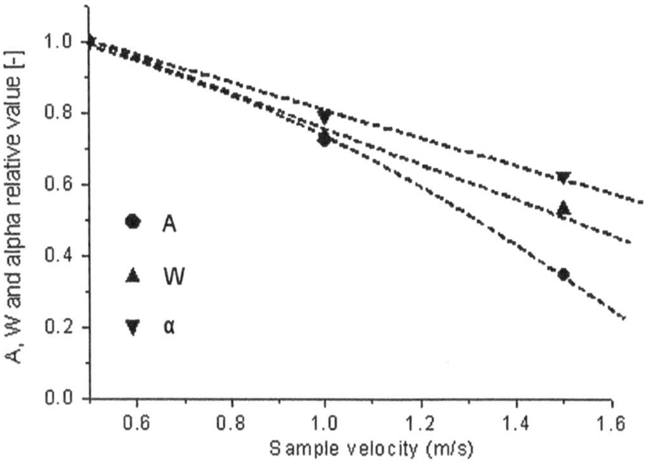

Fig. 9 Deposit profiles versus spray velocity

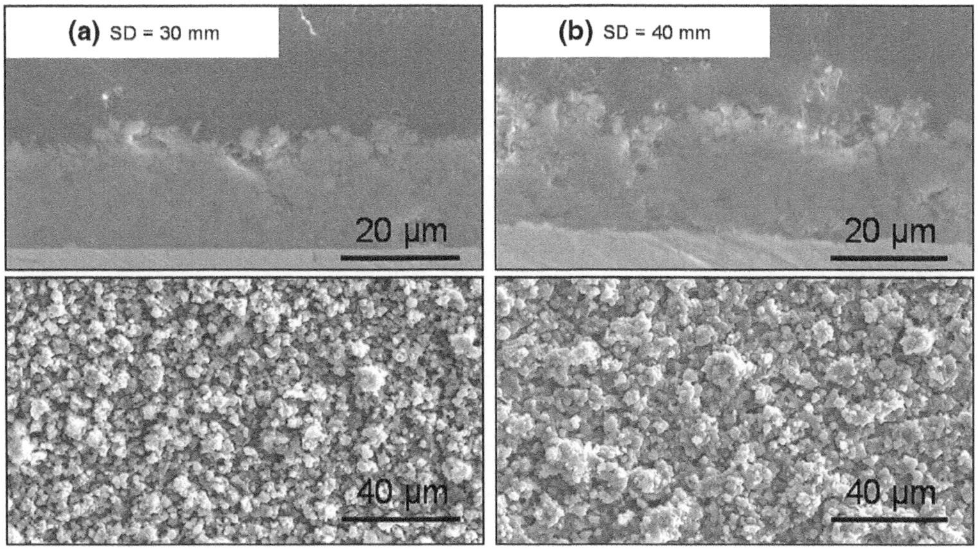

Fig. 8 Al_2O_3 SPS deposited bead architectures and surface morphologies at different spray distances

Peer Reviewed

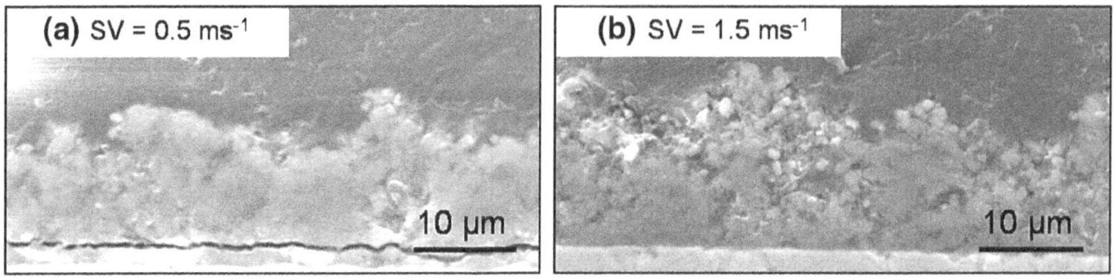

Fig. 10 Al_2O_3 SPS deposited bead architectures for different spray velocities

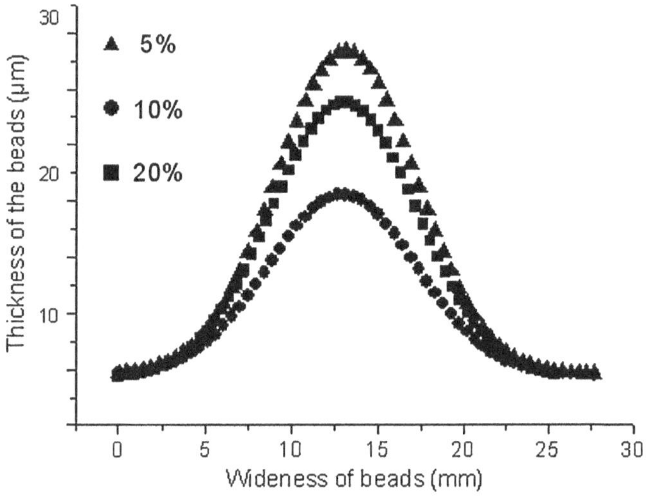

Fig. 11 Evolution of the spray bead profiles versus the powder mass percentage in suspension

extremum for MP = 10% (Fig. 11). At such a value, the dispersion angle is lowered, leading to a better treatment of the particles in the plasma jet. Therefore, the fraction of poorly treated particles is decreased and the coating density is increased consequently (Fig. 12). *A contrario*, for MP = 5 or 20%, the dispersion is increased and the fraction of poorly treated particles embedded into the spray bead is increased, leading to a porous coating architecture. MP becomes critical at high values (higher than 20%) and causes the layer delamination very likely due to the too high fraction of poorly treated particles embedded within the structure.

3.5 Comparison Between the Structures of Spray Beads and Coatings

The aforementioned results permit to clearly define some trends to adjust operating parameters to manufacture coatings with various pore levels.

At short spray distances (30-40 mm) and low powder mass percentage (5-10%), coatings are denser than those manufactured at longer spray distances (50 mm) or higher mass load (20%) (Fig. 13). By estimating the quantity of poorly treated particles embedded in the coating P (resolidified R and unmolten U particles) to the quantity of well melted particles M (molten particles), a qualitative ratio T = P/M can be defined as a function of operating parameters (plasma torch scan velocity, number of passes, spray distance, etc.).The higher the ratio, the higher the density of stacking defects and so the higher the pore level.

Moreover, the coatings manufactured at SD = 50 mm are very heterogeneous and porous whatever the other operating parameters, clearly demonstrating the criticity of the spraying distance on the coating microstructure. At last, for MP = 20%, the layers are porous and very poorly cohesive (some are even fractured). This is mostly due to the thermal stresses which develop during deposition consecutively to the high rate of poorly treated particles embedded into the coatings. Those particles are not removed by successive passes of the plasma torch in front of the substrate surface and consequently this generates a high density of stacking defects.

3.6 Temperature During Coating Manufacturing

During coating manufacturing, the average substrate temperature has been estimated to be about 500 °C. Even if the heat flux transmitted by the plasma flow to the substrate is fairly high at such short spray distance (a few MW m^{-2}), in situ sintering should not occur due to the too low temperature and relatively short duration at this temperature.

Nevertheless, temperature reaches very likely locally much higher values particularly at a 30-mm spray distance. Under such conditions, fast sintering mechanism could be achieved. Works are underway to identify such phenomena and to quantify them.

4. Conclusion

Several operating parameters including the spray distance, the number of passes, the spray velocity, and the powder mass percentage in suspension were varied to discriminate their effects upon sizes, shapes, and structures of spray patterns and structures of coatings.

The deposit shape and microstructure evolution have been linked to different operating conditions and have permitted to optimize the kinematics parameters to obtain dense or porous coating in a better reproducible way. The growth law and the average thickness deposited per pass have been quantified. The plasma torch scan velocity and

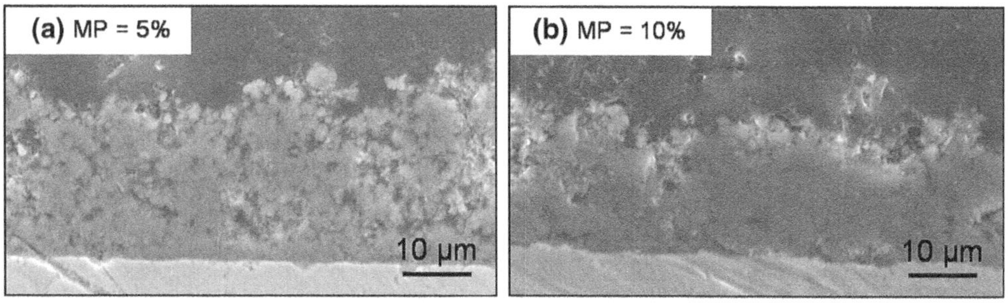

Fig. 12 Al_2O_3 SPS deposited bead architectures at different powder mass percentages in suspension

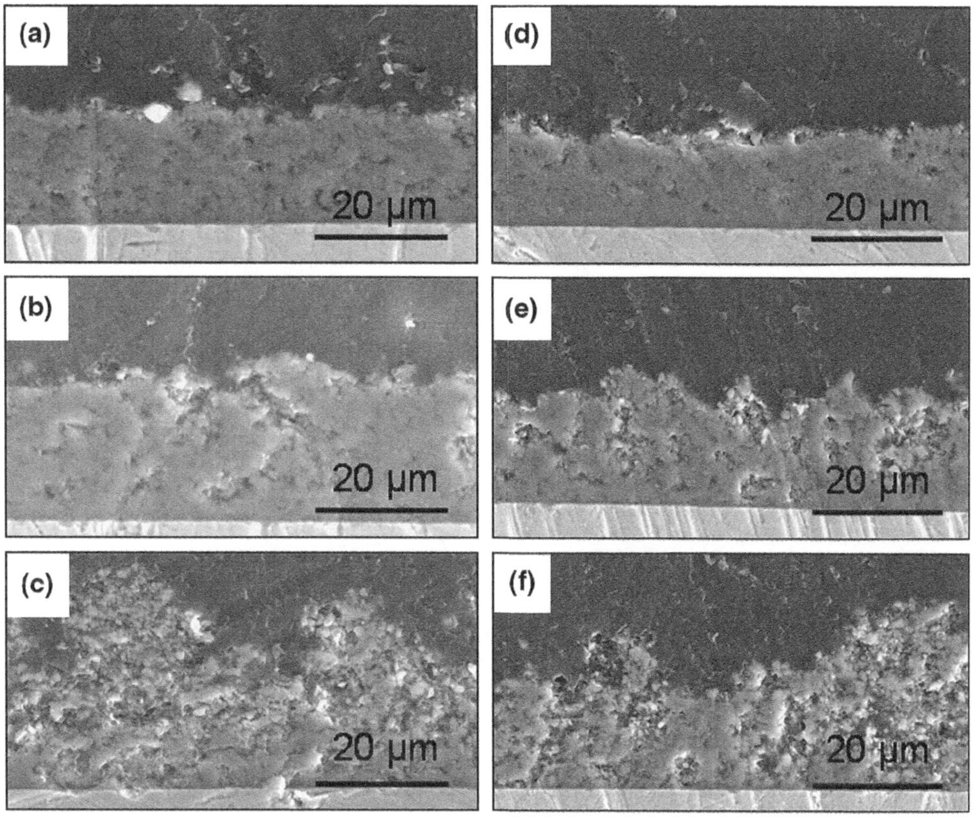

Fig. 13 Al_2O_3 SPS coating microstructures for MP = 5% at (a) 30 mm, (b) 40 mm, and (c) 50 mm, and for MP = 10% at (d) 30 mm, (e) 40 mm, and (f) 50 mm

especially the spray distance affect the microstructure of the coating and the deposition efficiency. An increase in those parameters leads to an increase in the pore level and a decrease in the layer cohesion. The powder mass load also plays a role but its effects are of lower amplitude.

This study demonstrates that besides the criticity of the fragmentation processes that are related to the processing of the liquid stream by the plasma jet bringing together the suspension properties, the injection configuration and the plasma flow thermodynamic properties, the kinematics and geometric parameters including the spray distance and the torch scan velocity and scanning step, are also very important in SPS process and permit to control the coating architecture.

References

1. R. Siegert, J.E. Döring, J.L. Marqués, R. Vaβen, D. Sebold, and D. Stöver, Denser Ceramic Coatings Obtained by the Optimization of the Suspension Plasma Spraying Technique, *Thermal Spray Solutions – Advances in Technology and Application*, DVS-Verlag GmbH, Düsseldorf, Germany, 2005, ISBN: 3-87155-792-7
2. R. Rampon, C. Filiatre, and G. Bertrand, Suspension Plasma Spraying of YPSZ Coatings for SOFC: Suspension Atomization and Injection, *J. Therm. Spray Technol.*, 2008, **17**(1), p 105-114
3. J. Jaworski, L. Pawlowski, F. Rondet, S. Kozerski, and A.L. Magner, Influence of Suspension Plasma Spraying Process Parameter on TiO_2 Coatings Microstructure, *J. Therm. Spray Technol.*, 2008, **17**(1), p 73-81
4. S. Bouaricha, J. Oberste-Berghaus, J.G. Legoux, C. Moreau, and T. Chraska, Suspension Plasma Spraying of Nano-Ceramics Using an Axial Injection Torch, *Thermal Spray Connects:*

Explore its Surfacing Potential!, E. Lugscheider, Ed., DVS-Verlag GmbH, Düsseldorf, Germany, 2005, ISBN: 3-87155-793-5

5. R. Etchart-Salas, "Atmospheric Plasma Spraying of Submicron-Sized Solid Particles in Suspension. Experimental and Analytical Approach of Involved Phenomena in Reproducibility and Quality of Coatings," Ph.D. Thesis, University of Limoges, France, 2007 (in French)
6. B.E. Gelfand, Droplet Breakup Phenomena in Flows with Velocity Lag, *Prog. Energy Combust. Sci.*, 1996, **22**, p 201-265
7. T. Watunabe and K. Ebihara, Numerical Simulation of Coalescence and Breakup of Rising Droplets, *Comput. Fluids*, 2003, **32**, p 823-834
8. C.S. Lee and R.D. Reitz, Effect of Liquid Properties on the Breakup Mechanism of High Speed Liquid Drops, *Atomization Sprays*, 2001, **11**, p 1-19
9. J. Fazilleau, C. Delbos, V. Rat, J.-F. Coudert, P. Fauchais, and B. Pateyron, Phenomena Involved in Suspension Plasma Spraying Part 1: Suspension Injection and Behavior, *Plasma Chem. Plasma Proc.*, 2006, **26**(4), p 371-391
10. P. Fauchais, M. Fukumoto, A. Vardelle, and M. Vardelle, Knowledge Concerning Splat Formation: An Invited Review, *J. Therm. Spray Technol.*, 2004, **13**(3), p 337-360
11. M.P. Planche, J.F. Coudert, and P. Fauchais, Velocity Measurements for Arc Jets Produced by d.c. Plasma Spray Torches, *Plasma Chem. Plasma Proc.*, 1998, **18**, p 263-283
12. J. Oberste-Berghaus, J.-G. Legoux, and C. Moreau, Injection Conditions and In-Flight Particle States in Suspension Plasma Spraying of Aluminia and Zirconia Nano-Ceramics, *Thermal Spray Connects: Explore its Surfacing Potential!*, E. Lugscheider, Ed., DVS-Verlag GmbH, Düsseldorf, Germany, 2005, ISBN: 3-87155-793-5
13. J.-F. Bisson and C. Moreau, Effect of Direct-Current Plasma Fluctuations on In-Flight Particle Parameters: Part II, *J. Therm. Spray Technol.*, 2003, **12**, p 256-265
14. R. Etchart-Salas, V. Rat, J.-F. Coudert, and P. Fauchais, *Thermal Spray 2007: Global Coating Solutions*, B.R. Marple, M.M. Hyland, Y.-C. Lau, C.-J. Li, R.S. Lima, and G. Montavon, Eds., CD-Rom, Pub. ASM International, Materials Park, OH, USA, 2007, ISBN: 0-87170-809-4
15. P. Fauchais, Understanding Plasma Spraying, *J. Phys. D: Appl. Phys.*, 2004, **37**, p R86-R108
16. C. Delbos, "Contribution to the Understanding of Ceramic (Y-PSZ, Perovskite, etc.) or Metallic (Ni, etc.) Particles Injection by a Liquid Carrier in a Plasma Jet to Manufacture Finely-Structured Coatings for SOFC," Ph.D. Thesis, University of Limoges, 2004 (in French)
17. O. Tingaud, A. Grimaud, A. Denoirjean, G. Montavon, V. Rat, J.-F. Coudert, and P. Fauchais, Effects of Operating Parameters on SPS Alumina Coatings Structures to Manufacture Functionally-Graded Layers, *Surf. Coat. Technol.* (in press)
18. Y.P. Wan, J.R. Fincke, S. Sampath, V. Prasad, and H. Herman, Modeling and Experimental Observation of Evaporation from Oxidizing Molybdenum Particles Entrained in a Thermal Plasma Jet, *Int. J. Heat Mass Transf.*, 2002, **45**, p 1007-1015
19. I. Ahmed and T.L. Bergman, Three-Dimensional Simulation of Thermal Plasma Spraying of Partially Molten Ceramic Agglomerates, *J. Therm. Spray Technol.*, 2000, **9**, p 215-224
20. X. Chen, Heat and Momentum Transfer Between a Thermal Plasma and Suspended Particles for Different Knudsen Numbers, *Thin Solid Films*, 1999, **345**(1), p 140-145
21. L. Xie, X. Ma, A. Ozturk, E.H. Jordan, N.P. Padture, B.M. Cetegen, D.T. Xiao, and M. Gell, Processing Parameter Effects on Solution Precursor Plasma Spray Process Spray Patterns, *Surf. Coat. Technol.*, 2004, **183**(1), p 51-61
22. F.-I. Trifa, G. Montavon, and C. Coddet, Model-Based Expert System for Design and Simulation of APS Coatings, *J. Therm. Spray Technol.*, 2007, **16**(3), p 128-139
23. E. Pfender, Thermal Plasma Technology: Where do We Stand and Where are We Going?, *Plasma Chem. Plasma Proc.*, 1999, **19**, p 1-31
24. F.-I. Trifa, G. Montavon, and C. Coddet, On the Relationships Between the Geometric Processing Parameters of APS and the Al_2O_3-TiO_2 Deposit Shapes, *Surf. Coat. Technol.*, 2005, **195**, p 54-69
25. A. Haddadi, R. Hamacha, A. Grimaud, P. Fauchais, and F. Nardou, Residual Stresses and Microstructure of Plasma Sprayed Zirconia Coatings, *High Temp. Mater. Processes*, 1998, **2**(3), p 327-337

JTTEE5 17:671–678
DOI: 10.1007/s11666-008-9219-8
1059-9630/$19.00 © ASM International

Peer Reviewed

High-Velocity Oxy-Fuel (HVOF) Suspension Spraying of Mullite Coatings

J. Oberste Berghaus and B.R. Marple

(Submitted May 1, 2008)

Mullite coatings ($3Al_2O_3 \cdot 2SiO_2$) were deposited by suspension thermal spraying of micron-sized ($D_{50} = 1.8$ μm) feedstock powders, using a high-velocity oxy-fuel gun (HVOF) operated on propylene (DJ-2700) and hydrogen fuels (DJ-2600). The liquid carrier employed in this approach allows for controlled injection of much finer particles than in conventional thermal spraying, leading to coatings with low porosity and fine and homogeneous porosity distribution, making this process potentially suitable for creating thin layers with low gas permeability. In-flight particle states were measured for a number of spray conditions of varying fuel-to-oxygen ratios and standoff distances and related to the resulting microstructure, stoichiometry, phase composition (EDS, SEM, XRD), and hardness ($VHN_{300g \cdot f}$) of the coatings. In an attempt to retain the crystalline phase in the coatings, HVOF operating conditions were varied to limit in-flight particle melting. However, fully crystalline coatings were only obtained by gradually heating the coating during deposition to temperatures above 400 °C.

Keywords HVOF suspension spraying, mullite coatings, suspension thermal spraying

1. Introduction

Mullite, an aluminosilicate of composition $3Al_2O_3 \cdot 2SiO_2$, has attracted considerable interest as a protective coating in high-temperature applications due to its high thermal and chemical stability, low thermal expansion coefficient, low thermal conductivity, and low oxygen permeability, associated with the sealing properties of silica. Over the entire temperature range of the crystalline phase, no polymorphic phase transitions occur, which results in particularly high thermal shock resistance and high strength (Ref 1-3). Mullite-based protective and thermal barrier coatings are applied in heat exchangers, gas turbines, and internal combustion engines, i.e., diesel engines, often in combination with SiC or Si_3N_4 thermal barriers by virtue of similar thermal expansion coefficients and chemical compatibility (Ref 4, 5). In particular, mullite coatings are often used in environmental barrier coating (EBC) systems as a basic protection against oxidative or water vapor-containing atmospheres at elevated temperatures (Ref 6, 7).

Plasma sprayed mullite is generally reported to contain a high portion of amorphous phase due to the rapid solidification of the molten droplets during the coating formation. Recrystallization during service at 700-1000 °C has been reported to cause cracking and delamination from the substrate due to the associated volume reduction of about 0.4% (Ref 1) By heating the substrate above the crystallization temperature (>980°) inside a furnace during spraying, desired crystallinity can be created directly in the spray process (Ref 4). This method is well suited for SiC substrates, but has limited use for metallic substrates, being subject to high-temperature oxidation. Other efforts are directed toward carefully limiting the thermal energy delivered to relatively large (>44 μm) crystalline feedstock powders by adjusting the thermal spray conditions, thereby fostering only partial melting of the particles and retaining above 80% of the original crystallinity in the coating (Ref 8, 9). Due to a low remaining amorphous content, the resulting recrystallization stresses are reported to be of less concern. To overcome some disadvantages inherently associated with the use of large feedstock particles for protective coatings, such as large coating thickness, coarse porosity, and high surface roughness, Siegert et al. (Ref 10) reports the use of a sol-gel feedstock with mullite precursors in a plasma spray process. The desired coating crystallinity and low porosity is attained by repeatedly passing over the layer with the plasma flame after deposition, thereby effectively reheating and in situ recrystallizing the coating. Similarly, the amorphous phase formation can somewhat be avoided in conventional thermal sprayed mullite coatings by hot spraying, i.e., maintaining high substrate temperature conditions during deposition, as illustrated by Vassen et al.

This article is an invited paper selected from presentations at the 2008 International Thermal Spray Conference and has been expanded from the original presentation. It is simultaneously published in *Thermal Spray Crossing Borders, Proceedings of the 2008 International Thermal Spray Conference*, Maastricht, The Netherlands, June 2-4, 2008, Basil R. Marple, Margaret M. Hyland, Yuk-Chiu Lau, Chang-Jiu Li, Rogerio S. Lima, and Ghislain Montavon, Ed., ASM International, Materials Park, OH, 2008.

J. Oberste Berghaus and **B.R. Marple**, Industrial Materials Institute, NRC, 75 de Mortagne Blvd., Boucherville, QC, Canada J4B 6Y4. Contact e-mail: jorg.oberste-berghaus@bekaert.com.

(Ref 6) for mullite coatings on siliconized carbon reinforced carbon substrates.

High-velocity oxy-fuel (HVOF) suspension spraying is an emerging technology that permits better feeding and spraying of finer particles than conventional thermal spraying (Ref 11, 12). In principle, thinner coatings (20-100 μm) with more refined microstructures and potentially smoother surface finishes are created. Feeding micron-sized or submicron particles into a thermal spray gun with conventional equipment is virtually impossible due to the strong powder agglomeration, which impedes dry powder flow. Suspending the small particles in a liquid carrier, which is then injected into the HVOF gun, alleviates this problem and allows for precisely controlled feeding rates. In the HVOF combustion chamber, the suspension is atomized into a fine mist and the suspension medium is evaporated and combusted along with the fuel, thereby concentrating the solid content into micro-sized particles, which are then projected toward a substrate. The high particle velocity of HVOF systems is conducive for creating dense, and possibly gas impermeable, coatings. Furthermore, the maximum flame temperature in HVOF is substantially lower (2500-3100 °C) than in plasma systems (>8000 °C), which may, in the case of small mullite particles, limit complete particle melting to retain some of the original crystallinity. The lower flame temperature may also be advantageous in avoiding segregation into silica- and alumina-rich phases and consequent silica evaporation and loss.

In this study, the potential of this technique for creating thin, dense, and finely structured mullite coatings with some remaining crystallinity is explored. The suspension feedstock is axially injected into the combustion chamber of an HVOF gun. In-flight particle velocity and temperature are measured for different operating conditions and standoff distances, and related to the resulting microstructure and phase composition of the coating. Different HVOF fuels, i.e., propylene and hydrogen, and different fuel-to-oxygen ratios are compared. Finally, in situ recrystallization by gradually heating the coating to high temperatures during deposition is investigated.

2. Experimental

Suspension spraying was implemented using an HVOF DJ-2700 (propylene-C_3H_6) and DJ-2600 (hydrogen-H_2) hybrid gun (Sulzer-Metco, Westbury, NY). The systems were equipped with internal injection/atomization modules inserted into the standard powder feeding ports, thereby intimately contacting the coaxially fed suspension droplets with the fuel inside the HVOF combustion chamber. The suspensions were delivered by a prototype Nanofeed Liquid Powder Feeder (Model 640, Northwest Mettech, North Vancouver, Canada). Flow-rate control, start-up, shutdown, and rinsing sequences were fully automated and PC controlled. The injection module and fuel distributor were continuously cooled with nitrogen at a flow rate of 15 slpm, also serving as atomizing gas.

Table 1 DJ-2700 and DJ-2600 HVOF spray parameters

Parameter	Value
Propylene flow (DJ-2700)	85, 75, 65, 55 and 50 slpm
Oxygen flow (DJ-2700)	279 slpm
Air flow (DJ-2700)	202 slpm
Hydrogen flow (DJ-2600)	684 and 450 slpm
Oxygen flow (DJ-2600)	231 slpm
Air flow (DJ-2600)	330 slpm
Carrier gas (N_2) flow	15 slpm
Suspension feed rate	50 mL/min (3.0 kg/h)
Spray distance	7.6-12.7 cm
Gun traverse speed	762 mm/s

In addition, coolant medium, such as nitrogen, water, or suspension, was delivered through the feeding port at all times during operation of the gun to prevent overheating and possible blockage of the internal components. The system was adapted to deliver the suspension feed against the combustion backpressure of up to 6.46 atm in the case of propylene and up to 7.48 atm for hydrogen. In an attempt to vary particle temperature at high velocities, the gun was operated with the two different fuels of varying flow rates, while maintaining constant air and oxygen flows, respectively, as summarized in Table 1.

In-flight particle states were measured with a commercial diagnostic system (AccuraSpray® G2 Tecnar, St-Bruno, PQ, Canada). The temperature measurement is based on two-color pyrometry, and the velocity is determined by a time-of-flight technique. At a short spray distance of 7.6 cm from the gun nozzle exit, measurements of particle temperature and velocity were impeded by the presence of compression/decompression zones (shock barrels) in the supersonic flame. Based on measurements further downstream at 7.6-15.2 cm from the nozzle, an indirect approach to rank the in-flight particle states for each set of spray parameters was adopted.

Microstructures of the coatings were observed by SEM (JEOL JSM-610) and FE-SEM (Hitachi S4700). The samples were prepared by standard metallographic methods. Porosity was assessed on the cross section by SEM (1000 X), using image analysis. Such magnification was judged most suitable to capture the finely structured porosity in the coatings. The intensity range and thresholds were standardized on reference materials and ten measurements were averaged per sample. Elemental composition was measured by energy dispersive spectroscopy (EDX) at 500X magnification, averaging three measurements per sample.

Vickers microhardness measurements were performed under a 300 g load for 20 sec on the polished cross section of the coatings and ten measurements were averaged per sample.

Phase analysis was carried out by XRD using a Brucker D8-Discovery diffractometer (Brucker AXS Inc. Madison) with Cu-K_α radiation at an acquisition of 0.01°/sec. In an attempt to compare the degree of crystallinity between the coatings, the ratio of peak areas between 20 and 50° to the total integrated area in this region, including the diffuse amorphous hump, was measured and denoted here as index of crystallinity (IC). Particle sizes and distribution were measured with a Coulter LS Particle size analyzer (Beckman Coulter Canada Inc., Mississauga, Canada) using a liquid module.

Fused and crushed feedstock powders of high-purity mullite (Chichibu MP40, Scimarec Co. Ltd., Tokyo, Japan) with a nominal particle size of 1-2 μm were used in this study. Powder morphology is shown in Fig. 1.

Suspensions were prepared with 5wt.% solids in a mixture of ethanol and 30wt.% ethylene glycol. Polyethylenimine (MWT 25,000 Alfa Aesar) was used as dispersing agent. The pH of the suspension was adjusted to neutral conditions using nitric acid. The addition of viscous ethylene glycol further prolonged the settling time of the suspension. The procedure stabilized the dispersion sufficiently to be compatible with the spray process.

The size distribution of the particles and agglomerates in suspension is shown in Fig. 2. The powder shows a relatively narrow size distribution with D_{50} = 1.8 μm, but submicron satellite particles as well as aggregates as large as 10 μm are also present.

Fig. 1 Micrograph of MP40 mullite feedstock

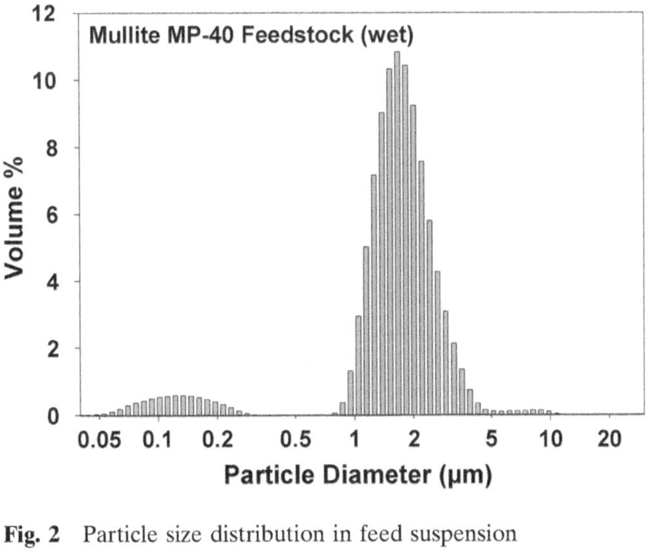

Fig. 2 Particle size distribution in feed suspension

Coatings were produced on mild steel substrates with dimensions of 25 × 75 × 12.5 mm. To increase coating adhesion, the substrate surface was grit-blasted with 54 grit Al_2O_3 particles prior to deposition. For most coatings, the surface temperature, as monitored with a pyrometer, was maintained at 250-300 °C during deposition by using forced air cooling. The substrate was preheated to approximately 200 °C by the HVOF flame in 10 passes immediately before deposition. Selected coatings were produced at an elevated surface temperature of up to 420 °C.

3. Results and Discussion

3.1 DJ-2700 Coatings

A typical microstructure of the coatings produced at a standard condition with propylene fuel (85 slpm) at a spray distance of 7.6 cm is shown in Fig. 3. Depending on the exact operating conditions, deposition rates varied from 6.3 to 7.5 um/pass. Deposition efficiencies were measured between 50 and 65% at spray distances up to

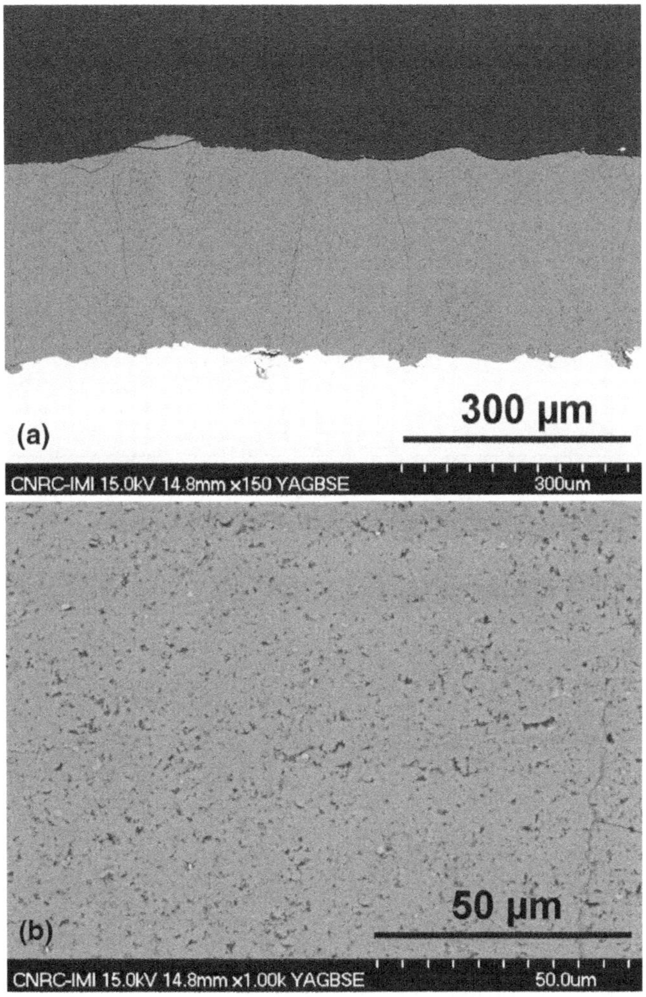

Fig. 3 Micrographs of cross sections of DJ-2700 coating (85 slpm C_3H_6 at SD 7.6 cm)

Peer Reviewed

12.6 cm. Typical coating thicknesses were between 110 and 130 μm. Coatings up to 260 μm in thickness were also produced. The initial layers of the coating filled the original surface features of the grit-blasted substrate, resulting in a final surface roughness of $R_a < 3.5$ μm, as measured with a profilometer.

The porosity of the sample shown in Fig. 3 is 1.4 ± 0.4%, but varied between 0.9% and 3.4% for coatings produced at the shortest spray distance (7.6 cm), depending on the exact fuel and gun conditions. The elemental composition of the DJ-2700 coatings (85 slpm C_3H_6) is summarized in Table 2. Besides a slight loss in oxygen, the stoichiometric ratio of Al to Si, corresponding to the mullite feedstock, is preserved in the coatings.

3.2 Spray Distance

In-flight particle states as a function of standoff distance are exemplified in Fig. 4 for this particular spray condition (85 slpm C_3H_6). The large error bars indicate significant uncertainty in the temperature measurement close to the gun nozzle exit. However, the temperature and velocity decrease monotonically from approximately 2500 °C and 850 m/s (at.7.6 cm), imparting some confidence to the measurement.

Coatings were produced at different standoff distances, using the same spray conditions. An increase in SD leads to a rise in porosity of up to 9%, as shown in Fig. 5. Correspondingly, the Vickers indentation hardness decreases from 623 $Hv_{0.3}$ to below 400 $Hv_{0.3}$ for the most porous coating. Figure 6 compares the XRD spectra of these coatings with the MP40 feedstock. The spectra show, besides the crystalline mullite peaks, a distinct hump in the region of 20-50°, indicative of a high amorphous content in the coatings.

An increase in porosity and the associated loss in hardness with lower particle temperature and velocity is somewhat expected. It is, however, less evident to observe the decrease in crystallinity, as suggested by the lowering of the peak heights in Fig. 6. Comparing the phase composition in terms of IC, a decrease from 37% at SD 7.6 cm to 17% at SD 10.2 cm and finally to 14% at SD 12.7cm is noted. It is unlikely that different rates of in situ recrystallization of the mullite coatings occurred in this case, since the maximum substrate temperature was carefully maintained at 230-250 °C for the three coatings. It is possible that the particle residence time in the heating combustion flame pays an important role. At the shortest standoff distance, unmolten particles may still be preserved, which deposit by virtue of the very high jet velocity. At larger spray distance, the particles may have had more time to fully melt and increase the amorphous content in the coating.

Table 2 EDX elemental composition

Sample	SD, cm	O, at.%	Al, at.%	Si, at.%
Feed MP40	...	59.0	30.1	10.8
85-P	7.6	55.8 ± 0.7	32.7 ± 0.5	11.5 ± 0.2
85-P	10.2	56.3 ± 0.6	32.4 ± 0.4	11.2 ± 0.2
85-P	12.7	56.3 ± 0.3	32.3 ± 0.4	11.4 ± 0.2

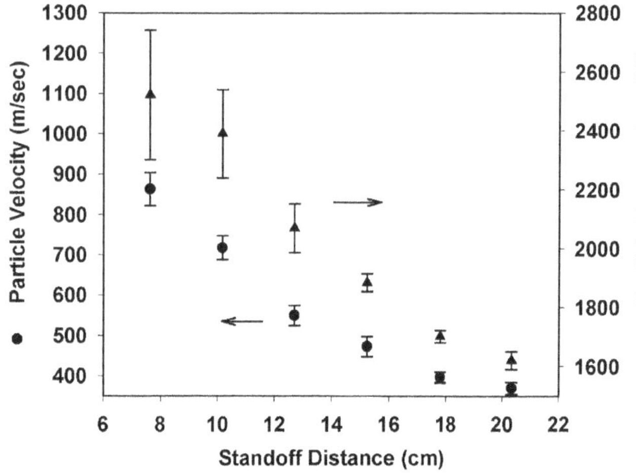

Fig. 4 Particle state versus standoff distance at 85 slpm propylene

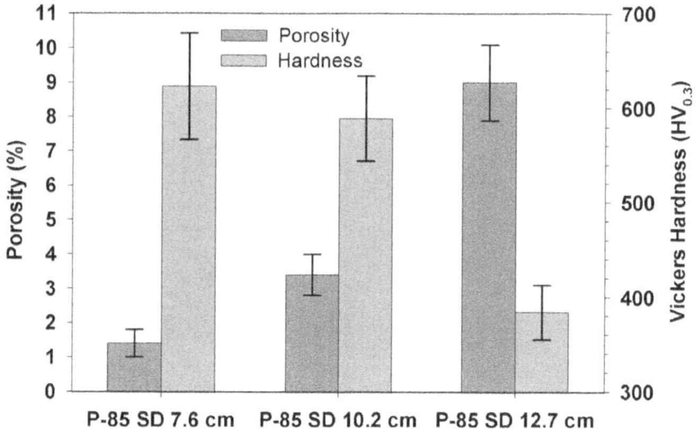

Fig. 5 Porosity and hardness versus standoff distance at 85 slpm propylene condition

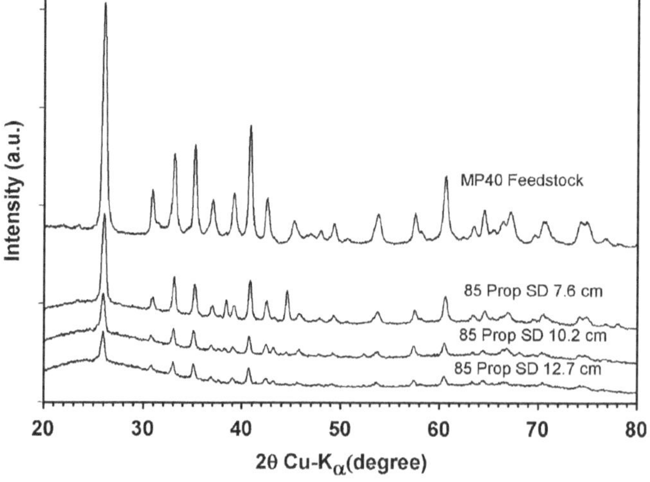

Fig. 6 XRD spectra of MP40 feedstock and coatings produced with 85 slpm propylene fuel

3.3 Fuel-to-Oxygen Ratio

Maximum particle temperatures in HVOF systems depend on, besides the torch design and feed conditions, the fuel-to-oxygen ratio (Ref 11). Figure 7 exemplifies the drop in particle temperature for decreasing propylene fuel rates from 85 to 50 slpm (at constant standoff distance). Replacement of the gun barrel resulted in higher temperature and velocity than reported above. In fact, during the course of the work, a high sensitivity of the coatings and particle states to the conditioning of the gun was noticed.

The characteristics of the corresponding coatings produced at SD of 7.6 cm are summarized in Fig. 8. In spite of the lower particle temperature and velocity with reduced fuel supply, a decrease in coating porosity and crystalline content can be discerned. It is again possible that residence time, and not maximum flame temperature, is a governing factor to limit particle melting to create more porous and crystalline coatings. In this case, high fuel ratios lead to high jet velocity and short dwell times, limiting particle melting, even though the flame temperature is elevated over a condition with less fuel. A trend in coating hardness cannot be identified within the measurement confidence. However, the highest hardness of 646 ± 33 $Hv_{0.3}$ was measured for a coating (Prop-85-hot) produced at much higher substrate temperature, gradually increasing to 410 °C during deposition. A significant densification and in situ crystallization has probably taken place, resulting in a crystalline coating, as measured on its top surface by XRD. The microstructure of this high-temperature coating shows distinct strata formation in its upper portion, composed of layers of spherical pores (Fig. 9a and b). This feature suggests that besides recrystallization, some remelting of the top surface in the last few deposition passes has occurred. A lower surface roughness of R_a 2.7 ± 05 μm, as compared to the coating produced at lower surface temperature (Prop-85) with R_a 3.1 ± 0.7 μm, also supports this idea. The microstructure of this high-temperature coating shows distinct strata formation in its upper portion, composed of layers of spherical pores (Fig. 9a and b). This feature suggests that besides recrystallization, some remelting of the top surface in the last few deposition passes has occurred. A lower surface roughness of R_a 2.7 ± 05 μm, as compared to the coating produced at lower

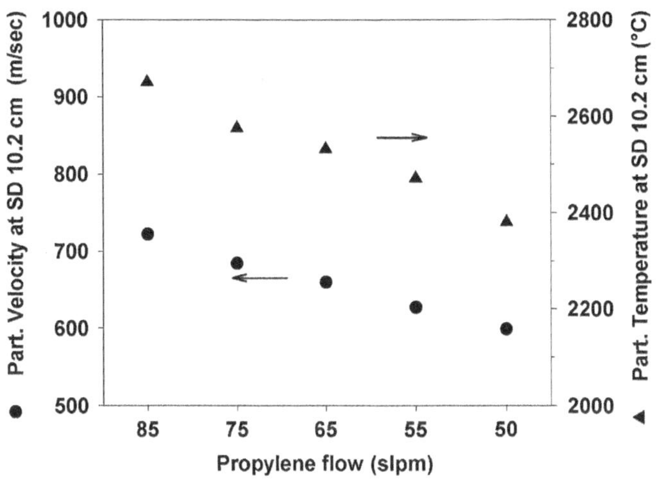

Fig. 7 Particle states for different propylene flows at SD 10.2 cm with new gun barrel

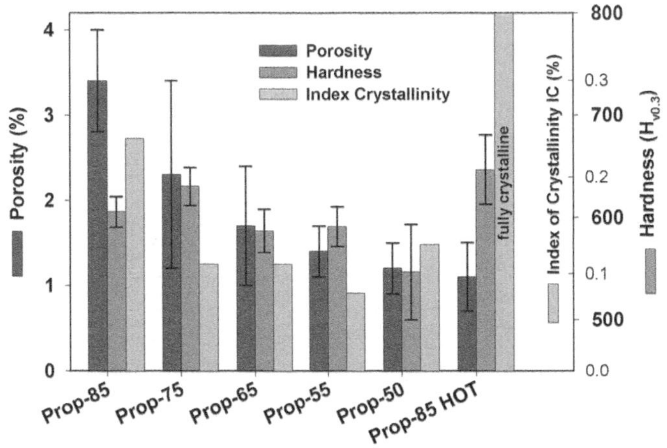

Fig. 8 Coating characteristics versus propylene rate

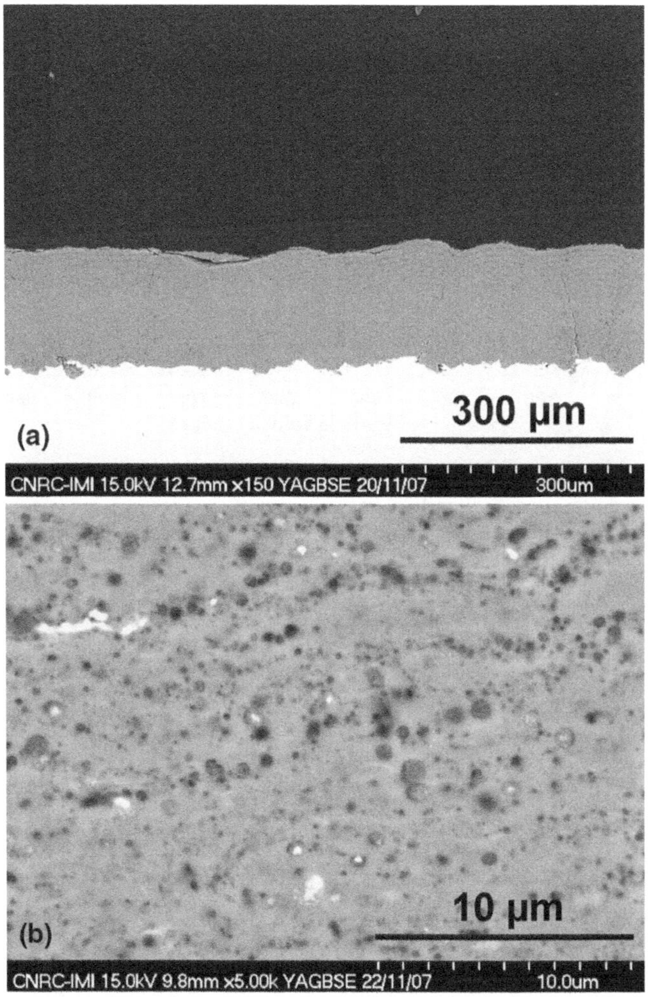

Fig. 9 Microstructure of coating produced at high surface temperature (Prop-85 Hot)

surface temperature (Prop-85) with R_a 3.1 ± 0.7 μm, also supports this idea. It should be noted that within the T-V operating space of the DJ-2700, particle temperature and velocity are strongly coupled and cannot be varied individually, as illustrated in Fig. 10. To retain some coating crystallinity, it was consequently difficult to limit particle melting by both a low flame temperature and a short dwell time at high particle velocity.

3.4 DJ-2600 Coatings

In an attempt to shift the operating space to lower particle temperatures, the system was operated with hydrogen fuel, as illustrated in Fig. 10. Coatings were produced at two different hydrogen flow rates of 684 and 450 slpm, keeping the other operating conditions constant. In this case, reducing the fuel supply from an over-stoichiometric to a close to stoichiometric fuel/oxygen ratio raised both the particle temperature and velocity as measured in Fig. 11.

The characteristics of the coatings produced at the shortest standoff distance (7.6 cm) are summarized in Fig. 12. Comparing samples H_2-684 and H_2-450, the coating becomes slightly denser, harder, and loses some of its remaining crystallinity. In addition, the deposition rate was found to increase from 5.7 to 7.7 μm/pass. Both coatings were produced at a maximum surface temperature of 300 °C.

Again, crystalline coatings, or at least coatings with crystalline top surfaces as measured by XRD, were obtained for both conditions by allowing the substrate temperature to gradually increase to 410 °C during deposition. The XRD spectra of the four coatings are shown in Fig. 13. Representative microstructures of samples produced at cold (H_2-450) and hot substrate temperatures (H_2-684 Hot) are shown in Fig. 14. Similar to the propylene coatings, strata composed of circular pores can be discerned in the top portion of the coating, indicating some surface remelting. It should be mentioned that when a high surface temperature was maintained during the entire deposition time, the coatings were found to spall from the substrate. The above microstructures and characteristics of the DJ-2600 coatings suggest that they

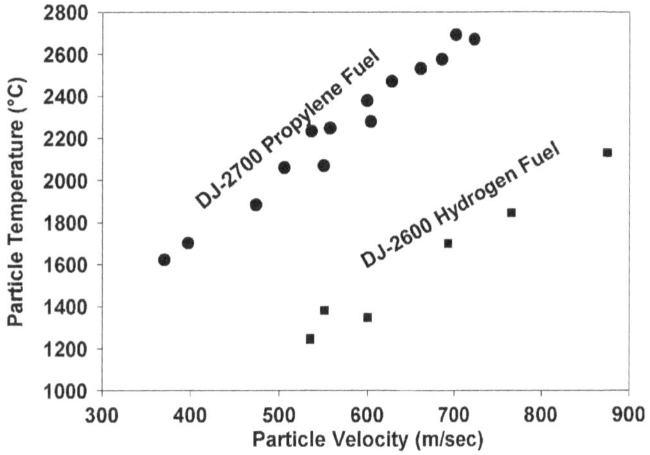

Fig. 10 DJ-2700 and DJ-2600 operating space in terms of particle temperature and velocity

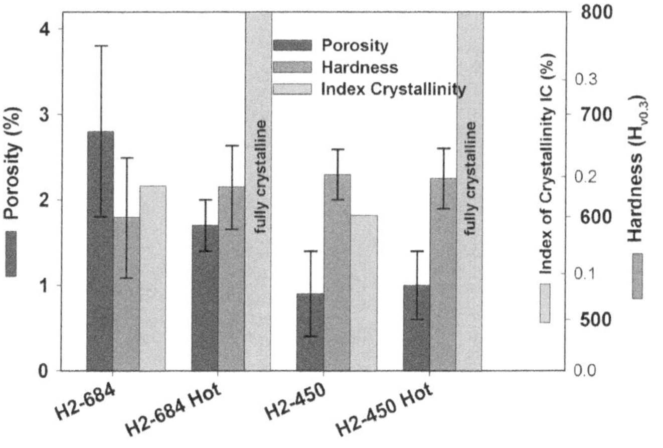

Fig. 12 Coating characteristics versus hydrogen feed rate

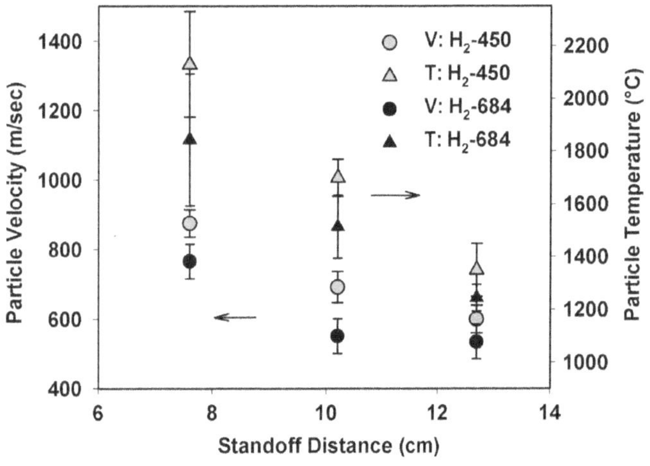

Fig. 11 DJ-2600 particle states versus SD

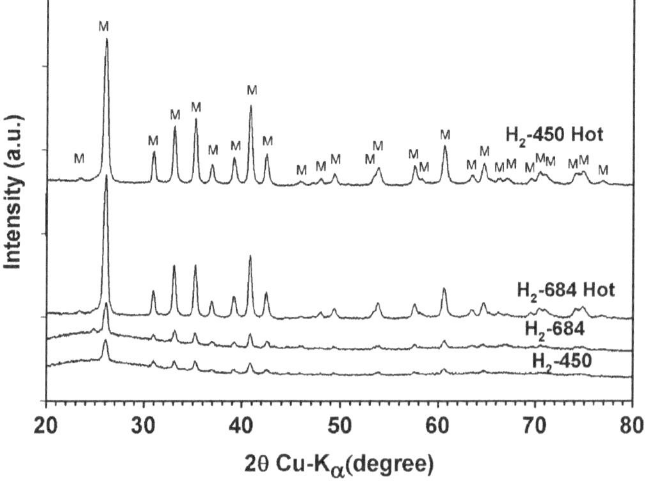

Fig. 13 XRD spectra of DJ-2600 coatings

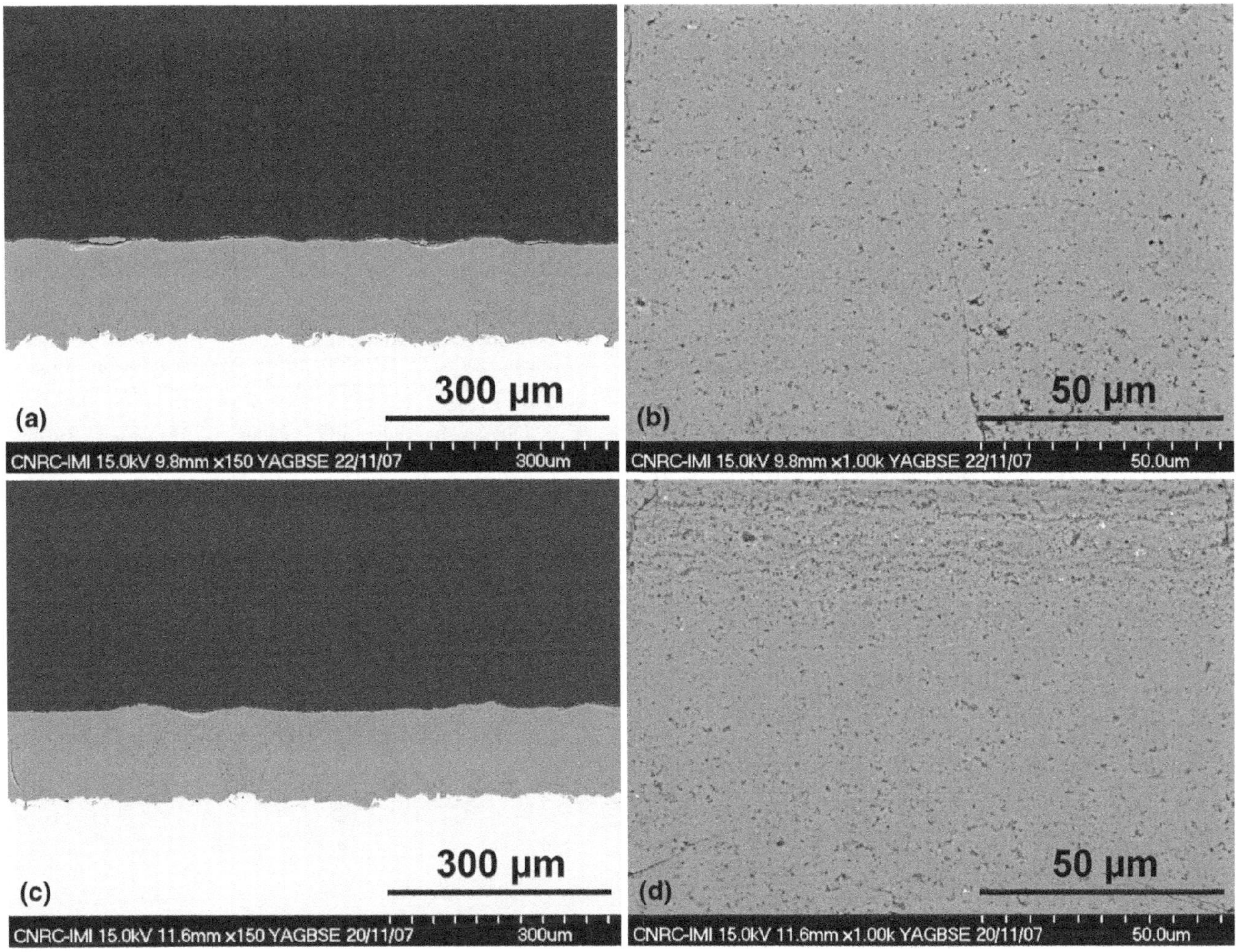

Fig. 14 Microstructures of DJ-2600 coating (a, b) H_2-450 (c, d) H_2-684 hot

are created from mostly molten particles, impacting at high velocity. A distinct difference from the coatings produced with propylene, in particular with respect to the crystalline content, cannot be discerned. It is possible that the background radiation emitted by the flame, which differs substantially between the DJ-2700 and the DJ-2600 jet, exaggerates the difference in particle temperature, as measured by the optical system. Furthermore, the information about the particle states at the short SD is subject to uncertainty since it is extrapolated from measurements further downstream. It cannot be excluded that at the short SD the particles are still in a process of heating, leading to the deposition of partially molten feedstock in some of the more crystalline DJ-2700 coatings.

Ultimately, the potential of HVOF over plasma spraying in terms of low flame temperature and the high velocity could not be fully exploited for suspension spraying of crystalline mullite coatings, since the particles experienced a dwell time long enough for complete melting in nearly all conditions investigated. However, allowing in situ crystallization by raising the substrate temperature during deposition can create fully crystalline coatings.

4. Conclusions

- The potential of HVOF with liquid suspension feed as a novel method to produce thin protective mullite coatings was explored. The lower flame temperature and high particle velocity, potentially limiting particle melting, make this technique a promising candidate.
- Mullite coatings with surface roughness $R_a < 3$ µm, with over 20% crystallinity and uniformly distributed and fine porosity below 2%, were produced with the DJ-2700 at high propylene fuel-to-oxygen ratios and at short standoff distance.
- Decreasing the flame temperature by reducing the fuel supply or changing to hydrogen fuel was not found to increase the crystalline content in the coatings. It was speculated that dwell time and not maximum flame temperature is the principal factor governing the degree of in-flight particle melting.
- Crystalline mullite coatings with porosities $\leq 2\%$ and Vickers hardness ≥ 640 $Hv_{0.3}$ were produced with DJ-2700 and DJ-2600 by gradually increasing the surface temperature above 400 °C during deposition.

Peer Reviewed

References

1. P. Rohan, K. Neufuss, J. Matějíček, J. Duvský, L. Prchlík, and C. Holzgartener, Thermal and Mechanical Properties of Cordierite, Mullite and Steatite Produced by Plasma Spraying, *Ceram. Int.*, 2004, **30**, p 597-603
2. F. Cipri, C. Bartuli, T. Valente, and F. Casdei, Electromagnetic and Mechanical Properties of Silica-Aluminosilcates Plasma Sprayed Composite Coatings, *Proceedings of the 2007 International Thermal Spray Conference*, *Global Coatings Solutions*, May 14-16 2007 (Beijing, China), ASM International, 2007, p 507-512
3. P. Ramaswamy, S. Seetharamy, K.B.R. Varma, and K.J. Rao, Thermal Shock Characteristics of Plasma Sprayed Mullite Coatings, *J. Therm. Spray Technol.*, 1998, **7**(4), p 497-504
4. K.N. Lee, R.A. Miller, and N.S. Jacoson, New Generation of Plasma-Sprayed Mullite Coatings on Silicon Carbide, *J. Am. Ceram. Soc.*, 1995, **78**(3), p 705-710
5. X.Q. Cao, R. Vassen, and D. Stoever, Ceramic Materials for Thermal Barrier Coatings, *J. Europ. Ceram. Soc.*, 2004, **24**, p 1-10
6. R. Vaβen and D. Stöver, Development of Thin and Gastight Ceramic Coatings by Atmospheric Plasma-Spraying, *Proceedings of the 2006 International Thermal Spray Conference, Thermal Spray: Building on 100 Years of Success*, May 15-18 2006 (Seattle, USA), ASM International, 2006
7. K.N. Lee, D.S. Fox, J.I. Eldridge, D. Zhu, R.D. Robinson, N.P. Bansal, and R.A. Miller, Upper Temperature Limit of Environmental Barrier Coatings Based on Mullite and BSAS, *J. Am. Ceram. Soc.*, 2003, **86**(8), p 1299-1306
8. P.H. McCluskey et al., "Plasma Sprayed Mullite Coatings on Silicon Based Ceramic Materials," U.S. Patent 6,468,648 B1, 2002
9. I.T. Spitsberg et al., "Method for Thermally Spraying Crack-Free Mullite Coatings on Ceramic Based Substrates," U.S. Patent 6,296,909 B1, 2001
10. R. Siegert et al., "Production of Gastight, Crystalline Mullite Layer by Using a Thermal Spraying Method," PCT Patent Application WO2006034674, 2006
11. A. Killinger, M. Kuhn, and R. Gadow, High-Velocity Suspension Flame Spraying (HVSFS), a New Approach for Spraying Nanoparticles with Hypersonic Speed, *Surf. Coat. Technol.*, 2006, **201**, p 1922-1929
12. J. Oberste Berghaus, J.G. Legoux, C. Moreau, F. Tarasi, and T. Chraska, Mechanical and Thermal Transport Properties of Suspension Thermal-Sprayed Alumina-Zirconia Composite Coatings, *J. Therm. Spray Technol.*, 2008, **17**(1), p 91-104

JTTEE5 17:679–684
DOI: 10.1007/s11666-008-9220-2

Synthesis and Preliminary Tests of Suspension Plasma Spraying of Fine Hydroxyapatite Powder

Roman Jaworski, Christel Pierlot, Lech Pawlowski, Muriel Bigan, and Maxime Quivrin

(Submitted April 30, 2008; in revised form September 8, 2008)

The synthetic hydroxyapatite (HA, $Ca_{10}(PO_4)_6(OH)_2$) is a very useful biomaterial for numerous applications in medicine, such as e.g., fine powder for suspension plasma spraying. The powder was synthesized using aqueous solution of ammonium phosphate ($H_2(PO_4)NH_4$) and calcium nitrate ($Ca(NO_3) \cdot 4H_2O$) in the carefully controlled experiments. The synthesized fine powder was characterized by X-ray diffraction (XRD) and scanning electron microscope (SEM). The powder was formulated into water and alcohol based suspension and used to carry out the initial tests of plasma spraying onto titanium substrate. The phase analysis of sprayed coating was made with the XRD.

Keywords biomedical coatings, hydroxyapatite, powder manufacturing, suspension plasma spraying

1. Introduction

Suspension plasma spraying is a new coating technique which enables obtaining relatively thin and finely structured coatings by the use of water (alcohol) based suspension of fine powders (Ref 1). The technique was reviewed recently (Ref 2) and can be used to obtain relatively thin ceramic coatings of TiO_2 (Ref 3, 4), yttria-stabilized zirconia (Ref 5), and hydroxyapatite (Ref 6-8). The latter is a bioceramic material used intensively as a plasma-sprayed coating material to promote osseointegration of the femoral stems (Ref 9, 10). Initial studies of suspension plasma sprayed HA made by our research group have been made by the use of commercial spray-dried powder (Ref 8). The powder crushed using a milling machine with zirconia balls. The present paper shows the way of synthesis of 600 g of HA starting from pure calcium nitrate and diammonium phosphate. The synthesis of HA by wet chemistry is well described in the literature and the reaction starting from phosphoric acid and calcium hydroxide were reported (Ref 6, 11). The present paper was inspired by the study of Hornez et al. (Ref 12). The synthesis was optimized in order to increase the productivity of the reaction by modifying concentration of calcium and phosphorus and volume the ammonium hydroxide in the reactor, and keeping under control the by-products of the reactions. The product of the reaction was subsequently calcined and crushed. The resulting fine powder was used to formulate water-based suspension and to carry out the initial test of plasma spraying.

This article is an invited paper selected from presentations at the 2008 International Thermal Spray Conference and has been expanded from the original presentation. It is simultaneously published in *Thermal Spray Crossing Borders, Proceedings of the 2008 International Thermal Spray Conference*, Maastricht, The Netherlands, June 2-4, 2008, Basil R. Marple, Margaret M. Hyland, Yuk-Chiu Lau, Chang-Jiu Li, Rogerio S. Lima, and Ghislain Montavon, Ed., ASM International, Materials Park, OH, 2008.

Roman Jaworski and **Lech Pawlowski**, Service of Thermal Spraying at Ecole Nationale Supérieure de Chimie de Lille, BP 90180, 8 avenue Mendeleïev, 59650 Villeneuve d'Ascq, France; and **Christel Pierlot**, **Muriel Bigan**, and **Maxime Quivrin**, LCOM, UMR CNRS 8009, ENSCL, BP 90108 F-59652 Villeneuve d'Ascq Cedex, France. Contact e-mail: lech.pawlowski@ensc-lille.fr.

2. Experimental Methods

2.1 Synthesis of Hydroxyapatite

The HA was synthesized using calcium nitrate, $Ca(NO_3)_2 \cdot 4H_2O$ (M = 236.15 g/mol) commercialized by Arcos Organics, purity of 99% and ammonium phosphate, $(NH_4)_2HPO_4$ (M = 132.06 g/mol) of Arcos Organics, purity of 99% and ammonium hydroxide, NH_4OH concentration of 28-30 mass%, (M = 35.95 g/mol) of Arcos Organics. The solution of each product in ionized water was made in the way to keep the ratio of concentrations (in mol/L) of [Ca]/[P] = 1.6. The pH of final solution was controlled by the ammonium hydroxide volume. The reaction of the synthesis is as follows:

$$6(NH_4)_2HPO_4 + 10Ca(NO_3)_2 + 8NH_4OH \rightarrow 6H_2O + 20NH_4NO_3 + Ca_{10}(PO_4)_6(OH)_2 \quad \text{(Eq 1)}$$

The procedure is shown schematically in Fig. 1 and includes the following stages:

- introduction of $Ca(NO_3)_2 \cdot 4H_2O$ solution in deionized water followed by heating and agitation;

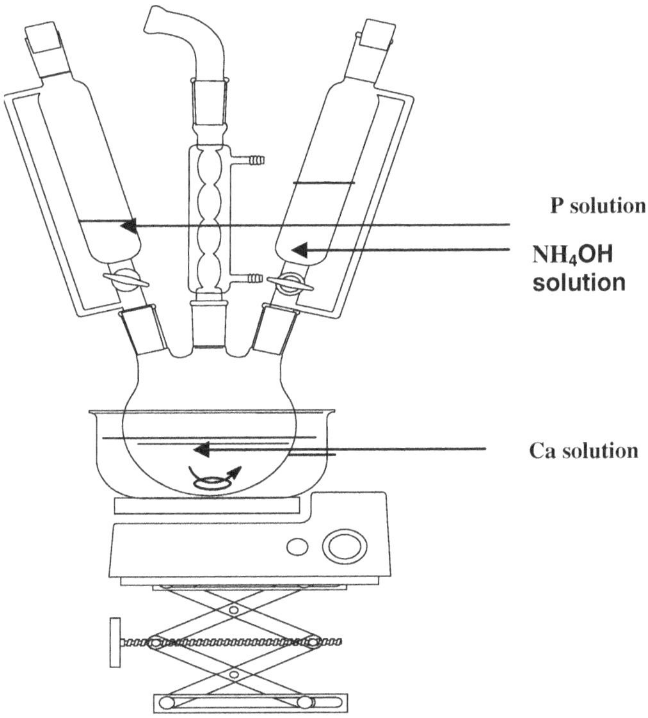

Fig. 1 Reactor of HA synthesis

- adding of NH_4OH at $T = 80$ °C and determination of pH (initial);
- adding of $(NH_4)_2HPO_4$ solution in deionized water in a "droplet by droplet way" during 15 min;
- 3-h heating and agitating and measuring pH (final);
- filtration and drying after 24 h of resting;
- calcinations at $T = 1000$ °C (for 4 h, 2 h of increasing of temperature up to maximum).

The experiments were designed in order to improve the productivity of reaction by using following variables:

- Concentration of [Ca] in mol/L, X_1: The lowest concentration was 0.5 and the highest was 4. The concentration of [P] was resulting from the condition [Ca]/[P] = 1.6 kept constant in all experiments.
- Volume of ammonium hydroxide in mL, X_2: The smallest volume was 10 mL and the greatest was 100 mL.
- Time of reaction in min, X_3: Two periods were used—15 min and 180 min.
- Time of resting, X_4: Two periods were tested—0 and 24 h.

First group of experiments were made in a small reactor (250 mL) and then a bigger one was used (1 L). The values of pH on the beginning of reaction were in the range of 11 to 12.5. The coefficient decreased after reaction to be in the range of 8.7 to 10. The optimized procedure of powder production enabled about 600 g of powder to be produced. The batches of 40 g after the optimized reaction and calcination were mixed together and ball milled using moliNEx system (Netzsch, Germany) with 1.85 mm diameter zirconia balls, and ethanol added as cooling medium. The obtained powder was formulated with distilled water 10 wt.% of the solid content.

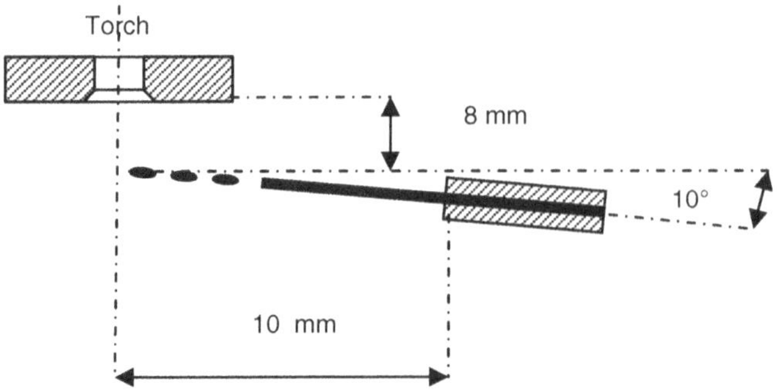

Fig. 2 Geometry of suspension injection to the plasma torch

2.2 Suspension Plasma Spraying

The suspension was fed to plasma jet using a pneumatic feeder described in details elsewhere (Ref 13) and an outlet having diameter of 0.75 mm in a way shown schematically in Fig. 2. Plasma torch used throughout all experiments was Praxair SG-100 mounted on 5-axis ABB IRB-6 industrial robot. The torch was alimented with 45 slpm of Ar and 8 slpm of H_2 and the electric power input was of 40 kW. The temperature at spraying was measured using pyrometer In 5 Plus of Impac. The pyrometer was about 90 cm from the sprayed substrates. The temperature variations at coatings deposition were very important, and it was decided to record only the maximal temperature. The torch trajectory over substrate is described elsewhere (Ref 3). There were 20 passes of torch above each substrate and after each five passes there was an interruption of 60 s for cooling. The process variables tested in this article are collected in Table 1. The substrates, being of metallic titanium, had size of 20 × 20 × 1 mm. These substrates were sand blasted using alumina grit with their sizes ranging from 250 to 500 μm at 4 bar of air pressure, directly prior to deposition. The angle of blasting was 90°.

2.3 Powders and Coatings Characterization Methods

The granulometry was tested using a laser device Master Sizer X of Malvern. The x-ray phase analysis of powder was made using a set-up Miniflex of Rigaku with Cu-Kα radiation and the Rigaku software. The synthesis of HA results in a formation of other phases, such as α-TCP, β-TCP (tricalcium phosphate $Ca_3(PO_4)_2$), and TTCP (tetracalcium phosphate, $Ca_4P_2O_9$). The percentage of the other phases with regard to hydroxyapatite were determined from the reference intensity ratio with the method described by Prevey (Ref 14) using the comparison of the strongest peaks' intensities. The XRD investigations of sprayed coatings were done by Bruker

Table 1 Variable spray parameters and coating thicknesses

Sample no.	Torch linear velocity, mm/s	Spray distance, cm	Maximal coating temperature at spraying measured with a pyrometer, °C	Coating thickness, μm
1	125	5	860	10
2	125	6	450	13
3	250	5	560	Very thin
4	250	6	290	33

Table 2 Experimental conditions

N°exp	[Ca], mol/L	[P], mol/L	Volume ammonium hydroxide, mL	Purity of obtained powder (a), % HA	Mass of obtained powder, g
1	1	0.6	20	100	5.0
2	3	1.8	20	46	13.5
3	1	0.6	60	98	5.0
4	3	1.8	60	65	14.4
5	0.57	0.35	40	100	2.8
6	3.4	2.1	40	60	16.2
7	2	1.2	11.7	85	9.9
8	2	1.2	68.3	98	10.0
9	2	1.2	40	99	9.9
10	2	1.2	40	100	9.8
11	2	1.2	40	100	9.8

(a) Obtained by the analysis of X-ray peaks intensities

apparatus type D8 with Cu-Kα radiation in the range of 2θ angles from 20° to 80° or 20° to 60°. The patterns were identified using Diffrac^{plus} Eva software which is based on the International Centre of Diffraction Data JCPDS-ICDD. Following peaks were considered:

- Hydroxyapatite, peak: 210, d = 3.08 Å, JCPDS 9-0432;
- α-TCP, peak: 441, d = 2.905 Å, JCPDS 9-348;
- β-CP, peak: 0.2.10, d = 2.88 Å, JCPDS 9-169;
- TTCP, peak: 040, d = 2.995 Å, JCPDS 25-1137;
- CaO, peak: 200, d = 2.405 Å, JCPDS 4-0777;

Amorphous calcium phosphate was identified as a halo with a maximum intensity at d = 2.97 Å. The percentage of phases was calculated as a ratio of areas of the strongest peak with regard to the area of the strongest peak of hydroxyapatite, according to French norm (Ref 15). The SEM investigations of powders and coatings were made using a Leo 982 scanning electron microscope working under low acceleration voltage.

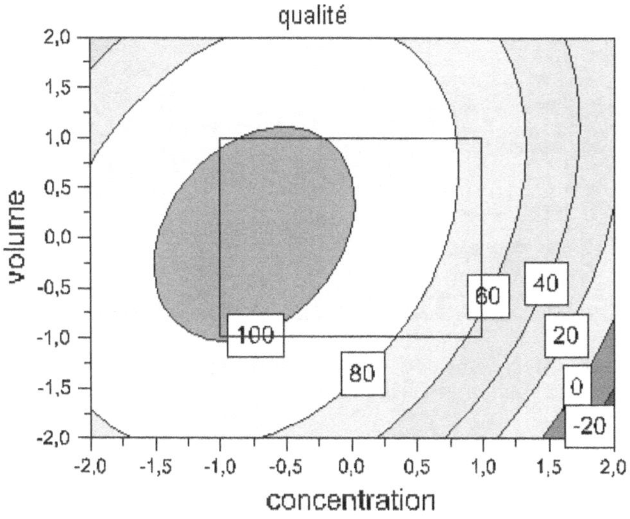

Fig. 3 Isopleths of constant purity of synthesized HA powder for two process variables: concentration of [Ca] and volume of ammonium hydrate (in normalized units)

3. Results

3.1 Powder Synthesis

After the selection of the best experimental conditions of synthesis in the small reactor, the optimization of synthesis in the big reactor was carried out. The optimization included two variables: concentration of calcium [Ca] and volume of ammonium hydroxide. The results of the experiments realized are collected in Table 2. Their design and the mathematical model are explained in detail elsewhere (Ref 16); the isopleths of constant purity of HA (Fig. 3) enabled us to find the optimal experimental conditions. Finally, optimal concentration of calcium nitrate was found to be about:

- 1.26 mol/L of calcium nitrate (and correspondingly 0.76 mol/L of diammonium phosphate);
- 41 mL of ammonium hydroxide.

A quantity of 600 g of HA powder was manufactured using the optimized conditions. The morphologies of the

Fig. 4 Morphologies of HA powder particles after synthesis and drying (a) and after calcinations (b)

powder after synthesis and crushing and after calcinations are shown in Fig. 4. The powder after synthesis is fine grained, the grains become larger after calcination. The final stage of the powder processing was crushing and the granulometry of final powder has the mean diameter, d_{VS}, varying between 1.0 and 4.4 μm and the distribution is shown in Fig. 5. There are two maxima in the distribution: the first one in the submicrometer range and the second one at about 30-40 μm. Most probably, the latter results from the agglomeration of fine particles.

3.2 Coatings

The sprayed coatings were very thin (Table 1) and their color turned to be dark. The phases' content of sprayed coatings is shown in Table 3. Among the HA decomposition phases, the α-TCP is most present in the sprayed followed by β-TCP and TTCP. CaO and amorphous

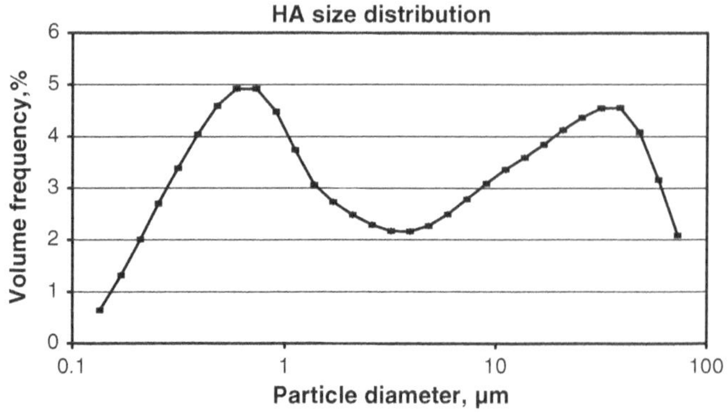

Fig. 5 Size distribution of synthesized HA powder particles

phosphate are only seldom present in sprayed coatings. The view of a typical surface of sprayed coatings is shown in Fig. 6.

4. Discussion

Suspension-plasma-sprayed HA coatings included the following phases: HA, α-TCP, β-TCP, TTCP, CaO, and amorphous calcium phosphates (ACP). The phase diagram shown in Fig. 7 for a low content of water allows, to a degree, predicting the phenomena at flight of HA particles. The formation of amorphous calcium phosphates (ACP) can be attributed the rapid cooling of melt. Similarly, α-TCP and TTCP formed on heating of HA and their presence in the coatings was explained at spraying of large particles by freezing of these phases at impact with substrate (or previously deposited coating), (Ref 18). Finally, the presence of HA can be explained by the fact that some particles, or their parts remain unheated or heated to the temperatures below the temperature of phase transformation to TTCP and α-TCP (see Fig. 7). The presence of β-TCP is unusual and the phase was not observed when sprayed with the same HA powder having coarse particles in water and/or onto substrate (Ref 19). The possible explanation is related to the small size of powder particles. The particles are a solid part of a water solution being injected into plasma jet. The jet of liquid is injected into plasma jet. Knowing that the size of suspension delivery outlet is 700 μm, the sizes of droplets must have diameter of a few hundreds of micrometers (see Fig. 2). It is possible that the jet of liquid was not injected perfectly in the middle of plasma jet, it could have remained on the periphery of the plasma jet and the fine particles inside get only heated without melting. If the jet were injected into the hot temperature region of plasma jet, the droplets of liquid would be accelerated in the plasma jet and disintegrated into smaller droplets. Simultaneously, water boiled and got evaporated. The solid particles inside any droplet get closer and may start to agglomerate by sintering in flight. Some of the solid droplets got molten in the initial part of their trajectory. Fauchais (Ref 2) estimated that the fine particles of 0.1 μm size gets molten after a distance of 2.5 mm in its trajectory

Table 3 Crystal phases' content in suspension plasma sprayed coatings obtained by the analysis the peaks on the x-ray diagrams

Sample no.	α-TCP/HA, %	β-TCP/HA, %	TTCP/HA, %	CaO/HA,%	Amorphous/HA, %
1	31	21	12	Absent	Absent
2	37	2	Absent	Absent	3
3	12	10	10	6	Absent
4	Absent	33	Absent	5	Absent

Fig. 6 SEM (secondary electrons) morphology of suspension plasma HA sample no. 1

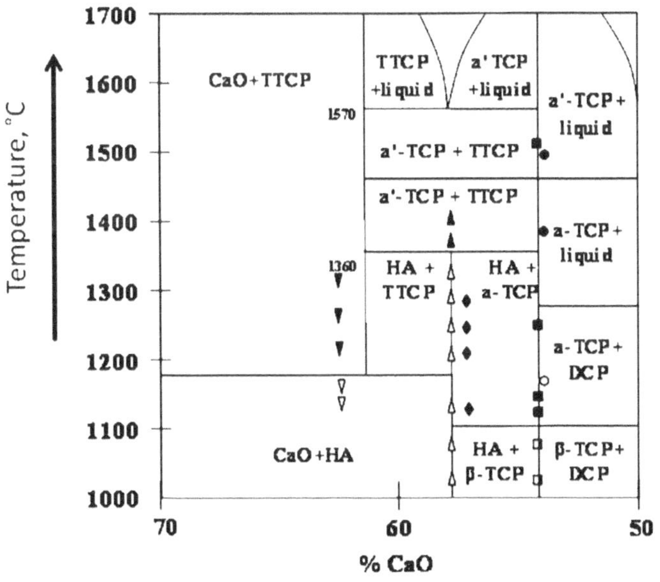

Fig. 7 Phase diagram of the system $CaO\text{-}P_2O_5$ at high temperature for the partial pressure of H_2O equal to 10 Torr (Ref 17)

in the plasma jet. In fact, the plasma gets colder because of evaporation of water. Subsequently, the fine particles may get agglomerated by sintering with other particles and/or evaporated. The agglomeration progresses until an impact with the substrate. Temperature of a few fine particles that remained on the periphery of the plasma jet may be greater than that of the transformation temperature of HA into α-TCP and TTCP. β-TCP could have been formed from α-TCP on cooling down. Similarly, CaO could have been transformed from TTCP. The cooling of particles being on the periphery of the jet is less intensive than that being in the center of the jet making possible such transformations. The β-TCP phase is present in all the sprayed coatings, and in the sample 4, all α-TCP transformed into β-TCP and all TTCP transformed into CaO (see Table 3). The spray conditions seem to be, however, far from optimal and it is hard to understand the influence of individual parameter onto the phase content.

5. Conclusions

The HA powder was synthesized using aqueous solution of ammonium phosphate ($H_2(PO_4)NH_4$) and calcium nitrate ($Ca(NO_3).4H_2O$) using a design of experiments. The synthesized fine powder was characterized by X-ray diffraction (XRD), scanning electron microscope (SEM). The powder was formulated into water-based suspension and used to carry out the initial tests of plasma spraying onto titanium substrate. The phase analysis of sprayed coating was made with the used of XRD and the phase content was related to the phenomena that could have happened in plasma jet. The process of suspension plasma spraying needs further optimization.

Acknowledgments

Mr. Marc Martel enabled us to realize the synthesis of HA powder in his laboratory, and the discussions with Dr. Hornez were helpful in initiating the synthesis of HA.

References

1. L. Pawlowski, The Science, Engineering of Thermal Spray Coatings, 2nd ed., Chichester, Wiley, 2008, p 101
2. P. Fauchais, Solution/Suspension Plasma Spraying of Finely Structured Coatings, *1st Scientific Workshop on Thermal Spraying of Suspensions*, Tours, France, 14 and 15 June 2007, available upon request
3. R. Jaworski, L. Pawlowski, F. Roudet, S. Kozerski, and A. Le Maguer, Influence of Suspension Plasma Spraying Process Parameters on TiO_2 Coatings Microstructure, *J. Therm. Spray Technol.*, 2008, **17**(1), p 73-81

4. R. Jaworski, L. Pawlowski, F. Roudet, S. Kozerski, and F. Petit, Characterization of Mechanical Properties of Suspension Plasma Sprayed TiO_2 Coatings Using Scratch Test, *Surf. Coat. Technol.*, 2008, **202**, p 2544-2553
5. K. Wittmann-Ténèze, K. Vallé, L. Bianchi, P. Belleville, and N. Caron, Nanostructured Zirconia Coatings Processed by PROSOL Deposition, *Surf. Coat. Technol.*, 2008, **202**, p 4349-4355
6. E. Bouyer, F. Gitzhofer, and M.I. Boulos, Suspension Plasma Spraying for Hydroxyapatite Powder Preparation by R.F. Plasma, *IEEE Trans. Plasma Sci.*, 1997, **25**(5), p 1066-1072
7. F. Gitzhofer, E. Bouyer, and M.I. Boulos, "Suspension Plasma Spraying," US Patent 5 609 921, 3 November 1997
8. R. Jaworski, C. Pierlot, R. Tomaszek, L. Pawlowski, Z. Znamirowski, and J. Zdanowski, Optimization of Dielectric Properties of Suspension Plasma Sprayed Hydroxyapatite Coatings, *Mat.-wiss. u. Werkstofftech.*, 2007, **38**, p 125-130
9. R. Heimann, Thermal Spraying of Biomaterials, *Surf. Coat. Technol.*, 2006, **201**, p 2012-2020
10. S.M. Best, A.E. Porter, E.S. Thian, and J. Huang, Bioceramics: Past, Present and for the Future, *J. Eur. Ceram. Soc.*, 2008, **28**(7), p 1319-1327
11. M. Markovic, B.O. Fowler, and M.S. Tung, Preparation and Comprehensive Characterization of Calcium Hydroxyapatite Reference Material, *J. Res. Natl. Inst. Stand. Technol.*, 2004, **109**, p 553-568
12. J.-C. Hornez, F. Chai, F. Monchau, N. Blanchemain, M. Descamps, and H.F. Hildebrand, Biological and Physico-Chemical Assessment of Hydroxyapatite (HA) with Different Porosity, *Biomol. Eng.*, 2007, **24**(5), p 505-509
13. L. Pawlowski, Finely Grained Nanometric and Submicrometric Coatings by Thermal Spraying: A Review, *Surf. Coat. Technol.*, 2008, **202**(18), p 4318-4328
14. P.S. Prevey, X-ray Diffraction Characterization of Crystallinity and Phase Composition in Plasma-Sprayed Hydroxyapatite Coatings, *J. Therm. Spray Technol.*, 2000, **9**(3), p 369-376
15. S 94-067, *Détermination qualitative et quantitative des phases étrangères présentes dans les poudres, dépôts et céramiques à base de phosphate de calcium*, AFNOR, Paris, France, 1993
16. R. Jaworski, C. Pierlot, L. Pawlowski, M. Bigan, and M. Martel, Design of the Synthesis of Fine HA Powder for Suspension Plasma Spraying, submitted for publication in *Surf. Coat. Technol.*
17. C. Santos, A. Pazo, F. Guitan, Water Vapour Pressure Influence on CaO-P_2O_5 System Phase Diagram, *Mater. Clinic. Appl.*, 1995, p 11-18
18. S. Dyshlovenko, L. Pawlowski, B. Pateyron, I. Smurov, and J.-H. Harding, Modelling of Plasma Particles Interactions and Coating Growth for Plasma Spraying of Hydroxyapatite, *Surf. Coat. Technol.*, 2006, **200**, p 3757-3769
19. S. Dyshlovenko, L. Pawlowski, P. Roussel, D. Murano, and A. Le Maguer, Relationship Between Plasma Spray Operational Parameters and Microstructure of Hydroxyapatite Coatings and Powder Particles Sprayed into Water, *Surf. Coat. Technol.*, 2006, **200**, p 3845-3855

JTTEE5 17:685–691
DOI: 10.1007/s11666-008-9259-0
1059-9630/$19.00 © ASM International

Peer Reviewed

Effective Parameters in Axial Injection Suspension Plasma Spray Process of Alumina-Zirconia Ceramics

F. Tarasi, M. Medraj, A. Dolatabadi, J. Oberste-Berghaus, and C. Moreau

(Submitted May 13, 2008; in revised form September 19, 2008)

Suspension plasma spray (SPS) is a novel process for producing nano-structured coatings with metastable phases using significantly smaller particles as compared to conventional thermal spraying. Considering the complexity of the system there is an extensive need to better understand the relationship between plasma spray conditions and resulting coating microstructure and defects. In this study, an alumina/8 wt.% yttria-stabilized zirconia was deposited by axial injection SPS process. The effects of principal deposition parameters on the microstructural features are evaluated using the Taguchi design of experiment. The microstructural features include microcracks, porosities, and deposition rate. To better understand the role of the spray parameters, in-flight particle characteristics, i.e., temperature and velocity were also measured. The role of the porosity in this multicomponent structure is studied as well. The results indicate that thermal diffusivity of the coatings, an important property for potential thermal barrier applications, is barely affected by the changes in porosity content.

Keywords alumina-yttria stabilized zirconia, effective parameters, nanocomposite, suspension plasma spray

1. Introduction

Suspension plasma spray (SPS) is an emerging process for spraying small feedstock particles with nano and/or a few micron size ranges, using a liquid carrier. These particles are mostly less than 5 μm in dimension and are difficult to feed using conventional equipment, since they tend to form larger aggregates that can cause blocking of the injection path. Injection of such particles into a plasma jet can be achieved with the help of liquid carriers. The resulting microstructure favors dense or finely porous structures as needed for solid oxide fuel cell electrolytes (Ref 1) or cathodes (Ref 2). However, more investigation is required for better controlling the range of porosity and other structural features obtainable with this process. SPS can generate unique and novel microstructure, not commonly seen in conventional thermal spraying (Ref 3) and understanding the role of the process parameters, which impart these microstructural features, has increasing importance.

In addition, the feedstock size and size distribution as well as the nature of the material play important roles. Using multicomponent feed materials increases the difficulty of producing uniform microstructures in the coatings, for example, mixtures of alumina and yttria-stabilized zirconia (YSZ). One reason is the effect of combination of the molten particle and substrate material on splat shape and morphology (Ref 4) that may continue in the following layers with alternating material (e.g., alumina or zirconia layers of the composite coating) and results in unexpected splat behavior as compared with a substrate of a single material. Meanwhile, substrate roughness is another important process parameter in the coating studies.

This study was conducted to identify the significant process parameters, involving the plasma torch, feed, and substrate in producing a sound and integrated microstructure. Such a structure is required for further studies of the characteristic properties of the coatings produced by the SPS process. The eutectic composition of alumina/8 wt.% YSZ was used in this study.

2. Experimental

2.1 Liquid Feed Material

To produce two different powder size ranges, 5 wt.% YSZ and 13 wt.% YSZ (Nano-Composite Powder, Inframat@Advanced Materials; Farmington, USA, size

This article is an invited paper selected from presentations at the 2008 International Thermal Spray Conference and has been expanded from the original presentation. It is simultaneously published in *Thermal Spray Crossing Borders, Proceedings of the 2008 International Thermal Spray Conference*, Maastricht, The Netherlands, June 2-4, 2008, Basil R. Marple, Margaret M. Hyland, Yuk-Chiu Lau, Chang-Jiu Li, Rogerio S. Lima, and Ghislain Montavon, Ed., ASM International, Materials Park, OH, 2008.

F. Tarasi, M. Medraj and **A. Dolatabadi**, Industrial and Mechanical Engineering Department, Concordia University, Sir George Williams Campus, 1515 St. Catherine W., Montreal, QC, Canada H3G 2W1; and **J. Oberste-Berghaus** and **C. Moreau**, NRC-IMI, Boucherville, QC, Canada. Contact e-mail: mmedraj @me.concordia.ca.

30-60 nm) nano-powders were proportionally mixed to produce 8 wt.% YSZ, which is the common thermal barrier coating (TBC) material in gas turbines and diesel engines. The resulting doped zirconia powders were then mixed with two different sizes of alumina powders. The alumina component was either nano size (Nanostructured & Amorphous Materials Inc. USA, Nominal size 27-43 nm) or micron size (Malakoff, TX, size 1.4 μm). The terms "Nano" and "Micron," respectively, will be used here for these mixtures. The mixture with the larger size powder was ball milled in a concentrated suspension of 60 wt.% solid for 24 h before dilution to the final solid concentration. This procedure ensured homogeneous mixing and stabilization of the suspension. The nano mixture was only milled for the same period for enhanced stability of the suspension. A weight ratio of 60/40 for the alumina/8 wt.% YSZ were prepared and suspended in ethanol at 10 and 30 wt.% concentrations. This resulted in four suspensions with different powder size ranges and solid contents. Suspension dispersion was done using 9 cc (poly-ethylene-imine) and 4.5 cc (nitric acid) (both with 10% concentration) for every 150 g of solids. Wet analysis of the suspension was performed by infrared Culter-Beckman particle size analyzer to determine the size of agglomerated solid particles in the suspension.

2.2 Methodology

The Mettech Axial III spray system (Northwest Mettech, North Vancouver, Canada) was used. This torch consists of three cathodes and three anodes that can reduce the average arc voltage fluctuations. The injection system is fully automated and PC controlled to ensure a constant feed rate, which can have significant effects on the spray process and the resulting coatings. The spray distance was kept constant at 50 mm, as short spray distances are commonly used in SPS (Ref 5, 6). In-flight particle characteristics (velocity and temperature) were monitored by the diagnostic system Accura-Spray G2 (Tecnar Automation, St. Bruno, Canada). This system measures particle temperature and velocity based on particle ensemble signals rather than the signal from individual particles. The measurements were taken at the substrate position.

Field emission scanning electron microscope FE-SEM (Hitachi S4700) was used to image the coating microstructure. Whereas SEM (Jeol JSM-610) was utilized in image analysis (IA) at 500× magnification and the average porosity was determined from measurements in 10 locations. Calibrations of brightness and contrast were selected between two reference materials, namely aluminum foil and the mounting material. This method is useful for large porosities in the coating; and is limited to the detection of the pores greater than 0.5 μm (Ref 7).

The vertical and horizontal cracks that are sometimes reported as a part of total porosity were individually assessed and counted per unit length or width of the coating. Five measurements were averaged per sample. The nature of the cracks can have significant influence on the properties of the coating. For example, it was observed that the planar defects parallel to the substrate are more influential on mechanical properties (Ref 8) as well as thermal diffusivity (Ref 8, 9), than the total porosity. Vickers microhardness test using 300 g loads for 15 s was done on selected samples. Thermal diffusivity was evaluated by the laser flash method (Ref 10).

The statistical method of Taguchi was used to evaluate the importance of seven selected variables in the SPS coating process, including feed, plasma torch condition, and substrate-related variables. The need to investigate a wide range of variables with the minimum number of experimental runs provoked the application of this design of experiment (Ref 11). The variables and their two selected levels are summarized in Table 1. Among the variables in Table 1, the levels for solid content, torch condition, and feed rate were selected based on the initial data from the equipment supplier, followed by the preliminary trial and errors by the authors for the favorable working condition. Additionally, the substrate roughness was in the range of what is used in the industry and the robot speed was changed from the medium to the maximum attainable speeds by the spray robot.

The microstructural features and in-flight particle characteristics on which the effects of variables are being studied were:

- Particle temperature, T_p, at impact distance (°C).
- Particle velocity at impact distance, V_p (m/s).
- Vertical cracks average spacing (μm).
- Horizontal cracks average spacing (μm).
- Porosity content in the crack-free area (%).
- Thickness per pass of deposition or deposition rate (μm/pass).

3. Results and Discussions

3.1 Suspension Agglomerate Size Distribution

The wet analyses of the agglomerate size in different suspensions diluted in ethanol were closely comparable

Table 1 Variables and their levels (values)

Variable		Low & high levels
A	Solid content in suspension	10 & 30 wt.%
B	Auxiliary gas	H_2 and He
C	Torch condition (total plasma gas flow, gas composition, torch current)	(245 slm, 75Ar/10N$_2$/15 auxiliary gas, 200 A) & (275 slm, 65Ar/15N$_2$/20 auxiliary gas, 240 A)
D	Injected feed rate	1.3 & 1.8 kg/h
E	Powder type	Nano & Micron
F	Substrate roughness	#24 & #60 mesh size of alumina for grit blasting
G	Spraying robot travel speed	0.6 & 2 m/s

regardless of the initial particle size or solid concentrations and are all in the range of 10 μm. That is, 0.2-3 μm in 30% solid of micron-sized particle suspension, 0.2-9 μm in 10% solid of the same powder, and 0.2-5 μm in nano-sized powder in 10% solid suspension.

3.2 Effects of the Variables

A summary of the results of the Taguchi evaluations are shown in bar chart graphs in Fig. 1. The first column for every variable shows the effect of its variation from the low level to high level as defined in Table 1. The second column for each variable is the standard error to help comparison of the significance of the effects with the error.

When changing the variable (e.g., solid weight%) from low to high level increases the specific structural parameter (e.g., vertical crack spacing) in the coating, the related column is shown on the positive side of the Y-axis and the decreasing effects of variables on the measured parameter are shown on the negative side of this axis.

By relating the particle temperature and velocity to the spray conditions, some general trends can be observed and will be discussed in the following sections.

3.2.1 Solid Concentration. Increasing the solid content from 10 to 30% in the suspension liquid, as observed in Fig. 1(a) and (b), has decreased both particle temperature and velocity. The coatings from this lower temperature and velocity have shown no considerable change in vertical crack density, but the spacing of horizontal cracks through the thickness increased slightly as presented in the diagrams of Fig. 1(c) and (d). Within the precision of the method used in these experiments the porosity content of the coatings was not affected by the solid concentration. It, however, shows to be the most effective factor on deposition rate, as shown in Fig. 1(e) and (f). The higher solid content in the liquid feed has produced thicker layers per pass, which can be translated to the higher rate of material deposition. It can also be reasoned that thicker splats are a result of lower T_p, causing higher droplet viscosity, and lower V_p, causing lower momentum for deformation, both of which could reduce the splat flattening.

3.2.2 Plasma Auxiliary Gas. Changing the plasma auxiliary gas from commonly used hydrogen to helium, was implemented to induce more porous microstructures for applications like thermal barrier coatings. Helium is

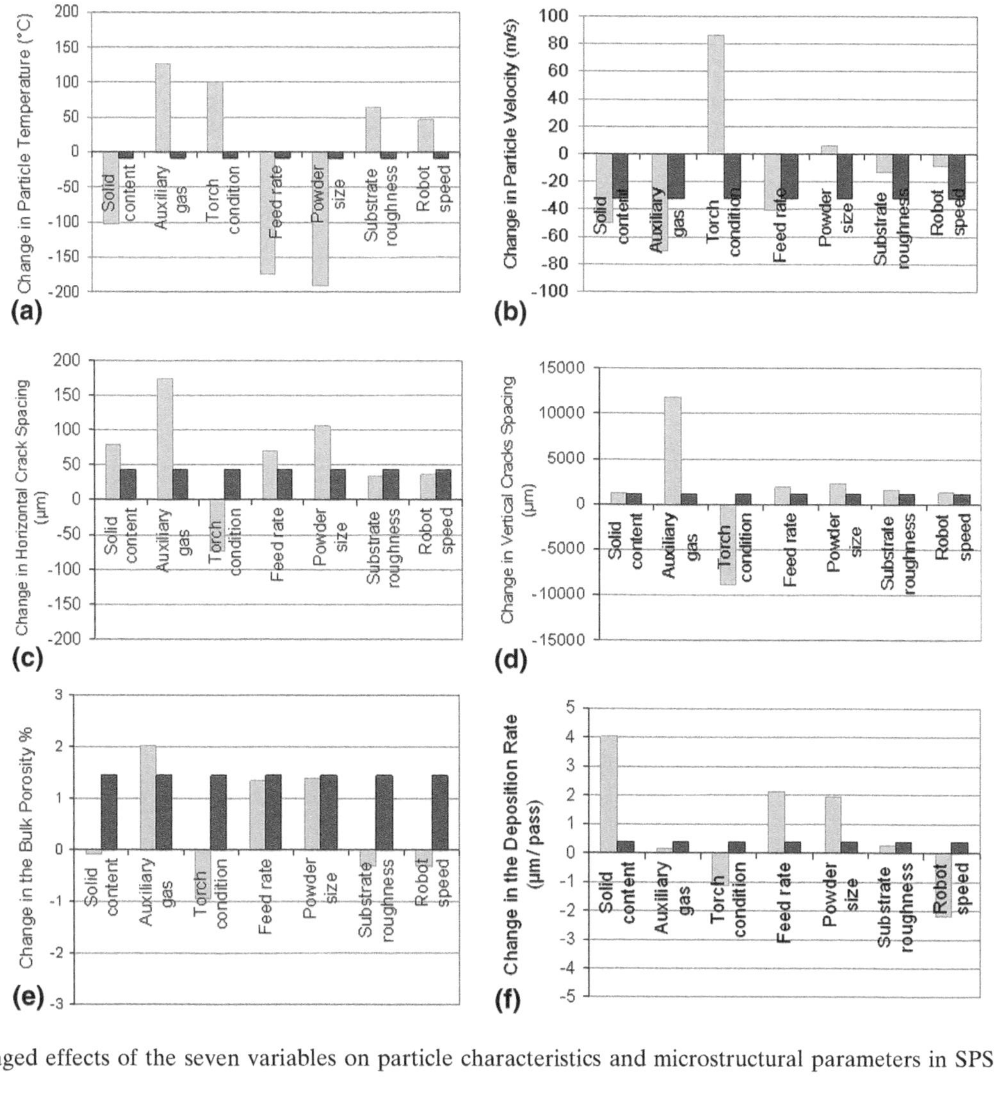

Fig. 1 The averaged effects of the seven variables on particle characteristics and microstructural parameters in SPS process

known to increase the plasma stability with its higher viscosity at high temperature [5]. It also has a higher conductivity than Argon and generally produces wider hot core area that promotes entrapment of larger number of small particles, which could otherwise escape from the particle jet without deposition. The observed effects of this change are summarized in Fig. 1(a) to (f).

Interestingly, changing from hydrogen to helium auxiliary gas has shown the most drastic effect on almost all of the measured parameters in this work. This experiment using the Taguchi method has shown that by replacing H_2 with He gas, the average particle temperature increases and the average velocity decreases. One by one comparison, however, gives some more information that may be extracted from Fig. 2. This figure shows that at the same plasma torch condition (in terms of total gas flow rate, plasma gases ratio, and arc current), He has dropped the resulting plasma power by 20-40 kW. This can be observed in two different plasma conditions of "245 slm total gas flow rate, $75Ar/10N_2/15$ auxiliary gas, 200 A current" and "275 slm total gas flow rate, $65Ar/15N_2/20$ auxiliary gas, 240 A current."

On the other hand, comparison between the two sets of experiments in Fig. 2 clearly shows that He auxiliary gas resulted in both higher velocity and temperature of the in-flight particles, even though only small differences of plasma power around 80-82 and 84-85 kW were recorded. This temperature increase, in spite of shorter heat exposure time at higher velocity, is a result of higher heat conduction by helium gas.

The SPS coatings produced within the range of variables in this experiment show a very dense microstructure. The porosity in crack-free areas ranges from a minimum of almost zero, produced with hydrogen auxiliary gas, to a maximum of 8% with helium. The two extreme microstructures of alumina/YSZ coatings are shown in the micrographs of Fig. 3(a) and (b). It is clear that while the high density of the coating in Fig. 3(c) causes the vertical microcracks to develop within the structure, the porous structure, especially in the case of distributed porosities as in Fig. 3(d), eliminates the microcracks. A comparison of the particle temperatures and velocities indicated on the micrographs as (T_p, V_p) shows that for high densities a high particle velocity is necessary.

3.2.3 Plasma Torch Condition. Changing the plasma condition from the low to high level, as described in Table 1 and based on the results in Fig. 1(a) and (b), raised the particle temperature and, to an even higher degree, the particle velocity. At higher particle temperature and velocity the density of both vertical and horizontal cracks increases. This may originate from formation of thinner splats that can more readily form vertical cracks during the cooling process. The horizontal cracks branch from the vertical cracks. In this way, the similar behavior from both types of cracks may also be justified. The porosity remains invariant and the deposition rate decreases slightly. A direct comparison, however, is difficult since the parameters of spray torch condition and auxiliary gas are not independent. To gain further insight into the effect of plasma power, deposition runs can be grouped into four ranges of plasma power of 56-57, 80-82, 84-85, and 116-118 kW. Accordingly, the effect of different plasma input powers on the particle characteristics is summarized in Fig. 4. It can be seen that an increase in plasma power generally increases the particle velocity. At similar plasma power (81-84 kW), particle velocity in

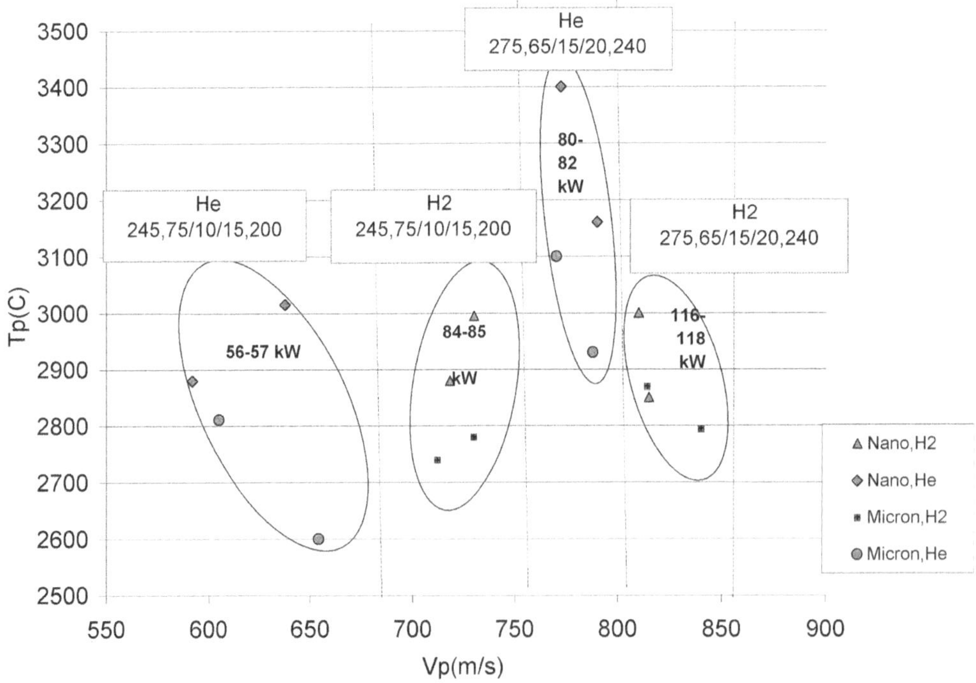

Fig. 2 The effect of auxiliary gas, powder size, torch condition, and plasma power on particle velocity and temperature

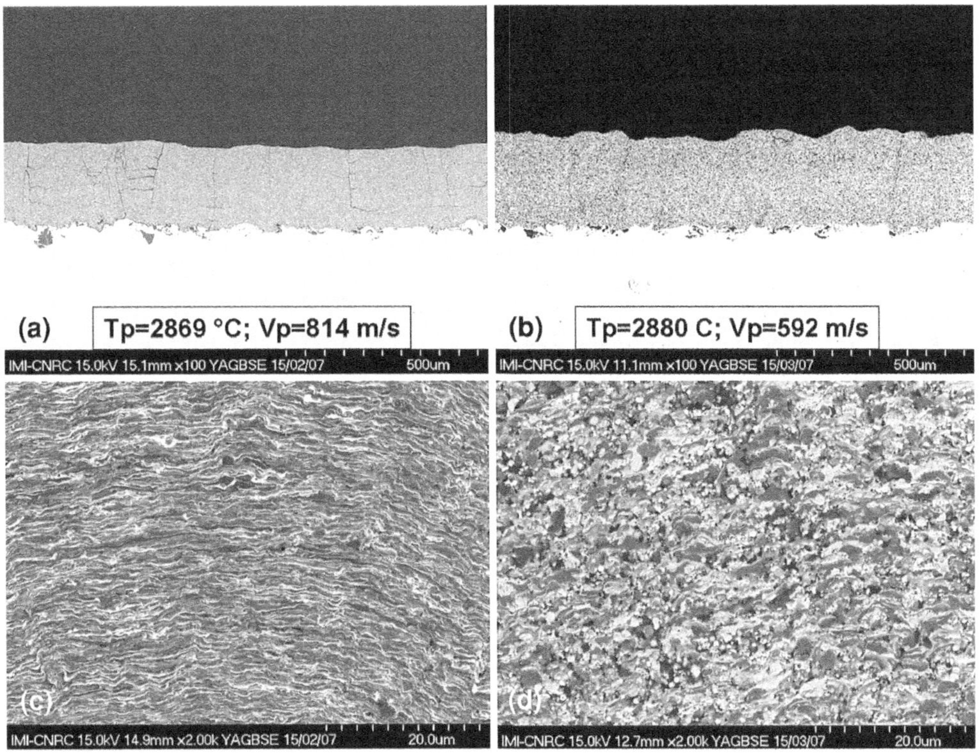

Fig. 3 60/40 wt.% alumina/zirconia suspension plasma-sprayed coatings (a, b) coatings at 100×. (c, d) Same coatings as (a, b), respectively at 2000×

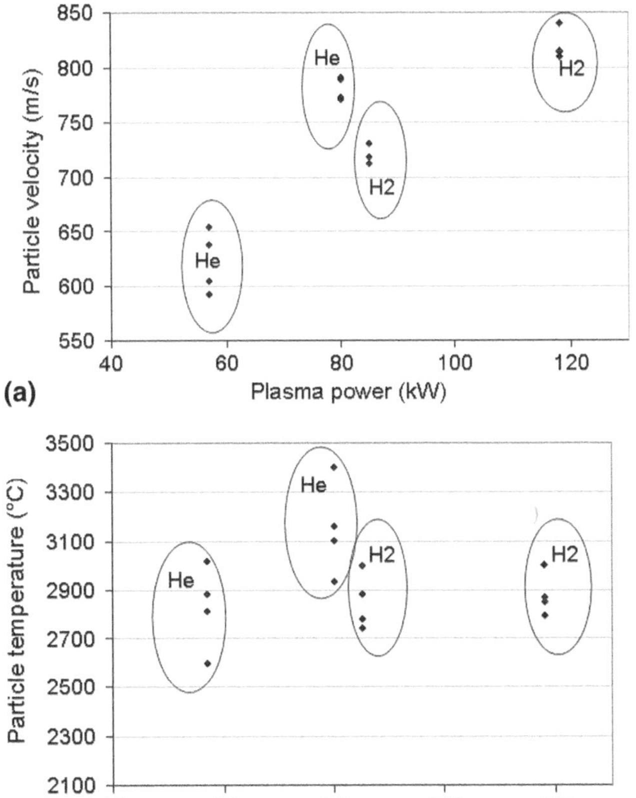

Fig. 4 Plasma power effect on (a) particle velocity and (b) particle temperature

helium exceeds that in hydrogen as shown in Fig. 4(a). Nonetheless, the highest particle velocities are attained with hydrogen.

As a result of the above condition, the average porosity increased with helium and both vertical and horizontal cracks spacing show an increasing trend presented in Fig. 1(c) to (f). Moreover, Fig. 1(e) shows that the deposition rate was indifferent to this change. The temperature of the particles, however, does not follow a definitive trend, as seen in Fig. 4(b). Generally, it was observed that the feed parameters are better tools for controlling the particle temperature than the torch operating parameters.

3.2.4 Feed Rate. Based on the Taguchi analysis in Fig. 1(a) to (d), increasing feed rate decreases both particle temperature and to a lesser extent particle velocity. As a result, the microcracks densities do not show any considerable variation. The porosity also did not get affected by this change, as represented in Fig. 1(e). Conversely, the deposition rate, shown in Fig. 1(f), increases remarkably. This characteristic of the process that can tolerate the increase of the feed rate and deposition rate without introducing more structural defects in the coating is promising for higher production rates.

It is noteworthy that the feed rate and the solid content have shown similar effects and may be interchangeably used in controlling the coating microstructure in the SPS process.

3.2.5 Powder Size Distribution. The change in the initial particle size range was done by changing nano- to micron-size alumina powders mixed with nano-size 8 wt.% YSZ.

This variation showed a recurring drop in particle temperature (T_p) in spite of the similar agglomerate size in the suspension; and no significant velocity drop occurred. The lower temperature from larger particle feedstock may be explained by the formation of dense particles within the plasma plume in comparison with hollow particles, which can result from nano-size suspension feedstock (Ref 12). A second reason for higher (T_p) can be that the nano particles that form loose aggregates are of considerably higher surface area thus showing lower energy barrier for melting than solid micron-sized particles in the aggregates. Experiencing the same velocity and spray distance, the nano aggregates, more rapidly melted, can rise to higher temperatures.

In the resulting coating microstructure the average porosity presented no change and while the density of vertical microcracks remained almost constant and the horizontal microcrack spacing increased remarkably. The lower microcrack densities observed with larger particles can also be justified by the lower T_p as the high particle temperatures can cause higher thermal stresses. The porosity content looks indifferent to the powder size variation. This is somewhat unexpected. Considering the large standard error, a possible underlying effect may not be captured. The limited sensitivity of the porosity measurement method also has restricted the observation of the smaller pores (nano-pores) that might have affected the results.

3.2.6 Substrate Roughness. Keeping in mind that the substrate roughness has no effect on in-flight particle characteristics, this experiment shows that its variation is one of the least effective parameters on the microstructural features, as seen in Fig. 1. Neither the porosity nor the deposition rate has changed, and even the microcrack densities (vertical/horizontal) have not been considerably altered by changing the substrate roughness taking into account the standard error in these measurements.

Additionally, the averages of coating roughness on the substrates blasted with #60 alumina grit ($R_a = 3\text{-}4$ μm) and #24 grit ($R_a = 6\text{-}7$ μm) substrates are almost the same, namely $R_a = 5.5$ and 6 μm, respectively. The absence of strict correlation between the initial substrate roughness and the resulting coatings is attributed to the small aggregate sizes comparable with the size of substrate roughness. The small particles at high velocity diffuse into the roughness asperities and after the first few runs of deposition eliminate the role of the substrate roughness. This independency of the roughness between the substrate and coating suggests that the coating roughness can be controlled by spray condition for various substrate roughnesses. On the other hand, changing the initial particle size from nano to micron-size powders causes a slightly more considerable effect on coating roughness, namely, from $R_a = 5.1$ to 6.3 μm.

3.2.7 Robot Travel Speed. Increasing the robot travel speed, again not having any effect on particle state, has neither changed the porosity nor the vertical/horizontal microcracks density. A drop in deposition rate is simply related to less mass deposition time at higher robot travel speed.

3.3 Particle Velocity, Porosity, and Hardness

A demonstration of the effect of particle velocity on the coating density indicated in Fig. 5(a) shows that coating density increases as a result of higher particle velocity that provides higher momentum for the splat flattening, as is expected. This also has resulted in enhanced intersplat bounding that has clearly improved the hardness of the resulting coating as illustrated in Fig. 5(b). This figure shows that with even less than 5% porosity, the coating hardness drops to less than one third of the highly dense structure.

3.4 Porosity and Thermal Diffusivity

A group of samples with a range of porosities ranging from the minimum to the maximum were evaluated for thermal diffusivity. The results are summarized in Fig. 6. A porosity of 0-8% in the crack-free area obtained in this experience has not altered the thermal diffusivity of the coating. The measured thermal diffusivities of the coatings are, however, comparable with that of the stabilized zirconia sprayed with the same process. It is noteworthy that the thermal conductivity of the 8 wt.% YSZ formed by this process was measured as 0.89 W/m °C. This value is in turn comparable to values reported in the literature (Ref 13) for air plasma-sprayed YSZ (0.9-1 W/m °C) and YSZ deposited by EB-PVD (Electron beam physical vapor deposition) in the order of 1.8-2 W/m °C.

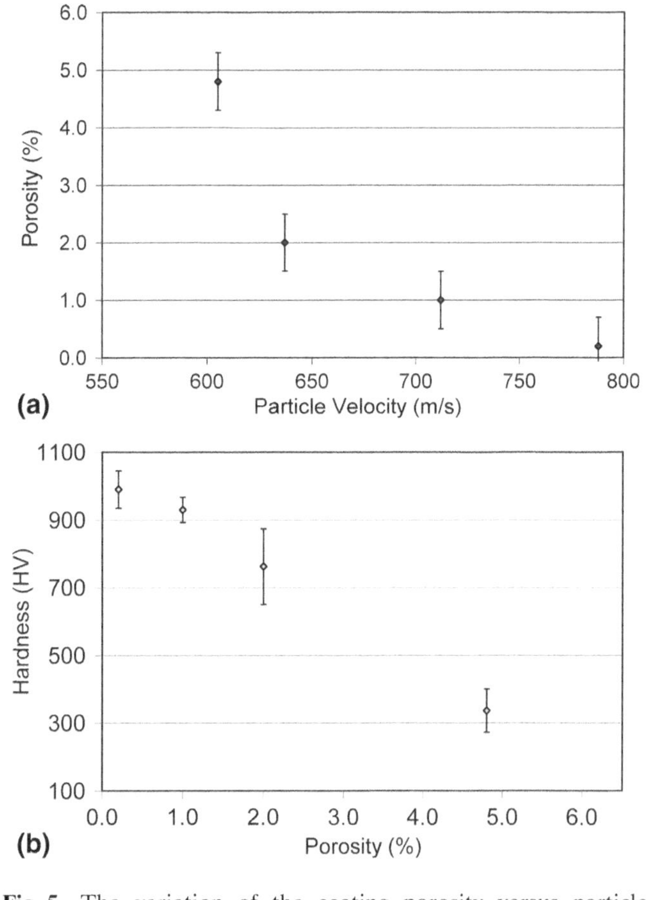

Fig. 5 The variation of the coating porosity versus particle velocity (a) and the coating hardness versus coating porosity (b)

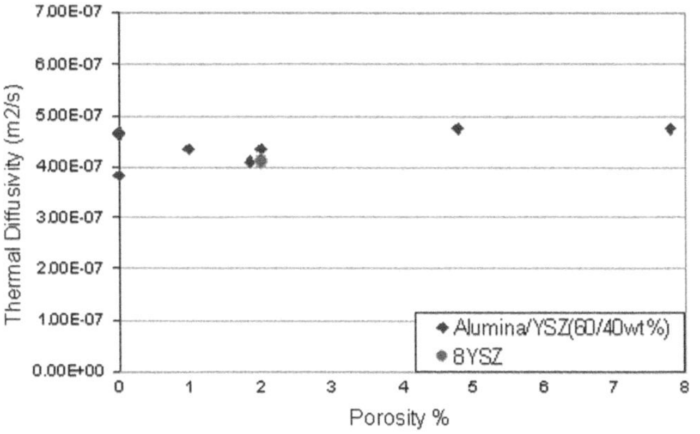

Fig. 6 The variation of thermal diffusivity with the porosity content

These results suggest that porosity may not be the most significant factor for thermal barrier applications. In this way the implementation of dense composite coating can promote the potentially better mechanical properties. It should be considered, in the thermal diffusivity versus porosity results, that the smaller pores of submicron sizes could not be detected in the image analysis method with magnification as low as 500×. However, it is proven that the roles of size, shape, and densities of the intersplat pores in thermal diffusivity are more important than the total porosity content (Ref 14).

4. Summary and Conclusion

In this study, the importance of some key variables in suspension plasma spraying of multicomponent alumina/stabilized zirconia coatings was evaluated.

It was observed that the variables that directly affect the particle velocity and temperature were the most influential on the microstructure. Considering the greater significance of particle velocity in this regard, variables with more prominent effect on V_p were of prime importance. The substrate and robot-related variables that do not play a role on any of T_p or V_p did not consequently affect the coating microstructure.

It was noticed that the plasma torch parameters, e.g., total gas flow rate and plasma gas composition, were most important in varying the particle velocity. On the other hand, particle temperature is more readily manipulated by feed stock parameters like solid content, particle size, and feed rate.

Helium auxiliary gas was successfully used as a tool to achieve a wider microstructural variety in the SPS coatings. It especially helped in increasing the porosity content in the coating. The porosity, however, increased at the expense of reducing the coating hardness.

Thermal diffusivity in SPS coatings for a multicomponent system of 60alumina/40YSZ is reasonably low and does not change with bulk porosity of up to 8%.

References

1. C. Delbos, C. Rat, C. Bonhomme, J. Fazilleau, J.F. Coudert, and P. Fauchais, Influence of Powder Size Distributions on Microstructural Features of Finely Structured Plasma Sprayed Coatings, *High Temp. Mater. Process.*, 2004, **8**, p 397-406
2. C. Monterrubio-Badillo, H. Ageorges, T. Hartier, J.F. Codert, and P. Fauchais, Plasma Spraying of a Perovskite Suspension for SOFC Cathodes, *Proceedings of the 2003 International Thermal Spray Conference: Advancing the Science & Applying the Technology*, B.R. Marple and C. Moreau, Eds., May 5-8, 2003 (Orlando, FL), ASM International, 2003, p 687-692
3. S. Siegmann, N. Margadant, A. Zagorski, and M. Arana-Antelo, Composition of Particle In-Flight-Characteristics and Coating Properties, *Proceedings of the 2003 International Thermal Spray Conference: Advancing the Science & Applying the Technology*, B.R. Marple and C. Moreau, Eds., May 5-8, 2003 (Orlando, FL), ASM International, 2003, p 1591-1598
4. C.J. Li, W.B. Wang, A. Ohmori, and K. Tani, Relationship Between the Microstructure and Properties of Thermally Sprayed Deposits, *Proceedings of the 15th Thermal Spray Conference*, C. Coddet, Ed., May 25-29, 1998 (Nice, France), ASM International, **1**, p 481-487
5. C. Delbos, J. Fazilleau, V. Rat, J.F. Coudert, P. Fauchais, and B. Pateyron, Phenomena Involved in Suspension Plasma Spraying Part 2: Zirconia Particle Treatment and Coating Formation, *Plasma Chem. Plasma Process.*, 2006, **26**, p 393-414
6. P. Fauchais, V. Rat, C. Delbos, J.F. Coudert, T. Chartier, and L. Bianchi, Understanding of Suspension DC Plasma Spraying of Finely Structured Coatings for SOFC, *IEEE Trans. Plasma Sci.*, 2005, **33**, p 920-930
7. S. Deshpande, A. Kulkarni, S. Sampath, and H. Herman, Application of Image Analysis for Characterization of Porosity in Thermal Spray Coatings and Correlation with Small Angle Neutron Scattering, *Surf. Coat. Technol.*, 2004, **187**, p 6-16
8. C. Li and A. Ohmori, Relationship Between the Microstructure and Properties of Thermally Sprayed Deposits, *J. Therm. Spray Technol.*, 2002, **11**, p 365-374
9. F. Cernuschi, P. Bianchi, M. Leoni, and P. Scardi, Thermal Diffusivity/Microstructure Relationship in Y-PSZ Thermal Barrier Coatings, *J. Therm. Spray Technol.*, 1999, **8**, p 102-109
10. A.S. Houlbert, P. Cielo, C. Moreau, and M. Lamontagne, Measurement of Thermal Diffusivity and Anisotropy of Plasma-Sprayed Coatings, *Int. J. Thermophys.*, 1994, **15**, p 525-546
11. G.-E.-P Box, W.-G Hunter, and J.-S. Hunter, *Statistics for Experimenters an Introduction to Design, Data Analysis, and Model Buildings*. John Wiley & Sons Inc., Canada, 1978
12. J. Oberste Berghaus, J.G. Legoux, C. Moreau, F. Tarasi, and T. Chraska, Mechanical and Thermal Transport Properties of Suspension Thermal Sprayed Alumina-Zirconia Composite Coatings, *J. Therm. Spray Technol.*, 2008, **17**(1), p 91-104
13. U. Schulz, C. Leyens, K. Fritscher, M. Peters, B. Saruhan-Brings, and O. Lavigne, Some Recent Trends in Research and Technology of Advanced Thermal Barrier Coatings, *Aerosp. Sci. Technol.*, 2003, **7**, p 73-80
14. R. Soltani, T.W. Coyle, J. Mostaghimi, R.S. Lima, and C. Moreau, Thermo-Physical Properties of Plasma Sprayed Yttria Stabilized Zirconia Coatings, *Surf. Coat. Technol.*, 2008, **202**(16), p 3954-3959

Peer Reviewed

JTTEE5 17:692–699
DOI: 10.1007/s11666-008-9227-8
1059-9630/$19.00 © ASM International

Peer Reviewed

Optimization of Solution Precursor Plasma Spray Process by Statistical Design of Experiment

Y. Wang and T.W. Coyle

(Submitted May 14, 2008; in revised form September 15, 2008)

The solution precursor plasma spray (SPPS) process, in which a solution precursor of the desired resultant material is fed into a plasma jet by atomizing gas or high pressure, was developed in the 1990s and has been studied extensively since then. Recently, it has been shown that the SPPS process is suitable for deposition of porous electrodes for solid oxide fuel cells (SOFC). High efficiency SOFC requires electrodes with 30-40% porosity. Due to the complexity of the SPPS process and the large number of processing parameters, it is difficult to investigate the effect of each parameter on the two important properties, i.e., coating porosity and deposition efficiency, separately. Design of experiments can use a small number of experimental runs to analyze the effect of each processing parameter on the properties of the fabricated product, after which the processing parameter combinations for fabricating a target product can be found. In this project, a small central composite design (CCD), a second-order statistical model, was used to analyze and optimize the SPPS process for Ni-YSZ anode deposition. The processing parameters investigated include: (1) Hydrogen flow rate, which determines arc voltage, (2) Current, (3) Solute flow rate, (4) Solution concentration, (5) Distance between nozzle and gun, and (6) Stand off distance. The effects of the selected processing parameters were analyzed, and the resultant model was used to select a combination of processing parameters, which produced a coating with the desired characteristics.

Keywords Electrode, Ni-YSZ anode, Solid oxide fuel cell (SOFC), Solution precursor plasma spray (SPPS)

1. Introduction

The solution precursor plasma spray (SPPS) process, in which a solution precursor was injected into a plasma jet to synthesize product materials and deposit coatings simultaneously, was developed in the 1990s and has been investigated extensively since then (Ref 1-5). Recently SPPS by direct current arc plasma spraying (DC-SPPS) has been used to produce coatings with high porosity (Ref 6, 7), suitable for use as solid oxide fuel cell (SOFC) electrode and sensor applications. Since a large number of processing parameters are involved in the SPPS process, it is difficult to establish relationships between the processing parameters and the properties of the deposited coatings. However, many applications require a particular property or combination of properties to be within a specific range, for example, high-quality electrodes for SOFC application require the porosity in the deposited coating to be in the range of 30-40%. In addition, to reduce cost and production time, the deposition efficiency of the SPPS process should be as high as possible. Therefore, deposition of coatings with one or several target properties and high-deposition efficiency is required from a practical SPPS process.

Statistical design of experiments is a mathematical method to establish the relations between the properties (responses) of a product and the processing parameters (factors) used in fabricating it, after which the processing parameter combination for making a target product could be found (optimization) (Ref 8, 9). The relations are obtained by measuring the properties of the product fabricated with different combinations of the processing parameters (the experimental design matrix) and employing statistical techniques to analyze the measured properties.

The classical approach of investigating the effect of each parameter on the final coatings one by one requires a large number of experimental runs because of the large number of processing parameters involved in the SPPS process. Design of experiments can use a small number of experimental runs to obtain a large amount of information about the effects of the processing parameters on the

This article is an invited paper selected from presentations at the 2008 International Thermal Spray Conference and has been expanded from the original presentation. It is simultaneously published in *Thermal Spray Crossing Borders, Proceedings of the 2008 International Thermal Spray Conference*, Maastricht, The Netherlands, June 2-4, 2008, Basil R. Marple, Margaret M. Hyland, Yuk-Chiu Lau, Chang-Jiu Li, Rogerio S. Lima, and Ghislain Montavon, Ed., ASM International, Materials Park, OH, 2008.

Y. Wang and **T.W. Coyle**, Centre for Advanced Coating Technologies, Department of Materials Science and Engineering, University of Toronto, Toronto, ON, Canada. Contact e-mail: youliang.wang@cnrc-nrc.gc.ca.

coating's properties. This approach has been widely employed to identify desirable process conditions for the deposition of thermal spray coatings, for example the vacuum plasma spray deposition of alumina, nickel-base alloys and tungsten carbide/cobalt cermet coatings (Ref 10), the high-velocity oxy fuel (HVOF) spraying of chromium carbide/nickel chromium coatings (Ref 11) and air plasma spray deposition of Ni-based superalloy (Ref 12). The application of statistical design of experiments to the development of thermal spray coatings has been reviewed recently (Ref 13).

In this article a small central composite design (CCD) (Ref 14), a second-order model, was used to investigate the effects of six processing parameters on coating porosity and deposition efficiency in the DC-SPPS process. The parameters investigated were: (1) Hydrogen flow rate, which determines arc voltage, (2) Current, (3) Solute flow rate, (4) Solution concentration, (5) Distance between nozzle and gun, (6) Stand off distance. The effects of the selected parameters were analyzed, and the SPPS process was successfully optimized.

2. DC-SPPS Setup, Solution Precursor, and Substrates

The DC-SPPS setup consists of a solution precursor feeder system, a solution atomizer-injector (nozzle), an atmospheric plasma spray system, and a sample holder (Ref 15). The solution precursor feeder system delivers measured quantities of the solution precursor to the nozzle fixed in front of the plasma torch, which produces atomized droplets of the precursor and injects them into the high-temperature plasma jet. A SG-100 torch (Miller Thermal) mounted on a robot was used. The solution precursor was made by adding distilled water to the mixture of $ZrOCl_2 \cdot 8H_2O$, $Y(NO_3)_3 \cdot 6H_2O$, and $Ni(NO_3)_3 \cdot 6H_2O$. The planned resultant material is 8YSZ-40 vol.%Ni (Ref 6, 15). Sintered partially stabilized zirconia discs with a diameter of 2 cm were used as substrates.

3. Optimization of SPPS Process for SOFC Anode Deposition

3.1 Small Central Composite Design

Central composite design (CCD) is the most popular response surface method (RSM) design (Ref 8, 9). A CCD has three groups of design points (1) two-level factorial or fractional factorial design points; (2) axial or star points; and (3) center points as shown in Fig. 1 Generally, the two level factorial points of the CCD are used to select critical factors first while the star points and center points are added to evaluate the curvature of the response surface and experimental error. Each factor in a CCD can have five levels. However, if the λ in Fig. 1 is set to 1, three levels for each factor can be used to get the well-known face-centered cubic array (FCC). When the number of factors is large, it is not necessary to run all combinations of factors. The factorial part of the design can be run in a fraction of the total number of available points. Standard CCDs may include any regular fractional factorial with a resolution of at least V, by which all the main effects and two factor interactions can be estimated.

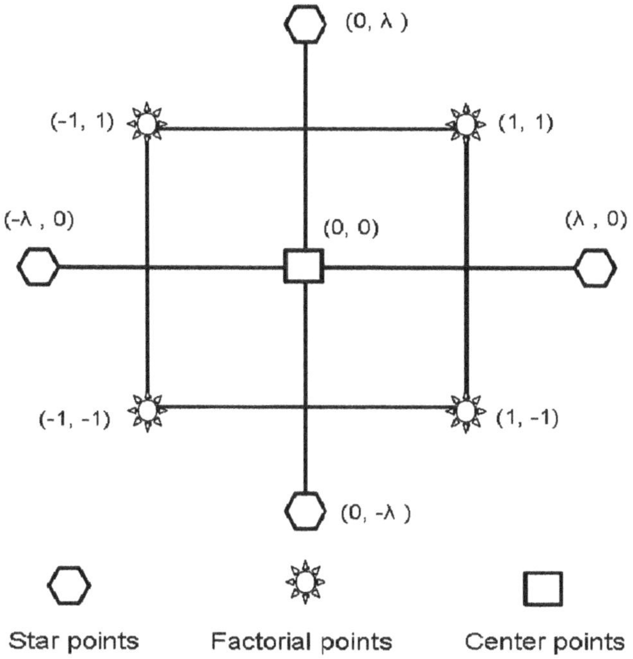

Fig. 1 Generation of a central composite design for two factors

If the experiment is expensive, small composite designs (SCDs), created by Hartley in 1959 (Ref 14), are alternatives to the standard CCDs. For the SCDs, the factorial part is not a regular fractional factorial of at least resolution V (Ref 14), but a special resolution III fractional factorial. Although by the resolution III factorial itself the main effects may be aliased with two factor interactions, the SCDs can give second-order models of the response. For *cubic* designs, all factors have three levels and the axial parts of the SCD have $\lambda = 1$.

3.2 Small Composite Design for Deposition of Ni-YSZ Anodes

All six processing parameters, including secondary plasma gas flow rate (HydrogenFR), which determines the gun arc voltage, gun current (I), solution precursor solute flow rate (SoluteFR), solution precursor concentration (SolutionC), distance between nozzle and gun (DistanceNG), and standoff distance (StandOD) were included in the small CCD. The design consisted of 34 experiments with 6 repeated center points as shown in Table 1. Each factor has three levels; therefore, the small central composite matrix is a face-centered (six dimensional) *cube*. The 34 deposition experiments are randomly distributed in the matrix to reduce the effect of systematic errors. In addition to the processing parameter values displayed in the matrix, the primary plasma gas (argon) flow rate was 62 l/min for all the runs; the atomizing gas pressure for different nozzle

Peer Reviewed

Table 1 Small central composite design matrix for Ni-YSZ anode deposition

Std	Run Unit	HydrogenFR L/min	I A	SoluteFR g/min	SolutionC M	DistanceNG cm	StandOD cm	EfficiencyR g	Porosity Area % ± std dev
6	1	3	570	7	0.5	0.3	8	0.134	47 ± 1.2
12	2	3	570	4	0.25	1.3	6	0.193	37 ± 2.1
19	3	2	600	5.5	0.375	0.8	7	0.071	35 ± 2.8
18	4	2	600	5.5	0.375	0.8	7	0.048	34 ± 0.3
2	5	3	630	7	0.25	1.3	6	0.149	33 ± 2.0
16	6	1	570	4	0.25	0.3	6	0.109	32 ± 1.6
7	7	1	630	7	0.25	0.3	6	0.071	30 ± 3.3
1	8	3	630	7	0.5	0.3	6	0.140	40 ± 2.4
8	9	3	630	4	0.25	1.3	8	0.057	29 ± 0.9
11	10	1	630	4	0.25	0.3	8	0.055	44 ± 2.5
20	11	2	600	5.5	0.375	0.8	7	0.064	34 ± 4.0
10	12	1	570	7	0.25	0.3	8	0.099	51 ± 1.5
13	13	1	570	4	0.5	1.3	6	0.152	35 ± 1.1
4	14	3	570	7	0.25	1.3	8	0.074	41 ± 1.8
14	15	1	570	7	0.5	1.3	8	0.252	47 ± 0.8
5	16	1	630	4	0.5	1.3	8	0.159	29 ± 0.7
9	17	3	570	4	0.5	0.3	6	0.137	42 ± 2.2
3	18	3	630	4	0.5	0.3	8	0.075	35 ± 2.1
15	19	1	630	7	0.5	1.3	6	0.180	43 ± 1.1
17	20	2	600	5.5	0.375	0.8	7	0.069	37 ± 1.6
34	21	2	600	5.5	0.375	0.8	7	0.072	36 ± 2.3
25	22	2	600	4	0.375	0.8	7	0.072	36 ± 0.4
23	23	2	570	5.5	0.375	0.8	7	0.079	38 ± 0.7
33	24	2	600	5.5	0.375	0.8	7	0.064	34 ± 2.8
22	25	3	600	5.5	0.375	0.8	7	0.076	29 ± 0.8
32	26	2	600	5.5	0.375	0.8	8	0.093	41 ± 0.4
26	27	2	600	7	0.375	0.8	7	0.117	33 ± 2.0
29	28	2	600	5.5	0.375	0.3	7	0.073	32 ± 2.1
30	29	2	600	5.5	0.375	1.3	7	0.099	33 ± 0.8
27	30	2	600	5.5	0.25	0.8	7	0.079	37 ± 1.8
24	31	2	630	5.5	0.375	0.8	7	0.093	34 ± 1.0
31	32	2	600	5.5	0.375	0.8	6	0.144	39 ± 1.6
28	33	2	600	5.5	0.5	0.8	7	0.106	40 ± 0.3
21	34	1	600	5.5	0.375	0.8	7	0.108	42 ± 2.5

positions was adjusted so that circular deposit footprints were generated on the substrates when the gun was held stationary. The duration of each deposition run was selected to compensate for the different amounts of solute consumed per minute due to the variations in solute flow rate levels, ensuring that the same total amount of solute was injected into the plasma jet over the course of each run.

Two responses were selected as the basis for the optimization. The porosity of the fabricated coating (Porosity) is one of the most important characteristics of the microstructure for SOFC performance and was found in the preliminary experiments to be the microstructural feature most sensitive to variations in deposition parameters. Deposition efficiency (EfficiencyR) was chosen because of its practical importance in production efficiency.

3.3 Experimental Data and Statistical Analysis

The 34 samples were deposited with a plasma gun scan speed 9.3 cm/s, scan length of 10 cm, and spray duration of 4 min, 5 min and 6 sec, and 7 min for the high, middle, and low solute flow rate levels, respectively. The deposition efficiency (EfficiencyR) in Table 1 is the weight gain (g) of the zirconia substrate discs. This is a relative measure valid only within this study since it depends on the substrate size, the amount of solute consumed, and other experimental conditions. The porosities (Porosity) in Table 1 were estimated by image analysis of SEM images of the polished cross sections of sintered and reduced coatings with the commercial software Clemex Vision PE 3.5. The polished cross sections for SEM observation were made by cutting the samples with a high-speed diamond saw (Clemex, Brillant 221), and then mounting the cut samples in a low-viscosity castable mounting resin using a vacuum impregnation setup. The analysis of the experimental data was done using Design Expert 7.0.2 (Ref 16). A detailed background of the data analysis procedures is described in Ref 8, 9.

3.4 Analysis Results

3.4.1 Quadratic Model for Deposition Efficiency. A six variable quadratic model including all 28 terms was fit to the 34 values of deposition efficiency (EfficiencyR) in Table 1. Tests on individual regression coefficients were done with $\alpha = 0.1$ in the t test statistic described in Ref 9. The terms found not to be significant at the 90% level were removed from the model equation by the backward elimination method (Ref 16). Ten terms were found to make significant contributions to the deposition efficiency

Table 2 ANOVA table for deposition efficiency (EfficiencyR)

Source	Sum of squares	df	Mean square	F value	*P*-value Prob > F	
Model	0.064	10	0.0064	28.8	<0.0001	Significant
A-HydrogenFR	0.0012	1	0.0012	5.6	0.0268	
C-SoluteFR	0.0023	1	0.0023	10.5	0.0035	
D-SolutionC	0.011	1	0.011	50.1	<0.0001	
E-DistanceNG	0.0098	1	0.0098	44.3	<0.0001	
F-StandOD	0.0042	1	0.0042	19.0	0.0002	
AF	0.0068	1	0.0068	30.8	<0.0001	
CD	0.0026	1	0.0026	11.7	0.0023	
CF	0.0043	1	0.0043	19.6	0.0002	
DF	0.0038	1	0.0038	17.22	0.0004	
F^2	0.017	1	0.017	79.0	<0.0001	
Residual	0.0051	23	0.00022			
Lack of fit	0.0047	18	0.00026	3.3	0.0917	Not significant
Pure error	0.00039	5	0.000078			
Cor total	0.069	33				

(EfficiencyR). The 10-term model for estimating deposition efficiency by coded and actual processing parameters (Ref 8, 9) can be expressed by Eq 1 and 2, respectively.

$$\begin{aligned}\text{EfficiencyR} = {} & 0.081 - (8.333 \times 10^{-3})\,\text{A} + 0.011\text{C} \\ & + 0.025\text{D} + 0.023\text{E} - 0.015\text{F} - 0.021\text{AF} \\ & + 0.013\text{D} + 0.017\text{CF} + 0.015\text{DF} + 046\text{F}^2 \end{aligned} \quad \text{(Eq 1)}$$

$$\begin{aligned}\text{EfficiencyR} = {} & +2.84 + 0.14\text{HydrogenFR} - 0.095\text{SoluteFR} \\ & - 1.04\text{SolutionC} + 0.059\ \text{DistanceNG} - 0.72\text{StandOD} \\ & - 0.021\text{HydrogenFR.StandOD} + 0.068\text{SoluteFR.} \\ & \text{SolutionC} + 0.011\text{SoluteFR.StandOD} \\ & + 0.12\text{SolutionC.StandOD} + 0.046\text{StandOD}^2 \end{aligned} \quad \text{(Eq 2)}$$

From the model equations it can be seen that all processing parameters are included among the linear terms of the model, except for arc current (B-I); only the factor distance between nozzle and gun (E-DistanceNG) does not appear among the 2 factor interaction terms; and standoff distance (F-StandOD) is the only variable with a significant second-order effect. Thus the model equation indicates that standoff distance (F-StandOD) is the most important processing parameter. The analysis of variance (ANOVA) Table 2 shows the effects of the different terms in the model equation. Since the F values (signal to noise ratio) for standoff distance (F-StandOD) squared, solution precursor concentration (D-SolutionC), distance between nozzle and gun (E-DistanceNG), and interaction of gun arc voltage (A-HydrogenFR) and standoff distance (F-StandOD) are the largest in the ANOVA Table 2, these four have the most important effects on deposition efficiency (EfficiencyR) in the selected experimental area.

3.4.2 Quadratic Model for Porosity. By the same analysis as for deposition efficiency, the equations for predicting porosity (Porosity) in coded and actual variables were found to be expressed by Eq 3 and 4, respectively.

$$\begin{aligned}\text{Porosity} = {} & +35.88 - 6.77\text{A} - 3.03\text{B} - 1.58\text{C} + 1.27\text{D} \\ & + 0.72\text{E} + 1.89\text{F} - 0.69\text{AB} - 0.79\text{AC} \\ & + 2.39\text{AD} - 1.90\text{AF} - 4.61\text{BF} + 1.45\text{CD} \\ & + 1.00\text{CE} - 6.33\text{DE} - 2.20\text{DF} - 2.31\text{EF} \\ & + 2.61\text{D}^2 - 3.41\text{E}^2 + 3.76\text{F}^2 \end{aligned} \quad \text{(Eq 3)}$$

$$\begin{aligned}\text{Porosity} & \\ = {} & -621.31 + 17.22\text{HydrogenFR} + 1.11\text{I} \\ & - 5.22\text{SoluteFR} + 104.62\text{SolutionC} \\ & + 140.26\text{DistanceNG} + 67.73\ \text{StandOD} \\ & - 0.03\text{HydrogenFR.I} - 0.52\text{HydrogenFR.SoluteFR} \\ & + 19.13\text{HydrogenFR.SolutionC} \\ & - 1.89\text{HydrogenFR.StandOD} - 0.16\text{I.StandOD} \\ & + 7.72\ \text{SoluteFR.SolutionC} \\ & + 1.67\text{SoluteFR.DistanceNG} \\ & - 126.60\ \text{SolutionC.DistanceNG} \\ & - 17.58\ \text{SolutionC.StandOD} \\ & - 5.78\text{DistanceNG.StandOD} + 166.80\text{SolutionC}^2 \\ & - 21.31\ \text{DistanceNG}^2 + 3.76\text{StandOD}^2 \end{aligned} \quad \text{(Eq 4)}$$

There are 19 terms out of the 28 which have statistically significant effects on the porosity (Porosity), and remain in the quadratic equation. All six processing parameters appear among the linear terms, but the distance between nozzle and gun (E-DistanceNG) and solute flow rate (C-SoluteFR) are not statistically significant items as shown in the ANOVA Table 3; they were included in the linear part of the model since these two factors are involved in the interaction and/or second-order terms of the model. Solution concentration (D-SolutionC), standoff distance (F-StandOD), distance between nozzle and gun (E-DistanceNG) and hydrogen flow rate (A-HydrogenFR) appear 6, 6, 5, and 5 times, respectively, in the equations; therefore, they are important processing parameters for controlling the porosity of fabricated coatings. The ANOVA Table 3 shows that the F values related to

Table 3 ANOVA table for porosity (Porosity)

Source	Sum of squares	df	Mean square	F value	*P*-value Prob > F	
Model	959.69	19	50.51	41.3	<0.001	Significant
A-HydrogenFR	91.57	1	91.57	74.9	<0.001	
B-I	165.01	1	165.01	135.1	<0.001	
C-SoluteFR	5.01	1	5.01	4.1	0.062	
D-SolutionC	28.88	1	28.88	23.6	<0.001	
E-DistanceNG	1.02	1	1.027	0.8	0.374	
F-StandOD	64.22	1	64.22	52.5	<0.001	
AB	7.58	1	7.58	6.2	0.025	
AC	10.10	1	10.10	8.2	0.012	
AD	10.16	1	10.16	8.3	0.012	
AF	57.69	1	57.69	47.2	<0.001	
BF	37.70	1	37.70	30.8	<0.001	
CD	33.54	1	33.54	27.4	<0.001	
CE	16.06	1	16.06	13.1	0.002	
DE	71.23	1	71.23	58.3	<0.001	
DF	77.29	1	77.29	63.3	<0.001	
EF	85.71	1	85.71	70.1	<0.001	
D^2	19.46	1	19.46	15.9	0.001	
E^2	33.33	1	33.33	27.3	<0.001	
F^2	40.62	1	40.62	33.2	<0.001	
Residual	17.09	14	1.22			
Lack of fit	9.53	9	1.05	0.70	0.697	Not significant
Pure error	7.56	5	1.51			
Cor total	976.79	33				

arc current (B-I), hydrogen flow rate (A-HydrogenFR), interaction of distance between nozzle and gun (E-DistanceNG) and standoff distance (F-StandOD), interaction of solution concentration (D-SolutionC) and standoff distance (F-StandOD), interaction of solution concentration (D-SolutionC) and distance between nozzle and gun (E-DistanceNG), and standoff distance (F-StandOD) are larger than those related to the other items. So these six items are critical for determining the porosity of the fabricated coatings.

3.4.3 Model Adequacy Checking. Model adequacy checks were made by residual analysis, predicted versus actual plots, testing for lack of fit, and R squared calculations. A residual e_i is the difference between the measured experimental response y_i and the reponse $\hat{y}_i$ predicted by the model equation, $e_i = y_i - \hat{y}_i (i = 1, 2 \ldots n)$. According to the basic assumptions of experimental design, the residuals should be normally distributed (Ref 8, 9), otherwise model inadequacy is indicated. A check of the normality assumption is usually done by constructing a normal probability plot of the residuals. If the residuals fall approximately on a straight line, the normal distribution of residuals is confirmed. Analysis indicates that the residuals are on the lines and concentrated on the center parts of the lines, showing that there are no problems with the residual normalities. Figure 2 is the normal plot of residuals for the model of deposition efficiency.

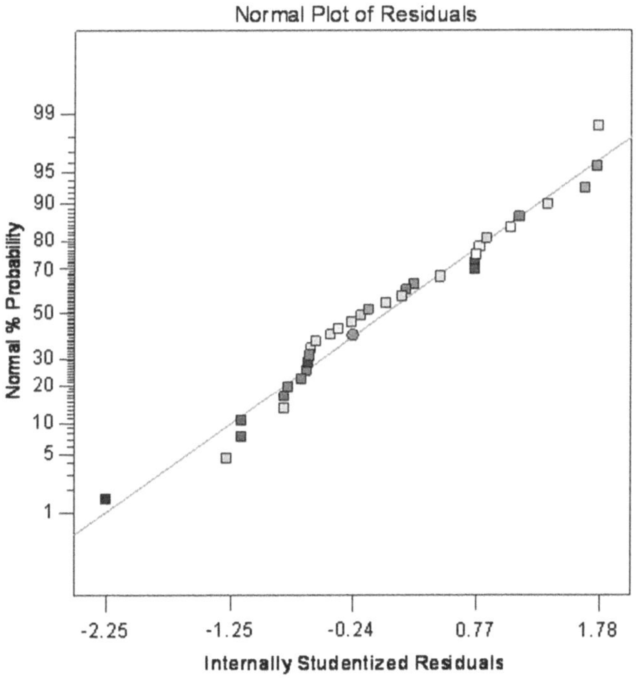

Fig. 2 Normal plot of residuals for model of deposition efficiency

The predicted versus actual plot is a graph of the measured versus the predicted response values. The data points should be split evenly by the 45-degree line or ideally all points should fall on the line. Figure 3 is the predicted versus actual plot for porosity (Porosity). The plot for deposition efficiency (EfficiencyR) has a similar distribution, demonstrating that the models appear to provide an excellent description of the effects of the various process parameters.

Testing for lack of fit is to examine if the model fits the data by calculating the lack of fit F value. The larger the F value, the more likely that the model does not adequately fit the data. The lack of fit F values for models of deposition efficiency (EfficiencyR) and porosity (Porosity) are

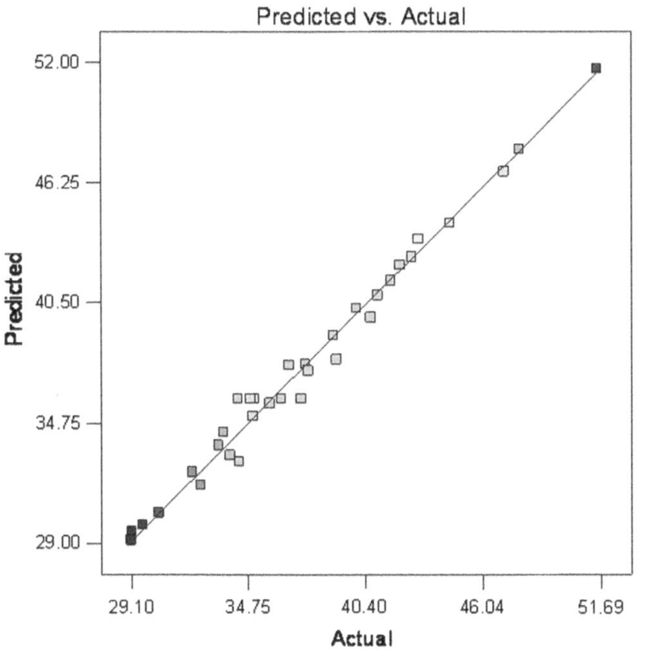

Fig. 3 Predicted versus actual plots for model of Porosity (Porosity)

	EfficiencyR	Porosity, %
Predicted	0.235	40
Measured	0.220	43

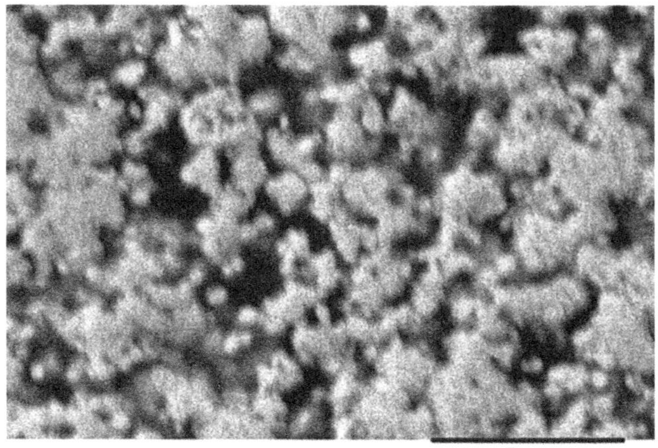

Fig. 4 SEM micrograph of a cross section of the coating deposited with optimized conditions

3.3 and 0.7 as shown in Table 2 and 3, respectively, which indicate that the models are acceptable.

R-Squared is the part of the variation about the mean that is explained by the fitted model. The calculations of R-Squared are described in Ref 8, 9. The R-Squared values for deposition efficiency (EfficiencyR) and porosity (Porosity) are 0.93 and 0.98, respectively. To overcome the increase in R-Squared with the addition of insignificant terms to the model, Adjusted R-Squared values are used. The Adjusted R-Squared values for deposition efficiency (EfficiencyR) and porosity (Porosity) are 0.89 and 0.96, respectively, meaning the two models are satisfactory.

3.5 Process Optimization

Within the experimental area in which NiO-YSZ deposits can be produced, the deposition process was optimized by the experiment design software (Design Expert 7.0.2) based on the analyses described in previous sections. This optimization process can also be done manually by using a desirability function (Ref 9). According to the microstructure requirements for a SOFC anode, the porosity (Porosity) target was 40%. The deposition efficiency (EfficiencyR) was desired to be maximum. The optimization analysis on the above conditions generated 59 processing parameter combinations and predicted responses as described in Ref 15. Considering the operational limits of the equipment, the parameter combination used to deposit a coating was: hydrogen flow rate = 1l/min; arc current = 610 A; solute flow rate = 6.9 g/min; solution concentration = 0.5 M; distance between nozzle and gun = 1.69 cm; standoff distance = 8 cm. Characterization of this coating yielded the following results:

Figure 4 is an SEM micrograph of a polished cross section of the coating deposited with the selected conditions, illustrating that the coating is very porous. The results above indicate that the optimized process parameters can be used to generate coatings with the desired porosity and at relatively high-deposition efficiency.

4. Understanding the Results of Statistical Design of Experiments

Relations between the processing parameters and the responses, i.e., the deposition efficiency (EfficiencyR) and the porosity (Porosity) as described by Eq 1 and 3, respectively, were obtained in the optimization of Ni-YSZ coatings. Although each response is a function of the processing parameters, it is very difficult to explain why each term takes the exact form it has in the functions. This difficulty arises from several considerations including the fact that the physical mechanism determining the relation between the response and each term in the function was not investigated directly and that some effects from higher order interactions may be confounded with the terms included in the models since only a fractional factorial design of experiment was used in the optimizations reported here. Therefore, in this section possible physical explanations for the effects described by only the linear terms in each model will be discussed based on the experimental observations and theoretical considerations. Other terms in the functions will be interpreted where possible. Since the equations with coded factors are independent of the numerical values of factors, they will be used as the basis of the discussion in the following subsections.

4.1 Deposition Efficiency

The deposition efficiency (EfficiencyR) is expressed by Eq 1 according to the analysis in the optimization of Ni-YSZ coating deposition. There are five terms, hydrogen flow rate (A), solution flow rate (C), solution concentration (D), distance between nozzle and gun (E) and standoff distance (F), in the linear part of the equation. The deposition efficiency (EfficiencyR) increases with the increase of solution flow rate (C), solution concentration (D), and distance between nozzle and gun (E). It decreases with the increase of hydrogen flow rate (A) and standoff distance (F).

A larger hydrogen flow rate (A) will increase the plasma gun voltage and generates a higher temperature and higher velocity plasma jet (Ref 17, 18). The higher temperature may vaporize more solute, decreasing the deposition efficiency (EfficiencyR), while the higher velocity would produce smaller droplets and smaller agglomerates as discussed in Ref 15. Small synthesized particles will follow the gas flow streams and may not impact onto the substrates, reducing the deposition efficiency (EfficiencyR) again. So the deposition efficiency (EfficiencyR) decreases with the increase of hydrogen flow rate (A).

As the solution flow rate (C) and solution concentration (D) increase, the mass density of synthesized material in the plasma jet would rise, which will result in the increase of probability of particle collisions. As a result, the agglomerates formed in the plasma jet would be larger (Ref 15), increasing the stopping distance and the deposition efficiency (EfficiencyR) as described in Ref 15. Therefore, the deposition efficiency (EfficiencyR) increases with the increase of solution flow rate (C) and solution concentration (D).

Increasing distance between nozzle and gun (E) is equal to decreasing plasma temperature and gas velocity. The lower temperature will vaporize less solute while the lower gas velocity will produce larger droplets and larger agglomerates as discussed in Ref 15. Therefore, increasing the distance between nozzle and gun (E) would increase deposition efficiency (EfficiencyR).

A longer standoff distance (F) allows a longer residence time of the synthesized materials and more probability of synthesized particle collisions, promoting the formation of larger agglomerates (Ref 15); therefore, a longer standoff distance (F) could increase deposition efficiency (EfficiencyR). However, a longer standoff distance (F) will also reduce the velocity of the synthesized particles, reducing the stopping distance and the deposition efficiency (EfficiencyR). The F term in Eq 1 (minus sign) indicates that the deposition efficiency (EfficiencyR) decreases with the increase of standoff distance (F), suggesting the velocity of synthesized particles is too low for impact onto the substrates (Ref 15).

4.2 Porosity

There are six terms in the linear part of Eq 3, which estimates the porosity (Porosity) of the fabricated Ni-YSZ coatings. However, the solution flow rate (C) and distance between nozzle and gun (E) are not significant. The porosity (Porosity) increases with the increase of solution concentration (D) and standoff distance (F) while it decreases when hydrogen flow rate (A) and gun current (B) increase.

According to the observations in Ref 15, there are two kinds of pores in the fabricated Ni-YSZ coatings, i.e., large pores between agglomerates and fine pores inside the agglomerates. The large pores, which account for a large fraction of the total porosity (Ref 15), are formed by stacking of the partially molten agglomerates.

A larger hydrogen flow rate (A) and gun current (B) will generate a higher temperature and velocity plasma jet (Ref 17, 18). The higher velocity would produce smaller droplets (Ref 19) and smaller agglomerates, which result in the formation of smaller pores between agglomerates. The higher temperature of the plasma jet would make the smaller agglomerates more molten and more deformable during impact onto the substrates, decreasing the porosity (Porosity). Therefore, the porosity (Porosity) decreases with increase of hydrogen flow rate and plasma gun current.

A higher solution concentration (D) increases the mass density of the particles in the plasma jet, resulting in more collisions of particles and larger and less molten agglomerates (Ref 15), promoting the formation of large pores between agglomerates. In addition, a lower concentration precursor tends to produce denser agglomerates (Ref 20). Therefore the porosity (Porosity) increases with the increase of solution concentration (D).

A longer standoff distance (F) increases the probability of synthesized particle collisions, forming larger agglomerates (Ref 15). Since the temperature of the plasma at impact is lower at longer standoff distance and the larger agglomerates have a higher heat capacity, the larger agglomerates would be relatively less molten, promoting the formation of coatings with higher porosity; therefore, the porosity (Porosity) increases with the increase of standoff distance (F).

A plasma gun operating at low power will generate lower temperature and velocity plasma gases, which promotes generation of larger agglomerates (Ref 15). Larger agglomerates would be less molten and less deformable during impact, therefore the low power coating would have larger porosity. This conclusion is consistent with the fact that there is a minus sign before the term describing the effect of the interaction between hydrogen flow rate (A), which determines gun voltage, and gun current (B) in Eq 3.

5. Conclusion

Quadratic models for predicting deposition efficiency and porosity were obtained by the small central composite experimental design technique. The model adequacy checking indicated that the models are statistically satisfactory. The models were used successfully to select process parameters which generated coatings with the desired porosity at a relatively high-deposition efficiency.

Physical-based explanations of the empirical relations between the processing parameters and the responses, i.e., the deposition efficiency (EfficiencyR) and the porosity (Porosity), were discussed with reference to experimental observations and theoretical considerations.

Acknowledgments

This project was financially supported by Natural Science and Engineering Research Council (NSERC) of Canada. The authors would like to express their gratitude to Dr. L. Pershin and Mr. T. Li in our laboratory for their kind help in the plasma spray experiments.

References

1. H. Wang, J. Williams, K.D. Vuong, C.Q. Shen, V. Wu, D.H. Lee, R.A. Condrate, and X.W. Wang, RF Plasma Fabrication of Nano-Scaled Ceramic Oxides for Energy Devices. *Proceedings of the 30th Intersociety Energy Conversion Engineering Conference. Orlando, July 30-August 4, 1995,* p 295-300
2. M. Gell, L. Xie, X. Ma, E.H. Jordan, and N.P. Padture, Highly Durable Thermal Barrier Coatings Made by the Solution Precursor Plasma Spray Process, *Surf. Coat. Technol.*, 2004, **177-178**, p 97-102
3. Y. Shan, T.W. Coyle, and J. Mostaghimi, Numerical Simulation of Droplet Breakup and Collision in the Solution Precursor Plasma Spraying, *J. Therm. Spray Technol.*, 2007, **16**(5-6), p 698-704
4. F. Gitzhofer, M.-E. Bonneau, and M. Boulos, Double Doped Ceria Electrolyte Synthesized by Solution Plasma Spraying with Induction Plasma Technology, *Thermal Spray 2001: New Surfaces for a New Millennium*. C.C. Berndt, D.A. Dhor, and E.F. Lugscheider, Eds., (Materials Park, Ohio, USA), ASM International, 2001, p 61-68
5. S.H. Miralai, D. Morvan, J. Amouroux, A. Naoumidis, H. Nickel, and R. Avni, Perovskite Thin Coating Deposition from Aqueous Nitrate Solutions of La and Mn in a Low Pressure Plasma Expanded Through a Nozzle, *12th International Symposium on Plasma Chemistry (ISPC 12)*. Vol. 2, Minneapolis, August 21-25, 1995, p 1999-2004
6. Y. Wang and T.W. Coyle, Solution Precursor Plasma Spray of Nickel-Yittia Stabilized Zirconia Anodes for Solid Oxide Fuel Cell Application, *J. Therm. Spray Technol.*, 2007, **16**(5-6), p 898-904
7. K. Chien and T.W. Coyle, Rapid and Continuous Deposition of Porous Nanocrystalline SnO_2 Coating with Interpenetrating Pores for Gas Sensor Applications, *J. Therm. Spray Technol.*, 2007, **16**(5-6), p 886-892
8. G.E.P. Box and N.R. Draper, *Empirical Model-Building and Response Surfaces*, John Wiley & Sons, Inc., New York, 1986
9. R.H. Myers and D.C. Montgomery, *Response Surface Methodology*, 2nd ed., John Wiley & Sons, Inc., New York, 2002
10. R. Kingswell, K.T. Scott, and L.L. Wassell, Optimizing the Vacuum Plasma Spray Deposition of Metal, Ceramic and Cermet Coatings Using Designed Experiments, *J. Therm. Spray Technol.*, 1993, **2**(2), p 179-186
11. C.J. Li, G.C. Ji, and Y. Wang, Effect of Powder Type on the Relationship Between Spray Parameters and Properties of HVOF Sprayed Cr_3C_2-NiCr Coatings, Thermal Spray: Meeting the Challenges of the 21st Century, *Proceedings of the 15th International Thermal Spray Conference*, C. Coddet, Ed., (Nice, France), ASM International, 1998, p 287-292
12. F. Azarmi, T.W. Coyle, and J. Mostaghimi, Optimization of Atmospheric Plasma Spray Process Parameters Using a Design of Experiment for Alloy 625 Coatings, *J. Therm. Spray Technol.*, 2008, **18**, p 1-12
13. C. Pierlot, L. Pawlowski, M. Bigan, and P. Chagnon, Design of Experiments in Thermal Spraying: A Review, *Surf. Coat. Technol.*, 2008, **202**, p 4483-4490
14. H.O. Hartley, Smallest Composite Designs for Quadratic Response Surfaces, *Biometrics*, 1959, **15**(4), p 611-624
15. Y. Wang, "Deposition of Solid Oxide Fuel Cell Electrodes by Solution Precursor Plasma spray," Ph. D. Thesis, University of Toronto, 2008
16. http://www.statease.com/software.html. Accessed 2006
17. M.P. Planche, J.F. Coudert, and P. Fauchais, Velocity Measurements for Arc Jets Produced by a DC Plasma Spray Torch, *Plasma Chem. Plasma Process.*, 1998, **18**(2), p 263-283
18. S. Semenov and B. Cetegen, Spectroscopic Temperature Measurement in Direct Current Arc Plasma Jets Used in Thermal Spray Processing of Materials, *J. Thermal Spray Technol.*, 2001, **10**, p 326-336
19. M. Pilch and C.A. Erdman, Use of Breakup Time Data and Velocity History Data to Predict the Maximum Size of Stable Fragments for Acceleration-Induced Breakup of a Liquid Droplet, *Int. J. Multiphase Flow*, 1987, **13**(6), p 741-757
20. A. Curav, T. Kodas, T. Pluym, and Y. Xiong, Aerosol Processing of Materials, *Aerosol Sci. Technol.*, 1993, **19**, p 411-452

JTTEE5 17:700–707
DOI: 10.1007/s11666-008-9249-2
1059-9630/$19.00 © ASM International

Peer Reviewed

Suspension HVOF Spraying of Reduced Temperature Solid Oxide Fuel Cell Electrolytes

Jörg Oberste Berghaus, J.-G. Legoux, C. Moreau, R. Hui, C. Decès-Petit, W. Qu, S. Yick, Z. Wang, R. Maric, and D. Ghosh

(Submitted April 29, 2008; in revised form August 19, 2008)

Metal-supported solid oxide fuel cells (SOFCs) composed of a $Ce_{0.8}Sm_{0.2}O_{2-\delta}$ (SDC) electrolyte layer and Ni-$Ce_{0.8}Sm_{0.2}O_{2-\delta}$ (Ni-SDC) cermet anode were fabricated by suspension thermal spraying on Hastelloy × substrates. The cathode, a $Sm_{0.5}Sr_{0.5}CoO_3$ (SSCo)-SDC composite, was screen-printed and fired *in situ*. The anode was produced by suspension plasma spraying (SPS) using an axial injection plasma torch. The SDC electrolyte was produced by high-velocity oxy-fuel (HVOF) spraying of liquid suspension feedstock, using propylene fuel (DJ-2700). The emerging technology of HVOF suspension spraying was explored here to produce thin and low-porosity electrolytes in an effort to develop a cost-effective and scalable fabrication technique for high-performance, metal-supported SOFCs. In-flight particle temperature and velocity were measured for a number of different gun operating conditions and standoff distances and related to the resulting microstructures. At optimized conditions, this approach was found to limit material decomposition, enhance deposition efficiency, and reduce defect density in the resulting coating, as compared to previous results reported with SPS. Produced button cells showed highly promising performance with a maximum power density (MPD) of 0.5 W cm^{-2} at 600 °C and above 0.9 W cm^{-2} at 700 °C, with humidified hydrogen as fuel and air as oxidant. The potential of this deposition technique to scale-up the substrate size to 50 × 50 mm was demonstrated.

Keywords HVOF suspension spraying, reduced temperature solid oxide fuel cells, suspension thermal spraying

1. Introduction

Solid oxide fuel cells (SOFCs) are highly efficient power converting devices combining low pollution and high fuel flexibility. The key obstacles for the widespread implementation of SOFC technology are the high component and overall manufacturing costs, which are compounded by the technical requirements of high performance, long-term stability, thermal cycling capability, and short startup time (Ref 1). It is increasingly recognized that for small-stationary SOFC units and mobile systems (1-5 kW), such as SOFC-based auxiliary power units (APU) for cars and trucks, metal-supported SOFCs operating at temperatures far below the traditional 1000 °C can offer many advantages over the conventional electrode or electrolyte-supported SOFCs (Ref 2). Reduced-temperature (RT) SOFC systems not only permit low-cost peripheral material (metal support and insulation), but also offer increased structural robustness, thermal stability, reduced degradation, and increased flexibility in design and assembly (Ref 3, 4).

Possible approaches to compensate for the increased resistance to ion transport at a lower operating temperature are to reduce the electrolyte thickness of the traditional yttria-stabilized zirconia (YSZ) layer to only a few microns and/or to use electrolyte materials of inherently higher ionic conductivity, such as, for example, scandia-stabilized zirconia (ScSZ) or samarium-doped ceria (SDC) (Ref 3).

Ultimately, to realize a high performance of metal-supported SOFCs, a thin, low-porosity, and fully crystalline electrolyte layer is essential to separate the fuel from the oxidant atmosphere (Ref 5, 6). Among the different fabrication technologies available, thermal spraying (TS) is gaining increasing attention due to its high deposition rates and scale-up capacity, as well as its potential to produce multiple SOFC components sequentially and directly on metallic substrates without the traditional and costly postsintering step (Ref 3, 7). However, significant challenges in TS for SOFC production remain, which are

This article is an invited paper selected from presentations at the 2008 International Thermal Spray Conference and has been expanded from the original presentation. It is simultaneously published in *Thermal Spray Crossing Borders, Proceedings of the 2008 International Thermal Spray Conference*, Maastricht, The Netherlands, June 2-4, 2008, Basil R. Marple, Margaret M. Hyland, Yuk-Chiu Lau, Chang-Jiu Li, Rogerio S. Lima, and Ghislain Montavon, Ed., ASM International, Materials Park, OH, 2008.

Jörg Oberste Berghaus, Bekaert Advanced Coatings NV, E3-Laan 75-79, Deinze 9800, Belgium; **J.-G. Legoux** and **C. Moreau**, Industrial Materials Institute, NRC, Boucherville, QC, Canada; and **R. Hui, C. Decès-Petit, W. Qu, S. Yick, Z. Wang, R. Maric,** and **D. Ghosh**, Institute for Fuel Cell Innovation, NRC, Vancouver, BC, Canada. Contact e-mail: jorg.oberste-berghaus@bekaert.com.

primarily related to high coating thickness, residual porosity/defects, and thermal stresses.

As a promising extension of conventional thermal spraying, the use of a liquid feed carrier in suspension plasma spraying (SPS) permits feeding and spraying of much finer particles to form thinner coatings with more refined microstructure and grain size (Ref 8-10). In their previous work, the authors reported a metal-supported fuel cell with a samarium-doped ceria (SDC) electrolyte and NiO-SDC anode fabricated by suspension plasma spraying using an axial injection plasma torch (Axial III, Northwest Mettech, North Vancouver, Canada). The cell showed promising performance at 650 °C with a MPD of 0.216 W cm^{-2} and an open cell voltage (OCV) of 0.768 V (0.94 V theoretical OCV). Its resistance to oxidation and to rapid thermal cycling was evaluated (Ref 11).

The small particles created in SPS carry little momentum and low thermal inertia. They generally have to be heated far above their melting point in the high-temperature plasma (T_{flame} > 8000 °C) to retain the melting temperature and high velocity further downstream at the impact on the substrate to form a dense coating. For ceria, this high superheating in the chemically reducing plasma can lead to an undesirable transformation of CeO_2 to Ce_2O_3, which is associated with a lowering of the melting and boiling points (Ref 12). The decomposition and resulting vaporization can limit deposition efficiency, electrochemical performance, and mechanical integrity of the electrolyte. Employing a much lower flame temperature of a high-velocity oxy-fuel (HVOF) system (2600-3200 °C) may alleviate these effects, under the condition that the ceria particles are still heated to at least their melting point (~2730 °C) (Ref 12). Maximum particle temperatures in HVOF depend on a number of factors, including the torch design, the fuel and fuel-to-oxygen ratio, the feedstock and feed rate, and the position in the flame. Using acetylene, a high-energy fuel with some safety restrictions, HVOF suspension spraying of alumina, titania, and zirconia ceramics was recently reported at the University of Stuttgart (Ref 13). Advantages of HVOF spraying of conventional feedstock powders for SOFC production were also demonstrated (Ref 5). The authors reported earlier on suspension HVOF spraying of ceramic composites using more standard fuels (Ref 10). In this paper, HVOF suspension spraying for the production of ceria-based electrolyte coatings and the role of operating parameters on the in-flight particle states and microstructure are discussed. A metal-supported fuel cell was produced and its electrochemical performance was evaluated.

2. Experimental Procedure

2.1 Spray Systems and Materials

Suspension spraying for the SDC coatings was implemented using an HVOF DJ-2700 hybrid gun (Sulzer-Metco, Westbury, NY) using propylene fuel (C_3H_6). The system was equipped with an internal injection/atomization module inserted into the standard powder feeding port, thereby intimately contacting the coaxially fed suspension droplets with the fuel inside the HVOF combustion chamber. The suspension was delivered by a prototype Nanofeed Liquid Powder Feeder (Model 640, Northwest Mettech). Flow-rate control, start-up, shut-down, and rinsing sequences are fully automated and PC controlled. The injection module and fuel distributor were continuously cooled with nitrogen. The system delivered the suspension feed against the combustion backpressure of up to 6.46 atm. HVOF operating conditions, used in this experiment, are summarized in Table 1.

Table 1 DJ-2700 HVOF spray parameters

Parameter	Value
Propylene flow	90, 85, 75, and 70 slpm
Oxygen flow	279 slpm
Air flow	202 slpm
Carrier gas (N_2) flow	15 slpm
Spray distance	7.6-12.7 cm
Gun traverse speed	762 mm/s
Suspension solid content	2.5-5 wt.%
Suspension feed rate	25-50 mL/min

In-flight particle states were measured with a commercial diagnostic system (AccuraSpray® G2 Tecnar, St-Bruno, PQ, Canada). The temperature measurement is based on two-color pyrometry, and the velocity is determined by a time-of-flight technique. Microstructures of the coatings were observed by FEG-SEM (Hitachi S4700). The samples were prepared by standard metallographic methods. Porosity was assessed on the cross-section microstructure images taken by SEM (5000×), and then using image analysis.

Phase analysis was carried out by XRD using a Brucker D8-Discovery diffractometer (Brucker AXS Inc. Madison, USA) with Cu Kα radiation at 0.01°/s. The crystallite size was approximated from the Sherrer equation at the principal diffraction lines.

For the electrolyte coatings, two different $Ce_{0.8}Sm_{0.2}O_{2-\delta}$ feedstock powders were used, namely, a micron-sized powder (d_{50}, 1.45 μm) with a specific surface area (SSA) of 5.37 m^2/g (Nextech Materials, Lewis Center, OH) and nanosized powders with SSA of 80-220 m^2/g (nGimat, Atlanta, GA), having a particle size of approximately 20 nm, which was estimated from the crystallite size as measured by XRD. Suspensions of initially 5 wt.% and then 2.5 wt.% solids were prepared in a mixture of ethanol and ethylene glycol. The nanosized powder was prone to form aggregates in the suspension. Wide-size distributions with aggregate ranging from 40 nm (lower detection limit) to 10 μm were measured with a Coulter LS Particle size analyzer (Beckman Coulter Canada Inc., Mississauga, CAN) using a liquid module in highly diluted aqueous suspensions. Polyethylenimine (MWT 25,000 Alfa Aesar, USA) was used as dispersing agent. It was found that the addition of viscous ethylene glycol further prolonged the settling time of the suspension. Most coatings were produced on steel substrates or functional anodes disks of 16 mm diameter (button cells). Electrolyte coatings were also produced on larger substrates with rectangular dimension of 50 × 50 mm. Surface temperatures, as monitored by a pyrometer, were controlled during deposition, using a specialized substrate holder.

2.2 Substrate and Anode

Tape cast and sintered Hastelloy × plates (Mott Corp., Farmington, CT) were used as metallic substrates. A substrate porosity of 16% was determined by image analysis.

NiO-SDC (50 wt.% NiO) anodes were produced by suspension plasma spraying using a torch with three converging plasma jets and an axial feed injection port (Axial III, Northwest Mettech). Relatively large NiO powders (d_{50}, 14.4 μm; SSA, 0.21 m^2/g) and micron-sized SDC powders (d_{50}, 1.2 μm; SSA, 8.9 m^2/g) (Nextech Materials, Lewis Center, OH) were used as feedstock. In an attempt to produce a relatively porous anode layer, while fully covering the substrate pores, spray conditions were selected to provide high particle velocities (700 m/s), but relatively low particle temperatures (2400 °C). These particle states were attained at a high plasma gas flow rate (275 slpm) and by using helium as the secondary gas at 30%, with argon and nitrogen as balance. Furthermore, the suspensions in 30 wt.% ethylene glycol in ethanol were fed at a high rate (70 mL/min), thereby imposing a high thermal load on the plasma flame. The anodes were produced at a spray distance of 62.5 mm with maximum surface temperature of 300 °C. Further details can be found elsewhere (Ref 14). After deposition, the NiO was reduced to Ni at 600 °C in hydrogen for 30 min, thereby inducing the first and possibly most significant volume change in the fine-grained anode. This additional process step was previously found to reduce crack formation in the electrolyte during SOFC testing (Ref 11). Large surface humps on the anode, which can sometimes occur in suspension spraying, were removed by mechanic polishing.

2.3 Cathode Deposition and Cell Testing

After electrolyte deposition, a $Sm_{0.5}Sr_{0.5}CoO_3$ (SSCo)-SDC (weight ratio 70:30) composite cathode was stencil printed on the half cell and fired *in situ* at 800 °C for 2h. To further reduce the anode, the cell was then maintained at 650 °C for 5h while gradually introducing hydrogen. The power generation characteristics were measured from 500 to 700 °C in 50 °C intervals. At each temperature, electrochemical performance and ac impedance measurements were performed twice using humidified hydrogen (3% H_2O) as fuel and air as oxidant. Before cathode deposition, the half cells were subjected to a helium leak test at 0.0689 bar differential pressure. Details on the testing setup for gas permeability and electrochemical performance are published elsewhere (Ref 11).

3. Results and Discussion

3.1 Initial Tests

Initial tests of HVOF suspension spraying with SDC revealed limited, if any, coating deposition. A typical example of a microstructure of these early coatings, using the larger micron-sized SDC feedstock, is shown in Fig. 1(a). A 5 wt.% SDC suspension in ethanol was fed at a feed rate of 50 mL/min. The propylene flow was 80 slpm and the spray distance 12 cm. The microstructure in the backscattered electron image features regions of gray contrast, which can be interpreted as pull-out and fracture surfaces created during the polishing step, indicative of limited material fusion. Individual particles can be discerned, which resemble unmolten feed particles. The flame enthalpy was likely insufficient to melt most of the micron-sized ceria particles. Similar observations are reported for high velocity suspension flame spraying of zirconia (Ref 13). Another example, using the much finer nanosized SDC feedstock dispersed as 5 wt.% in ethanol and fed at 50 mL/min, is shown in Fig. 1(b) and (c). This coating was deposited with 80 slpm propylene flow at a standoff distance of 10 cm on a stainless steel 430 substrate. Some isolated regions of dense material can be discerned. A higher degree of particle melting possibly occurred with the fine particles due to the higher surface area available for heat transfer and lower melting times of the smaller material volume. However, the coating is strongly fractured and the material appears to adhere only by virtue of mechanical anchoring between the roughness asperities of the substrate. Again, regions of gray contrast are indicative of limited material melting and fusion and limited splat contact due to a low flame enthalpy.

3.2 Spray Distance

In-flight particle states as a function of standoff distance are exemplified in Fig. 2 at the spray condition for the coating from the nanosized SDC in Fig. 1(b) and (c), i.e., 80 slpm C_3H_6. At the spray distance (SD) used in the initial test, i.e., 10 cm, the average particle temperature of 2550 °C was far below the melting point of ceria and the velocity was as high as 800 m/s. These values support the idea that most particles were indeed unmelted and that their impact at high velocity contributed to the cracking and material removal by grit blasting, consistent with the observed microstructure.

It is interesting to note that the maximum average particle temperature was measured far downstream from the gun nozzle exit (16 cm). At SD of 10 cm, the particles are still in a process of heating in the supersonic flame. This suggests that, in this example, they have not experienced temperatures higher than approximately 2650 °C during their in-flight history, including inside the HVOF combustion chamber. In principle, vaporization and decomposition due to superheating of the ceria particles, as it can occur in plasma spraying, are avoided by the inherently imposed upper temperature limit of the HVOF system. The diagnostic system indirectly provided evidence for less vaporization by capturing a meaningful signal, which was previously distorted in plasma spraying due to the spectral emission of evaporating ceria (Ref 14).

Using the on-line diagnostics, subsequent efforts were focused on increasing the particle temperature by adjusting the operating and feed conditions using the nanosized SDC feedstock.

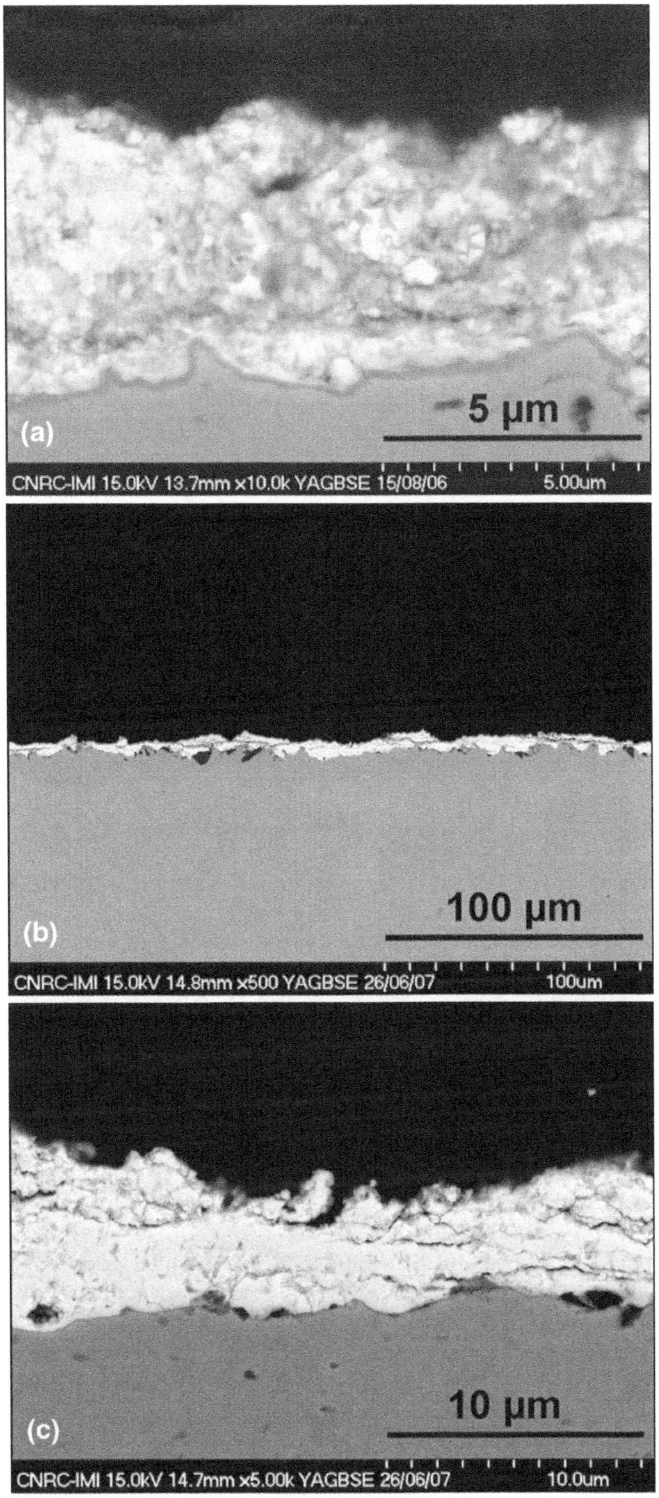

Fig. 1 Microstructure of initial HVOF SDC coating from (a) micron-sized and (b, c) nanosized feedstock

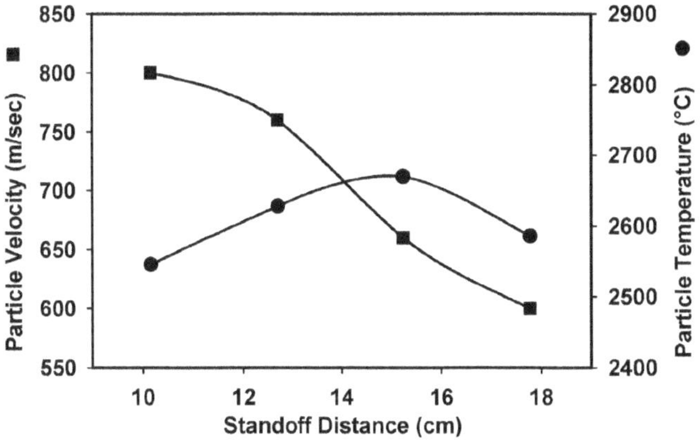

Fig. 2 Axial T-V profile at suboptimal conditions

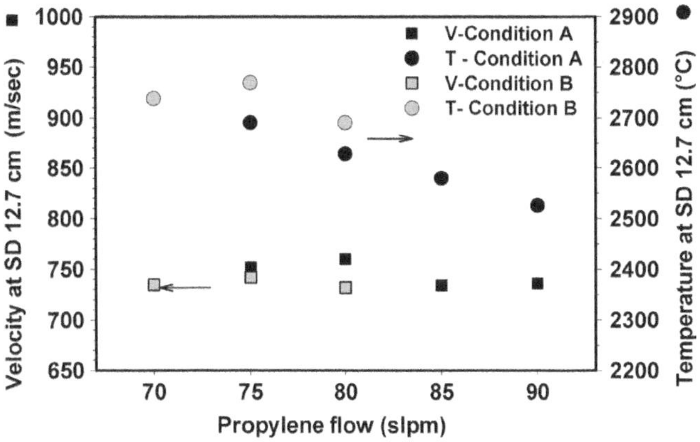

Fig. 3 Effect of propylene flow on particle T and V

3.3 Fuel-to-Oxygen Ratio

Figure 3 exemplifies the effect of propylene fuel rates at constant measurement distance (127 mm), oxygen flow, and two slightly different suspension feed conditions (A&B), i.e., 5 wt.% solid suspension in ethanol at 50 mL/min without (A) and with the addition of 50% ethylene glycol (B). While the particle velocity remained essentially constant, higher particle temperatures were attained at lower fuel rates, approaching those of the stoichiometric oxygen/fuel ratio of 4.5 to 1. Considering that both entrained shroud air and the suspension solvent can contribute to the combustion, the most efficient conditions may shift from stoichiometry. A maximum particle temperature was generally obtained with over-stoichiometric fuel supply, as in the example of condition B in Fig. 3.

3.4 Suspension Feed Rate

Figure 4 exemplifies the effect of the suspension feed rate at constant spray conditions (85 slpm C_3H_6, SD 127 mm) and suspension composition (5 wt.% solids in ethanol). The observed increase in particle temperature and velocity with decreasing feed loading is typical for suspension thermal spraying (Ref 9). In SPS, this is mainly associated with the lowering of the heat requirement for evaporation of the suspension medium. Such loading effect was expected to be less pronounced in HVOF suspension spraying, since the combustion of the ethanol solvent in intimate contact with the particles could partially compensate for heat loss by evaporation. In this

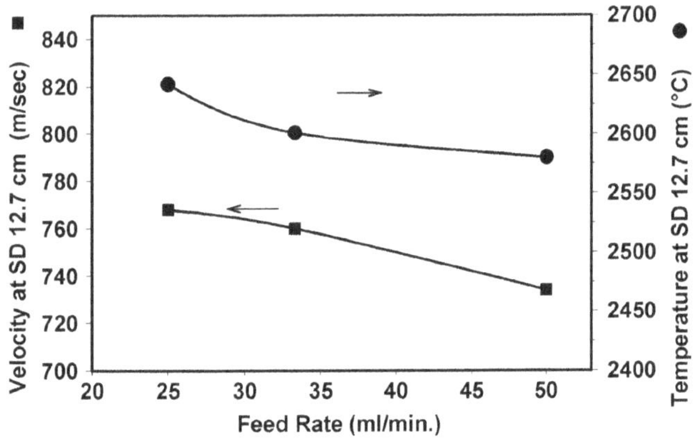

Fig. 4 Effect of feed rate on particle T and V

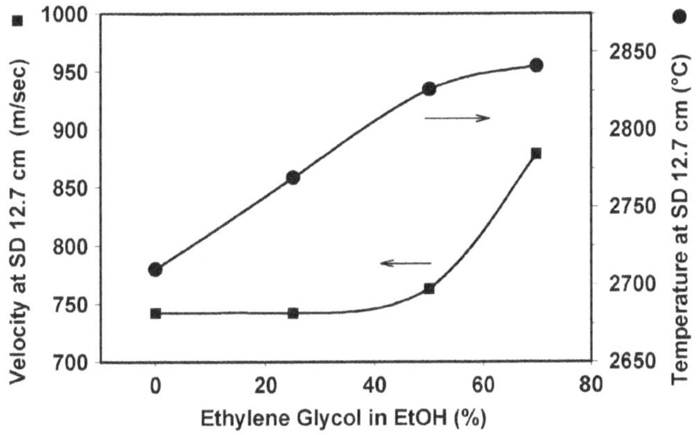

Fig. 5 Effect of ethylene glycol to ethanol ratio

case, however, the flame was already fuel rich and the ethanol may not have significantly contributed to the combustion heat.

3.5 Nature of Suspension Solvent

Figure 5 shows the effect of adding ethylene glycol (EG) to the ethanol-based suspension solvent at different mixing ratios, keeping the torch operating conditions (75 slpm C_3H_6, SD 127 mm) and feed conditions (5 wt.% solids at 50 mL/min) constant. The high boiling point glycol (B.P. 197.3 °C) has both a slightly higher vaporization enthalpy (63.9 kJ/mol) and a lower heat of combustion (19.8 kJ/g) than ethanol ($\Delta_{vap}H_{(25\ °C)}$ 42.32 kJ/mol, $\Delta_cH°$ 29.87 kJ/g). Increasing particle temperature and velocity with increased EG content was consequently somewhat surprising. A possible explanation may be an improved dispersion of the particles in the viscous glycol suspension ($\eta \sim 16.1$ mPa s) as compared to ethanol ($\eta \sim 1.2$ mPa s), which may stabilize the suspension to form smaller aggregates, which may then translate, in average, into smaller in-flight particles created during the atomization and solvent combustion step. In SPS the agglomerate distribution in the feed plays an important role in their transformation into ceramic droplets (Ref 8, 10). Smaller particulates, in turn, are more readily heated and melted and accelerated to the gas velocity, as explained earlier. Following this reasoning, lowering the solid content in the suspension from 5 to 2.5 wt.% may also improve the particle dispersion or, at least, slow down the formation of larger aggregates during processing by providing a longer mean-free path for particle/aggregate collisions. A lower solid content is further expected to decrease the heat load on the flame available for particle melting and heating. Based on the findings, optimized spray conditions were established for the nanosized SDC feedstock at slightly over-stoichiometric propylene flow, low suspension feed rate and solid loading and with high glycol content, as summarized in Table 2. At these conditions an average particle temperature close to 3000 °C was measured. The corresponding T-V profile as a function of SD is shown in Fig. 6.

Table 2 Optimized spray parameters for SDC electrolyte

Parameter	Value
Propylene flow	75 slpm
Oxygen flow	279 slpm
Air flow	202 slpm
Carrier gas (N_2) flow	15 slpm
Spray distance	12.7 cm
Gun traverse speed	762 mm/s
Suspension solid content	2.5 wt.%
Suspension feed rate	33 mL/min
Particle size	~20 nm SDC
Suspension composition	70/30 wt.% glycol/ethanol

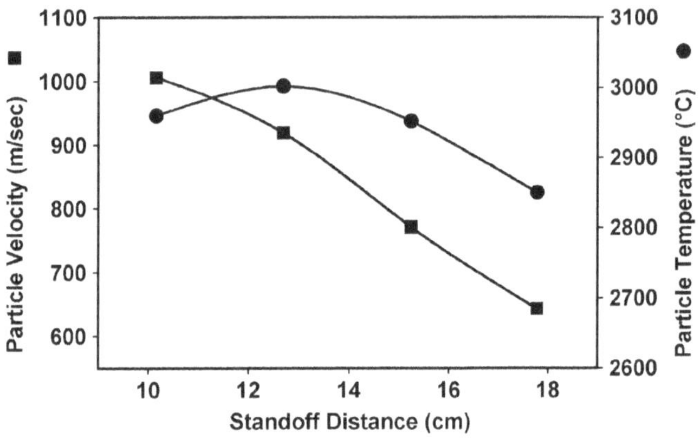

Fig. 6 Axial T-V profile at optimum spray conditions for SOFC production

3.6 As-Sprayed SOFC Half Cell

By positioning the substrate at the highest measured particle temperature (127 mm) at the optimized conditions, SDC electrolytes (~20 μm thick) were deposited on 50-60 μm thick anode layers supported by 16 mm diameter Hastelloy × substrates. A surface temperature of 450 °C was maintained. Figures 7 and 8 show cross-sectional micrographs of the as-sprayed cell, without the cathode. The electrolyte has a relatively dense microstructure (no visible open porosity and closed porosity

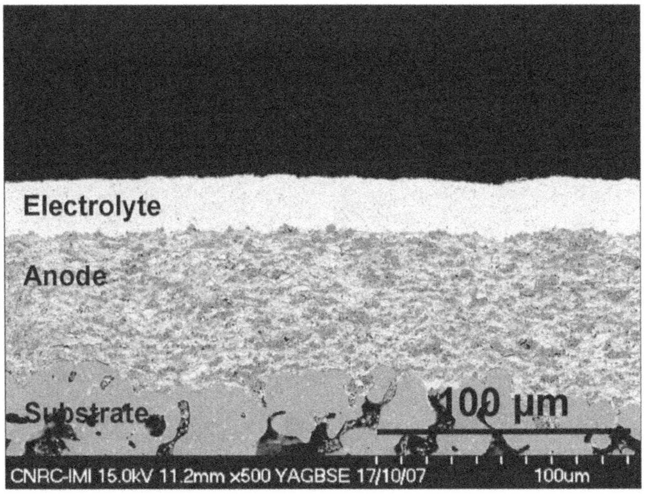

Fig. 7 Microstructure of HVOF sprayed half cell

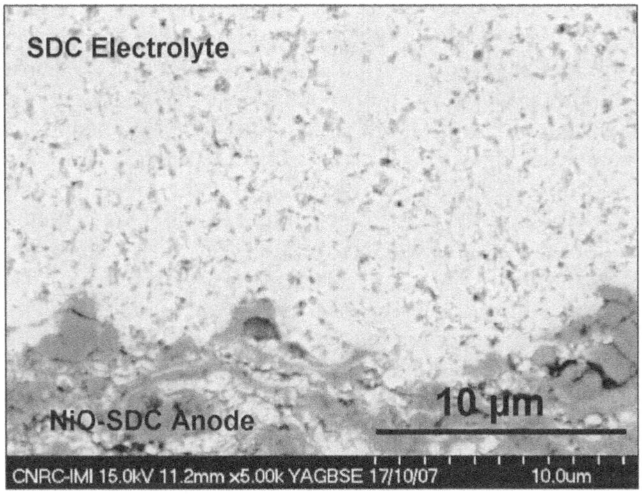

Fig. 8 As-sprayed SDC-NiO/SDC microstructure

<2% as estimated from cross-section SEM image analysis). A lamellar structure, typical for thermal sprayed coatings, cannot be discerned. The SDC layer also appeared well adherent to the anode. The backscattered electron image shows only few gray contrasted fracture and pull-out regions, which is indicative of a high degree of material fusion. The lack of gray regions may also suggest the absence of Ce_2O_3 phases. XRD analysis indicated that the crystal structure was exclusively doped cerianite (CeO_2) with a grain size of approximately 37 nm. It was noted that the electrolyte was virtually free of cracks or pinholes. Gas permeability below 0.1 slpm/cm^2 was determined under helium pressure (6.89 kPa). In thermal spraying, tensile stresses generally arise due to the material shrinkage during solidification and cooling of the molten droplets, often leading to transverse cracks. Using HVOF, these stresses may be partially compensated by the plastic deformation of unmolten or partially molten ceramic particles impacting at high velocity (Ref 5). At the

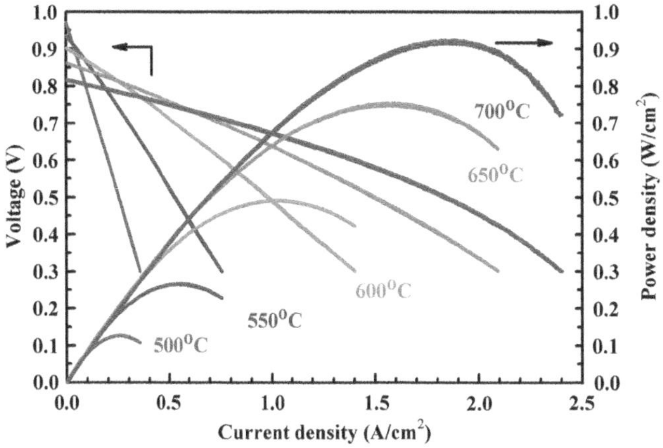

Fig. 9 Performance of HVOF sprayed SOFC

spray conditions used, a substantial portion of the particles were likely not fully molten and their impact at nearly 900 m/s may have affected the intrinsic stresses by introducing a compressive component. Furthermore, the top surface of the electrolyte was relatively smooth without the humps and defects often seen in SPS (Ref 10). The impacting colder particles may remove loosely attached material during deposition leading to this low surface roughness.

3.7 SOFC Cell Performance

After cathode deposition, the cell performance and impedance was tested from 500 to 700 °C, as described above. Figure 9 shows an open circuit voltage (OCV) of 0.97 V at 500 °C, nearly 90% of the theoretical OCV value of 1.08 V (Ref 15). The OCV then decreases with temperature due to the rise in electronic contribution to the SDC conductivity (Ref 4). The maximum power density increases continuously to 0.49 W cm^{-2} at 600 °C and up to 0.92 W cm^{-2} at 700 °C. This performance compares very favorably with the previous results of the authors (Ref 11), as well as other published results for metal-supported SOFCs (Ref 3, 7) in the reduced temperature range, of which the authors are aware. Figure 10 summarizes the area specific resistances (ASRs). Above 600 °C, the contribution of the polarization resistance to the total resistance is almost negligible. This suggests that catalytic activity of the electrodes is not a limiting factor. At lower temperatures, however, both ohmic and polarization resistance rise significantly. Ohmic resistance is generally attributed to restricted ion conductivity through the electrolyte and may be a consequence of the residual pores/imperfect splat contact in the coating, at least in comparison with fully sintered bulk-like ceramic (Ref 4). On the other hand, an imperfect contact between the electrolyte and the electrodes may also contribute to both polarization and ohmic resistance. Indeed, the interfaces proved to be of limited quality. Thermal cycling tests at a heating rate of 60 °C/min, i.e., heating the cell from 25 to 600 °C in 10 min multiple times, revealed a significant degradation at the cathode-electrolyte interface with a

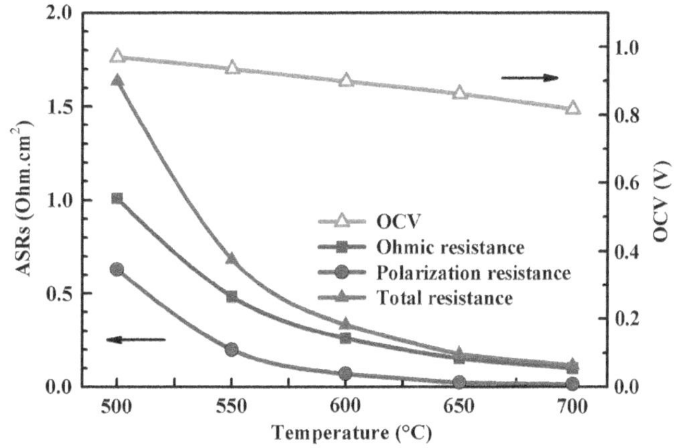

Fig. 10 Area specific resistances (ASR) and OCV

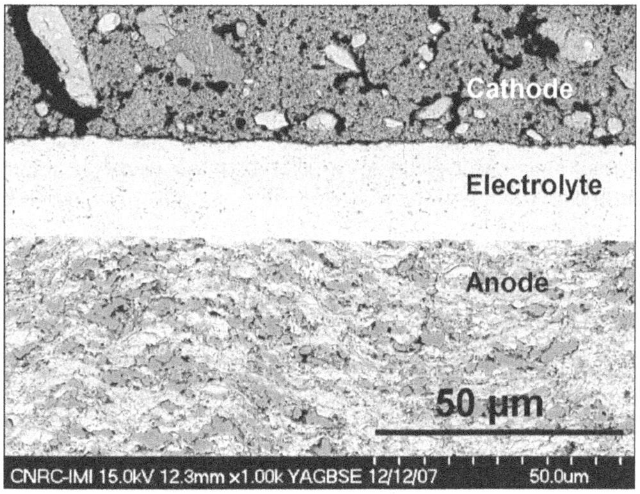

Fig. 11 SOFC after testing and thermal cycling

strong increase in overall cell resistance. The OCV remained stable throughout the tests. Further details will be presented elsewhere. Figure 11 shows the cell cross section after testing and thermal cycling. While the electrolyte appears unaltered, which is consistent with the constant OCV, a partial separation of the electrolyte from the cathode can be discerned. This degradation was previously observed in the SPS cells and was attributed to the mismatch in thermal expansion coefficient between the layers (Ref 11). Development of more suitable electrodes is in progress.

3.8 Short Stack Components (5 × 5 cm)

To assess the potential for planar SOFC stack production, half cells with rectangular dimensions of 50 × 50 mm were produced. Without optimizing for the deposition rate, the electrolyte coating was deposited in 16 min, consuming approximately 13.3 g of feedstock powder. This relatively high deposition rate, in comparison with the previous SPS tests conducted by the authors,

Fig. 12 Top-view of 50 × 50 mm half cell

may be allocated to the low flame temperature, which precludes feedstock evaporation. Figure 12 shows a top-view photograph of an as-sprayed functional electrolyte-anode-metal assembly. In spite of the relatively high surface temperature (440 °C) and the localized heating by the impinging spray jet, the samples did not show any distortion during or after spraying. Furthermore, no optical visible defects in the electrolyte coating were detected. Electrochemical characterization of the cell is in progress at the time of writing. Deposition at relatively large standoff distances (~13 cm) was made possible by the ongoing heating of the particles far outside the HVOF gun as they approach the substrate, see Fig. 2 and 6. This stands in contrast to SPS, where the small particles cool rapidly after exiting the torch and the substrate must consequently be placed in close proximity to the plasma (Ref 11). In HVOF suspension spraying, the thermal stresses and local overheating, a severe limitation for atmospheric pressure thermal spraying of SOFC electrolytes (Ref 3), promise to be manageable.

4. Conclusions

Suspension HVOF spraying (DJ-2700) for the production of ceria-based SOFC electrolytes was implemented. Using in-flight particle diagnostics, optimum spray parameters in terms of fuel-to-oxygen ratio, feedstock/suspension conditions, and standoff distance were established. The HVOF flame was found to limit evaporation and decomposition of the feedstock and to favorably affect the coating stresses, resulting in low-porosity, smooth, and virtually defect-free coatings of ~20 µm thickness. The flame characteristics allow for relatively large standoff distance to facilitate thermal management of the substrate. The cells showed high performance with a MPD of 0.49 W cm^{-2} at 600 °C and above 0.9 W cm^{-2} at 700 °C. Ongoing work focuses on improving component compatibility and long term stability.

Acknowledgments

The financial support by the National Fuel Cell and Hydrogen Program of the National Research Council of Canada is greatly acknowledged.

References

1. E. Ivers-Tiffée, A. Weber, and D. Herbstritt, Materials and Technologies of SOFC-Components, *J. Eur. Ceram. Soc.*, 2001, **21**, p 1805-1811
2. N.P. Brandon, D. Corcoran, D. Cummings, A. Duckett, K. El-Khoury, D. Haigh, R. Leah, G. Lewis, N. Maynard, T. McColm, R. Trezona, A. Selcuk, and M. Schmidt, Development of Metal Supported Solid Oxide Fuel Cells for Operation at 500-600 °C, *J. Mater. Eng. Perform.*, 2004, **13**(3), p 253-256
3. R. Henne, Solid Oxide Fuel Cells: A Challenge for Plasma Deposition Processes, *J. Therm. Spray Technol.*, 2007, **16**(3), p 381-402
4. X. Zhang, M. Robertson, C. Decés-Petit, Y. Xie, R. Hui, S. Yick, E. Styles, J. Roller, O. Kessler, R. Maric, and D. Ghosh, NiO-YSZ Cermet Supported Low Temperature Solid Oxide Fuel Cells, *J. Power Sources*, 2006, **161**, p 301-307
5. R. Gadow, A. Killinger, A. Candel Ruiz, H. Weckmann, A. Öllinger, and O. Patz, Investigation on HVOF Technique for Fabrication of SOFCs (Solid Oxide Fuel Cells) Electrolyte Layers, *Proceedings of the 2007 International Thermal Spray Conference, Global Coatings Solutions*, May 14-16 2007 (Beijing, China), ASM International, 2007, p 1053-1058
6. D. Stöver, Dag Hathiramani, R. Vaßen, and R.J. Damani, Plasma-Sprayed Components for SOFC Applications, *Surf. Coat. Technol.*, 2006, **201**, p 2002-2005
7. R. Hui, Z. Wang, O. Kessler, L. Rose, J. Jankovic, S. Yick, R. Maric, and D. Ghosh, Thermal Plasma Spraying for SOFCs: Applications, Potential Advantages, and Challenges, *J. Power Sources*, 2007, **170**, p 308-323
8. P. Fauchais, R. Etchart-Salas, C. Delbos, M. Tognonvi, V. Rat, J.F. Couder, and T. Chartier, Suspension and Solution Plasma Spraying of Finely Structured Layers: Potential Application to SOFCs, *J. Phys. D.: Appl. Phys.*, 2007, **40**, p 2394-2406
9. J. Oberste Berghaus, S. Bouaricha, J.-G. Legoux, C. Moreau, and T. Chráska, Suspension Plasma Spraying of Nano-Ceramics Using an Axial Injection Torch, *Proceedings of the 2005 International Thermal Spray Conference, Thermal Spray: Building on 100 Years of Success*, May 2-5, 2005 (Basel Switzerland), ASM International, 2005, p 1434-1440
10. J. Oberste Berghaus, J.G. Legoux, C. Moreau, F. Tarasi, T. Chraska, Mechanical and Thermal Transport Properties of Suspension Thermal-Sprayed Alumina-Zirconia Composite Coatings, *J. Therm. Spray Technol.*, 2008, **17**(1)
11. Z. Wang, J. Oberste Berghaus, S. Yick, C. Decès-Petit, W. Qu, R. Hui, R. Maric, and D. Ghosh, Dynamic Evaluation of Low-Temperature Metal-Supported Solid Oxide Fuel Cell Oriented to Auxiliary Power Units, *J. Power Sources*, 2008, **176**(1), p 90-95
12. S. Sodeoka, M. Suzuki, K. Ueno, H. Sakuramoto, R. Shibata, and M. Ando, Thermal and Mechanical Properties of ZrO_2-CeO_2 Plasma-Sprayed Coatings, *J. Therm. Spray Technol.*, 1997, **6**(3), p 361-367
13. A. Killinger, M. Kuhn, and R. Gadow, High-Velocity Suspension Flame Spraying (HVSFS), A New Approach for Spraying Nanoparticles with Hypersonic Speed, *Surf. Coat. Technol.*, 2006, **201**, p 1922-1929
14. J. Oberste Berghaus, J.-G. Legoux, C. Moreau, R. Hui, and D. Ghosh, Suspension Plasma Spraying of Intermediate Temperature SOFC Components Using an Axial Injection DC Torch, *Mater. Sci. Forum, Trans. Tech. Publisher, Switzerland*, 2007, **539-543**, p 1332-1337
15. T. Matsui, T. Kosaka, M. Inabe, A. Mineshige, and Z. Ogumi, Effects of Mixed Conduction on the Open-Circuit Voltage of Intermediate-Temperature SOFCs Based on Sm-Doped Ceria Electrolytes, *Solid State Ionics*, 2005, **176**, p 663-668

JTTEE5 17:708–714
DOI: 10.1007/s11666-008-9276-z
1059-9630/$19.00 © ASM International

Peer Reviewed

Effects of Silver Addition on Properties and Performance of Plasma Sprayed $La_{0.6}Sr_{0.4}Co_{0.2}Fe_{0.8}O_{3-\delta}$ Interconnect Layer

Soodong Park, S. Kumar, Hyuntaek Na, and Changhee Lee

(Submitted April 30, 2008; in revised form September 19, 2008)

Interconnect layers on stainless steel substrates (STS430) for solid oxide fuel cells (SOFC) were built up by atmospheric plasma spraying (APS) using spray dried $La_{0.6}Sr_{0.4}Co_{0.2}Fe_{0.8}O_{3-\delta}$ (LSCF) and blended LSCF/Ag composites. The microstructure and phase of each coating were analyzed using scanning electron microscopy (SEM) and x-ray diffraction (XRD) studies, respectively. Furthermore, bond strength, microhardness, performance in a thermal cycle test and in an oxidation test, and electrical conductivity were measured and compared. The coatings prepared from spray dried LSCF have higher porosity and more cracks within the splats and at intersplat boundaries. In contrast, the coatings prepared from LSCF/Ag had fewer cracks and less porosity due to the relatively high ductility of silver. After oxidation testing at 800 °C for 200 h, the weight change of the STS430 substrate and the LSCF and LSCF/Ag-coated alloys were found to be 0.06833, 0.01950, and 0.01656 mg/cm^2, respectively. Also the electrical conductivity of LSCF and LSCF/Ag coatings were higher than that of STS430 by two orders.

Keywords Ag, atmospheric plasma spraying, interconnect, $La_{0.6}Sr_{0.4}Co_{0.2}Fe_{0.8}O_{3-\delta}$, solid oxide fuel cell (SOFC)

1. Introduction

A solid oxide fuel cell (SOFC) is a device that directly converts chemical energy into electrical energy (Ref 1, 2). Among all components in a SOFC, the interconnect plays an important role. The primary function of the SOFC interconnect is to connect the anode of one cell to the cathode of the adjacent cell for an electrical series arrangement, thereby preventing the mixing of fuel and the oxidant gases. In general, the interconnect has to meet the following demands:

- High electronic and low ionic conductivities (electrical conductivity > 1 Scm^{-1})
- Chemical stability under the reducing and oxidizing conditions at the anode and cathode sides, respectively, of the cell at temperatures up to 1000 °C (range of oxygen activity 10^{-18} to 10^{-8} atm)
- Gas tightness
- Thermal expansion coefficient matching with that of other cell components
- Easy manufacture and low cost

In recent years, ferritic stainless steels have been developed as alternative to ceramic interconnects. However, ferritic stainless steel interconnects have some limitations such as the formation of insulating oxide scales during high-temperature operation. Chromium-based oxides tend to form on the surface of interconnects. The electrical conductivity of chromia at 800 to 1000 °C has been reported to be 1 to 5 × 10^{-2} Scm^{-1} (Ref 3, 4). Also, the vaporization of volatile chromium species from the metallic interconnect can lead to severe degradation of the electrical properties of SOFC. A description of the degradation mechanism was reported by Hilpert et al. (Ref 5). It is based on the vaporization of Cr_2O_3 on the interconnect surface as CrO_3(g) or $CrO_2(OH)_2$(g) as major gaseous species with chromium in the 6^+ oxidation state. The vapor species are reduced at the three-phase boundary by electrochemical reaction forming Cr_2O_3(s). The Cr_2O_3 or the reaction products formed inhibit the oxygen reduction necessary for the operation of the SOFC and lead to polarization losses (Ref 5, 6). Hence, ferritic stainless steel has to be protected by a coating layer suited for high-temperature SOFC operation, which can be produced for instance using APS (Ref 7).

This article is an invited paper selected from presentations at the 2008 International Thermal Spray Conference and has been expanded from the original presentation. It is simultaneously published in *Thermal Spray Crossing Borders, Proceedings of the 2008 International Thermal Spray Conference*, Maastricht, The Netherlands, June 2-4, 2008, Basil R. Marple, Margaret M. Hyland, Yuk-Chiu Lau, Chang-Jiu Li, Rogerio S. Lima, and Ghislain Montavon, Ed., ASM International, Materials Park, OH, 2008.

Soodong Park, S. Kumar, Hyuntaek Na, and **Changhee Lee**, Division of Advanced Materials Science & Engineering, Hanyang University, 17 Haengdang-dong, Seongdong-ku, Seoul 133-791, South Korea. Contact e-mails: chlee@hanyang.ac.kr, airtwenty4@hanyang.ac.kr, hai_kumars@rediffmail.com and nht6480@hanyang.ac.kr

Many different coating technologies have been applied to fabricate coating layers on metallic interconnects, including electrochemical vapor deposition (EVD), chemical vapor deposition (CVD), physical vapor deposition (PVD), tape casting, screen printing, sol-gel, and thermal spraying methods (Ref 8-13). Thermal spraying techniques, especially atmospheric plasma spraying (APS), seem to be economically attractive and effective techniques for industrial production of SOFCs due to its low cost, easy operation, high deposition efficiency, ability to coat large surfaces, wide selection of materials, and so forth. In this study, coating layers for SOFC interconnects were deposited using APS.

A high density of pores and cracks were obtained in the thermally sprayed ceramic coating layers produced in a previous study. Therefore, the cracks were interconnected through vertical microcracks in the individual splats within the coating, allowing gas or liquids to propagate throughout the coating. This leads to the generation of oxide scales on metallic interconnects during long-term operation of the SOFC. In this study, silver powder was added to the ceramic feedstock as a crack-reducing means.

Table 1 The APS process parameters

Plasma gas composition				
Ar, SCFH (a)	H_2, SCFH	Arc current, A	Distance, mm	Powder feed rate, g/min
100	5/10/15	500	100	10

(a) SCFH, standard cubic feet per hour (1 SCFH = 0.47 lpm)

2. Experiments

2.1 Atmospheric Plasma Spraying

Two kinds of coatings were deposited by an atmospheric plasma spray system (Sulzer Metco 9 MB, Switzerland) from different feedstock materials. A mixture of argon and hydrogen served as plasma process gas for these experiments, where argon was also used as powder carrier gas. The spraying distance between the substrate and the exit of the nozzle was maintained at 100 mm. The feedstock powder was vertically injected into the plasma jet center downstream of the exit with a feed rate of 10 g/min. In order to change the characteristics of the plasma jet, the hydrogen gas flow rate was changed while the other process parameters were kept constant. The processing parameters are tabulated in Table 1. The feedstock powders were spray deposited on grit-blasted STS430 stainless steel plates.

2.2 Feedstock Characterization

Prior to spraying, characterization of the blended powder feedstock was performed. Feedstock powders used for this study were spray dried $La_{0.6}Sr_{0.4}Co_{0.2}Fe_{0.8}O_{3-\delta}$ (LSCF) and blended LSCF/Ag. LSCF/Ag composite powder with 25 wt.% of Ag were produced via mechanical mixing using the mixer TURBULA (WAB, Willy A, Bachofen AG Maschinenfabrik, Basel, Switzerland) with a rotational speed of 30 rpm. The mixing time was 5 h. The characterization of the powder feedstocks were performed using scanning electron microscopy (SEM). Figure 1 shows SEM

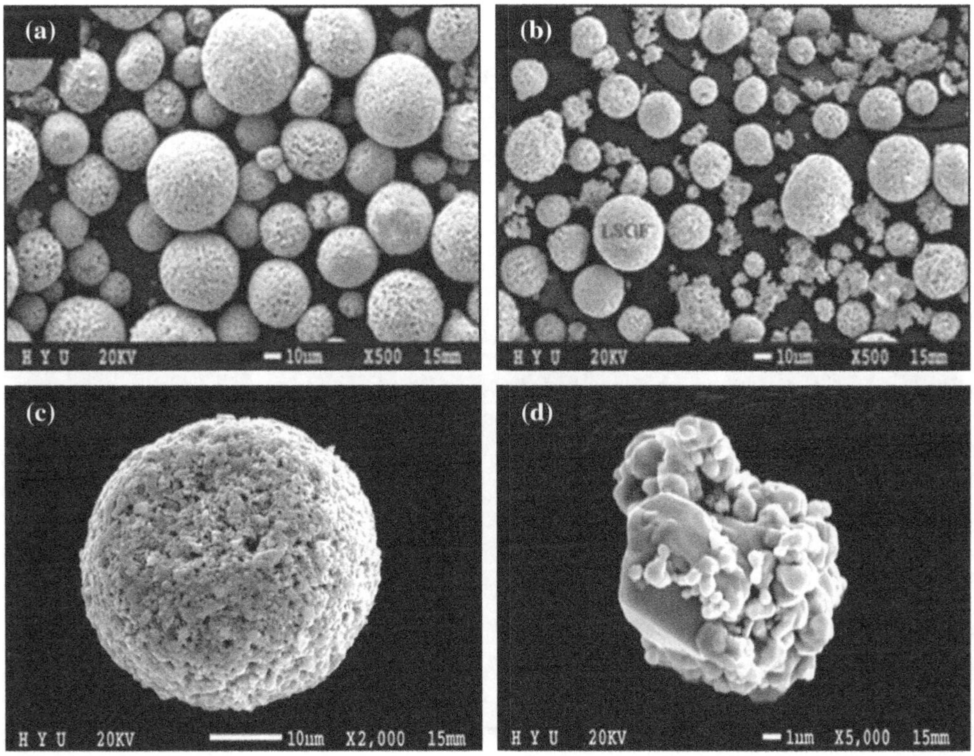

Fig. 1 Scanning electron images of feedstock materials. (a) $La_{0.6}Sr_{0.4}Co_{0.2}Fe_{0.8}O_{3-\delta}$. (b) LSCF/Ag. (c) Individual LSCF powder particle. (d) Individual silver powder particle

images of each feedstock powder. Figure 1(a) shows the SEM image of the spray dried LSCF powder. Figure 1(b) displays the SEM image of LSCF blended with Ag powder. Figures 1(c) and (d) are individual LSCF and silver powder particles, respectively. The morphology of LSCF and silver particles are nearly spherical and dendritic, respectively. The average particle sizes of LSCF and Ag powders are 30 to 40 μm and 10 to 40 μm, respectively. The characteristics of the feedstock are tabulated in Table 2.

2.3 Characterization of Coating

The microstructure of the coatings was observed through scanning electron microscopy (SEM) with a JEOL microscope, and the phase compositions of each condition were measured using a D/MAX-2500/PC x-ray diffractometer (XRD). The porosity of the coatings was measured using an image analysis method, and the area fraction of the porosity was calculated by Image Pro-Plus software. Cross sections of coatings were polished as smooth as a mirror. Up to 20 SEM images (magnification: 300×) of the cross sections of each sample were taken for the analysis. The software Image-Pro Plus can be used to estimate the percentage of open pores in a selected area. The software used for image analysis procedure is Image-Pro Plus 6.0 (Media Cybernetics, USA).

Table 2 Characteristics of the feedstock

Powder	Size, μm	Morphology	Density, g/cm^3	Melting point, °C
LSCF	−40 + 30	Spherical	2.30	...
Silver	−40 + 10	Irregular	10.49	961.93

Vickers microhardness was measured on polished surfaces using an applied load of 1.961 N. Bond strength was measured using Romulus Bond Strength tester. For the thermal cycle test, the samples were put in a furnace at 800 °C for 1 h and then quenched in water. Twenty continuous cycles were performed on each sample. Oxidation tests were performed in the same way. The samples were heat treated at 800 °C for 20 h and allowed to quench in air. After measuring the weight gain, the samples were put again in the furnace and the same process was done for 200 h with 20 h intervals. Samples of STS430 substrate coatings prepared from LSCF and LSCF/Ag blend were oxidized at 800 °C for 200 h for measuring the electrical conductivity using direct current (DC) quasi-four-probe method. Keithley 2430 Source Meter and 2000 Digital Multi-Meter were used for this measurement, and the samples were tested for 80 h at 800 °C. Samples used for this study were 5 × 5 × 1 mm. After oxidation at 800 °C for 200 h, the electrical conductivity of coated STS430 samples was measured.

3. Results and Discussion

3.1 Microstructure

Figure 2 shows the SEM images of the as-sprayed coatings prepared from LSCF (a) to (c) and LSCF/Ag powder (d) to (f) with different hydrogen gas flow rates. In the case of LSCF coatings, the porosity decreases and crack content is reduced with increasing hydrogen gas flow rate. However, the cracks could not be controlled for all

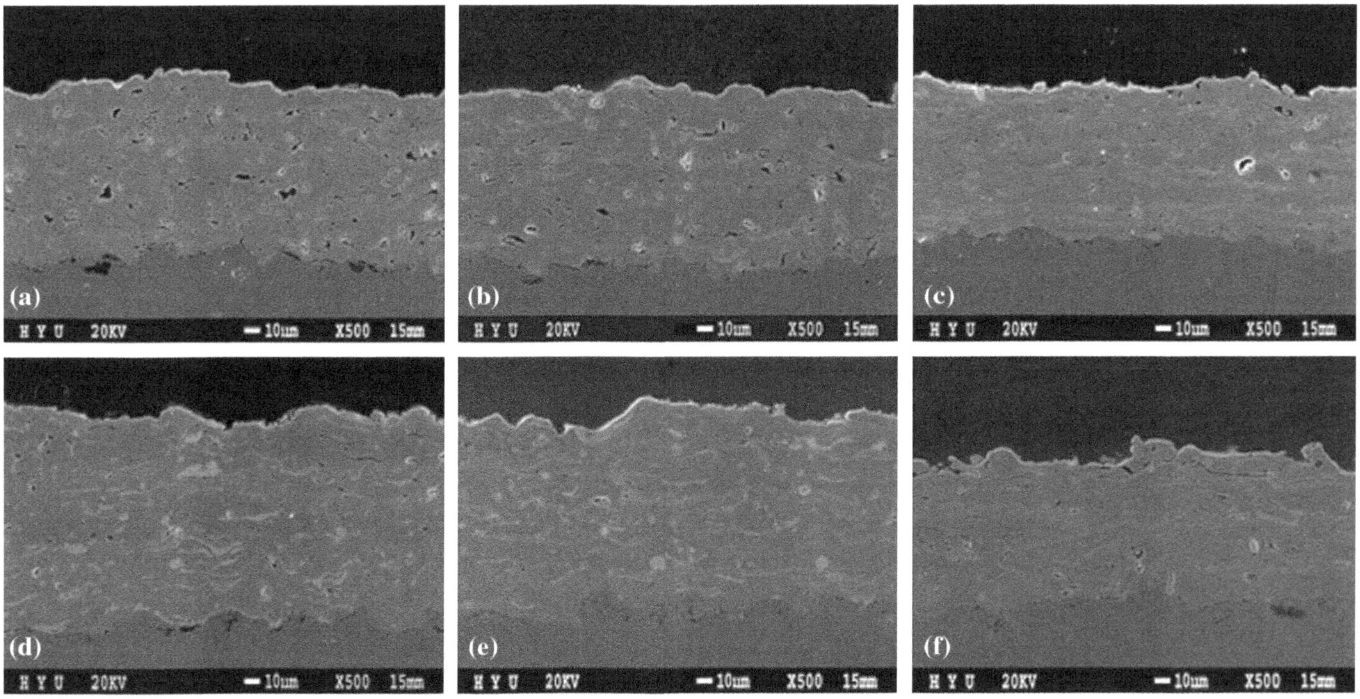

Fig. 2 Cross-sectional SEM images of LSCF and LSCF/Ag coatings at different hydrogen flow rates (SCFH). (a) 5H_2 LSCF. (b) 10H_2 LSCF. (c) 15H_2 LSCF. (d) 5H_2 LSCF/Ag. (e) 10H_2 LSCF/Ag. (f) 15H_2 LSCF/Ag

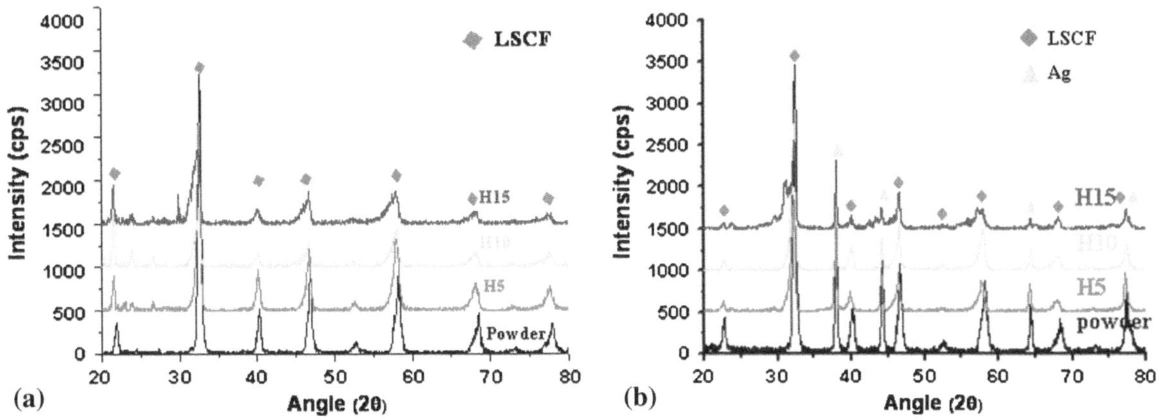

Fig. 3 X-ray diffraction patterns of feedstock and coatings, sprayed with different hydrogen content in the plasma. (a) LSCF and (b) LSCF/Ag

process parameters. Hence, silver was added to the LSCF as crack reducing agent. When silver is added, porosity increases with increasing hydrogen gas flow rate. Especially at a hydrogen flow rate of 15 SCFH, the silver content in the coating was reduced abruptly as consequence of increased silver evaporation. At 5 and 10 SCFH hydrogen gas flow rates, the quantity of silver remaining in the coating was 4.81 and 3.98%, respectively. In case of 15 SCFH, the Ag content was reduced to 1.06% because of the higher enthalpy of the plasma jet.

Figures 3(a) and (b) show the XRD patterns of plasma sprayed LSCF and LSCF/Ag coatings at different processing conditions. X-ray diffraction patterns indicate that peaks are almost similar for all conditions with two exceptions: with the high hydrogen content in the plasma (15 SCFH) the LSCF starts to decompose as seen by the new small peak left of the main peak at about 30°, and in the case of LSCF with Ag the peak intensity for silver is reduced because of high plasma enthalpy resulting in evaporation.

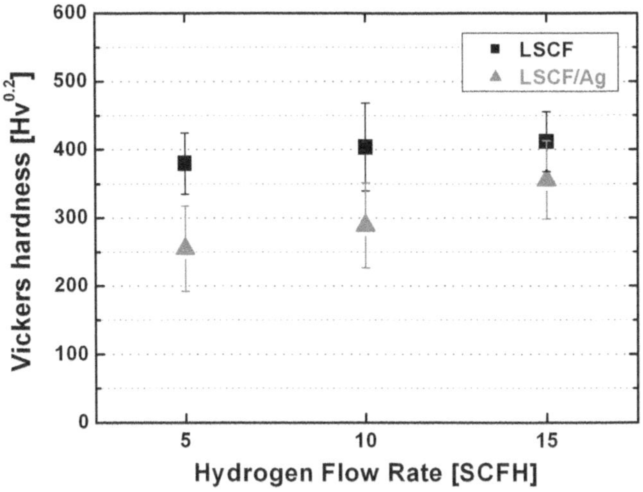

Fig. 4 Vickers microhardness values for coatings as a function of hydrogen gas flow rate prepared from LSCF and LSCF/Ag powders

3.2 Mechanical Properties

Figure 4 shows the microhardness results for the coatings prepared from different feedstock powders with different plasma parameters. From Fig. 4 it becomes clear that increasing hydrogen gas flow rate increases the microhardness values for both feedstocks. The microhardness values for coatings prepared from LSCF powder at 5, 10, and 15 SCFH hydrogen flow rates are 380, 404, and 412 HV, respectively. The values for LSCF with silver are 255, 289, and 356 HV, respectively. It is also clear that adding silver with LSCF reduces the microhardness values for all plasma parameters. However, the difference is diminished at higher hydrogen gas flow rates again as a result of increased silver evaporation.

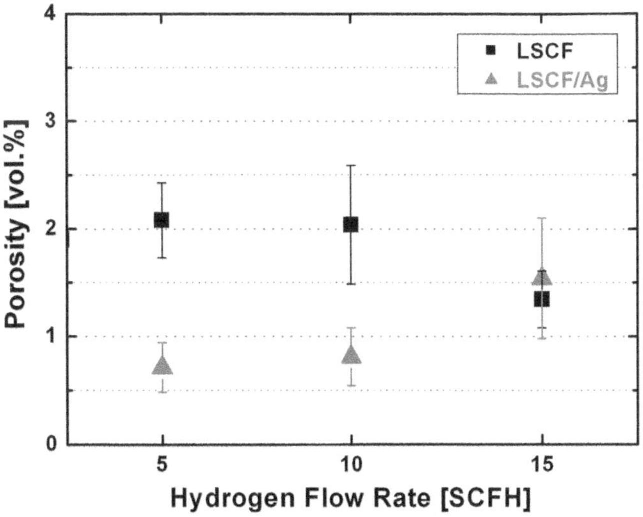

Fig. 5 Porosity of coatings as a function of hydrogen gas flow rate prepared from LSCF and LSCF/Ag powder

Figure 5 shows the porosity values for the coatings prepared with both powders. For coatings prepared from LSCF powder at 5, 10, and 15 SCFH, the porosity values are 2.08, 2.038, and 1.347%, respectively. The values for coatings prepared from LSCF with silver are 0.713, 0.813, and 1.542%, respectively. It is clear from the data that with increasing hydrogen flow rate the porosity decreases in the case of LSCF powder. The result is opposite for LSCF with silver. This effect is clearly due to the silver evaporation. However, the porosity value of the coating is lower for the powder with silver addition. Figure 6 shows the bond strength data of the coatings. The bond strength values for the coatings prepared from LSCF with silver were higher for all the applied plasma parameters.

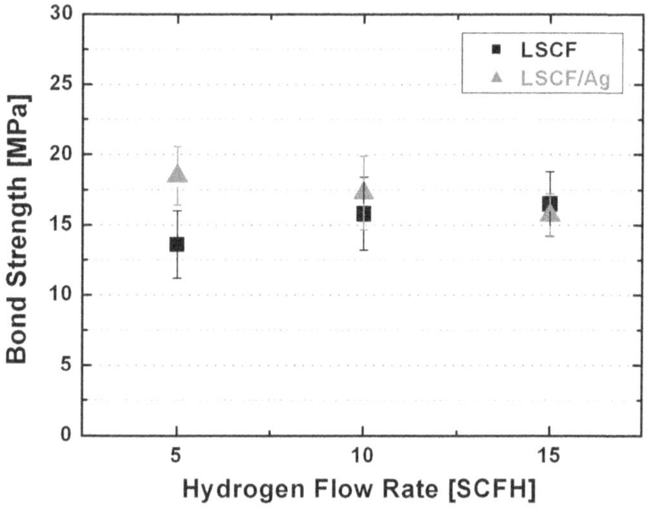

Fig. 6 Bond strength values of coatings as a function of hydrogen gas flow rate prepared from LSCF and LSCF/Ag powders

3.3 Thermal Cycle Test

Figures 7(a) and (c) show the SEM images of the coatings prepared at 5 SCFH of hydrogen from LSCF and LSCF with silver after thermal cycle testing. It is obvious that after testing coatings prepared from LSCF powder have many pores and cracks. The cracks are interconnected with each other, which leads to propagation effects. Coatings prepared from LSCF with silver show lower porosity and very few cracks. Figures 7(b) and (d) show the magnified images of the respective coatings. The silver layer can be seen in the coating where it acts as a crack-reducing agent.

3.4 Oxidation Test

Figure 8 shows the results of the oxidation test. In the figure, the weight gain caused by oxidation is illustrated for STS430 (the substrate), and for such substrates with a coating prepared from LSCF powder and from LSCF with silver, as well. It can be seen from the figure that the weight gain abruptly increased for the substrate material within 20 h. There was no such dramatic increase observable in the case of substrates with coatings prepared from LSCF and LSCF with silver. After 20 h, the weight gain was gradual for all the three samples. The change in weight for STS430 and substrates with coatings prepared from LSCF and LSCF with silver were 0.06833, 0.0195, and 0.01656 mg/cm^2, respectively. Hence, it is clear that addition of silver to the LSCF powder increases the resistance against oxidation of the coatings. Figures 9(a),

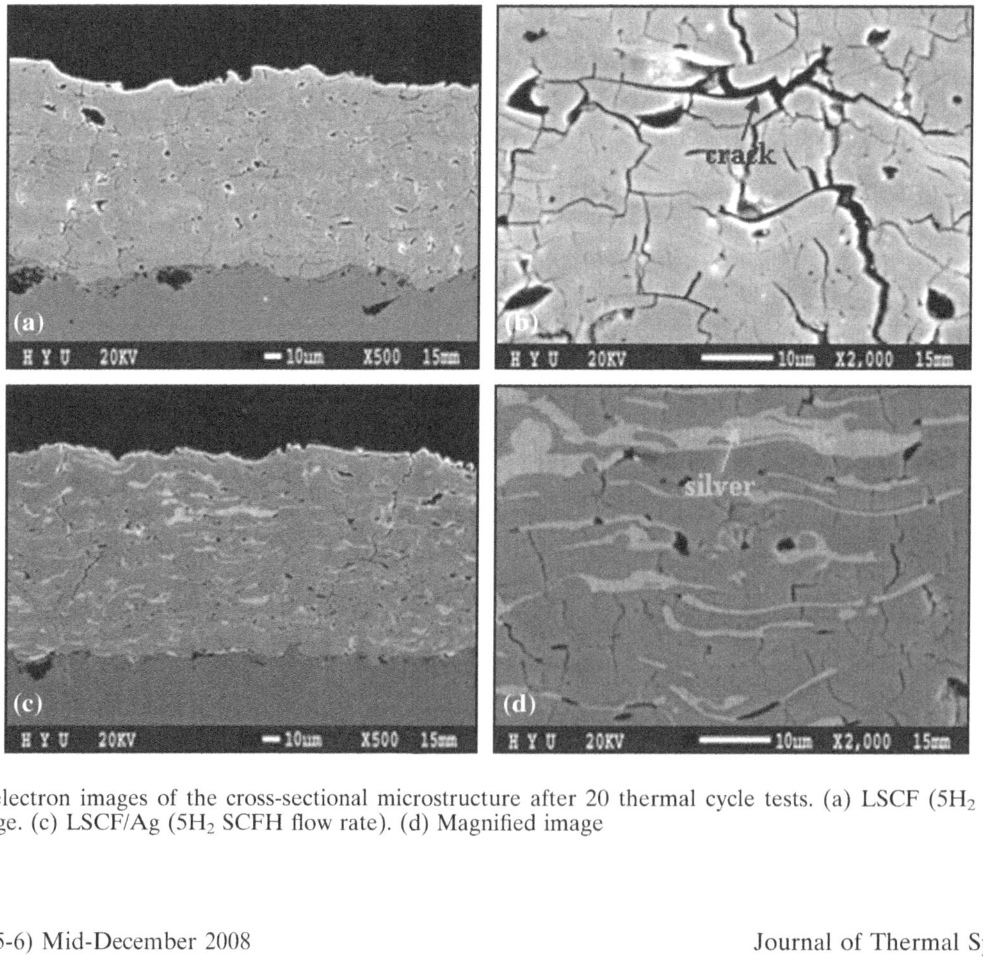

Fig. 7 Scanning electron images of the cross-sectional microstructure after 20 thermal cycle tests. (a) LSCF (5H_2 SCFH flow rate). (b) Magnified image. (c) LSCF/Ag (5H_2 SCFH flow rate). (d) Magnified image

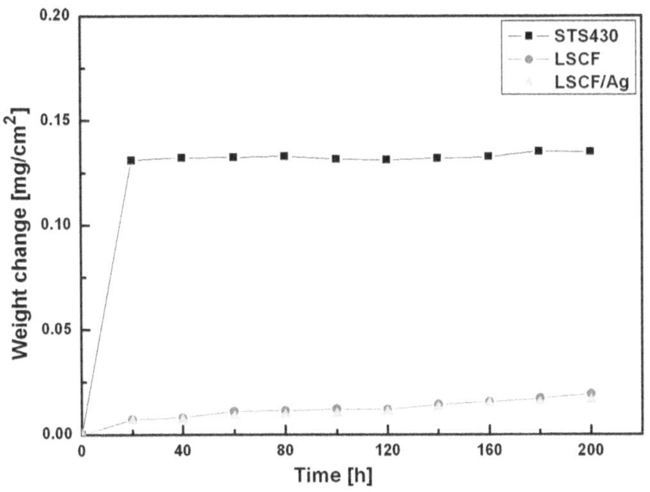

Fig. 8 Weight change with time during oxidation of STS430 alloy, alloy coated with LSCF and LSCF/Ag at 800 °C in air

(b), and (c) show the top view of the oxidized samples of STS430 and coatings prepared from LSCF and LSCF/Ag, respectively. Figures 9(d) to (f) show the x-ray diffraction results of oxidized and as-sprayed coatings. The uncoated substrate oxidized severely and formed chromium, iron, and manganese based oxides. However, coatings prepared from LSCF and LSCF/Ag show no significant phase changes after oxidation test.

3.5 Electrical Conductivity

Figure 10 shows the high-temperature transversal electrical conductivity of STS430 substrate and coatings prepared from LSCF and LSCF/Ag after oxidation at 800 °C for 200 h. It can be seen from the figure that for coatings, the conductivity values are much higher than for the STS430 substrate alone. Up to a certain time period, increasing time does not affect the conductivity value of the substrate. At the earlier state of the process, the oxides

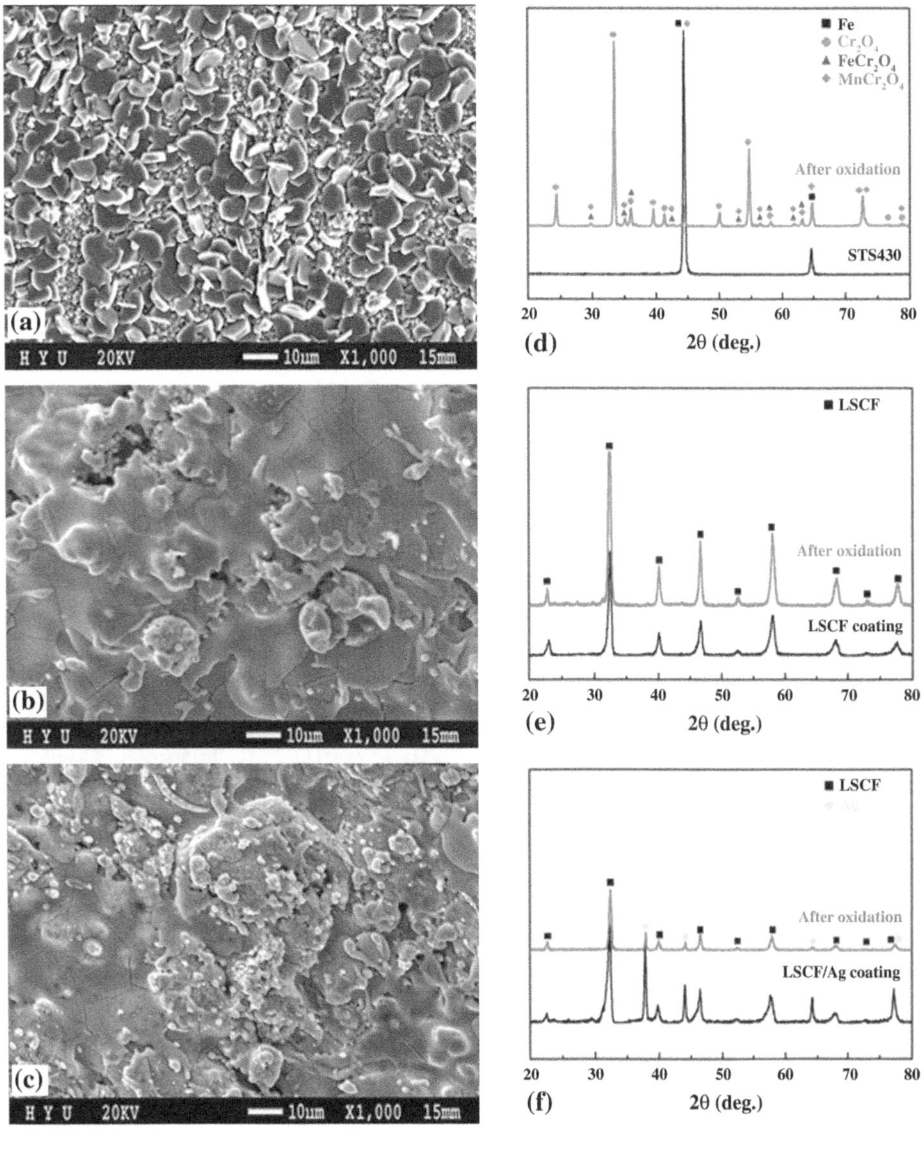

Fig. 9 Scanning electron microscopy images and x-ray patterns of STS430 alloy. (a) and (b), LSCF coated (c) and (d) and LSCF/Ag coated alloy (e) and (f), respectively, after the oxidation test

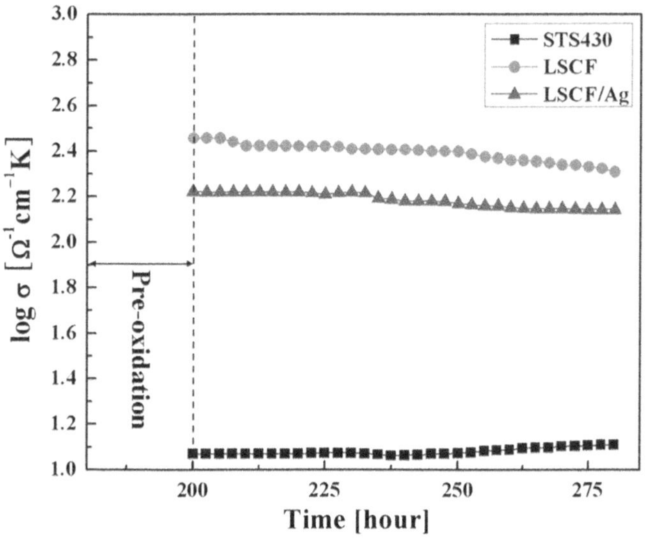

Fig. 10 Transversal electrical conductivity of STS430, and STS430 coated with LSCF and LSCF/Ag, respectively, measured at 800 °C in air

present in the substrate consist of Cr_2O_3, mostly. With increasing time, the formation of $(Mn,Cr)_3O_4$, which has a spinel structure, takes place. Hence the conductivity increases slightly. Coatings prepared from LSCF have a relatively higher conductivity than that of LSCF/Ag. LSCF has a perovskite structure in which the electrical conductivity is lattice defect dominated. Hence, the addition of silver to the LSCF decreases the electrical conductivity. The electrical conductivity after 200 h annealing time of the STS430 substrate was 1.08737 log $[\Omega^{-1}\ cm^{-1}\ K]$ and the same for STS430/LSCF and for STS430/LSCF/Ag were 2.36211 log $[\Omega^{-1}\ cm^{-1}\ K]$ and 2.15488 log $[\Omega^{-1}\ cm^{-1}\ K]$, respectively. The reduction of the electrical conductivity for LSCF and LSCF/Ag coatings between 200 and 280 h interval were 0.05203 log $[\Omega^{-1}\ cm^{-1}\ K]$ and 0.01016 log $[\Omega^{-1}\ cm^{-1}\ K]$, respectively. Hence for long-term application, LSCF in addition to silver seems to be advantageous.

4. Conclusions

Interconnect layers were built up by coating ferritic stainless steel substrates (STS430) by atmospheric plasma spraying (APS) using spray dried LSCF and blended LSCF/Ag powders. The coatings with silver addition were denser with fewer cracks and lower porosity than that of powders without silver. After oxidation testing for 200 h, the change in weight for substrates coated with LSCF and LSCF with silver were 0.0195 and 0.01656 mg/cm^2, respectively. The compactness and oxidation resistance at 800 °C of LSCF coatings can be significantly promoted by adding the ductile Ag. This is due to the effective barrier to inward oxygen migration by the tamping effect of Ag at the interparticle boundaries within the coating.

Acknowledgment

This work was financially supported by the Korea Science and Engineering Foundation (KOSEF) through the National Research Laboratory (NRL) program (No. 2006-02289).

References

1. N.Q. Minh and T. Takahashi, Selected Reference Relevant to Solid Oxide Fuel Cell Technology, *Sci. Technol. Ceram. Fuel Cells,* 1995, p 351-353
2. N.Q. Minh, A. Anumakakonda, B. Chung, R. Doshi, J. Ferrall, G. Lear, K. Montgomery, E. Ong, L. Schipper, and J. Yamanis, High-Performance, Reduced-Temperature SOFC Technology, *Fuel Cells Bull.*, 1999, **2**, p 9-11
3. A. Holt and P. Kofstadt, Electrical Conductivity and Defect Structure of Mg-Doped Cr_2O_3, *Solid State Ionics*, 1997, **100**, p 201-209
4. A. Holt and P. Kofstadt, Electrical Conductivity of Cr_2O_3 Doped with TiO_2, *Solid State Ionics*, 1999, **117**, p 21-25
5. D. Das, M. Miller, H. Nickel, and K. Hilpert, Chromium Evaporation from SOFC Interconnector Alloys and Degradation Process by Chromium Transport, *Europ. SOFC Forum*, 1994, **2**, p 703
6. K. Hilpert, D. Das, M. Miller, D.H. Peck, and R. Weiß, Chromium Vapour Species over SOFC Interconnect Materials and their Potential for Degradation Processes, *J. Electrochem. Soc.*, 1996, **143**, p 3642
7. R. Zheng, X.M. Zhou, S.R. Wang, T.L. Wen, and C.X. Ding, A Study of Ni + 8YSZ/8YSZ/$La_{0.6}Sr_{0.4}CoO_{3-\delta}$ ITSOFC Fabricated by Atmospheric Plasma Spraying, *J. Power Sources*, 2005, **140**, p 217-225
8. A.O. Isenberg, Energy Conversion via Solid Oxide Electrolyte Electrochemical Cells at High Temperatures, *Solid State Ionics*, 1981, **3-4**, p 431-437
9. J. Will, A. Mitterdorfer, C. Kleinlogel, D. Perednis, and L.J. Gauckler, Fabrication of Thin Electrolytes for Second-Generation Solid Oxide Fuel Cells, *Solid State Ionics*, 2000, **131**, p 79-96
10. D. Simwonis, H. Thulen, F.J. Dias, A. Naoumidis, and D. Stöver, Properties of Ni/YSZ Porous Cermets for SOFC Anode Substrates Prepared by Tape Casting and Coat-Mix Process, *J Mater. Process Technol.*, 1999, **92-93**, p 107
11. J. Van Herle, R. Ihringer, R. Vasquez Cavieres, L. Constantin, and O. Bucheli, Anode Supported Solid Oxide Fuel Cells with Screen-Printed Cathodes, *J. Europ. Ceram. Soc.*, 2001, **21**, p 1855-1859
12. W. Qu, J. Li, and D.G. Ivey, Sol-Gel Coatings to Reduce Oxide Growth in Interconnects Used for Solid Oxide Fuel Cells, *J. Power Sources*, 2004, **138**, p 162-173
13. O. Kwon, S. Kumar, S. Park, and C. Lee, Comparison of Solid Oxide Fuel Cell Anode Coatings Prepared from Different Feedstock Powders by Atmospheric Plasma Spray Method, *J. Power Sources*, 2007, **171**, p 441-447

JTTEE5 17:715–720
DOI: 10.1007/s11666-008-9225-x
1059-9630/$19.00 © ASM International

Peer Reviewed

Formation of NiAl Intermetallic Compound by Cold Spraying of Ball-Milled Ni/Al Alloy Powder Through Postannealing Treatment

Qiang Zhang, Chang-Jiu Li, Xiu-Ru Wang, Zhi-Liang Ren, Cheng-Xin Li, and Guan-Jun Yang

(Submitted May 13, 2008; in revised form July 23, 2008)

Ni/Al alloy powders were synthesized by ball milling of nickel-aluminum powder mixture with a Ni/Al atomic ratio of 1:1. Ni/Al alloy coating was deposited by cold spraying using N_2 as accelerating gas. NiAl intermetallic compound was evolved in situ through postspray annealing treatment of cold-sprayed Ni/Al alloy coating. The effect of annealing temperature on the phase transformation behavior from Ni/Al mechanical alloy to intermetallics was investigated. The microstructure of the mechanically alloying Ni/Al powder and NiAl coatings was characterized by scanning electron microscopy and x-ray diffraction analysis. The results show that a dense Ni/Al alloy coating can be successfully deposited by cold spraying using the mechanically alloyed powder as feedstocks. The as-sprayed alloy coating exhibited a laminated microstructure retained from the mechanically alloying powder. The annealing of the subsequent Ni/Al alloy coating at a temperature higher than 850 °C leads to complete transformation from Ni/Al alloy to NiAl intermetallic compound.

Keywords annealing, cold spraying, intermetallics, mechanical alloying, NiAl

1. Introduction

The NiAl intermetallic compounds, owing to their superior properties such as high-melting point, high-creep strength, low density, high corrosion, and oxidation resistance, are regarded as promising candidates for the development of the next generation high-performance high-temperature structural materials (Ref 1, 2). These characteristics have resulted in the widespread use of NiAl-based bond coats in TBC systems.

The diffusion process is generally utilized to fabricate NiAl on Ni-based superalloy by diffusing Al, either as a separate thermal treatment or combined with the aluminizing step. The aluminizing is performed using a pack cementation (Ref 3-5) or a chemical vapor deposition (CVD) process (Ref 6-8). Although these methods have successfully generated nickel aluminide layers, the processes are complex. These reaction-diffusion methods result in significant interdiffusion of the materials added to the surface and the underlying substrate alloy. This can compromise the coating if substrate alloy elements, such as W, Ta, Ti, or S, are able to diffuse from the substrate into the nickel aluminide layer (Ref 8, 9).

To address these difficulties, thermal spraying was employed to deposit NiAl coat using NiAl intermetallic as feedstocks (Ref 10-12). However, oxidation during spraying modified the compositions of materials. Cold spray emerged as a new coating process. A coating is formed through plastic deformation of spray particles in solid state during impact in cold spraying. The temperature of spray particles prior to impact is much lower than their melting point and spray materials experience little microstructure change, oxidation, or decomposition. Therefore, the deposition of many metals such as Ti (Ref 13), Zn (Ref 14), Cu (Ref 15), Fe/Si alloy (Ref 16), Fe/Al alloy (Ref 17), MCrAlY (Ref 18), and Ni/Al composite (Ref 19) have confirmed the feasibility to coating. However, it is difficult to directly deposit intermetallic compounds alloy due to their low-temperature brittleness. In previous report (Ref 17), cold spraying of Fe/Al alloy followed by postannealing has been employed to successfully fabricate FeAl intermetallic coating.

In this paper, fabrication of NiAl intermetallic compound coating was investigated through postannealing treatment of cold-sprayed Ni/Al alloy coating.

2. Experimental Materials and Procedures

The commercially available Ni (99.8 wt.%, 89 μm, Jinchuan Group Ltd., Jinchang, China) and Al (99.5 wt.%,

This article is an invited paper selected from presentations at the 2008 International Thermal Spray Conference and has been expanded from the original presentation. It is simultaneously published in *Thermal Spray Crossing Borders, Proceedings of the 2008 International Thermal Spray Conference*, Maastricht, The Netherlands, June 2-4, 2008, Basil R. Marple, Margaret M. Hyland, Yuk-Chiu Lau, Chang-Jiu Li, Rogerio S. Lima, and Ghislain Montavon, Ed., ASM International, Materials Park, OH, 2008.

Qiang Zhang, Chang-Jiu Li, Xiu-Ru Wang, Zhi-Liang Ren, Cheng-Xin Li, and **Guan-Jun Yang**, State Key Laboratory for Mechanical Behavior of Materials, School of Materials Science and Engineering, Xi'an Jiaotong University, Xi'an, Shaanxi 710049, P.R. China. Contact e-mail: licj@mail.xjtu.edu.cn.

Table 1 Cold spray conditions

Driving gas	N_2
Powders carrying gas	N_2
Accelerating gas pressure, MPa	2.0
Powders carrying gas pressure, MPa	2.5
Accelerating gas temperature, °C	300
Gun traverse speed over substrate, mm/s	100
Standoff distance, mm	20

74 μm, Youxinglian Nonferrous Metals Ltd., Beijing, China) powders were blended with a Ni/Al atomic ratio of 1:1. The powders were mixed with stainless balls of 6.4 mm in diameter at a ball to powder charge mass ratio of 30:1 and were milled at a rotation speed of 180 rpm in a modified attrition mill (Union Process 01-ST) for 8 h in liquid nitrogen atmosphere. Liquid nitrogen was continuously introduced into the cylinder of the mill during milling operation to ensure complete immersion of the powders in the nitrogen. The cryogenic milled powder was then milled in ethanol to regulate the particle size to be suitable for cold spraying. The milled powder was sieved to a size less than 50 μm and used as the feedstock for spraying deposition. The size distribution of the milled powder was determined by a laser diffraction sizer (MASTERSIZER 2000, Malvern Instruments Ltd., UK). The oxygen contents of the milled powder and corresponding coating were determined using oxygen determinator (Ro-316, LECO, USA). Chemical analysis of the as-sprayed coating was performed using inductively coupled plasma-optical emission spectrometer (ICP) (IRIS-Intrepid II XSP, PE, USA) and all analysis were conducted in triplicate.

The homemade cold spray system which consisted of nozzle, gas heater, and high-pressure gas supply unit was used for coating deposition tests. The setup of the system has been described in detail elsewhere (Ref 13). The particles were accelerated through a De Laval type of nozzle of 100 mm long with a circular cross section with an exit diameter of 6 mm and throat diameter of 2 mm. During cold spraying, all spray parameters were fixed, as shown in Table 1. In the present study, nitrogen gas was used as the driving gas and powders carrying gas. The accelerating gas was operated at a prechamber pressure of 2.0 MPa and a preheating temperature of 300 °C. The standoff distance from the nozzle to the substrate was 20 mm. During deposition, the spray gun was manipulated by a robot at a traverse speed of 100 mm/s relative to the Inconel 738 substrate. The as-sprayed coating was then annealed at different temperatures in a furnace in argon atmosphere at a heating rate of 10 °C/min.

The phase structure of Ni/Al alloy powder and as-sprayed coating and annealed coatings were analyzed by x-ray diffraction diffractometer (XRD) (XRD-6000, Shimadzu, Kyoto, Japan) using Cu-$K\alpha$ radiation. The cross-sectional microstructure and compositions of coatings were characterized by a scanning electron microscope (SEM, VEGA II-XMU, TESCAN, Czech) with an energy dispersive spectrometer (EDS).

Fig. 1 Typical morphology (a) and cross-sectional microstructure (b) of the milled powder for 8 h

3. Results and Discussion

3.1 Morphology and Microstructure of the Milled Powder

Figure 1 shows typical SEM images of the morphology and cross-sectional microstructure of the powders milled for 8 h. It can be noted in Fig. 1(a) that the as-milled powder presented an irregular morphology, although both Ni and Al starting powder presented a spherical morphology. The milled powder for spray has a mean particle size of 46.2 μm, which was suitable for cold spraying. The powder morphology is typical of milled powders and is generally attributed to the continuous welding and fracturing of particles caused by the collision of balls. It can be noted in Fig. 1(b) that a fine lamellar structure is present in the powder. Two distinguishable regions with different microstructural characteristics can be clearly observed.

The nitrogen, iron, and oxygen contamination were introduced into the powders as a consequence of the ball milling process. The contents of nitrogen, iron, and oxygen were 0.51, 0.14, and 0.75 wt.%, respectively. The presence of iron possibly resulted from the stainless steel balls and stainless steel cylinder. The nitrogen contamination is

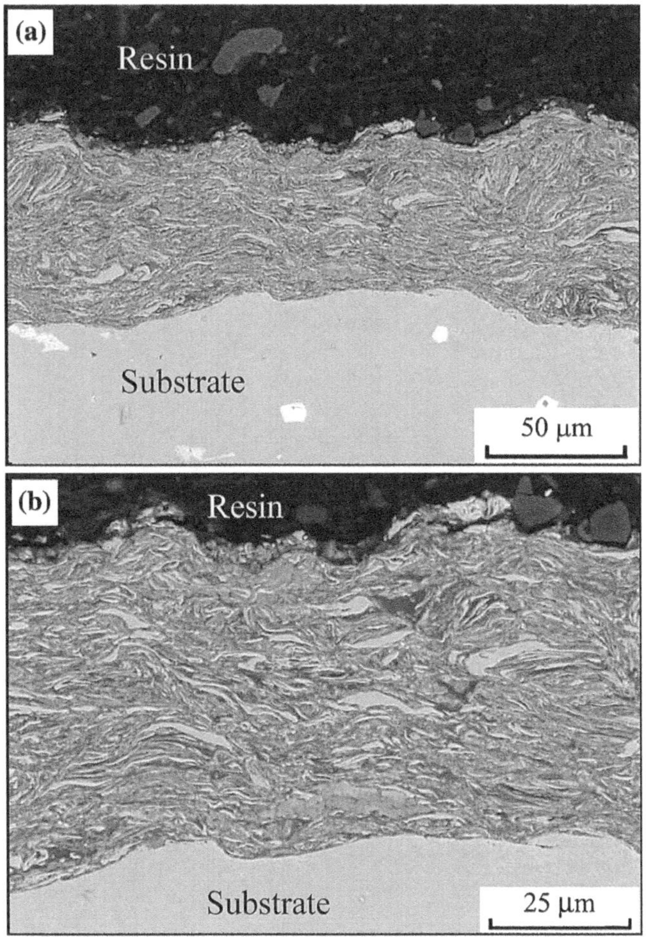

Fig. 2 Cross-sectional microstructure of cold-sprayed Ni/Al alloy coating observed at different magnifications

attributed to inclusion during milling from the milling environment. An increase of oxygen concentration is believed to be caused by the incorporation of the leaking air during milling or storage.

3.2 Microstructure of Cold-Sprayed Ni/Al Alloy Coating

Figure 2 shows typical SEM images of cross-sectional microstructure of cold-sprayed Ni/Al alloy coating using milled Ni/Al alloy powder. It was observed that the coating exhibited a dense microstructure and some apparent thick layers with a white contrast appeared on the coating microstructure. XRD patterns of the as-sprayed coating and feedstock powder are shown in Fig. 3. Only the peaks of nickel and aluminum were identified in XRD pattern of the ball-milled powder. It is clear that the XRD pattern of the coating is almost the same as that of the milled powder. This fact indicates that the coating and feedstock exhibited the same phase structure and no oxide was identified in the powder and the coating by XRD. The oxygen content of cold-sprayed coating was determined to be 0.78 wt.% using the oxygen determinator, which is similar to that of the powder. Therefore, the particles in flight were subjected to little

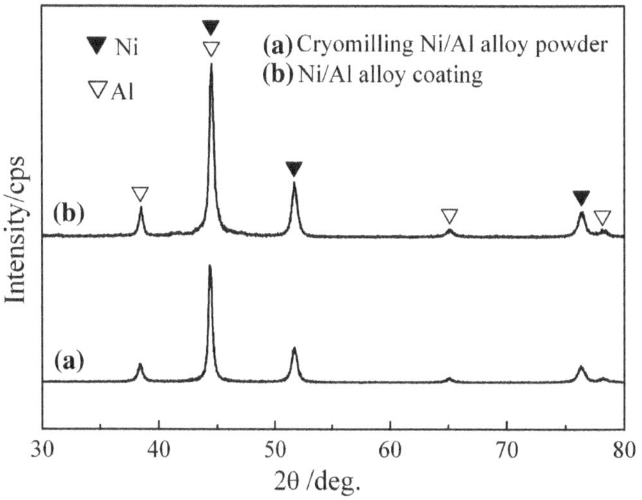

Fig. 3 XRD patterns of the feedstock powder fabricated by ball milling and the as-sprayed coating

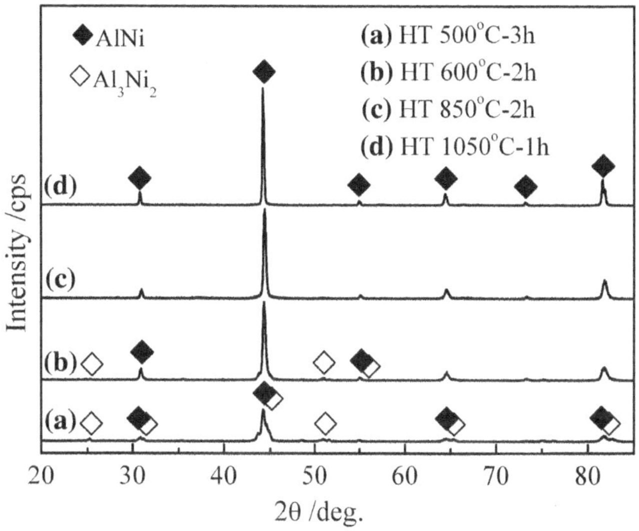

Fig. 4 XRD patterns of the Ni/Al alloy coatings annealed at different temperatures

oxidation. In cold spraying, the particle deposition takes place in a solid state. Consequently, the lamellar structure of the milled powder will be completely retained in the coating, giving a unique effect on the microstructure and properties of the cold-sprayed coating. According to EDS analysis of the coating, the thicker layer in a white contrast was a Ni-rich phase and the fine lamella was a Ni-Al solid solution with high-Al content.

3.3 The Evolution of the NiAl Intermetallic Compound Coating

The XRD patterns of annealed Ni/Al alloy coatings are shown in Fig. 4. After annealing treatment at 500 °C for 3 h, Ni and Al peaks completely disappeared and peaks corresponding to Ni_2Al_3 and NiAl appeared (Fig. 4a), compared to the as-sprayed NiAl alloy coating (Fig. 3b).

With annealing temperature rising to 600 °C, it can be found from Fig. 4(b) that NiAl became the main phase, only minor Ni_2Al_3 exists. As the temperature was raised to 850 °C, the diffraction peaks of Ni_2Al_3 disappeared completely, and only diffraction peaks of NiAl phase were present in the XRD pattern, as shown in Fig. 4(c). This fact indicates that the annealing at temperature of higher than 850 °C completely converts Ni/Al alloy to NiAl intermetallic compound. As the temperature reached 1050 °C, no additional reaction was detected, and the NiAl phase was present in the coating (Fig. 4d).

The evolution microstructures of the annealed Ni/Al alloy coatings are shown in Fig. 5. At an annealing temperature of 500 °C, the fine bright layer changed to large gray layers, while minor thicker bright layer was still present in the coating and the interface of the thick layer

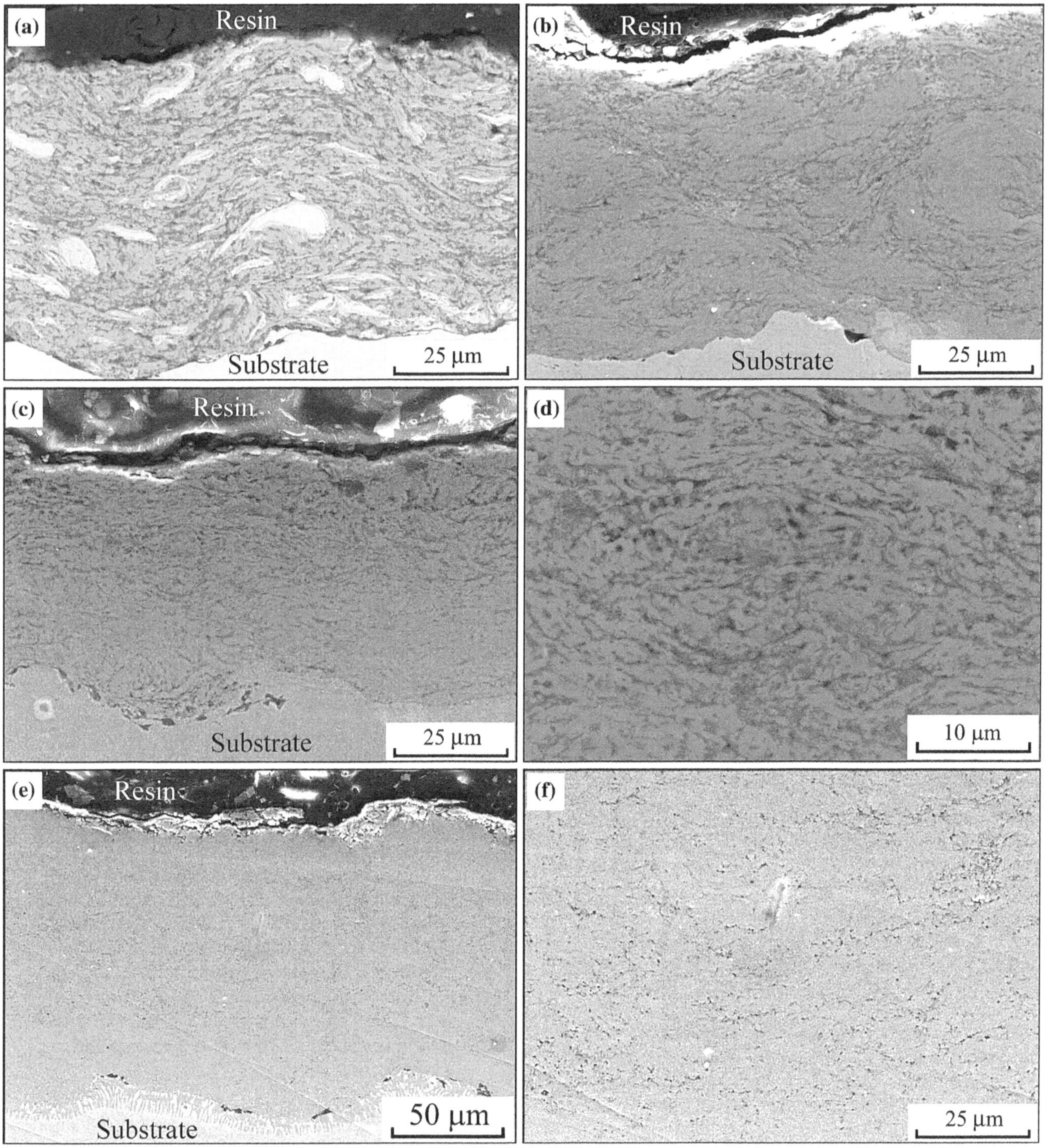

Fig. 5 Cross-sectional microstructure of Ni/Al alloy coatings annealed at different conditions: (a) 500 °C, 3 h; (b) 600 °C, 2 h; (c), and (d) 850 °C, 2 h; (e), and (f) 1050 °C, 1 h

was surrounded by the gray layer, as shown in Fig. 5(a). According to EDS analysis, in accord with the XRD data, it was clear that the NiAl phase was present in the coating with deep gray contrast and Ni_2Al_3 phase with gray contrast. Some large Ni lamellae with a bright contrast were still present in the core of the deformed particles. This fact is due to insufficient diffusion between Al and Ni. When the alloy coating was annealed at 600 °C, the thick bright layer disappeared; the gray layer and the deep gray layer co-exist, as shown in Fig. 5(b). Based on EDS and XRD analyses, the gray layer and the deep gray layer are Ni_2Al_3 and NiAl, respectively.

At 850 °C, it can be noted in Fig. 5(c) and (d) that NiAl became the dominant phase, which is reasonable by taking account of XRD results, although the Ni-rich and Al-rich NiAl phase detected by the EDS analysis presented in the coating. This fact indicates that the heterogeneity of the lamellar structure of the as-milled powder will influence the annealing time which is required to transform the Ni/Al composite alloy into NiAl intermetallics. Under the annealing condition of 1050 °C for 1 h, it is important to note in Fig. 5(e) and (f) that the coating contained a uniform NiAl structure with small amount of Al_2O_3 particulates dispersed and a fully dense NiAl was achieved. In addition, a diffusion layer was clearly observed at the interface between NiAl coating and substrate. These results clearly indicate that the reaction process is believed to be controlled via in situ solid-state diffusion. Zhu et al. (Ref 18) also reported that reactive processing of nickel-aluminide intermetallic compounds was controlled by solid-state diffusion. The phase formation sequence during the annealing of Ni/Al alloy coatings in the present study is well consistent with the published results (Ref 20-24).

4. Conclusions

A dense Ni/Al alloy coating was deposited by cold spraying using a mechanically alloyed powder as the feedstock. The as-sprayed coating exhibited a laminated microstructure retained from the mechanically alloying powder. The postspray annealing temperature influences significantly the in situ evolution of NiAl intermetallic compound in the Ni/Al alloy coating. The annealing of the subsequent Ni/Al alloy coating at a temperature higher than 850 °C leads to the complete transformation from Ni/Al alloy to NiAl intermetallic compound. The heterogeneity of the as-milled alloy powder microstructure may influence the annealing time needed to completely transform the alloy phase to the intermetallic phase. The high temperature annealing treatment also promotes the formation of a diffusion layer at the interface between the coating and substrate.

Acknowledgments

The present work is supported by National Basic Research Program of China (Grant No. 2007CB707702), National Science Fund for Distinguished Young Scholars (Grant No. 50725101).

Peer Reviewed

References

1. V.K. Sikka, J.T. Mavity, and K. Anderson, Processing of Nickel Aluminides and Their Industrial Applications, *Mater. Sci. Eng. A*, 1992, **153**(1-2), p 712-721
2. C.T. Liu and J.O. Stiegler, Ductile Ordered Intermetallic Alloys, *Science*, 1984, **226**, p 636-642
3. Z. Yu, D.D. Hass, and H.N.G. Wadley, NiAl Bond Coats Made by a Directed Vapor Deposition Approach, *Mater. Sci. Eng. A*, 2005, **394**(1-2), p 43-52
4. J. Angenete and K. Stiller, A Comparative Study of Two Inward Grown Pt Modified Al Diffusion Coatings on a Single Crystal Ni Base Superalloy, *Mater. Sci. Eng. A*, 2001, **316**(1-2), p 182-194
5. M. Chigasaki, K. Otaka, A. Okayama, K. Onisawa, and K. Soeno, Method of Producing Nickel Base Alloy Structure with NiAl Coating, U.S. Patent 4,486,245
6. B.M. Warnes and D.C. Punola, Clean Diffusion Coatings by Chemical Vapor Deposition, *Surf. Coat. Technol.*, 1997, **94-95**, p 1-6
7. Y. Zhang, W.Y. Lee, J.A. Haynes, I.G. Wright, K.M. Cooley, and P.K. Liaw, Synthesis and Cyclic Oxidation Behavior of a (Ni, Pt) Al Coating on a Desulfurized Ni-Base Superalloy, *Metall. Mater. Trans. A*, 1999, **30**(10), p 2679-2687
8. W.Y. Lee, Y. Zhang, I.G. Wright, B.A. Pint, and P.K. Liaw, Effects of Sulfur Impurity on the Scale Adhesion Behavior of a Desulfurized Ni-Based Superalloy Aluminized by Chemical Vapor Deposition, *Metall. Trans. A*, 1998, **29**(3), p 833-841
9. N.P. Padture, M. Gell, and E.H. Jordan, Thermal Barrier Coatings for Gas-Turbine Engine Applications, *Science*, 2002, **296**, p 280-284
10. W.Y. Hua, M.H. Li, and M. Fukumoto, Preparation and Properties of HVOF NiAl Nanostructured Coatings, *Mater. Sci. Eng. A*, 2008, **478**(1-2), p 1-8
11. J.A. Hearley, J.A. Little, and A.J. Sturgeon, The Effect of Spray Parameters on the Properties of High Velocity Oxy-Fuel NiAl Intermetallic Coatings, *Surf. Coat. Technol.*, 2000, **123**(2-3), p 210-218
12. Y. Wang, W. Chen, and L. Wang, Micro-Indentation and Erosion Properties of Thermal Sprayed NiAl Intermetallic-Based Alloy Coatings, *Wear*, 2003, **254**(3-4), p 350-355
13. C.J. Li and W.Y. Li, Deposition Characteristics of Titanium Coatings in Cold Spraying, *Surf. Coat. Technol.*, 2003, **167**(2-3), p 278-283
14. C.J. Li, W.Y. Li, and Y.Y. Wang, Formation of Metastable Phases in Cold-Sprayed Soft Metallic Deposit, *Surf. Coat. Technol.*, 2005, **198**(1-3), p 469-473
15. C.J. Li, W.Y. Li, Y.Y. Wang, G.J. Yang, and H. Fukanuma, A Theoretical Model for Prediction of Deposition Efficiency in Cold Spraying, *Thin Solid Films*, 2005, **489**(1-2), p 79-85
16. C.J. Li, W.Y. Li, W.H. Ma, and H. Fukanuma, Characterization of Microstructure of Nanostructured Fe-Si Coating Deposited by Cold Spraying, *Thermal Spray Solutions: Advances in Technology and Application*, Conference Proceedings, May 10-12 2004 (Osaka, Japan), DVS, Düsseldorf, 2004, p 371-377
17. H.T. Wang, C.J. Li, G.J. Yang, C.-X. Li, Q. Zhang, and W.-Y. Li, Microstructural Characterization of Cold-Sprayed Nanostructured FeAl Intermetallic Compound Coating and Its Ball-Milled Feedstock Powders, *J. Therm. Spray Technol.*, 2007, **16**(5-6), p 669-676
18. Q. Zhang, C.J. Li, C.X. Li, G.J. Yang, and S.C. Lui, Study of Oxidation Behavior of Nanostructured NiCrAlY Bond Coatings Deposited by Cold Spraying, *Surf. Coat. Technol.*, 2008, **202**(14), p 3378-3384
19. H.Y Lee, S.H. Jung, S.Y Lee, and K.H. Ko, Alloying of Cold-Sprayed Al-Ni Composite Coatings by Post-Annealing, *Appl. Surf. Sci.*, 2007, **253**(7), p 3496-3502
20. H.C. Zhu and R. Abbaschian, Reactive Processing of Nickel-Aluminide Intermetallic Compounds, *J. Mater. Sci.*, 2003, **38**(18), p 3861-3870

Peer Reviewed

21. L. Farber, L. Klinger, and I. Gotman, Modeling of Reactive Synthesis in Consolidated Blends of Fine Ni and Al Powders, *Mater. Sci. Eng. A*, 1998, **254**(1-2), p 155-165
22. L. Farber, Irena E.Y. Gutmanas, and A. Lawley, Solid State Synthesis of NiAl-Nb Composites from Fine Elemental Powders, *Mater. Sci. Eng. A*, 1998, **244**(1), p 97-102
23. A. Battezzati, P. Pappalepore, F. Durbiano, and I. Gallino, Solid State Reactions in Al/Ni Alternate Foils Induced by Cold Rolling and Annealing, *Acta Mater.*, 1999, **47**(6), p 1901-1914
24. A. Biswas and S.K. Roy, Comparrison between the Microstructural Evolutions of Two Modes of SHS of NiAl: Key to a Common Reaction Mechanism, *Acta Mater.*, 2004, **52**(2), p 257-270

JTTEE5 17:721–727
DOI: 10.1007/s11666-008-9245-6
1059-9630/$19.00 © ASM International

Peer Reviewed

Microstructure and Mechanical Properties of Low-Pressure Cold-Sprayed (LPCS) Coatings

Heli Koivuluoto, Juha Lagerbom, Mikko Kylmälahti, and Petri Vuoristo

(Submitted May 7, 2008; in revised form September 9, 2008)

In low-pressure cold spraying, compressed air is used as a process gas. The most important process parameters are temperature and pressure. In the Low-Pressure Cold Spraying (LPCS) system in this study, the maximum preheating temperature is 650 °C and pressure is 9 bar. Powders used in LPCS process contain alumina with metallic powders; therefore LPCS is the method to spray soft metallic coatings with ceramic hard phase for different application areas, e.g., thick coatings and coatings for electrical and thermal conduction and corrosion protection applications. The aim of this study was to investigate microstructure, denseness, and mechanical properties of LPCS Cu, Ni, and Zn coatings. LPCS coatings seemed to be dense according to Scanning Electron Microscope (SEM) studies but corrosion tests were needed to identify the existence of porosity. Through-porosity was observed in structures of the LPCS coatings. Bond strengths of LPCS Cu and Zn coatings were found to be 20-30 MPa, and hardness was high indicating reinforcement and work hardening.

Keywords copper, low-pressure cold spraying, mechanical properties, microstructure, nickel, zinc

1. Introduction

Cold spraying was developed in the former Soviet Union in the 1980s. Cold spraying is regarded as the latest development in the thermal spray techniques. The process is based on the utilization of significantly lower process temperatures with high particle velocities than those in other thermal spray methods. A coating is formed when powder particles impact at high velocities (high kinetic energy) on the substrate, deform, and adhere to substrate or to other particles. Moreover, good bonding between the cold-sprayed powder particles needs a high degree of plastic deformation upon the particle impact (Ref 1, 2). In low-pressure cold spraying (LPCS), preheating temperatures of the process gas (air) are between room temperature (RT) and 650 °C, and pressures are between 5 and 9 bar. Typically, compressed air is used in this method as the process gas to spray powder mixtures (Ref 3). Particle velocities are reported to be approximately from 350 to 700 m/s in the low-pressure cold spray process (Ref 4). Irissou et al. (Ref 5) have reported particle velocities of 580 m/s for Al_2O_3 particles (mean size 25.5 μm). In addition, Ning et al. (Ref 6) have updated mean particle velocities of 450 m/s for Cu particles (30 μm) (sprayed with helium). Furthermore, irregular particles have reportedly higher in-flight velocities compared to spherical particles (Ref 6) possibly explaining the utilization of dendritic particles in the some cases (e.g., Cu and Ni) in LPCS process.

Low-Pressure Cold-Sprayed (LPCS) is the method to spray metallic powders (e.g., Cu, Al, Ni, Zn) with an addition of ceramic powder in the blended spray powder. Different substrate materials, e.g., metals, ceramics, and plastics, can be used. The main functions of the ceramic addition are the activation of the sprayed surfaces and the cleaning of the nozzle of the gun. In addition to these, ceramic particles are affected by mechanical hammering of the substrate/sprayed layers or by the so-called shot peening via particle impacts (Ref 7). Ceramic addition in the powder also has a compacting effect during the impact, indicating influences on coating properties and improved deposition efficiency (Ref 8). Shkodkin et al. (Ref 8) have reported increased bond strength and coating density with increasing ceramic addition. The amount of ceramic particles in the sprayed coating is low compared to initial powder composition. Usually, coatings contain ceramic particles below 5% of the total amount of ceramic powder (Ref 8). This indicates the erosion occurrence of and also activation by ceramic particles (Ref 8, 9). A hard phase

This article is an invited paper selected from presentations at the 2008 International Thermal Spray Conference and has been expanded from the original presentation. It is simultaneously published in *Thermal Spray Crossing Borders, Proceedings of the 2008 International Thermal Spray Conference*, Maastricht, The Netherlands, June 2-4, 2008, Basil R. Marple, Margaret M. Hyland, Yuk-Chiu Lau, Chang-Jiu Li, Rogerio S. Lima, and Ghislain Montavon, Ed., ASM International, Materials Park, OH, 2008.

Heli Koivuluoto, Juha Lagerbom, Mikko Kylmälahti, and **Petri Vuoristo**, Department of Materials Science, Tampere University of Technology, P.O. Box 589, 33101 Tampere, Finland. Contact e-mail: heli.koivuluoto@tut.fi.

can be used for the purpose of reinforcement to strengthen the metallic matrix in LPCS process (Ref 10). Low-pressure cold spraying has many advantageous features, e.g., possibility to produce thick coatings, use in repair and shape modification of molds (Ref 3, 8). Low-pressure cold spraying or radial injection gas dynamic spray (RIGDS) reportedly have applications in the fields of aerospace, automotive, and rebuild industries (Ref 2).

Copper has good electrical and thermal conductivity (Ref 11); therefore it has been used often as a well-conducting coating material. Furthermore, nickel has good corrosion-resistance in many corrosion environments (Ref 12). Zinc is also a good corrosion resistant material for applications in which corrosion protection is provided by cathodic protection of steels. Zinc coatings act as a sacrificial anode protection (Ref 13). Zinc coatings are, therefore, mostly used only for corrosion protection, e.g., in car industry (Ref 3).

The aim of this study was to characterize microstructures of LPCS Cu, Ni, and Zn coatings and to get more information about the denseness of these coatings by corrosion studies. Moreover, mechanical properties (hardness and bond strength) on grit-blasted steel and copper substrates were investigated to get more information about coatings, and, as well as the influences of hard, ceramic phase on the coating properties.

2. Experimental Procedure

In this study, LPCS Cu, Ni, and Zn coatings were prepared at Tampere University of Technology with a DYMET 304 K equipment. Compressed air was used as a process gas. Spraying parameters are summarized in Table 1. DYMET equipment was installed into an x-y manipulator. A round (Ø 5 mm) tubular nozzle was used in this study. Substrates were grit-blasted (mesh 24, Al_2O_3 grits) steel and copper plates with the dimensions of 100 × 50 × 5 mm.

Three commercial powders were tested in LPCS process. Figure 1 shows the morphologies of copper powder K-01-01 (Cu + 50vol.-% Al_2O_3), nickel powder K32 (Ni + 50vol.-% Al_2O_3), and zinc powder K-0011 (Zn + 50vol.-% Al_2O_3) from OCPS (Obninsk Center for Powder Spraying). Copper and nickel powders were electrolytically prepared, having a dendritic structure, whereas zinc powder was atomized and possessed a spherical shape. With all these powder mixtures, the ceramic particles in the metal/ceramic powder blends were a fused and crushed alumina (Al_2O_3), having a blocky and irregular shape.

Table 1 Spraying parameters of LPCS coatings. Beam distance stands for distance between two adjacent spray beads

Spraying parameter	LPCS Cu	LPCS Ni	LPCS Zn
Pressure, bar	6	6	6
Preheating temperature, °C	540	540	440
Powder feed, g/min	20	15	24
Traverse speed, m/min	6	6	6
Beam distance, mm	1	1	1
Spraying distance, mm	10	10	10
Number of layers	2	7	2

LPCS coatings were characterized using a Philips XL30 Scanning Electron Microscope (SEM). The microstructures of LPCS coatings were studied from metallographic cross sections (unetched and etched structures). LPCS Cu coating was etched with an etching solution of 95 mL ethanol, 5 g $FeCl_3$ and 2 mL HCl. Denseness, and especially through-porosity of the LPCS coatings, was tested with corrosion tests, open-cell electrochemical potential measurements, and salt spray (fog) test. The electrochemical cell used in the open-cell potential measurements consisted of a tube, of diameter 20 mm and volume 12 mL, glued on the surface of the coating specimen. A 3.5 wt.% NaCl solution was poured into the tubes for

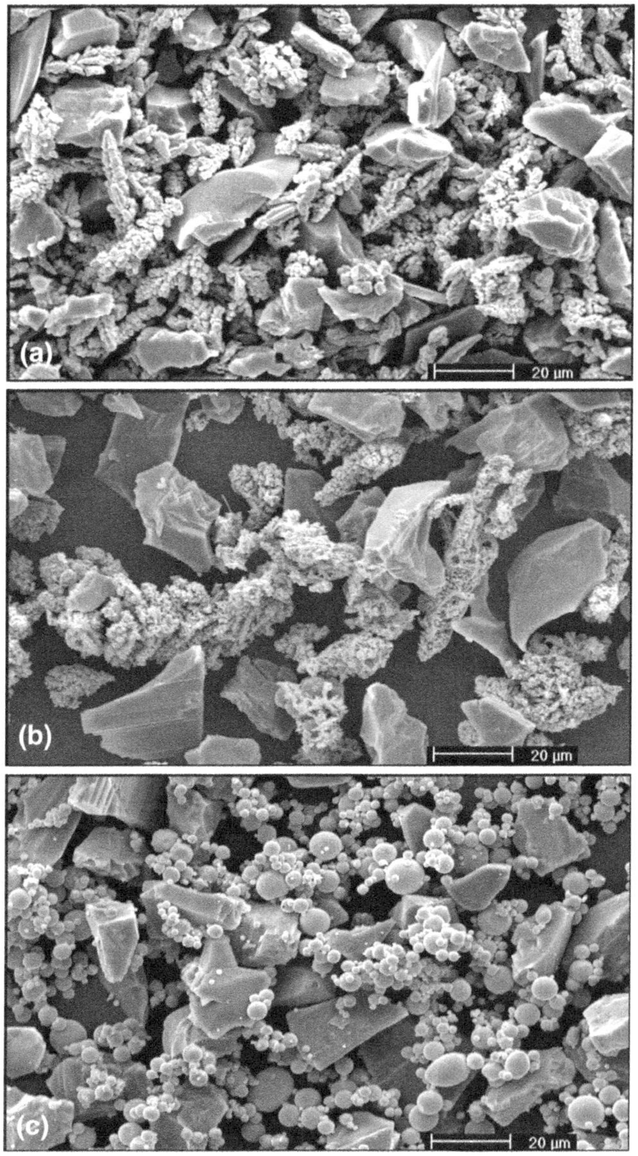

Fig. 1 Morphologies of powders (a) Cu + Al_2O_3, (b) Ni + Al_2O_3, and (c) Zn + Al_2O_3 used in LPCS process

nine-day measurements. Open-cell potential measurements were done with a Fluke 79 III true RMS multimeter. A silver/silver chloride (Ag/AgCl) electrode was used as a reference electrode. Salt spray fog test was done according to the ASTM B117 standard. A 5-wt.% NaCl solution was used with an exposure of 96 h, a temperature of 35-40 °C, a solution pH of 6.3, and a solution accumulation of 0.04 mL/cm^2 h. Samples were visually examined before, during, and after exposure.

Mechanical properties, hardness, and bond strength were also tested. Vickers hardness ($HV_{0.3}$) was measured as an average of ten measurements with a Matsuzawa MMT-X7 hardness tester. Bond strength values were determined according to the standard EN582 in a tensile pull test (Instron 1185 mechanical testing machine). Three measurements were carried out to calculate the average values of bond strength or in this case cohesive strength because of fracture planes (more details later).

3. Results and Discussion

The microstructural details, i.e., microstructure and denseness (existence of through-porosity) were investigated in this study. In addition, mechanical properties, hardness and bond strength, were obtained to gain more information about coating properties.

3.1 Microstructure of LPCS Coatings

Microstructures of LPCS Cu, Ni, and Zn coatings on grit-blasted steel substrates were investigated with SEM (SE and BSE images) analysis (Fig. 2, 3, and 4, respectively). In the microstructures of LPCS Cu, Ni, and Zn coatings, the detected black particles are Al_2O_3, arising from the powder mixture. The thickness of LPCS Cu coating was 200 μm. According to visual examination, LPCS Cu coating seemed to be dense without any noticeable pores (Fig. 2a). However, some oxidized areas (dark gray areas) between particle boundaries are detected in the coating structure (Fig. 2b, BSE image). LPCS Cu coating was etched to reveal the real cross-sectional microstructure. Powder particle boundaries cannot be noticed in the microstructure but the primary particle boundaries can be observed from the etched microstructure (Fig. 2c). It is possible that powder particle with dendritic structure breaks down during the particle impact on the surface of substrate or other particles. Moreover, flattened shape of primary particles indicates that high degree of plastic deformation occurred during spraying.

Figure 3(a) presents the microstructure of LPCS Ni coatings. The thickness of LPCS Ni coating was 290 μm. LPCS Ni coating was relatively dense according to visual observation. However, some open boundaries can be detected near the coating surface and some oxidized boundaries can be observed in the coating structure (Fig. 3b). Oxidation was mostly concentrated in the primary particle boundaries in the structure of LPCS Cu and Ni coatings, which is shown in Fig. 2(b) and 3(b).

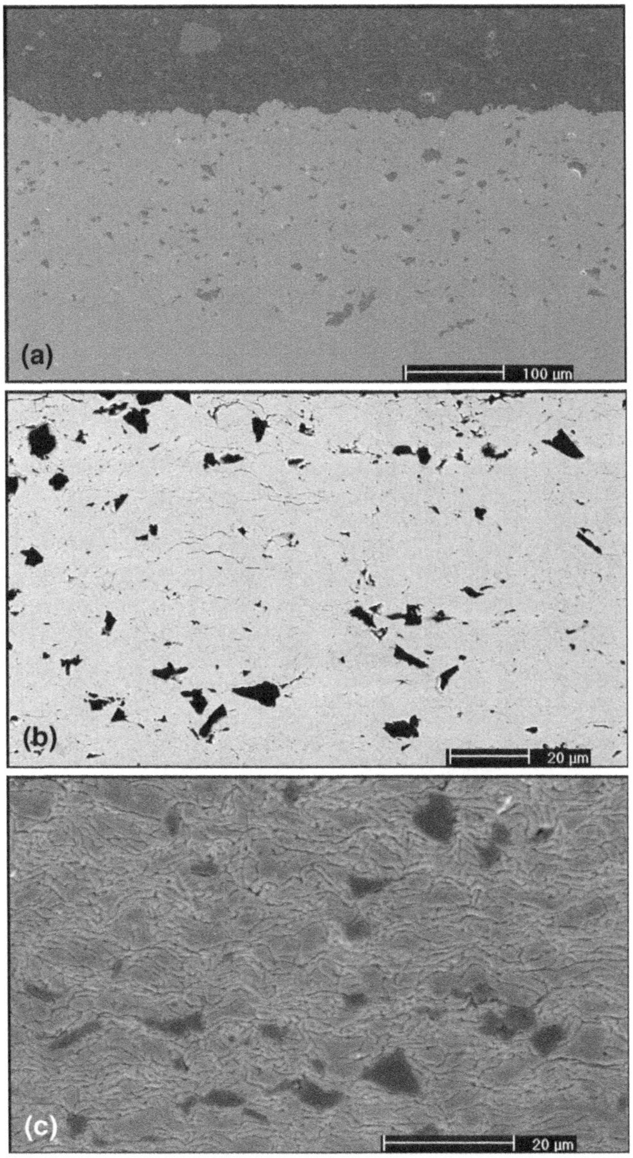

Fig. 2 LPCS Cu coating on grit-blasted steel substrate (a) SE image, (b) BSE image, and (c) SE image, etched microstructure

As-received powder particles had an oxidized layer on the surface caused by the electrolytic production method of the powder. Dendritic particles have large surface areas and, hence, more oxidized areas than on the surface of, e.g., atomized spherical particles. In addition, it is possible that powder particles oxidized during spraying because air was used as a process gas. Thus, oxidized boundaries in the coating structure were possibly originated both from oxidized layers in the powder particles and from oxidation during spraying (both during in-flight and on the substrate).

The detected black particles in the LPCS Zn coating also were Al_2O_3 particles on the gray metallic matrix (Fig. 4). The thickness of LPCS Zn coating was 150 μm. The structure of LPCS Zn coating by BSE image (SEM) differs from the structure of LPCS Cu and Ni coatings because of the different types (shape, production method)

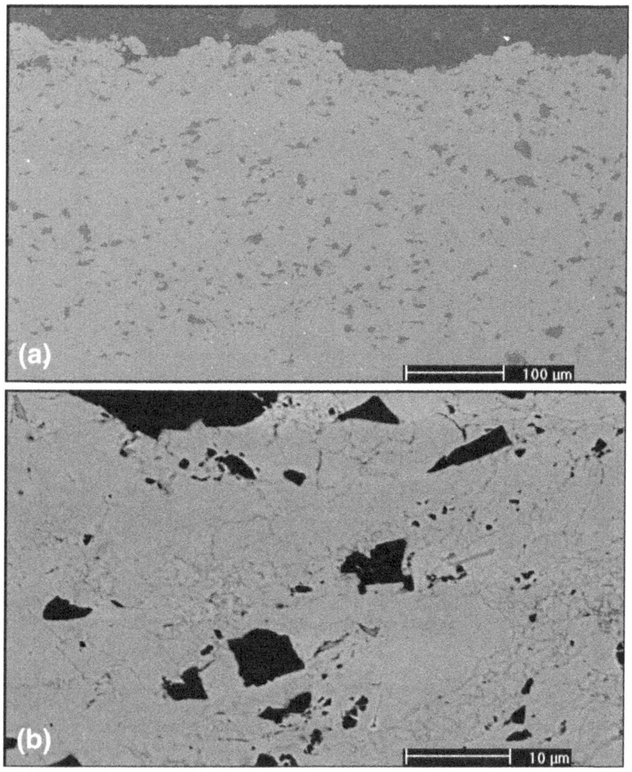

Fig. 3 LPCS Ni coating on grit-blasted steel substrate (a) SE image and (b) BSE image

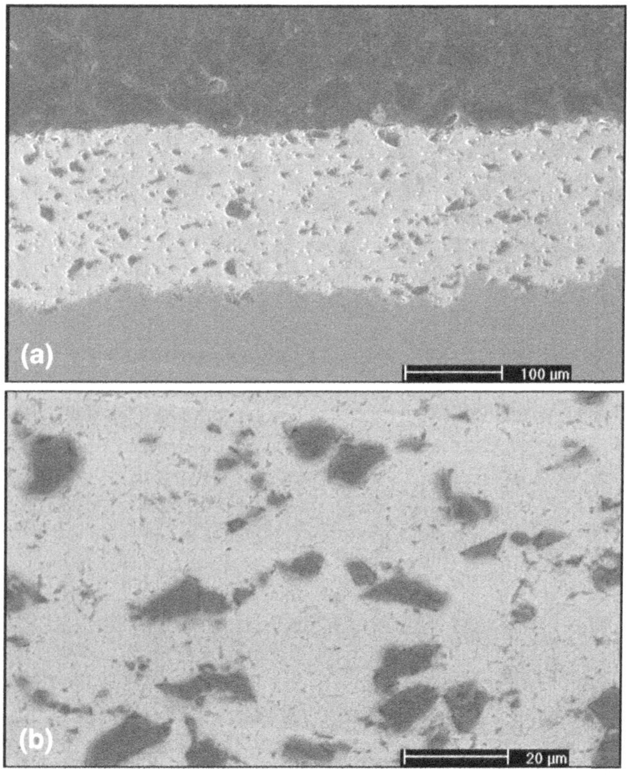

Fig. 4 LPCS Zn coating on grit-blasted steel substrate (a) SE image and (b) BSE image

of powders. Zinc powders were atomized, whereas nickel and copper were electrically prepared. Particle boundaries (initial state spherical particle shape) were imperceptible from unetched structure, indicating visually dense microstructure (Fig. 4b). Besides, it should be noticed that low amount of porosity in zinc coatings can be allowed because the steel substrate is in any case protected cathodically by the zinc coating. In summary, LPCS Cu, Ni, and Zn coatings were dense according to visual evaluation by SEM. Noticeable pores were not observed in the coating structures; nevertheless, some oxidized areas or boundaries can be detected especially in the LPCS Ni coating.

3.2 Denseness of LPCS Coatings

LPCS coatings were found to be quite dense according to SEM examinations. Furthermore, denseness was investigated with corrosion tests to get more information about existing through-porosity (open-porosity). The corrosion tests used were open-cell potential measurement and salt spray fog test.

Open-cell potentials of LPCS Cu and Ni coatings are presented in Fig. 5 and 6 as a function of exposure time. During the exposure, the behavior of the LPCS Cu coating was compared to the behavior of bulk Cu plate and

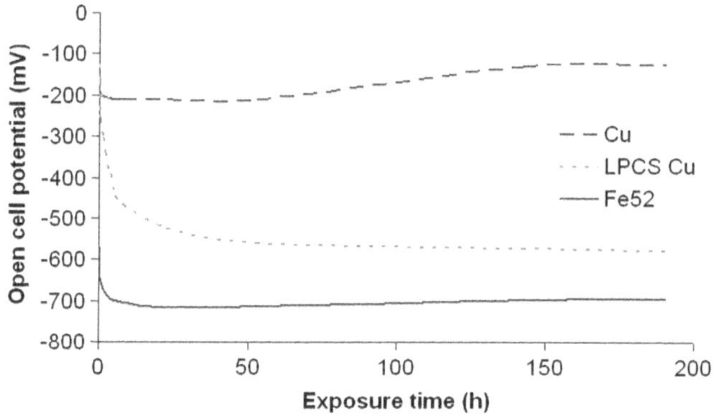

Fig. 5 Open-cell potential of Cu bulk material, LPCS Cu coating, and Fe52 substrate material as a function of exposure time (Ag/AgCl reference electrode)

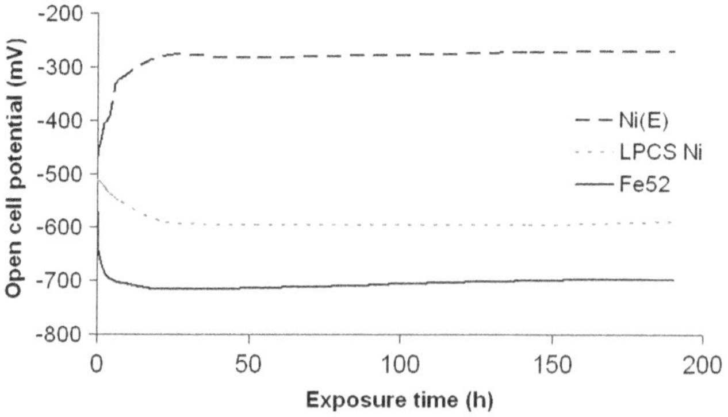

Fig. 6 Open-cell potential of electrolytically prepared Ni coating (Ni(E)), LPCS Ni coating, and Fe52 substrate material as a function of exposure time (Ag/AgCl reference electrode)

substrate material Fe52 plate to find out existing porosity. When open-cell potential curve of the coating is close to the curve of bulk material, it indicates denseness (impermeability) of the coating. However, if the value of the coating is closer to the substrate material, it reflects through-porosity, indicating the open way for exposure solution (NaCl in this study) from the coating surface to interface between coating and substrate. LPCS Cu coating was not fully dense because of existing through-porosity (Fig. 5). The LPCS Ni coating showed similar behavior as the LPCS Cu coating in the open-cell potential measurements. LPCS Ni coating contained also through-porosity (Fig. 6). Electrolytically prepared Ni coating was chosen as a reference sample in this study. In addition, corrosion products from the substrate material (iron oxide) were found by visual examination on the surfaces of LPCS Cu and Ni coatings after exposure. Open-cell potential of LPCS Zn coating was stabilized to the value of −1020 mV. Typically, zinc coating gives cathodic protection to steel substrate, behaving like a sacrificial anode. Because of cathodic protection, denseness (overall dense structure) is not critical for corrosion resistance in the case of zinc coatings.

In addition, denseness of LPCS Cu, Ni, and Zn coatings were tested with salt spray fog test. After exposure, coatings were visually examined. According to visual analysis, corrosion products from substrate material (iron oxide, rust) were detected on the surface of LPCS Cu and Ni coatings, indicating existence of through-porosity. Moreover, only white rust (corrosion product of zinc) was observed on the surface of LPCS Zn coating, indicating sacrificial behavior and at the same time corrosion protection.

According to corrosion tests (open-cell potential measurements and salt spray fog test) LPCS Cu and Ni coatings contained through-porosity. However, steel substrate was protected cathodically by LPCS Zn coating. The addition of ceramic particles has been reported to have an influence on the denseness of the LPCS Al coatings, indicating the possibility of spraying dense coatings with low-pressure cold spray equipment (Ref 5). This will need optimization of powder mixtures and spraying parameters for other coating materials to produce dense coatings with LPCS process.

3.3 Mechanical Properties of LPCS Coatings

3.3.1 Hardness of LPCS Coatings. Hardness of LPCS Cu, Ni, and Zn coatings on both steel and copper substrates is summarized in Table 2. Increased hardness of coatings indicates work hardening due to particle impacts. Somewhat higher hardness values of the LPCS coatings indicated a high level of plastic deformation and strain hardening. Moreover, it should be noticed that Al_2O_3 addition can also have influence on the hardness by increasing the values. The hard phase in the powder mixture was found to have an activation effect of the surface and also a hardening effect, which is seen in the hardness values. Substrate material did not have remarkable influence on the hardness of LPCS Cu, Ni, or Zn coatings.

The hardness of LPCS Cu coating was higher as compared to bulk material than in the cases of LPCS Ni coating. This clearly indicates a higher degree of plastic deformation. The impact may be much stronger in the case of LPCS Cu coating. The hardness of bulk Cu is reported in earlier studies to be $HV_{0.3}$ 90 (Ref 14), which is lower than the value of the LPCS Cu coating. The hardness of high-pressure cold-sprayed (HPCS) Cu and Ni coatings are $HV_{0.1}$ 147 and 238, respectively (Ref 14). The hardness of Ni and Zn bulk materials was $HV_{0.3}$ 111 (stand. dev. 9.0) and 43 (stand. dev. 1.8), respectively. HPCS coatings had higher hardness possibly arising from higher level of plastic deformation. However, it should be noticed that the production method of metallic powder is different and is therefore possibly affecting the hardness values as well. The hardness of LPCS Cu was closer to the value of HPCS Cu than LPCS Ni compared to HPCS Ni coating. This indicates higher level of deformation and more work hardening due to impacts in the case of LPCS Cu particles. In addition, high values of standard deviations (Table 2, stand. dev.) in the hardness measurements of LPCS Ni coatings reflect poor coating quality and very heterogeneous structure.

3.3.2 Bond Strength of LPCS Coatings. The bond strengths of LPCS Cu, Ni, and Zn coatings on both steel and copper substrates were measured with tensile pull tests. Fracture planes were inside the coating, meaning cohesive type fracture. In all the cases, the fracture was found nearer to the interface between coating and substrate but still inside the coating (Table 3).

LPCS Zn coating has significantly high bond strength on both steel and copper substrates, indicating good bonding between coating and substrate, and also between the sprayed particles. Adhesion of LPCS Zn was significantly higher compared to HPCS Zn coating, 13 MPa (Ref 14) on steel substrate arising from the reinforcement effect of Al_2O_3. In this study, all LPCS coatings had the weakest point in the coating. It can be supposed that

Table 2 Hardness ($HV_{0.3}$) of LPCS coatings

Coating	Substrate	Hardness, $HV_{0.3}$	Stand. dev.
Cu	Steel	105	5.3
Cu	Cu	104	3.6
Ni	Steel	119	21.4
Ni	Cu	124	31.7
Zn	Steel	57	0.8
Zn	Cu	57	1.1

Table 3 Bond strength of LPCS coatings

Coating	Substrate	Bond strength, MPa	Stand. dev.
Cu	Steel	20	3.6
Cu	Cu	23	4.1
Ni	Steel	8	3.6
Ni	Cu	9	0.5
Zn	Steel	33	4.7
Zn	Cu	38	0.7

Peer Reviewed

adhesion between coating and substrate was at least the value of the bond (cohesive type) strengths. The bond strength of LPCS Cu coating was also reasonable, 20-23 MPa, and higher than the value of HPCS Cu coating (process gas N_2) on steel substrate (8 MPa) (Ref 14). In the case of LPCS Ni coating, the bond strength was very low, indicating poor bonding between powder particles. Moderate values of bond strengths of LPCS coatings are caused by defects (e.g., pores and imperfect interparticle contacts) and also thin oxide layers between particles (Ref 15), which were also noticed in this study, especially with the LPCS Ni coating. LPCS Ni particles did not undergo as high a plastic deformation as LPCS Cu particles, and thus, the bonds between particles were very moderate due to the possibly stronger oxide layers of powder particles, making even weaker bonding without destroying the oxide layer during impact. On the other hand, ceramic phase has been reported elsewhere to affect adhesion between particles and substrates by increasing it (Ref 5). Lee et al. (Ref 16) have reported that ceramic addition in the powder increased the adhesion of cold-sprayed Al-Al_2O_3 coatings compared to pure Al coating. High adhesion is due to the size of craters between coating and substrate (like activation) (Ref 16). Cohesive type fracture after tensile tests can be explained by poor adhesion between metallic and ceramic particles (Ref 5).

4. Summary

LPCS Cu and Ni coatings were dense according to SEM analysis, whereas through-porosity was observed in the corrosion tests. Some oxidized boundaries and other microscale defects were detected as the weak points in the microstructure, indicating poor bonding and existence of small interparticle porosity. Although the hard phase (Al_2O_3) had a reinforcing effect, it apparently did not hammer the metallic matrix enough to form a dense coating structure (e.g., destroy oxide layers in the powder particle and primary particle boundaries). At least, visually dense and thick coatings can be sprayed with low-pressure cold spraying for applications where corrosion resistance is not crucial (e.g., for electrical and thermal conduction applications). Moreover, LPCS Zn coating was also visually dense and the open-cell potential was electronegative, indicating corrosion resistance behaving like sacrificial anode for the protection of steel substrates.

In this study, mechanical properties were reasonable instead of the bond strength of LPCS Ni coating. High hardness values measured indicated work hardening occurring during spraying. On the other hand, ceramic addition can affect the values. The weak points were at the bonding between particles, indicating cohesive failure type fracture. Therefore, adhesion between particles and substrates was not the most critical point (similar to thermally sprayed coatings).

Optimization of powders, substrates, their combinations, and, moreover, substrate pretreatments and hard phase addition will be investigated in future to produce dense and well-adhered coatings with low-pressure cold spray process.

Acknowledgments

The research was funded by Finnish Funding Agency for Technology and Innovation (TEKES) and a group of Finnish industrial companies.

References

1. C. Borchers, F. Gärtner, T. Stoltenhoff, H. Assadi, and H. Kreye, Microstructural and Macroscopic Properties of Cold Sprayed Copper Coatings, *J. Appl. Phys.*, 2003, **93**(12), p 10064-10070
2. R.G. Maev and V. Leshchynsky, Air Gas Dynamic Spraying of Powder Mixtures: Theory and Application, *J. Therm. Spray Tech.*, 2006, **15**(2), p 198-205
3. A. Kashirin, O. Klyuev, T. Buzdygar, and A. Shkodkin, DYMET Technology Evolution and Application, *Global Coating Solutions: Proceedings of the 2007 International Thermal Spray Conference*, B.R. Marple, M.M. Hyland, Y.C. Lau, C.-J. Li, R.S. Lima, and G. Montavon, Eds., May 14-16, 2007 (Beijing, China), ASM International, p 141-145
4. X.-J. Ning, J.-H. Jang, H.-J. Kim, C.-J. Li, and C. Lee, Cold Spraying of Al-Sn Binary Alloy: Coating Characteristics and Particle Bonding Features. *Surf. Coat. Technol.*, 2008, **202**(9), p 1681-1687
5. E. Irissou, J.-G. Legoux, B. Arsenault, and C. Moreau, Investigation of Al-Al_2O_3 Cold Spray Coating Formation and Properties, *J. Therm. Spray Tech.*, 2007, **16**(5-6), p 661-668
6. X.-J. Ning, J.-H. Jang, and H.-J. Kim, The Effects of Powder Properties on In-Flight Particle Velocity and Deposition Process during Low Pressure Cold Spray Process, *Appl. Surf. Sci.*, 2007, **253**(18), p 7449-7455
7. B.B. Djordjevic and R.G. Maev, SIMAT^{TM} Application for Aerospace Corrosion Protection and Structural Repair, *Building on 100 Years Success: Proceedings of the 2006 International Thermal Spray Conference*, B.R. Marple, M.M. Hyland, Y.C. Lau, R.S. Lima, and J. Voyer, Eds., May 15-18, 2006 (Seattle, Washington, USA), ASM International, 2006
8. A. Shkodkin, A. Kashirin, O. Klyuev, and T. Buzdygar, The Basic Principles of DYMET Technology, *Building on 100 Years Success: Proceedings of the 2006 International Thermal Spray Conference*, B.R. Marple, M.M. Hyland, Y.C. Lau, R.S. Lima, and J. Voyer, Eds., May 15-18, 2006 (Seattle, Washington, USA), ASM International, 2006
9. A. Shkodkin, A. Kashirin, O. Klyuev, and T. Buzdygar, Metal Particle Deposition Stimulation by Surface Abrasive Treatment in Gas Dynamic Spraying, *J. Therm. Spray Tech.*, 2006, **15**(3), p 382-386
10. H. Weinert, E. Maeva, and E. Leshchynsky, Low Pressure Gas Dynamic Spray Forming Near-net Shape Parts, *Building on 100 Years Success: Proceedings of the 2006 International Thermal Spray Conference*, B.R. Marple, M.M. Hyland, Y.C. Lau, R.S. Lima, and J. Voyer, Eds., May 15-18, 2006 (Seattle, Washington, USA), ASM International, 2006
11. R.N. Caron and J.T. Staley, Effects of Composition, Processing, and Structure on Properties of Nonferrous Alloys, Copper and Copper Alloys, *ASM Handbooks Online*, 20, Materials Selection and Design, ASM International, 1997
12. R.N. Caron and J.T. Staley, Effects of Composition, Processing, and Structure on Properties of Nonferrous Alloys, Nickel and Nickel Alloys, *ASM Handbooks Online*, 20, Materials Selection and Design, ASM International, 1997
13. R.N. Caron and J.T. Staley, Effects of Composition, Processing, and Structure on Properties of Nonferrous Alloys, Zinc and Zinc Alloys, *ASM Handbooks Online*, 20, Materials Selection and Design, ASM International, 1997

14. H. Mäkinen (Koivuluoto), J. Lagerbom, and P. Vuoristo, Mechanical Properties and Corrosion Resistance of Cold Sprayed Coatings, *Building on 100 Years Success: Proceedings of the 2006 International Thermal Spray Conference*, B.R. Marple, M.M. Hyland, Y.C. Lau, R.S. Lima, and J. Voyer, Eds., May 15-18, 2006 (Seattle, Washington, USA), ASM International, 2006
15. M. Beneteau, W. Birtch, J. Villafuerte, J. Paille, M. Petrocik, R.G. Maev, E. Strumban, and V. Leshchynsky, Gas Dynamic Spray Composite Coatings for Iron and Steel Castings, *Building on 100 Years Success: Proceedings of the 2006 International Thermal Spray Conference*, B.R. Marple, M.M. Hyland, Y.C. Lau, R.S. Lima, and J. Voyer, Eds., May 15-18, 2006 (Seattle, Washington, USA), ASM International, 2006
16. H.Y. Lee, S.J. Jung, S.Y. Lee, Y.H. You, and K.H. Ko, Correlation between Al_2O_3 Particles and Interface of Al-Al_2O_3 Coatings by Cold Spraying, *Appl. Surf. Sci.*, 2005, **252**, p 2892-2898

JTTEE5 17:728–735
DOI: 10.1007/s11666-008-9254-5
1059-9630/$19.00 © ASM International

Peer Reviewed

Characterization of Low-Pressure Cold-Sprayed Aluminum Coatings

K. Ogawa, K. Ito, K. Ichimura, Y. Ichikawa, S. Ohno, and N. Onda

(Submitted May 14, 2008; in revised form September 22, 2008)

Aluminum alloys are widely used as materials for engineering components of automobiles and airplanes because of their light weight and high corrosion resistance. However, cracks may develop sometimes in aluminum components, which have to be repaired by welding. It is difficult to weld aluminum components due to its high specific thermal conductivity and high coefficient of thermal expansion. The low-pressure cold-spray technique can be used instead of welding for repairing cracks. However, the effects of surface conditions on particle deposition and the mechanical properties of cold-sprayed coatings have not been investigated thus far. In this study, the effect of surface conditions focusing on active newly formed surface on aluminum particle deposition is studied and the mechanical properties of low-pressure cold-sprayed aluminum coatings are investigated by four-point bending tests. It is found that for efficient particle deposition it was necessary to obtain active newly formed surface of the substrate and particle surfaces by several impingements because the existence of inactive native oxide films has an adverse effect on the deposition. Furthermore, the strength of a cold-sprayed specimen is found to be higher than that of a cold-rolled specimen under compressive loading.

Keywords low-pressure cold spray, aluminum, bonding strength, characterization, deposition mechanism, four-point bending tests, newly-formed surface

1. Introduction

Surfaces are widely coated to obtain the required surface conditions and properties. In the past, many coatings used have been applied using the thermal spray processes. In a conventional thermal spray process, the coating material is heated to the molten or semimolten state and impinged onto a substrate surface. Because of heating, thermal spray coatings show various unfavorable phenomena, e.g., oxidation, phase transformation, etc. A new coating technique, the cold-gas dynamic-spray process, often simply called "cold spray," was initially developed in the mid-1980s (Ref 1-5). This technique is based on the high-velocity impinging (300-1200 m/s) of minute solid particles (generally 10-50 μm in diameter) on a substrate (Ref 4-6). In this process, the powder particles are accelerated by a supersonic gas jet at a high gas temperature, which is usually lower than the melting point of the powder material. Consequently, cold spray has solved the problems encountered in the thermal spray process (Ref 7). Many researchers have carried out studies on cold-sprayed coatings formed at high pressures of the order of a few megapascals (Ref 8-10). Recently, low-pressure cold-spray equipment, operating at a pressure typically less than 1 MPa, has drawn some researchers' attention (Ref 11), because, the equipment for low-pressure cold-spraying is considerably simpler and smaller than the high-pressure cold-spray equipment. As a result, the application field for cold spray has become wider, e.g., low-pressure cold-spray equipment can be used for on-site maintenance. In particular, aluminum alloys are widely used as materials for engineering components of automobiles and airplanes because of their light weight and high corrosion resistance. However, cracks may develop in aluminum components sometime, which have to be repaired by welding and conventional thermal spray techniques. Cold spraying can be used an alternative technique to repair the cracks in aluminum components.

During cold spraying, solid particles are accelerated to high velocities and impinged onto a substrate. The impinged particles deform the substrate surface and form a new surface. This process plays an important role in the deposition. The kinetic energy of the high-velocity particles is an important factor on which the impingement, deformation, and deposition mechanisms depend. The critical particle velocity of a spray material under the

This article is an invited paper selected from presentations at the 2008 International Thermal Spray Conference and has been expanded from the original presentation. It is simultaneously published in *Thermal Spray Crossing Borders, Proceedings of the 2008 International Thermal Spray Conference*, Maastricht, The Netherlands, June 2-4, 2008, Basil R. Marple, Margaret M. Hyland, Yuk-Chiu Lau, Chang-Jiu Li, Rogerio S. Lima, and Ghislain Montavon, Ed., ASM International, Materials Park, OH, 2008.

K. Ogawa, K. Ito, K. Ichimura, and **Y. Ichikawa**, Fracture and Reliability Research Institute, Tohoku University, Aoba 6-6-01, Aramaki, Aoba-ku, Sendai 980-8579, Japan; and **S. Ohno** and **N. Onda**, Tohoku Electric Power Co., Inc., 1-7-1, Honcho, Aoba-ku, Sendai 980-8550, Japan. Contact e-mail: kogawa@rift.mech.tohoku.ac.jp.

high-pressure cold-spray conditions, which is the threshold particle velocity for adhesion/separation, has been numerically studied (Ref 12-16). Assadi et al. reported that the critical velocity of aluminum was 620-675 m/s in the case of high-pressure cold spray (Ref 12). In our previous study (Ref 17), it has been found that the critical velocity of the aluminum under low-pressure cold spray was approximately 300 m/s. The difference between the both critical velocities was almost twofold. Hence, the deposition mechanisms in low-pressure cold spraying may differ from those in high-pressure cold spraying each other.

In this study, the effect of surface conditions on low-pressure cold spraying was investigated. Further, to evaluate the mechanical properties of cold-sprayed coatings, the interfacial strength of the coatings was determined by four-point bending tests.

2. Experimental

2.1 Spraying System and Materials Used

In this study, the commercial cold-spray facility DYMET403J (Obninsk Center for Powder Spraying, Russia) was used for spraying. Since this equipment uses compressed air as the carrier gas, only an air compressor is required. Moreover, the main unit has a small built-in gas heating system that is used to heat the working gas directly; hence, the equipment is small and lightweight. A round-shaped exit nozzle with a diameter of 4.8 mm was used. Compressed air at 0.6 MPa was used as the carrier gas. Pure aluminum powder (Al G-AT, Fukuda Metal Foil & Powder Corporation) was used. The oxygen content of the particle was approximately 0.2%. The averaged particle diameter of the powder used in the study is 17.3 μm. A cold-rolled plate of pure aluminum alloy (A1050) with dimensions of 40 × 150 × 3 mm was used as the substrate. Due to formation of thick coating, 40 m/s was the traverse speed of gun for deposition efficiency measurements and making specimens of four-point bending tests. On the other hand, due to evaluation of deposition behavior of an independent particle, the traverse speed was 500 m/s in the cases of evaluation of number of particle deposited, and of effects of roughness and oxide films for deposition.

2.2 Evaluation of Number of Particles Deposited

To investigate the particle deposition phenomenon in the early stage, the specimens were impinged with limited particles. At a high traverse speed and low powder feed rate, it can be realized to make very limited particles impinge on a mirror-polished aluminum substrate. The surface roughness Ra of the substrate was 0.2 μm. The number of particles deposited on the substrate in a spray area (1 × 5 mm), as shown in Fig. 1, was determined by the scanning electron microscope (SEM) observation.

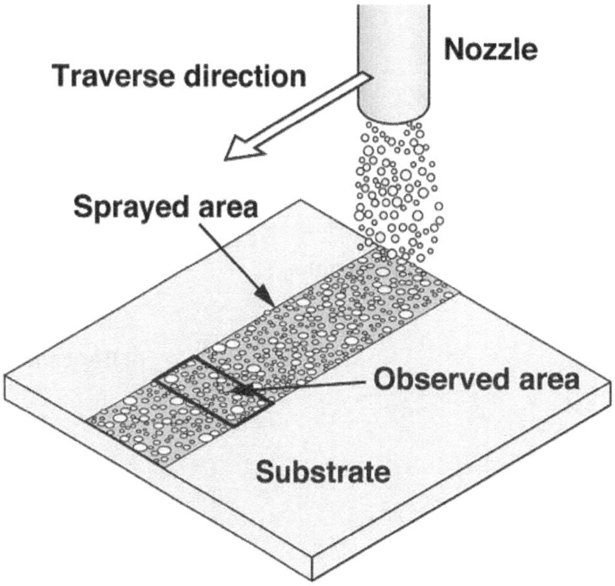

Fig. 1 Schematic illustration of measurement of number of particles deposited

2.3 Deposition Efficiency

The deposition efficiency (DE) of particles was measured as follows. Before spraying, the weights of the substrate (W_{s1}) and the feed stock powder (W_{p1}) were measured. Then, the weights of the sprayed substrate (W_{s2}) and the remaining feed stock powder (W_{p2}) were measured. Finally, the DE of particles was determined by using these measured values and Eq 1.

$$\mathrm{DE} = \frac{W_{s2} - W_{s1}}{W_{p2} - W_{p1}} \times 100\,(\%) \quad \text{(Eq 1)}$$

It was difficult to measure the weight of the remaining feed stock powder because some powder remained inside the spraying equipment. This powder loss may introduce a measurement error. Therefore, we measured the average weight difference (M) between the input feed stock powder and take off powder. Using M, Eq 1 was modified to obtain Eq 2.

$$\mathrm{DE} = \frac{W_{s2} - W_{s1}}{W_{p2} - W_{p1} - M} \times 100\,(\%) \quad \text{(Eq 2)}$$

Finally, the DE of particles was calculated using Eq 2.

2.4 Effect of Ra on Particle Deposition

To evaluate the effect of Ra on particle deposition, specimens with different Ra values, i.e., 0.2, 1.0, and 3.0 μm, were prepared. Then, the relationship between the number of particles deposited in the sprayed area (1 × 5 mm) and the number of traverses was investigated.

2.5 Effect of Native Oxide Film on Particle Deposition

Initial spraying can remove native oxide films. Therefore, to evaluate the effect of the formation of native oxide

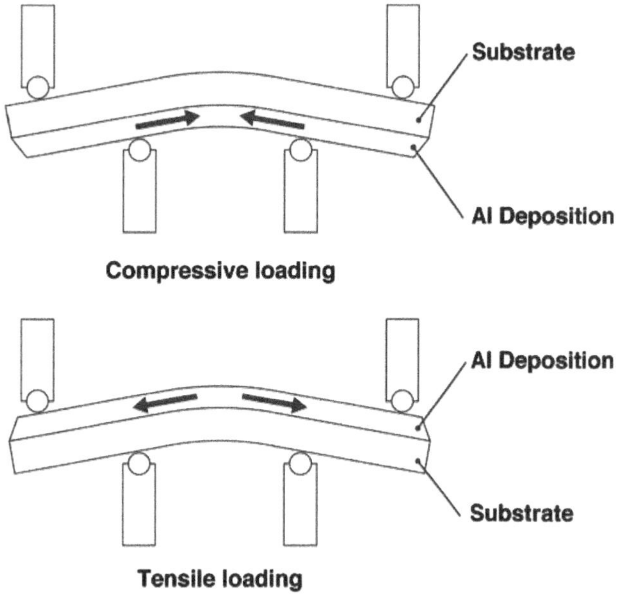

Fig. 2 Schematic illustrations of loading direction during four-point bending tests

films on particle deposition, specimens with different traverse intervals were prepared. The traverse intervals were set at 0.6, 60, 600, 1800, and 3600 s.

2.6 Four-Point Bending Tests

To evaluate the mechanical properties of the cold-sprayed coatings under tensile and compressive loadings, four-point bending tests were carried out.

Figure 2 shows the schematic illustration of the loading direction of the specimens under tensile and compressive loadings during the four-point bending tests. Under the four-point bending test, the maximum stress is always located on the surface. By cold spraying, 3 mm-thick pure aluminum coatings were formed on 3 mm-thick aluminum substrates. Then, a 1 mm-thick aluminum coating was removed by machining. Four-point bending specimens with dimensions of $5 \times 40 \times 5$ mm were prepared from the cold-sprayed specimens. Further, to compare bending force, cold-rolled aluminum specimens with a thickness of 5 mm were prepared. The specimens were subjected to four-point bending tests by using a servohydraulic testing machine with an inner span of 15 mm and an outer span of 34 mm at a crosshead speed of 0.6 mm/min. After the four-point bending tests, the fracture surfaces were observed by SEM.

3. Results and Discussion

3.1 Measurements of Number of Particles Deposited

Figure 3 shows the typical SEM images of the cold-sprayed specimens with different traverse numbers, and Fig. 4 shows the change in the number of particles deposited for different traverse numbers. In the case of the specimens with traverse numbers equal to 1 and 2, no particle deposition was observed on the substrates. For traverse numbers in the range of 6-12, the number of particles deposited increased slightly. This result indicates the presence of a time delay in particle deposition in low-pressure cold-spray deposition. The time delay causes changes in the substrate surface conditions, such as roughening and formation of a newly formed surface. The surface roughness influences elimination of the oxide film on the particle surface. And also, the formation of the newly formed surface, which was formed by several particle impingements, has higher surface energy, because of eliminating inactive oxide films. Therefore, it can be possible to improve bonding force. As a result of these changes, the DE of particles increases. The effects of surface roughness and newly formed surface are described in sections "Effect of Ra on Particle Deposition" and "Effect of Formation of Newly Formed Surface on Particle Deposition" in detail, respectively.

The time delay indicates that particle deposition by low-pressure cold spraying requires repeated impingements. Repeated impingements enhance particle deposition. Even though low-pressure cold spray exhibits low deposition efficiency, the velocity of particle deposition in low-pressure cold spray is lower than that in high-pressure cold spray. Further, the latter technique may not require repeated impingements because it has sufficient energy for making the newly formed surface due to the high-speed and high-pressure impingements. Accordingly, it is assumed that the difference in the critical velocities of low- and high-pressure cold sprays is due to the difference in the mechanism of active newly formed surface formation of the two techniques.

3.2 DE of Particles

Figure 5 shows the DE of low-pressure cold-sprayed aluminum coatings. The DE increases with the traverse number. At the early starting of the spraying, particle deposition was unstable. Therefore, the DE of less traverses may include influence of such unstable. However, after four traverses that have no effect of unstable, deposition gets saturated. Even for such case, the maximum spray efficiency of low-pressure cold spray is very low, approximately 7%, because it can be difficult to break oxide films and to form an active newly formed surface, due to low impingement force.

3.3 Effect of Ra on Particle Deposition

The relationship between the number of particles deposited and the traverse number for the specimens with different Ra values is shown in Fig. 6. From this figure, it is evident that the deposition in the specimens with Ra equal to 1.0 and 3.0 starts earlier than that of the specimen with Ra equal to 0.2. It is thought that Ra has a significant effect on deposition of coatings. However, the specimens with Ra equal to 1.0 and 3.0 show almost no difference in their coatings. From this result, it is assumed that particle depositions require a certain level of roughness.

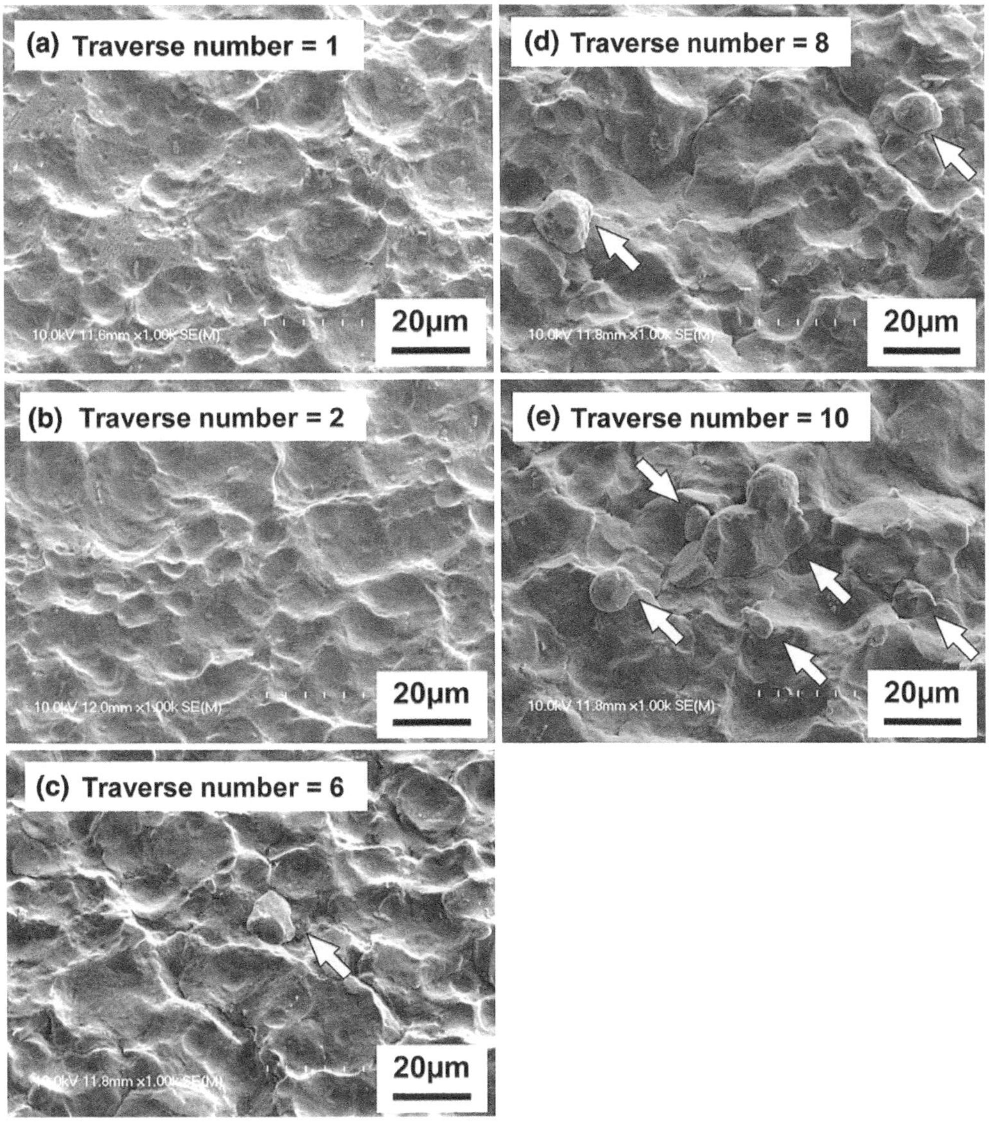

Fig. 3 Typical SEM images of substrate surfaces for different traverse numbers. (a) Traverse number = 1, (b) Traverse number = 2, (c) Traverse number = 6, (d) Traverse number = 8, and (e) Traverse number = 10. The white arrows indicate deposited particles

3.4 Effect of Formation of Newly Formed Surface on Particle Deposition

The relationship between the number of particles deposited and the traverse number for the specimens with different traverse intervals is shown in Fig. 7. Significant deposition is observed in the specimen with a short traverse interval. Thus, the removal of native oxide films, i.e., the formation of an active newly formed surface, is the key factor in the depositions. Further, the formation of native oxide films has an adverse effect on the deposition. Therefore, it is necessary to form a newly formed surface at the substrate surface for deposition. In the case of particle/particle deposition, it is also necessary to form a newly formed surface at the particle surface. With regard to the effect of newly formed surface, Tanaka et al. also reported that the formation of a newly formed surface directory improved joining of friction stir-welded aluminum/steel (Ref 18).

However, it was considered that the other factor such as the temperature rising during repeated scanning of the gun over the surface. Consequently, the temperature near the surface was measured by K-type thermocouple during the scanning of the gun, in the case of interval of 0.6 s. Schematic illustration of setup of the measurement is shown in Fig. 8. The result of the temperature measurement is shown in Fig. 9. As a result, only 2 °C temperature rising was observed. Therefore, it is thought that there is almost no temperature influence.

3.5 Four-Point Bending Tests

The results of the four-point bending tests are shown in Fig. 10. A comparison of the slopes of the plots of the cold-sprayed specimens and the cold-rolled aluminum specimens in the elastic region, shows that the former have higher elastic moduli as compared to the latter. In the case of compressive loading, the former showed a higher

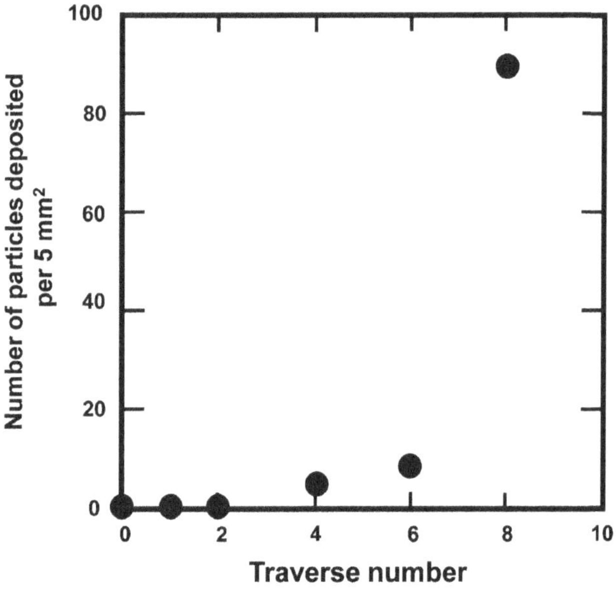

Fig. 4 Relationship between number of particles deposited and traverse number

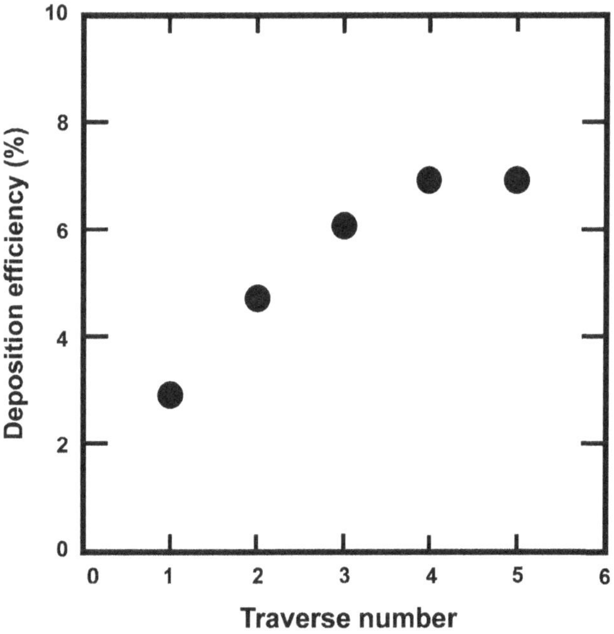

Fig. 5 Relationship between deposition efficiency and traverse number

strength as compared to the latter. In the case of tensile loading of the cold-sprayed specimens, the cold-sprayed coatings were found to be hardly ductile and fragment occurred after 0.2 mm displacement, leading to the unstable fractures.

The images of the specimens after the four-point bending tests are shown in Fig. 11. In the compressive loaded specimen, cracks were not observed in the deposited aluminum coating. On the other hand, in the tensile-loaded specimen, a vertical crack was suddenly initiated from the surface of the aluminum coating during loading,

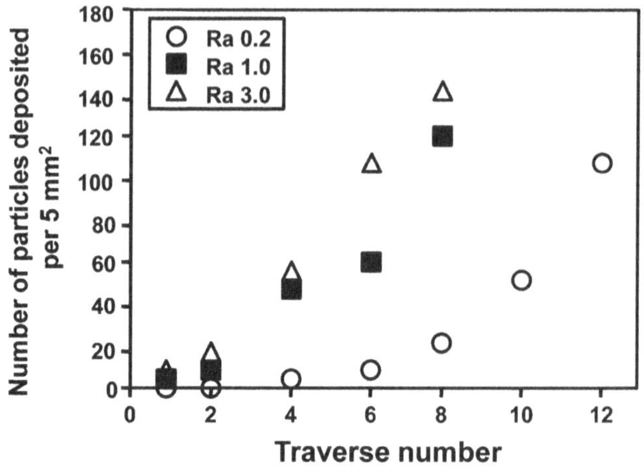

Fig. 6 Effect of surface roughness on particle deposition

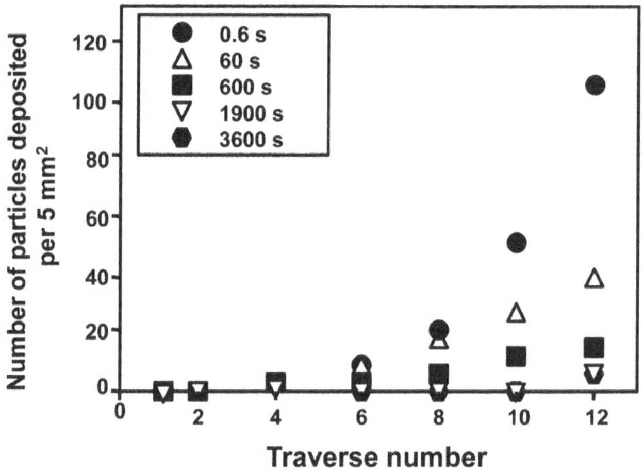

Fig. 7 Effect of traverse interval on deposition

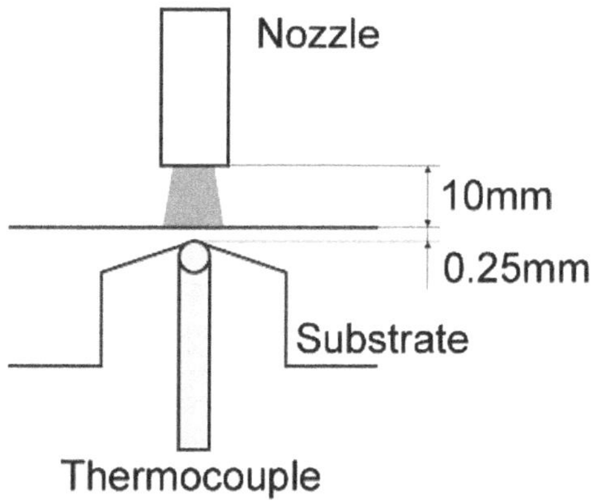

Fig. 8 Schematic illustration of temperature measurement

which propagated down to the interface between the aluminum coating and the substrate. Finally, a crack was formed along the horizontal direction. The point where the coating fractured (Fig. 10) possibly corresponded to the point at which the vertical crack initiated.

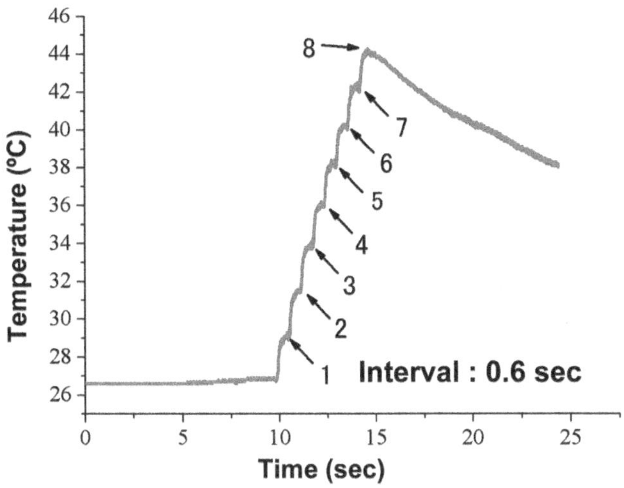

Fig. 9 Results of temperature measurement during scanning of the gun over the surface

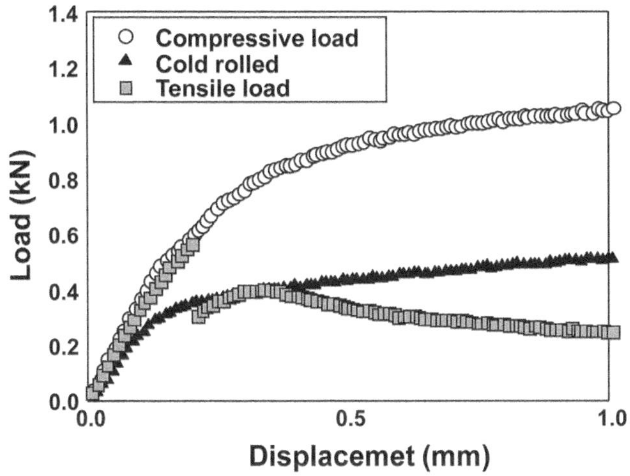

Fig. 10 Results of four-point bending tests

The sprayed particles are attracted toward the substrate or the deposited particles. A strong adhesion can be attained between the particles and the substrate and cohesion between the deposited particles. However, the cohesion between adjacent particles cannot be strong. Therefore, it is easy to initiate a vertical crack in the tensile loaded specimen. An SEM image of the fracture surface of the tensile loaded specimen after being subjected to the four-point bending test is shown in Fig. 12. In this figure, the deposited aluminum particles are indicated by arrows. Consequently, it is assumed that the cohesive strength between adjacent particles is weak. To improve the cohesive strength, the aluminum coatings were subjected to heat treatment.

To improve the strength of the specimens under tensile loading, they were subjected to heat treatment. The temperature for heat treatment of pure aluminum was set at 270 °C, on the basis of a study (Ref 19). The heat treatment time was evaluated by hardness tests. After 9 h of heat treatment, the hardness values of the specimens

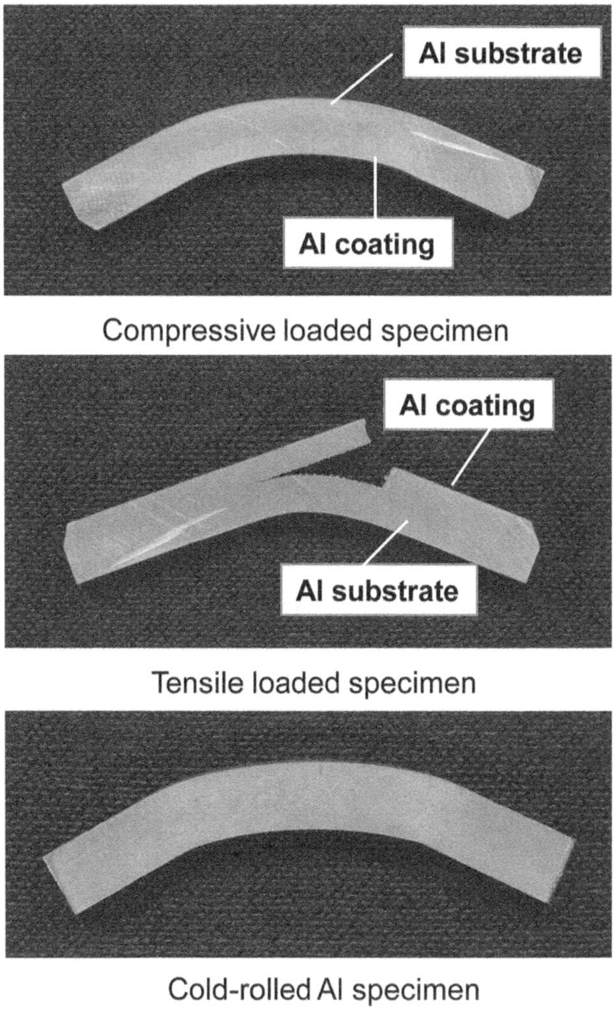

Fig. 11 Images of specimens after four-point bending tests

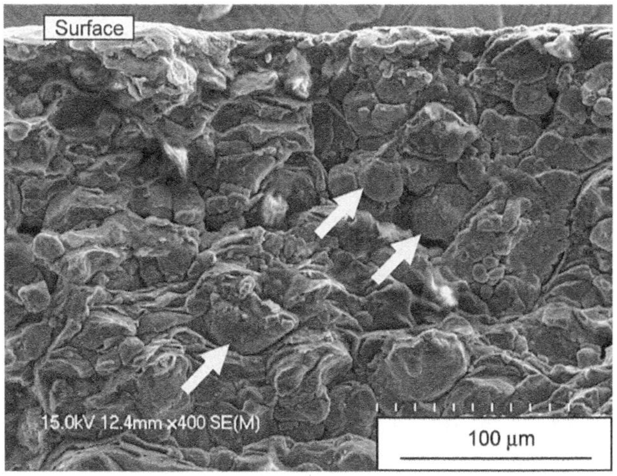

Fig. 12 Typical SEM image of fracture surface of tensile loaded specimen

became saturated. From this result, it was concluded that aluminum recrystallized after 9 h; therefore, the specimens were treated for 9 h.

Peer Reviewed

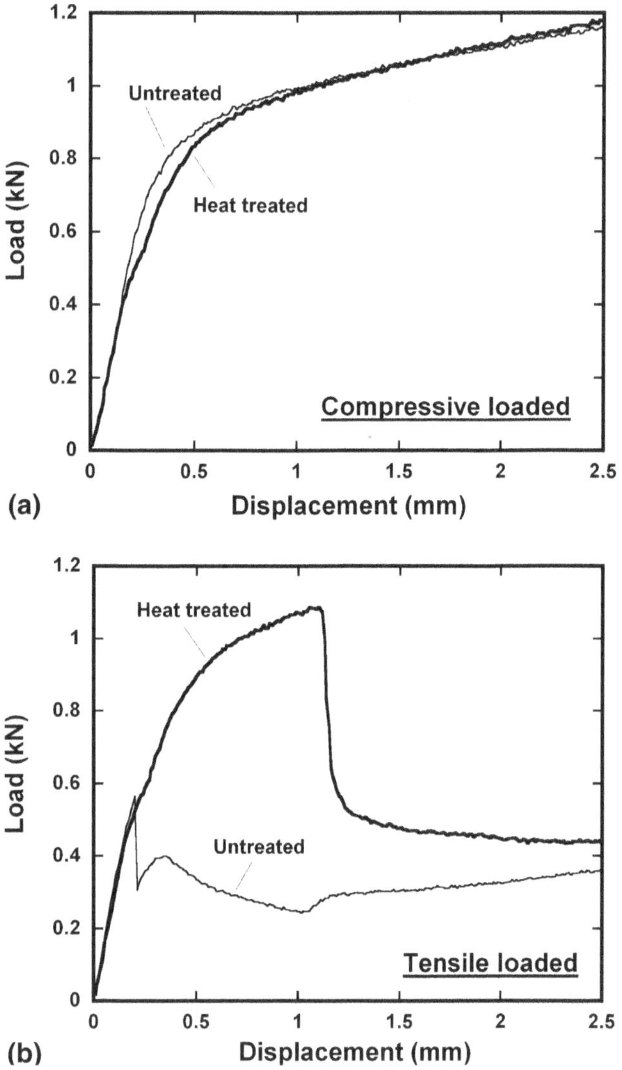

Fig. 13 Results of four-point bending tests of heat-treated specimens. (a) Compressive loaded and (b) Tensile loaded

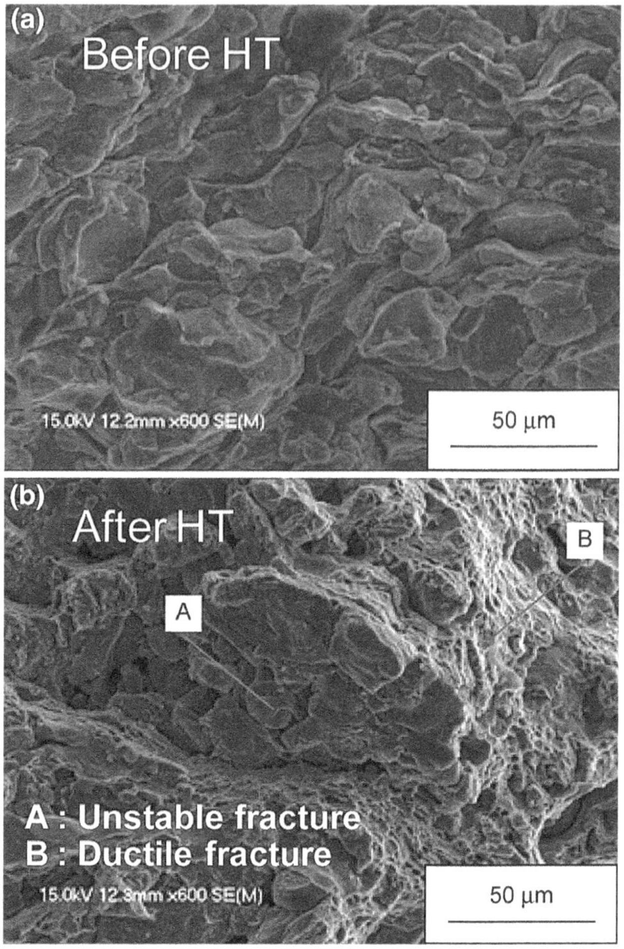

Fig. 14 Typical SEM images of fracture surface of heat-treated specimens after four-point bending tests

The results of the four-point bending tests of the heat-treated (HT) specimens are shown in Fig. 13. In the case of compressive loading, the untreated specimens showed higher strength as compared to the HT specimens because of the decrease in the hardness value after heating. On the other hand, in the case of tensile loading, the HT specimens showed significantly higher strength and displacement as compared to the untreated specimens; the strength of the former is two times higher than that of the latter. Moreover, the displacement of the HT specimens is five times higher than the untreated specimens.

The typical SEM images of the fracture surfaces of the tensile loaded specimens after being subjected to the four-point bending tests are shown in Fig. 14. In the case of the untreated specimens, few particles remain on the fracture surface, leading to the unstable fractures. On the other hand, in the HT specimens, due to the formation of ductile fracture surfaces, their strength and displacement were improved dramatically.

4. Conclusions

To evaluate the effects of surface conditions on aluminum deposition by the low-pressure cold-spray technique, the aluminum coatings on aluminum substrates have been characterized. The following main conclusions can be drawn:

1. In the case of low-pressure cold spraying, there is a time delay in particle deposition. This delay caused change in surface conditions, aiding particle deposition. During spraying, the substrate surface was exposed due to repeated impinging, resulting in the cleaning of the surface and formation of active newly formed surface.
2. The repeated impingements improve the deposition efficiency of particles. However, in the case of low-pressure cold spray, the deposition efficiency is very low, approximately 7%.
3. The repeated impingements enhance particle deposition. Hence, even though low-pressure cold spray exhibited low deposition efficiency, the velocity of particles deposited is lower than that in the case of

high-pressure cold spray. Further, the latter may not require repeated impingements because it has sufficient energy for formation of active newly formed surface due to the high-speed and high-pressure impingements. Accordingly, it is assumed that the difference in the critical velocities of the low- and high-pressure cold sprays is due to the difference in the mechanism of the active newly formed surface of the two techniques.

4. In the case of tensile loading, vertical cracks were formed in the aluminum coatings and the load value droped off sharply. However, the deposits have sufficient strength within the elastic region. Further, after heat treatment, the elongation of the specimens was improved.
5. Under compressive loading, the cold-sprayed specimens showed higher strength as compared to the cold-rolled aluminum specimen.
6. Under tensile loading, the heat-treated specimens showed significant improvement in strength and displacement. The strength of the HT specimens is two times higher than that of the untreated specimens. Moreover, the displacement of the former is five times longer than that of the latter.

Acknowledgments

This research was partially supported by the 21st COE Program "The Exploration of the Frontiers of Mechanical Science Based on Nanotechnology," Tohoku University, by the Industrial Technology Research Grant Program, 2008, of the New Energy and Industrial Technology Development Organization (NEDO), Japan, and by Japan-France Integrated Action Program SAKURA.

References

1. A.P. Alkhimov, V.F. Kosarev, and A.N. Papyrin, A Method of Cold Gas-Dynamic Deposition, *Sov. Phys. Dokl.*, 1990, **35**, p 1047-1049
2. A.P. Alkimov, V.F. Kosarev, N.I. Nesterovich, and A.N.Papyrin, Method of Applying Coatings, Russian Patent No. 1618778, 8 Sept 1990
3. A.P. Alkhimov, A.N. Papyrin, V.F. Kosarev, N.I. Nesterovich, and M.M. Shushpanov, Gas-Dynamic Spray Method for Applying a Coating, U.S. Patent No. 5302414, 12 April 1994
4. A.P. Alkhimov, A.N. Papyrin, V.F. Kosarev, N.I. Nesterovich, and M.M. Shushpanov, Method and Device for Coating, European Patent No. 0484533 B1, 25 Jan 1995
5. A.P. Alkimov, A.N. Papyrin, V.F. Kosarev, N.I. Nesterovich, et al., Gas Dynamic Spraying Method for Applying a Coating, U.S. Patent No. 5302414, April 12, 1994, Re-examination Certificate, 25 Feb 1995
6. F. Leonardi, J.M. Ginder, and R.C. McCune, Method of Manufacturing Electromagnetic Devices Using Kinetic Spray, U.S. Patent No. 6592935, 15 July 2003
7. A.N. Papyrin, Cold Spray Technology, *Adv. Mater.*, 2001, **159**, p 49-51
8. W. Kroemmer, P. Heinrich, and P. Richter, Cold Spraying—Equipment and Application Trends, *Proceedings of the International Thermal Spray Conference 2003*, B.R. Marple and C. Moreau, Eds., May 5-8, 2003 (Orlando, FL), ASM International, 2003, p 97-102
9. R.E. Blose, T.J. Roemer, A.J. Mayer, D.E. Beatty, and A.N. Papyrin, Automated Cold Spray System: Description of Equipment and Performance Data, *Proceedings of the International Thermal Spray Conference 2003*, B.R. Marple and C. Moreau, Eds., May 5-8, 2003 (Orlando, FL), ASM International, 2003, p 103-111
10. J. Karthikeyan, and C.M. Kay, Cold Spray Technology: An Industrial Perspective, *Proceedings of the International Thermal Spray Conference 2003*, B.R. Marple and C. Moreau, Eds., May 5-8, 2003 (Orlando, FL), ASM International, 2003, p 117-121
11. H.Y. Lee, S.H. Jung, S.Y. Lee, Y.H. You, and K.H. Ko, Correlation Between Al_2O_3 Particles and Interface of Al-Al_2O_3 Coatings by Cold Spray, *Appl. Surf. Sci.*, 2005, **252**, p 1891-1898
12. H. Assadi, F. Gartner, T. Stoltenhoff, and H. Kreye, Bonding Mechanism in Cold Gas Spraying, *Acta Mater.*, 2003, **51**, p 4379-4394
13. T. Schmidt, F. Gartner, H. Assadi, and H. Kreye, Development of a Generalized Parameter Window for Cold Spray Deposition, *Acta Mater.*, 2006, **54**, p 729-742
14. A.N. Paryrin, S.V. Klinkov, and V.F. Kosarev, Modeling of Particle–Substrate Adhesive Interaction Under the Cold Spray Process, *Proceedings of the International Thermal Spray Conference 2003*, B.R. Marple and C. Moreau, Eds., May 5-8, 2003 (Orlando, FL), ASM International, 2003, p 27-35
15. R.C. Dykhuizen, M.F. Smith, D.L. Gilmore, R.A. Neiser, X. Jiang, and S. Sampath, Impact of High Velocity Cold Spray Particles, *J. Therm. Spray. Technol.*, 1999, **8**, p 559-564
16. J. Wu, H. Fang, S. Yoon, H.J. Kim, and C. Lee, The Rebound Phenomenon in Kinetic Spraying Deposition, *Scr. Mater.*, 2006, **54**, p 665-669
17. Y. Ichikawa, K. Ito, K. Ogawa, and T. Shoji, Deposition Mechanisms of Low Pressure Type Cold Sprayed Aluminum Coatings, *Surface Modification Technologies XXI*, 2007, p 497-506
18. K. Tanaka, M. Kumagai, and H. Yoshida, Dissimilar Joining of Aluminum Alloy and Steel Sheets by Friction Stir Spot Welding, *J. Jpn. Inst. Light Metals*, 2006, **56**, p 317-322 (in Japanese)
19. M. Tsuchida, H. Yoshida, and K. Hirano, Heat Treatment of Aluminum Alloys, *Sumitomo Metal Giho*, 1990, **31**, p 126-143 (in Japanese)

JTTEE5 17:736–741
DOI: 10.1007/s11666-008-9256-3
1059-9630/$19.00 © ASM International

Peer Reviewed

Effect of Process Conditions on Microstructure and Corrosion Resistance of Cold-Sprayed Ti Coatings

Hong-Ren Wang, Bao-Rong Hou, Jun Wang, Qi Wang, and Wen-Ya Li

(Submitted May 13, 2008; in revised form September 21, 2008)

Ti and Ti alloys can be applied to steels as a protective coating in view of its excellent resistance to corrosive environment. Cold spraying, as a new coating technique, has potential advantages in fabrication of Ti coating in comparison with conventional thermal spraying techniques. In this study, Ti coatings were prepared on carbon steel substrates by cold spraying via controlling the process conditions. The microstructure of coatings was observed by SEM. The porosity of coatings was estimated by image analysis and the bond strength was tested for comparison of the process conditions. Potentiodynamic polarization and open-circuit potential (OCP) measurements were performed to understand the corrosion behavior of the coatings. The SEM examination shows that the coatings become more compact with the increases of pressure and temperature of driving gas. The potentiodynamic polarization curves indicate that the coating which has lower porosity has lower corrosion current. The polarization and OCP measurement reveal that cold-sprayed Ti coating can provide favorable protection to carbon steel substrate. The polishing treatment of coating surface polishes the rough outer layer including the small pores as well as decreases the actual surface area of the coating, leading to the considerable improvement of corrosion resistance.

Keywords cold spray, corrosion resistance, microstructure, process conditions, Ti coating

1. Introduction

Steels used in oceanic engineering are prone to corrosion and invalidation due to their lower corrosion resistance in marine environment. Titanium and its alloys have an excellent corrosion resistance in the severe marine environment and can be widely used for the protection of steel structures. So the corrosion resistance of steels can be improved by covering Ti coatings on their surfaces. Ti coatings are generally deposited by thermal spray process (Ref 1-6). However, owing to its high activity at elevated temperatures, thermal spraying of Ti in an air atmosphere usually results in a severely oxidized, unhomogeneous coating, which is brittle and porous. Therefore, in order to ensure the phase and chemical purity of Ti coatings, the spray process is usually conducted at a shrouded or low-pressure atmosphere, such as the shrouded plasma spray (APS/S), low-pressure plasma spray (LPPS). Other thermal spray technique is also used to produce dense Ti coatings such as modified HVOF spraying.

Cold spraying, as an emerging coating technique, has been developed to deposit high-quality metallic coatings. In this process, high pressure gas is introduced into a Laval type nozzle and produces a high-speed gas flow, and then spray particles are fed axially into the gas flow and accelerated to a high velocity (300-1200 m/s) and deposit through intensive plastic deformation upon impact on a substrate in a solid state and at a temperature well below the melting point of spray material (<600 °C) (Ref 7). According to lots of studies on cold spray since its invention, cold spraying is a suitable technique for production of most metals and their alloys (Ref 7-13), and even cermet coatings (Ref 14). It is generally accepted that there exists a critical velocity for a successful particle deposition. Only the particles have a velocity higher than this critical value can be deposited to form a coating. There are some reports concerned with the spray forming,

This article is an invited paper selected from presentations at the 2008 International Thermal Spray Conference and has been expanded from the original presentation. It is simultaneously published in *Thermal Spray Crossing Borders, Proceedings of the 2008 International Thermal Spray Conference*, Maastricht, The Netherlands, June 2-4, 2008, Basil R. Marple, Margaret M. Hyland, Yuk-Chiu Lau, Chang-Jiu Li, Rogerio S. Lima, and Ghislain Montavon, Ed., ASM International, Materials Park, OH, 2008.

Hong-Ren Wang and **Bao-Rong Hou**, Institute of Oceanology, Chinese Academy of Sciences, Qingdao 266071, People's Republic of China; **Hong-Ren Wang**, Graduate School of the Chinese Academy of Sciences, Beijing 100039, People's Republic of China; **Hong-Ren Wang, Jun Wang,** and **Qi Wang**, State Key Laboratory For Marine Corrosion and Protection, Luoyang Ship Material Research Institute (LSMRI), No.12A Jinhu Road, Qingdao 266071, Shandong Province, People's Republic of China; and **Wen-Ya Li**, Shaanxi Key Laboratory of Friction Welding Technologies, School of Materials Science and Engineering, Northwestern Polytechnical University, Xian 710072, People's Republic of China. Contact e-mail: wanghr@sunrui.net.

numerical simulation and experimental characterization, metal reactivity of cold spraying Ti coatings (Ref 15-19).

In the present study, the influence of process conditions, including gas type and its pressure and temperature on the microstructure and corrosion performance of cold-sprayed Ti coating were examined experimentally, compared with Ti bulk metal and carbon steel substrate, in order to optimize the process conditions and improve its anticorrosive property.

2. Materials and Experimental Procedures

2.1 Materials and Sample Preparation

Angular Ti powder sieved to the size from 5 to 45 μm was used as feedstock material, which was manufactured through hydrogenizing and dehydrogenating method. The composition of Ti powder is shown in Table 1. Carbon steel (Chinese trademark Q235A) plates were used as substrates. Prior to spraying, the substrate surface was sandblasted using 24 mesh alumina grits.

The cold spray system Cold Spray CS2000 was used to produce Ti coatings (Ref 13). The system included gas pressure regulators, gas preheater, powder feeder, and spray gun. The spray gun mainly consisted of a gas pre-chamber and a convergent-divergent accelerating nozzle. The nozzle had a throat diameter of 2 mm and an exit diameter of 6 mm. The length from the throat to the exit was 100 mm. The gas temperature and pressure in the pre-chamber were measured via the thermocouple and pressure gage mounted on the spray gun.

The driving gas used in cold spray process is pure N_2 or compressed air with a pressure up to 3 MPa, which is also used as the powder feeder gas synchronously. The controlled conditions of cold spray process applied to deposit different types of Ti coatings are listed in Table 2. In cold spraying process, the gun was controlled by a robot and the relative traverse speed of spray gun relative to substrate was 80 mm/s with a standoff distance of 15 mm.

2.2 Experimental Methods

The microstructure of Ti coating was characterized using an environmental scanning electron microscopy (ESEM, Philips XL-30) with voltage of 20 kV. The bond strength of coatings was tested using a WDW-20 tensile-testing machine at a rate of 10 mm/min with three parallel samples. The porosity of coatings was evaluated by image analysis with photoshop software.

The corrosion performance of Ti coatings was investigated in a 3.5wt.% NaCl solution at 30 °C using the as-sprayed and as-polished coatings, which were finished with Al_2O_3 abrasive paper from 220# to 1200# and polished with W2.5-μm silicon carbide abrasive on polishing cloth. The thickness of as-sprayed and as-polished Ti coatings was up to ~1 mm. The sample 2 Ti coatings were used for the open-circuit potential (OCP) measurement solely. The samples for polarization and OCP measurement were cut and sealed with epoxy to insulate the substrate and edges, only exposing a working coating surface of 10 × 10 mm. Samples of TA2 bulk metal and carbon steel substrate were prepared to the same dimension with a mechanical treatment surface. Samples for each polarization and OCP measurement had three parallel ones.

A Model 273A Potentiostat was used for potentiodynamic polarization measurement and a multicenter Potentiostat for OCP measurement. A saturated calomel electrode (SCE) with a Luggin capillary was used as a reference electrode, and a platinum wire twisted to a

Table 1 Composition of Ti powder

Element	O	Cl	H	N	C	Si	Fe	Ti
Composition, wt.%	0.33	0.03	0.024	0.056	0.01	<0.02	0.05	Bal.

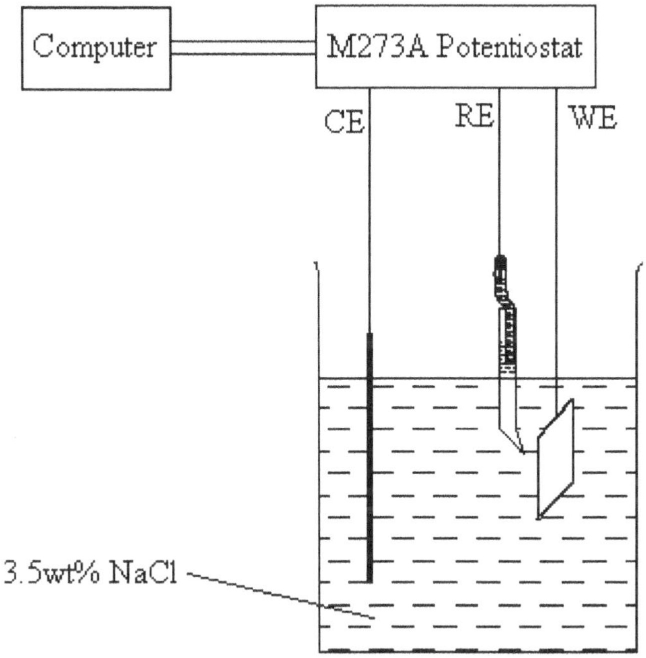

Fig. 1 Schematic diagram of electrochemical test system. CE, counter electrode; RE, reference electrode; WE, working electrode

Table 2 The controlled process conditions applied to deposit Ti coatings

Parameter	Sample 1	Sample 2	Sample 3	Sample 4	Sample 5	Sample 6
Driving gas	N_2	N_2	N_2	Compressed air	Compressed air	Compressed air
Temperature, °C	350 (T1)	500 (T2)	650 (T3)	500 (T2)	500 (T2)	500 (T2)
Pressure, MPa	2.0 (P1)	2.0 (P1)	2.0 (P1)	2.0 (P1)	1.8 (P2)	1.5 (P3)

multi-ring shape served as a counter electrode. Figure 1 shows the schematic diagram of the electrochemical test system. An initial delay time of 1 h was employed before polarization measurement. The polarization scan was started from 600 mV below the steady OCP to 600 mV over the OCP at a scanning rate of 10 mV/min. OCP measurement was carried out in 3.5wt.% NaCl solution for 30 days at an interval of 5-30 min according to the corrosion characteristic and experimental period.

3. Results and Discussion

3.1 Microstructure of Ti Coatings Using Different Operation Temperatures

Figure 2 shows the typical cross-sectional microstructure of cold-sprayed Ti coatings deposited using N_2 under a pressure of 2 MPa and temperatures of T1-T3. It is found that the interface of three kinds of coatings between the coating and substrate seems well bonded with little disfigurement. The average bond strength and porosity of coatings were listed in Table 3. It is found that the bond strength increases with the increase of gas temperature. And the density of coatings has also been enhanced obviously with increasing the gas temperature. The coating deposited under the temperature T3 has the most compact microstructure with porosity of 1.6%, and the coating under the temperature T1 has a relative porous structure with porosity of 10.3%.

It could be attributed to the increase of gas velocity due to the increase of gas temperature, which gives more kinetic energy to the particles. This enhances the plastic deformation during particle deposition. Further deformation of the previously deposited coating surface by the impacts of following particles is also obtained. The accumulative intensive deformation enhances the particle cohesion and leads to the formation of a dense coating (Ref 13).

3.2 Microstructure of Ti Coatings Using Different Operation Pressures

Figure 3 shows the typical cross-sectional microstructure of cold-sprayed Ti coatings deposited using compressed air under a temperature of T2 and pressures of P3 to P1. The bond strength and porosity were listed in Table 3. It can be found that the coating deposited under the pressure of P1 is denser with porosity of 10.0%. The coating becomes porous under the pressure of P2 with porosity of 14.8%, and the coating deposited under the pressure of P3 has the most porous microstructure with porosity of 17.5%. Table 3 shows that the coating bond strength increases with the increase of gas pressure.

It can be demonstrated that the density of coating could be improved by increasing the gas pressure. It could be attributed to the increase of gas velocity and dragging force due to the increase of gas pressure. The kinetic energy of particles is enhanced, leading to the formation of a denser coating (Ref 20).

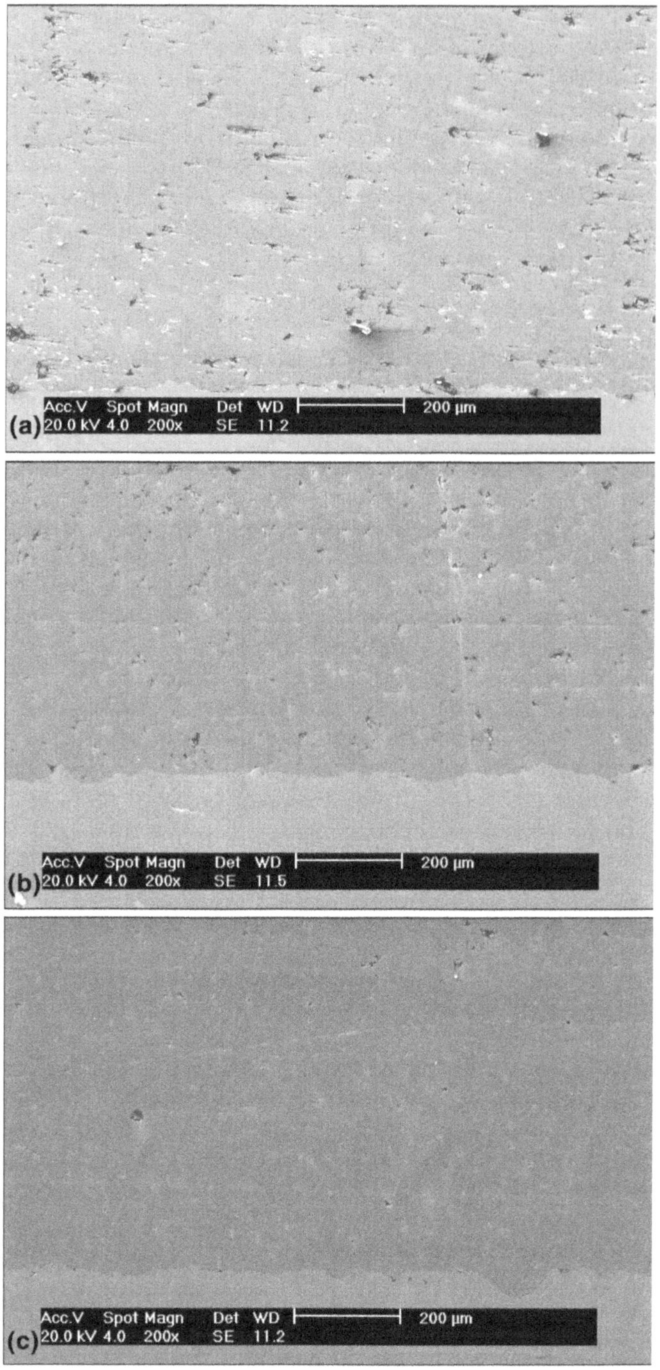

Fig. 2 Typical cross-sectional microstructure of cold-sprayed Ti coatings at the temperatures of (a) T1, (b) T2, and (c) T3 using N_2 under a pressure of P1

Table 3 Average bond strength and porosity of Ti coatings

Sample number	1	2	3	4	5	6
Bond strength, Mpa	9.4	11.2	16.1	10.3	10.0	8.7
Porosity, %	10.3	5.8	1.6	10.0	14.8	17.5

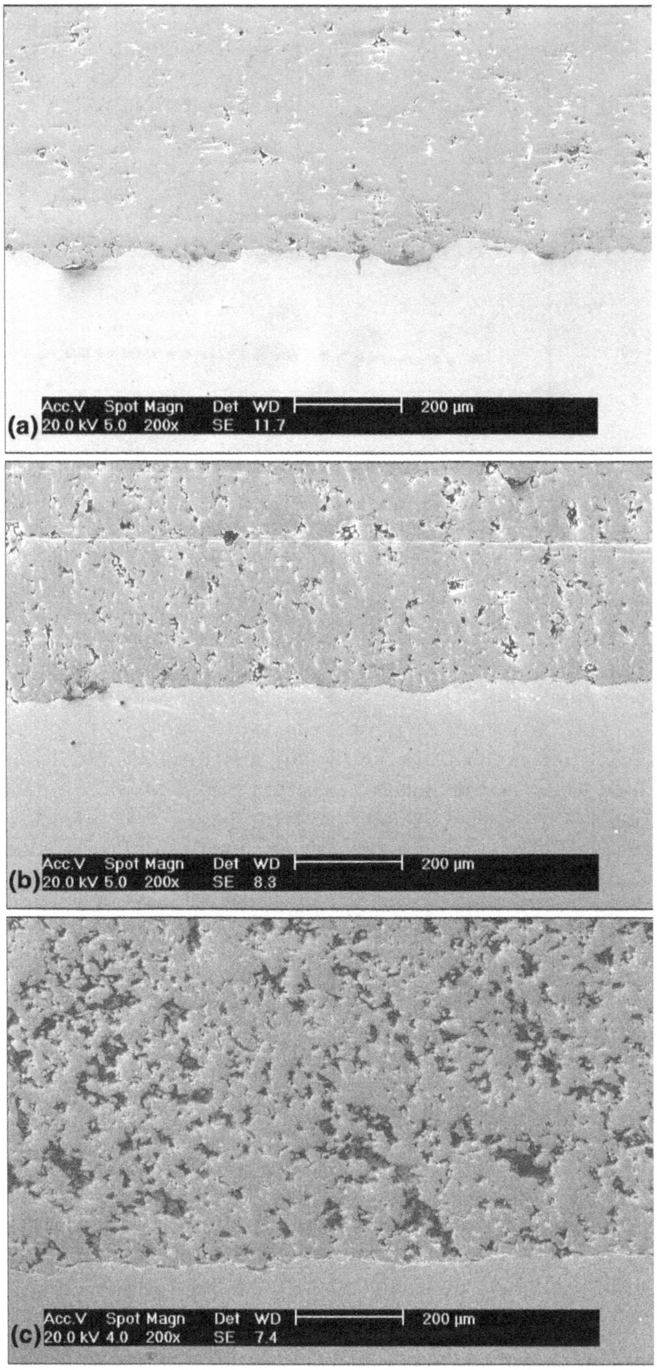

Fig. 3 Typical cross-sectional microstructure of cold-sprayed Ti coatings at the pressures of (a) P1, (b) P2, and (c) P3 using compressed air under a temperature of T2

3.3 Microstructure of Ti Coatings Using Different Driving Gas

Figure 2(b) and 3(a) show the comparison of the cross-sectional microstructure of different coatings using N_2 and compressed air as driving gas, respectively, under the same temperature and pressure. According Table 3, it is found that the coating deposited with N_2 has a much denser structure and higher bond strength than that deposited with air. The gas type has an evident influence on the

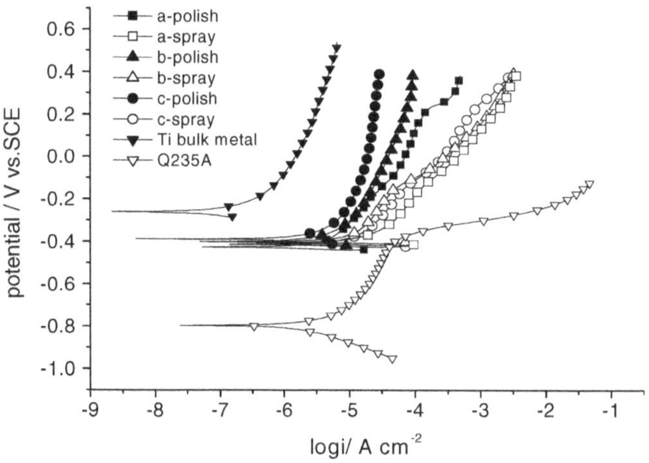

Fig. 4 Polarization curves of cold-sprayed Ti coatings at the pressures of (a) P3, (b) P2, and (c) P1 using compressed air under a temperature of T2 with TA2 bulk metal and Q235A substrate in 3.5wt.% NaCl

porosity of coating. The result is mainly attributed to the surface reactivity of Ti with oxygen. Besides particle velocity, the metal reactivity and oxide films at particle surfaces also take important roles on coating deposition process in cold spraying (Ref 17). The surface reactivity of Ti particle with oxygen in the adopted compressed air is evidently higher than that with entrained air by N_2 driving gas at the gun exit. This surface reaction could be helpful for the formation of a relative porous structure in compressed air (Ref 17-19).

3.4 Corrosion Resistance of Ti Coatings with Different Operation Conditions

Figure 4 shows the potentiodynamic polarization curves for the as-sprayed and as-polished Ti coatings using compressed air as driving gas under pressures of P3 to P1 at a temperature of T2. The as-polished coating has almost the similar polarization behavior with Ti bulk metal, but it has a larger polarization current due to their difference in density. The as-sprayed coatings have smaller polarization slope and larger polarization current compared with the as-polished coatings. It is clear that the polarization currents of coatings deposited under different pressures are different. The polarization current decreases with the increase of gas pressure, and the polarization current of as-sprayed surface is larger than the corresponding as-polished surface. This is because the polishing treatment peels the rough outer layer including small pores, as well as decreasing the actual surface area of the coating, leads to the considerable improvement of corrosion resistance.

As aforementioned, the coatings deposited under a lower pressure has more porous microstructure, the dissolved oxygen concentration is lower within the porous structure compared with the surrounding environment. Since Ti is a passive metal and its corrosion resistance depends on sufficient dissolved oxygen to keep its passivation in corrosive medium. The regions with lower dissolved oxygen concentration become the active corrosion

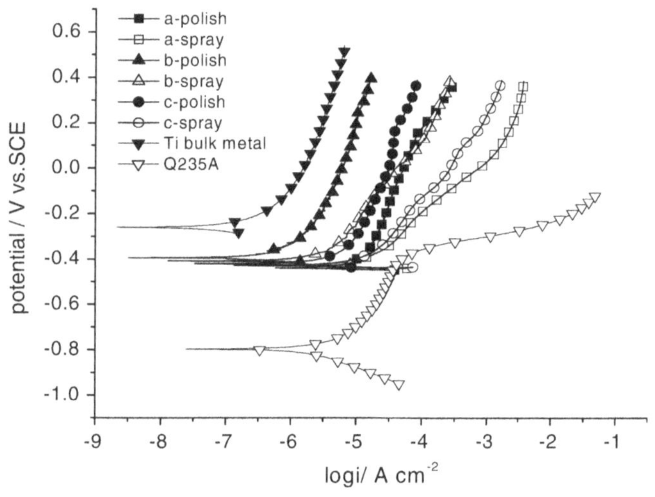

Fig. 5 Polarization curves of cold-sprayed Ti coatings at the temperatures of (a) T1, (b) T2, and (c) T3 using N_2 under a pressure of P1 with TA2 bulk metal and Q235A substrate 3.5wt.% NaCl

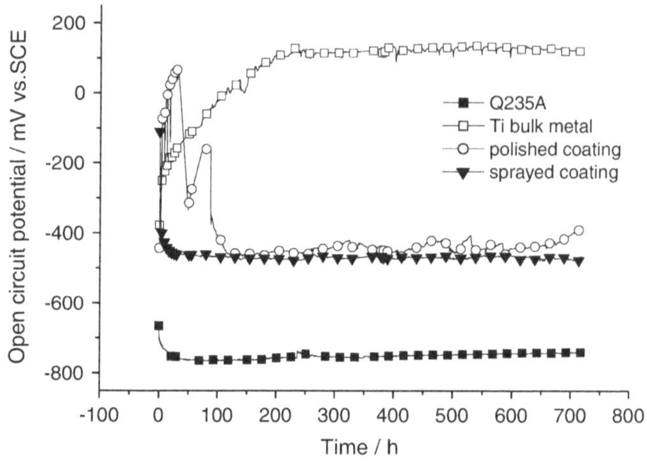

Fig. 6 Change of open-circuit potential versus immersion time of as-sprayed, as-polished sample-2 Ti coatings with TA2 bulk metal and Q235A substrate in 3.5wt.% NaCl

center and lead to the oxygen concentration difference corrosion (Ref 21).

Figure 5 shows the polarization curves of the as-sprayed and as-polished Ti coatings deposited using N_2 under different temperatures. There is obvious difference with polarization slope and current between the curves of coatings deposited under different temperatures. The as-polished coatings have higher corrosion resistance than the corresponding as-sprayed coatings.

According to Fig. 5, the coating deposited under the temperature T1 has the largest corrosion current density, and the coating deposited under the temperature T2 has the smallest one. The corrosion currents of coatings are correlative with the porosity and oxygen content on coating surface. The two factors influence the corrosion behavior of coating simultaneously and emulously. Due to the influence of the oxygen content to corrosion exceeds that of the density and keeps the dominant step, the corrosion current of coating deposited under T3 exceeds that of coating under T2.

According to the more negative corrosion potential and larger corrosion current of Q235A obtained from plorization curves in Fig. 4 and 5, the carbon steel substrate reveals much lower corrosion resistance than the coatings.

Figure 6 shows the change of OCP with immersion time for the as-sprayed and as-polished Ti coatings compared with TA2 bulk metal and Q235A substrate in 3.5wt.% NaCl. The OCP of TA2 bulk metal gradually shifts positively from −360 mV to 100 mV (SCE) in the initial stages of 200 h, subsequently achieves relatively steady OCP and finally keeps at 130 mV (SCE), indicating evident passivation occurs during immersion. The OCP of two coatings show negative shift due to initial corrosion. The OCP of as-sprayed coating achieves relatively steady state after 50 h and keeps at −470 mV (SCE), while the as-polished coating achieves steady OCP after 120 h and gradually shifts positively from −450 mV to −380 mV (SCE), indicating it is prone to achieve passivation than as-sprayed coating. The OCP of Q235A substrate keeps steadily at about −740 mV (SCE) due to its general corrosion.

The OCP of the coating was more negative than the TA2 bulk metal from 500 to 600 mV. It could be attributed to the porous surface structure of the coatings, which leads to a more active surface than that of TA2 (Ref 18). No evidence reveals that carbon steel substrate being corroded after 30 days immersion. It can be concluded from the OCP measurement that carbon steel substrate is favorable protected by coating.

4. Conclusions

For cold-sprayed Ti coatings, the gas pressure and temperature influence significantly the density of the deposited coatings. As the gas temperature increases, the porosity decreases. A higher gas pressure yields a lower porosity. The porosity of coating deposited using compressed air is relatively serious than coating deposited using N_2 under the same operation pressure and temperature due to the Ti reactivity with oxygen.

Potentiodynamic polarization results indicated that the coatings which had lower porosity had lower corrosion current. The polarization and OCP measurement reveals that cold-sprayed Ti coating has much higher corrosion resistance than carbon steel and can provide favorable protection to the substrate. The polishing treatment with coating surface peels the rough outer layer including the small pores, as well as decreasing the actual surface area of coating, leads to the considerable improvement of corrosion resistance.

Acknowledgment

This article is financially supported by the Foundation of China Shipbuilding Industry Corporation 03J2.2.1.

References

1. T. Kinos, S.L. Chen, P. Siitonen, and P. Kettunen, Corrosion Properties of Shrouded Plasma Sprayed Titanium Coatings, *Thermal Spraying—Current Status and Future Trends*, A. Ohmori, Ed. (Kobe, Japan), High Temperature Society of Japan, 1995, p 573-576
2. J. Kawakita, S. Kuroda, T. Fukushima, H. Katanoda, K. Matsuo, and H. Fukanuma, Dense Titanium Coatings by Modified HVOF Spraying, *Surf. Coat. Technol.*, 2006, **201**(3-4), p 1250-1255
3. E. Galvanetto, F.P. Galliano, F. Borgioli, U. Bardi, and A. Lavacchi, XRD and XPS Study on Reactive Plasma Sprayed Titanium-Titanium Nitride Coatings, *Thin Solid Films*, 2001, **384**(2), p 223-229
4. C.-E. Cui, Q. Miao, and J.-D. Pan, Ti/Cr Multi-Layer Coating on Magnesium Alloy AZ91 by Arc-Added Glow Plasma Depositing Technique, *Surf. Coat. Technol.*, 2007, **201**(9-11), p 5400-5403
5. J. Kawakita, H. Katanoda, M. Watanabe, K. Yokoyama, and S. Kuroda, Warm Spraying: An Improved Spray Process to Deposit Novel Coatings, *Surf. Coat. Technol.*, 2008, **202**(18), p 4369-4373
6. H.R. Salimijazi, T.W. Coyle, J. Mostaghimi, and L. Leblanc, Microstructure and Failure Mechanism in As-Deposited, Vacuum Plasma-Sprayed Ti-6Al-4V Alloy, *J. Therm. Spray Technol.*, 2005, **14**(2), p 215-223
7. A. Papyrin, A Cold Spray Technology, *Adv. Mater. Process*, 2001, **159**(9), p 49-51
8. T. Stoltenhoff, H. Kreye, and H.J. Richter, An Analysis of the Cold Spray Process and Its Coatings, *J. Therm. Spray Technol.*, 2002, **11**(4), p 542-550
9. R.C. McCune, W.T. Donlon, O.O. Popoola, and E.L. Cartwright, Characterization of Copper Layers Produced by Cold-Gas-Dynamic Spraying, *J. Therm. Spray Technol.*, 2000, **9**(1), p 73
10. T.H. Van Steenkiste, J.R. Smith, R.E. Teets, J.J. Moleski, D.W. Gorkiewicz, R.P. Tison, D.R. Marantz, K.A. Kowalsky, W.L. Riggs, P.H. Zajchowski, B. Pilsner, R.C. McCune, and K.J. Barnett, Kinetic Spray Coatings, *Surf. Coat. Technol.*, 1999, **111**(1), p 62
11. J. Karthikeyan, C.M. Kay, J. Lindeman, R.S. Lima, and C.C. Berndt, Cold Spray Processing of Titanium Powder, *Thermal Spray: Surface Engineering via Applied Research*, C.C. Berndt, Ed. (Materials Park, OH), ASM International, 2000, p 255-262
12. J. Karthikeyan, C.M. Kay, J. Lindemann, R.S. Lima, and C.C. Berndt, Cold Sprayed Nanostructured WC-Co, *Thermal Spray 2001: New Surfaces for a New Millennium*, C.C. Berndt, K.A. Khor, and E.F. Lugscheider, Eds., (Materials Park, OH), ASM International, 2001, p 383-387
13. C.-J. Li and W.-Y. Li, Deposition Characteristics of Titanium Coating in Cold Spray, *Surf. Coat. Technol.*, 2003, **167**(2-3), p 278-283
14. H.J. Kim, C.H. Lee, and S.Y. Hwang, Fabrication of WC-Co Coatings by Cold Spray Deposition, *Surf. Coat. Technol.*, 2005, **191**(2-3), p 335-340
15. T. Marrocco, D.G. McCartney, P.H. Shipway, and A.J. Sturgeon, Production of Titanium Deposits by Cold-Gas Dynamic Spray: Numerical Modeling and Experimental Characterization, *J. Therm. Spray Technol.*, 2006, **15**(2), p 263-272
16. R.E. Blose, Spray Forming Titanium Alloys Using the Cold Spray Process, *Thermal Spray Connects: Explore Its Surfacing Potential*, E. Lugscheider, Ed. (Düsseldorf, Germany), DVS, 2005 (in CD-ROM)
17. W.-Y. Li, C. Zhang, H.-T. Wang, X.-P. Guo, H.-L. Liao, C.-J. Li, and C. Coddet, Significant Influences of Metal Reactivity and Oxide Film of Powder Particles on Coating Deposition Characteristics in Cold Spraying, *Appl. Surf. Sci.*, 2007, **253**, p 3557-3562
18. H.R. Wang, W.Y. Li, J. Wang, Q. Wang, and L. Ma, Corrosion Behavior of Cold Sprayed Titanium Protective Coating on 1Cr13 Substrate in Seawater, *Surf. Coat. Technol.*, 2007, **201**(9-11), p 5203-5206
19. R.S. Lima, A. Kucuk, and C.C. Berndt, Deposition Efficiency, Mechanical Properties and Coating Roughness in Cold-Sprayed Titanium, *J. Mater. Sci. Lett.*, 2002, **21**, p 1687-1689
20. W.-Y. Li, C.-J. Li, Y.-Y. Wang, and G.-J. Yang, Effect of Parameters of Cold Sprayed Cu Particles on Its Impacting Behavior, *Acta Metall. Sin.*, 2005, **41**(3), p 282-286 (in Chinese)
21. J. Kawakita, T. Fukushima, S. Kuroda, and T. Kodama, Corrosion Behavior of HVOF Sprayed sus316L Stainless Steel in Sea Water, *Corros. Sci.*, 2002, **44**, p 2561-2581

JTTEE5 17:742–749
DOI: 10.1007/s11666-008-9258-1
1059-9630/$19.00 © ASM International

Peer Reviewed

Influence of Powder Porous Structure on the Deposition Behavior of Cold-Sprayed WC-12Co Coatings

Pei-Hu Gao, Yi-Gong Li, Chang-Jiu Li, Guan-Jun Yang, and Cheng-Xin Li

(Submitted May 13, 2008; in revised form September 9, 2008)

The limited deformation of hard cermet particles and impacted coating makes it difficult for conventional thermal spray powders to continuously build up on impact in cold spraying. In this study, three nano-structured WC-12Co powders with different porous structure and apparent hardness were employed to deposit WC-Co coatings on stainless steel substrate by cold spraying. The deposition characteristics of three powders of porosity from 44 to 5% were investigated. It was found that WC-Co coating is easily built-up using porous powders with WC particles bonded loosely and a low hardness. The microhardness of WC-12Co coatings varied from 400 to 1790 Hv with powders and spray conditions, which depends on the densification effects by impacting particles. With porous WC-Co powders, the fracture of particles on impact may occur and low deposition efficiency during cold spraying. The successful building up of coating at high deposition efficiency depends on the design of powder porous structure.

Keywords cold spraying, deposition behavior, hardness, porous structure, WC-12Co

1. Introduction

Carbide cermets are typical wear-resistant coating materials that are widely employed through thermal spraying. The carbide decomposition and dissolution into the binder occurs during thermal spraying of cermets, which deteriorates the performance of cermet materials. The carbide decomposition can be reduced to the minimum through limiting heating of spray powders. Therefore, high-velocity oxy fuel (HVOF) process becomes popular to deposit cermet coatings due to higher flame velocity but lower temperature compared with plasma spraying (Ref 1-10). With using a properly designed WC-Co powder, the decompositions can be reduced to a reasonable low degree (Ref 1). On the other hand, because of the necessity for sufficiently melting the binder alloy to increase the density of the deposit, the carbide dissolution into the molten binder is inevitable during thermal spraying (Ref 1-7). The dissolution becomes severe with the decrease of carbide particles owing to the increased specific surface area of smaller particles. Therefore, despite the limited dissolving time during HVOF, the dissolution can not be significantly avoided, especially when WC carbide particle size decreases to tens of nanometers (Ref 8-12).

With the development of cold spraying, it becomes possible to eliminate completely the decarburization of carbides through decompositions and dissolution because solid spray particles in a much low temperature are employed to deposit the coating (Ref 13-20). However, the limited deformation of both impacting hard cermet particles and the underlying cermet coating makes it difficult for the hard coating to continuously build up on impact in cold spraying using a conventional dense thermal spray powder. In previous papers (Ref 16-20), the deposition of WC-Co particles by cold spraying has been discussed intensively. It was speculated that the successful built-up of WC-Co coating requires the deformation of both the low part of impacting particle near the interface and the underlying WC-Co deposited previously. On the other hand, the upper part of the deposited particle retains its porous structure, which is kept as deformable under the impact of the next particle on its top. Such deformation requirement can easily be fulfilled using a porous powder. However, in such case, it can be considered that several requirements need to be fulfilled to deposit a dense WC-Co coating. The certain cohesion between WC particles bonded by the Co binder is required to avoid the splashing of small WC particles on impact. Then, the impacting particle acquires sufficient kinetic energy to force the

This article is an invited paper selected from presentations at the 2008 International Thermal Spray Conference and has been expanded from the original presentation. It is simultaneously published in *Thermal Spray Crossing Borders, Proceedings of the 2008 International Thermal Spray Conference*, Maastricht, The Netherlands, June 2-4, 2008, Basil R. Marple, Margaret M. Hyland, Yuk-Chiu Lau, Chang-Jiu Li, Rogerio S. Lima, and Ghislain Montavon, Ed., ASM International, Materials Park, OH, 2008.

Pei-Hu Gao, Yi-Gong Li, Chang-Jiu Li, Guan-Jun Yang, and **Cheng-Xin Li,** State Key Laboratory for Mechanical Behavior of Materials, School of Materials Science and Engineering, Xi'an Jiaotong University, Xi'an, Shaanxi 710049, P.R. China. Contact e-mail: licj@mail.xjtu.edu.cn.

underlying porous layer to deform sufficiently for full densification. However, excessive energy over that required for full densification may induce intensive cracking and erosion of the deposited layer. Therefore, the matching of porous structure with spray conditions is necessary. With certain porous WC-Co powder, the dense coating can be deposited only in a narrow window of spray conditions. Therefore, the investigations into the influences of porous structure of spray powder and substrate deformability on deposition behavior are essentially important issues to deposit a dense superhard cermet coating.

In this study, three porous WC-12Co powders with different porosity levels were used as feedstocks to deposit nanostructured WC-Co coatings on stainless steel substrate by cold spraying using helium as processing gas. The effect of powder porous structure on deposition characteristics was investigated. The dependence of successful build-up of WC-Co coating with a high deposition efficiency on the design of powder porous structure was discussed.

2. Experimental Materials and Procedures

The powder materials used for this work are three nanostructured WC-12Co powders. They are referred to as HP (high porosity), MP (middle porosity, Zhuzhou Cemented Carbide Group Corporation, Ltd., China), and LP (low porosity, InframatTM S7412, Inframat Corporation, USA) powders.

The cross-sectional microstructures of WC-12Co spray powders were characterized by scanning electron microscopy (SEM, VEGA II-XMU, TESCAN, Czech). The porosities of powders were determined using image processing from the cross section of three powders.

The home-made cold spraying system (CS2000) was employed to deposit the coatings in this study (Ref 21). A spray gun with a converging-diverging de Laval-type nozzle with a throat diameter of 2 mm was adopted. Helium gas was employed as both accelerating gas and powder feeding gas at a pressure of 2.0 and 2.5 MPa, respectively. The gas temperature in the prechamber was 640 °C. The standoff distance from the nozzle exit to the substrate surface was 20 mm. During deposition, the spray gun was manipulated by a robot at a traverse speed of 20 mm/s relative to the substrate for coating deposition and 800 mm/s for single particle deposition, respectively. Stainless steel with the surface sand-blasted was used as a substrate for coating deposition and the polished stainless steel was prepared for single particle deposition.

The microhardness of the powder was tested by a microhardness tester under a 20 g load for a loading time of 10 s owing to small size of the powders, while the microhardness of the coating was tested under a 300 g load for a loading time of 30 s.

The deposition efficiency was measured through the weight gain of the test specimen after spray torch traversed over the specimen and the weight of powders fed into the spray gun at a time interval that the gun took to traverse over the specimen. The deposition efficiency is defined as:

$$D_{\mathrm{E}} = \frac{\Delta m_{\mathrm{c}}}{\Delta m_{\mathrm{p}}}$$

where D_{E} is the deposition efficiency of powders, Δm_{c} is the weight increment of the substrate resulting from WC-Co deposition at a time interval that the spray gun traverses over the substrate, and Δm_{p} is the weight of WC-Co powders fed into the gun at the same time intervals.

3. Results

3.1 Characterization of the Feedstock

Figure 1 shows the cross-sectional microstructure of three WC-12Co powders. All powders exhibited a spherical morphology and porous microstructure. The measurement results showed that the porosity of the three powders changed from 44, 30 to 5%, and the microhardness of the three powders changed from 78, 548 to 1317 Hv, as shown in Table. 1.

3.2 Deposition Test of the Isolated Particles

Figure 2 shows typical particles deposited on stainless steel substrate by three powders, exhibiting the deformation degree on impact, for comparison. It can be clearly recognized that with the increase of powder porosity the powder particles deform more greatly and, however, the substrate deforms less. It was evident that the powder porosity influences significantly the deformation of particles on impact. Moreover, the deformation of the substrate under impact of particles is also influenced by the powder porosity. Therefore, using a porous WC-Co powder of certain porosity and cohesive strength would easily fulfill the requirements for the deposition of hard cermet coating by cold spraying.

3.3 Deposition Behavior and Microstructure of the WC-Co Coatings

Figure 3 shows the cross-sectional microstructure of the coatings prepared with HP, MP, and LP powders. Three coatings were deposited with the same spray parameters except spray passes. The coatings shown in Fig. 3(a) and (c), in the form of the free standing, were deposited by 2 passes using the HP and MP powders, respectively, while the coating in Fig. 3(e) was deposited by LP powder by five passes. Although the coatings deposited by HP and MP powders exhibited a dense microstructure, the lateral cracks were observed in the coating, especially in the region near the coating surface as shown in Fig. 3(d). With LP powder, only a thin coating was present on the stainless substrate surface with lateral cracks distributed on cross section as shown in Fig. 3(f). The lamellar structure was characterized by the lateral cracks in the coating. Compared with the much limited deformation of single particles as shown in Fig. 2(c), the particles in the coating evidently experienced remarkable deformation. Such effect

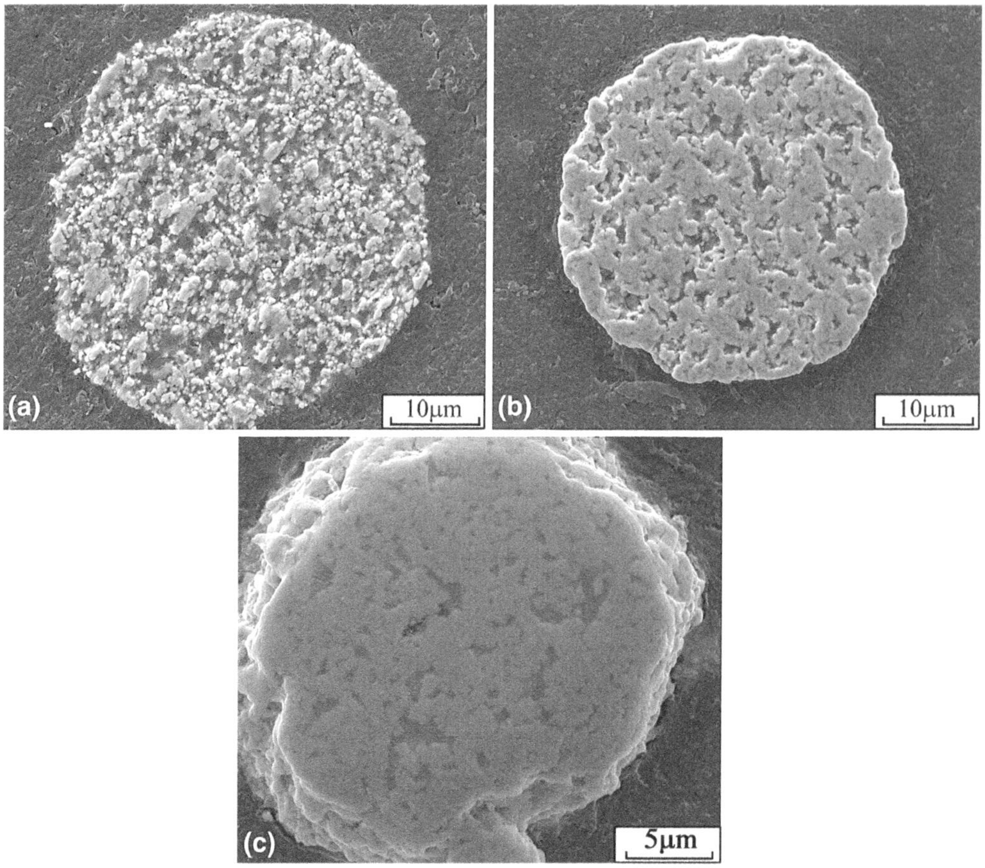

Fig. 1 Cross-sectional microstructure of three powders: (a) HP, (b) MP, and (c) LP

Table 1 Porosity and hardness of three powders

Powder type	HP	MP	LP
Porosity, %	44 ± 1	30 ± 1	5 ± 1
Microhardness, Hv	77.6 ± 2.3	548 ± 46	1317 ± 247

may be attributed to the forced deformation by multi-impacts by spray particles.

Figure 4 shows the microstructure of the coatings deposited by the MP powder by 1 and 10 passes. The coating deposited by 2 passes is shown in Fig. 3(c) and (d). The measurements yielded a thickness of 200, 330, and 350 μm for the coatings deposited by 1, 2, and 10 passes, respectively. It was clearly observed that although the coating thickness increased with the increase of spray pass, the thickness increment of each pass tended to decrease with the increase of spray pass. Moreover, cracks were observed in the coatings, as shown in Fig. 3(d), Fig. 4(b) and (d). The detailed examination revealed that the apparent crack density decreased with the increase of spray pass. The possible reason for crack density reduction with increasing spray pass is the densification through successive impacts of spray particles. With the cracks near the surface region, the spalling by peening effect of hard particle may eliminate the cracks and is also responsible for the reduction of the deposition efficiency.

The deposition efficiency measurements yielded 18.1, 19.2, and 7.1% at spray passes 1, 2, and 10 for the MP powder, respectively, as shown in Fig. 5. It is clear that the deposition efficiency till 2 passes changed less significantly, while it tended decrease with the increase of overlaying spray pass. These results are consistent with those observed from the coating thickness change with spray pass. During cold spraying, the densification through successive impacts of spray particles would make the hardness of the deposited WC-Co to increase gradually. Therefore, the deformability of the previously deposited WC-12Co layer would be decreased comparably. The spray particles would impact on the hard surface of the deposited layer. As a result, the DE value of MP powder decreased enormously with the increase of the spray pass, which was consistent with the thickness increment tendency. LP powder had the highest microhardness and the lowest deformability among the three powders. Once a layer of LP powders embedded into the stainless steel substrate and a thin layer of WC-Co coating covered over the substrate, the hardness of the newly formed substrate, i.e., WC-Co coating, would increase intensively. The further deposition of LP powders, behaviors just like that of MP powder deposition, experienced many spray passes. Both the LP powder and newly formed substrate were hard to deform, which led to a much low DE value. On the other hand, the HP powder had higher porosity and deformability than the MP powder.

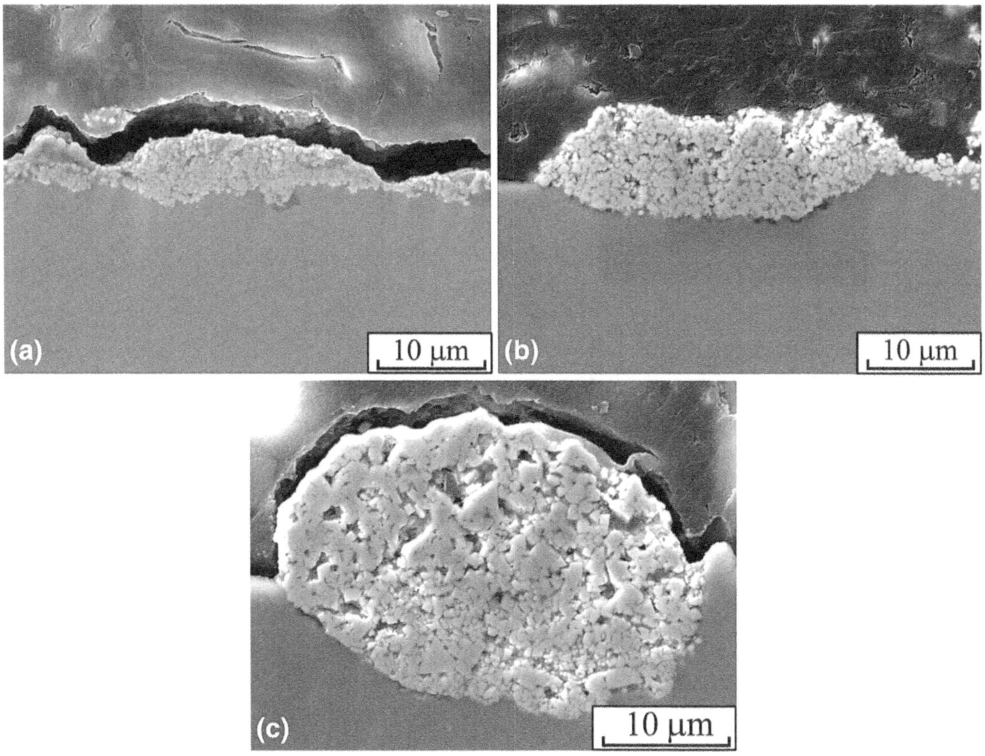

Fig. 2 Cross-sectional microstructure of the deposited single particles using different powders: (a) HP, (b) MP, and (c) LP

During deposition of the HP powder, even though the deposited WC-Co layer exhibited high hardness and poor deformability, the HP powder could deform intensively and flatten onto the deposited layer to build up a layer of coating. However, due to the relatively weak cohesion, the HP powder deformed intensively on impact and partially disintegrated, which resulted in rebounding off of a fraction of powder material as shown in Fig. 2(a). Accordingly, the deposition using the HP powder yielded a lower DE value than the MP powder.

3.4 Microhardness of the Coating

The microhardness of WC-Co coating deposited by the HP powder is 1790 ± 23 Hv. It can be seen from Table 2 that the microhardness of the WC-Co coatings deposited by MP powders increased with the increase of the spray pass. The microhardness of the coatings deposited at 1, 2, and 10 passes were 400, 670, and 1340 Hv, respectively. Such increase is possibly attributed to the densification effect caused by tamping of particle jet impacting on the coating surface. Taking account of the high coating hardness and low deposition efficiency using the HP powder compared with that of the MP powder, it is evident that the densification effect through tamping of impacting hard particle is essential to deposit a dense nanostructured coating using porous powder.

3.5 Deposition Test at a Fixed Position

It is clear that under the same spray conditions the nanostructured WC-Co coatings were deposited using HP and MP powders. The spalling of the deposited layer makes it difficult to deposit a well-cohered WC-Co coating using the LP powder. However, it was found that the dense deposit can be obtained when spray gun was held stationary during deposition using all three powders. Figure 6 shows the microhardness of WC-Co deposits obtained using three powders for 5 s. The microhardness of the cold-sprayed three nanostructured WC-Co deposits is comparable despite initial hardness of starting powders. This fact suggests that some other factors such as the change of the deformability of the deposit may be important for continuous deposition of WC-Co cermet coating by cold spraying using a relatively dense powder.

4. Discussion

The three porous WC-12Co powders in the present study showed different microhardness and different porosity values. The hardness of the powder decreases with the increase of porosity, which increases deformability of the powder. The HP powder had the highest porosity and the greatest deformability among the three powders. From the results of the single-particle deposition test, it was clearly observed that the HP powder deformed more intensively and flattened on the substrate surface. The MP powder deformed greatly but not as intensively as the HP powder. On the other hand, the LP powder had the highest microhardness and could embed into the substrate easily, but it presented much less deformation. Such fact suggests that the LP powder may not be

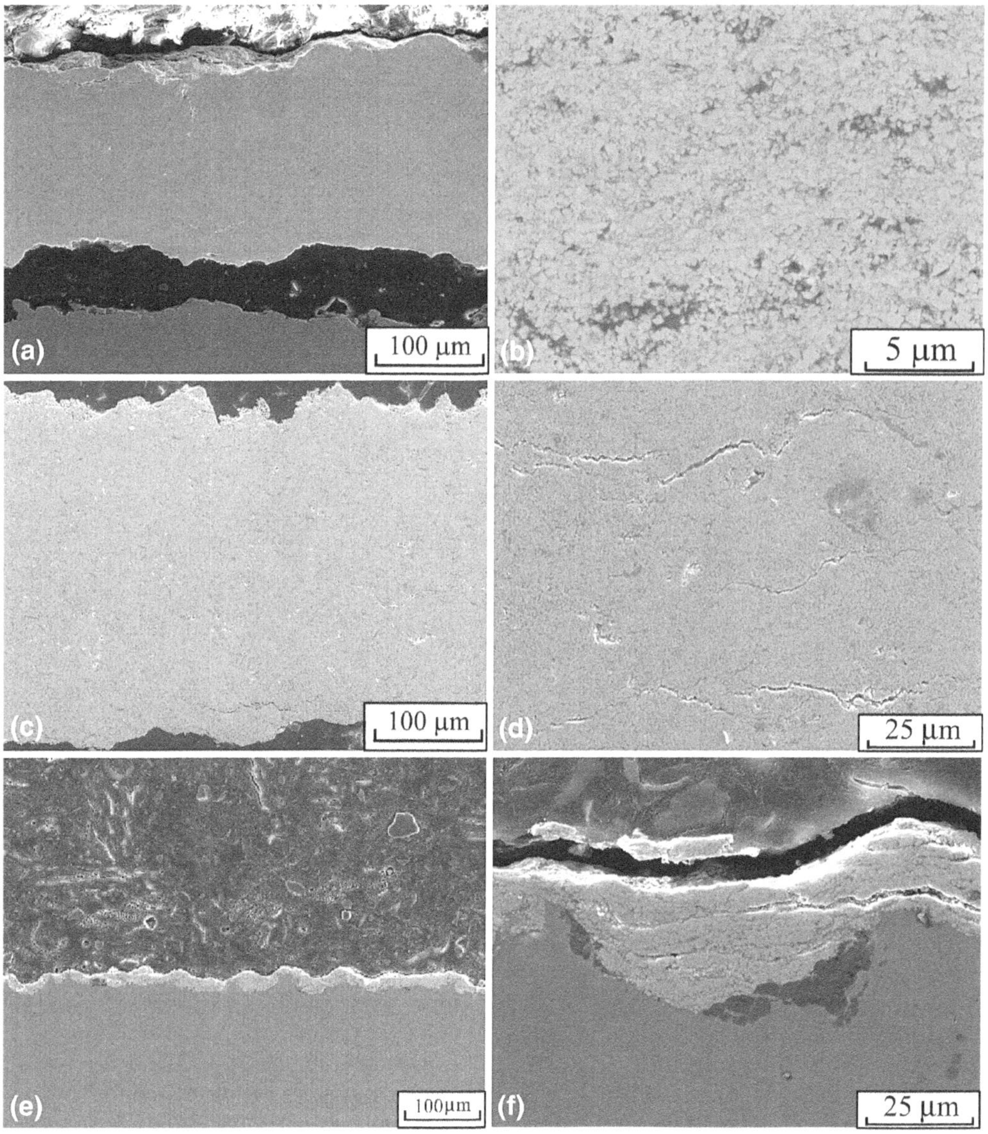

Fig. 3 Typical microstructure of WC-Co coatings by three powders: (a, b) HP, (c, d) MP, and (e, f) LP

favorable to continuous build up of WC-Co coating. This was consistent with the coating deposition test result as shown in Fig. 3. Kim et al. (Ref 17, 18) reported that the nanostructured WC-Co of high hardness comparable to bulk materials was deposited through preheating of spray powder. The preheating of spray powder caused softening effect and consequently enhanced the deformability of the particle on impact. It can be considered that realizing the continuous coating built-up using the porous WC-Co powder in the present study resulted from the similar enhanced deformability effect to that reported by Kim et al.

From the coatings prepared with HP and MP powders at the same deposition parameters, it could be found that the deposited coatings exhibited different deposition efficiency and microhardness. Moreover, it seemed that a higher deposition efficiency was associated with a lower microhardness of the coating. It can be considered that the difference in the deposition efficiency with two powders is mainly associated with the integrity of the particles on impact. The fracture or collapse of the HP powder on impact possibly is one reason that led to the low deposition efficiency. On the other hand, the low deposition efficiency may contribute more accumulative densification effect by multi-tamping. As a result, the low deposition efficiency was possibly associated with an increased coating hardness.

With the MP powder, it was found that the spray pass influenced significantly the deposition efficiency and coating hardness. During cold spraying, the substrate was changed with the deposition of WC-Co particles. Once a WC-Co layer is deposited on the initial substrate, spray particles will impact on the surface of the deposited layer; therefore, the previously deposited WC-12Co layer acts as a new substrate. Moreover, the densification through successive impacts of spray particles would harden the deposited WC-Co. As a result, the hardness of the deposited WC-Co would be increased intensively. The

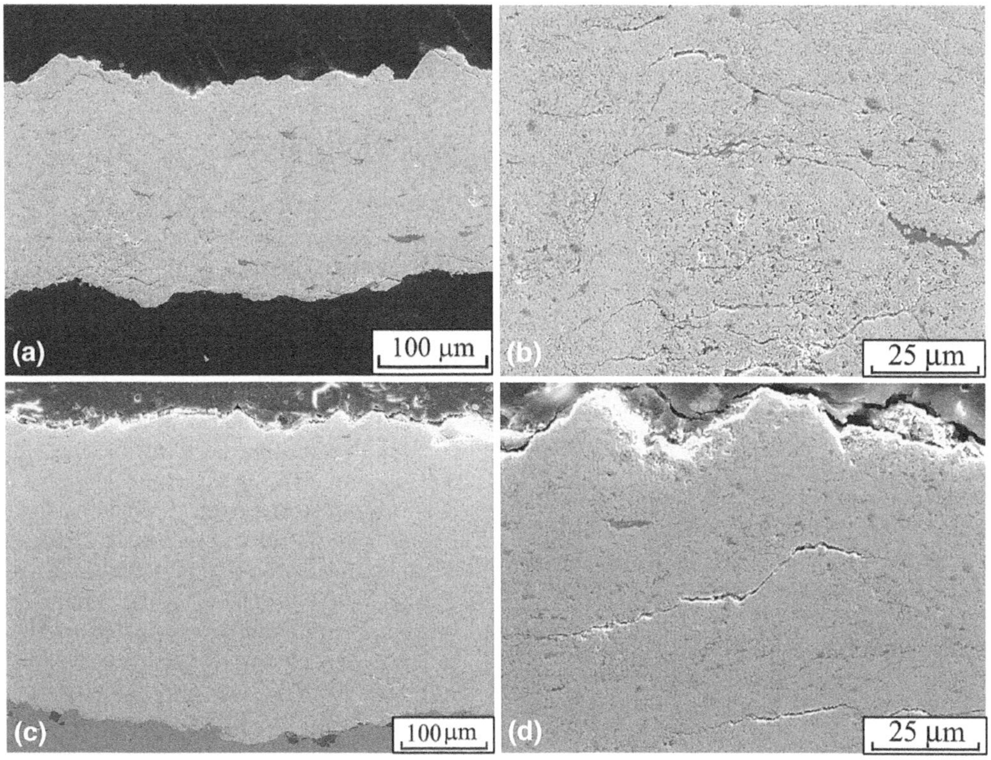

Fig. 4 Microstructure of WC-Co coatings deposited with MP powder by different passes: (a, b) 1 pass and (c, d) 10 passes

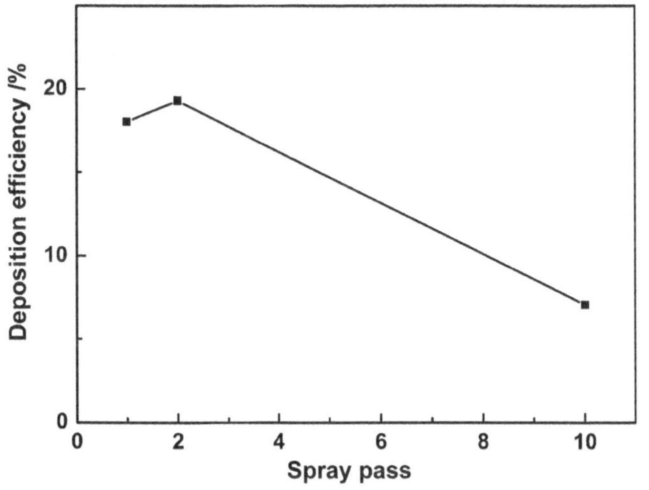

Fig. 5 Deposition efficiencies of MP powder at different spray passes

Table 2 Effect of spray pass on the coating thickness and coating hardness using MP powder

Pass number	Thickness, μm	Hardness, $HV_{0.3}$
1	200	400 ± 11
2	330	670 ± 38
10	350	1340 ± 78

deformability of the deposited WC-Co would become gradually poorer. Therefore, with increasing overlying spray passes, the deposition efficiency was evidently

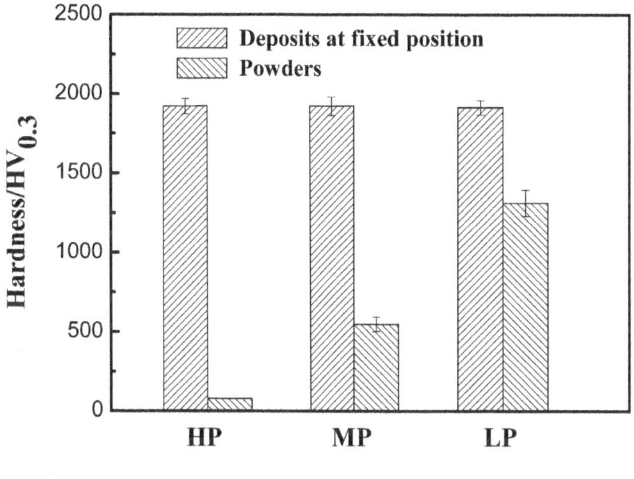

Fig. 6 Microhardness of the nanostructured powders and cold-sprayed deposits at fixed position

decreased and the hardness of the coating was increased. The previous study on the Ti coating deposition clearly revealed that the top porous layer can be gradually densified by tamping effect of the later coming particles (Ref 21). With porous WC-Co powders, the limited deformation of the particle on impact leads to the deposition of partial porous layer. The accumulative impacts by particles jet seem effective to densify the coating. As a result, the coating hardness can be significantly increased. However, as the cracks were observed especially near the surface region, it can be considered that the adequate particle impact pressure is required to eliminate the cracks through densification. The excessively high impact pressure may cause the propagation of the preexisted cracks and subsequently spalling of the deposited layer as clearly observed during deposition using LP powder. To avoid the spalling on impacts of spray particles, the sufficient cohesion within the deposited layer is necessary. Therefore, using an adequately designed porous spray powder, controlling of dynamic impact pressure, and particle cohesion through spray conditions would be essential for continuous build-up of a cermet coating with high hardness.

Moreover, besides the deformability of spray powder, low density of the powder with high porosity may also be beneficial to WC-Co particle deposition. High powder porosity in the same powder size would correspond to light weight under the same spray conditions, and the lighter particles would be accelerated to a higher velocity. Correspondingly, powders with higher porosity would be accelerated to a higher velocity which would be more easily deposited onto the substrate than the powders with lower porosity.

It was found that the traverse speed of spray gun influences significantly the deposition behavior of WC-Co. When the traverse speed was reduced to zero, that is, performing deposition at a fixed position, dense WC-Co deposits, with the hardness values comparable to those of bulk materials, were continuously built up. This is possibly due to the improved deformability of deposited WC-Co top layer resulting from the intensive heating of gas stream. Recently, it has been reported (Ref 22) that under the stationary nozzle operation a significant heat transfer from the supersonic gas jet to the substrate occurs and the heat transfer could bring out a significant temperature increasing of the substrate, especially near the deposited top layer surface. Such temperature increasing of the deposited WC-Co layer would decrease the surface hardness of WC-Co (Ref 23, 24). As a result, the deformability of the WC-Co substrate on which WC-Co impacts will be enhanced. Therefore, the transient softening of the top surface of WC-Co layer on which spraying particles impacts during spraying also influences significantly the built-up of the hard WC-Co coating. At a very low traverse speed, even to zero (i.e., at a fixed position), the deposited particles were kept at a high temperature so that the deformability of the previous deposited particles was kept at a higher level. The high deformability of the surface deposit layer leads to the continuous deposition of dense WC-12Co of high hardness.

The hardness test of the WC-Co deposits sprayed at the fixed position yielded a comparable mean hardness of about 1900 Hv despite powder microstructure. Those hardness numbers are comparable to those of bulk materials, which is consistent with that reported by Kim et al. (Ref 17, 18) However, it is clear that the hardness of the coatings built up with continuous traversing was significantly influenced by the powder structure and gun traverse speed. It is clear that with the decrease of traverse speed gun and increase of powder porosity the hardness of cold-sprayed WC-Co coating was increased. Those facts clearly indicated that the deformation of both the deposited layer and impacting particle are responsible for the deposition of hard coating.

5. Conclusions

WC-Co coating can be successfully built-up using the porous powders with WC particles bonded more loosely. The build-up of the coating was mainly contributed by the deformation of porous powders on impact. It was found that the deposition efficiency was significantly influenced by the porosity levels of the powders. Moreover, the deposition efficiency tended to decrease with the increase of coating thickness. The densification effect resulting from the successive tamping of the later impacting particles increased the coating hardness and, on the other hand, decreased the deposition efficiency. The high intensive impact with harder particles can cause spalling rather than densification of the deposited layer. Therefore, designing a porous WC-Co powder with adequate hardness is one important approach to deposit hard WC-Co cermet coating by cold spraying through controlling the deformability of spray powder. The deposition test at a fixed position suggested that the enhancement of the deformation of the deposited top layer at the impact of spray particle benefits the deposition of dense and hard WC-Co coating.

Acknowledgments

This work is supported by the National Natural Science Foundation of China (NSFC) (No. 50571080), the National Science Fund for Distinguished Young Scholars (No. 50725101), and R&D Project of Shaanxi Province (No. 2005KW-31).

References

1. C.-J. Li, A. Ohmori, and Y. Harada, Effect of Powder Structure on the Structure of Thermally Sprayed WC-Co Coatings, *J. Mater. Sci.*, 1996, **31**(3), p 785-794
2. P.K. Aw and B.H. Tan, Study of Microstructure, Phase and Microhardness Distribution of HVOF Sprayed Multimodal Structured and Conventional WC–17Co Coatings, *J. Mater. Process. Technol.*, 2006, **174**(1-3), p 305-311
3. V. Richter and M.V. Ruthendorf, On Hardness and Toughness of Ultrafine and Nanocrystalline Hard Materials, *Int. J. Refract. Metal Hard Mater.*, 1999, **17**(1-3), p 141-152
4. C.-J. Li, H. Yang, and H. Li, Effect of Gaseous Conditions on the Characteristics of HVOF Flame and Structure and Property of WC-Co Coatings, *Mater. Manuf. Process.*, 1999, **14**(3), p 383-395
5. J. Subrahmanyam, M.P. Srivastava, and R. Sivakumar, Characterization of Plasma-Sprayed WC-Co Coatings, *Mater. Sci. Eng.*, 1986, **84**, p 209-214
6. J. Voyer and B.R. Marple, Sliding Wear Behavior of High Velocity Oxy-fuel and High Power Plasma Spray-Processed Tungsten Carbide-Based Cermet Coatings, *Wear*, 1999, **225-229**(1), p 135-145
7. M.E. Vinayo, F. Kassabji, J. Guyonnet, and P. Fauchais, Plasma Sprayed WC-Co Coatings: Influence of Spray Conditions (Atmospheric and Low Pressure Plasma Spraying) on the Crystal Structure, Porosity, and Hardness, *J. Vac. Sci. Technol. A*, 1985, **3**(6), p 2483-2489

8. H. Lovelock, Powder/Processing/Structure Relationships in WC-Co Thermal Spray Coatings: A Review of the Published Literature, *J. Therm. Spray Technol.*, 1998, **7**(3), p 357-373
9. V. Ramnnath and N. Jayaraman, Characterization and Wear Performance of Plasma Sprayed WC-Co Coatings, *Mater. Sci. Technol.*, 1989, **5**(4), p 382-388
10. D.A. Stewart, P.H. Shipway, and D.G. McCartney, Microstructural Evolution in Thermally Sprayed WC–Co Coatings Comparison Between Nanocomposite and Conventional Starting Powders, *Acta Mater.*, 2000, **48**(7), p 1593-1604
11. Y.-C. Zhu, K. Yukimura, C.-X. Ding, and P.-Y. Zhang, Tribological Properties of Nanostructured and Conventional WC-Co Coatings Deposited by Plasma Spraying, *Thin Solid Films*, 2001, **388**(1-2), p 277-282
12. Y.-Y. Wang, C.-J. Li, J. Ma, and G.-J. Yang, Effect of Flame Conditions on the Abrasive Wear Performance of HVOF Sprayed Nanostructured WC-12Co Coatings, *Trans. Nonferr. Metal Soc. China*, 2004, **14**(Special 2), p 72-76
13. A. Papyrin, Cold Spray Technology, *Adv. Mater. Process.*, 2001, **159**(9), p 49-51
14. T. Stoltenhoff, H. Kreye, and H.J. Richter, An Analysis of the Cold Spray Process and Its Coatings, *J. Therm. Spray Technol.*, 2002, **11**(4), p 542-550
15. B. Jodoin, F. Raletz, and M. Vardelle, Cold Spray Modeling and Validation Using an Optical Diagnostic Method, *Surf. Coat. Technol.*, 2006, **200**(14-15), p 4424-4432
16. C.-J. Li, G.-J. Yang, P.-H. Gao, J. Ma, Y.-Y. Wang, and C.-X. Li, Characterization of Nanostructured WC-Co Deposited by Cold Spraying, *J. Therm. Spray Technol.*, 2007, **16**(5-6), p 1011-1020
17. H.-J. Kim, C.-H. Lee, and S.-Y. Hwang, Superhard Nano WC-12%Co Coating by Cold Spray Deposition, *Mater. Sci. Eng. A*, 2005, **391**(1-2), p 243-248
18. H.-J. Kim, C.-H. Lee, and S.-Y. Hwang, Fabrication of WC-Co Coatings by Cold Spray Deposition, *Surf. Coat. Technol*, 2005, **191**(2-3), p 335-340
19. M. Yandouzi, E. Sansoucy, L. Ajdelsztajn, and B. Jodoin, WC-Based Cermet Coatings Produced by Cold Gas Dynamic and Pulsed Gas Dynamic Spraying Processes, *Surf. Coat. Technol.*, 2007, **202**(2), p 382-390
20. R.S. Lima, J. Karthikeyan, C.M. Kay, J. Lindemann, and C.C. Berndt, Microstructural Characteristics of Cold Sprayed Nanostructured WC-Co Coating, *Thin Solid Films*, 2002, **416**(1-2), p 129-135
21. C.-J. Li and W.-Y. Li, Deposition Characteristics of Titanium Coating in Cold Spraying, *Surf. Coat. Technol.*, 2003, **167**(2-3), p 278-283
22. E. Irissou, J.-G. Legoux, C. Moreau, and A.N. Ryabinin, How Cold is Cold Spray? An Experimental Study of the Heat Transfer to the Substrate in Cold Gas Dynamic Spraying, *Thermal Spray Crossing Borders: Conference Proceedings of International Thermal Spray Conference & Exposition 2008*, E. Lugscheider, Ed., June 2-4, 2008 (Maastricht, The Netherlands), DVS, 2008, p 625-631
23. M. Lee, High Temperature Hardness of Tungsten Carbide, *Metall. Mater. Trans. A*, 1983, **14**(8), p 1625-1629
24. Yu.V. Milman, S. Chungunova, V. Goncharuck, S. Luyckx, and I.T. Northrop, Low and High Temperature Hardness of WC-6 wt%Co Alloys, *Int. J. Refract. Metal Hard Mater.*, 1997, **15**(1-3), p 97-101

JTTEE5 17:750–756
DOI: 10.1007/s11666-008-9271-4
1059-9630/$19.00 © ASM International

Peer Reviewed

Development of WC-Co Coatings Deposited by Warm Spray Process

Pornthep Chivavibul, Makoto Watanabe, Seiji Kuroda, Jin Kawakita, Masayuki Komatsu, Kazuto Sato, and Junya Kitamura

(Submitted June 24, 2008; in revised form September 19, 2008)

The high-velocity oxy-fuel (HVOF) process is commonly used to deposit WC-Co coatings. There are some problems with this process; especially the decomposition and decarburization of WC during spraying make a coating brittle. To suppress such degradation, the warm spray (WS) process was applied to deposit WC-Co coatings, which is capable of controlling the flame temperature in the range of 500-2000 °C. The microstructure and phases of the deposited coatings were characterized by using SEM and XRD, and the mechanical properties such as hardness, fracture toughness, and wear properties were also investigated. WS process successfully suppressed the formation of the detrimental phases such as W_2C and W, which are usually observed in HVOF coatings. The WS coatings showed the similar trend of the hardness variation for Co content with a sintered bulk material. Improvement of toughness and wear behavior was also observed in WS coatings.

Keywords coating, HVOF, warm spray, WC-Co

1. Introduction

WC-Co cermet coatings have been used to enhance the wear resistance of various engineering components in a variety of industrial environments. Many thermal spraying techniques, such as atmospheric plasma spraying (APS) and high-velocity oxy-fuel (HVOF) spraying, can be applied to deposit WC-Co coatings; however, the properties of such coatings strongly depend on the spraying technique. Compared to other spraying techniques, HVOF spraying is one of the best methods for depositing conventional WC-Co cermets, because the higher velocities and lower temperatures experienced by the powder results in less decomposition of the WC during spraying (Ref 1). Therefore, coatings with higher amount of retained WC and lower porosity are expected. However, when compared to sintered bulk WC-Co, for which the sintering atmosphere has been carefully controlled, HVOF-sprayed WC-Co coatings suffer from decomposition and decarburization during spraying process leading to a formation of undesirable phases such as W_2C, W, and amorphous or nanocrystalline Co-W-C phase (Ref 2, 3). Many studies have shown the effects of these phases on the properties of the coating (Ref 3-7). Chivavibul et al. (Ref 5) systematically studied the effects of both carbide size and Co content on the mechanical properties of the coatings. They found that the hardness and fracture resistance behaviors of the coating were different from those of the bulk materials. This discrepancy could be explained by the hardening of the binder phase due to the dissolution of WC and the formation of amorphous/nanocrystalline phases. The hardness of the binder phase estimated from the experimental data was ranged from 1000 to 1300 Hv, and significantly higher than that of the bulk binder phase (490-660 Hv).

Many efforts have been made to deposit WC-Co coatings at lower process temperature than HVOF to suppress decarburization and decomposition of WC phases (Ref 8-14). Jacob et al. successfully deposited WC-Co and WC-Co-Cr coatings using high-velocity air fuel (HVAF) process (Ref 12, 13). These coatings showed no detrimental phase from decomposition or oxidation during deposition. Compared to HVOF coatings, the HVAF WC-Co-Cr coating showed the improved hardness and wear resistance. On the other hand, HVAF WC-Co coating showed lower hardness than HVOF and there was no comparative result of wear property. Recently, the success in using cold spray technique to deposit various metallic materials (Ref 15, 16) has invoked a great interest to apply this technique to deposit WC-Co coatings. The earlier work by Lima et al. (Ref 8) showed that the nano WC-Co coatings could be deposited, but the thickness of the coating was less than 10 μm. The nano WC-Co

This article is an invited paper selected from presentations at the 2008 International Thermal Spray Conference and has been expanded from the original presentation. It is simultaneously published in *Thermal Spray Crossing Borders, Proceedings of the 2008 International Thermal Spray Conference*, Maastricht, The Netherlands, June 2-4, 2008, Basil R. Marple, Margaret M. Hyland, Yuk-Chiu Lau, Chang-Jiu Li, Rogerio S. Lima, and Ghislain Montavon, Ed., ASM International, Materials Park, OH, 2008.

Pornthep Chivavibul, Makoto Watanabe, Seiji Kuroda, Jin Kawakita and **Masayuki Komatsu**, Composites and Coatings Center, National Institute for Materials Science, Ibaraki, Japan; and **Kazuto Sato** and **Junya Kitamura**, Fujimi Incorporated, Gifu, Japan. Contact e-mail: Watanabe.Makoto@nims.go.jp.

coatings with thickness over 100 μm were successfully deposited by Kim et al. (Ref 9, 10) and Li et al. (Ref 14). These coatings have extremely high hardness (1820-2050 Hv) and show no detrimental phase formation in the coatings. Although the overall performances of these coatings such as toughness and wear properties are still unknown, it indicates the capability of cold spray to fabricate WC-Co coatings, which have a similar microstructure as the sintered bulk materials.

Kawakita et al. (Ref 17, 18) has developed a new process named *warm spray* process (WS). This process has a mixing chamber between the combustion chamber and the powder feed ports. The combustion is generated within the chamber and then it is mixed with nitrogen to control the temperature of the gas. Compared to the cold spray deposition, this technique provides a capability to control the temperature of sprayed particles in a wide range. The success of using this technique to deposit various materials (Ref 17-20) such as titanium and metallic glass has been reported. Recently, Watanabe et al. (Ref 11, 18) demonstrated a capability of this technique to deposit WC-12Co coating. The coating showed no detrimental phases and had the same level of hardness as the sintered bulk material and HVOF coating of the same composition. However, the systematic study of this process is still required on the application of WC-Co coatings.

In the present investigation, a comparative study of WS and HVOF coatings is performed to verify the capability of WS process as a new candidate for cermet coating. The nano-sized WC-Co powders with various Co contents were sprayed by both WS and HVOF techniques. Phase distribution and microstructure of the coatings were examined by X-ray diffraction and scanning electron microscope (SEM), respectively. Hardness, fracture toughness, and abrasive wear properties were measured. The relation between the microstructural features and the mechanical properties are discussed based on the experimental results.

2. Experimental Procedure

2.1 Materials and Spraying Process

Three commercially available WC-Co powders with an average carbide size of 0.2 μm and different cobalt contents (12, 17, and 25 wt.%) were used in this work. The powders were manufactured by spray drying of slurries containing WC and Co particles, followed by light sintering, crushing, and classification. The particle size of the powders ranged from 15 to 45 μm. Examples of the morphology and cross section of WC-12Co and WC-25Co powder are shown in Fig. 1. All powders have typically spherical shape. The WC-12Co powder shows dense packing of fine carbide, while WC-25Co powder shows more loosely packed structure with higher amount of pores. The Co distribution in all the powder is not uniform. Many Co pools can be clearly observed in WC-25Co powder.

A commercial HVOF spraying equipment (JP5000, Praxair Technology Inc., USA) was used to spray these powders onto carbon steel (0.45% C) substrates. The WS process was performed by adding the mixing chamber between the combustion chamber and powder feed port of the HVOF equipment. Details of the spraying equipment were given elsewhere (Ref 17, 18). Table 1 shows the spraying conditions of HVOF and WS processes. The thickness of the coatings was approximately 300 μm.

2.2 Coating Characterization

X-ray diffraction (XRD) was conducted for the powders and as-fabricated coatings with Cu Kα radiation at 40 kV and 300 mA. Cross sections of the coatings were obtained by embedding a specimen in cold-mounting resin followed by grinding and polishing to a 1 μm finish and examined by scanning electron microscopy (SEM).

Microhardness tests were carried out with a 300 g load and a dwell time of 15 s. At least ten measurements were made for each sample. The fracture toughness of the coatings was determined by the indentation method with a 10 kg load and a dwell time of 15 s. Both the crack lengths and Vickers diagonals were measured by an optical microscope. The fracture toughness, K_C, was determined under an assumption that the cracks generated from an indentation were radial cracks possessing the Palmqvist geometry by the following equation (Ref 21):

$$K_C = 0.0193(H_v D)\left(\frac{E}{H_v}\right)^{2/5}(a)^{-1/2}, \qquad \text{(Eq 1)}$$

where H_v is the Vickers hardness, E is the Young's modulus, D is the half-diagonal of the Vickers indentation, and a is the indentation crack length. At least five indentations were made for each sample. The Young's modulus of the coatings was assumed to be 300 GPa for all the samples.

Abrasive wear of the coatings was evaluated using a Suga-type wear tester (two-body abrasive wear tester). The test was performed by the reciprocating motion of the test specimen scratching on a rotating wheel under a load of 29.4 N. The wheel had a diameter of 50 mm and was covered with 180 mesh SiC paper (~78 μm). To eliminate the effect of surface condition, all test specimens were polished before the test. The weight loss of the test specimen was measured after each run of 400 cycles, which corresponds to 25.6 m sliding distance, and the tests were performed up to three runs for each specimen. The average weight losses of all the samples were converted to volume losses by using density values of the bulk materials and then normalized with that of the JIS-SS400 carbon steel.

3. Results and Discussion

3.1 Coating Microstructure

Examples of BSE images of HVOF and WS coatings are shown in Fig. 2. The image of HVOF and WS WC-12Co are shown in Fig. 2(a) and (b), respectively. Both revealed a dispersion of fine carbide particles in binder

Peer Reviewed

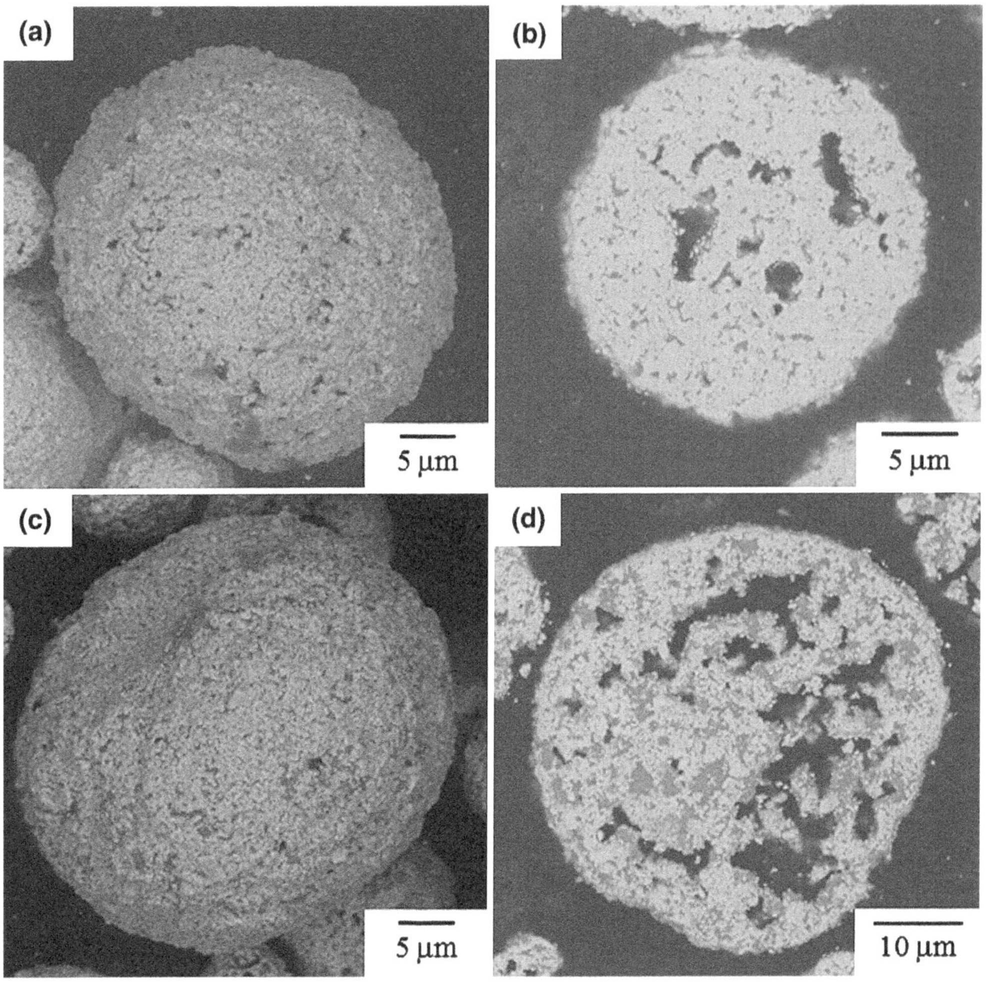

Fig. 1 SEM observation of powders (a), (b) WC-12Co and its cross section, (c) and (d) WC-25Co and its cross section

Table 1 Spray conditions for WC-Co coatings

	WS	HVOF
Barrel length, mm	203	152
Spraying distance, mm	200	380
Fuel, L/min	0.38	0.38
Oxygen, L/min	778	944
Nitrogen, L/min	500	...
Powder feed rate, g/min	80	80
Powder feed gas	Nitrogen	Nitrogen

layers. Pores were observed in both coatings. In some area of the HVOF WC-12Co coatings, lack of splat-splat bonding was also observed as shown in Fig. 2(a). A clear difference of microstructure between HVOF and WS coating can be observed in coating with higher Co content. The HVOF WC-25Co coating (Fig. 2c) showed carbides distribution in binder layers with different contrasts. These bright and dark binder layers were W-rich and Co-rich regions, respectively. The bright binder layer was not observed in the WS WC-25Co coating (Fig. 2d). Moreover, the carbide fractions in HVOF coatings were much lower than those in WS coatings. The shape of carbides in HVOF coating became rounder than that of WS coatings. The loss of carbide and change of carbide morphology in HVOF coatings could be attributed to the dissolution of WC into the Co binder, while such reaction can be suppressed by WS process. This process provides a deposition of feedstock powder in solid state, therefore the microstructure of the coating results from the stacking of particles under impact. This means that the microstructure of the starting powder strongly influences the coating microstructure. This can be confirmed by comparing the cross sections of WC-25Co powder (Fig. 1d) to that of the coating (Fig. 2b). Co pools could be observed in both the powder and the coating.

Examples of XRD patterns of HVOF and WS coatings are shown in Fig. 3(a) and (b), respectively. The results of WC-12Co feedstock powder is also given for a comparison. While the original powder contained only the peaks of Co and WC, HVOF coatings showed the crystalline peaks of WC, W_2C, W, and a broad shoulder at the angle of the Co peak, indicating that the binder has attained amorphous or nanocrystalline form. There is a tendency for amorphous peak to increase as the Co content increases, while W_2C tended to decrease. In the case of WS coatings, only the peaks of WC and Co phase can be clearly recognized for the 17 and 25% Co contents, while a

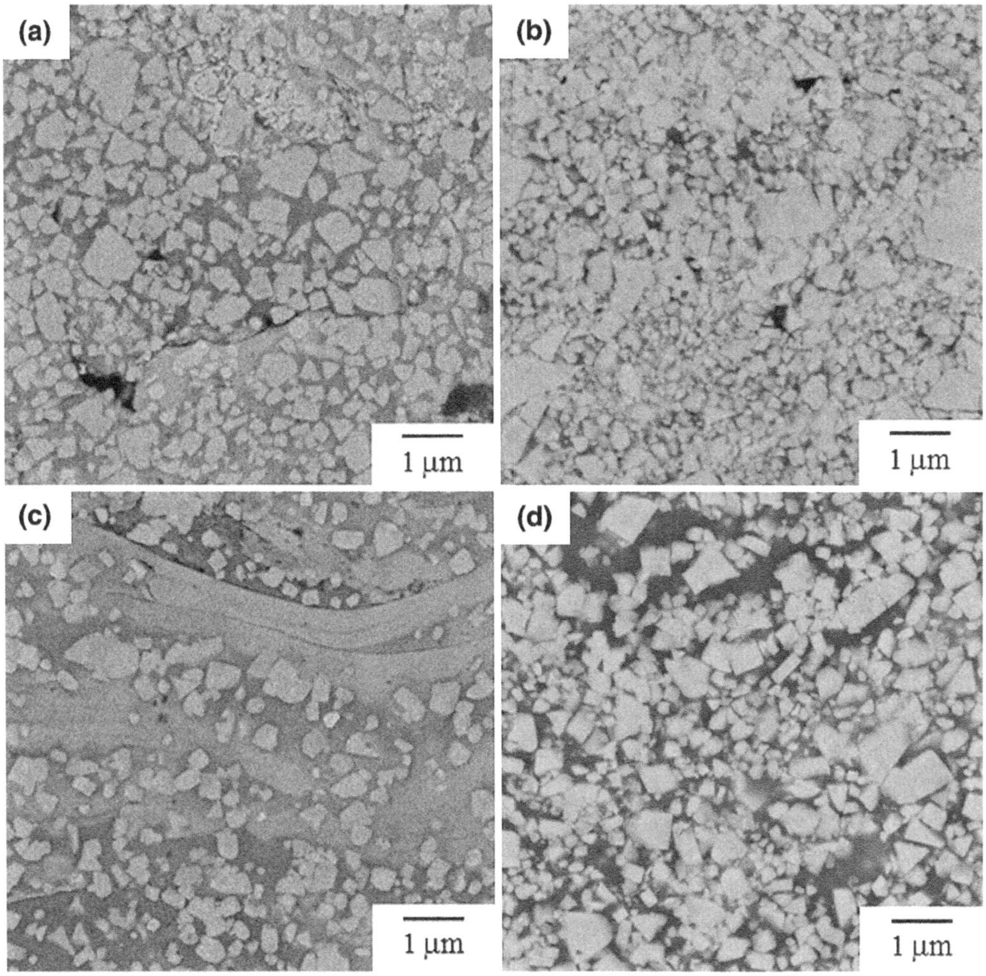

Fig. 2 SEM observations of (a) HVOF WC-12Co, (b) WS WC-12Co, (c) HVOF WC-25Co, and (d) WS WC-25Co coatings

very small peak of W_2C is observed in the 12% Co content. This meant that WS process can significantly suppress the decomposition and oxidation during deposition. However, it was noted from Fig. 2(b) that the Co peaks of WS coatings were broadened and shifted to lower angles. One possible explanation can be the deformation of Co phases due to high-velocity impact of particle during deposition. Large deformation of soft Co binder and constraining by the surrounding carbides can generate high residual compressive stress in the binder, and thus only the Co peak shows a substantial peak shift. For broadening of the peak, since the high-speed impact of solid particle induces the dynamic recrystallization and leads to the formation of nano-order polycrystalline structure (Ref 22), the formation of such nanocrystalline microstructure possibly occurred in the coatings.

3.2 Mechanical Properties of the Coating

The microhardness and fracture toughness values of the HVOF and WS coatings, as a function of Co content, are shown in Fig. 4. The values of a sintered bulk material are also given for a comparison (Ref 23). Generally, the hardness of a bulk material decreases with increasing cobalt content. The hardness of both HVOF and WS coatings also decreased with increasing Co content, but the gradients of the trend lines reveal clear difference between them. While the WS coatings had softer than the bulk for all Co contents, a trend of hardness variation was similar to the bulk. In contrast, the hardness of HVOF coatings only shows slight variation with Co content and the gradient of the line was quite different with the ones of the bulk and WS coatings. Based on the XRD results, the HVOF coatings contained new phases such as W_2C, W, and amorphous or nanocrystalline binder phase. Although the WC contents in HVOF coatings were lower than those in the bulk and WS coatings (e.g., Fig. 2d), the formation of such new phases affected the hardness significantly. The hardness of W_2C and amorphous binder are reported to be approximately 3000 and 1000-1300 Hv, respectively, which are higher than WC (Hv = 1300-2300) and Co binder (Hv = 490-660) (Ref 5). While the hardness of the bulk material and the WS coatings depends on the amount of WC content, one of the HVOF coatings depends more on the properties of binder resulting in relatively small variation for Co content.

When the data of WS coatings is compared with the bulk, it shows 10-20% lower values for 12-25% Co content. This may be due to the existence of the microstructural

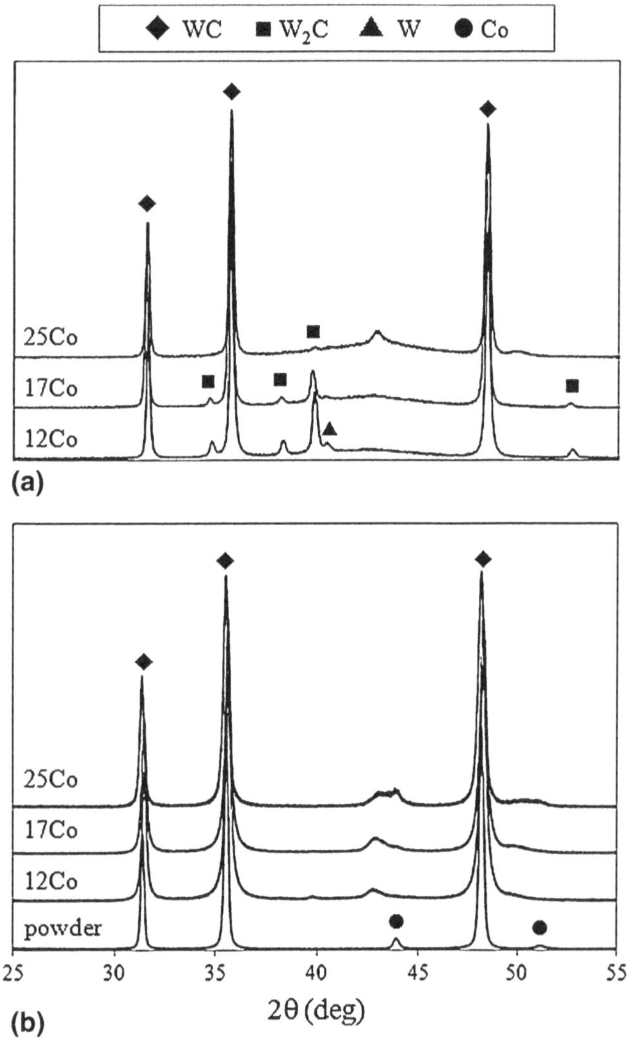

Fig. 3 XRD results of (a) HVOF and (b) WS coatings. The result of WC-12Co powder coatings is given for a comparison

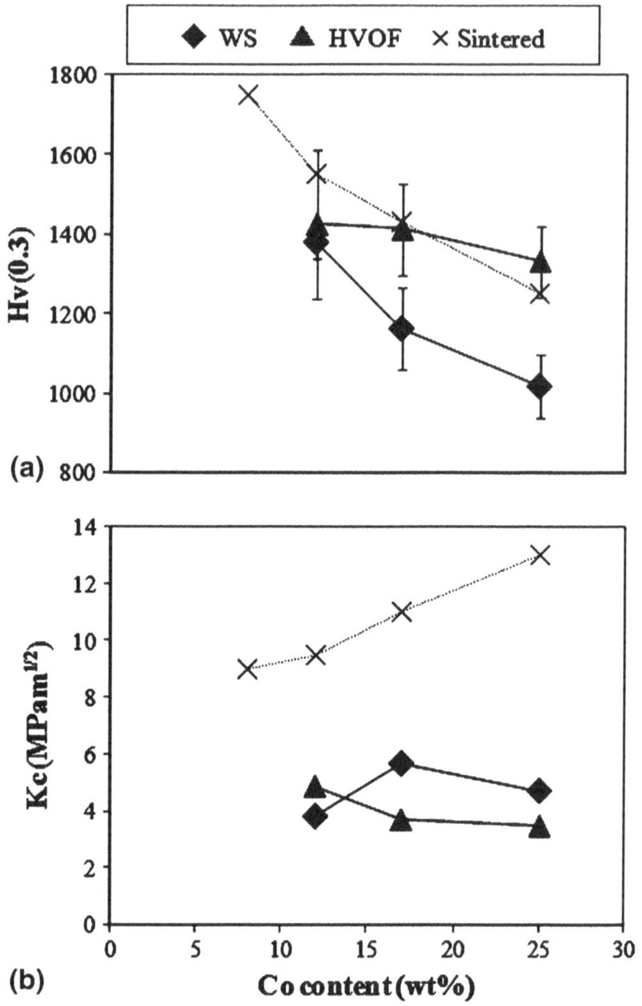

Fig. 4 (a) Hardness and (b) fracture resistance as a function of cobalt content. Values of the sintered materials were given for a comparison (Ref 23)

defects such as pores, which can be recognized in Fig. 2(b) as black dots. Although the adhesion mechanism of a WC-Co particle in WS deposition is yet to be studied, it is hypothesized that Co binder would work as a kind of glue with significant deformation upon impact. Thus, if there were not enough amounts of Co on the surface of a particle locally, for example, due to nonuniform distribution of Co in one particle, those regions could be poorly bonded. More detail experiments are required to minimize the gap between a bulk material and a WS coating.

In Fig. 4, the fracture toughness of both HVOF and WS coatings ranged from 4 to 6 $MPam^{1/2}$. Figure 5 shows the cross-sectional images of the crack configurations in WC-17Co coatings after the test deposited by (a) HVOF and (b) WS, respectively. The drastic reduction in crack length can be recognized in WS sample indicating the effectiveness of suppression of the formation of brittle W_2C and amorphous binder phases for crack resistance. These values, however, are still lower than those of bulk WC-Co (9-14 $MPam^{1/2}$). Figure 6(a) shows SEM photograph of the crack path generated from indentation test of HVOF WC-25Co coating. The crack preferentially propagated through the W-rich binder regions (white-colored area) indicating lower fracture toughness of these regions. This observation showed that the microstructure of the binder plays an important role in the mechanical properties of the HVOF coatings. Although WS coatings showed no changes in phase composition as mentioned in previous section, the increase of toughness with increasing Co content in WS coatings was very small as compared with bulk materials. This might be attributed to their inhomogeneous microstructure. Figure 6(b) shows SEM photograph of the crack path generated from fracture toughness indentation test of WS WC-25Co coating. In contrast to the HVOF coating, crack propagated through the WC-WC interface and the region which might be splat-splat boundary. Another point noted from Fig. 6(b) is that there are Co pools in the coating indicating the poor distribution of Co content. This also gave an effect on the fracture toughness properties by causing nonuniform stress distribution.

Abrasive wear behaviors of HVOF and WS coatings as a function of Co content are shown in Fig. 7. To avoid any

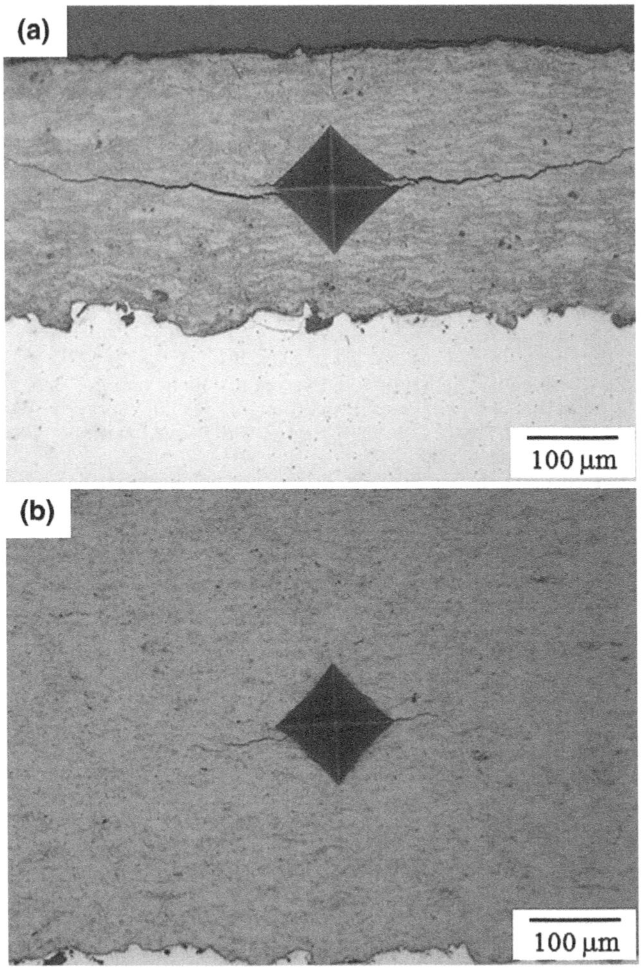

Fig. 5 Cracks generated by the indentation fracture toughness test (load: 10 kgf) in WC-17Co coatings by (a) HVOF and (b) WS depositions. Drastic reduction of crack length can be recognized in the WS coating

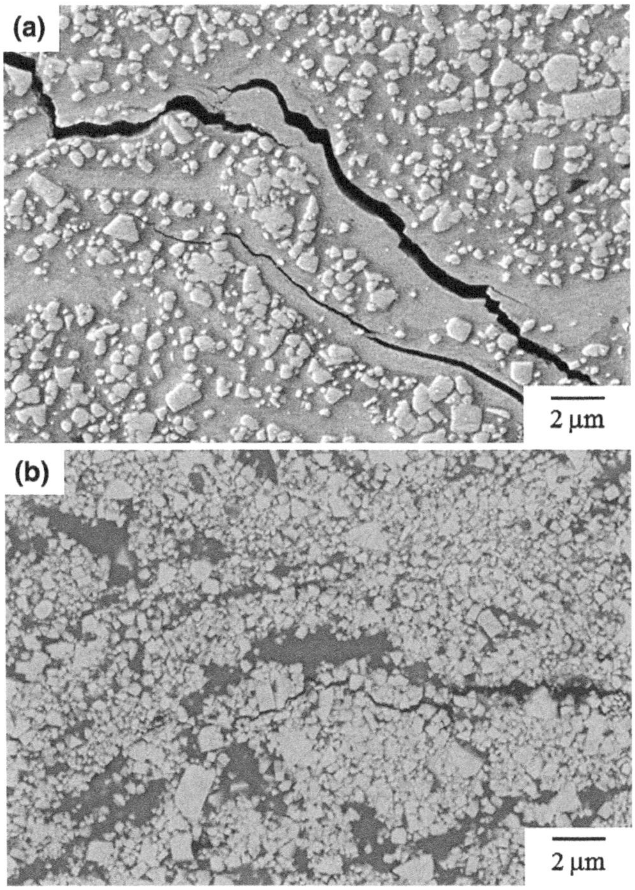

Fig. 6 SEM observation of cracks generated from fracture toughness test of WC-25Co deposited by (a) HVOF and (b) WS techniques. Arrows indicate the region which might be splat-splat boundary

effect of variation in the quality of SiC paper from different lots, data were presented as volume loss normalized with that obtained from JIS-SS400 steel. The volume wear loss increased with increasing Co content in both HVOF and WS coatings. The WS WC-12Co coatings showed great improvement of wear property compared to the HVOF one; however, higher volume wear losses were observed in the coatings with higher Co contents. This behavior was related to the phase distribution and microstructure of the coatings. Considering WC-12Co, HVOF and WS coatings showed the same hardness value while XRD results of HVOF coatings showed higher amount of W_2C and W. These brittle phases and the amount of defects in HVOF coatings lead to higher wear volume loss than that observed in WS coating. For higher Co contents, WS coatings contain a number of Co pools (Fig. 2d and 6b) while the binder phases in the HVOF were hardened by the dissolution and decarburization of WC (Fig. 2c and 6a). Such Co pools in WS coatings, presumably softer than the binder in HVOF coatings, could be preferentially removed during the wear test. This may

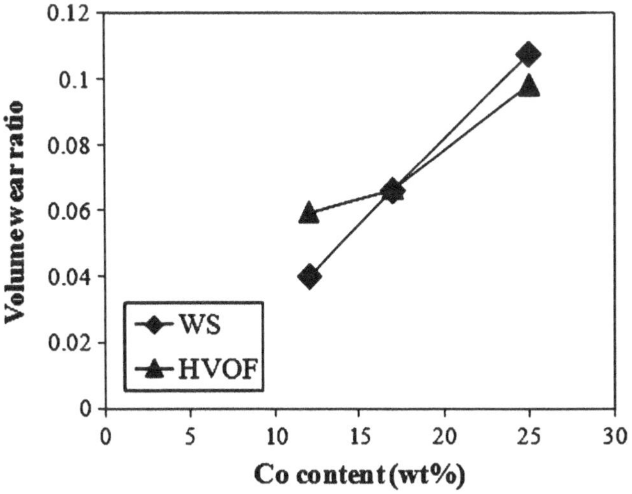

Fig. 7 Suga wear ratio as a function of Co content

be a reason for higher wear loss for 25% Co in WS coatings compared to HVOF ones. A plot of reciprocal of volume wear ratio as a function of coating hardness is shown in Fig. 8. The reciprocal volume wear ratio was

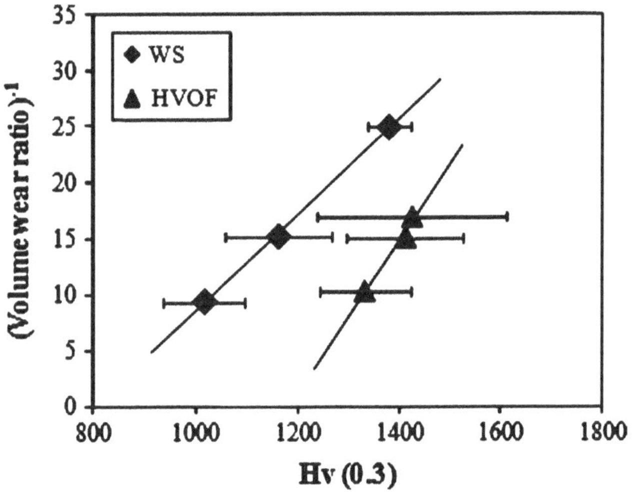

Fig. 8 Reciprocal of Suga wear ratio as a function of coating hardness

chosen because it likely shows the influence of hardness in a linear fashion. Fitting curves showed good linear relation for both HVOF and WS coatings. For a given hardness value, WS coating showed better wear performance than HVOF coating. This improvement was related to the microstructure and phase distribution of the WS coating. One important factor was the difference of binder properties between HVOF and WS coatings. Detailed studies for further improvement in mechanical properties of WS coatings are required.

4. Conclusions

WC-Co coatings with 0.2 μm carbide size and different cobalt contents were obtained by HVOF and warm spraying of three commercial powders. The microstructure, hardness, fracture toughness, and wear behavior of the coatings were investigated.

Characterization of the coatings revealed different mechanical properties between WS and HVOF coatings. The hardness-Co content relation in WS coatings was similar to that in sintered bulk materials, while different behavior was observed in HVOF coatings. Fracture toughness tended to increase with increasing Co content in WS coatings, while HVOF coatings showed the opposite trend. The improvement in wear property was also observed in WS coatings. These could be explained by the microstructure and phase distribution in the coating. The WS technique showed capability to deposit WC-Co coating without dissolution and decarburization of WC.

References

1. V.V. Sobolev, J.M. Guilemany, and J. Nutting, *High Velocity Oxy-Fuel Spraying*. Maney Publishing, UK, 2004
2. J.M. Guilemany, J.M. de Paco, J. Nutting, and J.R. Miguel, Characterization of the W2c Phase Formed During the High Velocity Oxygen Fuel Spraying of a Wc + 12 Pct Co Powder, *Metall. Mater. Trans. A*, 1999, **30**(8), p 1913-1921
3. C. Verdon, A. Karimi, and J.L. Martin, A Study of High Velocity Oxy-Fuel Thermally Sprayed Tungsten Carbide Based Coatings. Part 1: Microstructures, *Mater. Sci. Eng. A*, 1998, **246**(1-2), p 11-24
4. Y. Ishikawa, J. Kawakita, S. Sawa, T. Itsukaichi, Y. Sakamoto, M. Takaya, and S. Kuroda, Evaluation of Corrosion and Wear Resistance of Hard Cermet Coatings Sprayed by Using an Improved HVOF Process, *J. Therm. Spray Technol.*, 2005, **14**(3), p 384-390
5. P. Chivavibul, M. Watanabe, S. Kuroda, and K. Shinoda, Effects of Carbide Size and Co Content on the Microstructure and Mechanical Properties of HVOF-Sprayed WC-Co Coatings, *Surf. Coat. Technol.*, 2007, **202**(3), p 509-521
6. M. Watanabe, A. Owada, S. Kuroda, and Y. Gotoh, Effect of Wc Size on Interface Fracture Toughness of WC-Co HVOF Sprayed Coatings, *Surf. Coat. Technol.*, 2006, **201**(3-4), p 619-627
7. Y. Ishikawa, S. Kuroda, J. Kawakita, Y. Sakamoto, and M. Takaya, Sliding Wear Properties of HVOF Sprayed WC-20%Cr3c2-7%Ni Cermet Coatings, *Surf. Coat. Technol.*, 2007, **201**(8), p 4718-4727
8. R.S. Lima, J. Karthikeyan, C.M. Kay, J. Lindemann, and C.C. Berndt, Microstructural Characteristics of Cold-Sprayed Nanostructured WC-Co Coatings, *Thin Solid Films*, 2002, **416**(1-2), p 129-135
9. H.J. Kim, C.H. Lee, and S.Y. Hwang, Superhard Nano WC-12%Co Coating by Cold Spray Deposition, *Mater. Sci. Eng. A*, 2005, **391**(1-2), p 243-248
10. H.J. Kim, C.H. Lee, and S.Y. Hwang, Fabrication of WC-Co Coatings by Cold Spray Deposition, *Surf. Coat. Technol.*, 2005, **191**(2-3), p 335-340
11. M. Watanabe, C. Pornthep, S. Kuroda, J. Kawakita, J. Kitamura, and K. Sato, Development of WC-Co Coatings by Warm Spray Deposition for Resource Savings of Tungsten, *J. Jpn. Inst. Metals*, 2007, **71**, p 853-859
12. L. Jacobs, M.M. Hyland, and M. De Bonte, Comparative Study of WC-Cermet Coatings Sprayed via the HVOF and the HVAF Process, *J. Therm. Spray Technol.*, 1998, **7**(2), p 213-218
13. L. Jacobs, M.M. Hyland, and M. De Bonte, Study of the Influence of Microstructural Properties on the Sliding-Wear Behavior of HVOF and HVAF Sprayed WC-Cermet Coatings, *J. Therm. Spray Technol.*, 1999, **8**(1), p 125-132
14. C.J. Li, G.J. Yang, P.H. Gao, J. Ma, Y.Y. Wang, and C.X. Li, Characterization of Nanostructured WC-Co Deposited by Cold Spraying, *J. Therm. Spray Technol.*, 2007, **16**, p 1011-1020
15. L. Ajdelsztajn, B. Jodoin, and J.M. Schoenung, Synthesis and Mechanical Properties of Nanocrystalline Ni Coatings Produced by Cold Gas Dynamic Spraying, *Surf. Coat. Technol.*, 2006, **201**(3-4), p 1166-1172
16. T.H. Van Steenkiste, J.R. Smith, and R.E. Teets, Aluminum Coatings via Kinetic Spray with Relatively Large Powder Particles, *Surf. Coat. Technol.*, 2002, **154**(2-3), p 237-252
17. J. Kawakita, S. Kuroda, T. Fukushima, H. Katanoda, K. Matsuo, and H. Fukanuma, Dense Titanium Coatings by Modified HVOF Spraying, *Surf. Coat. Technol.*, 2006, **201**(3-4), p 1250-1255
18. S. Kuroda, J. Kawakita, M. Watanabe, and H. Katanoda, Warm Spraying—A Novel Coating Process Based on High-Velocity Impact of Solid Particles, *Sci. Technol. Adv. Mater.*, 2008, **9**(3), p 033002
19. J. Kawakita, S. Kuroda, S. Krebs, and H. Katanoda, In-Situ Densification of Ti Coatings by the Warm Spray (Two-Stage HVOF) Process, *Mater. Trans.*, 2006, **47**(7), p 1631-1637
20. J. Kawakita, N. Maruyama, S. Kuroda, S. Hiromoto, and A. Yamamoto, Fabrication and Mechanical Properties of Composite Structure by Warm Spraying of Zr-Base Metallic Glass, *Mater. Trans.*, 2008, **49**(2), p 317-323
21. M.M. Lima, C. Godoy, J.C. Avelar-Batista, and P.J. Modenesi, Toughness Evaluation of HVOF WC-Co Coatings Using Non-Linear Regression Analysis, *Mater. Sci. Eng. A*, 2003, **357**(1-2), p 337-345
22. K. Kim, M. Watanabe, J. Kawakita, and S. Kuroda, Grain Refinement in a Single Titanium Powder Particle Impacted at High Velocity, *Scr. Mater.*, 2008, **59**(7), p 768-771
23. G.S. Upadhyaya, *Cemented Tungsten Carbides: Production, Properties, and Testing*. Noyes Publications, Westwood, NJ, 1998

JTTEE5 17:757–761
DOI: 10.1007/s11666-008-9257-2
1059-9630/$19.00 © ASM International

Peer Reviewed

Microstructure and Wear Performance of Arc Sprayed Fe-FeB-WC Coatings

Ding-Yong He, Bin-You Fu, Jian-Min Jiang, and Xiao-Yan Li

(Submitted May 11, 2008; in revised form September 13, 2008)

Two Fe-FeB-WC coatings were deposited on the Q235 steel substrate by arc spraying. The microstructure and the abrasive wear performance of the coatings were characterized by x-ray diffraction (XRD) and scanning electron microscope (SEM). The wear mechanisms of the coatings were examined. It was found that Fe-Cr alloy and Fe_2B are present in the coating as the main phases. The results showed that adding hard particle powders could obviously increase the hardness and wear resistance of the coatings. The average microhardness of the coatings was about 870 to 920 $HV_{0.1}$. The coatings exhibited excellent abrasive wear resistance, being 3.3 to 4.8 times higher than that of arc sprayed 3Cr13 coating.

Keywords abrasive wear resistance, arc spraying, Fe-FeB-WC coating

1. Introduction

Thermally spraying cermet coatings have emerged as a viable solution for a wide range of wear-resistance applications to improve the service life of machine components, due to their high hardness and good wear resistance (Ref 1). In order to enhance the wear resistance of many types of engineering components, WC, TiC, and Cr_3C_2 particulates were commonly used as hard phases in the cermet. The wear resistance of FeNiCrB coatings, especially for abrasive wear, can be greatly increased by adding refractory carbides into the FeNiCrB matrix. Compared with cermet systems involving other carbides, the coatings containing WC are superior in hardness and wear resistance. WC coupled with Co, CoNi, Ni, or FeNiCrB matrix has been the preferred material for industrial applications used to reduce wear because of the combined effect of the hardness of carbides and toughness of the metal matrix (Ref 2). Hence, many research efforts have been made to prepare the cermet coatings containing WC by different methods (Ref 3-10), including high-velocity oxyfuel (HVOF), plasma spraying (PS), and detonation gun spraying (DGS). Although industrial applications of these methods, especially HVOF, have become very popular because of the excellent wear performance of the resultant coating, arc spraying of wear-resistant coatings has been widely used for many industrial applications where good wear and corrosion resistance are highly required. The major advantages of arc spraying are low cost, high efficiency, and simple operation.

In order to improve the wear performance of arc sprayed coating, one possible solution is to add wear-resistant hard particles into the arc sprayed metallic alloy matrix. The development of the cored wire makes it possible to deposit hard phase containing cermet coating by arc spraying. In this study, cobalt-coated tungsten carbide (WC/12Co), nickel-coated tungsten carbide (WC/12Ni), and FeB alloy powders were used as fillers in the cored wire to produce Fe-FeB-WC composite coatings. However, the previous study showed that it was difficult to deposit a dense coating. In this paper, Ni powder was added in the filler materials to modify deposition behavior and coating microstructure. The microstructure and phase compositions of the coatings were investigated. The wear performance and mechanisms of the coatings were also investigated.

2. Experimental

Cored wire with a diameter of 2.0 mm was used to produce Fe-FeB-WC coatings. The sheath of the cored wire was 304L stainless steel. FeB ($FeB_{18}C_{0.5}$-A: 81.5% Fe, 18.0% B, 0.5% C; 150-250 μm, crushed; Liaoyang International Boron Alloys Co., Ltd. (LIB), China), WC/12Co (KF-60: 88% WC, 12% Co; 15-45 μm, agglomerated and sintered; Beijing General Research Institute of Mining & Metallurgy, China), and WC/12Ni (KF-60-1: 88% WC, 12% Ni; 15-45 μm, agglomerated and sintered;

This article is an invited paper selected from presentations at the 2008 International Thermal Spray Conference and has been expanded from the original presentation. It is simultaneously published in *Thermal Spray Crossing Borders, Proceedings of the 2008 International Thermal Spray Conference*, Maastricht, The Netherlands, June 2-4, 2008, Basil R. Marple, Margaret M. Hyland, Yuk-Chiu Lau, Chang-Jiu Li, Rogerio S. Lima, and Ghislain Montavon, Ed., ASM International, Materials Park, OH, 2008.

Ding-Yong He, Bin-You Fu, Jian-Min Jiang, and **Xiao-Yan Li,** College of Materials Science and Engineering, Beijing University of Technology, Beijing 100124, China. Contact e-mail: dyhe@bjut.edu.cn.

Table 1 Composition of the filler materials

	Element, wt.%						
Coatings	Fe	Cr	Ni	B	W	Co	C
W1	70.12	12.67	7.14	4.05	5.07	0.72	0.23
W2	70.12	12.67	7.86	4.05	5.07	···	0.23

Beijing General Research Institute of Mining & Metallurgy, China) thermal spray powders were employed as the filler materials in the cored wire. The mass ratio of the 304L sheath to filler powders inside the sheath was 7:3. The coatings were sprayed onto degreased and grit-blasted Q235 steel substrates ($56 \times 25 \times 5$ mm^3) using two cored wires by arc spraying system (JZY-250, Beijing Jiazhiyuan Scientific & Trading Co., Ltd., China) under the optimized spray parameters. The parameters were: arc voltage, 32 V; arc current, 180 A; air pressure, 0.55 MPa; spray distance, 100 mm. Two coatings were deposited on Q235 steel, coatings W1 and W2. The thickness of both coatings was approximately 1.0 mm. The nominal composition of the coatings is listed in Table 1.

The microhardness of the coatings was measured using a microVickers hardness tester (MicroMet 1, Buehler). The measurements were carried out on the midplane of coating cross section with a 100 g load and 15 s dwell time. A given microhardness value was an average of 10 measurements.

The abrasive wear test was carried out using a modified wet sand rubber wheel (Model MLS-225, Zhangjiakou Chengxin Balance Testing Machineries Company, China). The specimen dimensions were $57 \times 25 \times 6$ mm^3, and the coatings in a thickness of 1.0 mm were used. The silica in a sharp, angular morphology was used as an abrasive and was sieved into a particle size range of 224 to 355 μm prior to the test. A load of 100 N was applied on the contact surface between the coating and the rubber wheel, and the rubber wheel peripheral speed was 240 rpm, The specimen was immersed in the abrasive mortar with the weight ratio of W_{silica}:W_{water} = 3:2. The tests were performed for a set number of revolutions with a preabrading of 1000 cycles and a refining abrading of 2000 cycles. The specimens were cleaned ultrasonically in the acetone bath for 3 to 5 min before and after the experiment. The weight loss of the specimen before and after abrasion was measured by a 0.1 mg BS224S electronic balance. An arc sprayed 3Cr13 coating was also deposited for comparison of the relative wear resistance of the hard phase contained coatings. The relative wear resistance (σ) of the coatings was calculated using:

$$\sigma = \frac{m_{3Cr13}}{m_{Coating}} \qquad \text{(Eq 1)}$$

where m_{3Cr13} and $m_{Coating}$ represent the wear mass loss of 3Cr13 coating and W1 or W2 coatings, respectively.

After abrasive wear test, a square specimen of 5×5 mm^2 was cut from the central part of the abrasive track. The wear track was examined by a FEI Quanta 200 FEG scanning electron microscopy (SEM) under the secondary electron (SE) images model. Composition analysis was also performed in the SEM by using a Noran energy-dispersive x-ray analysis (EDX) system, with a thin-window detector capable of analyzing elements with atomic number down to carbon. Cross-section microstructures of the coatings were observed by an optical microscope. The phases of the coatings were characterized by the x-ray diffraction (XRD) (D8 ADVANCE, BRUKER/AKS, Germany) using Cu Kα target at 40 kV and 20 mA. The porosity of the coatings was evaluated through image analysis of cross-sectional microstructure of the coatings. A total of 10 pictures of each coating were taken by an optical microscope at a magnification of 200× to evaluate the average porosity levels.

3. Results

3.1 Microstructural Characterization

The x-ray diffraction patterns of feedstock powders used as the fillers are shown in Fig. 1. In WC/12Co powder, WC, Co, and Co_3W_3C were detected, while in WC/12Ni powder, only WC phase was detected. Only FeB phase was detected in FeB powder. X-ray diffraction patterns of the coatings W1 and W2 are shown in Fig. 2. From the patterns, both coatings were mainly composed of Fe-Cr alloy with a fraction of Fe_2B and Fe_3O_4, and the broadening of the peak was observed. Probably the matrix material was presented in amorphous/crystalline form, presumably because of the high cooling rates (typically more than 10^6 K/s) occurring in such deposition processes upon impact of molten particles on the target material. For the coating W1, peaks from WC and α-W_2C were recognized, and the C-deficient phases W_2C formed as a result of oxidation and decarburization during the coating deposition process. However, a distinct Co peak was not present in the coating in agreement with the results reported (Ref 11), which suggested that Co in WC/Co based thermal spray coatings was present as an amorphous phase. The XRD pattern of coating W2 is similar to that of the coating W1, but distinct WC and α-W_2C peaks were not present; it may have formed because most of the original WC/12Ni particle was decarburized during the coating deposition process and the residual WC particles were in an amount that could not be detected by XRD.

Figure 3 shows the cross-sectional microstructure of the as-sprayed coatings at the unetched condition. As indicated in Fig. 3, both coatings consist of flattened platelike lamellae oriented parallel to the substrate. Some pores were also present in the coatings. The measured porosity of the coatings W1 and W2 was 2.1% and 3.2%, respectively. Based on the analyzing results of microstructures and XRD patterns, the phases can be identified in both coatings, and the white metal matrix is Fe-Cr and the particles in a deeper gray contrast corresponds to Fe_2B hard phase, and the area in a dark contrast corresponds to pores. It is clear that a significant fraction of Fe_2B was distributed in the coating. However, it was confirmed that

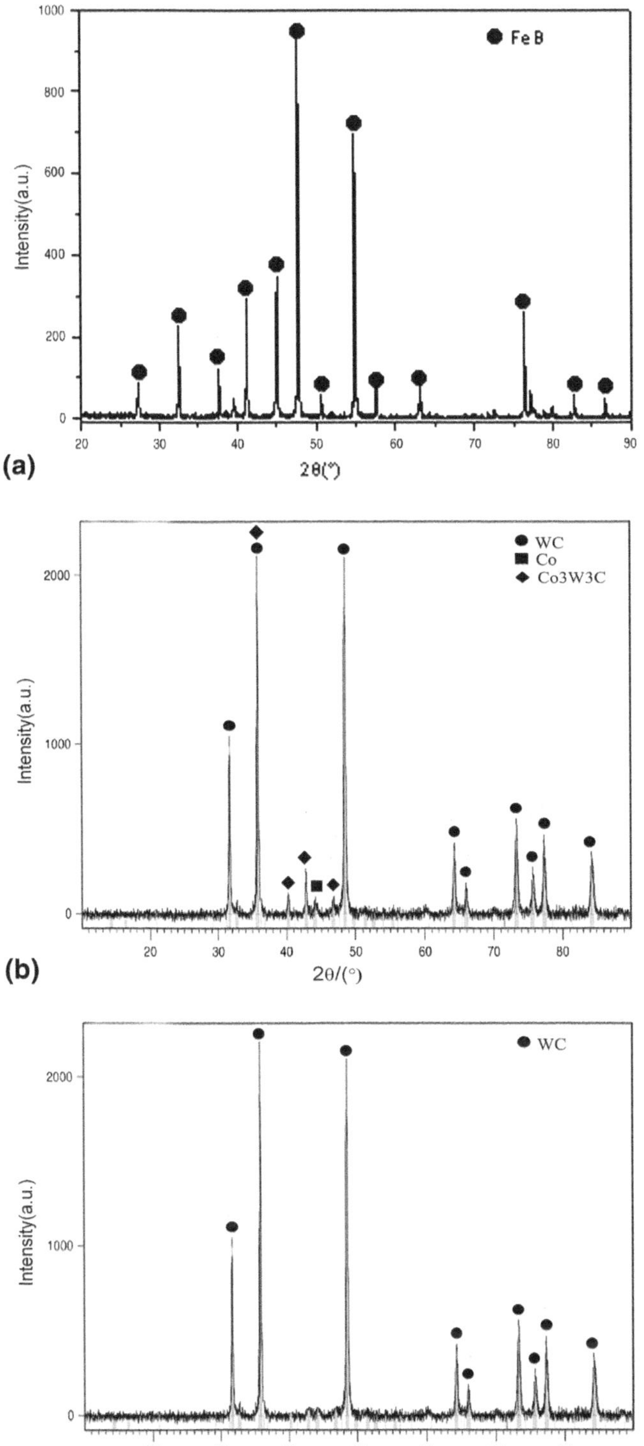

Fig. 1 XRD spectrum of feedstock powders used as fillers in cored wires. (a) FeB. (b) WC/12Co. (c) WC/12Ni

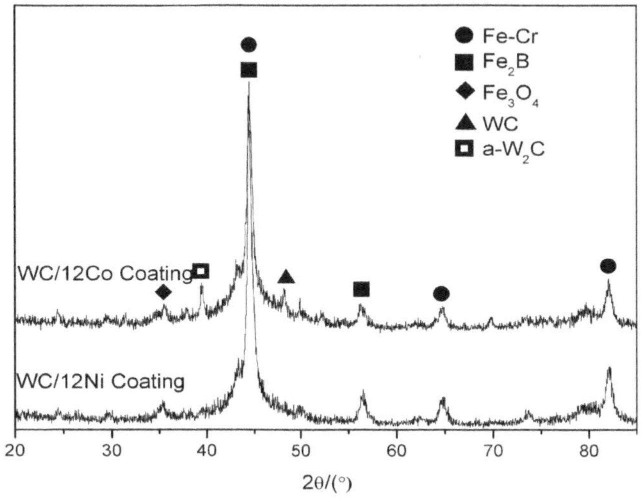

Fig. 2 XRD spectrum of the as-sprayed coatings

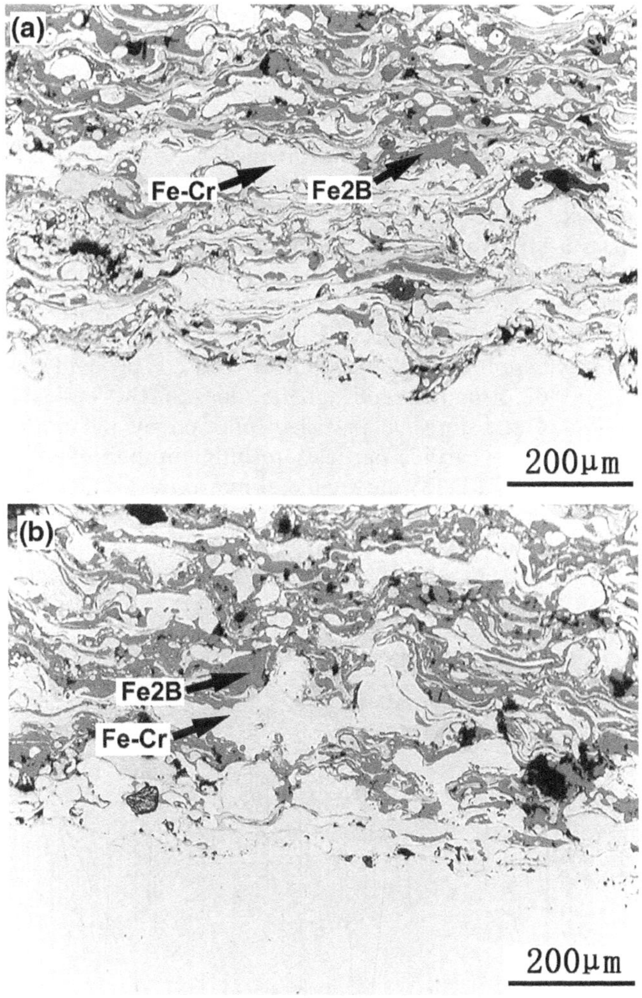

Fig. 3 Cross-sectional microstructure of the as-sprayed coatings. (a) Coating W1. (b) Coating W2

few carbides were recognized from the cross-sectional microstructure of the coatings. This possibly resulted from the fact that the substantial FeB particles were melted during spraying, while most WC particles were not melted and rebounded off after impact on the substrate. Therefore, as-hard phases in the filler materials in the cored wire designed for arc spraying, FeB can be more easily retained in the coating than WC.

3.2 Wear Properties of Coatings

The hardness and relative wear resistance of the as-sprayed coatings are shown in Fig. 4. The microhardness values are the average of 10 measurements taken in the cross section of the coatings. Microhardness measurements on coatings yielded a value of 920 $HV_{0.1}$ for the coating W1 and 872 $HV_{0.1}$ for the coating W2. The average mass loss of the coating W1 was 0.1243 g, while the average mass loss of the coating W2 was 0.1794 g. As shown in Fig. 4, the coating W1 exhibited an excellent abrasive wear resistance that was 4.8 times higher than that of the arc sprayed 3Cr13 coating. Although the abrasive wear resistance of the coating W2 was not good as the coating W1, its wear resistance was still 3.3 times higher than that of the arc sprayed 3Cr13 coating.

Worn surface morphologies were examined under SEM to understand the wear behavior of the as-sprayed coatings. The locations for EDS analysis are shown in Fig. 5, and the contents of different elements at corresponding locations are shown in Table 2. As shown in Table 2, composition analysis of points A and B confirmed the inclusion of the primary carbide WC in the coatings.

The morphology of the abrasive worn surface of the as-sprayed coatings are shown in Fig. 5. As shown in Fig. 5(a), there are numerous plastic furrows on the worn surface of the 3Cr13 coating. It was believed to result from a typical microcutting mechanism. For the coating W1, some WC particles have come out in some regions of the coating surface (point A). It was suggested that the wear mechanism of the cermet coating was by selective removal of the binder caused probably by plastic deformation and fatigue due to the repeated action of the abrasive particles followed by the undermining of the carbide particles, resulting in their eventual pullout (Ref 12, 13). Fe-Cr metal matrix was softer than the WC and Fe_2B hard phases and thus prone to plastic deformation. WC and Fe_2B grains in the coating were held by the metallic phase. When the metallic phase was

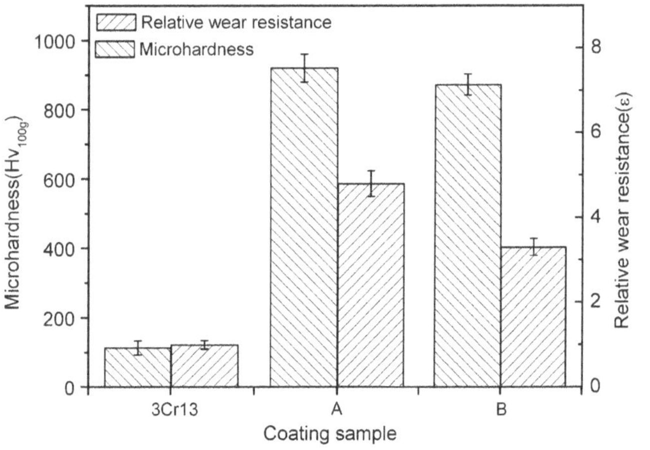

Fig. 4 Microhardness and relative wear resistance of the as-sprayed coatings

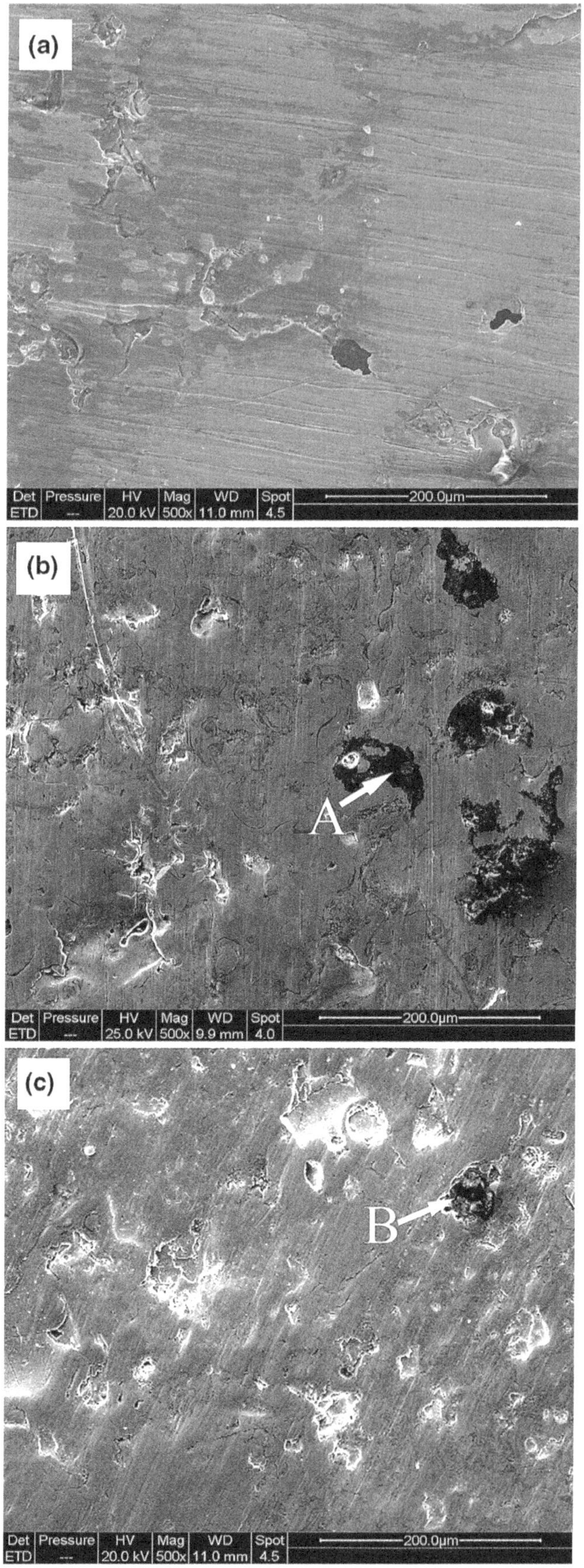

Fig. 5 Worn surface morphologies of the as-sprayed coatings. (a) 3Cr13 coating. (b) Coating W1. (c) Coating W2

Table 2 Composition of the different points in as-sprayed coatings

Points	Element, wt.%				
	Fe	Cr	Ni	W	C
A	1.12	0.89	...	93.84	4.15
B	2.67	2.13	0.92	91.02	3.26

removed from the coating, the WC and Fe_2B grains automatically come out. A similar type of grain removal has been reported elsewhere (Ref 12). For the coating W2, the mechanism of wear was similar with that of the coating W1. The high porosity and little carbide content may possibly be the reason for lower hardness and wear resistance. It was easier for cracks to initiate from defective regions such as pores than in a normal smooth surface; under the abrasion conditions, microcracks initiate and propagate along the interface between hard phase and the matrix. The propagation of cracks cause flaking off of the lamellae.

4. Conclusions

The results of the present study on the arc sprayed Fe-FeB-WC coatings can be summarized:

- The coatings prepared by arc spraying using cored wires exhibited a typical lamellar structure. The majority of phases in both coatings were Fe-Cr and Fe_2B. A trace of WC and α-W_2C was recognized from the XRD pattern of coating W1, but was not recognized in coating W2. The results indicated that it was easier to deposit an alloy-based coating containing FeB than one containing WC by arc spraying using cored wire with hard phase fillers.
- The average microhardnesses of two arc sprayed Fe-FeB-WC coatings were 920 $HV_{0.1}$ and 872 $HV_{0.1}$, higher than that of an arc sprayed 3Cr13 coating. The coatings W1 with W2, coating W1 exhibited better abrasive wear resistance, which was 3.3 to 4.8 times than that of the 3Cr13 coating. The results showed that a wear-resistant coating can be deposited by arc spraying using a cored wire of hard filler materials.

Peer Reviewed

References

1. F. Rastegar and D.E. Richardson, Alternative to Chrome: HVOF Cermet C for High Horse Power Diesel Engines, *Surf. Coat. Technol.*, 1997, **90**(1-2), p 156-163
2. V. Stoica, R. Ahmed, and T. Itsukaichi, Influence of Heat-Treatment on the Sliding Wear of Thermal Spray Cermet Coatings, *Surf. Coat. Technol.*, 2005, **199**(1), p 7-21
3. P.K. Aw and B.H. Tan, Study of Microstructure, Phase and Microhardness Distribution of HVOF Sprayed Multi-Modal Structured and Conventional WC-17Co Coatings, *J. Mater. Process. Technol.*, 2006, **174**(1-3), p 305-311
4. Q. Yang, T. Senda, and A. Hirose, Sliding wear behavior of WC-12%Co coatings at elevated temperatures, *Surf. Coat. Technol.*, 2006, **200**(14-15), p 4208-4212
5. H.S. Sidhu, B.S. Sidhu, and S. Prakash, Mechanical and Microstructural Properties of HVOF Sprayed WC-Co and Cr_3C_2-NiCr Coatings on the Boiler Tube Steels Using LPG as the Fuel Gas, *J. Mater. Process. Technol.*, 2006, **171**(1), p 77-82
6. H. Du, C. Sun, W.G. Hua, T.G. Wang, J. Gong, X. Jiang, and S.W. Lee, Structure, Mechanical and Sliding Wear Properties of WC-Co/MoS_2-Ni Coatings by Detonation Gun Spray, *Mater. Sci. Eng. A*, 2007, **445-446**, p 122-134
7. G.C. Ji, C.J. Li, Y.Y. Wang, and W.Y. Li, Microstructural Characterization and Abrasive Wear Performance of HVOF Sprayed Cr_3C_2-NiCr Coating, *Surf. Coat. Technol.*, 2006, **200**(24), p 6749-6757
8. Y. Ishikawa, S. Kuroda, J. Kawakita, Y. Sakamoto, and M. Takaya, Sliding Wear Properties of HVOF Sprayed WC-20%Cr3C2-7%Ni cermet coatings, *Surf. Coat. Technol.*, 2007, **201**, p 4718-4727
9. A.K. Maiti, N. Mukhopadhyay, and R. Raman, Effect of Adding WC Powder to the Feedstock of WC-Co-Cr Based HVOF Coating and Its Impact on Erosion and Abrasion Resistance, *Surf. Coat. Technol.*, 2007, **201**(18), p 7781-7788
10. T. Sahraoui, N.E. Fenineche, G. Montavon, and C. Coddet, Structure and Wear Behaviour of HVOF Sprayed Cr_3C_2-NiCr and WC-Co Coatings, *Mater. Des.*, 2003, **24**(5), p 309-313
11. S. Luyckx and C.N. Machio, Characterization of WC-VC-Co Thermal Spray Powders and Coatings, *Int. J. Refract. Met. Hard Mater.*, 2007, **25**(1), p 11-15
12. J.K.N. Murthy and B. Venkataraman, Abrasive Wear Behaviour of WC-CoCr and Cr3C2-20(NiCr) Deposited by HVOF and Detonation Spray Processes, *Surf. Coat. Technol.*, 2006, **200**(8), p 2642-2652
13. T. Sudaprasert, P.H. Shipway, and D.G. McCartney, Sliding Wear Behaviour of HVOF Sprayed WC-Co Coatings Deposited with Both Gas-Fuelled and Liquid-Fuelled Systems, *Wear*, 2003, **255**(7-12), p 943-949

JTTEE5 17:762–773
DOI: 10.1007/s11666-008-9252-7
1059-9630/$19.00 © ASM International

Peer Reviewed

Erosion, Abrasive, and Friction Wear Behavior of Iron Aluminide Coatings Sprayed by HVOF

J.M. Guilemany, N. Cinca, J. Fernández, and S. Sampath

(Submitted May 3, 2008; in revised form September 22, 2008)

Iron aluminides have been proposed lately as promising materials for wear applications. Many authors have focused their investigations on the friction behavior of FeAl coatings, emphasizing the role of this intermetallic phase as a new matrix to embed ceramic particles and replace the extensively studied WC-Co cermet system for high temperature. However, few of these studies deal with the evaluation of the different tribological properties and their relationship with the coating microstructure. In the present study, the near stoichometric Fe40Al was successfully sprayed by means of high velocity oxy-fuel using different particle size distribution and the tribological behavior was assessed through solid particle erosion, abrasive and dry sliding tests. The wear mechanisms of the deposited coatings are discussed with regard to the observed results. In addition, oxidized samples were tested to evaluate the role of the oxide top layer; also, the powder was previously annealed to produce a coating with an almost fully ordered FeAl structure.

Keywords erosion and abrasion resistance, friction and wear, HVOF microstructures, HVOF spray parameters, wear, wear mechanisms

1. Introduction

One of the current challenges in materials science is to find versatile materials able to withstand good resistance for different service performance situations. A combination of good mechanical properties, corrosion, oxidation, and wear resistance is usually difficult to reach.

Currently, intermetallic compounds (IMCs) are being investigated as a good compromise for the above requirements. They are regarded as advanced materials, attractive due to their unique properties: high melting points, great strength but poor ductilities. In these respects, they resemble ceramic materials. Unlike ceramics, however, they have a metallic lustre and conduct heat and electricity well. Thus, they have been exploited not only for their mechanical properties but their chemical, electrical, magnetic, optical, and semiconducting properties as well (Ref 1).

Although IMCs have been studied and employed over decades, their interest is relatively new and has been motivated by the demand of finding some sort of materials capable of substituting superalloys, especially for jet engine development (Ref 2, 3); their particular structure is the reason of their high stability. Whereas conventional alloys are defined by atoms linked with relatively weak metallic bonds, IMCs have either partly ionic/covalent bonds or, having entirely metallic bonds, the atoms take up preferred positions within the crystal lattice. Because of such atom arrangement, IMCs tend to have a much lower self-diffusion coefficient than do disordered alloys and so their structure can be maintained to high temperatures (Ref 4). Nevertheless, their long-range ordered superlattices reduce dislocation mobility and diffusion processes leading to their inherent embrittlement. It has been observed that some can even keep such order up to their melting point, otherwise they disorder below a certain critical temperature (T_c) at which an order-disorder transformation might occur.

Contemporary efforts have been addressed to aluminides of transition metals, i.e. Fe, Ni, Ti, Co, and Nb; these compounds possess relatively low densities, high specific modulus, high melting points, strength and excellent oxidation and corrosion resistance because of the formation of a protective alumina layer (Ref 5, 6). However, since the first investigations about intermetallics carried out by Kumakov et al. (1916), their low ductility has hampered their fabrication. To overcome such a handicap, many attempts have been carried out in the last decades to

This article is an invited paper selected from presentations at the 2008 International Thermal Spray Conference and has been expanded from the original presentation. It is simultaneously published in *Thermal Spray Crossing Borders, Proceedings of the 2008 International Thermal Spray Conference*, Maastricht, The Netherlands, June 2-4, 2008, Basil R. Marple, Margaret M. Hyland, Yuk-Chiu Lau, Chang-Jiu Li, Rogerio S. Lima, and Ghislain Montavon, Ed., ASM International, Materials Park, OH, 2008.

J.M. Guilemany, N. Cinca and **J. Fernández**, Thermal Spray Centre, University of Barcelona, Barcelona, Spain; and **S. Sampath**, Center for Thermal Spray Research, Stony Brook, New York. Contact e-mail: nuria.cinca@ub.edu.

enhance their ductility, such as reduction of grain size (Ref 7) or addition of certain alloying elements (Ref 8, 9).

Their difficulty to fabricate as structures has stimulated their use as coatings. In a previous study, the oxidation behavior of high velocity oxy-fuel (HVOF) sprayed FeAl coatings was already reported (Ref 10). Such aluminide (B2 ordered structure, CsCl-type) has a density of 5.56 g/cm^3, excellent sulphidation resistance, and good strength up to 600 °C (Ref 6, 11). Its applications include, among others, molten salt systems for chemical air separation, automotive exhaust systems, immersion heaters, heat exchangers, catalytic conversion vessels, chemical production systems, and coal conversions systems. Since some of the previous applications involve removal of material due to contact damage, few studies have recently evaluated their wear response either in bulk or coating form. Friction has been the most investigated mechanism and delamination was found to be the predominant cause for volume loss in coatings (Ref 12-14). The wear resistance was further improved by reinforcement with hard particles (Ref 13, 15). The presence of such particles has also been reported to affect the abrasive and erosion properties even for NiAl (Ref 15-17).

The other two wear mechanisms, mainly abrasion and erosion, have been studied for FeAl and NiAl coatings (Ref 18, 19). Alman et al. (Ref 14, 16) compared the wear rates of Fe-Al based alloys with those of other metals and ceramics and concluded that their resistance was as much as that encountered in typical austenitic stainless steels. Moreover, the higher the Al content, the further the wear rate decreased relating to bulk hardness of the compound.

Ordering in these compounds is also induced by heat treatments to yield B2 (36-48 at.% Al) or $D0_3$ (interpenetrating FCC lattices, composition range 23-36 at.% Al and from room temperature to 550 °C) ordered structures (Ref 20), which result in lower wear rates compared to these non-long-range ordered iron aluminides (Ref 21, 22).

The present study intends to assess all the above wear responses for FeAl intermetallic coatings based on different microstructures. A main comparison is reported concerning two different microstructures obtained by using different particle size distribution; secondly, the effect in lattice ordering is assessed by previously annealing the feedstock powder and, finally, thermal oxidation is intended to evaluate how surface modification would affect to wear properties.

2. Experimental Procedure

The FeAl powder alloy namely Grade 3, provided by CEA-DTEN (*C*ommissariat à l'*E*nergie *A*tomique – *D*epartement de *T*echnologie des *E*nergies *N*ouvelles, Grenoble) and Mecachrome (Vibraye, France) (Ref 23), was used to produce the FeAl coatings onto low alloyed carbon steel substrate. The powder with an initial composition of Fe40Al-0.05Zr (at.%) + 50 ppm B was prealloyed by gas atomization and subsequently ball-milled. Boron and zirconium were introduced as alloying elements to improve ductility in the material (Ref 9, 24, 25) and yttria (ODS, oxide dispersed strengthening) was added during the milling step in order to keep fine grain size after hot consolidation and forming so that mechanical strength (Hall-Petch mechanism) and room temperature ductility (reduction of the environmental embrittlement caused by water vapor at RT thanks to grain size reduction down to 1 μm) are improved (Ref 26-28).

CEA-DTEN was the first supplier and it was afterwards transferred to Mecachrome. Both suppliers provided the same powder with the same composition but with the respective powder distributions as observed in Fig. 1. The whole CEA-DTEN powder distribution was used to produce the coating which will be referenced as

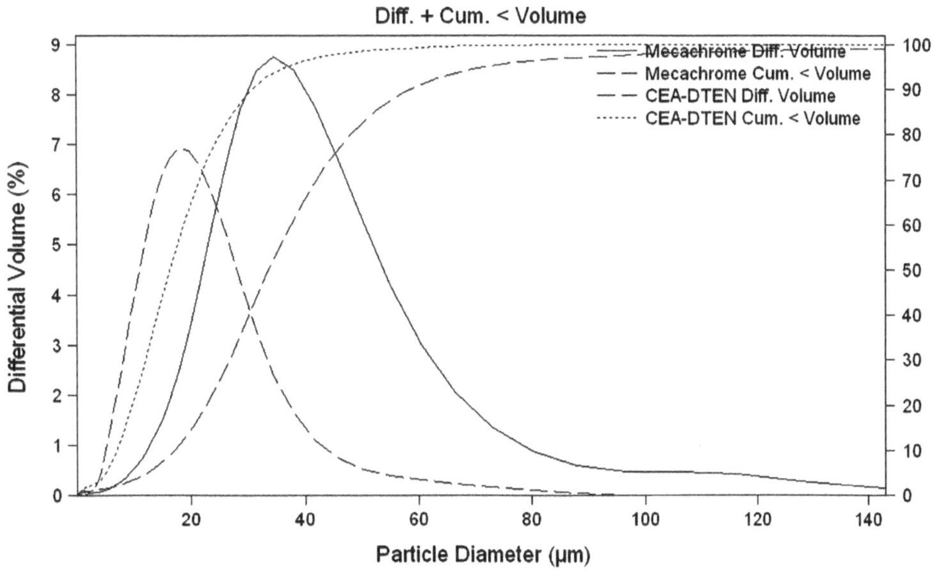

Fig. 1 Powder size distribution

Peer Reviewed

Table 1 Thermal spraying parameters

	FeAl
Oxygen flow rate, L/min	189
Propylene flow rate, L/min	87
Carrier gas, L/min	305
Oxygen/fuel ratio	2.874
Feeding rate, g/min	20
Spraying distance, mm	250

CEA, while from the 40-60 μm fraction, the one referenced as MEC40-60 was obtained. Furthermore, a third coating was produced by spraying the 40-60 fraction size being annealed at 700 °C for 10 min in order to induce the ordering in the intermetallic structure and then evaluate the properties of the resulting coating compared to that of MEC40-60 (this will be referenced as $MEC40\text{-}60_{ANN}$). The annealing treatment was carried out in a controlled atmosphere furnace with a continuous argon flux.

A Diamond Jet Hybrid gun model 2600/2700 was employed for powder deposition, and Table 1 presents the spraying parameters. Such parameters were previously optimized with the intention to achieve the less oxidation content as possible (Ref 29).

The CEA coating was oxidized at 900 °C for 72 h (referenced as CEATT). The selection of the temperature for the treatment was based on a previous study where the formation of an adherent oxide scale was shown (Ref 10). The idea came from the work carried out by Xia et al. (Ref 30) who claimed that the average wear rate is reduced 16 times when the bulk intermetallic is compared with an untreated specimen. Both the microhardness of the as-sprayed and the oxidized coatings were investigated using a Vickers indentation tester MATSUZAWA MXT-a at 200 gf load for 15 s in the cross sections. At least 20 indentations were performed for average.

For the microstructural cross sections and top surfaces characterization, a JEOL scanning electron microscope (SEM) was used with energy dispersive spectroscopy (EDS) for microanalysis. X-ray diffraction (XRD) scans helped to identify the phases present after the spraying process. In addition, the "image analysis" option within the Matrox Inspector 1.71 software was employed for the oxidation and porosity evaluation. This was done by taking a micrograph of the overall coating cross section; at least 10 pictures were analyzed like this for each coating.

2.1 Abrasive Tests

According to ASTM G65-00, the Rubber Wheel test consists of a rubber wheel rotating at a constant speed of 139 rpm on the specimen while silica particles (particle size in the range of 0.4-0.8 mm) flow by gravity between the two contacting surfaces. The samples were tested in their as-sprayed state.

A normal load of 50 N is applied and the mass loss is measured up to 30 min. The abrasion wear rate W ($mm^3\ N^{-1}\ m^{-1}$) is calculated as:

$$W = \frac{\Delta m}{t}\frac{1}{N}\frac{1}{\rho} \qquad \text{(Eq 1)}$$

where Δm is the average mass loss (g), t is the test time (min), N is the normal load, and ρ ($g\ cm^{-3}$) is the FeAl powder density calculated as follows: a known mass of powder is introduced into a 25 mL flask and is filled with a high wettable liquid with a known density, here cyclohexanone.

$$\rho = \frac{m_p}{\left(25 - \frac{m_T - m_p}{\rho_{liq}}\right)} \qquad \text{(Eq 2)}$$

where m_p is the mass powder introduced, m_T is the total mass (powder + cyclohexanone), and ρ_{liq} is the cyclohexanone density at working temperature.

This methodology for calculating the density has been considered a much better approximation to the real value of the material than vibrated and tap densities referenced according to ASTM B-212 and ASTM B-527, respectively. Although it is evident that both coatings do not have the same density, powder density has just been used as approximation.

2.2 Erosion Tests

The erosion tests have been performed at room temperature using a grit-blasting apparatus. The Al_2O_3 abrasive particles whose average size is about 50 μm leave the nozzle at the velocity of 1 g/min controlled by selecting the air pressure at 10 psi (regularly checked by weighting the alumina collected for 1 min). The nozzle tube was 1.5 mm inner diameter and 50 mm long.

The sample's dimensions were 20 × 20 × 5 mm and, before testing, they were polished down to 1 μm. The distance from the specimens surface to nozzle end was 10 ± 1 mm and the angle between the nozzle axis and the specimens surface was 90°. The mass loss was obtained by weighting on an analytical precision balance every 30 s and the test was stopped when the substrate was reached.

The average erosion value of the coatings was calculated by dividing erosion rate ($g\ min^{-1}$) by the abrasive flow rate. It is then normally divided by the specimen density (powder density, $g\ cm^{-3}$) and the average value is then reported as $mm^3\ g^{-1}$.

2.3 Dry Sliding Behavior

Friction tests were carried out with a ball-on-disc system according to ASTM G99-03, where an 11 mm diameter WC-6Co ball was frictioning the polished coating surface (below 0.8 Ra). A temperature around 25 °C and the humidity (below 20%) were controlled during the test time into the closed camera where the experiments were performed. A sliding speed of 0.11 m/s, track diameter of 16 mm, and sliding distance of 1000 m were constant for all the tests. The variation of friction coefficient with sliding distance was registered at 5 N for the FeAl coatings and a comparison with the wear behavior of the steel substrate and the thermal treated sample is given.

According to the standard, the friction coefficient (FC) is given as the average of the values obtained for the last 200 m.

The volume loss was obtained by means of white light interferometry (SWLI). The interferometric surface analysis microscope (Zygo New View 100) uses SWLI to image and measure tested surfaces and provide surface analysis without contacting it. Depths up to 100 μm, with 0.1 nm resolution and 0.3 nm RMS repeatability, are imaged independently of the objective magnification; lateral resolution is 500 nm. SWLI is a technique based on the constructive or destructive interferences of two light beams, one going to a reference and the other to the sample, which leads to the obtention of a 3D image with the profile of the wear track.

The wear rate K ($mm^3\ N^{-1}\ m^{-1}$) is calculated as:

$$K = \frac{V}{ws}, \qquad \text{(Eq 3)}$$

where V is the worn volume, w (N) is the normal load, and s (m) is the total distance.

The as-sprayed coatings were characterized by SEM and microhardness indentation. The worn surfaces were also examined using SEM-EDS.

3. Results and Discussion

3.1 Coating Deposition and Characterization

Figures 2(a) and (b) show the SEM and XRD morphology of the powder. X-ray presents the typical fundamental lines of the FeAl pattern (h + k + l = even), being only exhibited when the structure is disordered as, otherwise, the superlattice lines (h + k + l = odd) would also appear; this was already foreseen as result of the high plastic deformation induced by ball-milling. It is even worth noting that the peak broadening comes from the finer grain size and microstrain introduced during milling. By contrast, when the powder is annealed, the X-ray presents the superlattice peaks conserving a slight broadening, which means that after the annealing treatment the structure has been partly reordered (Fig. 2b).

Using different particle distributions, a great difference was observed on the resulting microstructures (Fig. 3a and b). Whereas MEC40-60 mainly consists of partially melted particles with some porosity and an oxide layer at the interparticle zones, CEA is well featured by partially and fully melted particles with some light and dark-grey contrasts at the intersplats boundaries. These have been characterized as iron-rich and oxide zones by EDS, respectively; the presence of Al-depleted regions suggests that during spraying, oxidation has occurred to the powder to form the Al-rich spinel $FeAl_2O_4$ oxide leaving Fe-rich areas at the lamellar boundaries. XRD also confirms the former results (Fig. 4); the additional peaks also identified as FeAl correspond to the superlattice lines; some authors have attributed their presence to the ordering within the

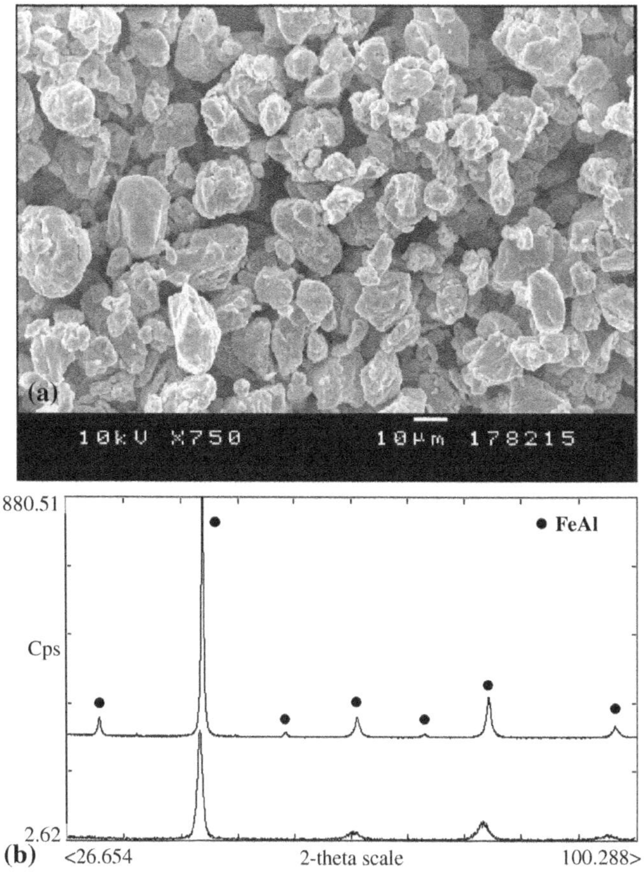

Fig. 2 (a) Morphology and (b) X-ray of the iron aluminide powder grade 3 as-supplied and annealed at 700 °C

melted areas (Ref 31). The difference in both microstructures can be easily observed also by the examination of the fracture sections (Fig. 5 and 6). Figure 5 illustrates the CEA coatings being built up by melted splats aggregation, while Fig. 6(a) and (b) do not exhibit such morphology and one can see two features: one where a particle has been torn off, and the other where the particle still remains in that part of the coating. Coating thicknesses, as well as porosity and oxide contents, are presented in Table 2; they point out the difference between the two coatings and the features that have been discussed. $MEC40\text{-}60_{ANN}$ is not illustrated here as it shows an analogous structure as MEC40-60.

By contrast, oxidized CEA presents a microstructure as shown in Fig. 7(a), where it can be seen that a very thin layer has been formed on the top. Although not being pure alumina as commonly found for bulks, the oxide characteristics are good enough to provide a good corrosion resistance; otherwise an easy oxygen inward diffusion would have produced an increase in oxidation kinetics. Such good performance has been associated to the presence of Al_2O_3 nuclei among the Fe_2O_3 nodules that prevent their growth. The X-ray spectrum indicates the presence of the before mentioned phases (Ref 10) (Fig. 7b).

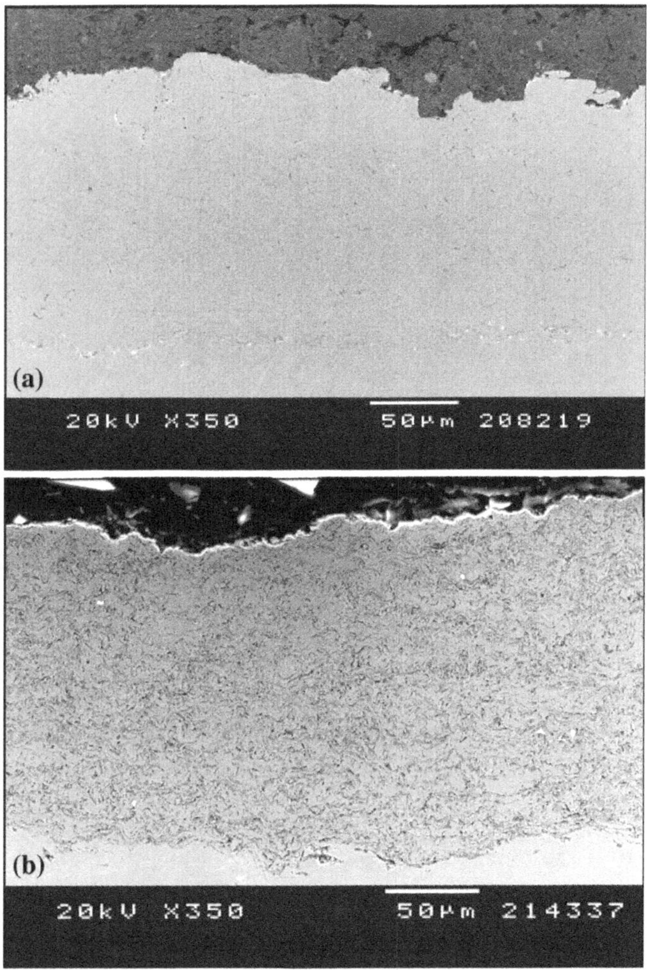

Fig. 3 SEM cross sections of (a) MEC40-60 and (b) CEA

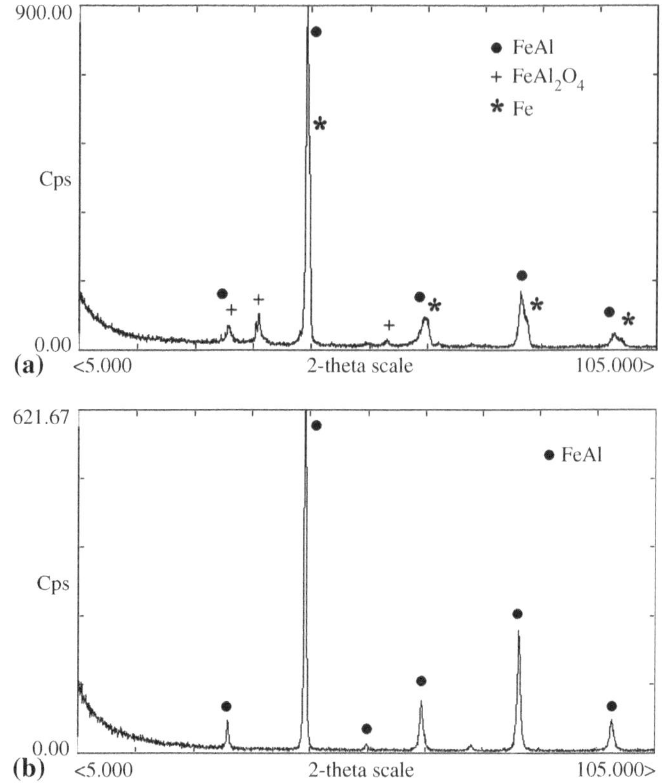

Fig. 4 XRD of (a) CEA and (b) MEC40-60 coatings respectively

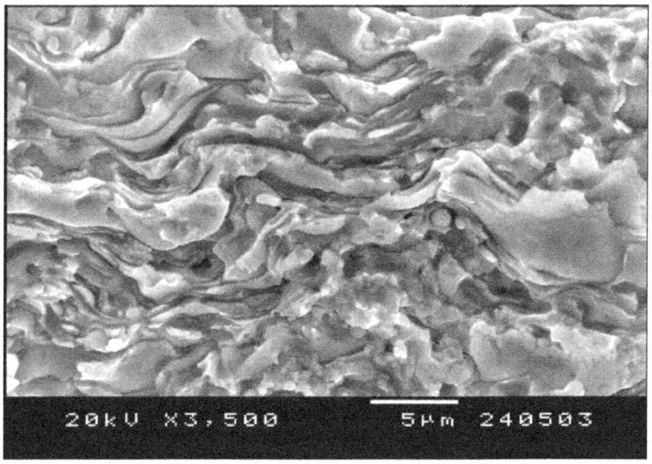

Fig. 5 Fracture section of the CEA coating

Regarding the microhardness values (Table 3), it can be seen that despite having different porosity levels, the overall presence of oxidation in CEA produces an increase in hardness as well as an embrittlement compared to MEC40-60. In addition, it might be associated to the higher occurrence of intersplat boundaries; in CEA, the indentation has more probabilities to include representative features of the overall coating structure, whereas in MEC40-60 the indentation might be performed inside an unmelted particle.

The comparison between the hardness values of MEC40-60 and MEC40-60_{ANN} can be discussed as follows: as a consequence of not having the stoichiometric 50:50 FeAl and even more, having a milled strained structure, the lattice must present constitutional defects; according to Jordan and Deevi (Ref 32), vacancy hardening can ameliorate hardness and strength of iron aluminides but is detrimental to room temperature ductility. Although the cooling rate during the deposition process might also have a considerable influence on the vacancy concentration (Ref 33-35), the main difference between the two coating structures is the fact that annealing the feedstock powder might have eliminated most of the defects produced by milling and has ordered the lattice; therefore, from this point of view, there has been a strain release and the logical conclusion would be that lowering vacancy defects should lead to a less hard deposit. Nevertheless, the reason why this does not occur must find answer assuming that some oxidation might have occurred during the annealing process, even though it was carried out in argon atmosphere as some powder particles seemed to have a different color tonality.

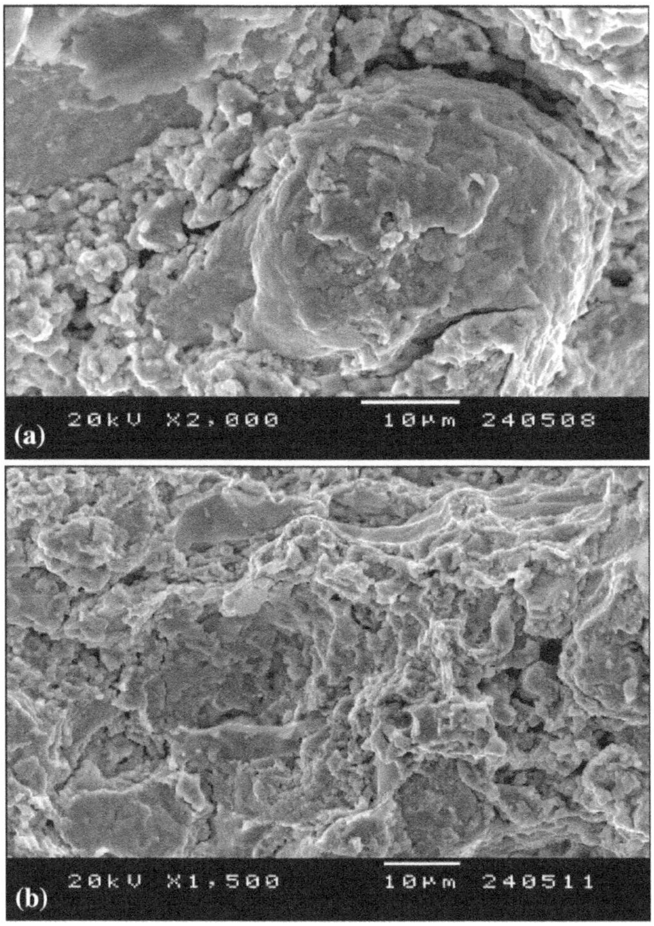

Fig. 6 (a, b) Different zones of the fracture section of the MEC40-60 section

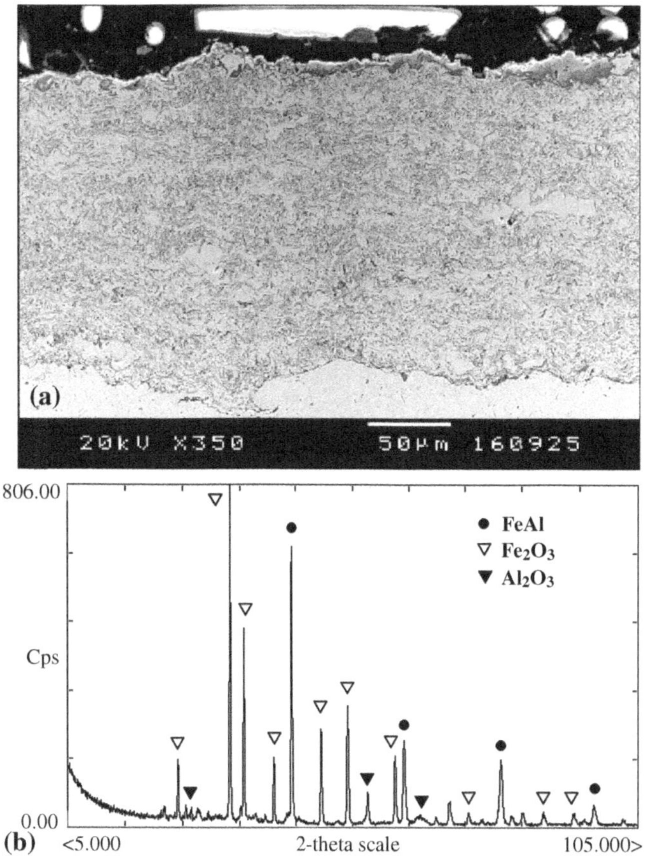

Fig. 7 (a) SEM cross section and (b) X-ray of CEATT

Table 2 Coating features

	Thickness	% Oxidation	% Porosity
CEA	162 ± 6	13.4 ± 2.6	1.8 ± 0.1
MEC40-60	147 ± 9	4.4 ± 0.6	0.3 ± 0.1

Furthermore, with regard to CEATT, looking at the standard deviation, HV differences between CEA and CEATT are not really significant.

3.2 Abrasive Tests

The present work examines a three-body abrasive wear, where silica particles are introduced and flow by gravity between the two surfaces (rubber rotating wheel and the as-sprayed coating). This kind of wear involves plastic deformation whenever the coating hardness is lower than $H_a/1.2$, where H_a is the hardness of silica particles; otherwise, the abrasive particles would smash (Ref 36). In the case of all studied coatings, $H_a/H_c > 1.2$ (hard abrasion) was observed to cause plastic scratching.

In a previous study, the abrasive behavior was already discussed for the as-called CEA coating (Ref 29). Compared to the others, CEA exhibits the highest abrasive wear resistance, which is assumed to be due to its highest hardness (Table 4). Hence, the harder the coating is, the less damage is produced by abrasive silica particles as a harder coating reduces wear by preventing ploughing on both a macro and micro-scale. Abrasive wear has also been found to be dependent on crystal structure and orientation; Alison (Ref 37) showed that cubic metals wear at about twice the rate of hexagonal metals, which was attributed to the lower work-hardening rate of the hexagonal metals. This reasoning serves also to discuss that the ordered FeAl lattice (MEC40-60_{ANN}) is less work-hardened than the disordered structure (MEC40-60).

Figure 8 shows the wear rate along the test time. The noticeable rapid initial decrease in the oxidized CEA is attributed to removal of the oxide surface layer; an examination of both damaged surfaces, that of the as-sprayed CEA and the oxidized one, indicates that the as-sprayed coating shows the grooves the hard silica particles have produced during the test, whereas the oxidized coating shows the cracks produced on the oxide scale. Such scale is not enough homogeneous to protect the coating against ploughing. Moreover, the fractured oxide debris can produce a more aggressive three-body abrasive effect.

The additions of hard particles such as TiC or TiB_2 in a Fe_3Al or FeAl matrix have been shown to reduce the wear rate by an order of magnitude. Such improvement has been also claimed by some authors (Ref 38, 39) who focused on

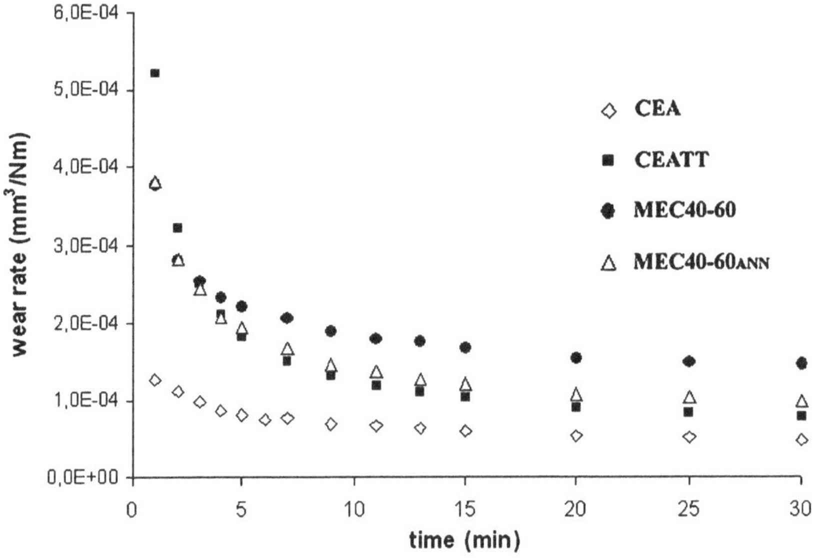

Fig. 8 Mass loss (g/min) versus time

Table 3 Microhardness of intermetallic FeAl coatings

	HV_{200}
CEA	434 ± 48
CEATT	426 ± 23
MEC40-60	377 ± 36
$MEC40\text{-}60_{ANN}$	415 ± 21

Table 4 Abrasive wear rates

	Wear rate, $mm^3\ N^{-1}\ m^{-1}$
CEA	3.7×10^{-5}
CEATT	5.2×10^{-5}
MEC40-60	1.2×10^{-4}
$MEC40\text{-}60_{ANN}$	6.7×10^{-5}

the comparison of the use of this intermetallic instead of usual Co as matrix. Interestingly, they stated that WC-40%FeAl exhibits similar wear resistance to that of commercial WC-10%Co. Particularly, FeAl ordered matrix is attractive for wear resistance in high-temperature environments.

However, there is not an extensive study on the abrasion resistance of thermally sprayed FeAl coatings and even in the case of their comparison to other deposited materials, the conditions they have been tested with, are not always comparable.

What can be actually compared are the present results with that of other materials tested with the same equipment and following the same standard. NiCr-based coatings exhibit wear rates of the same magnitude as MEC40-60, whereas CEA and CEATT are one order of magnitude lower because of the higher hardness (Ref 40). WC-12Co, for instance, showed a wear rate in the order of 10^{-6} $mm^3\ N^{-1}\ m^{-1}$ (Ref 41).

3.3 Erosion Behavior

With regard to test variables, in abrasive wear, the amount of material removed depends on the normal load pressing the particles against the surface and on the distance slid. In erosion, however, the extent of damage mainly depends on the number and mass of individual particles striking the surface, and on their impact velocity. As in the case of abrasion, mechanisms by which a metallic material is removed can involve both plastic deformation and brittle fracture; they can either depend on erodent particle sizes, velocities, impact angles, or environments. Hence, if plastic deformation dominates, the maximum wear occurs at low angles ("ductile behavior"), while at high impact angles erosion by brittle fracture is more rapid ("brittle behavior"). Striking at low angles with rounded particles normally yields to surface ploughing, whereas using angular particles produces cutting (Ref 42). At 90°, the damage normally occurs by cracking and chipping of surface material; in the case of coatings, this is favored by low strength of inter-lamella.

Figure 9 shows one of the damaged surfaces, which is clearly associated to the eroding effect of angular shaped particles where the material is displaced forming rims or lips. It is worth noting that such a mechanism, known as *cutting*, is normally encountered in ductile mechanisms (more attributed for metals at low impingement angles). However, the angular dependence of erosion is not a characteristic of the material alone, but depends also on the conditions of erosion. Thus, there can be alloys that erode in a "brittle behavior", that is, alloys of high hardness and low ductilities that show their maximum erosion rates at normal incidence. The terms "brittle" and "ductile" in erosion context must be then used with caution (Ref 42). Wang et al. (Ref 18, 43) also claimed that; in their results, the eroded surfaces also appear to result from a ductile mechanism. Few studies report their results on erosion performance of FeAl coatings; nevertheless,

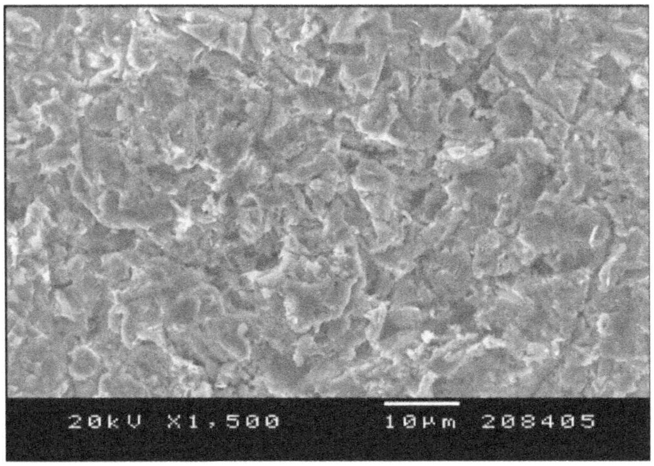

Fig. 9 Erosion surface morphology

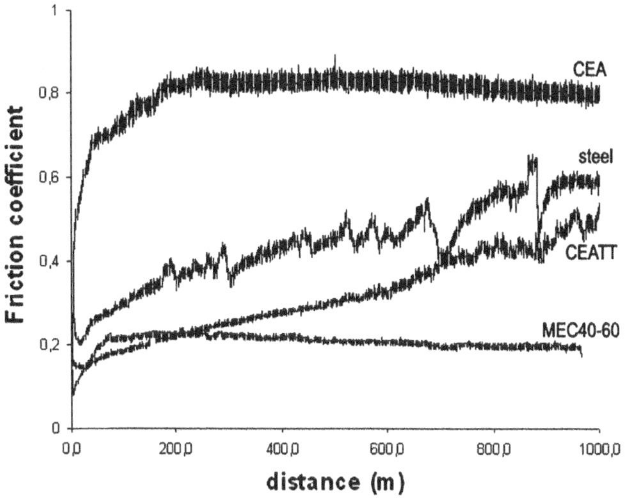

Fig. 10 Friction coefficient versus sliding distance

Table 5 Wear rates of the eroded surfaces

	CEA	MEC40-60
Wear rate, $mm^3\ g^{-1}$	2.42×10^{-4}	2.09×10^{-4}

those carried out on NiAl intermetallics also support the former evaluation (Ref 18, 44). The addition of ceria has shown to ameliorate significantly the erosion resistance (Ref 43).

Table 5 shows the wear rates having MEC40-60 slightly higher erosion resistance than CEA. This cannot be understood from the reasoning of hardness material as in abrasive wear. There is normally an inverse relationship between wear and hardness. In the most simple way, assuming a single particle striking the surface with mass (m), the work done by the retarding force will be equal to the initial kinetic energy of the particle (v_0), resulting in a final indentation volume (V) (Ref 42):

$$V = \frac{mv_0^2}{2H} \qquad \text{(Eq 4)}$$

Even more complex theoretical models predict that erosion rates mainly involving plastic deformation should be inversely proportional to the hardness of the material. However, when using H in the equations, the fact that the material at the worn surface is strain-hardened by plastic flow is ignored, as well as the consequence that its hardness will generally be greater than that of the unstrained material; therefore, better correlation is found between wear resistance and the hardness of the worn surface. It is assumed then that as result of the different microstructures the material is more or less strain hardened (Ref 42). It might be also said that the cohesion between splats in the CEA specimen is not as high as in MEC40-60, maybe because of the presence of much more porosity and oxide embrittlement. It can be compared with difficulty to other investigations because of the different parameters set chosen in each study.

Before choosing the ones established in the present work, there were tested those corresponding to G76-04 ASTM standard but they were discarded for being so aggressive for the coatings.

3.4 Dry Sliding Behavior

Figure 10 shows the variation of friction coefficient versus sliding distance at a normal load of 5 N for the steel substrate and the different coating systems as well. While the FC of the non-coated sample increases continuously during the entire test, the FC of the CEA coating shows the following trend: in the early minutes, there is a sharp increase and then it rapidly reaches a steady state at about 0.7, which persists during the rest of the test. However, the oxidized sample exhibits a different behavior slightly increasing during the whole test and showing some fluctuations at the end. MEC40-60 and MEC40-$_{60ANN}$ show, by contrast, surprisingly much lower friction coefficient. A common trace of all these plots is the initial raise, which is associated to the high adhesive contact between the WC-Co counterpart material and the coating material; the slight decrease afterwards may be caused by the hardening effect under the compressive stress the material undergoes.

In the case of similar microstructures but different hardness, lower friction coefficients would have been achieved with the hardest coating. Lower friction indicates that lower shear has taken place and therefore contact area is much smaller. This can be explained by the examination of the wear tracks.

Figures 11-13 show the common features of damaged coatings and Table 6 gives the magnitudes of their respective widths as well as the volume loss and the calculated wear rates. SWLI had not enough resolution to detect MEC40-60 track's depth.

The magnification in Fig. 11 shows the plate-like debris on the worn CEA surface consisting of ploughed portions caused by adhesive mechanism and posterior delamination. EDS analysis indicated that oxidation takes place on the fresh coating when the ball slides over the deposit

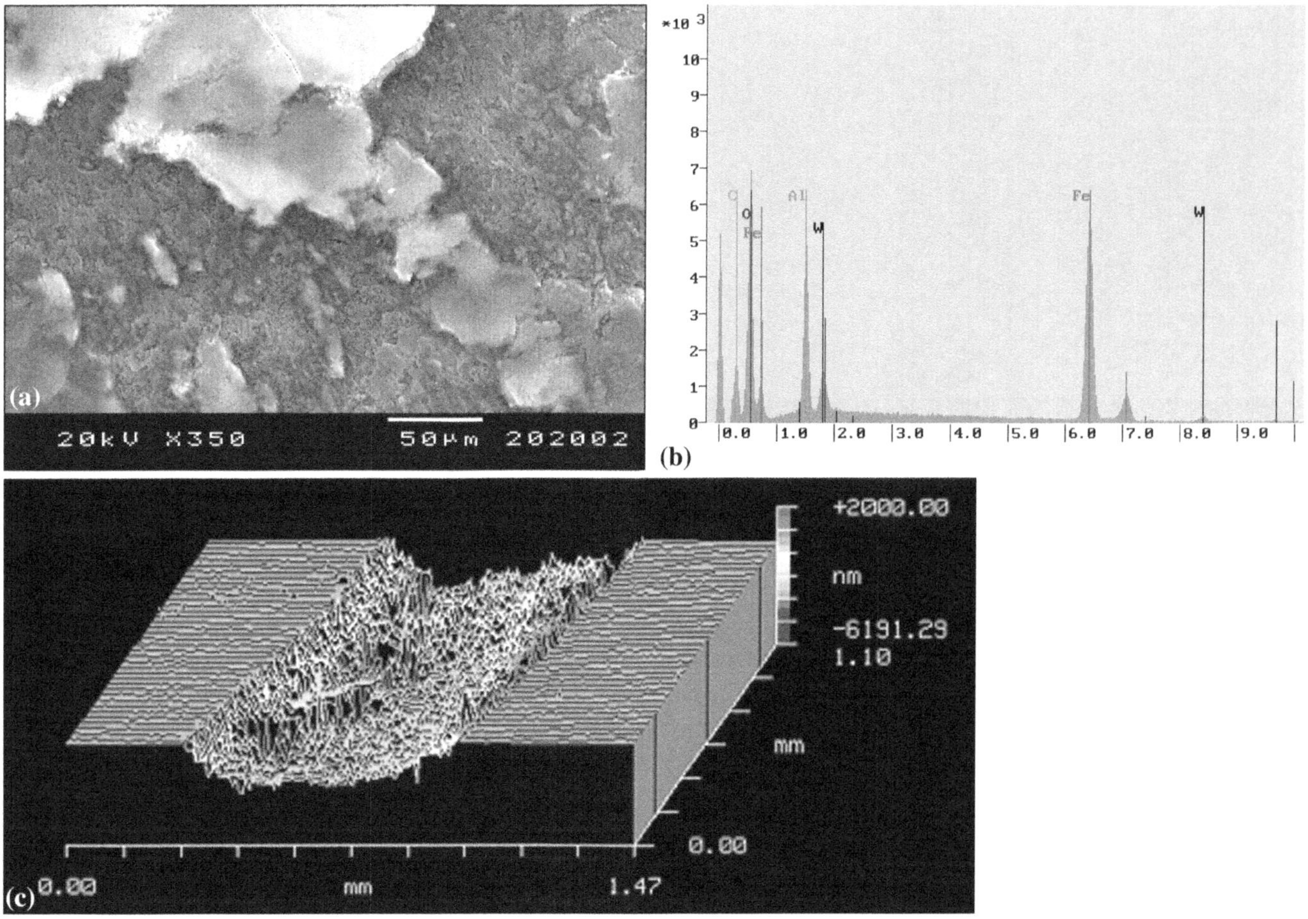

Fig. 11 (a) Wear debris after the ball-on-disk test on the CEA sample, (b) EDS analysis on the plate-like debris, and (c) wear track profile

(Fig. 11b). Such oxidation may be explained by the temperature rise during friction contact. Moreover, the debris observed in the laterals of the wear track could also play some role on constraining laterally and making easier the deformation of the central layer. The debris also entrapped between the two surfaces could have also caused more severe damage in CEA.

Friction coefficients are calculated as an average value of the last 200 m. Although there is a fluctuation for CEA TT and steel substrate, according to the standard, these would then result in 0.50_6 and 0.59_1, respectively. In the other cases, that value is more reliable as a uniform stage is reached.

Unlike CEA, which shows delamination as also reported by other authors (Ref 45), the as-thermally treated and MEC40-60 exhibit completely different wear mechanisms. Figure 12 corresponds to the wear track of the oxidized sample which is covered by an oxide layer of several microns as shown in the cross section in Fig. 12(b) (this is the reason why volume increment is expressed as a positive value, because it is above the surface plane). The thin and brittle layer corresponds to alumina as demonstrated by EDS. It could be thought that it could have a lubricant role as well as the oxidation debris in CEA, but it seems that the continuous break-away of such oxide layer, being alternatively formed and detached from the fresh coating, produces the non-uniformity in the friction coefficient. Xia et al. (Ref 30) claimed that a uniform FC value was maintained in their oxidized samples along the whole test. Nevertheless, as they underwent the study in compacted FeAl grade 3 specimens, the wear mechanism was not the same as that of a coating such as CEA. FC graphics are more similar to MEC40-60 because that had a microstructure similar to a bulk sintered material.

Sliding behavior has also been evaluated by Xu et al. (Ref 12, 13) with FeAl HVAF (High Velocity Air Fuel) coatings, which exhibited a microstructure similar to that of CEA with melted particles and oxidation in the interlamellas. After performing pin-on discs at 3, 5, and 7 N, they concluded that when load is increased, friction leads to lower FC owing to the presence of a similar oxide film as that encountered in CEA. However, what these authors did not try was to change spraying conditions in order to ameliorate wear response. In the present study, the MEC40-60 coating significantly shows an improvement. MEC40-60 possessed the lowest friction coefficient, no delamination was found but the wear track was identified by some remaining oxidation at both sides (Fig. 13). Splat delamination observed in CEA is commonly associated to the weak bonding between lamellas. Also, porous, oxides-sprayed material interfaces, and other imperfections behave as stress concentrators that serve as trigger factors to yield crack initiation. Therefore, it is difficult to predict how a thermal sprayed deposit will behave at friction dynamic contact and often complex simulations are needed; only hypothesis can be made depending on the microstructure obtained. Hence, the absence of these many defects in MEC40-60 is the key for improvement. Here, the annealing treatment of the powder seems not to have a strong influence on the friction behavior, indicating that the wear mechanism is more influenced by the microstructure rather than the hardness of the material itself.

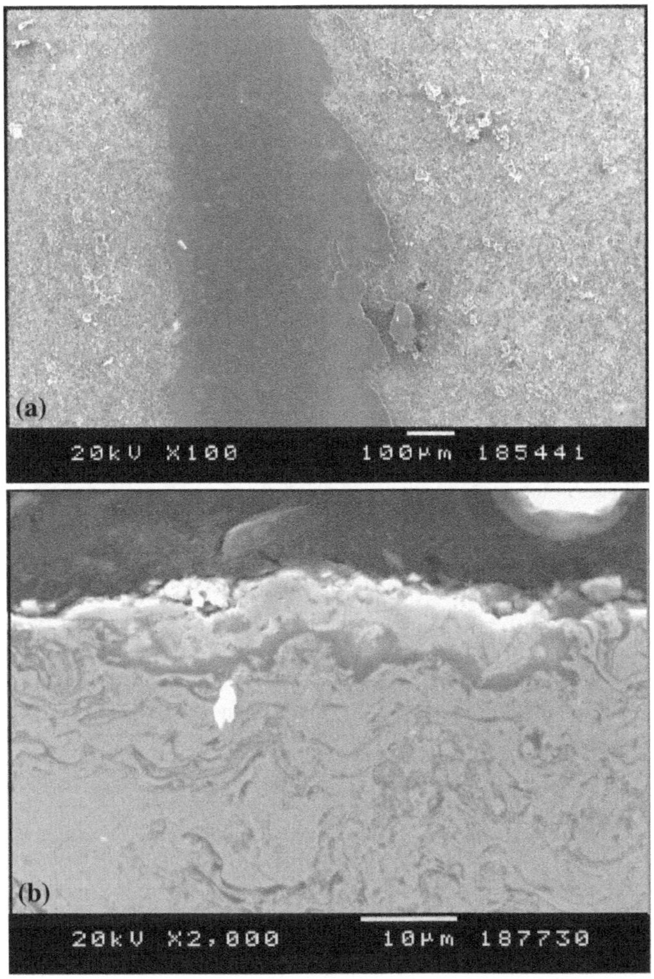

Fig. 12 (a) Wear track observed on the oxidized CEA coating and (b) cross section of the oxide layer formed after the sliding test

As already discussed in terms of erosive and abrasive behavior, sliding behaviour has also been compared with other HVOF-coating materials tested under the same conditions. Thus, when volume loss is compared (mm^3 normalized for the whole perimeter) with a cermet WC-metal coating, the damage produced in CEA is as high as two order of magnitudes more (Ref 15, 17). More similar friction coefficients were obtained with NiCr-based coatings (Ref 16).

4. Conclusions

The results of the present work can be summarized as follows:

- Two very different coatings were obtained from the same powders just modifying the particle size for spraying. This resulted in different microstructures: CEA is dominated by almost fully melted particles surrounded by iron-rich and oxide phases, while MEC40-60 is similar to a bulk material where particles have been compacted.
- Different hardness coatings yield to different abrasive wear rates, where CEA showed the higher abrasive resistance.
- The difference between MEC40-60 and MEC40-60_{ANN} hardness has been associated to the internal oxides rather than to the lower vacancy defects produced during annealing the powder. Both, milling conditions from the as-atomized powder and quenching effect

Fig. 13 Wear track of the MEC40-60 coating

Table 6 Wear track features

	CEA	CEATT	MEC40-60	MEC40-60_{ANN}
Friction coefficient	0.706	0.506	0.192	0.190
Track width, μm	579 ± 8	361 ± 15	377 ± 26	375 ± 20
ΔV, mm^3	-3.3×10^{-2}	$+1.6 \times 10^{-2}$	...	...
Wear rate, $mm^3\ N^{-1}\ m^{-1}$	5.9×10^{-6}	3.6×10^{-6}	...	...

Peer Reviewed

produced during the spraying process, have been identical for both powders; thus, knowing that stress release would reduce hardness, oxidation during the annealing treatment must have enhanced the hardness.

- Erosion tests demonstrated that iron aluminides possess a ductile behavior even when the impingement angle is 90°, which is normally usual of brittle mechanisms.
- Ball-on-disc tests were analyzed in terms of different microstructures and it was stated that CEA was worn by delamination, whereas MEC40-60 almost does not suffer any damage. The first one exhibits FC more than three times higher than the second one.
- Oxidation of the coating was unfavorable for its wear properties. The formation of the alumina layer when frictioning would have displayed a good role if it would not have been so brittle that its fracture left fresh coating areas ready to be damaged further.
- A good goal of this work has been the idea of previously annealing the powder; $MEC40\text{-}60_{ANN}$ shows an analogous microstructure as that of MEC40-60 but higher hardness and abrasion resistance while maintaining the same friction coefficient trend.

Acknowledgments

N. Cinca would like to thank to the Generalitat de Catalunya—project 2005SGR00310—and the Ministerio de Educación y Ciencia for the project MAT2006-06025 for the economical support and the researcher grant with reference number AP-2004-2453.

References

1. J.H. Westbrook, Applications of Intermetallic Compounds, *MRS Bull.*, 1996, **21**(5), p 26-28
2. K. Savolainen, J. Mononen, R. Ilola, and H. Hänninen, *Materials Selection for High Temperature Applications*, Helsinki University of Technology
3. R. Tanaka, Research and Development of Ultra-High Temperature Materials in Japan, *Mater. High Temp.*, 2000, **17**, p 457-464
4. P. Adeva, Materiales alternativos de las superaleaciones: compuestos intermetálicos, *Revista de la Asociación Española de Científicos*, 1999, **1**, p 1-6
5. S.C. Deevi and V.K. Sikka, Nickel and Iron Aluminides: An Overview on Properties, Processing, and Applications, *Intermetallics*, 1996, **4**, p 357-375
6. C.T. Liu and J.O. Stiegler, *Ordered Intermetallics. Properties and Selection: Nonferrous Alloys and Special Purpose Materials*, 10th ed., Metals Handbook, ASM International, 1990
7. H. Skoglund, M. Knutson, and B. Karlsson, Processing of Fine-Grained Mechanically Alloyed FeAl, *Intermetallics*, 2004, **12**, p 977-983
8. S.C. Deevi, V.K. Sikka, and C.T. Liu, Processing, Properties and Applications of Nickel and Iron Aluminides, *Prog. Mater. Sci.*, 1997, **42**, p 177-192
9. C.T. Liu, E.P. George, P.J. Maziasz, and J.H. Scneibel, Recent Advances in B2 Iron Aluminide Alloys: Deformation, Fracture and Alloy Design, *Mater. Sci. Eng. A*, 1998, **258**, p 84-98
10. J.M. Guilemany, N. Cinca, S. Dosta, and C.R.C. Lima, High Temperature Corrosion of Fe-40Al Coatings, *Intermetallics*, 2007, **15**, p 1384-1394
11. G. Sauthoff, State of Intermetallics Development, *Mater. Corr.*, 1996, **47**, p 589-594
12. B. Xu, Z. Zhu, W. Zhang, and S. Ma, The Microstructure and High-Temperature Sliding Wear Behavior of Fe-Al Coating Produced by High Velocity Arc Spraying, *Thermal Spray 2004: Advances in Technology and Application*, ASM International, May 10-12, 2004, Osaka, Japan, 2004, p 1129
13. B. Xu, Z. Zhu, S. Ma, W. Zhang, and W. Liu, Sliding Wear Behavior of Fe-Al and Fe-Al/WC Coatings Prepared by High Velocity Arc Spraying, *Wear*, 2004, **257**, p 1089-1095
14. J.A. Hawk and D.E. Alman, Abrasive Wear of Intermetallic-Based Alloys and Composites, *Mater. Sci. Eng. A*, 1997, **239-240**, p 899-906
15. M. Shi-ning, H. Jun-zhi, C. Xue-rong, L. Chang-qing, L. Qian, and J. Hai, Erosion Resistance of Arc Spray Composite Coatings at High Temperature on Boiler Tubes of Pulverixed Coal Fired Power Station, *Thermal Spray 2003: Advancing the Science and Applying the Technology*, B.R. Marple and C. Moreau, Eds., May 5-8, 2003, Orlando, FL, ASM International, 2003, Vol. 1: p 845; Vol. 2, p 864
16. D.E. Alman, J.A. Hawk, J.H. Tylczak, C.P. Dogan, and R.D. Wilson, Wear of Iron-Aluminide Intermetallic Alloys and Composites by Hard Particles, *Wear*, 2001, **251**, p 875-884
17. B. Wang and S.W. Lee, Erosion-Corrosion Behaviour of HVOF $NiAl\text{-}Al_2O_3$ Intermetallic-Ceramic Coating, *Wear*, 2000, **239**, p 83-90
18. Y. Wang and M. Yan, The Effect of CeO_2 on the Erosion and Abrasive Wear of Thermal Sprayed FeAl Intermetallic Alloy Coatings, *Wear*, 2006, **11-12**, p 1201-1207
19. J.A. Hearley, J.A. Little, and A.J. Sturgeon, The Erosion Behaviour of NiAl Intermetallic Coatings Produced by High Velocity Oxy-Fuel Thermal Spraying, *Wear*, 1999, **233-235**, p 328-333
20. R.A. Varin, Intermetallics: Crystal Structures, *Encyclopedia of Materials: Science and Technology*, 1st. ed., K.H.J. Buschow, Ed., Elsevier Science Publ. Co., 2001, p 9913, ISBN: 0-08-0431526
21. M. Boufenghour, A. Hayoune, and D. Hamana, Study of the Ordered Structures in Fe-Al Alloys Using Dilatometric and Calorimetric Analysis, *J. Mater. Sci.*, 2004, **39**, p 1207-1212
22. S. Suriñach, X. Amils, S. Gialanella, L. Lutterotti, and M.D. Baro, Kinetics of Reordering in a Nanograined FeAl Alloy, *Mater. Sci. Forum*, 1997, **235-238**, p 415-420
23. http://www.mecachrome.com/
24. D.A. Muller, S. Subramanian, P.E. Batson, J. Silcox, and S.L. Sass, Structure, Chemistry and Bonding at Grain Boundaries in Ni_3Al-I. The Role of Boron in Ductilizing Grain Boundaries, *Acta Mater.*, 1996, **44**(4), p 1637-1645
25. M.M. Rico, J.M. Greneche, and G.A. Perez, Effect of Boron on Structural and Magnetic Properties of the $Fe_{60}Al_{40}$ System Prepared by Mechanical Alloying, *J. Alloys Compd.*, 2005, **398**, p 26-32
26. E. Artz, R. Behr, E. Göhring, P. Grahle, and R.P. Mason, Dispersion Strengthening of Intermetallics, *Mater. Sci. Eng. A*, 1997, **234-236**, p 22-29
27. S. Launois and A. Fraczkiewicz, Environmental Effect on Mechanical Behavior of Reinforced FeAl (40at%Al) Alloy, *J. Phys. IV*, 1996, **6**(2), p C2.223-C2.228
28. G. Ji, T. Grosdidier, N. Bozzolo, and S. Launois, The mechanisms of microstructure formation in a nanostructured oxide dispersion strengthened FeAl alloy obtained by spark plasma sintering, *Intermetallics*, 2007, **15**(2), p 108-118
29. J.M. Guilemany, N. Cinca, C.R.C. Lima, and J.R. Miguel, Studies of Fe40Al Coatings Obtained by High Velocity Oxy-Fuel, *Surf. Coat. Technol.*, 2006, **201**, p 2072-2079
30. J. Xia, C.X. Li, and H. Dong, Thermal Oxidation Treatment of B2 Iron Aluminide for Improved Wear Resistance, *Wear*, 2005, **258**, p 1804-1812
31. J. Gang, J. Morniroli, and T. Grosidider, Nanostructures in Thermal Spray Coatings, *Scr. Mater.*, 2003, **48**, p 1599-1604
32. J.L. Jordan and S.C. Deevi, Vacancy Formation and Effects in FeAl, *Intermetallics*, 2003, **11**, p 507-528
33. H. Skoglund, M. Knutson Wedel, and B. Karlsson, Thermal Vacancy Hardening in Fe(100-x) Alx ($28 < x < 40$), *Intermetallics*, 2007, **15**, p 1127-1131
34. M. Kupka, Can Vacancies be the Main Reason of FeAl Alloys Hardening? *J. Alloys Compd.*, 2007, **437**, p 373-377

35. X. Amils, J. Nogués, S. Suriñach, M.D. Baro, M.A. Muñoz-Morris, and D.G. Morris, Hardening and Softening of FeAl During Milling and Annealing, *Intermetallics*, 2000, **8**, p 805-813
36. K. Holmberg and A. Matthews, *Coatings Tribology Properties, Techniques and Applications in Surface Engineering*, Elsevier Science Publ. Co., Amsterdam, 1994, p 442, ISBN-10: 0444888705 and ISBN-13: 978-0444888709
37. P.J. Alison and H. Wilman, The different behaviour of hexagonal and cubic metals in their friction, wear and work hardening during abrasion, *Brit. J. Appl. Phys.*, 1964, **15**, p 281-289
38. A.Y. Mosbah, D. Wexler, and A. Calka, Abrasive Wear of WC-FeAl Composites, *Wear*, 2005, **258**, p 1337-1341
39. M. Ahmadian, D. Wexler, T. Chandra, and A. Calka, Abrasive Wear of WC-FeAl-B and WC-Ni3AL-B Composites, *Int. J. Refract. Met. Hard Mater.*, 2005, **23**, p 155-159
40. J.M. Guilemany, M. Torrell, and J.R. Miguel, Study of the HVOF Ni-Based Coatings Corrosion Resistance Applied on Municipal Solid-Waste Incinerators, *J. Therm. Spray Technol.*, 2008, **17**(2) p 1059–9630
41. J.M. Guilemany, S. Dosta, and J.R. Miguel, The Enhancement of the Properties of WC-Co HVOF Coatings Through the Use of Nanostructured and Microstructured Feedstock Powders, *Surf. Coat. Technol.*, 2006, **201**, p 1180-1190
42. I.M. Hutchings, *Tribology: Friction and Wear of Engineering Materials*, Metallurgy and Materials Science Series, ISBN 0-340-56184-x
43. Y. Wang, W. Chen, and L. Wang, Micro-Indentation and Erosion Properties of Thermal Sprayed NiAl Intermetallic-Based Alloy Coatings, *Wear*, 2003, **254**, p 350-355
44. B.J. Johnson, F.E. Kennedy, and I. Baker, Dry Sliding Wear of NiAl, *Wear*, 1996, **192**, p 241-247
45. L. Prchlik and S. Sampath, Effect of the Microstructure of Thermally Sprayed Coatings on Friction and Wear Response Under Lubricated and Dry Sliding Conditions, *Wear*, 2007, **262**, p 11-23

JTTEE5 17:774–781
DOI: 10.1007/s11666-008-9273-2

Peer Reviewed

Diamond-Bronze Coatings for Grinding Applications

W. Tillmann, E. Vogli, and J. Nebel

(*Submitted May 23, 2008; in revised form September 19, 2008*)

Grinding applications for the machining of stone and concrete require composite tools where large diamonds are perfectly embedded into a metallic matrix. With the detonation flame spraying process, it is possible to manufacture these super abrasive composites. Excellent embedment of the voluminous super abrasive particles into the matrix coating material can be realized to produce high quality composite layers for grinding applications of stone and concrete. In this paper, different diamond gradings as well as different volume contents of diamond in matrix are compared. Especially, the influence of particle size on its implantation efficiency is investigated and the influence of process and substrate temperature is analyzed. The thermal sprayed grinding tools are evaluated with respect to their microstructure as well as their grinding abilities. Compared to sintered diamond-bronze samples, the results of an adaptively designed grinding test for the machining of concrete are presented and analyzed.

Keywords detonation spraying, diamond-bronze, grinding tool, superabrasive composite, substrate preheating

1. Introduction

Diamond-based grinding tools are typically used for the machining of stone or concrete. Due to the high hardness and abrasiveness of the machined minerals, metal-bonded composites provide the best performance. Sintering, brazing, and electroplating are the main technologies for the industrial production of these metal-bonded diamond tools (Ref 1-4).

But sintering as well as brazing is characterized by a high procedural effort and a lot of production steps (Ref 5). Complex component geometries cannot be accomplished and the repairing of worn out grinding tools is usually neither possible nor economically justifiable. Detrimental effects on the environment mainly restrict the electroplating process (Ref 6).

Thermal spraying as an alternative production method has not yet been established. But the possibility to produce complex-shaped tools and to repair damaged surfaces is of high technological interest. Minor procedural complexity and fewer production steps characterize the coating process. Furthermore, it allows a production of the tool under atmospheric environmental conditions. Additionally, the spraying process is fast and custom-made products can be produced easily. No additional production-tool costs (e.g., pressing molds) are necessary to change the grinding tools design. Summarized, detonation flame spraying has an interesting potential for producing these super abrasive coatings.

However, the manufacturing of diamond composite layers by means of thermal spraying presents some challenges which need to be scientifically investigated. For example, diamonds are quite temperature-sensitive. In oxygen atmosphere, they react at elevated temperatures to graphite, carbon monoxide, and carbon dioxide (Ref 7-12). Therefore, the coating process has to ensure a low thermal influence induced into the sprayed diamonds.

If grinding tools for the machining of mineral materials are supposed to be produced by thermal spraying, the desired hard material has to be perfectly embedded in the metallic matrix. Additionally, it is to be assured that the diamond is distributed homogeneously within the functional layer to achieve afterward a constant cutting performance during the grinding. Moreover, the manufacturing process needs to guarantee the production of dense composites without oxidizing the matrix metal and decreasing the MMC ductility.

The main challenge for the production of grinding tools by thermal spraying is the voluminous diamond particle size. A good stone grinding performance requires the embedding of diamonds with a large grain size. Depending on the grinding task and the mineral material 150-400 μm diamonds are used for sawing, cutting and coarse grinding applications (Ref 4, 12).

This article is an invited paper selected from presentations at the 2008 International Thermal Spray Conference and has been expanded from the original presentation. It is simultaneously published in *Thermal Spray Crossing Borders, Proceedings of the 2008 International Thermal Spray Conference*, Maastricht, The Netherlands, June 2-4, 2008, Basil R. Marple, Margaret M. Hyland, Yuk-Chiu Lau, Chang-Jiu Li, Rogerio S. Lima, and Ghislain Montavon, Ed., ASM International, Materials Park, OH, 2008.

W. Tillmann, E. Vogli, and **J. Nebel**, Institute of Materials Engineering, TU Dortmund University, Leonhard-Euler-Str. 2, 44227 Dortmund, Germany. Contact e-mail: wolfgang.tillmann@udo.edu.

Hence, the diamonds are not plastically forming or fusing with the matrix bronze form fit joints are the exclusive bonding mechanism. Form fits are generated by the injection of the diamonds displacing the matrix metal during the impact. A successful implantation of the voluminous diamonds demands the movement of a high amount of matrix material. More than half of the diamond volume is needed to be fixed after the impact.

Due to the requirements mentioned above the diamond impregnation by means of thermal spraying can only be carried out at high particle velocity and moderate spraying temperatures. Detonation flame spraying is well-suited for the application of the MMC-layers.

2. Experimental

2.1 Base Materials

Diamond and CuSn 85/15 were applied as base materials to produce composite coatings by detonation flame spraying. In order to fulfill the cutting demands for the grinding of concrete or stone the diamonds must have a certain size. Different grain sizes were investigated regarding spraying efficiency and grinding abilities (Table 1).

The diamond quality SDB 1055 (Element Six GmbH, Hanau, Germany) is commonly employed in grinding and sawing tool production. Fine nickel alloy inclusions are distributed throughout the cubo-octahedral particle due to the catalytic synthesis manufacturing process (Ref 12). Bronze powder with a grain-size of 45-90 μm (GTV GmbH, Betzdorf, Germany) was used as matrix material. SE-micrographs of a 240 μm diamond and the bronze powder are given in Fig. 1.

Table 1 Used diamond gradings (SDB 1055)

μm	U.S.-Mesh
300-420	40-50
250-300	50-60
210-250	60-70
180-210	70-80
150-180	80-100

Cylindrical C45 steel billets (1.0503) with a diameter of 16 mm and a length of 40 mm were taken as substrate. They were coated at their grid blasted tip.

In grinding tests CEM III A 42.5 N cement disks with a diameter of 105 mm and a thickness of 10 mm were machined, which were produced at the Department of Building Materials (TU Dortmund University, Germany).

To verify the results of the grinding test, sintered diamond-bronze tools of the same geometry were compared with the thermally sprayed tools. Detailed information for the production of the sintered tools is given in Ref 5.

2.2 Spraying of Diamond-Bronze Superabrasives

The investigations were carried out using a detonation flame spraying equipment (Surface Advanced Technology Inc., Warsaw, Poland). Two different powder-feeders (Elektroplasma Inc., CA, USA; Single-10C, Plasma-Technik AG, Switzerland) were used to insert super abrasives and matrix material independently. The detailed setup is described in Ref 13.

C45 substrates were grit blasted with Al_2O_3 with a grain size of 1180-1700 μm before coating. Three bar blasting pressure and a blasting distance of 100 mm in an angle of 70° result in a substrate roughness of $Rz = 73.5 \pm 8.7$ μm and $Ra = 14.4 \pm 0.9$ μm.

Before coating the substrates were heated by induction. The induction equipment (HG 3002, Himmelwerk GmbH, Tübingen, Germany) worked with a frequency of 2 MHz and a maximum power of 3 kW. The induction coil was designed with seven windings and an inner diameter of 28 mm.

With this configuration, preheating of the cylindrical steel substrates could be rapidly achieved. A temperature of 923 K (650 °C) at the substrate tip was reached in 2 min. The temperature before and during the coating process was controlled by a thermocouple, which was fixed at the tip of the steel sample right under the coated surface.

Fig. 1 SE-micrographs. (a) 240 μm diamond (SDB 1055). (b) 45-90 μm CuSn 85/15 matrix powder

Peer Reviewed

It was determined, that the sample temperature deviates max. 10% during the process. Starting with 923 K, it was 863 K after coating.

2.3 Coating Characterization

Vickers hardness was measured on an universal hardness tester (Wolpert, Hahnkolb GmbH, Stuttgart, Germany) relating DIN EN ISO 6507-1. Warm hardness tests were carried out on the self-designed Vickers vacuum hardness tester of the Chair of Materials Technology (University of Bochum, Germany).

Optical microscopy (Axiophot, Zeiss, Germany) with analytic module (KS 300, Zeiss, Germany) was used to characterize the coating morphology and hard-particle content and distribution. Additionally, SEM analyses were conducted (JXA-840, Jeol GmbH, Japan).

Roughness and wear measurements on grinding pins and cement disks were carried out by 3D analysis. They were realized on an Alicona Infinitive Focus optical surface measurement system (Infinitive Focus, Alicona Imaging GmbH, Austria). Additionally, the implantation of 180-210 μm and 300-420 μm sized diamonds into metal materials with different hardness was investigated by means of 3D surface analysis. After detonation spraying, the diamonds without matrix feedstock into Pb, Al, AlMgLi, AlMg, Cu, and CuZn the diamonds were mechanically removed out of the surface. By gauging minimal 10 indentations sinks per sample the average diamond implantation depth and the volume of the displaced material was measured.

2.4 Grinding Performance Test

A grinding test was built up which assured controlled and comparable grinding parameters and results. The assembly is given in Fig. 2.

The concrete disk was rotating with 60 rpm. The pin was pressed on with a contact force of 20 N without using a lubricant or additional cooling. Grinding dust was continuously removed by an external vacuum device.

The weight loss of pin and disk was measured after a running distance of 2500, 5000, 7500, 20,000, and 35,000 m.

3. Results and Discussion

3.1 Substrate Heating

It was shown in previous works (Ref 5, 13, 14) that the implantation efficiency of impacting hard particles could be affected positively by heating the substrate material during the spraying process. For copper as matrix material, an optimum was confirmed at 623 K (350 °C). Without significantly oxidizing, the copper it was softening the layer.

Compared to pure copper the oxidation affinity of CuSn 85/15 is much lower. Respectively, the oxidation velocity at higher temperatures is decelerated. Higher process and substrate temperatures are possible without sensibly oxidizing the matrix material.

Even though the oxidation affinity of bronze allows higher temperatures, the spraying of the diamond-bronze composite is additionally affected by the temperature sensitiveness of the diamond. Mostly the diamond deterioration is induced through high process temperatures. For example, diamond reacts under high temperature in oxidizing conditions to CO_2. However, the temperature range for this deterioration varies in literature sources depending on the investigated diamond size and quality. Under atmospheric conditions the deterioration temperature of diamond differs from 773 K (500 °C) to 1273 K (1000 °C) (Ref 7-12). However, the literature sources confirm a low deterioration velocity up to temperatures of 1023 K (750 °C) (Ref 10, 15).

The durability of SDB 1055 diamonds at elevated temperatures is only known under inert conditions. Element six guarantees a thermal resistance of SDB 1055 of min. 1373 K (1100 °C) with a small reduction in the diamonds strength (Ref 12).

Although, the behavior of SDB 1055 in oxidizing atmospheres is only partially investigated no affecting

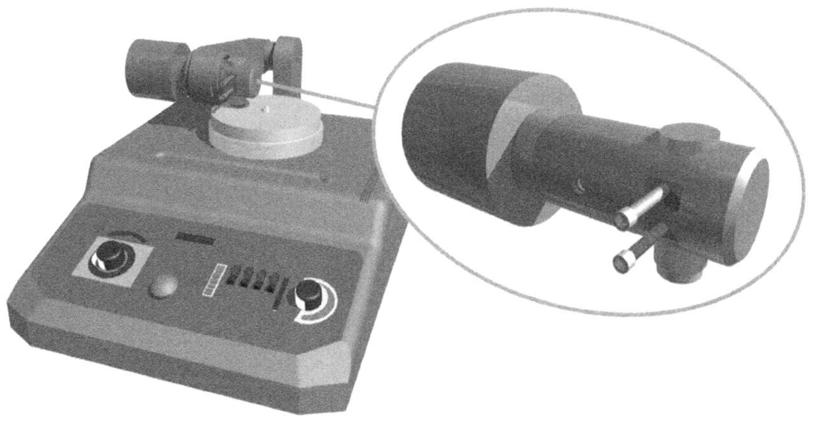

Fig. 2 Grinding test setup

deterioration can be assumed at temperatures below 1023 K (Ref 10, 15) regarding to the short thermal spraying time.

3.2 Spraying Parameter Optimization

The spraying parameters of the diamond-bronze composites were optimized by variation the oxygen and propane flow rate, ignition frequency, powder feed rates, spray distance, and preheating of substrate.

The value range is given in Table 2. Additionally, the optimal detonation flame spraying parameters are specified.

The experiments have shown that compared to copper (Ref 13), the diamond-bronze layer could be sprayed at much higher temperatures without sensibly oxidizing the matrix material. Due to higher process temperatures, the bronze-layer hardness is reduced. Hence, the implantation of the diamond particles is improving.

But if the spraying temperatures are too high, the bronze layer quality is getting affected negatively. High porosity, oxidation, and poor bonding were detected by choosing spraying parameters which are typically used for spraying ceramic materials ($O_2 = 72$ L/min, $C_3H_8 = 4.2$ L/min, $f_{det} = 2$ Hz). Furthermore, some diamond particles deteriorate and crack regarding the strong thermal influence and the shock wave of the detonation (Fig. 3).

A much better independent variable was confirmed by substrate heating and colder spraying conditions. Induction heating permitted high temperatures of 923 K (650 °C) during the spraying process. A short preheating period of 2 min reduced the oxidation of the sample. The continuous impact of matrix and diamond particles through the spraying process predominantly avoided the oxidation of the layer.

Table 2 Detonation spraying parameters for the diamond-CuSn 85/15 composites

	Parameter range	Optimized parameters
Substrate		
Material	1.0503	
Dimension, mm	Ø 16 × 40	
Sprayed material (1)		
Analysis	CuSn 85/15	
Size, μm	45-90	
Feed-rate, g/min	10-30	15
Sprayed material (2)		
Analysis	Diamond SDB 1055	
Size, μm	150-420	
Feed-rate, g/min	0.2-10	0.2
Spraying parameters		
O_2, L/min	29-72	58
C_3H_8, L/min	1.7-4.2	2.6
Deton. frequency, Hz	2-4	2
Distance, mm	100-220	100
Substrate temperature		
Pre-heating	20-650	650
During spraying	80-600	600
Others		
Layer thickness, μm	380-2200	~1300

3.3 Microstructure

The microstructure of the optimized detonation flame sprayed diamond-bronze composite is given in Fig. 4. In the cross section, it is visible that the diamonds are well embedded. Almost no voids in the layer or in the interface between the substrate and coating lead to a good bonding and durability of the composite. Further, the bronze oxidation is very low. Only few oxides were found on the bronze grain boundaries. The low porosity and homogeneity of the layer as well as the perfect mechanical embedment is more obvious in the SE-micrograph (Fig. 5). The picture shows the same section of the sample.

Even though, diamond deterioration and cracking were observed by spraying with "strong" conditions ($O_2 = 72$ L/min, $C_3H_8 = 4.2$ L/min, $f_{det} = 2$ Hz, Fig. 3) no diamond damage was detected in the coatings with optimized spraying parameters. Considering the results of previous spraying experiments with Al_2O_3-Cu and SiC-Cu (Ref 13), it was assumed that diamond cracking cannot be avoided completely due to the high pressure during the detonation

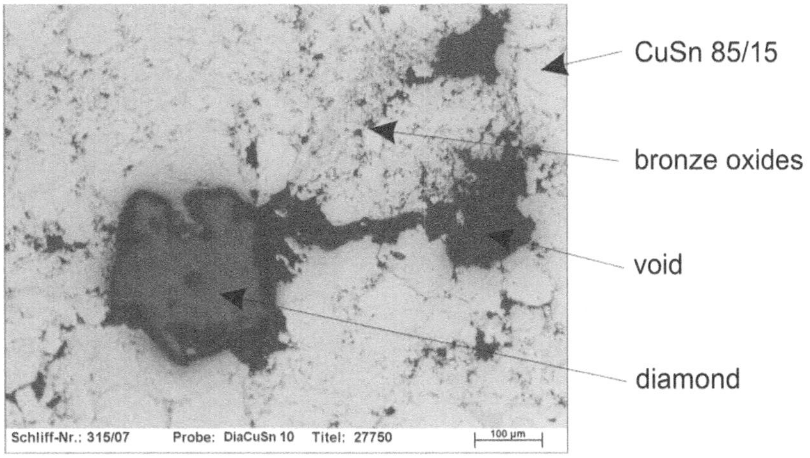

Fig. 3 Diamond-CuSn 85/15 layer before optimization

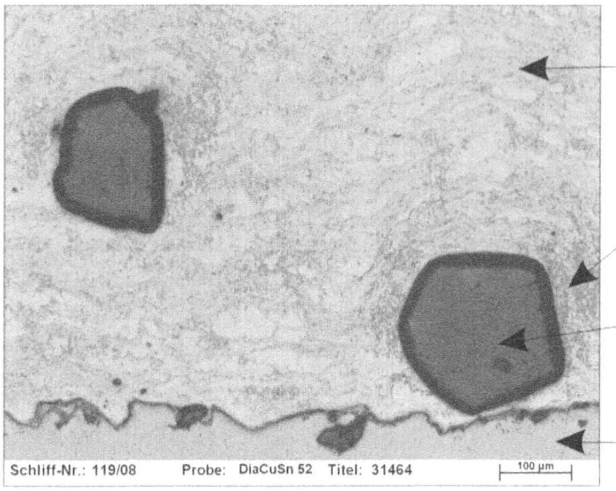

Fig. 4 Optical micrograph of detonation sprayed diamond-bronze composite after process optimization

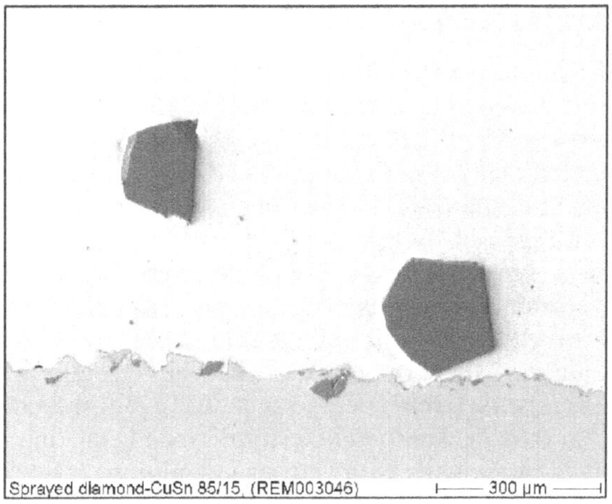

Fig. 5 SEM of detonation sprayed diamond-bronze layer

Fig. 6 SEM of sintered diamond-bronze segment

or the impact on the substrate. But due to the diamonds' higher fracture toughness compared to Al_2O_3 and SiC, smooth surface (lower notch effect), optimized spraying conditions, and substrate heating no cracking was observed in all layer cross sections.

3.4 Comparison to Sintered Composite

To evaluate the layer morphology of the sprayed composite a cross section of a sintered diamond-CuSn 85/15 segment is added (Fig. 6).

Related to the sprayed composite the sintered compound shows a higher porosity. A dense structure is useful for a good grinding performance because it provides mechanical stability. Furthermore, a low porosity increases the wear resistance and durability of the tool. In this case, the detonation sprayed composite quality is indeed better. The embedding quality of the diamond into the bronze matrix is nearly identical in both cases. No diamond deterioration or cracking was detected.

3.5 Deposition and Implantation Efficiency

High deposition efficiency (DE) is an aim of thermal spraying to reduce costs and attain economical profit. DE is strongly affected by the spraying parameters such as powder materials, particle sizes, spraying velocity, particle temperature, substrate geometry, and the substrate temperature.

3.5.1 Bronze Deposition Efficiency. Typically kinetic spraying processes show deposition rates of about 35-55% for Al, Zn, Cu, CuSn, and Ni (Ref 16-20). For the cold spraying of bronze powder, Shin et al. determined a DE of 40% (Ref 16).

Compared to cold spraying, detonation flame spraying offers analogical process velocities in a wider range of possible particle temperatures.

Without preheating the substrate material, DE for the bronze powder was verified to 47% in our experiments. With the same optimized spraying parameters (Table 2) and heating the substrate to 923 K (650 °C) during the spraying process, the DE was slightly increased to 56%.

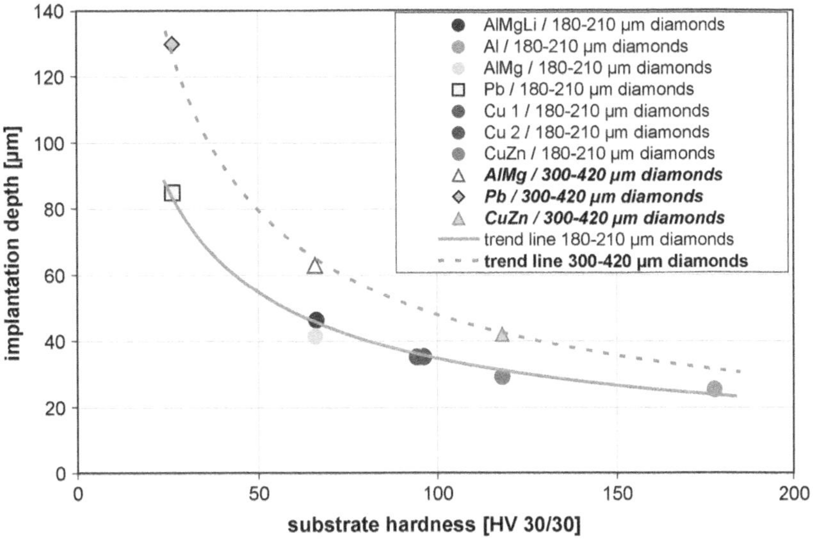

Fig. 7 Measured implantation depth of detonation sprayed diamonds into different substrate materials with dissimilar hardness

3.5.2 Diamond Implantation Efficiency. In the case of MMCs, the bonding mechanism of the hard material is entirely different to the metal matrix powder. When the process gases, propane and oxygen, are ignited in the spraying gun, the simultaneously fed bronze and diamond powders are heated and accelerated to high velocities.

During the impact of the bronze and diamond powder on the sample surface, their kinetic energy is transformed to heat and deformation of the matrix layer. Clamped bronze splats are building up the layer, whereas the hard and strong diamonds inject into it. Displacing the weaker metallic matrix, they build up form fit joints in the material. Plastic deformations of the diamonds are missing in this case. The DE of the diamonds is mainly dependent on the grain size, shape, velocity, and mass of the hard particle as well as the hardness of the matrix layer. Shin et al. confirmed a DE of max. 20% for kinetically sprayed diamond particles (5 µm) into a bronze matrix (Ref 16). Kim et al. specified the DE of uncoated diamonds with a size of 28-35 µm kinetically sprayed into Fe with 46%. For Ti-coated diamonds (53-63 µm), they determined a DE of 24% (Ref 21).

For significantly larger diamond particles, the DE is much lower. In the spraying experiments, it was identified that effective form fit joints of the cubo-octahedral diamond require the implantation of approximately half of the diamond volume. Because the surface of the diamond is very smooth, the DE decreases by an implantation depth of less than the half grain size. Under a critical injection depth of a quarter of the grain size the DE comes to zero. Impacted diamonds rebound from the substrate or are rejected during the next shock wave of the detonation.

With constant spraying parameters and diamond grain size, the depth of diamond implantation is only influenced by the hardness and elasticity of the substrate/matrix material. Figure 7 shows the implantation of SDB 1055 diamonds with a grain size of 180-210 µm and 300-420 µm diamonds into materials with different hardness. They were sprayed without substrate preheating and without bronze powder (additional parameters, see Table 2).

The experimental approach, relating the matrix hardness to the diamond implantation depth, underlines the challenge in realizing diamond-bronze composites (Fig. 7).

With a bronze hardness of 206 HV0.025/30 at room temperature the corresponding implantation depth of a 200 µm SDB1055 diamond is about 20 µm. As it was shown in the spraying experiments, it is not bonded to the matrix. Relating to the smooth surface of the diamond, the rebound rate was quantified to 100%.

By heating the sample to 923 K during the spraying process, the bronze matrix is getting significantly softer.

Because of the short time of the spraying and the high temperatures, the exact hardness for the sprayed bronze matrix could not be determined during the coating process. However, the layer hardness was measured afterward through vacuum warm hardness tests. At a temperature of 923 K, a hardness of 17 HV0.1/20 was determined.

Because of a much lower heating rate of 5 K/min in the vacuum chamber and a dwell-time of 60 min during the hardness measurements, the exposure time at high temperatures is much longer than in the real thermal spraying process. It can be assumed that reduced soft-annealing effects during the spraying lead to a bronze layer hardness of approximately 50 HV, which was deduced from the implantation results.

For sawing, cutting and coarse grinding of hard concrete material a diamond volume fraction of 10-20 vol.% in the layer is needed (Ref 4, 12, 22). But this concentration could not be achieved in our spraying experiments. By applying the optimized parameters with a 923 K heated substrate (Table 2) and a diamond powder feedstock of 0.2 g per spraying experiment, ~5 vol.% was the highest reached diamond concentration for diamonds with a grain size of 150-180 µm. The DE was confirmed to be ~3.6%.

Coherent to the implantation results given in Fig. 7, the DE and volume concentration in the layer reduced by employing diamond with larger size. For 180-210 μm diamonds, a concentration of ~2.7 vol.% was verified in the layer. With a feedstock of 0.2 g diamonds, the DE was calculated to be ~1.7%.

For 250-300 μm diamonds, a volume concentration of ~1.6% and a DE of ~0.9% were determined. Even with optimized detonation spraying parameters (Table 2) and sample heating of 923 K, the diamond implantation was limited to the size of 300 μm. For the larger grading of 300-420 μm only a small amount of diamonds could be detected in the coating.

Experiments with a diamond feedstock of 1 g showed nearly the same diamond quantity in the layer. For instance the volume concentration of diamonds in the coating was confirmed to be 3 vol.% for 180-210 μm diamonds. Due to the higher amount of feedstock material the DE was ~75% lower in this case.

3.6 Grinding Tests

The grinding performance of the diamond-bronze layers was verified in grinding tests. Detonation sprayed diamond-bronze composites with diamond gradings of 180-210 μm and 250-300 μm were compared to a detonation sprayed pure bronze layers and to sintered diamond-CuSn 85/15 tools (diamond size, 300-420 μm, Ref 5, Fig. 8).

As expected, the pure bronze layers had no grinding abilities (Fig. 8a). Hence, the wear of the CuSn 85/15 coated pin is much higher than the wear of the concrete disk.

With a diamond content of 2 vol.%, the grinding performance was improved (Fig. 8b). The mass loss at the sprayed composite layer reduces to more than 50% compared to the bronze layer without reinforcement.

By employing 3 vol.% diamonds, the grinding pin wear was further reduced (Fig. 8c). Although diamonds with smaller sizes (180-250 μm) were used in these tests, the concrete disk wear remains almost the same.

In Fig. 8(d), the grinding result of the sintered pin is exposed (Ref 5). A typical grinding composite with 10 vol.% SDB 1055 diamonds of 300-420 μm in CuSn 85/15 was used.

The higher amount of diamonds results in significant lower wear rates of the grinding tool. But in relation to the sprayed grinding tools, the machining performance on cement remained almost the same.

4. Conclusion

In this research work, it was confirmed that the hardness of the matrix material has a big influence on the

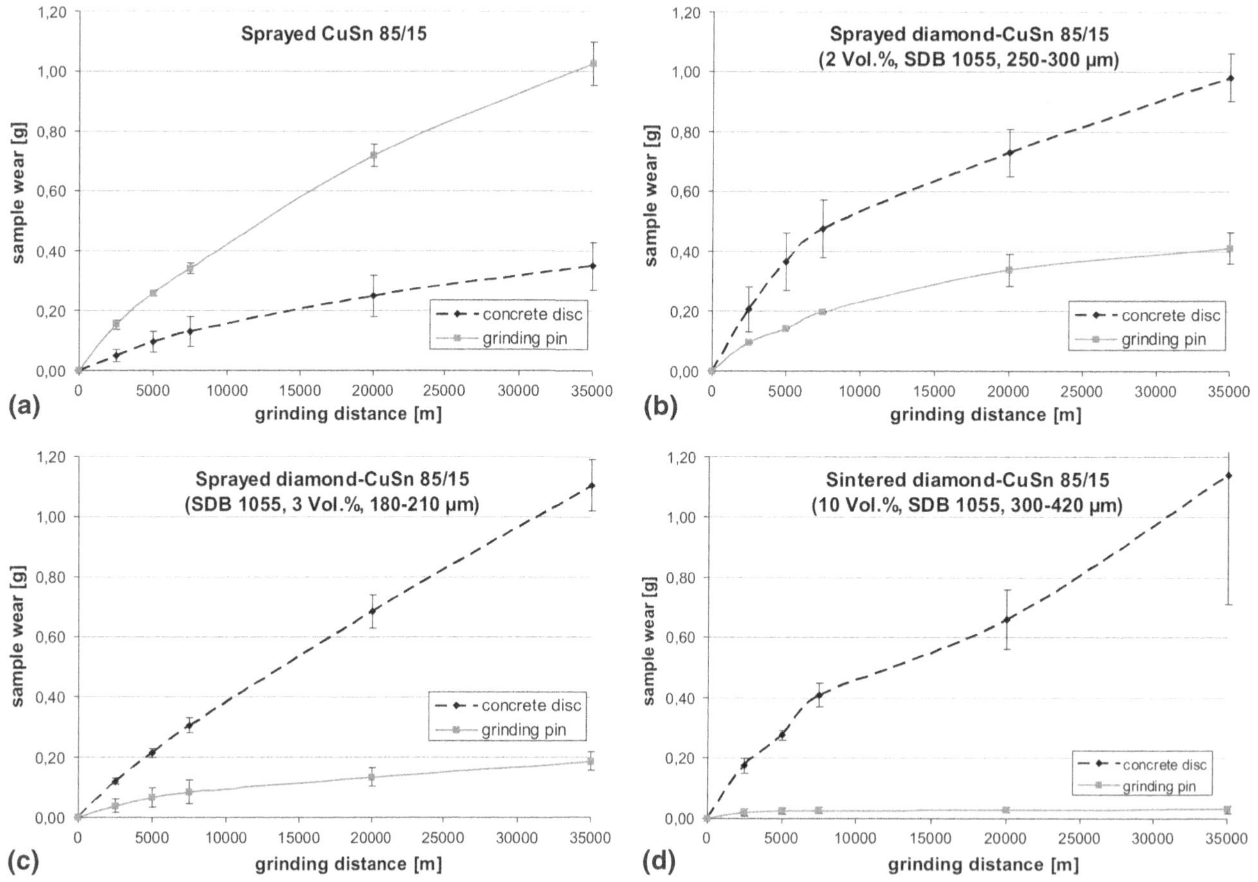

Fig. 8 Results of cement grinding. (a) Detonation sprayed CuSn 85/15 (Ref 5). (b) Detonation sprayed CuSn 85/15 with 2 vol.% diamonds (250-300 μm) (Ref 5). (c) Detonation sprayed CuSn 85/15 with 3 vol.% diamonds (180-210 μm). (d) Sintered CuSn 85/15 with 10 vol.% diamonds (300-420 μm) (Ref 5)

implantation rate of impacted diamonds. Substrate heating during the spraying process is necessary to soften the matrix material and to increase the implanting efficiency. The optimal substrate and matrix material temperature is therefore a compromise of bronze oxidation velocity and rate, as well as of diamond deterioration temperature. For this spraying setup, a substrate temperature of 923 K (650 °C) provided the best conditions.

With the employed experimental setup and substrate preheating diamond implantation was successful for particle sizes of 150-300 μm. The highest DE was reached with the smallest diamond grading (150-180 μm), which was determined to ~3.6%. It was figured out, that the DE decreases significantly for larger diamond sizes. Even with a diamond size of 180-210 μm the DE reduces to ~2.7%. Above a critical diamond size of 300 μm, the implantation into CuSn 85/15 was not possible. Additionally, it was observed that nearly the same amount of diamonds was implanted no matter whether a diamond feedstock of 0.2 or 1 g was used. This effect indicates the interaction of the diamonds during spraying. Impacting diamonds collide with injected diamonds on the tools surface and rebound or partially dissolve implanted particles.

The sprayed layers showed a dense structure. Excellent embedment of the voluminous diamonds into a bronze matrix was achieved leading to high quality composite layers. Furthermore, diamonds are not damaging during the detonation spraying process or in the heated layer.

Compared to sintered diamond-CuSn 85/15, the microstructure of the sprayed composites gave comparable results in diamond bonding quality and composite density. Only the bronze oxidation was slightly higher in the sprayed layers. The convenience and flexibility of the detonation spraying process coupled with the excellent layer microstructure and promising machining performance results indicate the future potential of sprayed diamond-bronze grinding tools.

Acknowledgments

These investigations of the thermal spraying of diamond composites for grinding applications were founded by the German Research Foundation. Thanks to the (DFG) for the financial promotion. Special thanks to A. Botscher and W. Piotrowski for their spraying assistance, M. Sander for manufacturing the concrete disks, K. Bambauer for his help in measuring the warm hardness of sprayed bronze, C. Kronholz for the production of sintered grinding tools, and J. Vahrenholt for the SEM analysis.

References

1. J. Konstanty, Powder Metallurgy of Diamond Tools, *Mater. Sci. Forum*, 2007, **534-536**, p 1121-1124
2. J.-M. Borel and B. Gartner, Effects of Varied CuSn-Alloying Systems as Selected Bonding Material in Hot Pressing Processing Diamond Tools, *PM World Congress*, (Granada, Spain), European Powder Metallurgy Association, October 18-22, 1998, p 213-216, ISBN 1899072098
3. W. Tillmann, M. Gathen, and M.A. Osmanda, Interfaces in Joints and Composites, Designing of Interfacial Structures in Advanced Materials and their Joints, *Solid State Phenom.*, 2007, **127**, p 25-30
4. H. Huang, H.M. Zhu, and X.P. Xu, Grinding of Granite with Three Kinds of Diamond Tools, *Key Eng. Mat.*, 2004, **259-260**, p 146-150
5. W. Tillmann, M. Gathen, E. Vogli, J. Nebel, and C. Kronholz, Innovative Herstellverfahren kupferasierter Werkstoffverbunde für Schleifanwendungen (Innovative Manufacturing Processes for Copper-Based Composites for Grinding Applications), *Metal – Int. J. Metall.*, 2007, **61**(11), p 746-751 (in German)
6. R. Zeller, B. Aumüller, K. Schutte, and H.W. Bergmann, Plasmaspritzen von Diamantwerkzeugen (Plasma Spraying of Diamond Tools), *Int. Congr. Adv. Mater. Process. Appl.*, 1996, **5**, p 373-377 (in German)
7. W. Schatt and K.-P. Wieters, *Pulvermetallurgie—Technologien und Werkstoffe (Powder Metallurgy—Technologies and Materials)*, (Berlin, Heidelberg), Springer, 1994, 454 p (in German), ISBN 978-3-54062-143-0
8. G.J. Bullen, Auswirkung von Temperatur und Bindungszusammensetzung auf synthetische Diamantkörnungen bei der Herstellung metallgebundener Diamantwerkzeuge (Effect of Temperature and Matrix Material on Synthetik Diamonds During the Manufacturing of Metal-Bonded Diamond Tools), *IDR*, 1975, **9**, p 223-227 (in German)
9. E. Riedel, *Anorganische Chemie (Anorganic Chemistry)*, vol 5, (Berlin, New York), Walter de Gruyter Verlag, 2002, 937 p (in German), ISBN 3-11-017439-1
10. V.V. Ogorodnik, E.A. Pugach, and G.G. Postolova, Application of DTA-DTG-TG Complex Thermal Analysis to Study Diamond Oxidation Processes, *Thermocinima Acta*, 1985, **93**, p 705-708
11. X.P. Xu and H. Huang, Thermal Study on Rock Grinding, Part 1-3, *Key Eng. Mat.*, 2003, **205**, p 147-154
12. Data sheet SDB 1055, http://www.e6.com (18.02.2008)
13. W. Tillmann, E. Vogli, and J. Nebel, Development of Detonation Flame Sprayed Cu-Base Coatings Containing Large Ceramic Particles, *J. Therm. Spray Technol.*, 2007, **16**(5-6), p 751-758
14. W. Tillmann, E. Vogli, and J. Nebel, Detonationsflammspritzen hartstoffhaltiger Funktionsschichten (Detonation Flamespraying of Functional Surfaces), *Materialwissenschaft und Werkstofftechnik*, 2007, **6**, p 481-488 (in German)
15. O. Beffort, S. Vaucher, and F.A. Khalid, On the Thermal and Chemical Stability of Diamond during Processing of Al/Diamond Composites by Liquid Metal Infiltration, *Diamond Relat. Mater.*, 2004, **13**, p 1834-1843
16. S. Shin, C. Lee, K.T. Han, and H.J. Kim, Effect of Process Parameters on the Deposition Characteristics of a Bronze-20 Vol.% Diamond Composite in Kinetic Spraying, *Therm. Spray – Global Coating Solutions (ITSC 2007)*, (Beijing, China), May 14-16, 2007, p 123-127
17. N. Sinha, B.J. York, A. Hosangadi, and S.M. Dash, First Principles-Based Computational Fluid Dynamic (CFD) Model of Thermal Spray Deposition Process, *8th National Thermal Spray Conference (NTSC 1995)*, (Houston, TX), September 11-15, 1995, p 199-206
18. X. Yang and S. Eidelman, Numerical Analysis of a High Velocity Oxygen Fuel Thermal Spray System, *J. Therm. Spray Technol.*, 1996, **5**(2), p 175-184
19. B.R. Marple and R.S. Lima, Process Temperature/Velocity Hardness Wear Relationships for High-Velocity Oxyfuel Sprayed Nanostructured and Conventional Cermet Coatings, *J. Therm. Spray Technol.*, 2005, **14**(1), p 67-76
20. M. Fukumoto, H. Wada, K. Tanabe, M. Yamada, E. Yamaguchi, A. Niwa, M. Sugimoto, and M. Izawa, Effect of Substrate Temperature on Deposition Behavior of Copper Particles on Substrate Surfaces in the Cold Spray Process, *J. Therm. Spray Technol.*, 2007, **16**(5-6), p 643-650
21. H.J. Kim, D.H. Jung, J.H. Jang, and C.H. Lee, Assessment of Metal/Diamond Composite Coatings by Cold Spray Deposition, *Therm. Spray – Building on 100 Years of Success (ITSC 2006)*, (Seattle, WA), May 15-18, 2006
22. J. Konstanty, T.W. Kim, and S.B. Kim, Resistance to Abrasive Wear of Materials Used as Metallic Matrices in Diamond Impregnated tools, *Mater. Sci. Forum*, 2007, **534-536**, p 1125-1128

JTTEE5 17:782–787
DOI: 10.1007/s11666-008-9222-0
1059-9630/$19.00 © ASM International

Peer Reviewed

High-Temperature Erosion of HVOF Sprayed Cr_3C_2-NiCr Coating and Mild Steel for Boiler Tubes

Guan-Jun Yang, Chang-Jiu Li, Shi-Jun Zhang, and Cheng-Xin Li

(Submitted May 13, 2008; in revised form July 19, 2008)

The comparison of the high-temperature erosion behavior of a High-velocity oxyfuel (HVOF) sprayed Cr_3C_2-NiCr coating with mild steel for circulating fluidized bed boiler tubes was investigated. Results showed that the erosion rate of the mild steel at 800 °C was four times that at 300 °C at an erosion angle of 30°. However, the erosion rate of the HVOF sprayed Cr_3C_2-NiCr coating was not influenced by the temperature in the range of 300-800 °C. It was found that the erosion resistance of HVOF sprayed Cr_3C_2-NiCr coating was more than three times higher than that of the mild steel at 700-800 °C. In addition to the ploughing on the coating surface, the cracking along splat interfaces in the coating was clearly observed on the cross-sectional microstructure. The results indicate that the erosion performance of the HVOF sprayed Cr_3C_2-NiCr coating is controlled by the cohesion between splats and can be further enhanced by improving splat cohesion.

Keywords boiler tube, Cr_3C_2-NiCr coating, high-temperature erosion, HVOF, mild steel

1. Introduction

Circulating fluidized bed (CFB) boilers are becoming more widely used in coal-fired power plants because of their advantages of high efficiency in the combustion of a wide variety of solid fuels, in-bed sulfur capture, and relatively low NO_x emission (Ref 1-3). The erosive wear resulting from the hard coal gangue is the main cause of failures in economizer boiler tubes at high temperature (Ref 2, 3). The improvement of the erosive wear resistance of the boiler tubes is therefore extremely important to elongate the continuous-operation time. The size of the coal gangue particles in CFB is on the millimeter scale, and the particle velocity is from several to tens of meters per second (Ref 4, 5). Remedial actions used to combat the erosion wastage include design modifications, use of horizontal plates welded on the wall, and use of wear-resistant materials in the area at risk (Ref 6-8).

High-velocity oxyfuel (HVOF) sprayed Cr_3C_2-NiCr coatings are one of the most important candidates for protection of materials from high-temperature wear and erosion and have been successfully used to protect pulverized coal-fired boiler tubes (Ref 6-8). Many papers investigate wear and erosion performance of HVOF sprayed Cr_3C_2-NiCr and other cermet coatings (Ref 8-24). It was recognized that the erosion is much more complex than abrasive wear in which the wear resistance is predominantly influenced by the hardness of the coatings (Ref 15-24). The erosion behavior depends on the response behavior of coatings to the impact of erosive particles at high velocity (Ref 8-14). Microcutting, gouging, carbide peeling out, and splat flaking have been observed in the erosion of the HVOF sprayed Cr_3C_2-NiCr coatings at room temperature (Ref 14). Microcutting and gouging can be partially understood from the viewpoint of abrasive wear. For example, since the wear resistance increases with the increase of carbide content and carbide hardness and the decrease of carbide size and carbon loss (Ref 17-22), the erosion resistance can also be improved by similar approaches (Ref 14). In addition, it is of interest to note that the erosion of the HVOF sprayed Cr_3C_2-NiCr coatings at room temperature was significantly influenced by the spalling off of lamellae from the interlamellar interface (Ref 14). Furthermore, the wear resistance also depends on the powder feedstock characteristics. For instance, the wear resistance of the coating deposited with sinter-crushed powder was much higher than that with a blended powder of alloy powder and carbide powder in the work of Ref 24. This is because the coating deposited with the blended powder presented poorly bonded,

This article is an invited paper selected from presentations at the 2008 International Thermal Spray Conference and has been expanded from the original presentation. It is simultaneously published in *Thermal Spray Crossing Borders, Proceedings of the 2008 International Thermal Spray Conference*, Maastricht, The Netherlands, June 2-4, 2008, Basil R. Marple, Margaret M. Hyland, Yuk-Chiu Lau, Chang-Jiu Li, Rogerio S. Lima, and Ghislain Montavon, Ed., ASM International, Materials Park, OH, 2008.

Guan-Jun Yang, Chang-Jiu Li, Shi-Jun Zhang, and **Cheng-Xin Li**, State Key Laboratory for Mechanical Behavior of Materials, School of Materials Science and Engineering, Xi'an Jiaotong University, Xi'an, Shaanxi 710049, P.R. China. Contact e-mail: ygj@mail.xjtu.edu.cn.

cracked carbides and a coarse level of inhomogeneity (Ref 24). The high-temperature erosion of the HVOF sprayed Cr_3C_2-NiCr coating by the erosive particles at relatively high velocity of more than 30 m/s was extensively reported (Ref 8-10), and the spray powder in the work was blended powder with NiCr alloy powder and Cr_3C_2 carbide powder. However, the high-temperature erosion of the coatings deposited with sinter-crushed powder is less well known.

The objective of this paper was to investigate the high-temperature erosion behavior of HVOF sprayed Cr_3C_2-NiCr coatings deposited with sinter-crushed powder and compare that with the typical mild steel for boiler tubes.

2. Experimental Procedures

2.1 Materials

A sinter-crushed Cr_3C_2-25%NiCr powder (Langli, Luoyang, China) was employed for HVOF spraying. The powder size ranged from 30 to 37.5 μm. A mild steel, which is often used to fabricate boiler tubes, was used as a substrate for the deposition of Cr_3C_2-NiCr coating. The composition of the mild steel is shown in Table 1. The bare steel was also subjected to the erosion test for comparison.

2.2 Deposition of the Coating

Cr_3C_2-NiCr coatings were deposited with a CH-2000 HVOF spray system (Xi'an Jiaotong University), the detailed description of which was given elsewhere (Ref 25). The system is operated using propane and oxygen to generate supersonic flame. Propane was used as fuel gas and operated at a pressure of 0.4 MPa and a flow of 1300 L/h. Oxygen was operated at a pressure of 0.55 MPa and a flow of 13,000 L/h. Nitrogen gas was used as a powder feed gas. The spray distance was 180 mm.

2.3 Characterization of the Coating

The topographic morphology and cross-sectional microstructures of the coating and the mild steel were examined by scanning electron microscopy (SEM) (Quanta 200, FEI, the Netherlands). Erosion of the coating and the mild steel was tested using the experiment setup as shown in Fig. 1. Alumina grit with a nominal size of 250 μm was employed as the erodent. The feed rate of erodent was 60 g/min. The erosive particles were accelerated to a velocity of 12 m/s by a modified HVOF spray torch. The impact angles during test were fixed at 30° and 90°, respectively. The temperature of the testpieces including the coating and the mild steel plate during erosion test was set to 300, 500, 600, 700, and 800 °C, respectively. To eliminate the effect of surface conditions, all testpieces were polished before the erosion test. The erosion rate of the coating and the mild steel was evaluated by the weight loss divided by the weight of erosive particles. Under each test condition, three testpieces were used and the erosion rate was the average of three tests.

Table 1 Composition of the mild steel for boiler tube

Element	Content, %
C	0.17-0.24
Si	0.17-0.37
Mn	0.35-0.65
S	≤0.035
P	≤0.035

3. Results

3.1 Microstructure of the HVOF Sprayed Cr_3C_2-25%NiCr Coating

Figure 2 shows the cross-sectional microstructure of the HVOF sprayed Cr_3C_2-NiCr coating. The coating thickness was about 600 μm. The microstructure at a higher magnification shown in Fig. 2(b) revealed that the carbide particle size was several micrometers. The carbide particles were uniformly distributed in the coating. However, some pores were observed on the microstructure. The pores may have resulted from the pulling out of weakly bonded particles during the sample polishing process. It was found from Fig. 2 that the coating was well adhered to the substrate.

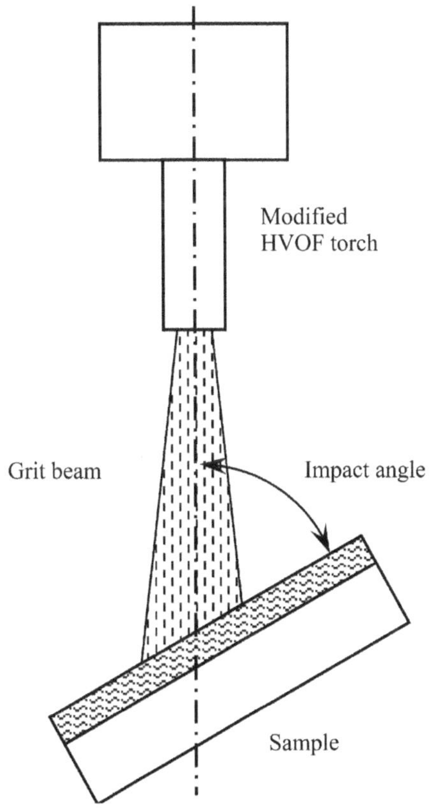

Fig. 1 Experimental setup for the high-temperature erosion test

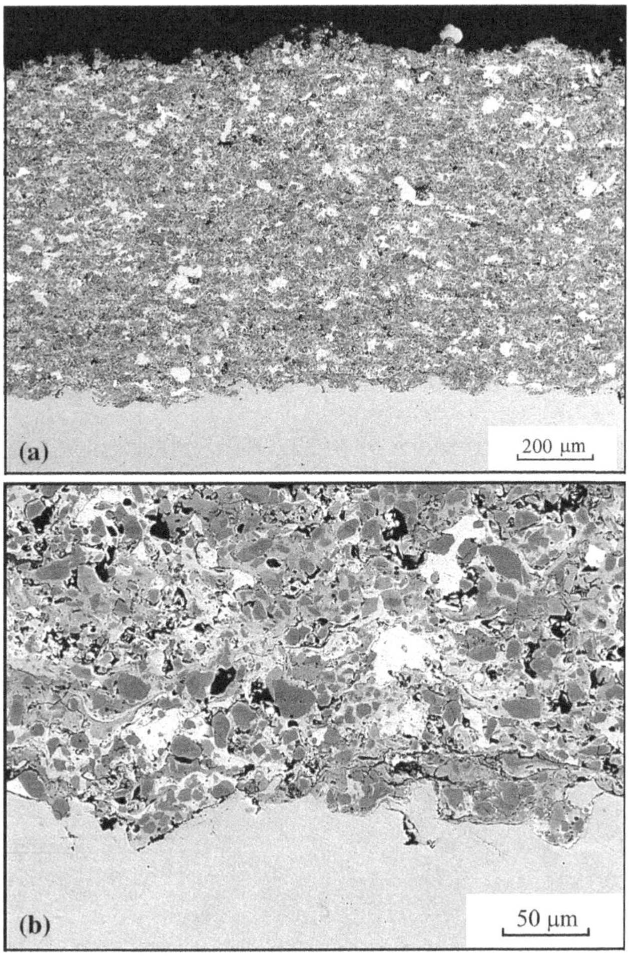

Fig. 2 Cross-section microstructure of the HVOF sprayed Cr_3C_2-NiCr coating. (a) Low magnification. (b) High magnification

3.2 Erosion of the Coating and Mild Steel

Figure 3 shows the effect of sample temperature on the erosion rate of the mild steel. The erosion rate was significantly increased with the increase of the temperature at the impacting angles of both 30° and 90°. The erosion rate at 30° was 0.2 mg/g at a temperature of 300 °C. The erosion rate at 800 °C became more than four times higher than that at 300 °C. At the impact angle of 90°, the erosion rate at 800 °C was nearly seven times higher than that at 300 °C. According to the erosion theory of ductile metallic materials, the erosion rate of metals becomes the maximum at impact angle range from about 15 to 20° (Ref 26). It will decrease with the increase of impact angle. It is clear that the present test results at the temperatures lower than 300 °C follow the classical theory. However, it is clearly observed that with the increase of test temperature the relative erosion rate at 90° with respect to that at 30° becomes large. This is possibly attributed to poor oxidation resistance of the mild steel.

Figure 4 shows the effect of coating temperature on the erosion rate of the Cr_3C_2-NiCr coating. The effect was very different from that for the mild steel. This is possibly attributed to the difference in the change of oxidation resistance of the Cr_3C_2-NiCr coating from the mild steel with the increase of temperature. The erosion rate at 30° was 0.25 mg/g and independent of coating temperature in the range from 300 to 800 °C. The erosion rate at 90° slightly decreased with the increase of the test temperature. Moreover, at a temperature of 300 °C, the erosion rate of the coating at 90° was two times that at 30°. However, at 800 °C, the erosion rate at 90° became comparable to that at 30°.

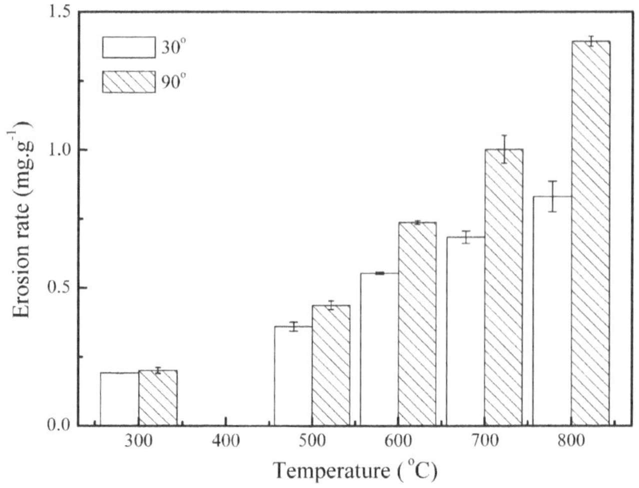

Fig. 3 Effect of erosion temperature on the erosion rate of mild steel

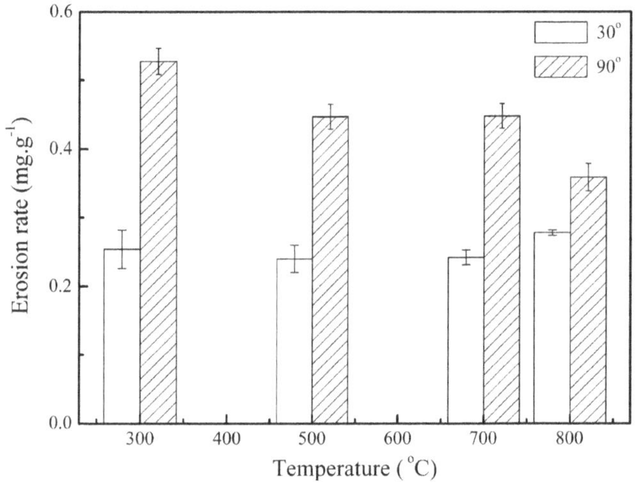

Fig. 4 Effect of erosion temperature on the erosion rate of the HVOF sprayed Cr_3C_2-NiCr coating

4. Discussion

Table 2 shows the erosion rate of the mild steel relative to that of the HVOF sprayed Cr_3C_2-NiCr coating as a function of test temperature. It can be clearly seen that the relative erosion resistance of the coating increases with the

Table 2 Comparison of the erosion rate of the coating and the mild steel

	$Y_{steel}/Y_{coating}$	
Test temperature, °C	**At 30°**	**At 90°**
300	0.39	0.77
500	1.0	1.5
700	2.2	2.9
800	3.8	3.0

increase of test temperature. These results indicate that the coating exhibits a higher relative erosion resistance at higher test temperature. Although the relative erosion resistance of the coating at a test temperature of 300 °C is only 0.39-0.77 at the impact angles of 30° and 90°, respectively, it becomes larger than one at test temperatures higher than 500 °C. The relative wear resistance at 800 °C reaches to 3.8 and 3.0 at the impact angles of 30° and 90°, respectively.

It was evidently found that the erosion rate of the HVOF sprayed Cr_3C_2-NiCr coating decreases with the increase of test temperature at an impact angle of 90°. This may be attributed to the composite characteristics of the coating. For ductile materials such as mild steel, the increase of test temperature leads to a decrease of hardness and therefore results in a significant decrease in erosion resistance. For the coating, the high content of hard carbide particles enables the coating to maintain a comparably high hardness even at high temperature. Furthermore, the apparent ductility of the coating is improved by the enhanced ductility of the NiCr alloy matrix at high temperature. When the coating is impacted by the erosive particles, the good ductility of the matrix allows the impacted local area of the coating to deform to prevent the surging of the highly localized stress that may result in cracking of the coating. Therefore, the erosion rate of the HVOF sprayed Cr_3C_2-NiCr coating decreases with the increase of erosion temperature.

Figure 5 shows the topographical morphologies of the mild steel and the HVOF sprayed Cr_3C_2-NiCr coating after erosion test at an impact angle of 30° and a temperature of 800 °C. It can be clearly seen from Fig. 5(a) that there are large ploughs in a width of tens of micrometers on the worn surface of the mild steel. This is attributed to a typical ductile erosion mechanism of the mild steel against alumina erosive particles. Small ploughs were also observed in the surface of the coating shown in Fig. 5(b). The width of these small ploughs is <2-5 μm. Therefore, the ploughs on the coating surface are much smaller than those on the mild steel surface. When the erosive particles impact on the coating surface at a low angle, the edge of an erosive particle will plough the surface. On account of the existence of hard carbide particles in the coating, only a small plough can be formed on the coating surface.

Figure 6 shows the cross-sectional morphology of the HVOF sprayed Cr_3C_2-NiCr coating after erosion test at an impact angle of 30° and a temperature of 800 °C. A large crack marked with a large white arrow can be clearly observed near the coating surface. This is attributed to the impact of the erosion particles on the coating surface.

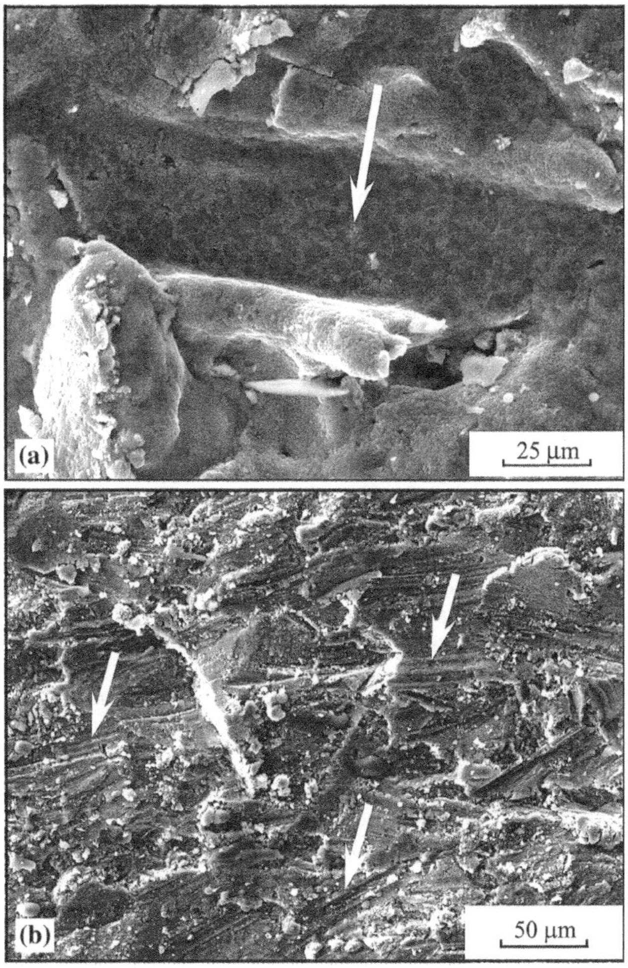

Fig. 5 Topographical morphology of the mild steel (a) and the HVOF sprayed Cr_3C_2-NiCr coating (b) after erosion test at impact angle of 30° and temperature of 800 °C

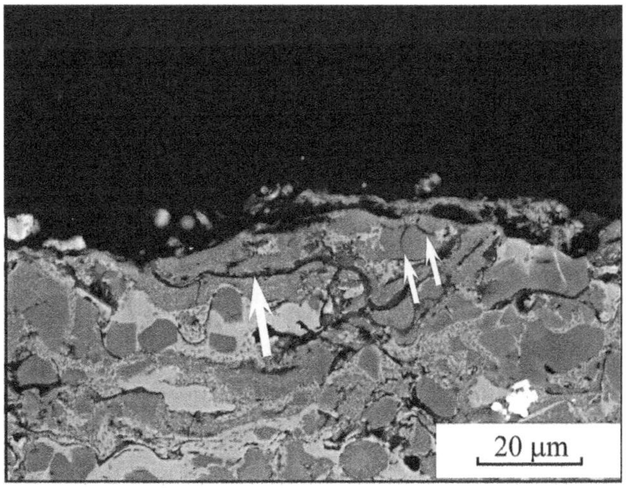

Fig. 6 Cross-section microstructure of the HVOF sprayed Cr_3C_2-NiCr coating after erosion test at impact angle of 30° and temperature of 800 °C

HVOF sprayed Cr_3C_2-NiCr coatings are composed of splats, which are formed through spreading of solid-liquid particles with a NiCr matrix in molten state and Cr_3C_2 particles in solid state (Ref 17, 18). The adjacent two splats are bonded partially, and significant nonbonded area exists between the two splats (Ref 14, 17, 18). The partial bonding between splats was also supported by the fact that the cohesive strength of the HVOF sprayed Cr_3C_2-NiCr coating was measured as 72 MPa (ASTM C 633-79), which is much less than the strength of the bulk cermet. The nonbonded area will act as a precrack, which propagates under the impact of erosive particles on the coating surface. Consequently, before the splat is completely ploughed by erosive particles, the crack propagation originating from the nonbonded interface area may lead to the flaking of one splat or several splats. In addition, cracks in carbide particles marked with two small white arrows are also observed in Fig. 6. The small carbides resulted from the cracking of the original large carbide particles. The new surface of small carbides was in the form of free surface, which was produced by the cracking of carbide particles. Therefore, these small particles were held less by the NiCr matrix. These particles are easily eroded off by the impact of erosion particles. Therefore, such cracking may also promote the erosion of the coating by removal of carbides by impact.

Although there are three mechanisms that cause the mass loss of the coating under erosive wear, splat flaking will be dominant over carbide cracking and ploughing. Therefore, the erosive resistance of the HVOF sprayed Cr_3C_2-NiCr coating will predominantly depend on the extent of splat flaking. The results suggest that the erosion performance of the HVOF sprayed Cr_3C_2-NiCr coating is controlled by the cohesion between splats in the coating and can be further enhanced by improving splat cohesion.

5. Conclusions

The erosion performance of the HVOF sprayed Cr_3C_2-NiCr coating and a mild steel at temperatures from 300 to 800 °C was investigated using alumina grit as erodent. Results showed that the erosion rate of the mild steel increased with the increase of the test temperature at the erosion angles of both 30° and 90°. The erosion rate of the mild steel at 800 °C was four times higher than that at 300 °C at an erosion angle of 30°. However, the erosion rate of the HVOF sprayed Cr_3C_2-NiCr coating at an erosion angle of 30° was not influenced by the test temperature in the range of 300-800 °C. It was found that the erosion resistance of HVOF sprayed Cr_3C_2-NiCr coating was more than three times higher than that of the mild steel at 700-800 °C. The mass loss of the coating was attributed to splat flaking and ploughing on the surface. The results suggest that the erosion performance of the HVOF sprayed Cr_3C_2-NiCr coating was controlled by the cohesion between splats in the coating and could be further enhanced by improving splat cohesion.

References

1. A. Gungor and N. Eskin, Two-Dimensional Coal Combustion Modeling of CFB, *Int. J. Therm. Sci.*, 2008, **47**, p 157-174
2. V.H. Hidalgo, F.J.B. Varela, A.C. Menendez, and S.P. Martinez, A Comparative Study of High-Temperature Erosion Wear of Plasma Sprayed NiCrBSiFe and WC-NiCrBSiFe Coatings under Simulated Coal-Fired Boiler Conditions, *Tribol. Int.*, 2001, **34**, p 161-169
3. V. Kain, K. Chandra, and B.P. Sharma, Failure of Carbon Steel Tubes in a Fluidized Bed Combustor, *Eng. Failure Anal.*, 2008, **15**, p 182-187
4. C. Yang and X. Gou, Dynamic Modeling and Simulation of a 410 t/h Pyroflow CFB Boiler, *Comp. Chem. Eng.*, 2006, **31**, p 21-31
5. J.S. Leszczynski, Z. Bis, and W. Gajewski, Evaluation of Structure and Particle Velocity Distribution in Circulating Fluidised Beds, *Powder Technol.*, 2002, **128**, p 22-35
6. M. Sawa and J. Oohori, Application of Thermal Spraying Technology at Steelworks, *Thermal Spraying: Current Status and Future Trends*, A. Ohmori, Ed., May 22-26 1995 (Kobe, Japan), High Temperature Society of Japan, 1995, 37 p
7. Y. Fukuda and M. Kumon, Application of High Velocity Flame Spraying for the Heat Exchanger Tubes in Coal Fired Boilers, *Thermal Spraying: Current Status and Future Trends*, A. Ohmori, Ed., May 22-26 1995 (Kobe, Japan), High Temperature Society of Japan, 1995, 107 p
8. B.Q. Wang, Chromium-Titanium Carbide Cermet Coating for Elevated Temperature Erosion Protection in Fluidized Bed Combustion Boilers, *Wear*, 1999, **225-229**, p 502-509
9. B.-Q. Wang and A. Verstak, Elevated Temperature Erosion of HVOF Cr_3C_2/TiC-NiCrMo Cermet Coating, *Wear*, 1999, **233-235**, p 342-351
10. K.J. Stein, B.S. Schorr, and A.R. Marder, Erosion of Thermal Spray MCr-Cr_3C_2 Cermet Coatings, *Wear*, 1999, **224**, p 153-159
11. S.J. Matthews, B.J. James, and M.M. Hyland, Microstructural Influence on Erosion Behaviour of Thermal Spray Coatings, *Mater. Charact.*, 2007, **58**, p 59-64
12. I. Hussainova, Some Aspects of Solid Particle Erosion of Cermets, *Tribol. Int.*, 2001, **34**, p 89-93
13. J. Vicenzi, D.L. Villanova, M.D. Lima, A.S. Takimi, C.M. Marques, and C.P. Bergmann, HVOF-Coatings Against High Temperature Erosion (300 °C) by Coal Fly Ash in Thermoelectric Power Plant, *Mater. Des.*, 2006, **27**, p 236-242
14. G.-C. Ji, C.-J. Li, Y.-Y. Wang, and W.-Y. Li, Erosion Performance of HVOF-Sprayed Cr_3C_2-NiCr Coatings, *J. Therm. Spray Technol.*, 2007, **16**(4), p 557-565
15. M. Factor and I. Roman, Use of Microhardness as a Simple Means of Estimating Relative Wear Resistance of Carbide Thermal Spray Coatings: Part 2. Wear Resistance of Cemented Carbide Coatings, *J. Therm. Spray Technol.*, 2002, **11**(4), p 482-495
16. J.K.N. Murthy and B. Venkataraman, Abrasive Wear Behaviour of WC-CoCr and Cr_3C_2-20(NiCr) Deposited by HVOF and Detonation Spray Processes, *Surf. Coat. Technol.*, 2006, **200**, p 2642-2652
17. C.-J. Li, Y.-Y. Wang, G.-J. Yang, A. Ohmori, and K.A. Khor, Effect of Solid Carbide Particle Size on Deposition Behavior, Microstructure and Wear Performance of HVOF Cermet Coatings, *Mater. Sci. Technol.*, 2004, **20**, p 1087-1096
18. G.-C. Ji, C.-J. Li, and Y.-Y. Wang, Microstructural Characterization and Abrasive Wear Performance of HVOF Sprayed Cr_3C_2-NiCr Coating, *Surf. Coat. Technol.*, 2006, **200**, p 6749-6757
19. R. Schwetzke and H. Kreye, Microstructure and Properties of Tungsten Carbide Coatings Sprayed with Various High-Velocity Oxygen Fuel Spray Systems, *J. Therm. Spray Technol.*, 1999, **8**(3), p 433-439
20. J. He and E.J. Lavernia, Precipitation Phenomenon in Nanostructured Cr_3C_2-NiCr Coatings, *Mater. Sci. Eng. A*, 2001, **301**, p 69-79
21. M. Roy, A. Pauschitz, J. Bernardi, T. Koch, and F. Franek, Microstructure and Mechanical Properties of HVOF Sprayed Nanocrystalline Cr_3C_2-25(Ni20Cr) Coating, *J. Therm. Spray Technol.*, 2006, **15**(3), p 372-381

22. S. Matthews, M. Hyland, and B. James, Microhardness Variation in Relation to Carbide Development in Heat Treated Cr_3C_2-NiCr Thermal Spray Coatings, *Acta Mater.*, 2003, **51**, p 4267-4277
23. S. Matthews, M. Hyland, and B. James, Long-Term Carbide Development in High-Velocity Oxygen Fuel/High-Velocity Air Fuel Cr_3C_2-NiCr Coatings Heat Treated at 900 °C, *J. Therm. Spray Technol.*, 2004, **13**(4), p 526-536
24. S. Wirojanupatump, P.H. Shipway, and D.G. McCartney, The Influence of HVOF Powder Feedstock Characteristics on the Abrasive Wear Behaviour of Cr_xC_y-NiCr Coatings, *Wear*, 2001, **249**, p 829-837
25. C.-J. Li, H. Yang, and H. Li, Effect of Gas Conditions on HVOF Flame and Properties of WC-Co Coatings, *Mater. Manuf. Process.*, 1999, **14**, p 383-395
26. P.A. Engel, Impact Wear of Materials. Elsevier, Amsterdam, 1976

JTTEE5 17:788–795
DOI: 10.1007/s11666-008-9255-4
1059-9630/$19.00 © ASM International

Peer Reviewed

Effect of the Structural Scale of Plasma-Sprayed Alumina Coatings on Their Friction Coefficients

G. Darut, H. Ageorges, A. Denoirjean, G. Montavon, and P. Fauchais

(Submitted May 8, 2008; in revised form September 18, 2008)

The objective of this study is to compare the tribological properties of alumina coatings with two different structural scales, a micrometer-sized one manufactured by atmospheric plasma spraying and a sub-micrometer-sized one manufactured by suspension plasma spraying. Coating architectures were analyzed and their friction coefficients in dry sliding mode measured. Sub-micrometer-sized structured coatings present a lower friction coefficient than micrometer ones, thanks to their higher cohesion and smaller characteristic structural feature sizes.

Keywords alumina, friction, plasma spraying, sub-micrometer-sized feedstock, suspension plasma spraying, structural scale, tribology, wear

1. Introduction

Emissions of greenhouse gases (GHGs) related to anthropogenic activities are responsible for climate global warming (CGW) (Ref 1). In its last released report (Ref 2), the Intergovernmental Panel of Climate Change estimates that carbon dioxide emissions resulting from fossil fuel use are mainly responsible for GHGs (~56% of GHGs). In Western European countries, about 30% of GHGs from fossil fuel use result from car and truck transportation. Reducing those GHGs to mitigate CGW to 2 °C over this century is absolutely required to limit to the maximum possible extent impacts associated with global average temperature change. Until the deployment on a large scale of sustainable energy sources such as solid oxide-fuel cells (SOFCs) or proton exchange membrane fuel cells (PEMFCs) at horizon 2030 (Ref 3), fossil energy will remain a major source of GHGs. Reducing those emissions requires improving the engine efficiency for energy saving.

Friction is responsible for about 50% of energy loss in an automotive engine. Reducing friction within an engine is hence a major challenge that has been pursued for numerous years by major automotive companies in collaboration with technical centers.

The development 25 years ago of Nicasil-type coatings, a composite electrolytic layer made of a dispersion of micrometer-sized SiC particles embedded in a Ni matrix, constituted the first effort to meet this challenge (Ref 4). Nowadays, low carbon alloyed steel layers are applied by thermal spraying (Ref 5, 6). The improvement of new sliding layers is done nowadays following three complementary directions:

1. The decrease in the material architecture scale to reach sub-micrometer-sized or nanometer-sized scales. Numerous developments aim hence at manufacturing composite micrometer-sized layers with embedded nanometer-sized particles by electrolytic (Ref 7) or thermal spraying routes or at manufacturing layers structured at the nanometer scale (Ref 8). In these both cases, published works have demonstrated significant improvements in coatings properties (wear resistance, toughness, etc.) (Ref 9-12).
2. The replacement of the metallic matrix by a ceramic matrix in order, on the one hand, to increase the local operating temperature and, on the other hand, to limit the emission of heavy metal dusts resulting from abrasion during sliding (Ref 13).
3. Less lubricated systems since synthetic oils, the most used ones in recent engines, are synthesized from fossil fuels. For example, the annual consumption of engine oil in France was 341,824 tons in 2006 (decrease of 4% compared to the previous year) to be added to several hundreds of tons of additives (Ref 14). This represents a challenge in recycling

This article is an invited paper selected from presentations at the 2008 International Thermal Spray Conference and has been expanded from the original presentation. It is simultaneously published in *Thermal Spray Crossing Borders, Proceedings of the 2008 International Thermal Spray Conference*, Maastricht, The Netherlands, June 2-4, 2008, Basil R. Marple, Margaret M. Hyland, Yuk-Chiu Lau, Chang-Jiu Li, Rogerio S. Lima, and Ghislain Montavon, Ed., ASM International, Materials Park, OH, 2008.

G. Darut, H. Ageorges, A. Denoirjean, G. Montavon, and **P. Fauchais**, SPCTS-UMR CNRS 6638, Faculty of Sciences, University of Limoges, 123 Avenue Albert Thomas, 87060 Limoges Cedex, France. Contact e-mail: ghislain.montavon@unilim.fr.

those products after usage (278,820 tons of used engine oils recycled in 2006 in France (Ref 14)). Decreasing their consumption would reduce their environmental impact.

In this context, thermal spraying appears as a pertinent process since it permits processing of a wide range of feedstock, in terms of composition (ceramic, alloys, polymers), architecture (composite powders), and sizes (from micrometer-sized to nanometer-sized).

The primary objective of this work is to estimate the effect of the structural scale of plasma-sprayed coatings on some of the tribological properties in the dry mode, particularly the friction coefficient.

Alumina was selected as a demonstrative material. Indeed, alumina is used for wear-resistant layers and it is not a material suitable for friction layers (as it exhibits poor friction coefficients whatever the counter material nature). The objective here is to discriminate the effect of the structural scale rather than developing layers with low friction coefficients. In order to study the behaviors of structures exhibiting two significantly different scales, two plasma spray processes were selected: atmospheric plasma spraying (APS) was implemented to manufacture alumina coatings at a micrometer-sized scale while suspension plasma spraying (SPS) was implemented to manufacture alumina coatings at a sub-micrometer-sized scale. There were almost two orders of magnitude difference in the scale of the studied structures due to the selected feedstock average diameters (d_{50} of the particle size distributions), taken as reference to quantify the structural scale.

SPS is a technology that permits the deposition of layers thinner and finer than those resulting from conventional APS. Compared to conventional coatings, those manufactured from suspensions exhibit quite interesting features such as the absence of lamella boundaries and cracks and porous microstructures with void dimension of the same order or lower than the one of the feedstock (Ref 15, 16). In this process, a stabilized suspension, made of a liquid, solid particles, and a dispersant, is injected within the plasma flow. The liquid is very quickly vaporized and the individual particles, or the particle agglomerates, depending on the average size of the solid feedstock, are heated and simultaneously accelerated toward the substrate surface where they impact, spread, and solidify, analogously in a first approximation to larger particles, to form a layer (Ref 17, 18). Compared to plasma spraying of micrometer-sized particles (i.e., conventional plasma spraying), SPS exhibit several major differences (Ref 19):

1. SPS exhibits a more pronounced sensitivity to arc root fluctuations, requiring specific operating parameters in order to operate the spray gun in its take over mode to avoid inhomogeneous processing of the suspension by the plasma which induces heterogeneity in the coating structure.
2. The spray distance in SPS is shorter (30-40 mm) since small particles decelerate much faster than bigger ones.
3. In SPS, a higher thermal flux is transmitted from the plasma flow to the substrate, between 5 and 10 times than the heat flux transmitted in conventional plasma spraying (from 7 to 15 MW m^{-2} compared to 1.5 to 2 MW m^{-2}, average values).

2. Experimental Protocols

2.1 Spray Process and Related Parameters

Table 1 summarizes the operating parameters for both APS and SPS processes.

SPS layers were sprayed implementing a d.c. stick-cathode plasma torch developed at SPCTS equipped with a 5-mm nozzle internal diameter and operated with an arc current intensity of 600 A with a binary Ar-He (30-30 slpm) plasma gas mixture. The injection system was made of a stainless steel injector at the tip of which a calibrated diaphragm with an internal diameter of 150 μm was mounted. The suspension to be sprayed was stored in pressurized tanks and, according to the air pressure applied in the tank, the velocity of the suspension at the injector exit was 25 m s^{-1}. The suspension was made of α-Al_2O_3 P152 SB (d_{50} = 0.3 μm) feedstock supplied by Alcan (Saint-Jean de Maurienne, France), with a mass percentage of 10% in pure ethanol. The spray velocity (relative velocity between the substrate and the plasma torch) was 1 m s^{-1} and the scanning step 10 mm per pass, whereas the spray distance (SD) was 30 or 40 mm.

APS layers were sprayed with the same spray torch equipped with a 7-mm internal diameter nozzle. In this case, the plasma forming gas mixture was Ar-H_2 (46-14 slpm) and the arc current intensity 600 A. Powders were injected perpendicularly to the plasma jet axis with argon as carrier gas through a 1.8-mm internal diameter injector. The feedstock was α-Al_2O_3 typified Medipure (d_{50} = 36 μm) fused and crushed powder supplied by Medicoat (Delle, France).

For both processes, specimens for microstructural characterizations were button-type substrates made of low carbon steel 25 mm in diameter and 20 mm in thickness.

Table 1 Main spray operating parameters for APS and SPS

Operating parameters	APS	SPS
Anode internal diameter, mm	7	5
Primary plasma forming gas flow rate, slpm	46 (Ar)	30 (Ar)
Secondary plasma gas flow rate, slpm	14 (H_2)	30 (He)
Arc current intensity, A	600	600
Plasma mass enthalpy, MJ kg^{-1}	14	12
Feedstock carrier	Ar (gas)	Et-OH (liquid)
Feedstock carrier flow rate, slpm	5	n.a.
Injector internal diameter, mm	1.80	0.15
Spray distance (SD), mm	110	30/40
Spray velocity, m s^{-1}	1	1
Scanning step, mm per pass	6	10

Peer Reviewed

Specimens for tribological characterizations were made of the same low carbon steel 50 mm in diameter and 5 mm in thickness. Substrates were polished to reach a specular surface finishing (i.e., average roughness ranging from 0.3 to 0.4 μm) for the SPS process and grit-blasted for APS in order to get an average roughness (R_a) of about 5 μm and a peak-to-valley ratio (R_z) of about 30 μm (R_z). Prior to spraying, they were degreased by immersion in ethanol with ultrasonic treatment.

2.2 Coating Characterization Techniques

Samples were cross-sectioned using a diamond saw, mounted in epoxy rings (vacuum impregnation), and polished using conventional polishing protocols consisting of pre-polishing using SiC papers and polishing using diamond slurries. Polished coating cross sections and fracture surfaces were observed using scanning electron microscopy (SEM) and field-emission scanning electron microscopy (FESEM). Image analysis was implemented to quantify the porosity level and crack network orientation (Ref 20).

A ball-on-disk (BOD) tribometer (CSM Instruments, Lausanne, Switzerland) was used to quantify the coating friction coefficients. A previous study established reference operating parameters as follows: α-Al_2O_3 (sintered) ball of 6 mm in diameter, 2 N load, relative speed of 0.1 m s^{-1}, and sliding distance of 1500 m. The test was operated in the dry mode and the wear scraps (i.e., third body) were constantly removed by an air jet located behind the contact point. Since as-sprayed layers exhibited significantly different surface roughnesses (both R_a and R_z, see Sect 3.1), coating surfaces were polished prior to BOD tests until the average surface roughness was in the order of a micrometer for all of them (see Sect 3.2). The surface polishing was performed using conventional metallographic polishing protocols.

The wear profile and roughness were measured with a Dektak IIA surface profiler (Veeco Instruments, Woodbury, NY) which has a tip radius of 10 μm.

3. Results and Discussion

3.1 Coating Architectures

The average thickness of APS coatings was about 330 μm, whereas the thickness of SPS layers was 12 μm for SD = 30 mm and 15 μm for SD = 40 mm. The average roughness of as-sprayed APS coatings was about 5.07 μm ± 0.19; the average of SPS coatings only about 1.6 μm ± 0.10.

Figure 1 displays the major difference in structural scales for both APS- and SPS-sprayed coatings. The average flattening ratio of lamellae forming the APS coating is estimated to vary from 3 to 4—a common value encountered for this type of process—whereas it is estimated to vary from 1 to 2 for SPS coatings that exhibit a granular-type structure (to be compared to the conventional layered structure of APS coatings). In the two processes, the predominant bounding mechanism of the

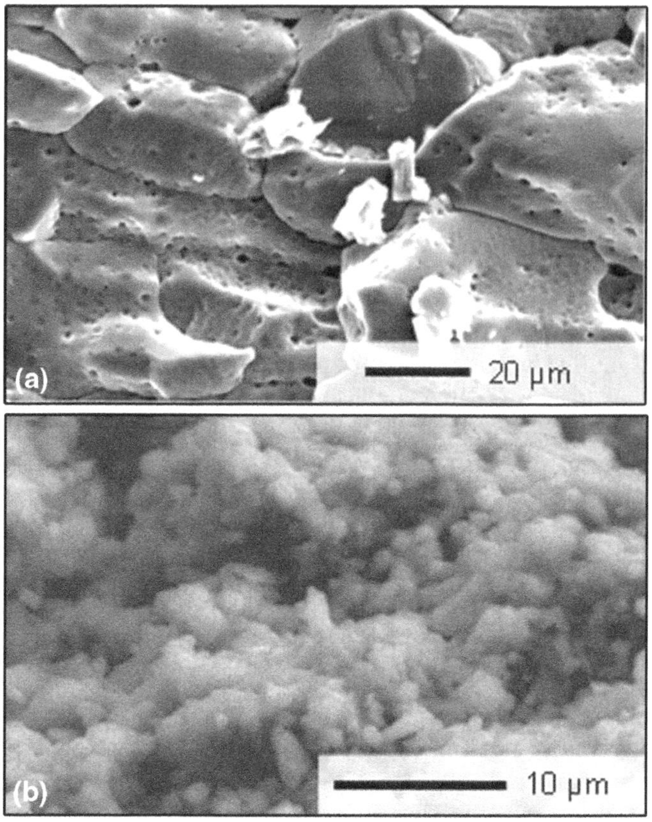

Fig. 1 Fracture surfaces of (a) micrometer-sized (APS) and (b) sub-micrometer-sized (SPS) Al_2O_3 coatings

coating to the substrate is very likely mechanical anchorage. Mechanical anchorage is promoted when the substrate surface average roughness is of the same order than the lamella thickness, a few micrometers (2-4 μm) in APS coatings and a few tenths of micrometers (0.2-0.4 μm) in SPS. Consequently, coatings are manufactured by APS onto grit-blasted substrate surface and by SPS onto polished substrate surface. In the peculiar case of SPS coatings, an additional bounding mechanism could result from the development of an interphase at the substrate/coating interface made of a layer of mixed oxides developing onto the substrate surface consecutively to the high heat flux imparted by the plasma flow. Indeed, Valette et al. (Ref 21) have demonstrated in conventional plasma spraying that an interphase could be promoted by oxidation resulting in an epitaxial solidification of alumina lamellae onto an iron oxide layer itself epitaxied onto the steel substrate. Works are underway to more specifically address this mechanism in SPS.

Figure 2 displays the pore network architecture of APS and SPS coatings. They differ by their pore level: about 4% for SPS layers to be compared to about 9% for APS ones. Besides, the pore network architecture differs also since it seems to be made exclusively of voids for SPS layers whereas it is made of voids (80 vol.%) and cracks (20 vol.%), the latter consisting of interlamellar delaminations perpendicular to the spray direction (80 vol.%) and intra-lamellar cracks parallel to the spray direction (20 vol.%) for APS layers.

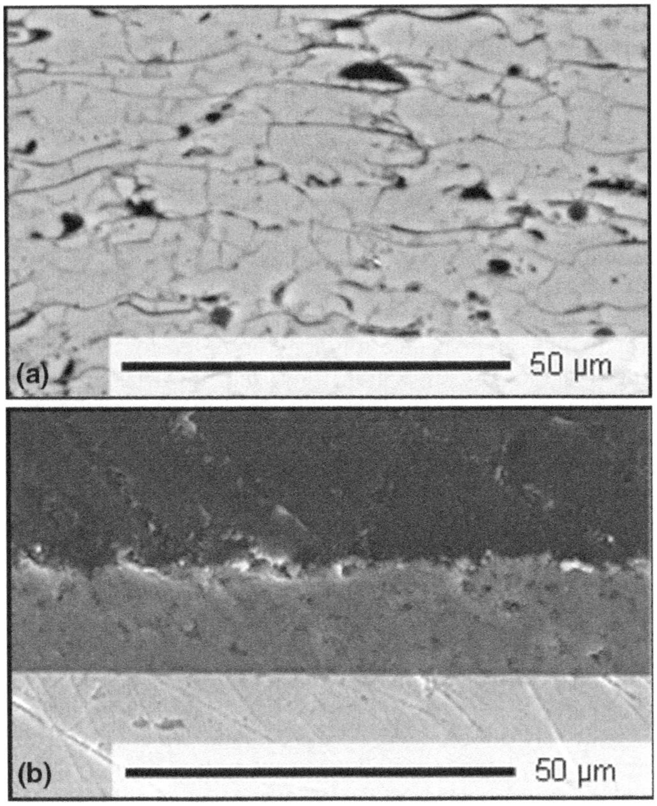

Fig. 2 Polished cross sections of (a) micrometer-sized (APS) and (b) sub-micrometer-sized (SPS) Al_2O_3 coatings presented at the same magnification

3.2 Coating Friction Coefficient

As mentioned in Sect 2.2, the differences in as-sprayed coating roughnesses required that they be polished prior to BOD tests. Tables 2 and 3 display measured values after polishing (average value and associated standard deviation of 10 randomly located measurements). R_a is the arithmetic mean of the absolute departures of the roughness profile from the mean line, R_z the maximum peak-to-valley height at the mean line, R_{sk} the skewness, R_{ku} the kurtosis, R_{Sm} the mean spacing between profile peaks at the mean line, and R_{HTp} the height of the bearing ratio of the complete profile (Ref 22).

The values differ by about 50% for R_a and R_z between SPS and APS coatings due to a polishing protocol not fully optimized at this stage. The mean spacing, R_{Sm}, is almost identical for coatings manufactured with both processes. Polished surface topologies differ mostly by their skewness and kurtosis. Polished APS coating surfaces are made of more valleys than peaks (i.e., negative skewness) while polished SPS surfaces have almost symmetrical topologies corresponding to skewness values close to zero. Polished SPS coating topologies exhibit a distribution of heights closer to the normal distribution (i.e., corresponding to a kurtosis equal to 3) compared to polished APS coating surfaces. Those characteristics lead to a bearing ratio higher for polished APS coatings compared to SPS ones.

Table 2 Roughness criteria for polished specimens

Surface parameters	APS	SPS	
		SD = 30 mm	SD = 40 mm
R_a, µm	0.71 ± 0.10	1.45 ± 0.24	1.42 ± 0.15
R_z, µm	5.77 ± 0.77	9.39 ± 1.69	8.31 ± 0.85
R_{sk}, –	−1.46 ± 0.33	0.67 ± 0.19	−0.09 ± 0.13
R_{ku}, –	7.32 ± 1.64	3.56 ± 0.77	2.55 ± 0.24
R_{Sm}, µm	57.86 ± 8.13	40.68 ± 2.28	39.68 ± 3.51
R_{HTp}, µm	1.37 ± 0.23	2.99 ± 0.54	3.16 ± 0.34

Table 3 Roughness criteria for polished specimens

Surface parameters	SPS		
	SD = 30 mm	SD = 40 mm	SD = 50 mm
R_a, µm	1.82 ± 0.14	2.57 ± 0.23	3.11 ± 0.27
R_z, µm	11.53 ± 1.02	15.26 ± 1.30	17.83 ± 1.25
R_{sk}, –	0.67 ± 0.21	0.28 ± 0.16	0.,19 ± 0.20
R_{ku}, –	3.36 ± 0.65	2.77 ± 0.23	2.64 ± 0.21
R_{Sm}, µm	48.90 ± 3.26	54.06 ± 4.85	62.61 ± 7.41
R_{HTp}, µm	3.86 ± 0.35	5.46 ± 0.56	6.64 ± 0.72

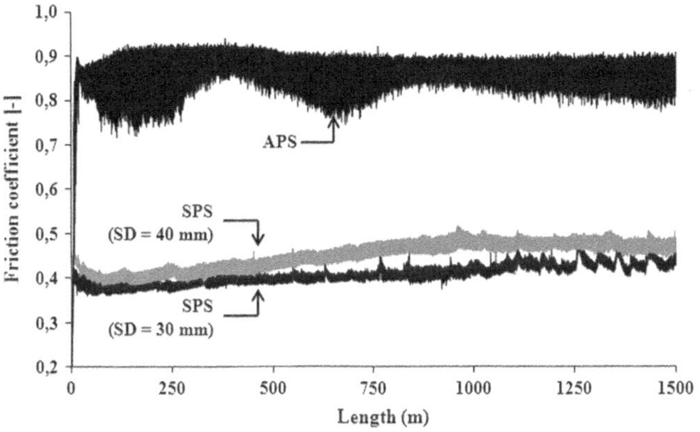

Fig. 3 Friction coefficient versus sliding distance for Al_2O_3 micrometer-sized (APS) and sub-micrometer-sized (spray distance = 30 and 40 mm) coatings (BOD, α-Al_2O_3 ball of 6 mm in diameter, 2 N load, relative speed of 0.1 m s^{-1})

Nevertheless, and in a first approximation, such differences do not play a significant role, except in the accommodation stage of the friction curves, since the roughness descriptor values vary in narrow ranges.

Figure 3 displays the evolution of coating friction coefficients versus sliding distance of micrometer-sized (APS) and sub-micrometer-sized (spray distance of 30 and 40 mm) structural scale layers. Table 4 summarizes friction coefficient average values for several ranges of sliding distance. The structural scale plays obviously a relevant role on the friction coefficient: it evolves from an average value of about 0.8 for APS coating (micrometer scale) to about 0.4-0.5 for SPS coatings (sub-micrometer scale), clearly demonstrating the benefit on tribological behavior provided by finely structured structures over larger ones.

Considering APS coatings, voids, interlamellar delaminations, and intralamellar cracks lead to weakened lamella boundaries. When a contact load is applied on the coating and because of a reduced mechanical strength, lamellae are exfoliated generating a rough surface topology. As a result, the friction force is high. Considering SPS coatings and because of a smaller grain size and the absence of apparent delaminations and cracks (Fig. 1), particles are strongly linked together. This improves the resistance to crack propagation and lamella exfoliation, as published works have shown (Ref 23, 24). Moreover, when lamella exfoliation occurs, changes in surface topology are lower in magnitude. This results in a lowered friction coefficient. This is confirmed by the lower fluctuations in friction coefficients (noise) imparted by surface heterogeneities for SPS coatings compared to APS coatings, thanks to their finer and more homogeneous structures. Concerning SPS coatings, coatings sprayed with a 30-mm spray distance show a slightly lower friction coefficient compared to the ones sprayed with a 40-mm spray distance. Figure 4 displays the evolution of coating friction coefficient versus sliding distance curves for three spray distances of SPS layers. The average friction coefficients are given in Table 5. As mentioned previously, the longer the spray distance, the lower the friction coefficient. This can be directly related to their porous architecture and cohesive strength, a shorter spray distance increasing the layer cohesion while reducing the stacking defect density, as shown by Tingaud et al. (Ref 25) (Fig. 5). In fact, for the largest spray distance, lack of cohesion leads to layer destruction at about 750 m.

Additional tests at a sliding distance of 3000 m, shown in Fig. 6 (instead 1500 m for the previous experiments), demonstrated that SPS coatings sprayed at a 30-mm distance behaved identically; that is, the friction coefficient remains equal to 0.4 with no clear evidence of wear according to Fig. 7.

Considering wear mechanisms, grooves, cracks, and a relatively rough surface topology observed for APS coatings (Fig. 8a and b) demonstrates that they have encountered severe wear. In fact, the wear mechanism is well known in this case. Lamella delamination begins with the appearance of cracks and their propagation between two lamellae. Another relevant mechanism is inter-lamellar delamination along a weak interface between successive lamellae (Ref 26), particularly when inter-lamellar delaminations develop due to particle rapid solidification (Ref 27). Of course, splat delamination is also facilitated by the presence of defects such as unmolten or semi-molten particles, voids, cracks, etc. For SPS coatings, the wear track presents no grooves nor cracks and the surface is relatively smooth (Fig. 8c and d). One explanation is the absence of lamella exfoliation but only a plastic deformation as depicted in Fig. 9. Wear occurs only at the

Table 4 Average friction coefficients for several ranges of sliding distances

Sliding distance range, m	Friction coefficient, –		
	APS	SPS (SD = 40 mm)	SPS (SD = 30 mm)
100-400	0.9	0.4	0.4
400-900	0.9	0.4	0.4
900-1500	0.9	0.5	0.4
1500-3000	n.a.	n.a.	0.4

Table 5 Average friction coefficients for several ranges of sliding distances

Sliding distance range, m	Friction coefficient, –		
	SPS (SD = 30 mm)	SPS (SD = 40 mm)	SPS (SD = 50 mm)
0-250	0.30	0.30	0.35
250-750	0.30	0.35	0.40
750-1000	0.30	0.35	n.a.
1000-1500	0.30	0.40	n.a.

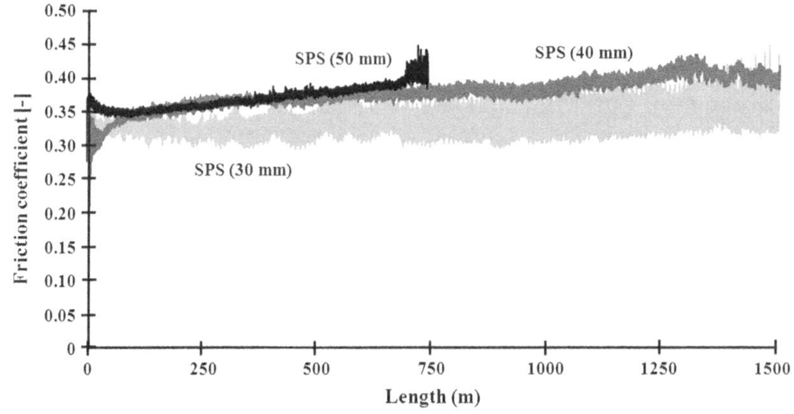

Fig. 4 Friction coefficient versus sliding distance for Al_2O_3 sub-micrometer-sized (spray distance = 30, 40, and 50 mm) coatings (BOD, α-Al_2O_3 ball of 6 mm in diameter, 2 N load, relative speed of 0.1 m s^{-1})

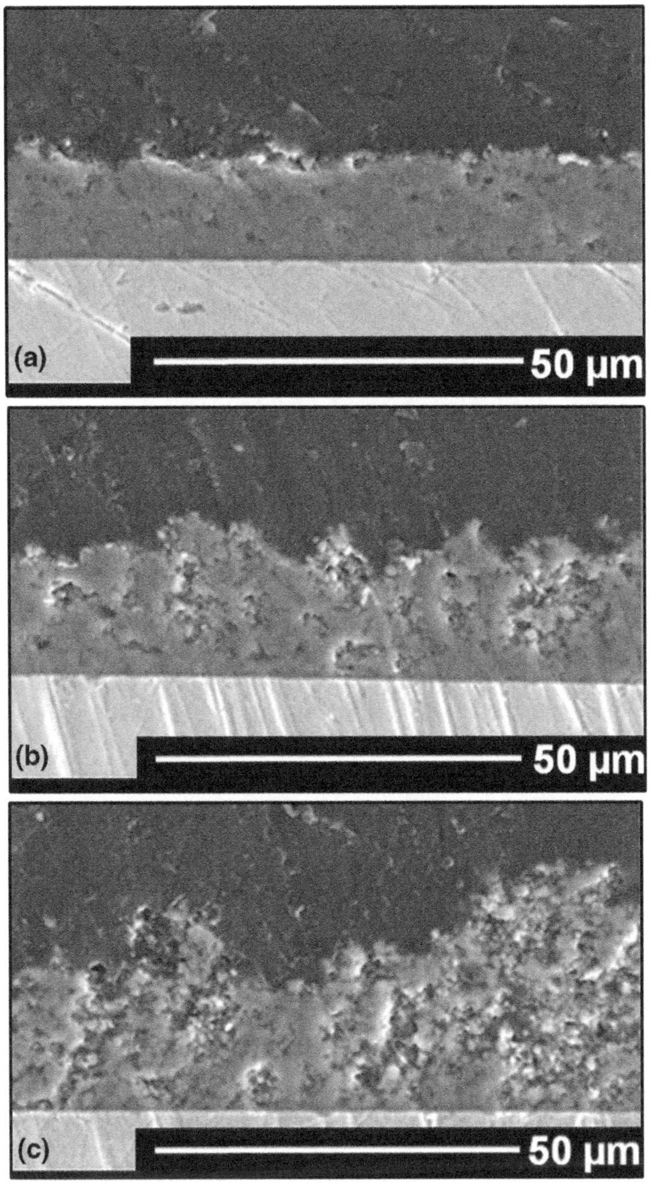

Fig. 5 Polished cross sections of sub-micrometer-sized Al_2O_3 (SPS) coatings sprayed at (a) 30 mm, (b) 40 mm, and (c) 50 mm

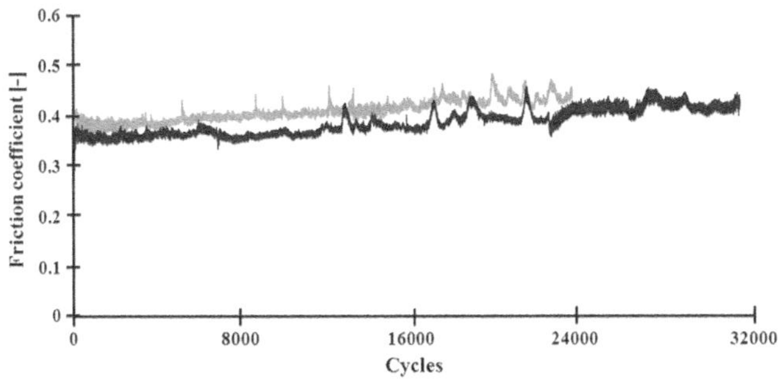

Fig. 6 Friction coefficient versus sliding distance for Al_2O_3 sub-micrometer-sized SPS (spraying distance of 30 mm) coatings for a sliding distance of 1500 and 3000 m (BOD, α-Al_2O_3 ball of 6 mm in diameter, 2 N load, relative speed of 0.1 m s^{-1})

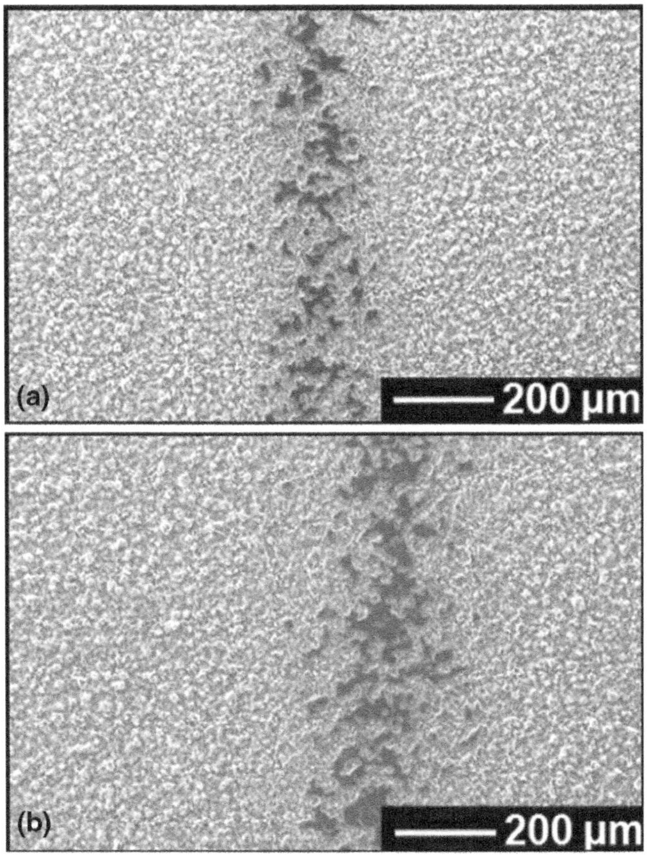

Fig. 7 Wear tracks resulting from BOD test with sub-micrometer-sized SPS (spraying distance of 30 mm) coatings for a sliding distance of (a) 1500 m and (b) 3000 m (α-Al_2O_3 ball of 6 mm in diameter, 2 N load, relative speed of 0.1 m s^{-1})

asperities, which would account very likely for the high cohesion strength between neighbor lamellae (Ref 28, 29). A published study points out that plastic deformation can permit to dissipate energy from strain (Ref 24). Besides, Lima et al. (Ref 23) have demonstrated that a thinner structure was characterized by a lower level of residual stresses due to a lower mismatch thermal expansion between coating and substrate (2-3 times). Indeed, Fig. 10 presents profiles of wear tracks for the three considered structures. The wear rate of APS coatings (micrometer-sized architecture) is significantly higher than the one of SPS coatings (sub-micrometer-sized architecture): almost 0 for SPS and 2.93×10^5 μm^3 N^{-1} m^{-1}. These results corroborate researches made on "nano effect" (Ref 30-32). At this stage of the experiments, no clear evidence of SPS wear has been identified as the track depth corresponds to the characteristic dimension of the surface roughness.

The behavior of SPS alumina sub-micrometer-sized coatings could be very likely extended for these specific test conditions to other oxides, such as ZrO_2 for example, until no transfer between coating and counter material (ball) occurs. The behavior of metallic coatings would be very likely very different due to such a transfer.

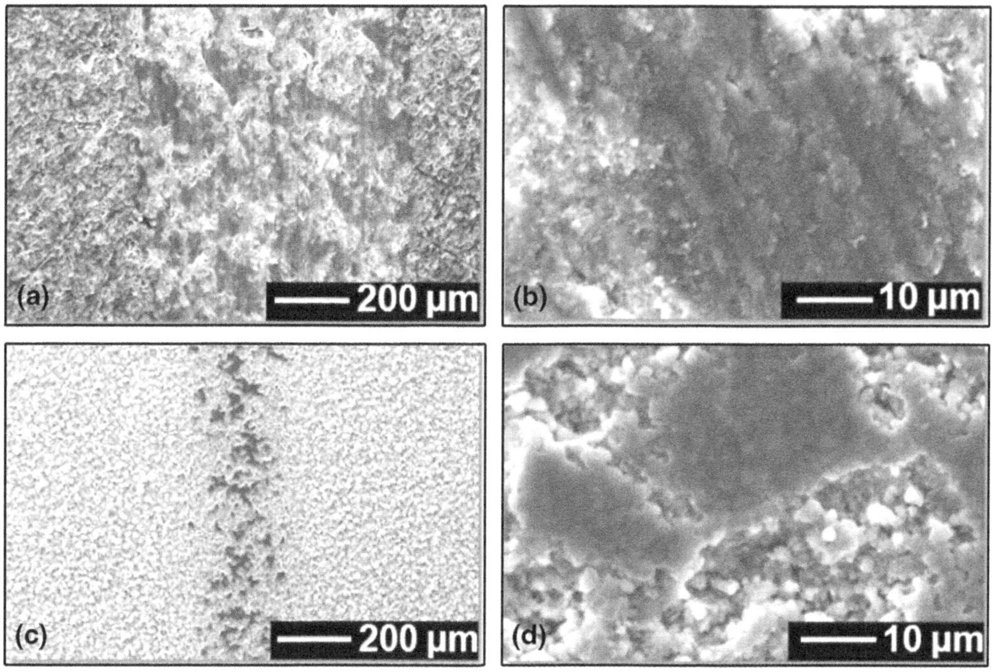

Fig. 8 Wear tracks for Al_2O_3 (a, b) micrometer-sized (APS) and (c, d) sub-micrometer-sized (SPS, spray distance = 30 mm) coatings (1500 m sliding distance following conditions specified in Sect 2.2)

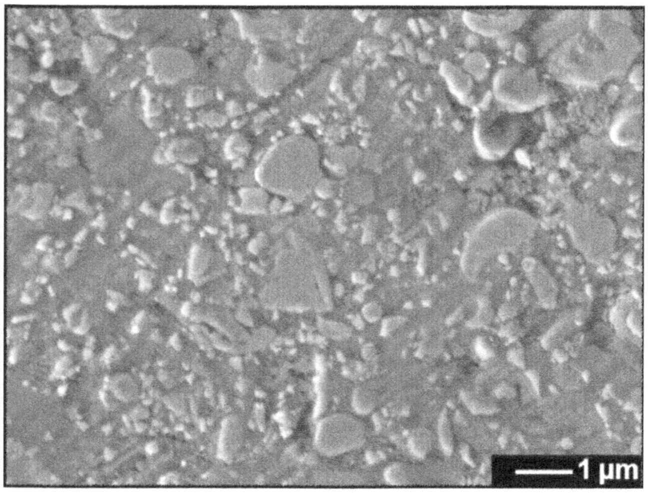

Fig. 9 Wear tracks (FESEM) of Al_2O_3 sub-micrometer-sized (SPS, spray distance = 30) coatings (1500 m sliding distance following conditions specified in Sect 2.2)

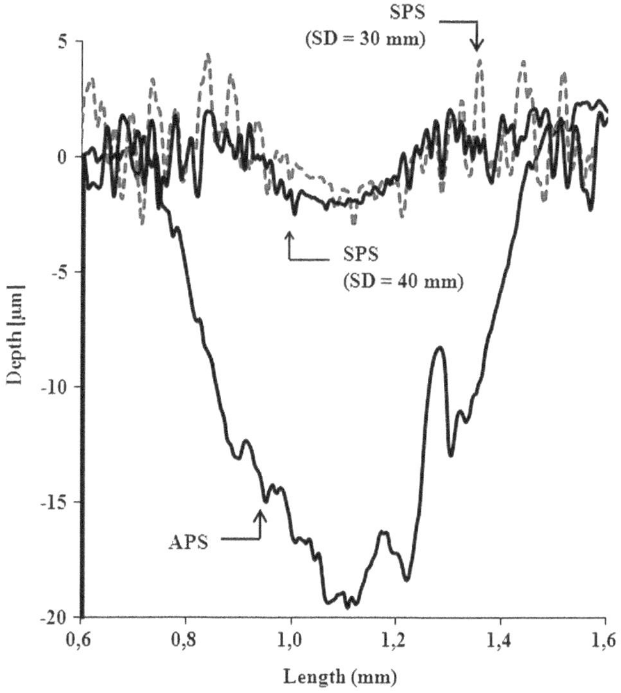

Fig. 10 Wear profiles for Al_2O_3 micrometer-sized (APS) and sub-micrometer-sized (SPS, spray distance = 30 and 40 mm) coatings (1500 m sliding distance following conditions specified in Sect 2.2)

4. Conclusions

This study highlights the improvement brought by the decrease in plasma spray coating structural scale on the coating friction coefficient.

The friction coefficient is decreased by a factor of two when the characteristic scale is reduced by two orders of magnitude. Accordingly, the wear resistance of SPS coatings is very significantly higher than APS coatings, thanks to a better toughness of the layer when shown by track view with plastic deformation. When considering different SPS coating architectures (resulting from different spray distances), the higher the cohesion, the lower the friction coefficient and the wear.

Acknowledgments

This work was conducted within the frame of the French FCE-NANOSURF consortium (Mecachrome, Frechin, CRT Plasma Laser, Cilas, CEA, CITRA) that is granted by the French Ministry and Industry and local governments of Région Centre and Limousin, the financial supports of which are acknowledged by the authors. The authors gratefully acknowledge the valuable help of Olivier Tingaud, SPCTS – UMR CNRS 6638.

References

1. Kyoto Protocol to the United Nations Framework Convention on Climate Change, United Nations, 1998
2. Intergovernmental Panel on Climate Change, Climate Change 2007, released November 17, 2007, Valencia, Spain
3. Y. Fukushima, M. Shimada, S. Kraines, M. Hirao, and M. Koyama, Scenarios of Solid Oxide Fuel Cell Introduction into Japanese Society, *J. Power Sources*, 2004, **131**, p 327-339
4. E. Broszeit, Mechanical, Thermal and Tribological Properties of Electro- and Chemodeposited Composite Coatings, *Thin Solid Films*, 1982, **2**(95), p 133-142
5. G. Barbezat, Coating for the Working Surface of the Cylinders of Combustion Engines and a Method of Applying such a Coating, US Patent 6,548,195B1, 2003
6. G. Barbezat, Surface Layer Forming a Cylinder Barrel Surface, a Spraying Powder Suitable Therefore and a Method of Creating such a Surface Layer, US Patent 6,578,539B2, 2003
7. Q. Feng, T. Li, H. Yue, K. Qi, F. Bai, and J. Jin, Preparation and Characterization of Nickel Nano-Al_2O_3 Composite Coatings by Sediment Co-Deposition, *Appl. Surf. Sci.*, 2008, **254**(8), p 2262-2268
8. R.S. Lima and B.R. Marple, Thermal Spray Coatings Engineered from Nanostructured Ceramic Agglomerated Powders for Structural, Thermal Barrier and Biomedical Applications: A Review, *J. Therm. Spray Technol.*, 2007, **16**(1), p 40-63
9. M. Gell, E.H. Jordan, Y.H. Sohn, D. Goberman, L. Shaw, and T.D. Xiao, Development and Implementation of Plasma Sprayed Nanostructured Ceramic Coatings, *Surf. Coat. Technol.*, 2001, **146-147**, p 48-54
10. E.H. Jordan, M. Gell, Y.H. Sohn, D. Goberman, L. Shaw, S. Jiang, M. Wang, T.D. Xiao, Y. Wang, and P. Strutt, Fabrication and Evaluation of Plasma Sprayed Nanostructured Alumina-Titania Coatings with Superior Properties, *Mater. Sci. Eng. A*, 2001, **301**(1), p 80-89
11. H. Chen, Y. Zhang, and C. Ding, Tribological Properties of Nanostructured Zirconia Coatings Deposited by Plasma Spraying, *Wear*, 2002, **253**(7-8), p 885-893
12. Y.-C. Zhu, K. Yukimura, C.-X. Ding, and P.-Y. Zhang, Tribological Properties of Nanostructured and Conventional WC-Co Coatings Deposited by Plasma Spraying, *Thin Solid Films*, 2001, **388**(1-2), p 277-282
13. J.H. Ouyang and S. Sasaki, Effects of Different Additives on Microstructure and High-Temperature Tribological Properties of Plasma-Sprayed Cr_2O_3 Ceramic Coatings, *Wear*, 2001, **249**, p 56-67
14. Coll, Used Lubricants. Pub. Agence de l'Environnement et de la Maîtrise de l'Energie, Angers, France, p 12 (in French)
15. P. Fauchais, V. Rat, C. Delbos, J.-F. Coudert, T. Chartier, and L. Bianchi, Understanding of Suspension DC Plasma Spraying of Finely Structured Coatings for SOFC, *IEEE Trans. Plasma Sci.*, 2005, **33**(2), p 920-930
16. H. Kassner, R.O. Siegert, D. Hathiramani, R. Vassen, and D. Stoever, Application of Suspension Plasma Spraying (SPS) for Manufacture of Ceramic Coatings, *J. Therm. Spray Technol.*, 2008, **17**(1), p 115-123
17. C. Delbos, "Contribution to the Understanding of Ceramic (Y-PSZ, perovskite, etc.) or Metallic (Ni, etc.) Particles Injection by a Liquid Carrier in a Plasma Jet to Manufacture Finely-Structured Coatings for SOFC," Ph.D. Thesis, University of Limoges, 2004 (in French)
18. J. Oberste Berghaus, J.-G. Legoux, C. Moreau, F. Tarasi, and T. Chráska, Mechanical and Thermal Transport Properties of Suspension Thermal-Sprayed Alumina-Zirconia Composite Coatings, *J. Therm. Spray Technol.*, 2008, **17**(1), p 91-104
19. O. Tingaud, R. Etchart-Salas, V. Rat, J.-F. Coudert, H. Ageorges, A. Grimaud, A. Denoirjean, P. Fauchais, G. Montavon, N. Caron, and S. Alexandre, *Suspension Plasma Spraying of Alumina Coatings: Process and Coating Structure*, Kyoto, Japan, Aug 26-31, 2007
20. G. Antou, G. Montavon, F. Hlawka, A. Cornet, and C. Coddet, Characterizations of the Pore-Crack Network Architecture of Thermal Sprayed Coatings, *Mater. Charact.*, 2004, **53**(5), p 361-372
21. S. Valette, G. Trolliard, A. Denoirjean, and P. Lefort, Iron/Wustite/Magnetite/Alumina Relationships in Plasma Coated Steel: A TEM Study, *Solid State Ionics*, 2007, **178**(5-6), p 429-437
22. ISO 4287:1997 Standard. Surface Texture: Profile Method – Terms, Definitions and Surface Texture Parameters
23. R.S. Lima and B.R. Marple, From APS to HVOF Spraying of Conventional and Nanostructured Titania Feedstock Powders: A Study on the Enhancement of the Mechanical Properties, *Surf. Coat. Technol.*, 2006, **200**(11), p 3428-3437
24. X. Lin, Y. Zeng, S.-W. Lee, and C. Ding, Characterization of Alumina-3 wt.% Titania Coating Prepared by Plasma Spraying of Nanostructured Powders, *J. Eur. Ceram. Soc.*, 2004, **24**(4), p 627-634
25. O. Tingaud, A. Grimaud, A. Denoirjean, G. Montavon, V. Rat, J.-F. Coudert, and P. Fauchais, Effects of Operating Parameters on SPS Alumina Coatings Structures to Manufacture Functionally-Graded Layers, *Surf. Coat. Technol.* (in press)
26. P.P. Psyllaki, M. Jeandin, and D.I. Pantelis, Microstructure and Wear Mechanisms of Thermal-Sprayed Alumina Coatings, *Mater. Lett.*, 2001, **47**(1-2), p 77-82
27. D.I. Pantelis, P. Psyllaki, and N. Alexopoulos, Tribological Behaviour of Plasma Sprayed Al_2O_3 Coatings Under Severe Wear Conditions, *Wear*, 2000, **237**(2), p 197-204
28. N.B. Dahotre and S. Naya, Nanocoatings for Engine Application, *Surf. Coat. Technol.*, 2005, **194**(1), p 404
29. H. Hahn, P. Mondal, and K. Padmanabhan, Plastic Deformation of Nanocrystalline Materials, *Nanostruct. Mater.*, 1997, **9**, p 603
30. X. Lin, Y. Zeng, C. Ding, and P. Zheng, Tribological Behaviour of Nanostructured Al_2O_3-3 wt% TiO_2 Coating Against Steel in Dry Sliding, *Tribol. Lett.*, 2004, **17**(1), p 19-26
31. J. Zhang, J. He, Y. Dong, X. Li, and D. Yan, Microstructure and Properties of Al_2O_3-13%TiO_2 Coatings Sprayed Using Nanostructured Powders, *Rare Met.*, 2007, **26**(4), p 391-397
32. Y. Wang, S. Jiang, M. Wang, S. Wang, T.D. Xiao, and P.R. Strutt, Abrasive Wear Characteristics of Plasma Sprayed Nanostructured Alumina/Titania Coatings, *Wear*, 2000, **237**(2), p 176-185

JTTEE5 17:796–804
DOI: 10.1007/s11666-008-9266-1
1059-9630/$19.00 © ASM International

Peer Reviewed

Fast Regime Fluidized Bed Machining (FR-FBM) of Thermally Sprayed Coatings

Massimiliano Barletta, Gianluca Rubino, Giovanni Bolelli, and Luca Lusvarghi

(Submitted June 9, 2008; in revised form September 15, 2008)

Finishing of thermally sprayed metallic, ceramic, and cermet coatings is required to meet tolerances and requirements on surface roughness in most industrial applications. Conventional machining is a costly and time-consuming process, and is difficult to automate. Therefore, this study investigates and develops a new technique highly amenable for automation: fast regime—fluidized bed machining (FR-FBM). Atmospheric plasma sprayed TiO_2, Cr_2O_3, and HVOF-sprayed WC-17%Co and Tribaloy-800 coatings, deposited on AISI 1040 steel substrates, were subjected to FR-FBM treatment. The effects of the leading operational parameters, namely, abrasive size, jet pressure, and processing time, were evaluated on all coatings by using a two/three-levels full factorial design of experiments. The FR-FBM treated surfaces were observed by FE-SEM and their surface finishing was evaluated by contact profilometry. Significant improvements in surface finishing of all the machined thermally sprayed coatings can always be detected, with FR-FBM being able to guarantee the precision and to ensure the closest geometrical tolerances.

Keywords automatic machining, atmospheric plasma spraying, fluidized bed, HVOF

1. Introduction

The conception of highly stressed components requires a fine assessment of materials, design, and manufacturing techniques. Whenever components withstanding dynamic solicitations have to be developed, their surface plays a crucial role, because damages often start in the outermost layers at notches, pores, fatigue slip steps, and inclusions (Ref 1). In this respect, the various techniques for surface treatment are assuming a remarkable interest in several industrial fields. Among them, the application of surface overlay coatings by thermal spray techniques is spreading over a wide range of applications, often including the manufacturing of high performance components (Ref 2). Nonetheless, reprocessing of the as-deposited thermally sprayed coatings is often necessary to meet surface finishing requirements and, above all, to ensure the closest geometrical tolerances (Ref 3). Nowadays, conventional reprocessing techniques are still costly and time consuming and are rather troublesome to automate (Ref 2, 3). Besides, conventional machining solutions are often manual, thus displaying poor or, at least, unstable results and strict dependency on the capability of skilled technicians (Ref 3, 4).

This is, therefore, the context in which the present study moves to analyze the possibility to apply fast regime—fluidized bed machininig (FR-FBM) for surface finishing of a wide range of thermally sprayed coatings by atmospheric plasma sprayed (APS) and HVOF. In particular, the influence of FR-FBM operational parameters on the resultant surface finishes was thoroughly investigated according to a design of experiments (DOE)-assisted experimental schedule. Furthermore, the morphological characteristics of the finished surfaces were examined by combined usage of SEM and contact-mode profilometry. Finally, variations in tribological properties before and after FR-FBM were evaluated by pin-on-disk tests. In the light of experimental findings, FR-FBM was found to significantly improve the finishing of thermally sprayed coatings, hence guaranteeing superior visual appearance, high machining accuracy, and good precision.

2. Experimental

2.1 Material

Four different thermally sprayed coatings were selected for the present test, namely, atmospheric plasma sprayed TiO_2 (deposition parameters in Ref 4, thickness 168 ± 16 μm), APS Cr_2O_3 (parameters in Ref 5, thickness 330 ± 21 μm), HVOF-sprayed WC-17%Co (thickness

This article is an invited paper selected from presentations at the 2008 International Thermal Spray Conference and has been expanded from the original presentation. It is simultaneously published in *Thermal Spray Crossing Borders, Proceedings of the 2008 International Thermal Spray Conference*, Maastricht, The Netherlands, June 2-4, 2008, Basil R. Marple, Margaret M. Hyland, Yuk-Chiu Lau, Chang-Jiu Li, Rogerio S. Lima, and Ghislain Montavon, Ed., ASM International, Materials Park, OH, 2008.

Massimiliano Barletta and **Gianluca Rubino**, Dipartimento di Meccanica ed Aeronautica, 'La Sapienza' Università di Roma, Via Eudossiana, 18, 00184 Rome, Italy; and **Giovanni Bolelli** and **Luca Lusvarghi**, Dipartimento di Ingegneria dei Materiali e dell'Ambiente, Università degli Studi di Modena e Reggio Emilia, Via Vignolese, 905, 41100 Modena, Italy. Contact e-mail: Barletta@ing.uniroma2.it.

230 ± 15 μm) (Ref 6), and Tribaloy-800 (thickness 402 ± 14 μm) (Ref 7), deposited on AISI 1040 steel substrates. Slabs 10 × 10 × 5 mm^3 were subjected to FR-FBM treatment.

Unsieved alumina powders (Smyris Abrasivi, Como, Italy) were used as abrasive media for finishing purposes. Abrasive sizes were 46, 80, and 180 mesh size (MS). The different sized powders were all characterized by a rather angular shape factor of about 0.67, this being the guarantee of the presence on them of a large number of 'self-sharpening cutting tools' (i.e., the sharp edges and the tips of the angular abrasive particle).

2.2 Apparatus

FR-FBM was used for easy-to-automate finishing of the thermally sprayed coatings. The description of FR-FBM system is detailed elsewhere (Ref 8-10).

Figure 1 shows a sketch of the system, which consists of an air feeding system (mostly, a blower), an air plenum chamber to ensure the proper air feeding to the fluidization column and its distribution across the column, a porous plate distributor, which allows the fluidization air to pass through while it retains the powders to be fluidized, a fluidized bed column in which the fluidization takes place, a couple of cyclones, a static or rotating sample holder, and, then, controls, tubes, valves, and all the necessary fittings.

Purified air is forced through the porous plate distributor (sintered bronze, 8 mm thick, average porosity of 70 μm) and acts to suspend the abrasive powders (typically, Al_2O_3) preloaded inside the fluidized bed column (40 × 40 × 1200 mm^3, AISI 304). The substrate to be finished is held inside the fluidization column by using a shaft (3 mm diameter, AISI 304). A direct current electrical motor is used to rotate the sample inside the fluidized abrasives, hence changing the impact conditions between the abrasives and the surface to be finished and ensuring fresh abrasive is repeatedly delivered toward the target.

During FR-FBM, abrasives and fines produced because of the heavy machining conditions tend to raise up through the fluidized bed column and they are progressively elutriated from the top of the column as a result of the fluidization in fast regime. To prevent the rapid emptying of the fluidized bed column and ensure the continuity of FR-FBM, a cyclone is located downstream the fluidized bed column to collect the elutriated abrasives and resupply them at the bottom of the column itself. A second cyclone is, then, used to recover the fines produced during the heavy machining, thus avoiding their unsafe dispersion to the environment.

2.3 Procedure

The effects of leading FBM operational parameters, namely, abrasive size, jet pressure, and processing time, on the surface finish of the thermally sprayed coatings were evaluated using a full factorial DOE. Table 1 summarizes the experimental schedule.

Table 1 Design of experiments

Mesh size (MS), μ/in	Abrasives diameter, μm	Pressure, Pa	Time, min
46	~354	4×10^5	2
80	177	4×10^5	2
180	~88	4×10^5	2
46	~354	4×10^5	5
80	177	4×10^5	5
180	~88	4×10^5	5
46	~354	7×10^5	2
80	177	7×10^5	2
180	~88	7×10^5	2
46	~354	7×10^5	5
80	177	7×10^5	5
180	~88	7×10^5	5

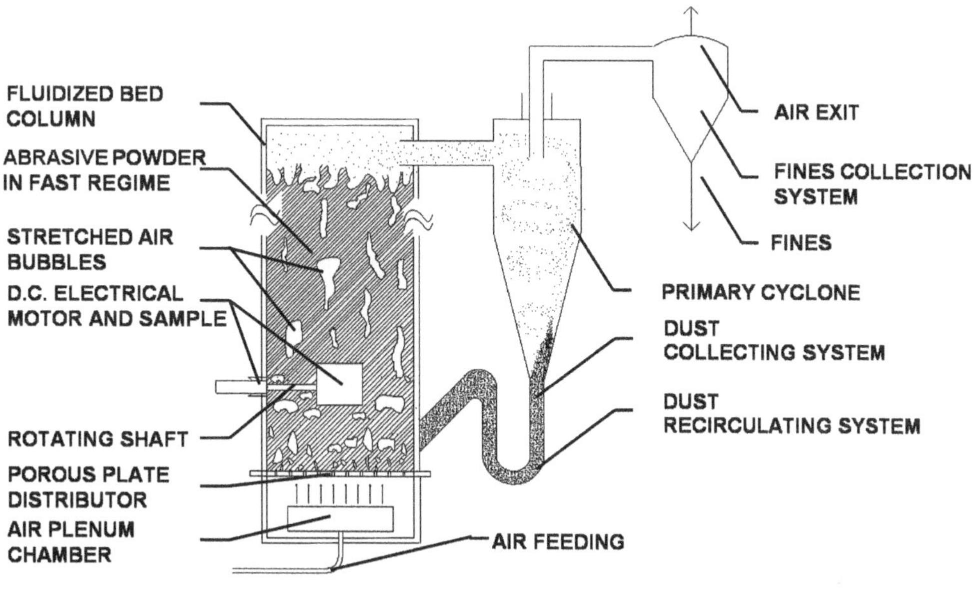

Fig. 1 The FR-FBM system

Mass loss after FR-FBM was calculated by using a Sartorious BP 211-D weighing instrument (0.01 mg resolution). Each workpiece was measured several times, and if the difference between the two following measurements failed to agree within 0.5 mg, the measurements were repeated again until agreement within this range was attained in successive determinations. The change in thickness was also carefully evaluated by using a Mitutoyo Absolute Digimatic Micrometers Series 227 (±1 μm resolution). A map of nine measurements equally spaced over the machined coatings were performed. This way, the evenness of the coating removal process, and thus the suitability of FR-FBM in obtaining tight tolerances, was carefully checked.

Surface roughness measurements were carried out using a contact-mode profilometer (Taylor-Hobson CLI 2000). TalyMap software release 3.1 was employed to work out the amplitude, spacing, and hybrid roughness parameters. For such purpose, wide enough areas of 4 × 4 and 7 × 7 mm^2 with a lateral resolution of 1 μm along the measurement direction and, respectively, of 4 and 100 μm along the perpendicular direction were measured. Finally, a Field Emission Scanning Electron Microscope (FE-SEM Leo Supra 35) was employed to obtain high resolution images of the machined surfaces at various levels of magnification.

3. Results and Discussion

The interactions between the fluidized abrasives and the surface to be finished were first analyzed by evaluating the mass loss and the material removal of the thermally sprayed coatings varying the FR-FBM operational parameters. In particular, machining conditions, which made use of abrasives MS 46, were first considered as it was supposed that the biggest abrasives produced the massive effects and the largest dimensional changes in the machined substrates.

Figures 2(a) and (b) reports the mass loss and the material removal, respectively. Both mass loss and material removal were found to increase drastically whenever higher jet pressure was set. Furthermore, if their trends are looked into, it is possible to note how very large reductions in the samples mass and in their thickness are induced by an increase in jet pressure, while slighter reduction was induced by an increase in the machining time. However, no strict relationship between the mass loss and the reduction in the coating thickness can be supposed, as the mass loss is very much influenced by the embedding phenomena of abrasive debris inside the outermost layers of the thermally sprayed coating as a result of the heavy machining conditions. This way, the measurement of the mass variation is altered by the presence of abrasive residuals into the machined surface, as the evidence of SEM image in Fig. 3 shows and in agreement with what is reported in the pertinent literature for similar process conditions (Ref 8-10).

It is also worth mentioning how the variance of coating thickness is very low (Fig. 2b) although measurements

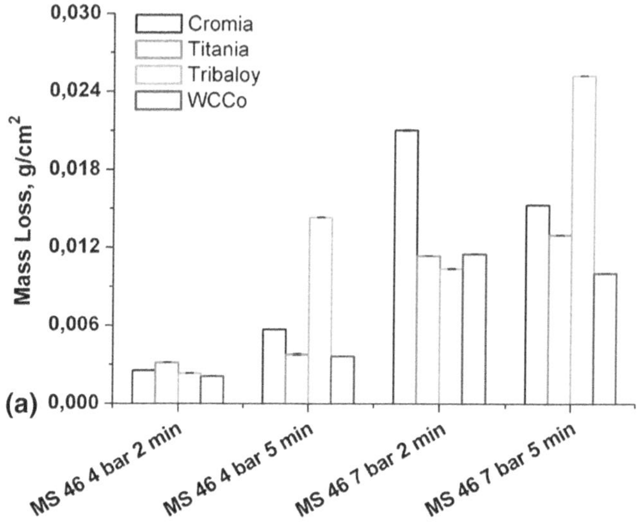

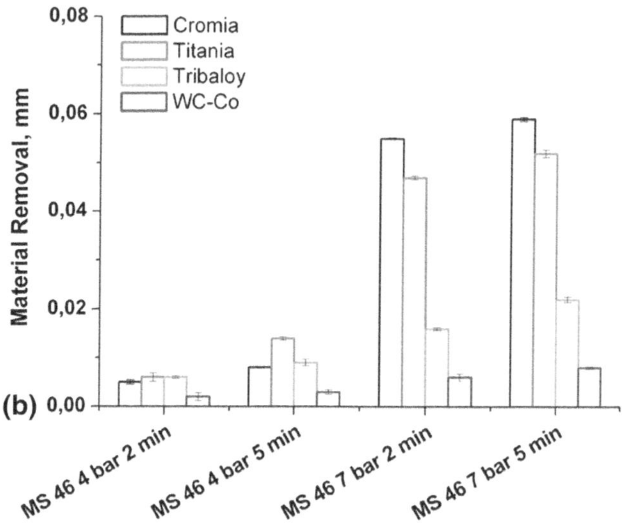

Fig. 2 (a) Mass loss and (b) material removal of thermally sprayed coatings after FR-FBM

were taken on nine different locations equally spaced over the treated samples. This suggests that FR-FBM can be a very promising technique to evenly machine thick and hard coatings and, thus, to ensure the closest geometrical tolerances. Further test on more complicated sample geometries would however be required to ascertain the evenness of the FR-FBM treatment and its capability to respect the dimensional tolerances.

Noticeable differences in the variation of samples mass and thickness characterize the investigated coating materials. Indeed, the different coating materials behave differently as a result of their specific hardness and toughness (Ref 4-7). WC-Co-based coatings, which present high hardness coupled with satisfactory toughness (Ref 5, 6), tend to be very difficult to machine, thus leading to minimum material removal during FR-FBM. Indeed, this favorable combination of material properties makes these coatings highly resistant to particle erosion (Ref 11, 12), a process similar to that occurring in FR-FBM. However,

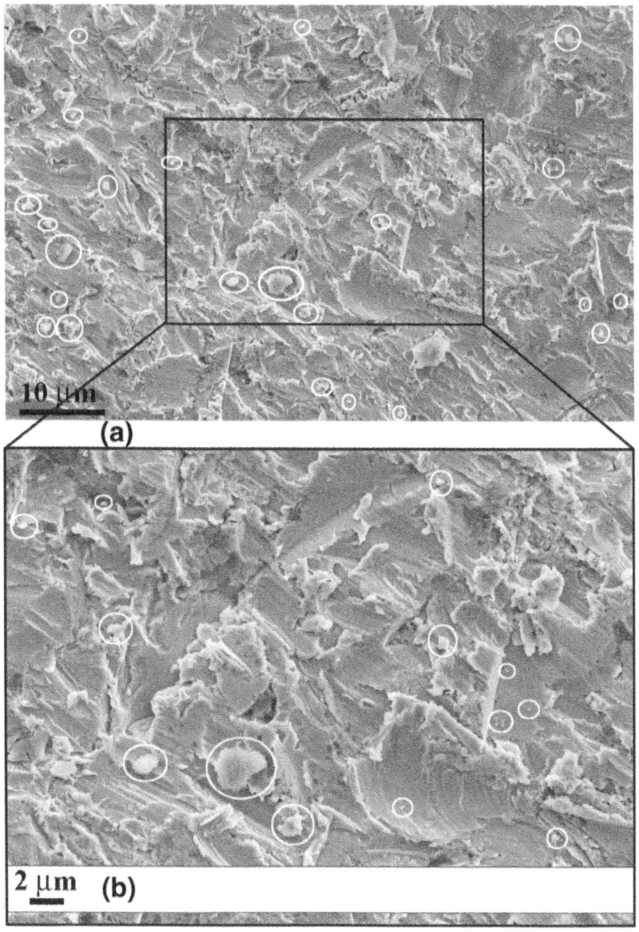

Fig. 3 (a) Abrasive splinters clung onto a TiO_2 coating after FR-FBM and (b) Higher magnification of the embedded abrasive particles

WC-Co coatings are rather sensitive to the changes in FR-FBM operational parameters. In fact, material removal can increase up to four times setting the jet pressure and the machining time at their maximum.

SEM micrographs (Fig. 4 and 5) reveal that material is removed from the as-deposited WC-Co coating surface (which exhibits a rough morphology, Fig. 4(d), due to the incomplete flattening of particles impinging in a semisolid state, as shown in Ref 13, 14 by a combination of ductile grooving and carbide fracture (Fig. 6), probably caused by erodent particles impinging at low or high (close to 90°) angles, respectively, consistently with erosion mechanisms described in Ref 11, 12.

Conversely, APS-Cr_2O_3 coatings do exhibit very large amount of material removal at any time the jet pressure is set at 7 bar. APS-Cr_2O_3 is the hardest among the present coatings, but it is also rather brittle (i.e., poor toughness) (Ref 5): this compromises its resistance to the more energetic impacts of the fluidized abrasives, thus determining severe mass loss via brittle fracture phenomena, a well-known erosive wear mechanism for APS ceramics (Ref 15). Indeed, on the FR-FBM-processed surface of Cr_2O_3, the gently curved morphology of sprayed lamellae is still well recognizable (compare the circled area in

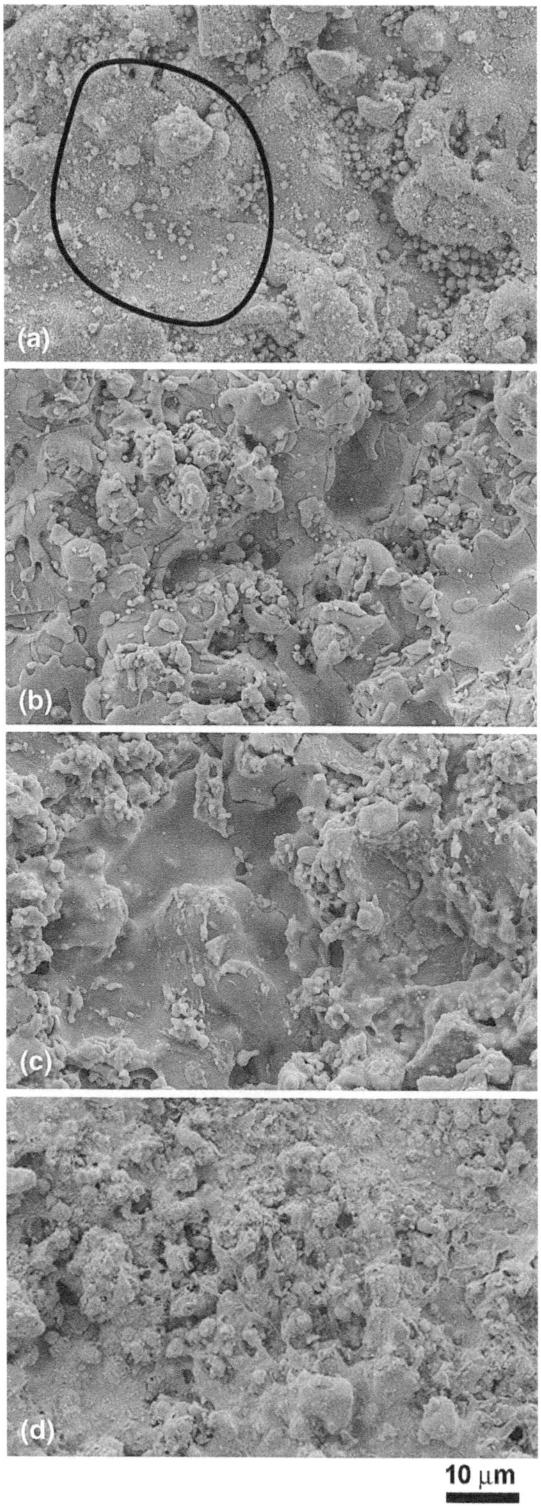

Fig. 4 SEM images of the thermally sprayed coating before FR-FBM (MS 80, jet pressure 7 bar, and machining time 5 min): (a) Cr_2O_3 as-deposited (the *circle* indicates a clearly recognizable lamella); (b) TiO_2 as-deposited; (c) Tribaloy-800 as-deposited; and (d) WC-17%Co as-deposited

Fig. 5a to the lamella shown in Fig. 4a), indicating that material was removed by cracks running preferentially along interlamellar boundaries (Fig. 5a, arrows indicate

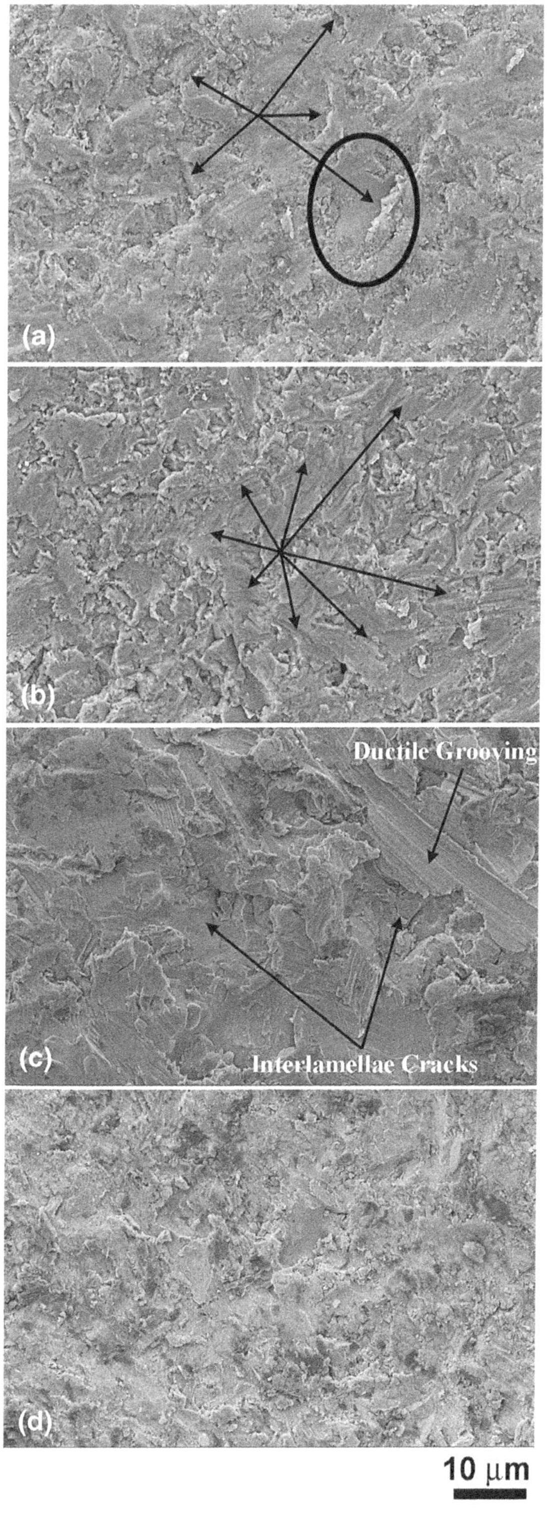

Fig. 5 SEM images of the thermally sprayed coatings after FR-FBM (MS 80, jet pressure 7 bar, and machining time 5 min): (a) Cr_2O_3 finished (*arrows*: interlamellae cracks; *circle*: a portion of a lamella uncovered by detachment of the above layers); (b) TiO_2 finished (*arrow*: ductile grooving); (c) Tribaloy-800 finished, and (d) WC-CO 17% finished

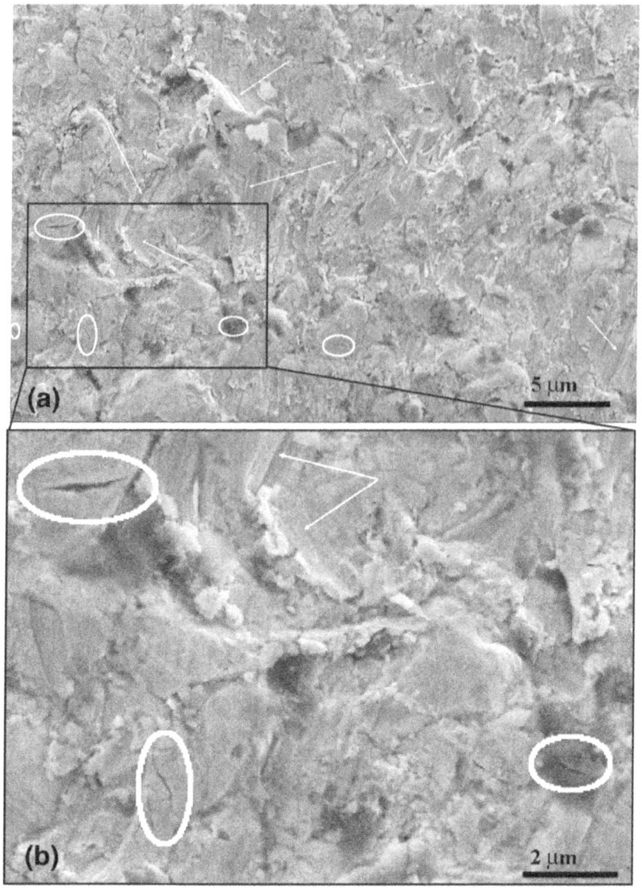

Fig. 6 (a) SEM image of ductile grooving (i.e., *white arrows*) and carbide grains fracture (i.e., *white circle*) onto WC-Co coating after FR-FBM and (b) higher magnification of ductile grooving and carbide grains fracture

some recognizable cracks). Indeed, such clear tendency to anisotropic crack propagation in this coating was formerly proven by indentation tests (Ref 5); moreover, an analogous wear mechanism was noted in dry particle abrasion tests (Ref 5). However, Cr_2O_3-based coatings also present the largest difference in behavior when machined at high and low jet pressures. In fact, at low jet pressure, they are less sensitive to the machining action of the abrasives as a result of their high hardness. At high jet pressure, the prominent effect of their rather poor toughness can increase the material removal rate by more than one order of magnitude.

TiO_2-based coatings behave differently. Their hardness ($HV_{0.1} = (10.31 \pm 1.11)$ GPa, Ref 4) is lower than that of Cr_2O_3-based coatings ($HV_{0.1} = (12.52 \pm 1.26)$ GPa, Ref 5), but their toughness is higher. Indeed, even though the quantitative reliability of indentation fracture toughness measurements on APS ceramics was criticized in Ref 4 by some of the authors, a useful qualitative ranking can be obtained: APS-Cr_2O_3 exhibited K_{Ic} to be (1.67 ± 0.67) MPa $m^{1/2}$, while for APS-TiO_2, using the data collected in Ref 4, it is computed to be (2.11 ± 0.41) MPa $m^{1/2}$, which confirms its better toughness. Therefore, TiO_2-based coatings present the highest material removal after FR-FBM at low pressure as a result of the very low hardness. At high pressure, TiO_2-based coatings keep on showing rather large material removal, but their good toughness prevents them to behave worse than the Cr_2O_3-based coatings. Indeed, the material removal mechanism of APS-TiO_2 is significantly different from that of Cr_2O_3. Indeed, ductile grooves appear in the former (Fig. 5b), witnessing the occurrence of microscale plastic deformation (microcutting and/or microploughing), which is in strict agreement with the report by Liu et al. (Ref 16) on abrasive machining of thermally sprayed coatings. They indicated that material removal by plastic grooving occurs below a critical depth-of-cut value, while brittle fracture prevails above that value. The critical depth-of-cut is given by the relation:

$$d_C = \beta \frac{E/HV}{(HV/K_{IC})^2}$$

where β is a constant and E, HV, and K_{IC} are elastic modulus, Vickers hardness, and toughness, respectively. Using the values listed above, $E = 129$ GPa for TiO_2 (Ref 4) and $E = 132$ GPa for Cr_2O_3 (Ref 17), the ratio between the critical depth-of-cut on Cr_2O_3 and that on TiO_2 is estimated to be 0.36. Thus, it is clearly explained why, under the same conditions, the former coating is much more prone to undergo brittle fracture.

Tribaloy coatings are softer but tougher. For this reason, they display intermediate machining behavior between the WC-Co-based coatings, which undergo lower surface modification due to superior hardness, and the Ti and Cr oxides, which undergo deeper alteration. In particular, remarkable grooving, together with limited interlamellar failure, is noted (Fig. 5c).

The evaluation of surface finishing of thermally sprayed coatings was performed by evaluating the variations in roughness parameters before and after FR-FBM. Figure 7 reports the trend of the average roughness R_a. In as-deposited condition, the various coatings possess different roughness values, since the particles clearly possess different behaviors, both in-flight and upon impact, due to different thermophysical material properties, different particle size distributions, and different spray conditions (including torch type and parameter setting). Useful indications are achieved by evaluating the relative improvement in roughness values after FR-FBM.

The data are in good agreement with that mentioned previously about the mass loss and material removal of the different thermally sprayed coatings after FR-FBM. In particular, TiO_2-based coatings present the most significant variations in R_a. Abrasives MS 46 and 80 lead to improvements of three to four times in R_a, which approach values as low as 1.6-1.7 μm. Indeed, remarkable plastic grooving helps cutting and removing the large surface asperities existing in as-deposited condition due to the lamellar structure, thus making the surface smoother and more homogeneous. Analogously, tribaloy coatings display very large improvements in R_a, with values reduced by about 250%, when the largest abrasives and high operating pressure are employed, since, under these conditions, a very efficient surface machining by plastic grooving is also obtained. Instead, under less energetic machining conditions (MS 180), the mechanical properties of the tribaloy coatings are high enough to make them

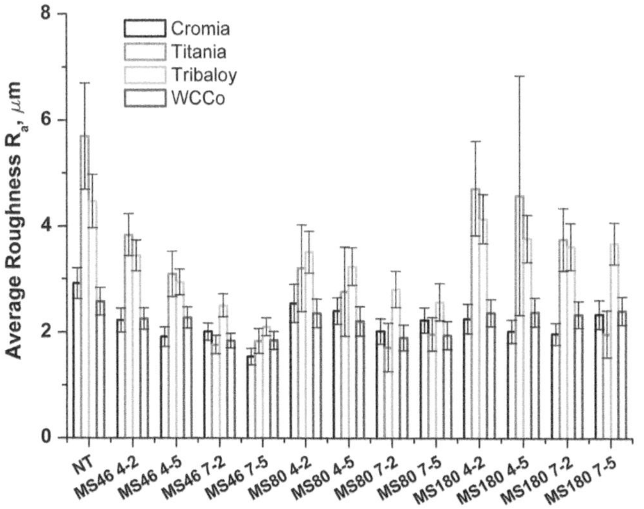

Fig. 7 Average roughness R_a of thermally sprayed coatings after FR-FBM (Note: NT = nontreated coatings)

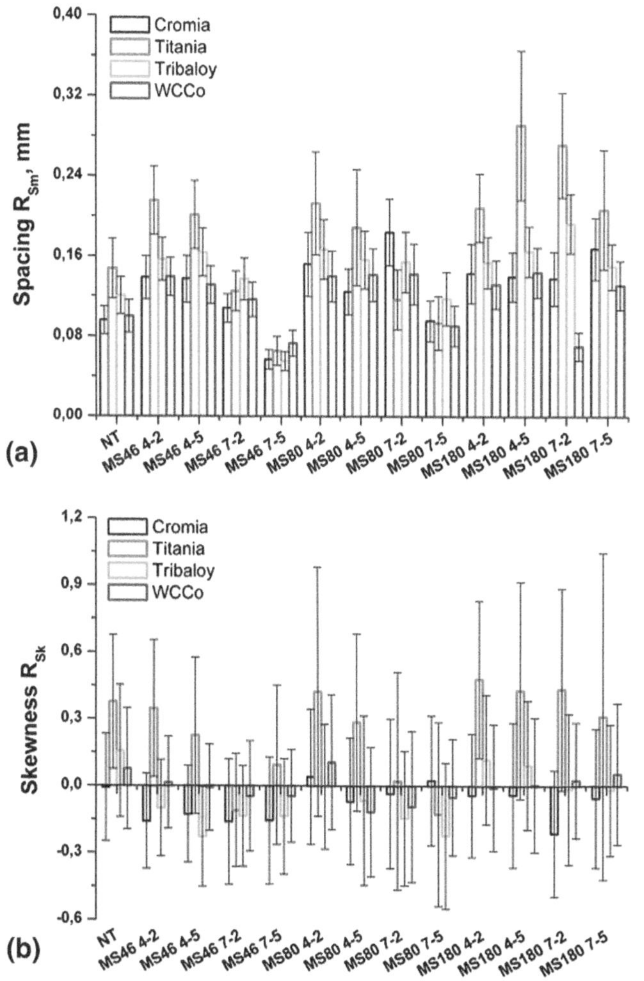

Fig. 8 (a) Spacing R_{Sm} and (b) skewness R_{sk} of thermally sprayed coatings after FR-FBM (Note: NT = non-treated coatings)

Peer Reviewed

almost insensitive to the FR-FBM. In fact, in that operating condition, the improvements in R_a are very small, whatever the setting of the jet pressure and of the machining time. Cr_2O_3- and WC-Co-based coatings display the lowest variations in R_a, with improvements in surface finish in the range of 30-40%, when the more energetic machining conditions (MS 46 or 80, higher jet pressure) are set. Such result can be interpreted as follows. Concerning WC-17%Co, its high hardness and the toughness reduce, to a larger extent, the machining capability of the FR-FBM. Concerning Cr_2O_3, under less intense machining conditions, the high hardness can also reduce the finishing effects, consistently with the low material loss; instead, under more energetic conditions, the onset of brittle fracture wear prevents substantial roughness improvements, because new lamellae are continuously being uncovered, as described earlier.

Figure 7 also reveals that R_a approaches very similar values for all the thermally sprayed coatings, whenever effective machining conditions are set. In fact, abrasives MS 46 or 80 joined with a jet pressure of 7 bar and a machining time of 2 or 5 min lead to R_a in the range of 1.6-2.4 μm, apart from the starting roughness and the coatings material. Accordingly, a hypothesis of a sort of asymptotic finishing conditions, typical of the FR-FBM technique, can be definitely formulated. This result is in agreement with what was found in a previous work of one of the authors, which usefully identifies an asymptotic finishing condition for two different aluminum alloys submitted to a fluidized bed of abrasives in bubble regime (Ref 18). However, these observations reveal that some improvement can still be expected for the more ductile coatings (APS TiO_2 and HVOF tribaloy), while little improvements are expected on HVOF WC-17%Co and on APS Cr_2O_3, the latter being possibly the most unsuitable coating for this kind of machining process.

Figures 8(a) and (b) reports spacing R_{Sm} and skewness R_{sk} of the thermally sprayed coatings before and after FR-FBM. R_{Sm} shows the lowest values at any time abrasives MS 46 and 80 are joined with high jet pressure.

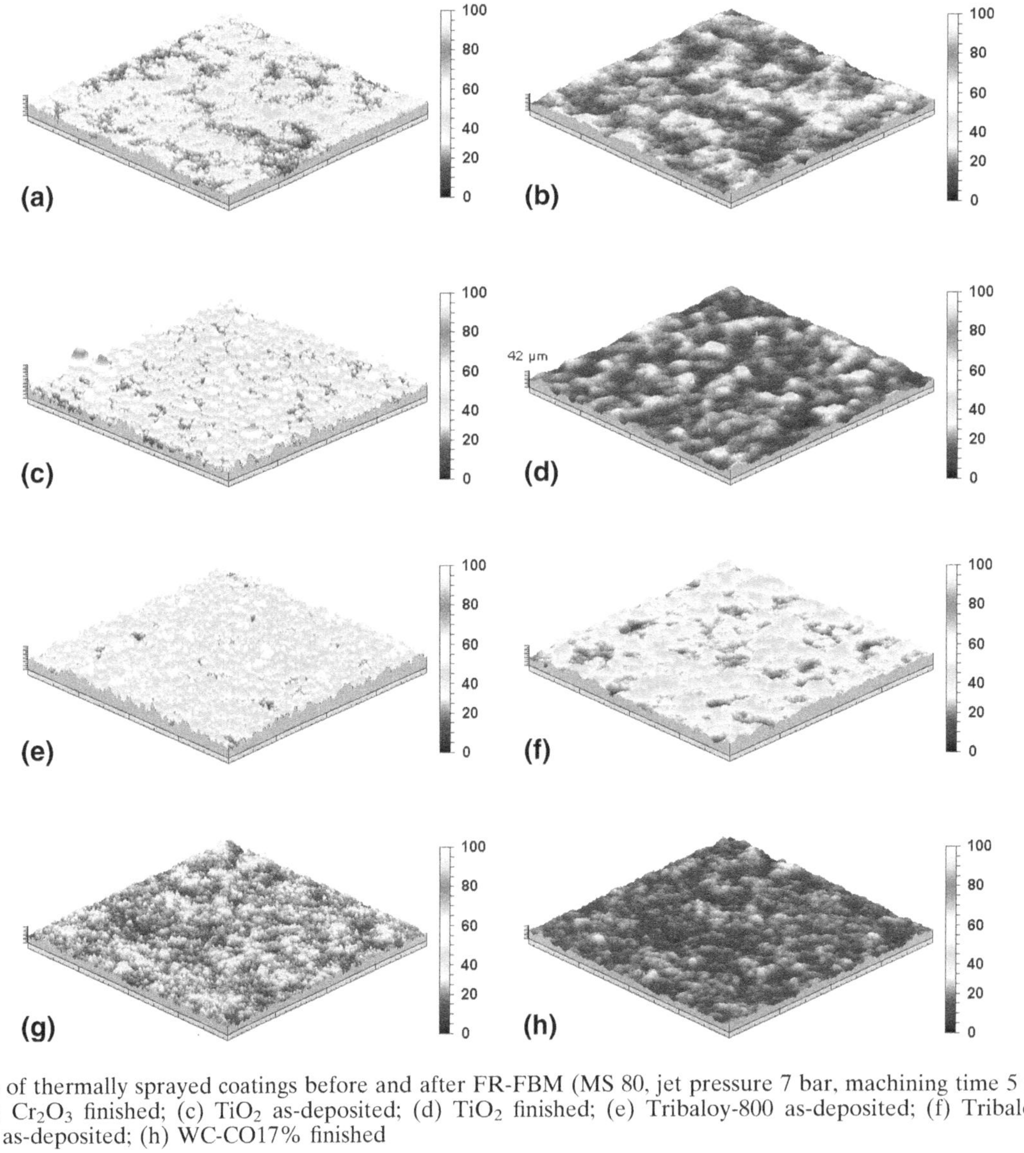

Fig. 9 3D maps of thermally sprayed coatings before and after FR-FBM (MS 80, jet pressure 7 bar, machining time 5 min): (a) Cr_2O_3 as-deposited; (b) Cr_2O_3 finished; (c) TiO_2 as-deposited; (d) TiO_2 finished; (e) Tribaloy-800 as-deposited; (f) Tribaloy-800 finished; (g) WC-CO17% as-deposited; (h) WC-CO17% finished

This result is ascribable to the presence of widespread microcraters all over the finished surface due to the high energetic impacts of the fluidized abrasives projected at high speed toward the surface to be finished. This way, the finished surface acquires a typical texture due to the characteristics of the FR-FBM.

This can be also the reason why all the materials finished in such FR-FBM conditions (MS 46 or 80, jet pressure of 7 bar, machining time of 2 or 5 min) approach nearly the same value of the R_a apart from the starting surface morphology.

The trend of skewness reveals further crucial aspects about the FR-FBM of the thermally sprayed coatings.

R_{Sk} turns negative at any time sufficiently energetic machining is performed. R_{Sk} negative means surface profile with prevalently upward material, that is, rather smooth surface. As seen, the result of the repeated actions of the abrasives all over the surface can result in a significant change in the overall morphology of the as-deposited thermally sprayed coatings even at a large scale. The achievable variations are displayed in the 3D maps reported in Fig. 9, which report an overview of the status of the thermally sprayed coatings before and after FR-FBM. As can be clearly appreciated, a significant smoothening of the surface finish can always be claimed whatever the coatings material, with an apparent flattening of the coarser morphological features, which initially spread all over the starting surface morphology, and the corresponding establishment of a quite smooth surface finish.

4. Conclusions

The present study deals with an application of FR-FBM for finishing of different thermally sprayed coatings. From the examination of the experimental findings, the following conclusions can be drawn:

(i) Machining capability of FR-FBM depends on the mechanical properties of the thermally sprayed coatings and, in particular, on the hardness at low jet pressure and on the toughness at high jet pressure. The best surface finish improvements are achieved when ductile grooving mechanisms occur. Therefore, harder and tougher materials are troublesome to machine. Analogously, excessively brittle materials, prone to undergo brittle fracture, are difficult to process.

(ii) Significant improvement in surface finish can be detected for TiO_2 and Tribaloy-800 coatings, which are know to present moderate hardness and sufficient toughness. Indeed, machining takes place by ductile grooving. In such cases, improvement in average roughness from 250 to 400% can be claimed.

(iii) The surface finishing of the thermally sprayed coatings, matter of the present investigation, always improves, thus approaching an average roughness of from 1.6 to 2.4 μm independently of the starting morphology and the material properties.

(iv) FR-FBM is a first step toward the automatization of the finishing process of thermally sprayed coatings, which, at the present, are only hand-machined by skilled technicians with recursive time- and cost-consuming procedures.

To conclude, FR-FBM can be a valid and easy-to-automate alternative to the handmade finishing techniques of the thermally sprayed coatings. Further research and optimization is certainly needed to optimize processing conditions and surface finishing of the tested coatings. In particular, future analyses will regard the application of FR-FBM to the finishing of complex-shaped components.

Acknowledgments

The authors wish to thank Mr. Ceccarelli (Università di Roma 'Tor Vergata') for the setup of the apparatus. They also wish to acknowledge Mr. Bassani and Mr. Pighetti (Università di Modena e Reggio Emilia) for the preparation of the samples.

References

1. C.M. Verpoort and C. Gerdes, *Shot Peening, Theory and Applications*, IITT International, Gorunay-Sur-Marne, France, 1989, p 11-70
2. T. Bernecki, *Handbook of Thermal Spray Technology*, J.R. Davis, Ed., 14-35, 2004 (Materials Park, OH), ASM International, 2004
3. Z.W. Zhong, Z.F. Peng, and N. Liu, Surface Roughness Characterization of Thermally Sprayed and Precision Machined WC–Co and Alloy-625 Coatings, *Mater. Charact.*, 2007, **58**, p 997-1005
4. G. Bolelli, V. Cannillo, L. Lusvarghi, F. Pighetti Mantini, E. Gualtieri, and C. Menozzi, An FIB Study of Sharp Indentation Testing on Plasma-Sprayed TiO_2, *Mater. Lett.*, doi:10.1016/j.matlet.2007.09.022
5. G. Bolelli, V. Cannillo, L. Lusvarghi, and T. Manfredini, Wear Behaviour of Thermally Sprayed Ceramic Oxide Coatings, *Wear*, 2006, **261**, p 1298-1315
6. G. Bolelli, V. Cannillo, L. Lusvarghi, and S. Riccò, Mechanical and Tribological Properties of Electrolytic Hard Chrome and HVOF-Sprayed Coatings, *Surf. Coat. Tech.*, 2006, **200**, p 2995-3009
7. G. Bolelli, V. Cannillo, L. Lusvarghi, M. Montorsi, F. Pighetti Mantini, and M Barletta, Microstructural and Tribological Comparison of HVOF-Sprayed and Post-treated M-Mo-Cr-Si (M = Co, Ni) Alloy Coatings, *Wear*, 2007, **263**, p 1397-1416
8. M. Barletta and V. Tagliaferri, Development of an Abrasive Jet Machining System Assisted by Two Fluidized Beds for Internal Polishing of Circular Tubes, *Int. J. Mach. Tool Manuf.*, 2006, **46**(3-4), p 271-283
9. M. Barletta, S. Guarino, G. Rubino, and V. Tagliaferri, Progress in Fluidized Bed Assisted Abrasive Jet Machining (FB-AJM): Internal Polishing of Aluminium Tubes, *Int. J. Mach. Tool Manuf.*, 2007, **47**(3-4), p 483-495
10. M. Barletta, D. Ceccarelli, S. Guarino, and V. Tagliaferri, Fluidized Bed Assisted Abrasive Jet Machining (FB-AJM): Precision Internal Finishing of Inconel 718 Components, *J. Manuf. Sci. E-T ASME*, 2007, **129**, p 1-15
11. H.M. Hawthorne, B. Arsenault, J.P. Immarigeon, J.G. Legoux, and V.R. Parameswaran, Comparison of Slurry and Dry Erosion Behaviour of Some HVOF Thermal Sprayed Coatings, *Wear*, 1999, **225-229**, p 825-834

Peer Reviewed

12. P. Kulu, I. Hussainova, and R. Veinthal, Solid Particle Erosion of Thermal Sprayed Coatings, *Wear*, 2005, **258**, p 488-496
13. S. Kamnis, S. Gu, T.J. Lu, and C. Chen, Computational Simulation of Thermally Sprayed WC–Co Powder, *Comput. Mater. Sci.*, 2008, **43**(4), p 1172-1182
14. W. Trompetter, M. Hyland, D. McGrouther, P. Munroe, and A. Markwitz, Effect of the Substrate Hardness on Particle Morphology in High Velocity Thermal Spray Coatings, *J. Therm. Spray Technol.*, 2006, **15**, p 663-669
15. R. Westergård, N. Axén, U. Wiklund, and S. Hogmark, An Evaluation of Plasma Sprayed Ceramic Coatings by Erosion, Abrasion and Bend Testing, *Wear*, 2000, **246**, p 12-19
16. X. Liu, B. Zhang, and Z. Deng, Grinding of Nanostructured Ceramic Coatings: Surface Observations and Material Removal Mechanisms, *Int. J. Mach. Tool Manuf.*, 2002, **42**, p 1665-1676
17. G. Bolelli, L. Lusvarghi, T. Manfredini, F Pighetti Mantini, R Polini, E Turunen, T Varis, and S-P Hannula, Comparison Between Plasma- and HVOF-Sprayed Ceramic Coatings. Part I: Microstructure and Mechanical Properties, *Int. J. Surf. Sci. Eng.*, 2007, **1**(1), p 38-61
18. M. Barletta, A New Technology in Surface Finishing: Fluidized Bed Machining (FBM) of Aluminium Alloys, *J. Mater. Process. Tech.*, 2006, **173**, p 157-165

JTTEE5 17:805–811
DOI: 10.1007/s11666-008-9246-5
1059-9630/$19.00 © ASM International

Peer Reviewed

Development and Investigation on New Composite and Ceramic Coatings as Possible Abradable Seals

U. Bardi, C. Giolli, A. Scrivani, G. Rizzi, F. Borgioli, A. Fossati, K. Partes, T. Seefeld, D. Sporer, and A. Refke

(*Submitted June 25, 2008; in revised form August 18, 2008*)

To improve gas turbine performance, it is possible to decrease back flow gases in the high-temperature combustion region of the turbo machine by reducing the shroud/rotor gap. Thick and porous thermal barrier coating (TBC) systems and composite CoNiCrAlY/Al_2O_3 coatings made by air plasma spray and composite NiCrAlY/graphite coatings made by laser cladding were studied as possible high-temperature abradable seal on shroud. Oxidation and thermal fatigue resistance of the coatings were assessed by means of isothermal and cyclic oxidation tests. Tested CoNiCrAlY/Al_2O_3 and NiCrAlY/graphite coatings after 1000 h at 1100 °C do not show noticeable microstructural modification. The oxidation resistance of the new composite coatings satisfied original equipment manufacturer (OEM) specifications. Thick and porous TBC systems passed the thermal fatigue test according to the considered OEM procedures. According to the OEM specifications for abradable coatings, the hardness evaluation suggests that these kinds of coatings must be used with abrasive-tipped blades. Thick and porous TBC coating has shown good abradability using tipped blades.

Keywords abradable coatings, air plasma spray, laser cladding

1. Introduction

New generation of gas turbine engines concentrate on performance gains, in particular by developing new materials that allow operating under much more severe conditions: superalloys and protective coatings are a permanent field of interest for engine designers (Ref 1). Gas turbine efficiency can be improved by increasing working temperatures, and also by improving dynamic sealing between rotating and stationary parts. One type of improved sealing being incorporated into turbines is abradable seals, to reduce the blade-tip clearances. Abradable materials must be able to withstand high-temperature erosion and oxidation, as well as to provide easy and controlled abrasion (Ref 2). The locations of abradable coatings include seals over compressor blades and seals over unshrouded first stage turbine blades and shrouded second and third stage turbine blades for industrial gas turbines (Ref 3). The commercial abradable coatings for compressor shrouds consist of polymer, AlSi-Polymer, metal matrix with solid lubricant, and MCrAlY with polyester, which withstand temperatures lower than 900 °C; abradable coatings for turbine shrouds are produced by using MCrAlY with polyester and ceramics that withstand temperatures greater than 900 °C. This kind of high-temperature abradable coatings must be used with blades tipped with abrasive coatings (Ref 4).

This work is focused on the development of abradable coatings to be applied on the first stage of industrial land base gas turbines where the temperatures are at the highest (Ref 5) for this type of engine. Composite metal-ceramic coatings were developed to combine the high-temperature oxidation (HTO) resistance of ceramic materials with the good mechanical properties of metal alloys, which should avoid the problems related to ceramic brittleness. Composite CoNiCrAlY/Al_2O_3 coatings made by air plasma spray (APS) and composite NiCrAlY/Graphite coatings made by laser cladding were produced and tested as a possible high-temperature abradable seal on shroud. The ceramic filler particles embedded in the

This article is an invited paper selected from presentations at the 2008 International Thermal Spray Conference and has been expanded from the original presentation. It is simultaneously published in *Thermal Spray Crossing Borders, Proceedings of the 2008 International Thermal Spray Conference*, Maastricht, The Netherlands, June 2-4, 2008, Basil R. Marple, Margaret M. Hyland, Yuk-Chiu Lau, Chang-Jiu Li, Rogerio S. Lima, and Ghislain Montavon, Ed., ASM International, Materials Park, OH, 2008.

U. Bardi, Dipartimento di Chimica, Università di Firenze, Via della Lastruccia 3, Sesto Fiorentino, 50019 Firenze, Italy; **U. Bardi** and **C. Giolli**, Consorzio Interuniversitario Nazionale per la Scienza e Tecnologia dei Materiali (INSTM), via Benedetto Varchi 59, 50132 Firenze, Italy; **C. Giolli, A. Scrivani,** and **G. Rizzi**, Turbocoating S.p.A., via Mistrali 3, 43010 Rubbiano di Solignano, Italy; **F. Borgioli** and **A. Fossati**, Dipartimento di Ingegneria Civile, Università di Firenze, via S. Marta 3, 50139 Firenze, Italy; **K. Partes** and **T. Seefeld**, Bremer Institut für angewandte Strahltechnik BIAS, Klagenfurterstr. 2, 28359 Bremen, Germany; **D. Sporer**, Sulzer Metco (Canada) Inc., Wies 21a, 6677 Schattwald, Austria; and **A. Refke**, Sulzer Metco AG (Switzerland), Rigackerstr. 16, 5610 Wohlen, Switzerland. Contact e-mail: carlogiolli@turbocoating.it.

metal matrix should improve the abradability due to their brittleness as it occurs for metal matrix with solid lubricant (Ref 4). Thick and porous thermal barrier coating (TBC) systems were also produced by APS without using plastic filler (Ref 6, 7) in the sprayed powder. The oxidation resistance of composite coatings and thermal fatigue resistance of thick and porous TBC systems were assessed according to original engine manufacturer (OEM) specifications by means of isothermal and cyclic oxidation tests, and their abradability was evaluated.

2. Experimental

2.1 Materials and Deposition Process

Hastelloy X (HX) samples (Ni-22Cr-19Fe-9Mo-2Co-1W), in the form of metal squared substrates (25 × 25 mm; thickness: 3 mm), were coated by APS. AMDRY 995 (Co-32Ni-21Cr-8Al-0.5Y) and Al_2O_3 powders were used. In the first stage, plasma spray parameters were adapted to the CoNiCrAlY powder through preliminary tests to obtain an average and homogeneous porosity of the coating of 16%. The resulting coating hardness was found to be 86.5 ± 1.5 HR 15Y. In fact, it is believed that abradable coatings benefit by a fairly high porosity, but this point has to be balanced against oxidation resistance (Ref 8). By using these selected spraying parameters, further spraying tests with both alumina and CoNiCrAlY powders were carried out. In the first tests, finer alumina, with an average size of −22 + 5 μm, was sprayed with a CoNiCrAlY powder in the ratio of 40% by weight of ceramic (Sample 1). Another set of samples (Sample 2) was obtained by spraying a coarser alumina powder, with an average size of −90 + 45 μm, again with a proportion of 40 wt.% of ceramic. In a third set of samples (Sample 3), a spheroidical ceramic powder, with an average size of −90 + 45 μm, in the same ratio, was used in order to determine whether the powder shape affects the coating properties.

Squared substrates (25 × 25 mm; thickness: 3 mm) of HX were also coated with a mixture of NiCrAlY and Ni/Graphite powders by laser cladding. Gas atomized AMDRY 9625 (Ni-22Cr-10Al-1.0Y) powder was used with a grain size distribution of −75 + 45 μm. The grain size distribution of the Ni/Graphite (60/40) powder (Durabrade™ 2200) was −115 + 45 μm. The powders were mixed inline with an Y-connecter (Sample 4). The laser cladded surface was generated using a solid-state Nd:Yag laser "Trumpf HL 4006 D" with a maximum output power of 4 kW. Powder was applied by a pneumatic powder feeding device in combination with an off axial nozzle. The laser power was 1200 W and the scan velocity was 1.2 m/min. The powder feed rate was 18 g/min for the MCrAlY and 2 g/min for the Ni/Graphite. The track offset was found to be optimal at 0.3 mm to realize the most constant clad height and a homogeneous dilution at the interface. The specimen (25 × 25 × 3 mm) was clad all over the upper (25 × 25 mm) surface.

A new APS process for making thick and porous TBC systems with a total thickness of up to 2 mm and a porosity level of the top coat of up to 30% without using plastic filler in the sprayed powder was developed (Ref 6). Three different kinds of TBC systems were coated by APS on circular substrates (diameter: 25 mm; thickness: 3 mm) of HX with the same bond coat of AMDRY 995 (average thickness of 250 μm) and with a top coat of yttria partially stabilized zirconia powder (Amperit 827.7 produced by H.C. Starck) (average thickness of about 1.8 mm) and low (17 ± 1%) (Sample 5), medium (21 ± 1%) (Sample 6), and high (29 ± 1%) (Sample 7) porosity levels of the top coat as described in previous papers [6, 7].

2.2 High-Temperature Oxidation Test and Furnace Cycling Test

The HTO tests were carried out in a Carbolite furnace (model CWF 13/5) with a maximum operating temperature of 1300 °C and a heating rate of 15 °C/min. Isothermal oxidation tests were performed at 1100 °C for 300 and 1000 h. Furnace cycle tests (FCTs) were performed using a test equipment consisting of an isothermal static air furnace (type: 3 zone split tube; T_{max}: 1300 °C), a specimen tray in Hastelloy X positioned on a vertical elevator, and a circular tube for the forced cooling of specimens when the elevator is lowered. Each thermal cycle in the FCT consisted of a 5-min heat up to the steady-state temperature, a 45-min soak at the steady-state temperature, and a 10-min forced air cool down. According to the OEM specifications, the chosen steady-state temperature was 1150 °C (Ref 5), and the minimum requested number of thermal cycles to pass the thermal cycling fatigue test was 250.

2.3 Microstructural and Phase Analysis

Samples before and after the oxidation tests were prepared for micrographic investigations by vacuum impregnation with epoxy, and then sectioned, ground, and finely polished (Ref 9). Microscopy analysis was performed by means of a metallographic optical microscope Zeiss Axiovert 100A with an image analyzer based on gray contrast, to determine powder morphology, coating microstructure, coating thickness, and coating porosity. Micrographic examinations and microprobe analysis were performed by ESEM QUANTA 200 FEI with EDAX-ZAF Quantification (Standardless).

X-ray diffraction analysis was performed by means of a Philips PW1050/37 diffractometer in Bragg-Brentano configuration (Cu Kα radiation generated at 40 kV and 25 mA). Diffraction patterns were analyzed by means of the program MAUD using the Rietveld method (Ref 10).

2.4 Hardness and Abradability Tests

Coating hardness was measured on a HR 15Y scale with a 12.5 mm steel ball and 15 kg major load. Before performing the tests, the coatings were carefully hand sanded with 60-grit SiC paper. According to the OEM specifications for abradable coatings, the recommended hardness must be about 55-65 HR 15Y (for use with untipped blades) and 90-95 HR 15Y (for use with abrasive-tipped blades).

AISI 304 stainless steel samples (35 × 100 × 4 mm) were coated with the studied systems all over one 35 × 100 mm surface to be used as samples for abradability rig testing. In the function of the incursion rate, the test can last between 0.5 s and 10 min. So, it is possible to use stainless steel samples instead of nickel superalloy because this does not affect the test results (Ref 11). The tests were performed according to Sulzer Metco specifications. The rig consisted of a rotor, a movable specimen stage, and a heating device as shown in Fig. 1.

The infrastructure of the test equipment is similar to a large turbine balancing facility. For the high-temperature blade tests, a nickel superalloy disc was employed. The disc had 60 blade slots, 4 of which were furnished with a special attachment for dummy blades. Tipped c-BN and untipped dummy blades were used for the tests. The blade mass determines the maximum attainable blade tip velocity. Blade tip velocities up to 420 m/s were possible with Sulzer dummy blades. The blade tip velocities used were 250, 350, and 410 m/s (Ref 5, 11). All tests were conducted with an incursion rate in a range of 5-500 μm/s (Ref 5, 11).

The shroud specimen surfaces were heated by means of a high-velocity burner. A thermal gradient similar to that present in an engine was attained. The test temperature was about 1100 °C. The abradable specimen has fine thermocouples immediately below the surface to ensure that abradable surface temperature, and not flame radiation effects, is recorded. The thermocouples lie on the center line forming a cross. The blade height change was measured using a jig, which allowed the measurement of the blade height from a reference point. The jig was rigged with a digital micrometer. The setup was calibrated with a premeasured calibration blade prior to each measurement series. The measurement error was lower than 2 μm. The blade wear was evaluated as a percentage of a total incursion. Positive values describe wear of the blade whereas negative ones show transfer from the shroud. Therefore, a value of 100 exhibits no incursion into the coating and the total blade wear; on the other hand, a value of 0 indicates 100% incursion into the coating without any blade wear. The test conditions are reported in Table 1 for each kind of analyzed sample.

3. Results and Discussion

3.1 Coatings Characterization

Figure 2(a) shows the optical micrographs of a CoNiCrAlY coating sprayed with coarser alumina powders (Sample 1). Alumina is embedded in the metal matrix (30 ± 3% as evaluated by image analysis) as an interrupted layer. The alumina deposit between the passages of the torch is evident (dark lines). The different densities of the materials probably result in different behaviors during the flight in the plasma jet. Figure 2(b) shows the optical micrograph of a CoNiCrAlY coating sprayed with a finer alumina powder injected separately (Sample 2). These samples show a good homogeneity in the alumina distribution in the metal matrix in comparison to other samples. The alumina amount is lower in comparison with the other samples (22 ± 3%). Figure 2(c) shows the optical micrograph of a CoNiCrAlY coating sprayed with a spheroidical alumina powder (Sample 3). In this coating, alumina is grouped into clusters (30 ± 3%) and not uniformly dispersed in the metal matrix. Hardness values for all the samples are in a range of 84-98 HR 15Y. The presence of alumina resulted in a light increase of the

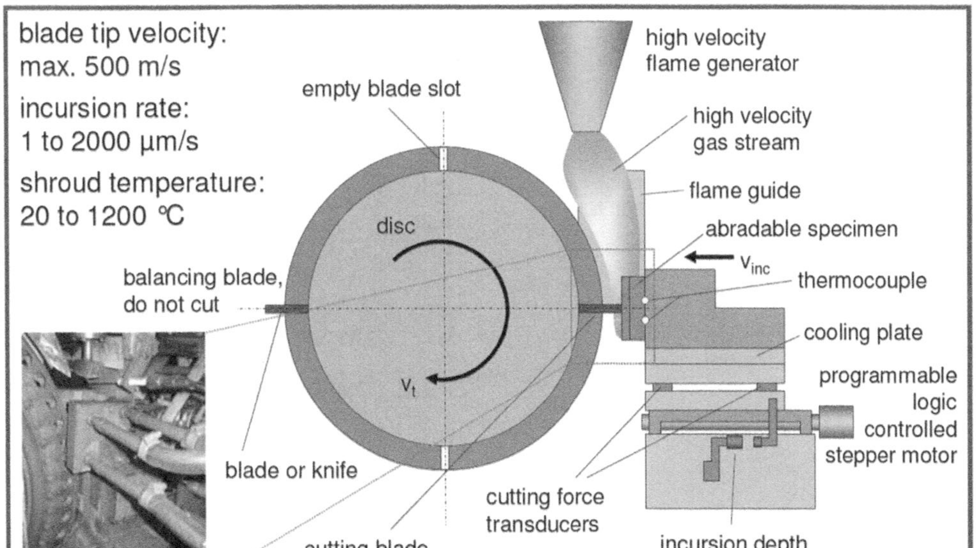

Fig. 1 Schematic of the test rig for abradability testing

Table 1 Abradability test conditions and results

Test No.	Abradable type	Remarks	Sample No.	Blade material and tipping	RPM	Velocity, m/s	Inc. rate, μm/s	Shroud wear track length, mm	Inc. depth, mm	Blade height change, mm	Total incursion depth, mm	Blade wear as % of incursion
1	MCrAlY/Alumina		2	IN718 none	9912	410	5	10.0	0.032	−0.41	0.442	92.8
2			2	IN718 cBN	9912	410	5	34.5	0.377	−0.16	0.537	29.8
3			2	IN718 cBN	9912	410	500	42.0	0.559	0.08	0.559	−14.3
4			2	IN718 cBN	8461	350	50	39.0	0.482	−0.22	0.702	31.4
5	MCrAlY/Graphite		4	IN718 none	9912	410	5	8.0	0.020	−0.43	0.450	95.5
6			4	IN718 cBN	9912	410	5	50.0	0.792	0.05	0.792	−6.3
7			4	IN718 cBN	9912	410	500	25.5	0.198	−0.72	0.918	78.4
8			4	IN718 cBN	8461	350	50	27.0	0.231	−0.04	0.271	14.8
9			4	IN718 cBN	6044	250	5	37.0	0.433	−0.20	0.633	31.6
10	Thick, porous TBC	Low porosity	5	IN718 cBN	6044	250	5	27.0	0.231	−0.70	0.931	75.2
11			5	IN718 cBN	8461	350	50	38.5	0.469	−0.36	0.829	43.4
12			5	IN718 cBN	9912	410	500	34.0	0.366	−0.49	0.856	57.2
13		Medium porosity	6	IN718 cBN	6044	250	5	58.0	1.066	−0.06	1.126	5.3
14			6	IN718 cBN	8461	350	50	51.5	0.840	−0.01	0.850	1.2
15			6	IN718 cBN	9912	410	500	51.5	0.840	−0.03	0.870	3.4
16		High porosity	7	IN718 cBN	6044	250	5	42.0	0.559	−0.03	0.589	5.1
17			7	IN718 cBN	8461	350	50	50.0	0.792	−0.01	0.802	1.2
18			7	IN718 cBN	9912	410	500	51.0	0.824	−0.01	0.834	1.2

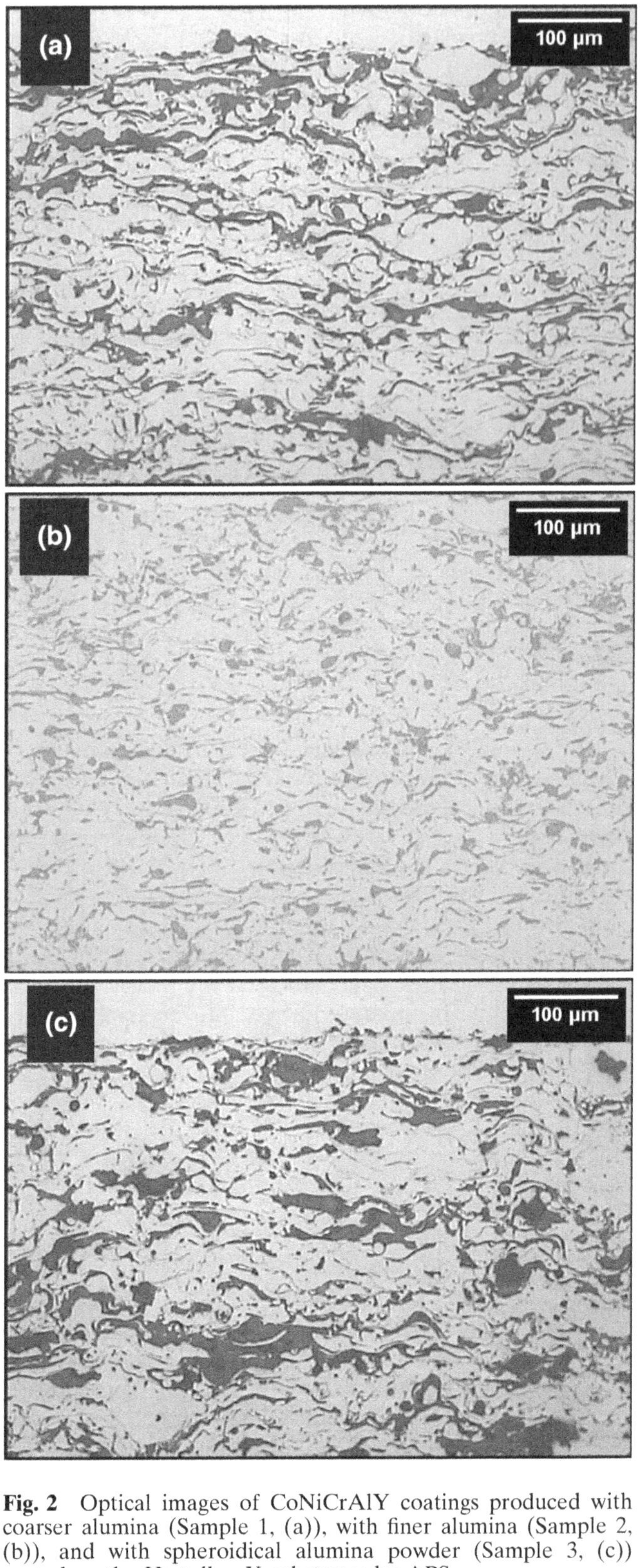

Fig. 2 Optical images of CoNiCrAlY coatings produced with coarser alumina (Sample 1, (a)), with finer alumina (Sample 2, (b)), and with spheroidical alumina powder (Sample 3, (c)) coated on the Hastelloy X substrates by APS

hardness, due to the embedding of the alumina in the metal matrix. Image analysis indicates that the porosity level is not higher than in the CoNiCrAlY coating without alumina. The average thickness of the manufactured

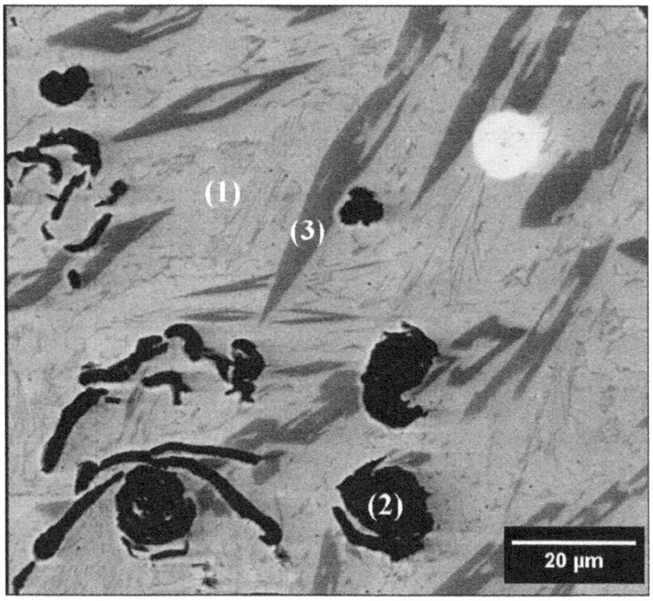

Fig. 3 Sample 4 (AMDRY 9625/graphite) as-coated: (1) Metal matrix, (2) graphite nodules, and (3) brittle phases (source BIAS)

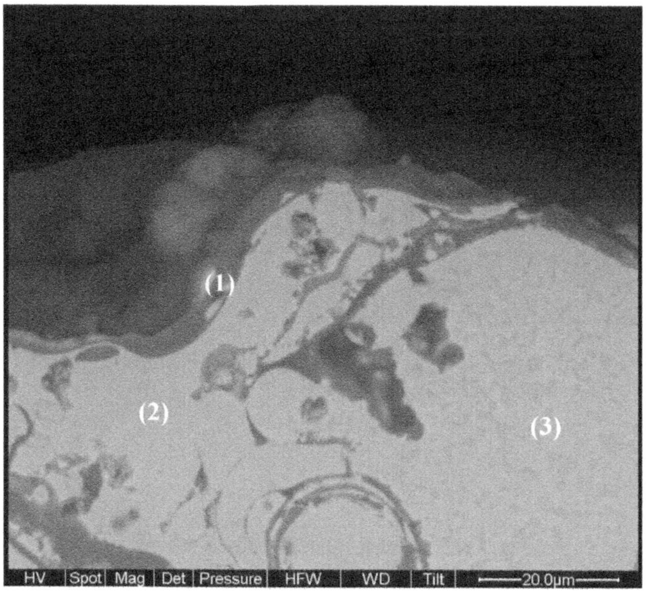

Fig. 4 Sample 2 after 1100 °C for 1000 h: 3000× BSE micrography of sample in section shows: (1) Al_2O_3 protective layer, (2) Metal matrix of CoNiCrAlY depleted in Al, and (3) Metal matrix

samples is about 940 ± 45 µm. Figure 3 shows the back scattered electron (BSE) image of NiCrAlY with added graphite after the laser cladding deposition process (Sample 4). The coating shows a multiphase structure. X-ray diffraction analysis shows the presence of a γ-Ni (c.f.c.) solid solution, a nickel-aluminum intermetallic phase, γ′-Ni_3Al (c.f.c.), and graphite (hex.); small amounts of nickel, Ni_3C (rhomb.), and chromium, Cr_3C_2 (orth.) and Cr_7C_3 (orth.), carbides are also detected. Very small amounts of oxides, α-Al_2O_3 (alumina type; rhomb.), Y_2O_3 (cubic), Cr_2O_3 (rhomb.), and Ni(Cr, Al)$_2$O$_4$ (cubic), are also observed, probably due to surface oxidation. The SEM/EDAX analysis (Fig. 3) suggests the presence of a metallic matrix (the lighter phase, Fig. 3(1)) with embedded graphite nodules (the black ones, Fig. 3(2)) and precipitates rich in chromium (the gray ones, Fig. 3(3)), which being brittle phases should improve the abradability of the metal matrix. The thickness of the coating is about 1000 ± 40 µm. The resulting hardness value is about 86 ± 5 HR 15Y.

TBC systems with different top coat porosity show an average thickness of about 1800 µm for all the sprayed coatings. As shown in some previous papers, three kinds of samples were made with different top coat porosity levels (Ref 6, 7). Hardness (HR 15Y) values were found to be 87 ± 4 for Sample 5, 84 ± 2 for Sample 6, and 83 ± 4 for Sample 7.

3.2 High-Temperature Oxidation and Thermal Fatigue Behavior

According to OEM specifications, composite coatings must support HTO tests while TBCs must withstand FCT. Samples 1, 2, and 3 coated with CoNiCrAlY and alumina showed a similar behavior when oxidized at high temperatures. A micrograph of the cross section of Sample 2 near the surface after 1000 h at 1100 °C is shown in Fig. 4.

After the oxidation a continuous dark layer formed as shown in Fig. 4. EDAX analysis suggests the formation of an Al_2O_3 layer (Fig. 4(1)). It is also possible to observe the presence of a beta phase at the bottom of the micrograph (Fig. 4(3)) and the presence of a beta phase-depleted zone (Fig. 4(2)) between the zone rich in aluminum and the alumina layer as EDAX analysis confirmed. CoNiCrAlY coatings typically exhibit a two-phase microstructure β + γ. The presence of γ increases the ductility of the coating thereby improving the thermal fatigue resistance. As for β-NiAl coatings, high-temperature exposure resulted in a depletion of Al in the zones near the interface of both the TGO (thermally grown oxide) and the substrate by interdiffusion. As the amount of Al decreases, the β-phase tends to dissolve. For this reason, this phase is often described as an aluminum reservoir, and the coating life is often measured in terms of β depletion (Ref 12). The large amount of Al in the AMDRY 995 metal matrix allowed the formation of a protective alumina layer, which slowed down the diffusion of the oxygen through the coating. Figure 5 shows the back scattered electron (BSE) micrograph of the cross sections of the AMDRY 9625 cladded with graphite after 1100 °C for 1000 h. Figure 5 shows the presence of two layers (a lighter one, Fig. 5(1) and a darker one Fig. 5(2)) at the top of the coating. X-ray diffraction analysis suggests that the oxide layers consist essentially of aluminum oxide, α-Al_2O_3, and spinel type oxide, Ni(Cr,Al)$_2$O$_4$; very small amounts of Cr_2O_3 and Y_2O_3 were also observed. It has to be pointed out that graphite was not detected in the depth explored by the x-ray beam, while the other matrix phases were still observable. Figure 5 also shows the presence of chromium-rich phases and also the absence of the graphite nodules as observed by x-ray diffraction analysis. No relevant variation in the microstructure after oxidation was observed in comparison with the as-coated samples. As mentioned previously, according to the applied OEM specifications, the minimum requested number of thermal cycles to pass the thermal cycling fatigue test is 250 cycles. All the tested samples passed the FCTs despite the different levels of the top coat porosity (Ref 6, 7).

3.3 Abradability Test Results

The abradability test results are summarized in Table 1. The test on an Inconel 718 bare blade at a low incursion rate (Test 1) produced a very high blade wear and transfer of blade material to the shroud. Similar results to the MCrAlY/graphite tests were produced against a c-BN-tipped blade producing partial cutting/smearing of the shroud in the same condition (Test 2) with removal of the tip. Transfer of shroud material to the blade tip, as indicated by a negative overall wear value, arose at maximum incursion conditions with cutting/smearing of the shroud (Test 3). Testing at mid-incursion conditions (50 μm/s) produced localized blade wear and rupture of the shroud (Test 4).

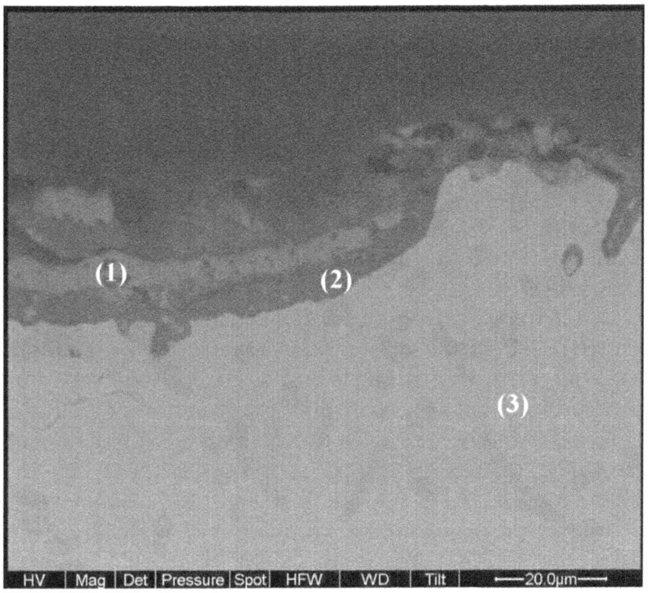

Fig. 5 Sample 4 (AMDRY 9625/graphite) after 1000 h at 1100 °C: (1) $NiAl_2O_4$ layer, (2) Al_2O_3 protective layer, and (3) Metal matrix with embedded brittle phases

The test against an Inconel 718 bare blade at an incursion rate of 5 μm/s (Test 5) produced a very high blade wear and transfer of blade material to the shroud. By contrast, a c-BN-tipped blade produced cutting/smearing of the shroud under the same conditions (Test 6). At maximum incursion condition, the blade was bent by the incursion process. Transfer of shroud material to the blade tip happened at minimum incursion conditions with cutting/smearing of the shroud (Test 6). Testing at mid-incursion conditions produced localized blade wear and rupture of the shroud (Test 8).

Abradability tests were also performed on thick and porous TBC against a c-BN-tipped blade. The sample with low top coat porosity showed very high blade wear and transfer of blade material to the shroud (Tests 10 and 12). For mid-incursion conditions (Test 11), it was possible to observe a lower blade wear. The samples with medium and high top coat porosity in the three different test conditions (Tests 13-18) all show a slight blade wear, with only little difference between them. Figure 6 shows a comparison of the rub paths produced in the low and high porosity thick, porous TBCs.

Composite coatings (Samples 2 and 4) resulted in being too tough, showing cutting/smearing of the shroud or/and localized blade wear and rupture of the shroud coating during the abradability tests. The TBC with a low porosity level (Sample 5) resulted to be too dense to have good abradability characteristics. Thick and porous TBC systems with medium (Sample 6) and high porosity (Sample 7) levels of the top coat showed a good abradability against tipped blades. The tests showed a low blade wear and no shroud coating rupture.

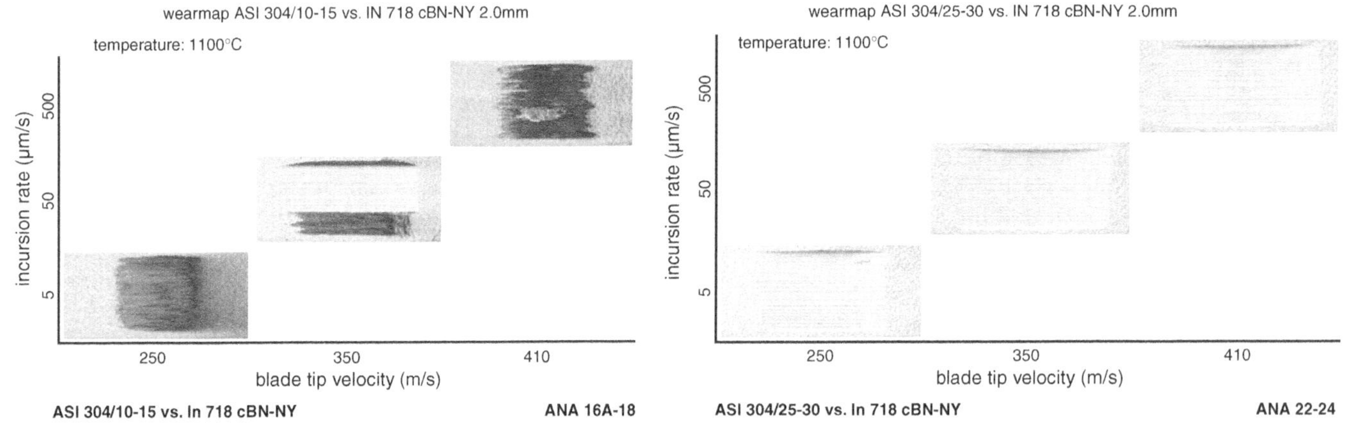

Fig. 6 Appearance of rub paths produced by three different test conditions in low porosity (left) and high porosity (right) thick TBCs with c-BN-tipped blades. Dark rub path appearance is indicative of blade material transfer to the coating

4. Conclusions

New composite metal-ceramic coatings were developed and tested as possible high-temperature resistance abradable materials for the first stage of industrial gas turbines. APS deposition process was used to obtain high-temperature resistant composite CoNiCrAlY/Al_2O_3 coatings. Composite CoNiCrAlY/Graphite coatings were produced by laser cladding. The HTO behavior of the composite coatings was studied. CoNiCrAlY/Al_2O_3 and CoNiCrAlY/graphite coatings after 1000 h at 1100 °C do not show relevant microstructural modifications. The oxidation resistance of the new composite coatings satisfied OEM specifications. The newly developed thick and porous TBC systems passed the thermal fatigue test according to the considered OEM procedures, regardless of the top coat porosity. Hardness Rockwell 15Y measurements suggest that all the produced coatings must be used with abrasive-tipped blades. MCrAlY/graphite and MCrAlY/alumina shrouds were generally too resistant to cutting or resulted in an extensive material transfer to the c-BN-tipped blades. Thick and porous TBC coatings with top coats having medium and high porosity levels have shown good abradability behavior against c-BN-tipped blades.

Acknowledgment

The present work was supported by ABRANEW project RS0098 (FP5 Programme).

References

1. Y. Tamarin, *Protective Coatings for Turbine Blades*, ASM International, Materials Park, OH, 2002
2. M. Dorfman, U. Erning, and J. Mallon, Gas Turbines Use 'Abradable' Coatings for Clearance-Control Seals, *Seal. Technol.*, 2002, **2002**(1), p 7/8
3. 2003 NASA Seal/Secondary Air System Workshop (2003), Nov 5-6, Ohio Aerospace Institute, Cleveland, OH
4. F. Ghasripoor, R. Schmid, and M. Dorfman, Abradables Improve Gas Turbine Efficiency, *Mater. World*, 1997, **5**(6), p 328-330
5. M.P. Boyce, *The Gas Turbine Engineering Handbook*, 2nd ed., Gulf Professional Publishing, Houston, TX, 2001
6. A. Scrivani, G. Rizzi, and C.C. Berndt, Enhanced Thick Thermal Barrier Coatings that Exhibit Varying Porosity, *Mater. Sci. Eng. A*, 2008, **476**(1-2), p 1-7
7. A. Scrivani, G. Rizzi, U. Bardi, C. Giolli, M. Muniz Miranda, S. Ciattini, A. Fossati, and F. Borgioli, Thermal Fatigue Behavior of Thick and Porous Thermal Barrier Coatings Systems, *J. Therm. Spray Technol.*, 2007, **16**(5-6), p 816-821
8. H.B. Guo, R. Vaβen, and D. Stöver, Thermophysical Properties and Thermal Cycling Behavior of Plasma Sprayed Thick Thermal Barrier Coatings, *Surf. Coat. Technol.*, 2005, **192**(1), p 48-56
9. J.R. Davis, et al., *Handbook of Thermal Spray Technology*, ASM International, Materials Park, OH, 2005
10. L. Lutterotti, S. Matthies, and H.R. Wenk, MAUD (Material Analysis Using Diffraction): A User Friendly JAVA Program for Rietveld Texture Analysis and More, *Proceedings of the 12th International Conference on Textures of Materials (ICOTOM-12)*, J.A. Szpunar, Ed., NRC Research Press, Ottawa, 1999, **1**, p 1599-1604
11. R. Schmid and A.R. Nicoll, Advance in Abradable Coatings for Gas Turbines, *Proceedings of International ASME Gas Turbine and Aeroengine Congress and Exposition*, June 13-16, 1994 (The Hague, Netherlands), 1994
12. R. Sivakumar and B.L. Mordike, High Temperature Coatings for Gas Turbine Blades: A Review, *Surf. Coat. Technol.*, 1989, **37**(2), p 139-160

Peer Reviewed

JTTEE5 17:812–817
DOI: 10.1007/s11666-008-9253-6
1059-9630/$19.00 © ASM International

Peer Reviewed

HVOF-Sprayed Nylon-11 + Nanodiamond Composite Coatings: Production & Characterization

A. Stravato, R. Knight, V. Mochalin, and S.C. Picardi

(Submitted July 2, 2008; in revised form August 22, 2008)

High velocity oxy-fuel (HVOF) combustion spraying has previously been shown to be a viable method for depositing polymer and polymer/ceramic composite coatings. The addition of hard particulate reinforcing phases to soft polymeric matrices should improve their durability and sliding wear performance. Nanosized diamond is an ideal reinforcing phase, owing to its high hardness, low coefficient of friction, and desirable thermal properties. Composite coatings comprising a Nylon-11 matrix reinforced with nanodiamonds have been successfully produced by HVOF. An important challenge is preserving the structure of the nanoparticles after thermal spray deposition and achieving their uniform dispersion within the polymeric matrix. Raman spectroscopy and x-ray diffraction were used to confirm the presence and retention of nanodiamonds after HVOF deposition. Understanding of the role of process parameters, including the content of reinforcing phase in the matrix and powder preparation route is necessary. Nanoindentation studies demonstrated an improvement in creep behavior and recovery of the HVOF Nylon-11/nanodiamond composites subjected to deformation.

Keywords functionalization, HVOF spraying of polymers, nanocomposites, nanodiamonds

1. Introduction

Nanoreinforced polymer composites have attracted increased attention because they exhibit unique properties compared to their conventional microreinforced counterparts. Ultradispersed (UDD) or nanocrystalline diamond, often referred to as "nanodiamond" (ND), is an ideal and desirable reinforcing phase due to its nanometer size and unique properties of diamond such as extreme hardness, high Young's modulus, high electrical resistivity, and high thermal conductivity. ND powder is produced at high pressure under nonequilibrium conditions by detonation synthesis from carbon-containing explosives with negative oxygen balance such as a mixture of TNT and RDX (Ref 1). Several applications of ND have recently been explored, including the deposition of wear- and corrosion-resistant metal coatings, additives to cooling fluids, lubricants, and polishing compositions for ultraflat optical or magnetic components (Ref 2, 3).

A key technical challenge in processing these composites is achieving a uniform dispersion and distribution of the nanofiller within the polymer matrix. Another challenge is to improve the interfacial bonding between the nanodiamond particles and polymer matrix. Previous results (Ref 4-6) demonstrated that the high velocity oxy-fuel (HVOF) combustion spray technique can be used to deposit polymers and polymer matrix nanocomposites incorporating ceramic phases such as silica and alumina without any need for volatile solvents. The size and morphology of the nanoparticles strongly affect their rate of heating and acceleration and, hence, the efficiency of deposition as well as structure and properties of the coating. Nanosized powders tend to agglomerate and can easily obstruct the feeding system, adhere to the walls of the nozzle, and be lost during spraying, resulting in a loss of material, and hence money. To avoid these problems, appropriate attention must be paid to preparation of the feedstock powders and optimization of the spraying parameters.

This paper reports on the first successful HVOF deposition of Nylon-11/nanodiamond composite coatings, preserving the structure of the nanodiamonds in the deposit microstructure. Nylon 11 was chosen because of its industrial significance and potential use in the powder coating industry. It is also one of the very few synthetic polymers produced from a natural source—the precursor for Nylon-11 is 11-aminoundecanoic acid monomer, typically manufactured from castor oil.

This article is an invited paper selected from presentations at the 2008 International Thermal Spray Conference and has been expanded from the original presentation. It is simultaneously published in *Thermal Spray Crossing Borders, Proceedings of the 2008 International Thermal Spray Conference*, Maastricht, The Netherlands, June 2-4, 2008, Basil R. Marple, Margaret M. Hyland, Yuk-Chiu Lau, Chang-Jiu Li, Rogerio S. Lima, and Ghislain Montavon, Ed., ASM International, Materials Park, OH, 2008.

A. Stravato, Dipartimento Ingegneria Meccanica e Industriale, Università degli Studi "Roma Tre", via della Vasca Navale 79, 00146, Rome, , Italy; **R. Knight** and **V. Mochalin**, Department of Materials Science and Engineering, Drexel University, 3141 Chestnut Street, Philadelphia, PA; and **S.C. Picardi**, Nanoblox, Inc., Boca Raton, FL. Contact e-mail: knightr@coe.drexel.edu.

2. Experimental Approach

2.1 Materials Selection

Semicrystalline Nylon-11 (Polyamide-11) powder, available commercially as Rilsan® D, "French Natural ES" (donated by Arkema, Inc., King of Prussia, PA) with nominal 80 μm particle size (designated D80), was selected as the feedstock for the experiments.

Nanodiamond (ND) powder (UD90 grade) produced by detonation synthesis was supplied by NanoBlox, Inc., (Boca Raton, FL). The as-produced powder was thoroughly characterized previously (Ref 7) using Raman spectroscopy, TEM, XANES, and FTIR. The ND particles have a diamond sp^3 core, covered by diverse functional groups such as -COOH, -OH, hydrocarbon chains, and surrounded by amorphous and graphitic carbon shells incorporating metal impurities. The as-produced UD90 was purified by the manufacturer through a multistage acid treatment with nitric and sulfuric acids. Additional oxidative purification in air at 425 °C was used to selectively remove the non-diamond carbon phases around the diamond cores of the nanoparticles (Ref 7). Air oxidation was followed by a treatment with boiling aqueous HCl to remove metal impurities, which become accessible after the oxidation of amorphous and graphitic carbon. This treatment, while removing metals and metal oxides from the surface, increases the number of surface carboxylic groups. IR spectra of as-received and purified ND (Fig. 1) reveal an increased number of oxygen-containing COOH, $=(CO)_2O$ (anhydro-), and OH groups. The higher content of ionogenic oxygen-containing surface functional groups results in a better aqueous dispersion stability of the oxidized HCl-treated nanodiamond as compared to as-received and oxidized nanodiamond at pH 7-9.

2.2 Feedstock Powder Preparation

Nylon-11 powder was dry ball-milled together with as-received UD90 nanodiamond powder for 48 h in a Norton Ball Mill using zirconia balls. The milling procedure mechanically embedded the hard reinforcement particles into the surface of the Nylon-11 powder particles. Composite powders with 2.5 to 10 vol.% (7 to 25 wt.%) nanodiamond loadings were produced for subsequent HVOF spray consolidation, as described below.

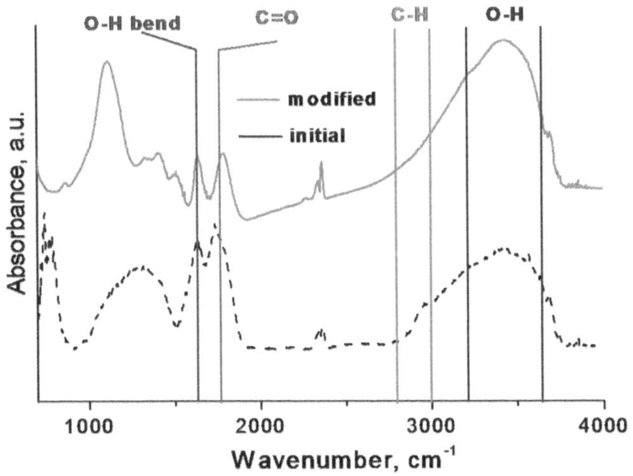

Fig. 1 IR spectra of UD90 powder before and after the air oxidation and HCl treatment

2.3 Thermal Spray Experiments

Spraying of the composite feedstock powders was carried out using a Jet Kote II® HVOF combustion spray system (Deloro Stellite, Inc., Goshen, IN) at oxygen and hydrogen flow rates of 0.0024 and 0.0039 m^3/s (300 and 500 scfh), respectively, and a spray distance of 200 mm (8 in.). Powders were fed using a Praxair Model 1207 volumetric powder feeder (Praxair Surface Technologies, Inc., Indianapolis, IN) with nitrogen as a powder carrier gas at a flow rate of 0.5×10^{-4} m^3 s^{-1} (60 scfh). A 76 mm long by 6 mm ID (3 × ¼ in.) spray nozzle was used. The materials were deposited using a gun traverse speed of ~0.06 m s^{-1}, with external compressed air substrate cooling (410 kPa or 60 psi) with a 6 mm step size between passes. Coatings were deposited onto steel substrates, grit blasted prior to spraying using 50-mesh angular alumina grit from a Trinco model 24/BP2 (Fraser, MI) grit blast cabinet. The substrates were preheated with the HVOF jet to ~180 °C before spraying.

2.4 Coating Characterization

Cross-sectional samples of the sprayed coatings were prepared using standard metallographic techniques: sectioning, mounting, and polishing with 400 and 600 grit SiC papers and 5 μm alumina powder.

Raman spectra were recorded using a Renishaw Model 1000 spectrometer (Renishaw, UK) with an excitation wavelength of 325 nm (He-Cd laser) in a backscatter configuration. Each Raman spectrum was obtained as an average of three accumulations recorded from three different spots on the sample with 300 s accumulation time. To minimize the thermal destruction of the samples caused by the UV laser, the samples were immersed in deionized water during exposure to the laser. Spectra were analyzed using the Renishaw *Wire 2.0* software.

X-ray diffraction (XRD) analysis was performed using a Siemens D500 x-ray powder diffractometer (Bruker AXS, Inc., Madison, WI) (Cu Kα, $\lambda = 1.54056$ Å) with a step size of 0.02° (2θ) and a hold time of 1 s. Results were analyzed using the MDI Jade 7 analysis software (Materials Data, Inc., Livermore, CA). Samples for XRD analysis were sprayed on glass substrates, rather than steel, to eliminate x-ray peaks from the steel determined to be close to the expected peaks of diamond.

Room temperature mechanical properties of the sprayed polymer/nanodiamond deposits were studied by nanoindentation using a Nanoindenter XP system (MTS Corp., Oak Ridge, TN) equipped with a continuous stiffness measurement (CSM) attachment. Each sample cross section was indented 10 times using a conical indenter with a 13.5 μm radius spherical tip, at 4000 nm maximum indentation depth with an allowable drift rate of 0.1 nm/s.

Using the Oliver and Pharr model (Ref 8), the contact depth h_c and hence the contact radius a were determined according to the following equations:

$$h_c = h - \varepsilon \cdot (P/S)$$

$$a = \sqrt{2h_c R - h_c^2}$$

where h is the total displacement, S the contact stiffness measured by the CSM, and ε a geometric constant that is 0.75 for a spherical indenter. From Hertz's theory E_r can be calculated as

$$E_r = \frac{\sqrt{\pi} \cdot S}{2\sqrt{A}}$$

where E_r is the reduced modulus, E_r is also given by

$$E_r = \frac{1 - \nu_i^2}{E_i} + \frac{1 - \nu_s^2}{E_s}$$

where i and s refer to the indenter and specimen, respectively.

The hardness H is defined as the indentation load divided by the projected contact area $A = \pi a^2$

$$H = P/A$$

Further details can be found in Ref 8.

3. Results and Discussion

X-ray diffraction (XRD) and Raman spectroscopy were used to confirm the retention and presence of nanodiamonds in the coatings after HVOF spraying. XRD patterns of a 13 wt.% ND-Nylon-11 coating sprayed onto glass are shown in Fig. 2. The peaks at 20.5 and 23.5° are related to the polymer. Curve b (red) shows the XRD diffraction pattern of the nanocrystalline diamond powder (Ref 9, 10). The peaks of ND were present in the XRD spectra of the coating (curve a, blue) confirming the presence of diamond in the deposited coating.

Additional confirmation of the presence of nanodiamonds in the HVOF-sprayed coatings was obtained via UV Raman spectroscopy. The UV Raman spectrum of a typical ND powder exhibits the G band of sp^2 carbon at ~1610 cm^{-1} originating from amorphous carbon, carbon onions, and fullerene-like shells, the disorder-induced double-resonance D band around 1400 cm^{-1}, and the diamond peak at ~1325 cm^{-1} that is downshifted compared to the normal bulk diamond peak at 1332 cm^{-1} (Ref 7, 11). In Fig. 2 the D band is weak and not seen. The Raman spectrum of Nylon-11 overlaps with the spectrum of ND (curve *a* in Fig. 3); moreover, the polymer shows high fluorescence level under UV laser excitation. Therefore, to eliminate the contribution of polymer, the coating was dissolved in a 1:1 (by volume) mixture of formic acid and dichloromethane (Ref 12). The Raman spectra were collected after rinsing the resulting powder several times with fresh solvent (Fig. 3b). The peak at 1325 cm^{-1} shown in Fig. 3(b), observed after the polymer dissolution, was assigned to the ND, thus confirming the presence of the diamond phase in the sprayed coating. The low intensity of the G-band (~1600 cm^{-1}) in Fig. 2(b) led to the conclusion that the HVOF process did not result in any significant graphitization of the nanodiamond.

Figure 4 shows the typical loading-hold-unloading nanoindentation curves for coatings with different nanodiamond contents. When the load is released from the indenter, the material attempts to regain its original shape,

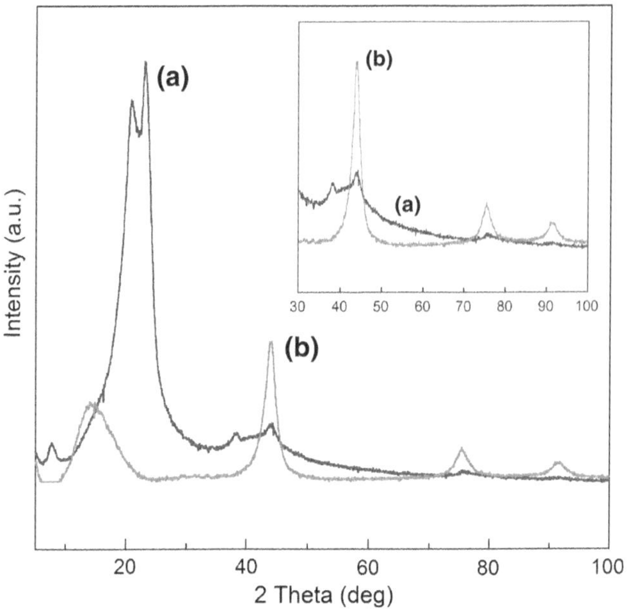

Fig. 2 X-ray spectra of (a) Nylon-11/nanodiamond coating and (b) nanodiamond powder. The inset region shows an enlargement of the three primary peaks of nanodiamond

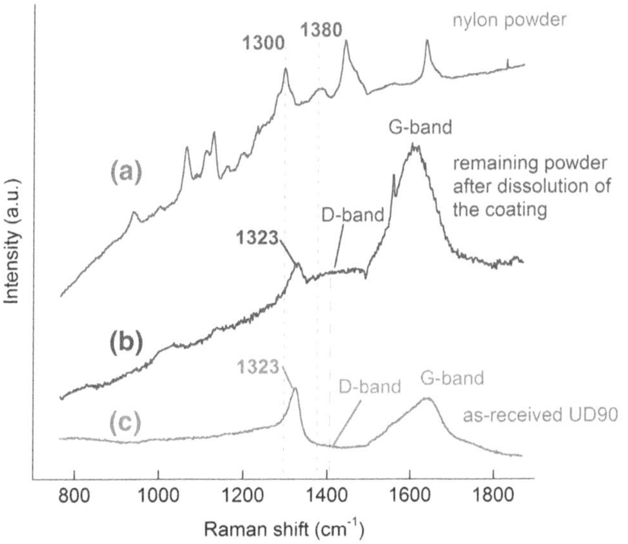

Fig. 3 Raman spectra of coating on a glass substrate (a); powder collected after the dissolution of the polymer in the coating (b); and ND powder (c)

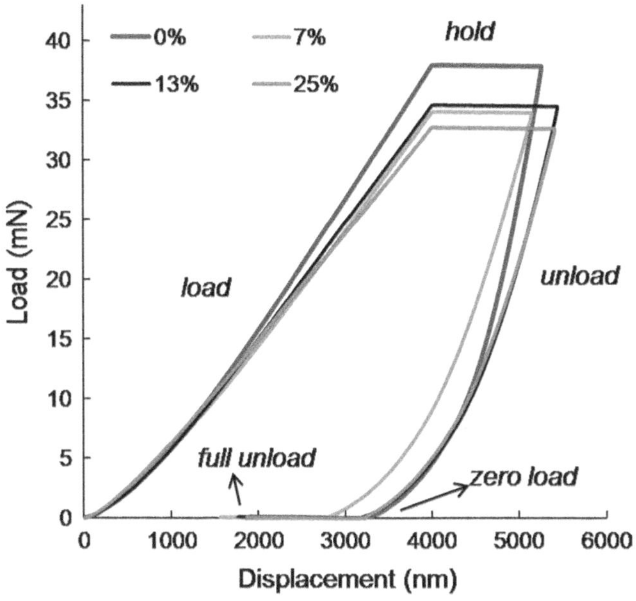

Fig. 4 Loading-hold-displacement curves for HVOF-sprayed Nylon-11/nanodiamond coatings with varying nanodiamond content

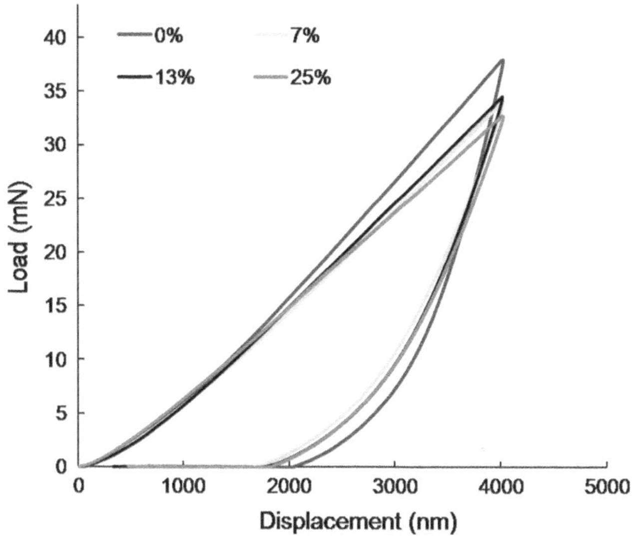

Fig. 5 Load-displacement curves for HVOF-sprayed Nylon-11/nanodiamond coatings with varying nanodiamond content

Table 1 Elastic recovery and average creep rate of the coatings

Nanodiamond loading, wt.%	Recovery at zero load, %	Recovery at full unload, %	Average creep rate, nm/s
0	49 ± 4	82 ± 5	1.21 ± 0.03
7	56 ± 12	88 ± 7	1.12 ± 0.03
13	56 ± 14	93 ± 4	1.41 ± 0.03
25	50 ± 22	87 ± 8	1.32 ± 0.04

but usually the recovery is not 100%, indicating a level of plastic deformation. After the complete removal of the load, the material slowly continued to recover the original shape with a delay due to the relaxation of the polymer chains, giving origin to different recovery values at zero load (removal of the load) and at full unloading (complete relaxation of the elastic strain in the material). To better visualize this recovery behavior, Fig. 5 shows the indentation curves for the coatings without the holding segment. The recovery at zero load and the recovery at full unloading are reported in Table 1. The data are the mean of 10 indents for each sample and are presented as a percentage of the total depth (Recovery * 100/total depth).

The presence of nanodiamond is expected to influence the viscoelastic behavior of the polymer. The nanoparticles in the matrix limit the mobility of the polymer chains and conformation changes. The chains are confined at nanoscale. The presence of nanodiamonds reduces the dissipation of energy during the deformation process so that the coatings with higher nanodiamond contents tend to return more energy at the end of the unloading cycle. This resulted in a slight trend of decrease of the residual depth compared to that for the pure polymer. Thus, the mean values of recovery at zero load and recovery at full unload are systematically higher for samples with nanodiamond as compared to the neat polymer (Table 1). The large values of standard deviations for both these parameters are due to the surface roughness and common problems with precise definition of the sample surface in nanoindentation.

The length of the creep zone at the maximum load divided by time (the average creep rate) is reported in Table 1. There was a slight decrease in the creep rate for the sample with 7 wt.% of ND. The hardness and modulus, calculated from the continuous stiffness measurement (CSM), are reported in Fig. 6. Overall, the improvement in the mechanical properties of the composites, measured by nanoindentation, was not as significant as expected. One of the reasons may be a poor distribution/dispersion of the nanodiamonds in the final coatings. This is a well-known common issue in the realization of the nanocomposites (Ref 13) and still is to be addressed.

Another key property of sprayed coatings is their adhesion to the substrate. For the HVOF coatings produced in this work, their adhesion to the substrate was significantly improved upon addition of the nanodiamond. Figure 7 compares optical images of a coating with 7 wt.% of nanodiamond and a pure Nylon-11 coating sprayed using the same parameters. Small regions of the coatings were peeled off using a razor blade. The sprayed pure Nylon-11 coating was readily peeled off whereas the Nylon-11/nanodiamond composite coating could not be peeled off at all without scratching the underlying metallic substrate, indicating significantly improved adhesion in the latter case.

It is known that many properties of nanoparticles are determined by their surface chemistry (Ref 14). The surface of the nanoparticles governs their dispersion/agglomeration behavior in solvents, interaction with the environment, biocompatibility, etc. Therefore, the full control and manipulation of the surface chemistry is

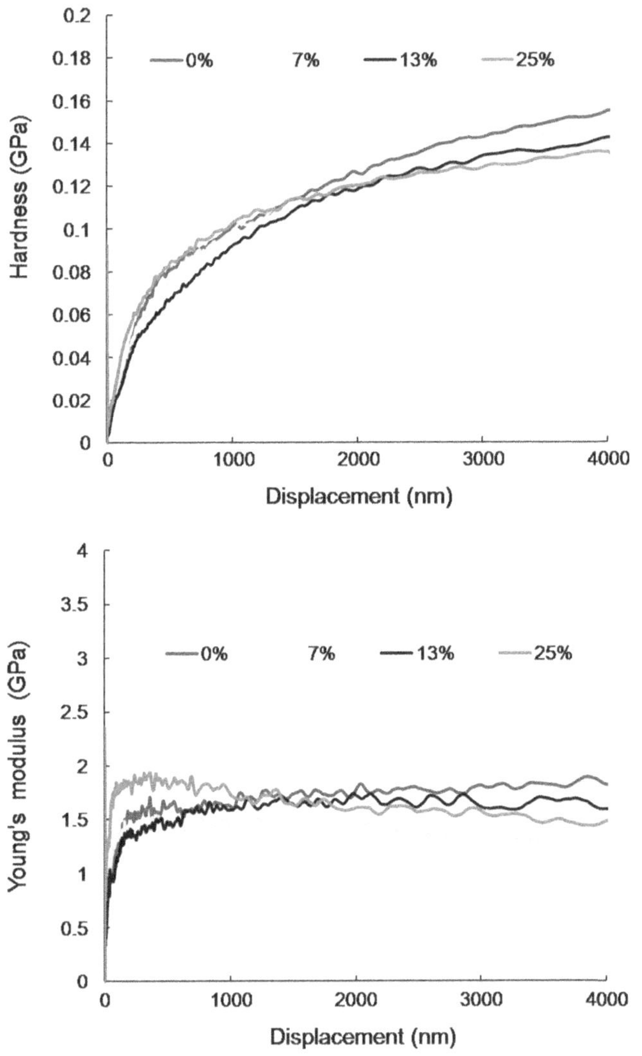

Fig. 6 Hardness (upper) and elastic modulus (lower) versus displacement for HVOF-sprayed Nylon-11/nanodiamond coatings with varying loadings of nanodiamond

important for optimal design of nanodiamond-polymer composites. Having the ability to purify the nanodiamond and control its surface chemistry allowed conversion of the majority of the surface functionalities to -COOH groups (Ref 7). Subsequent treatment with a dilute aqueous HCl, which is a standard step to remove metal impurities, increases the content of surface carboxylic groups through conversion of anhydrides into COOH. Carboxyl groups at the surface of ND are believed to interact with the nitrogen and oxygen atoms of the amide groups in the backbone of the Nylon-11 chain through the formation of hydrogen bonds. The formation of hydrogen bonds results in stronger interactions between the polymer and nanodiamond particles and more uniform distribution of the particles within the polymer matrix.

Scanning electron microscopy images of the surface of the polymer covered with oxidized HCl-treated UD90 (9 wt.% of ND) and as-received UD90 (13 wt.% of ND) are shown in Fig. 8. Despite the lower content of nanodiamonds, the oxidized HCl-treated UD90 shows a better coverage of the surface and a higher level of deagglomeration compared to the as-received one. Therefore, it is believed that substitution of the as-received UD90 with the oxidized and HCl treated UD90 will result in further improvement in properties of HVOF ND-containing Nylon-11 coatings.

4. Summary and Conclusions

This work has demonstrated the feasibility of producing polymer matrix nanocomposites incorporating nanodiamond as the reinforcing phase via thermal spray. X-ray diffraction and Raman spectroscopy confirmed the presence of nanodiamonds in the sprayed deposits. Qualitative assessment indicated that coating adhesion was improved through the addition of the nanodiamond to the Nylon-11.

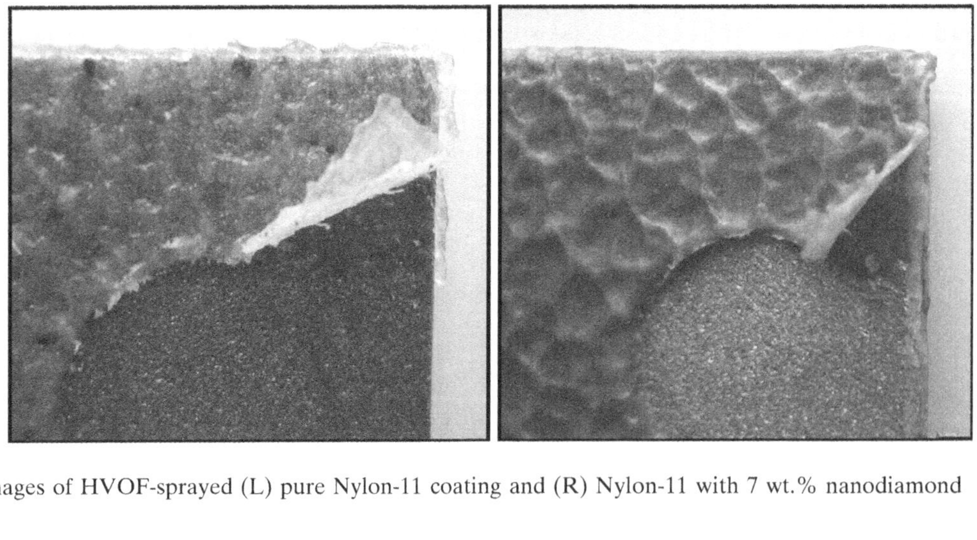

Fig. 7 Optical images of HVOF-sprayed (L) pure Nylon-11 coating and (R) Nylon-11 with 7 wt.% nanodiamond

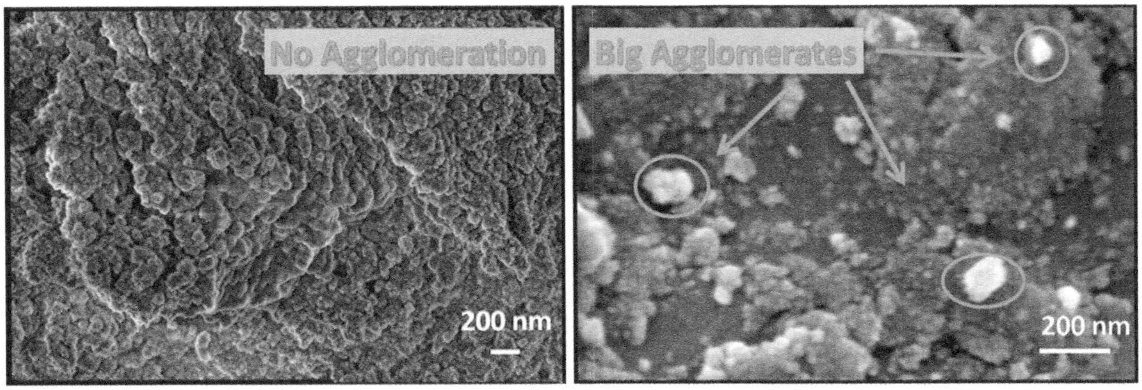

Fig. 8 SEM image of the surface of Nylon-11 covered by oxidized and HCl-treated UD90 (L) and by UD90 (R) as-received

Feedstock preparation is an important aspect and further work is clearly needed to investigate and optimize the process for better dispersion and covalent bonding of the nanodiamond particles to the polymer matrix and for characterizing the improvement in properties as a function of nanodiamond content.

These results indicate significant promise for the HVOF-produced nanodiamond-polymer composites and should lead to future applications for these materials. Nanodiamonds may also be used in place of, or in addition to, more conventional reinforcing phases, such as WC or Cr_3C_2, where their extreme hardness, low coefficient of friction, high thermal conductivity, and modest cost (~$1 per gram) may enable more environmentally friendly and improved wear resistant coatings to be developed.

Acknowledgments

The authors would like to acknowledge NanoBlox®, Inc. for supplying the nanodiamond powders and Arkema, Inc. for donating the Nylon-11 powders used in this work. The authors are grateful for the assistance of Mr. Dustin Doss, Ms. Dee Breger, Dr. Zhorro Nikolov, and the Centralized Research Facilities (CRF) at Drexel University. The authors would also like to thank K. Behler and S. Picchiotti for the helpful discussions.

References

1. V.V. Danilenko, Specific Features of Synthesis of Detonation Nanodiamonds, *Combust. Explo. Shock Waves*, 2005, **41**(5), p 577-588
2. G. Post, V.Y. Dolmatov, V.A. Marchukov, V.G. Sushchev, M.V. Veretennikova, and A.E. Sal'ko, Industrial Synthesis of Ultradisperse Detonation Diamonds and Some Fields of Their Use, *Russ. J. Appl. Chem.*, 2002, **75**(5), p 755-760
3. V. Livramento, J.B. Correia, N. Shohoji, and E. Ōsawa, Nanodiamond as an Effective Reinforcing Component for Nano-Copper, *Diam. Relat. Mater.*, 2007, **16**, p 202-204
4. E. Petrovicova, R. Knight, L.S. Schadler, and T.E. Twardowski, Nylon-11/Silica Nanocomposite Coatings Applied by the HVOF Process I. Microstructure and Morphology, *J. Appl. Polym. Sci.*, 2000, **77**, p 1684-1699
5. E. Petrovicova, R. Knight, L.S. Schadler, and T.E. Twardowski, Nylon-11/Silica Nanocomposite Coatings Applied by the HVOF Process II. Mechanical and Barrier Properties, *J. Appl. Polym. Sci.*, 2000, **78**, p 2272-2289
6. N. Ivosevic, R.A. Cairncross, and R. Knight, Melting and Degradation of Nylon-11 Particles during HVOF Combustion Spraying, *Proceedings of the 2007 International Thermal Spray Conference*, (ITSC-2007), (Beijing), ASM International, 2007, p 820-825
7. S. Osswald, et al., Control of sp^2/sp^3 Carbon Ratio and Surface Chemistry of Nanodiamond Powders by Selective Oxidation in Air, *J. Am. Chem. Soc.*, 2006, **128**(35), p 11635-11642
8. A.C. Fischer-Cripps, A Review of Analysis Methods for Submicron Indentation Testing, *Vacuum*, 2000, **58**, p 569-585
9. V.Y. Shevchenko, A.E. Madison, and G.S. Yur'ev, Structure of Nanodiamonds, *Glass Phys. Chem.*, 2006, **32**, p 261-266
10. Y. Gogotsi, T.K. Klaus, G. Nickel, and M.E. Zvanut, Hydrothermal Behavior of Diamond, *Diam. Relat. Mater.*, 1998, **7**, p 1459-1465
11. A.C. Ferrari and J. Robertson, Raman Spectroscopy of Amorphous, Nanostructured, Diamond-like Carbon, and Nanodiamond, *Phil. Trans. R. Soc. Lond. A*, 2004, **362**, p 2477-2512
12. K. Behler, M. Havel, and Y. Gogotsi, New Solvent for Polyamides and its Application to the Electrospinning of Polyamides 11 and 12, *Polymer*, 2007, **48**, p 6617-6621
13. S.G. Advani, *Processing and Properties of Nanocomposites*, World Scientific Publishing, Hackensack, NJ, 2007
14. V.Y. Dolmatov, Detonation Synthesis Ultradispersed Diamonds: Properties and Applications, *Russ. Chem. Rev.*, 2001, **70**(7), p 607-626

JTTEE5 17:818–823
DOI: 10.1007/s11666-008-9228-7
1059-9630/$19.00 © ASM International

Peer Reviewed

Electrically Conductive Flame Sprayed Aluminum Coatings on Textile Substrates

Joël Voyer, Peter Schulz, and Martha Schreiber

(Submitted May 15, 2008; in revised form September 3, 2008)

In this study, electrically conductive and flexible aluminum coatings using powder and wire flame spraying were successfully deposited onto diverse textiles. The influences of various process parameters and fabric materials on the electrical conductivity and microstructure of the metal-fabric composites were investigated. Preliminary results show that to obtain excellent coating surface conductivity values a specific coating quantity of higher than 20 mg/cm^2 is required. After further optimization of the spraying parameters, very good specific surface conductivities (~500 S_A) could be obtained even with reduced coating quantities. Through an adequate parameter optimization a reduction in the specific coating quantity was also achieved while high conductivity values were retained. In addition, when the coating quantity was reduced, the flexibility of the fabric substrates was better conserved. This investigation illustrates that optimized electrically conductive composites with flexible fabric substrates can be produced without any preliminary thermal or chemical fabric specifications.

Keywords composite materials, flame spray, influence of spray parameters

1. Introduction

Electrically conductive flexible fabric materials represent a new approach for the production of various integrated applications, e.g., wearable electronics, lighting, or communication in medical techniques (Ref 1-3).

To produce composite materials based on copper and textiles, chemical processes are presently being used. However, these techniques have two significant disadvantages: they are expensive and they are potentially dangerous for the environment because they consist of immersing the part to be covered into acid or basic electrolytes. Other physical processes, such as CVD, are high-temperature processes that can significantly overheat the fabric substrate. Furthermore, these processes produce only very thin coatings, which are barely conductive when applied to textiles. Therefore, due to the above-mentioned limitations for the chemical and physical processes, thin metallic copper sheets are usually used to produce such composites. However, these sheets are liable to crease and tear. Currently no process exists for the manufacture of sufficiently conductive aluminum-fabric materials composites.

For this study, the flame spray process was chosen to produce aluminum-coated fabric composites. The main advantages of this process are its low operational cost and its versatility, which can easily be scaled up for mass production (Ref 4). The use of two different flame spray processes (powder and wire) enables a comparison of the influence of the feedstock materials on the composites properties.

The combination of the high particle temperatures involved in the flame spraying of metallic materials and the extremely high heat sensitivity of the fabrics represents an interesting challenge in the production of such metal-textile composites. To protect the fabrics while spraying metallic materials onto them, an adequate cooling system must be developed, which does not cause physical damages to the textile.

2. Materials, Coatings, and Experimental Procedure

Two existing conventional flame spray gun apparatus were used for this study: a Metco 5P-II powder gun and a GTV-12E wire gun. Conventional aluminum powder (−90 + 45 μm) and aluminum wire (Ø 3.2 mm) were used throughout this work; details are reported in Ref 5. The powder mostly consists of blocky particles without undesired sharp edges, resulting in reasonable flowability without significant pulses.

Three different polyester fabrics, labeled 5248, 5911, and "XX" were used. Fabrics 5248 and 5911 are similar,

This article is an invited paper selected from presentations at the 2008 International Thermal Spray Conference and has been expanded from the original presentation. It is simultaneously published in *Thermal Spray Crossing Borders, Proceedings of the 2008 International Thermal Spray Conference*, Maastricht, The Netherlands, June 2-4, 2008, Basil R. Marple, Margaret M. Hyland, Yuk-Chiu Lau, Chang-Jiu Li, Rogerio S. Lima, and Ghislain Montavon, Ed., ASM International, Materials Park, OH, 2008.

Joël Voyer and **Peter Schulz**, ARC-Leichtmetallkompetenzzentrum Ranshofen GmbH, RanshofenAustria; and **Martha Schreiber**, Cellstrom GmbH, EisenstadtAustria. Contact e-mail: joel.voyer@arcs.ac.at.

except that 5911s structure is much looser, making its density (fibers/cm^2) lower than that of 5248. Fabric 5911 also has two significantly different rough sides. The third fabric, labeled "XX," has an even more open mesh structure combined with an uniform side structure. The properties of the different textiles are described elsewhere (Ref 5).

For the first optimization step, standard flame spray parameters for aluminum were used. These spray parameters were described in a previous work (Ref 5). In this first optimization procedure, only the standoff distance and the substrate cooling were varied for different amounts of deposited materials. For the experiments without any substrate cooling, the standard standoff distance for powder and wire spraying was chosen. Afterward, for the optimization of the process parameters, an adequate substrate cooling system was used, which enabled a standoff distance reduction of 250 mm for the powder flame process and of 300 mm for the wire process from the standard spray distance value. These reduced distances in conjunction with the use of the cooling system did not induce any thermal damage to the textile substrates (Ref 5). The A4-sized fabric substrates were coated using conventional gun movements, parallel to the substrate surface, using a 6-axis ABB robot.

The coating quantity (mg Al/cm^2 fabric) of the sprayed samples was determined by weighing the fabric samples before and after the spraying process and dividing this weight by the area of the sprayed sample.

The electrical surface conductivity of the sprayed samples was determined via a 4-point measurement method with a Keithley 2010 multimeter. The detailed electrical surface conductivity measurement procedure was documented previously (Ref 5). The conductivity measurement errors as well as the spread in the data were estimated to be 10% of the measured value.

A scanning electronic microscope (SEM) was used to investigate the surface and the microstructure of the coatings as well as the interface between the coatings and the fabrics. For the microstructure and interface investigations, the samples were embedded and cross sections were prepared using standard metallographic procedures.

3. Results and Discussion

For the first experiments performed in this study, coatings were produced with standard spraying parameters without any cooling system and without any optimization to identify the value of the coating quantity at which the surface conductivity had a value higher than 0 S_A, and this value was determined to be 20 mg/cm^2. Based on these results, to save material and time, it was decided to use a coating quantity of 20-25 mg/cm^2 as the target value for further experiments, where the spraying parameters were optimized. Figure 1 shows the electrical conductivity values as a function of the coating quantity for fabrics 5248 and 5911 for the first series of experiments, where no substrate cooling was deployed.

Samples produced using fabric 5248 possess a higher surface conductivity than do those based on fabric 5911. This is assumed to be due to a better interconnection between the individual sprayed splats for fabric 5248, which has a tightly woven structure, compared to fabric 5911, which has a more three-dimensional and looser structure. In addition, independent of the fabrics used, using of wires as raw materials produces samples with higher surface conductivity than the powder sprayed samples. This is believed to be due to the higher kinetic energy and more uniform melting of the particles involved in the wire flame spray process (Ref 6). The surface morphology of the samples produced onto the 5248 fabric using powder or wire as raw material is shown in Fig. 2 and 3. As a first observation, it is obvious that their microstructures consist of individual particles with a large number of nonmolten or partially molten particles. This is impressive since these samples, particularly the wire sprayed sample, possess a relatively good electrical conductivity.

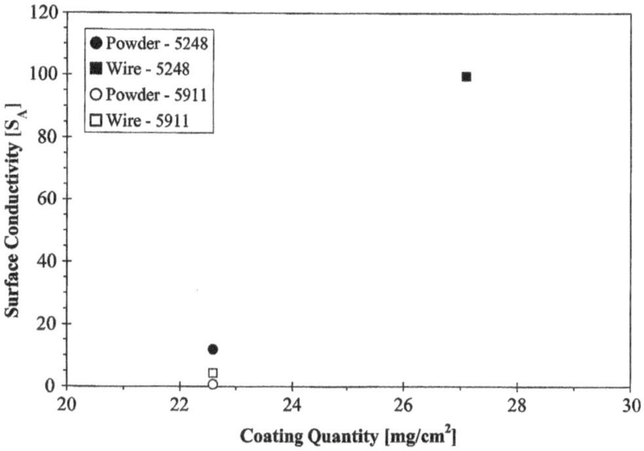

Fig. 1 Electrical conductivity as a function of the Al coating quantity for fabrics 5248 and 5911 using powder and wire as raw materials and without substrate cooling

Fig. 2 Morphology of the surface of the Al powder sprayed polyester 5248 without substrate cooling

Therefore, it may be concluded that even if only limited contacts and interconnections between the individual splats exist, a relatively good absolute sample surface conductivity may be achieved. Cross sections of the samples indicate that penetration of aluminum into the fabric thickness is minimal (Ref 5).

The second series of experiments were performed using two air amplifiers to cool the 5911 fabric surface during the spraying process. The 5911 fabric was chosen because of its three-dimensional structure, which improved the surface conductivity with optimized spray parameters, and also because the initial conductivity was extremely low (see Fig. 1), rendering increases in the conductivity value easily observable. As expected, substrate cooling made it possible to decrease the standoff distance significantly, as shown in Fig. 4 for the powder and in Fig. 5 for the wire as raw material. In these figures, the standoff distance value 0, hereafter called SDP or SDW (SDP: standard distance for powder; SDW: standard distance for wire), represents the standard distance without substrate cooling. The other distances are shorter than this standard distance by the values indicated. It is worth noting that for the results shown in Fig. 4 and 5, the coating quantity ranged between 23 and 32 mg/cm^2 (Ref 5). Apart from producing sprayed fabric samples without causing thermal damages, the reduced standoff distance generated by using an adequate cooling system has a significant influence on the electrical conductivity as well as on the microstructure of the aluminum coatings, as shown in Fig. 6-8. It is obvious that using an adequate cooling system allows to significantly reduce the standoff distance, resulting in an increase in the surface electrical conductivity without inducing any damage on the fabrics. The fabrics were visually observed after the spray process to identify the presence or the absence of any thermal damage. This was afterward completed by observations of the cross-sections of the samples using SEM.

Fig. 3 Morphology of the surface of the Al wire sprayed polyester 5248 without substrate cooling

For the powder sprayed samples with comparable coating quantities, the increase in surface conductivity with reduced spray distance is assumed to be due to a more uniform and better molten particle state on impact with the fabric surface. The optimum standoff distance was defined as the distance that gave the best surface conductivity without inducing any fabric damages and is equal to SDP-200 mm for the powder spray process.

For the wire sprayed samples, the optimal standoff distance is SDW-300 mm; at shorter distances (SDW-350 mm), the electrical conductivity is significantly reduced and the fabrics were damaged.

The standoff distance was further reduced for the wire sprayed (<SDW-350) and for the powder sprayed samples (<SDW-200), but these shorter distances induced either enormous damages on the fabrics for the shortest distances (easily seen by the naked eye) or light damages, which were sometimes only visible by microscopy. The surfaces of samples SDP-200 and SDW-300 are shown in Fig. 6 and 7 and their cross sections in Fig. 8. It is clearly visible that the textile fibers are completely covered on both sides of the fabric and the coatings consist of better

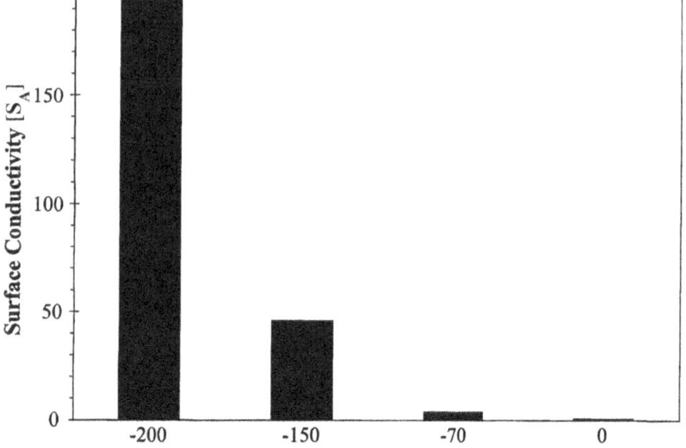

Fig. 4 Results of the standoff distance optimization using substrate cooling, fabric 5911, and powder as raw material

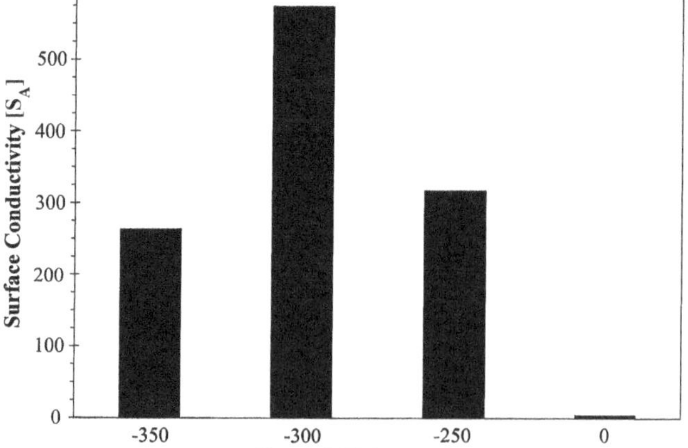

Fig. 5 Results of the standoff distance optimization using substrate cooling, fabric 5911, and wire as raw material

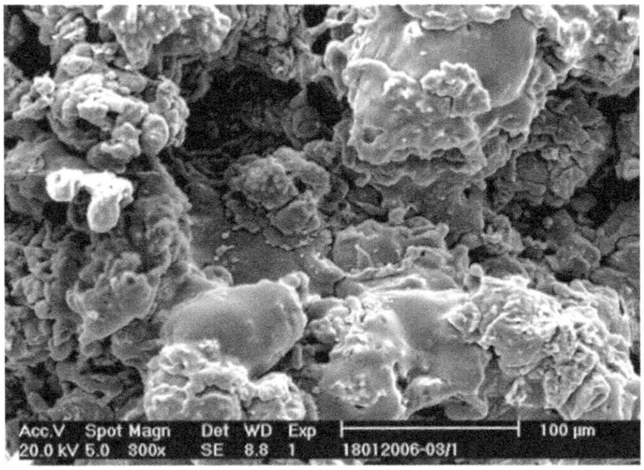

Fig. 6 Morphology of the surface of the Al powder sprayed polyester 5911 with substrate cooling at a standoff distance of SDP-200 mm

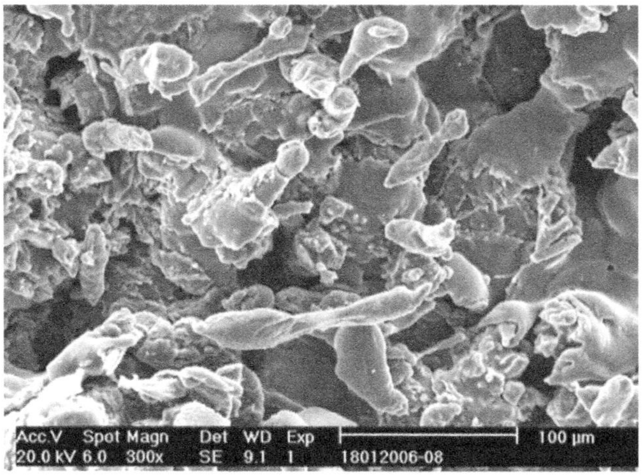

Fig. 7 Morphology of the surface of the Al wire sprayed polyester 5911 with substrate cooling at a standoff distance of SDW-300 mm

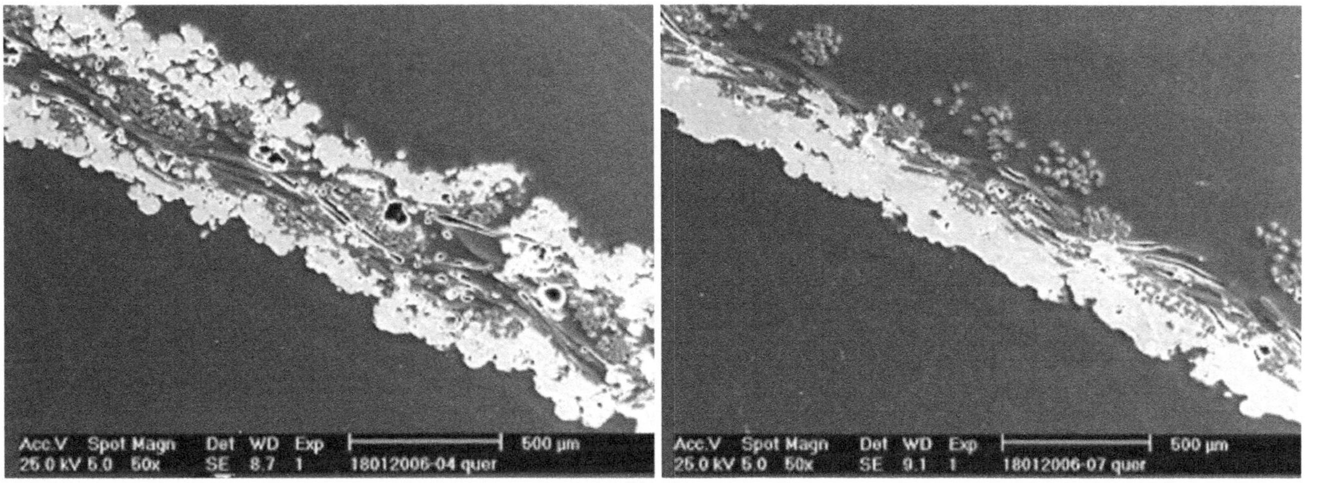

Fig. 8 Cross sections of the Al powder sprayed (left, sample SDP-200 mm, double-sided) and Al wire sprayed (right, sample SDW-300 mm) polyester 5911 using the optimal standoff distance and with substrate cooling

molten spray particles, which is better than when no cooling was used. For the wire sprayed sample (SDW-300 mm), an excellent and uniform aluminum distribution with surface electrical conductivity values of up to 500 S_A was obtained, as shown in Fig. 7. The coating microstructure is independent of the fabric side used, as shown by the cross sections of the double-sided sample SDP-200 mm in Fig. 8(a), but some cavities are present in the center of the fabric material and contacts between the two Al coatings may be observed at only a few locations. Through the wire spray process, the SDW-300 mm coating is considerably flatter and more uniform, showing a very good interconnected structure, as shown in Fig. 8(b).

The fabric structure has a significant influence on the electrical conductivity of the samples, as shown in Fig. 9 and 10. For a fixed coating quantity, smooth fabric sides possess higher surface electrical conductivity values than rough fabric sides, which is believed to be due to their lower specific surface values. This influence is only significant for the wire sprayed samples. It was also observed that more open fabric structures induce higher surface electrical conductivity values, due to the metallic contact between each coated side of the fabric material (see Fig. 10). A thin, light composite with a surface conductivity value interesting for the desired application (which is described later in this paper) using a dense fabric structure with an Al coating quantity of 13 mg/cm^2 is shown in Fig. 11.

An interesting application of electrically conductive aluminum coatings deposited onto polyester fabric materials is their deployment as positive electrodes in lithium-ion batteries. Figure 12 shows the typical structure of such a battery.

The replacement of the aluminum foil through an aluminum-coated fabric material as well as the copper foil through a copper-coated textile has several advantages in a battery structure, as shown in Fig. 12. Through the intrinsically three-dimensional structure of the fabric

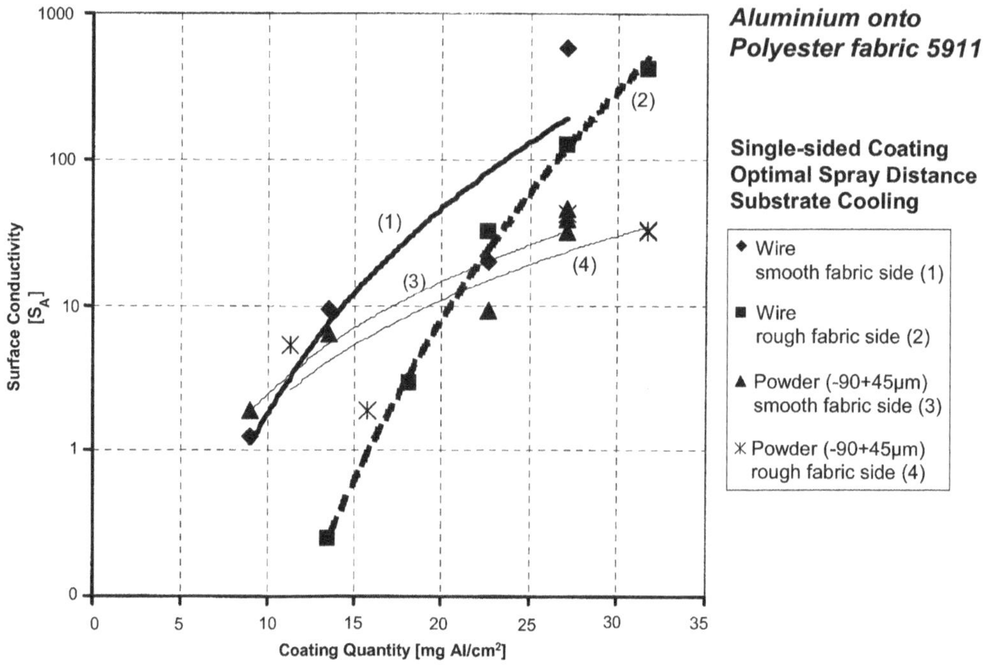

Fig. 9 Relationship between the coating quantity and the surface conductivity of aluminum-coated fabric sides and the influence of the smooth or rough textile side

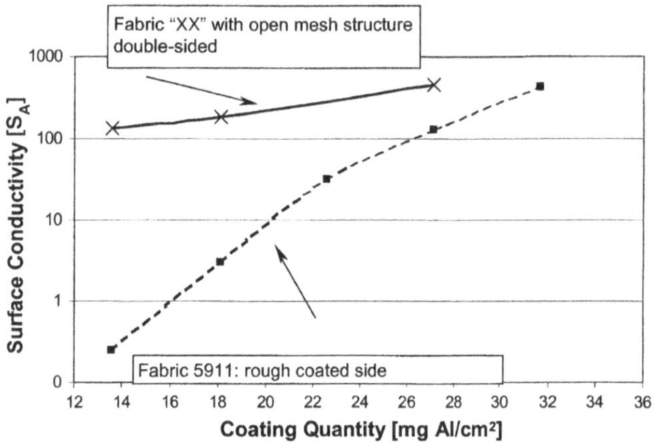

Fig. 10 Relationship between the coating quantity and the surface conductivity of aluminum-coated fabric sides: Influence of a dense textile structure in comparison to an open structure in double-sided coated fabric materials

Fig. 11 Morphology of the surface of an Al wire sprayed double-sided "XX" polyester fabric using the optimal standoff distance (13 mg Al/cm^2 per side)

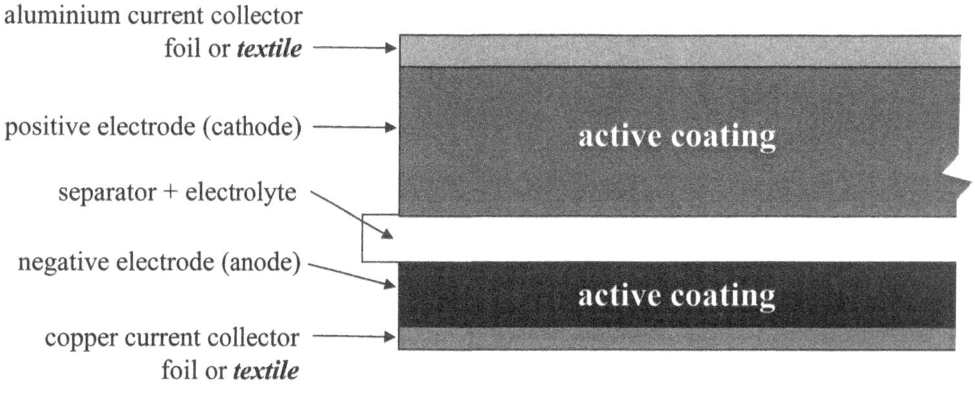

Fig. 12 Schematic setup of a lithium-ion battery

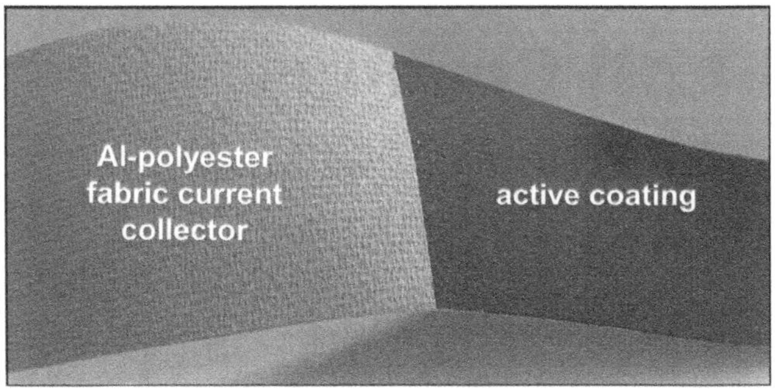

Fig. 13 Flexible lithium-ion cathode based on flame sprayed aluminum coatings onto polyester fabrics

materials, the elasticity and strength of the electrode rise, rendering feasible the production of flexible and soft batteries. Figure 13 shows a cathode designed for lithium-ion batteries. This cathode was produced using an aluminum flame sprayed fabric combined with an electrically active ceramic layer. Such produced batteries have already been successfully tested. In a near future, it is planned to upscale the process to produce aluminum-coated textile fabrics using a continuous flame spray process.

4. Conclusions

Electrically conductive powder and wire flame sprayed aluminum coatings were successfully deposited onto various polyester textile materials. Coatings produced using wire as raw material have a much better coating morphology than those produced using powder as starting material. By optimizing the spray parameters, electrical surface conductivity values of up to 500 S_A could be achieved. The fabric texture has a strong influence on the coating quantities necessary to obtain a good surface conductivity. Typical coating quantities lie around 20 mg/cm^2 for one-sided coated samples by using optimized spray parameter and an adequate cooling system.

However, by using a specific double-sided coated fabric, this quantity may be decreased to 15 mg/cm^2 or less and still produce coatings with good electrical surface conductivity values. To obtain these high conductivity values coupled with low coating quantities, an adequate cooling system for the substrate (air amplifiers) is essential. The shorter spray distances associated with such cooling systems enable a better particle melting, which in turn produces a more homogeneous coating structure, while retaining the flexibility of the fabric.

Successful preliminary test results for Al-polyester composites used as current collectors show that application-specific electrically conductive composites with adjustable specific surface conductivity values and microstructures can be produced on flexible fabric substrates without inducing any thermal or chemical damages to the fabric material.

It is intended in a near future to decrease the quantity of deposited aluminum while maintaining an adequate surface conductivity of the coatings for the application, which means finding a compromise between flexibility and conductivity.

Acknowledgments

The authors would like to thank the Austrian National Foundation for funding this research work in the frame of the project Austrian Light Weight Structures.

References

1. D.W. Carroll, "Wearable Personal Computer System Having Flexible Battery Forming Casing of the System", US Patent 5572401, 1996
2. Z.S. Nakad, "Architectures for e-Textiles", PhD Thesis, Virginia Polytechnic Institute and State University, Blacksburg, Virginia, USA, 2003
3. R. Service, Patterning Electronics on the Cheap, *Science*, 1997, **278**(5337), p 383-384
4. E. Lugscheider, *Handbuch der thermischen Spritztechnik* (Thermal Spray Handbook) (DVS-Verlag GmbH, 2002), 221 p (in German)
5. J. Voyer, P. Schulz, and M. Schreiber, Conducting Flame Sprayed Al-Coatings on Textile Fabrics, *J. Therm. Spray Technol.,* 2008, **17**(4), in press
6. *Handbook of Thermal Spray Technology*, J.R. Davis, Ed., ASM International, 2005, 338 p

JTTEE5 17:824–830
DOI: 10.1007/s11666-008-9235-8
1059-9630/$19.00 © ASM International

Peer Reviewed

Flame Sprayed Al-12Si Coatings for the Improvement of the Adhesion of Composite Casting Profiles

Joël Voyer, Christian Peterlechner, and Ulf Noster

(Submitted May 19, 2008; in revised form July 24, 2008)

In this study, flame sprayed Al-12Si coatings were produced on the surface of inlays (aluminum profiles) of composite castings parts. The aim was to enhance the strength between the joining partners inlay and cast. Due to the high surface roughness and the presence of pores in the coatings, combined with the formation of an intermetallic phase at the interface, the adhesion of flame sprayed inlays could be enhanced by a factor of 2 compared to blank inlays and by a factor of 1.3 when compared to sand-blasted inlays. However, results also show that gaps are present, mostly at the interface between the inlays and the flame sprayed coatings, and these gaps have a negative effect on the joining strength of the composite casting parts. Therefore, optimizing the adhesion of the coating on the Al profiles via an improvement in both the sand-blasting and the flame spraying parameters would be beneficial for further enhancement of the adhesion of composite casting parts.

Keywords aluminum composites, fatigue and fracture, flame spray, influence of properties

1. Introduction

One important advantage of thermal spray processes is the possibility of using lightweight substrates with specially optimized surface properties. For instance, today thermal spray processes are widely used in industry to protect the underlying substrate against a wide range of attacks: wear, corrosion, erosion, high-temperature, electrical current, and so forth (Ref 1-8). By adequately optimizing the spraying parameters, the coating properties can be tailored to meet the application requirements. However, various new fields of application for thermal spray processes, for example in joining of different materials, are gaining widespread recognition (Ref 9).

Flame spraying is one of the oldest, least expensive, and easiest thermal spray process used to produce coatings. In this technique, oxygen and acetylene are used to produce a flame having a temperature of approximately 3000 °C, in which the coating material to be sprayed is injected.

A major disadvantage of the flame spray process is its relatively low particle speed, which accounts for relatively high surface roughness and porosity values in the sprayed coatings compared to other thermal spray processes. However, in the present study, this is considered an advantage since a rough coating surface enables the anchoring of the fluid metal and a highly porous coating can eventually be impregnated by the liquid metal. Thus, it is assumed that both surface roughness, and porosity may contribute to an increase in the adhesion of the components during composite casting.

This article reports preliminary results on aluminum-magnesium (Al-Mg) compound cast joinings, where the solid aluminum part was flame sprayed with Al-12Si to enhance the interface properties of the cast part.

2. Materials and Experimental Procedure

2.1 Aluminum Profile

The aluminum profiles used for these composite casting experiments were made of EN AW 6060 wrought Al alloy. This alloy was chosen because it is one of the most often used alloys for producing profiles in the extrusion industry due to its extraordinary ability to deform plastically and its good machinability and weldability.

As shown in Fig. 1, the Al alloy profiles were rectangular, with dimensions of 25 × 25 × 55 mm, and the wall thickness was 4 mm.

To investigate the effective adhesion enhancement produced by using flame sprayed coatings, a total of three

This article is an invited paper selected from presentations at the 2008 International Thermal Spray Conference and has been expanded from the original presentation. It is simultaneously published in *Thermal Spray Crossing Borders, Proceedings of the 2008 International Thermal Spray Conference*, Maastricht, The Netherlands, June 2-4, 2008, Basil R. Marple, Margaret M. Hyland, Yuk-Chiu Lau, Chang-Jiu Li, Rogerio S. Lima, and Ghislain Montavon, Ed., ASM International, Materials Park, OH, 2008.

Joël Voyer and **Ulf Noster**, ARC-Leichtmetallkompetenzzentrum Ranshofen GmbH, Ranshofen, Austria; and **Christian Peterlechner**, Framag Industrieanlagenbau GmbH, Frankenburg, Austria. Contact e-mail: joel.voyer@arcs.ac.at.

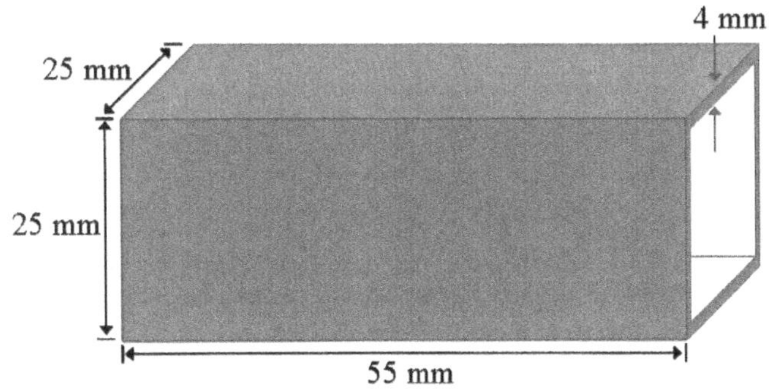

Fig. 1 Geometry of the Al alloy profiles

different Al part surface treatment procedures were compared:

- Blank (no treatment, reference samples)
- Sand-blasted
- Flame sprayed Al-12Si coatings

On the blank samples, no further treatment was performed except for cleaning and degreasing the surface with alcohol before performing the casting experiments. For the sand-blasted samples, the procedure consisted of sand blasting all four surfaces using a standard blasting device with an air pressure of 5 bars and a standoff distance of approximately 15 cm. Following this surface roughening, the parts were cleaned and degreased using alcohol before the casting experiments were performed. The procedure used for the flame sprayed samples is described thoroughly in the following section.

2.2 Flame Spraying

Al-12Si powder was used to produce the flame sprayed coatings. This powder was produced using a gas atomization process, and therefore it featured a spheroidal morphology. Its size distribution was −90 +45 μm.

For the flame sprayed samples, the procedure consisted of sand blasting a 30 mm long section on each of the four profile surfaces, followed by the aforementioned cleaning procedure. The sand blasting of only a 30 mm long section of the profiles was based on the fact that the overlapping length of the casting on the Al profiles in the ceramic form is equal to 30 mm, and therefore there is no need to sand blast the entire profile length. Due to the high affinity of aluminum and its alloys for forming a thin oxide film on their surfaces, no preheating of the samples was performed before the spraying process. In addition, the spraying procedure was initiated as soon as possible (<2 min) after the sand-blasting procedure to minimize the oxidation of the profiles surfaces. The 30 mm long sand-blasted sections of the Al profiles were then sprayed in an *x-y* pattern using a robot. A total of 10 Al profiles were simultaneously flame sprayed; the flame spray parameters used are listed in Table 1.

During the spraying of the profiles, two air compressors were used to cool down the sprayed samples and prevent any surface overheating. After the first side of the Al profiles was sprayed, the spray gun was stopped and the 10 Al samples were manually turned by 90° to expose a new uncoated surface, and the spray procedure was repeated. This was performed until all four sides of the Al profiles were coated. Successfully flame sprayed Al-12Si coatings with a length of 30 mm on Al profiles are shown in Fig. 2.

Table 1 Flame spray parameters used for the production of the Al-12Si coatings

Spray parameter	Value
Oxygen flow rate, L/h	1415
Acetylene flow rate, L/h	880
Powder flow rate, g/min	38
Standoff distance, mm	65
Horizontal speed, m/min	30
Vertical step, mm	8
Number of layers	4
Coating thickness, μm	250-275

Fig. 2 Flame sprayed Al-12Si coatings on Al profiles with a coating overlapping length of 30 mm

2.3 Composite Casting

The magnesium alloy used for the casting experiments was AZ91-F. The identification label AZ shows that aluminum and zinc are the main alloying elements, and the label F tells us that the alloy did not undergo any further heat treatments following the casting process. This alloy exhibits high strength in the as-cast condition, but at elevated temperatures its strength and creep resistance are rather moderate compared to aluminum alloys.

The casting experiments were performed using a squeeze casting equipment. The casting process is quite similar to the classic die-casting process: a small defined melt quantity is inserted into the casting chamber, and then a hydraulic piston pushes the melt into the cavity with high pressure and speed. The filling of the cavity occurs in a laminar manner. Prior to the casting experiments, the Al alloy profiles with their different surface treatments (blank, sand blasted, or flame sprayed) were installed in a ceramic casting form. To prevent or to minimize Mg infiltration into the ceramic form, which can cause severe adhesion problems and prevents the cast parts from being released from the form, a special procedure was performed, the description of which is beyond the scope of this paper and is not described here (Ref 10).

Peer Reviewed

Figure 3 shows the ceramic form with the glued Al alloy profiles before its installation in the squeeze-casting machine and the composite casting parts after their release from the casting form.

Before performing any tests or characterization, the composite casting parts were cleaned and separated from the overcast material. Thus, for each part, two composite casting test samples were obtained.

2.4 Tensile Testing

In order to quantify the adhesion strength of the composite castings interface, quasi-static tensile tests were performed using a Zwick tensile testing machine. Due to the chosen samples geometry, the tensile tests could be performed without designing and deploying an additional sample holder, but rather by placing the composite casting samples directly in the clamp devices of the tensile testing machine. During tensile testing, the applied force and the displacement of the machine were recorded. The tests were stopped when the Al alloy profiles were pulled out of the Mg castings, and the maximal force was defined as the interface adhesion strength.

2.5 Samples Characterization

The characterization of the interface of the composite casting samples for each surface treatment variation (blank, sand blasted, or flame sprayed) was performed using optical microscopy. The composite casting samples were prepared using standard metallographic procedures to characterize the following features:

- Interface between the Al profile and the Mg cast (blank and sand-blasted samples)
- Interface between the Al profile and the flame sprayed coating (flame sprayed samples) before and after the casting experiments
- Interface between the flame sprayed coating and the Mg-cast (flame sprayed samples) after the casting experiments
- The coating microstructure (flame sprayed samples) before and after the casting experiments.

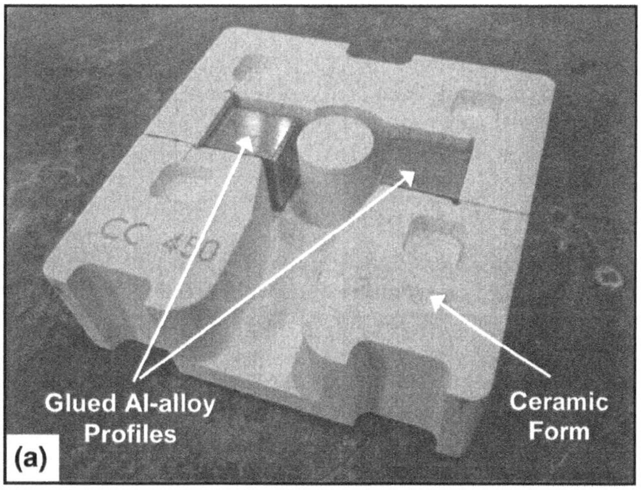

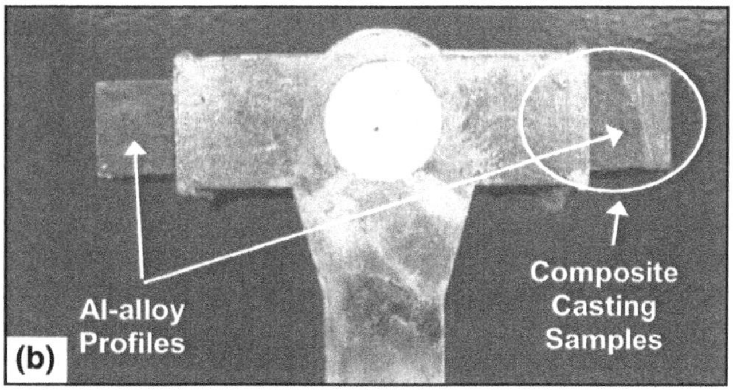

Fig. 3 Ceramic form with the glued Al alloy profiles before its installation in the squeeze casting machine (a) and the composite casting parts after their release from the casting form (b)

3. Results and Discussion

3.1 Profile Surface Characterization Prior to the Casting Experiments

In this section, the surface properties of the different profiles (blank, sand blasted, and flame sprayed) before the casting experiments are presented.

3.1.1 Blank Profile Surface. The topography measurements using a bifocal microscope provided a value of approximately 1 μm (average of three independent measurements) for the average surface roughness.

3.1.2 Sand-Blasted Profile Surface. The topography measurements using the same equipment and procedure as described previously for the blank profiles gave an average surface roughness of 5.7 μm.

3.1.3 Flame Sprayed Profile Surface. Figure 4 shows a cross-section micrograph of the flame sprayed Al-12Si coating on the Al-alloy profile surface before the casting experiments. The coating microstructure is laminar with

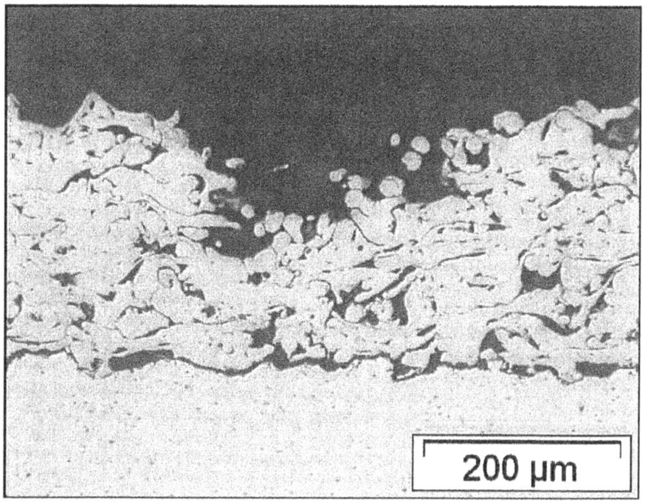

Fig. 4 Cross-section micrograph of the flame sprayed Al-12Si coating on the Al alloy profile surface

unmelted particles (round white particles), pores (black areas), and embedded oxides between Al-12Si splats (tiny lines) and is typical of thermally sprayed coatings. The average coating thickness is approximately 300 μm. The surface of the flame sprayed samples is much rougher than the surface of the sand-blasted profile, and the average of the surface roughness measurements is about 20 μm, which is three to four times higher than the value measured for the sand-blasted surface profile.

3.2 Profile Surface Characterization after the Casting Experiments

In this section, the surface properties of the different profiles (blank, sand blasted, and flame sprayed) after the casting experiments are presented.

3.2.1 Blank Profile Surface. Figure 5 shows a cross section of a composite casting blank profile. The overall interface between the blank profile and the casting is good. For example, it was observed that the corners of the profile are very well bonded to the casting, but as shown in Fig. 5, at larger distances from the corners, a gap at the interface can be observed and its maximum width, which is approximately at the centre of the profile, is equal to about 2 μm.

The gap is believed to have been initiated during the cooling stage of the casting process, due to different cooling rates of the Al and Mg parts, which induces tensile stresses and may result in interface delamination.

Figure 6 shows the diffusion zone present between the blank profile surface and the Mg casting on the inner side, where no gap was present. The top right-hand corner of Fig. 6 is the Mg casting, and the low left-hand corner is the blank profile. It can be easily measured from Fig. 6 that the diffusion zone has a total thickness of approximately 100 μm. The elemental distribution measurement shown in Fig. 7 shows that the concentration of both elements decreases constantly with the distance from their respective sources and that two different phases exist, one of which has an interlocked needle structure. However, no major differences in the phase composition could be identified between the two zones. One of these phases is believed to be $Al_{12}Mg_{17}$ (γ-phase), which is an intermetallic phase and solidifies at a temperature far below Al and Mg and is believed to be also an initiator for the formation of the gap at the interface.

3.2.2 Sand-Blasted Profile Surface. Figure 8 shows micrographs of the cross section of a composite casting sand-blasted profile. A diffusion zone, which is present along the total length of the profile interface, has been identified and no gaps are observed, verifying good adhesion between the sand-blasted profile and the Mg casting. However, pores and oxides are embedded in the diffusion zone.

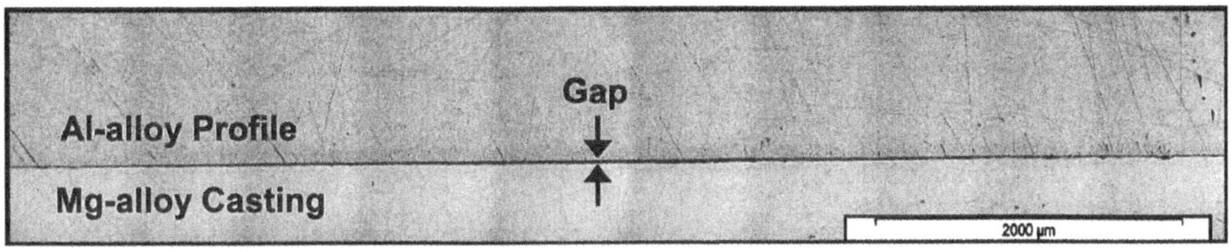

Fig. 5 Cross-section micrographs of a composite casting blank profile

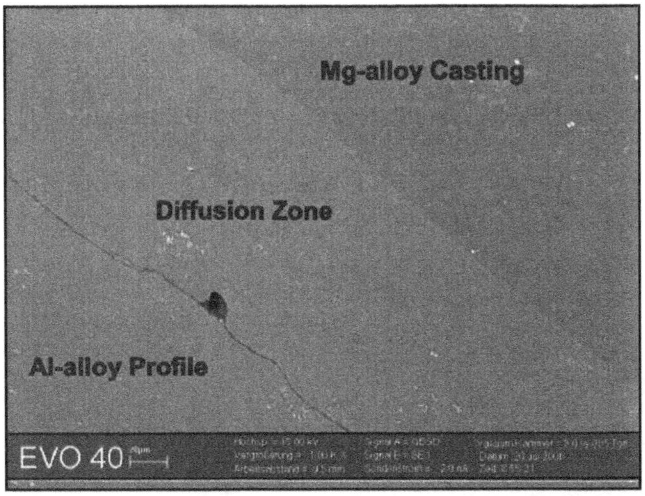

Fig. 6 Micrograph of the intermetallic phase zone of the blank profile interface, where no gap was present

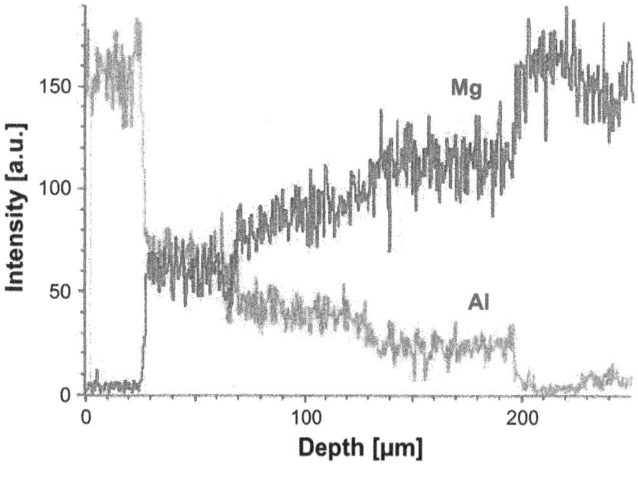

Fig. 7 EDX analysis of the blank profile interface

Peer Reviewed

Figure 9 shows micrographs of the interface between the sand-blasted profile and the Mg casting. The top right-hand corner of Fig. 9 is the Al profile and the low left-hand corner is the Mg casting. It is again clearly visible that a diffusion zone exists. It can be easily measured from Fig. 9 that the diffusion zone has a total thickness of approximately 100 μm. Elemental distribution measurements have shown that the concentration of both elements decreases constantly with the distance from their respective sources. Extended EDX measurements (not shown in this paper) have shown that the upper part of the diffusion zone consists of a homogeneous mixture of Al and Mg, while the lower part consists of large Al phases embedded in the Mg matrix. As described previously, one of these phases is believed to be $Al_{12}Mg_{17}$ (γ-phase), which is an intermetallic phase and solidifies at a temperature far below Al and Mg and is also believed to be an initiator for the formation of the gap at the interface.

3.2.3 Flame Sprayed Profile Surface. Figure 10 shows a micrograph of the cross sections of the interface between the flame sprayed Al profile surface and the Mg casting. It can be seen that the corners show very good adhesion to the casting, but in the middle of the profile, a gap has formed between the coating and the casting (depending on the chosen parts geometry). At several locations, it can be observed that the Mg casting has impregnated the flame sprayed Al-12Si coating for a few micrometers. More detailed observations on magnified micrographs (not shown in this paper) have shown that the adhesion zone could be divided in four different zones: Al profile, flame sprayed coating, diffusion zone, and Mg casting.

Figure 11 shows longitudinal micrographs of this interface. The coating thickness was measured to be

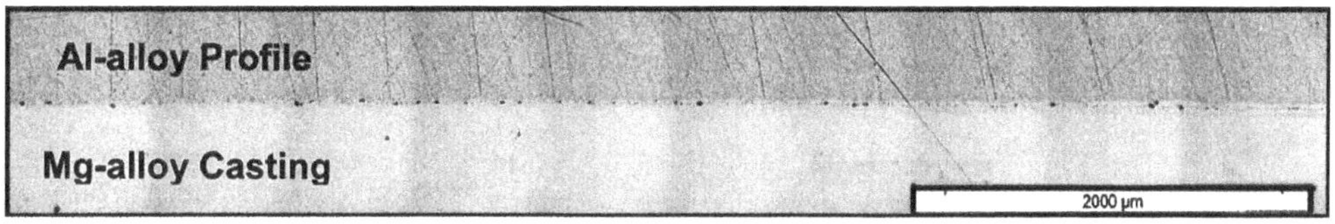

Fig. 8 Cross-section micrographs of a composite casting sand-blasted profile interface

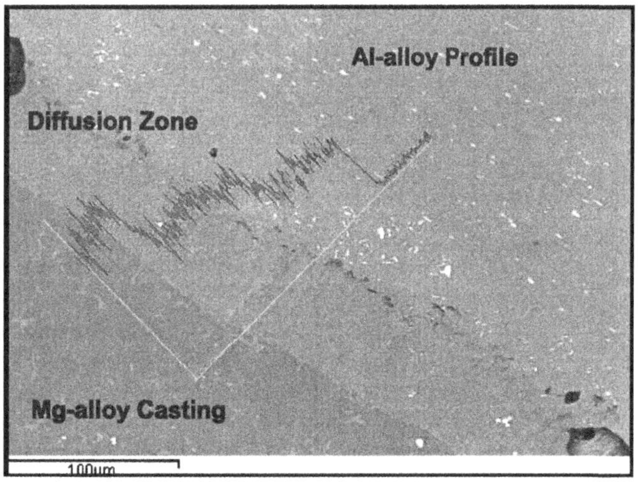

Fig. 9 Micrograph of the diffusion zone and EDX analysis of the sand-blasted profile interface

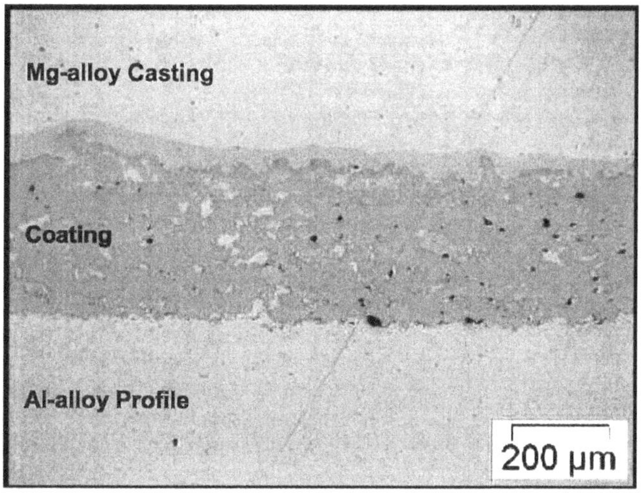

Fig. 11 Micrographs of a cross section of a composite casting flame sprayed profile interface

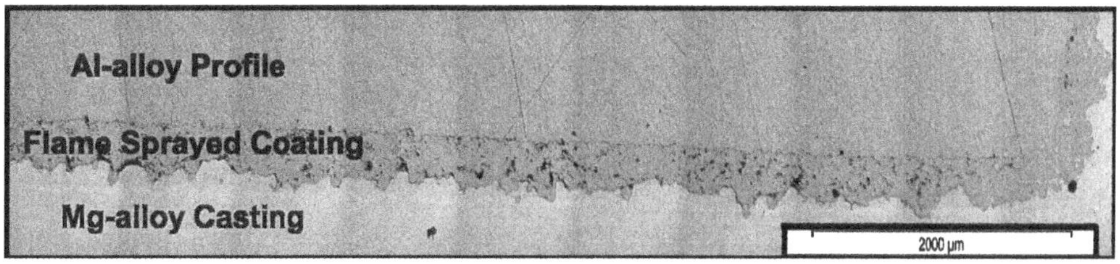

Fig. 10 Cross-section micrographs of a composite casting flame sprayed profile interface

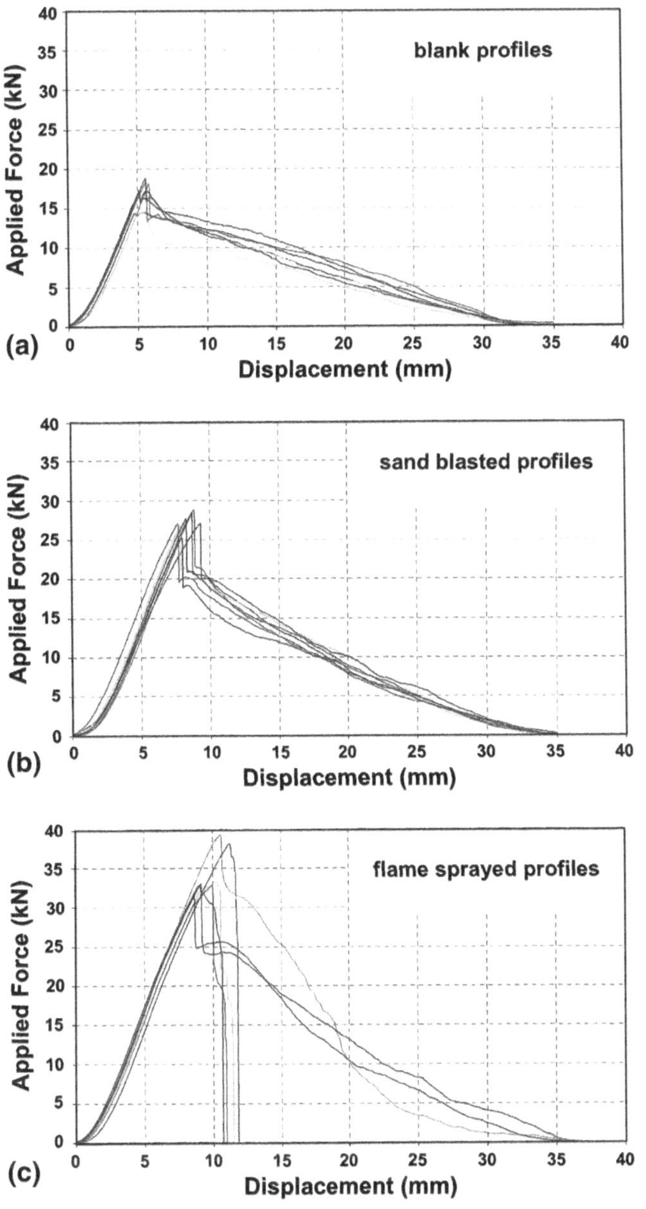

Fig. 12 Graphs showing the applied force as a function of the displacement for profiles having a blank (a), sand-blasted (b), or flame sprayed surface (c)

approximately 300 μm and possesses good adhesion throughout its interface with the Al profile and with the Mg casting. In addition, a distinct diffusion zone with a thickness of approximately 40 to 50 μm and a partial impregnation of the Mg casting into the flame sprayed coating can be observed. As reported earlier, the flame sprayed coating is partially impregnated from the Mg casting, and this could be observed on the EDX analysis by isolated Mg peaks in the coating.

As reported in the tensile test section, the flame sprayed coating has a strong tendency to adhere to the Mg casting; that is, the fracture mainly occurs between the Al alloy profile and the flame sprayed coating.

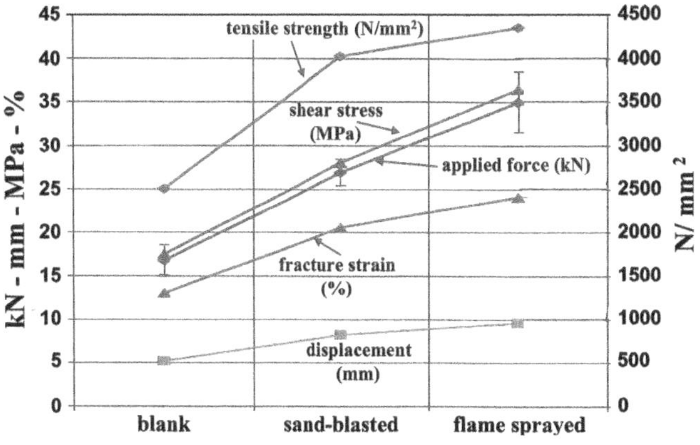

Fig. 13 Graph showing the average values of the applied force (kN), displacement (mm), shear stress (MPa), fracture strain (%) (all four on the left axis), and the average value of the tensile strength (N/mm^2) (on the right axis)

3.3 Tensile Tests

Figure 12 illustrates the results of the tensile tests as graphs of the applied force as a function of the displacement of the tensile machine crosshead for the profiles with a surface which was either blank, sand blasted, or flame sprayed. For almost all the samples, the displacement of the crosshead was equal or higher than the overlapping length of the casting. For the samples for which this displacement was smaller, the Mg casting broke away from the profile quite rapidly. Apart from these samples, independently of the experiments and of the side from which the sample was taken, the tensile results are quite reproducible, as can be seen in Fig. 12. For all the composite casting samples, by comparing the average maximum applied force, it can be observed that the blank profile surfaces have the lowest applicable force (15-18 kN), followed by the sand-blasted profiles (25-28 kN) and that the flame sprayed profiles surfaces possess the highest applicable force (30-40 kN) (comparison of Fig. 12a, b, and c). Therefore, the applied force seems to be proportional to the average surface roughness of the profile surfaces; that is, the roughest surface (flame sprayed profiles) may withstand the highest applied force.

In addition, the presence of a diffusion zone in the interface between sand-blasted profile surfaces and Mg casting may explain the increase in the applicable force compared to the blank profile. This diffusion zone enables the formation of a better adhesion between the Mg casting and the profile. For the flame sprayed samples, the benefit of such a diffusion zone is also present, but partial impregnation of the Al-12Si coating by the Mg casting is also assumed to be responsible for the further increase of the applied force when comparing these samples to the sand-blasted profile surfaces.

From Fig. 13, it can be easily observed that the applied force, displacement, shear stress, fracture strain, and tensile strength increase steadily when the profile surface is changed from a blank to a sand-blasted and finally to a flame sprayed surface. In fact, the applied force withstood by the composite casting samples is 60% higher when comparing the sand-blasted to the blank profile surface, and this applied force is 30% higher when comparing the flame sprayed profile to the sand-blasted samples. As discussed previously, this is believed to be due to the higher average surface roughness of the flame sprayed profiles but also to the presence of a diffusion zone combined with a partial impregnation of the Al-12Si coating by the Mg casting.

4. Conclusions

The use of a flame sprayed Al-12Si coating to enhance the adhesion strength between an Al-alloy profile (solid) and a Mg alloy (as cast) in composite casting samples shows impressive and successful results. Due to the fact that the flame sprayed coating surface has a higher surface roughness combined with the presence of a diffusion zone and a partial coating impregnation by the Mg casting, the latter being due to the intrinsic porous nature of typical flame sprayed coatings, the interface properties are significantly enhanced. In fact, the adhesion of flame sprayed composite castings increases by a factor of 2 compared to blank castings and by a factor of 1.3 when compared to sand-blasted castings.

However, delaminations are always present in the composite castings, and this is believed to be caused by high cooling rates and different temperature distributions along the profile surface and also to the chosen geometry. This could be eliminated by optimizing the cooling rate during the casting process and by optimizing the composite casting samples geometry by finite element programs (Ref 11, 12). However, the presence of such a gap indicates that part of the overlapping length does not participate in the overall adhesion strength of the sample.

The microscopic investigations as well as the tensile test results show that the adhesion between the flame sprayed coating and the Mg casting is much better than the adhesion between the coating and the Al profile. Therefore, an optimization of the adhesion of the flame sprayed coating to the profile would help to increase the strength of such composite casting samples. This could be achieved by optimizing the sand-blasting parameters and improving the flame spray parameters.

References

1. K. Ghosh, T. Troczynski, and A.C.D. Chakleder, Al-SiC Coatings by Plasma Spraying, *J. Therm. Spray Technol.*, 1998, **7**(1), p 78-86
2. P. Kulu and T. Pihl, Selection Criteria for Wear Resistant Powder Coatings under Extreme Erosive Wear Conditions, *J. Therm. Spray Technol.*, 2002, **11**(4), p 517-522
3. A.H. Dent, S. DePalo, and S. Sampath, Examination of the Wear Properties of HVOF Sprayed Nanostructured and Conventional WC-Co Cermets with Different Binder Phase Contents, *J. Therm. Spray Technol.*, 2002, **11**(4), p 551-558
4. B.R. Marple, J. Voyer, M. Thibodeau, D.R. Nagy, and R. Vaßen, Hot Corrosion of Lanthanum Zirconate and Partially Stabilized Zirconia Thermal Barrier Coatings, *J. Gas Turbines Power*, 2006, **128**(1), p 144-152
5. M.F.O. Schiefler, F. Gärtner, J. Voyer, A. Kirsten, H. Kreye, and A.J.A. Buschinellei, Protection of Steel Components Against Marine Corrosion by Thermally Sprayed Anodic Coatings, *Thermal Spray 2003: Advancing the Science and Applying the Technology*, B.R. Marple and C. Moreau, Eds., May 5-8, 2003 (Orlando, FL), ASM International, 2003, Vol 1, 845 pages; Vol 2, 864 pages
6. M.F.O. Schiefler, J. Voyer, F. Gärtner, and X. Qi, Corrosion Behaviour of High Velocity Combustion Wire Sprayed Coatings, *International Thermal Spray Conference 2002,* E. Lugscheider and C.C. Berndt, Eds., March 4-6, 2002 (Essen, Germany), DVS Deutscher Verband für Schweißen, 2002, 1061 pages
7. A. Nusair Khan and J. Lu, Thermal Cyclic Behavior of Air Plasma Sprayed Thermal Barrier Coatings Sprayed on Stainless Steel Substrates, *Surf. Coat. Technol*, 2007, **201**, p 4653-4658
8. J. Voyer, P. Schulz, and M. Schreiber, Conducting Flame Sprayed Al-Coatings on Textile Fabrics, *J. Therm. Spray Technol.,* 2008, **17**(4), in press
9. F.W. Bach, K. Möhwald, B. Drößler, and D. Kolar, Thermal Spray Joining—Soldering and Filling of Aluminium Substrates under Atmospheric Conditions by a Combined Thermal Spray/Fusing Technique, *Thermal Spray 2006:Building on 100 Years of Success,* B.R. Marple, M.M. Hyland, Y.C. Lau, R.S. Lima, and J. Voyer, Eds., May 15-18, 2006 (Seattle, WA), ASM International, 2006, on CD-ROM
10. C. Peterlechner, "Machbarkeit von Aluminium Magnesium Gusshybriden in Automotiven Rahmenstrukturen" ("Feasibility of Aluminium Magnesium Cast Composites in Automotive Frame Structures"), Diplomarbeit (Diploma Thesis), Hochschule Mittweida, University of Applied Sciences, 2006 (in German)
11. R.D. Bitsche, U. Noster, C. Peterlechner, and F.G. Rammerstorfer, Simulation von kraft- und formschlüssigen Hybridgussverbindungen als Designgrundlage (Simulation of Force and Form Closure Connections for Cast Composites as a Base Design), in *4. Ranshofener Leichtmetalltage 2006,* H. Kaufamnn, P.J. Uggowitzer, and U. Noster, Eds., October 17-18 2006 (Salzburg, Austria), LKR Verlag, 2006, p 275-286 (in German)
12. R.D. Bitsche, U. Noster, and C. Peterlechner, Fügepartner schlüssig zu Leichtbaukomponenten verbinden: Simulation von Hybridguss-Verbindungen (Joining Lightweight Partner Components: Simulation of Hybrid Cast Connections), *Lightweight-design,* 2008, **02-08,** p 30-34 (in German)

JTTEE5 17:831–837
DOI: 10.1007/s11666-008-9239-4
1059-9630/$19.00 © ASM International

Peer Reviewed

New Generation Perovskite Thermal Barrier Coating Materials

W. Ma, M.O. Jarligo, D.E. Mack, D. Pitzer, J. Malzbender, R. Vaßen, and D. Stöver

(Submitted June 13, 2008; in revised form August 6, 2008)

Advanced ceramic materials of perovskite structure have been developed for potential application in thermal barrier coating systems, in an effort to improve the properties of the pre-existing ones like yttria-stabilized zirconia. Yb_2O_3 and Gd_2O_3 doped strontium zirconate ($SrZrO_3$) and barium magnesium tantalate ($Ba(Mg_{1/3}Ta_{2/3})O_3$) of the ABO_3 and complex $A(B'_{1/3}B''_{2/3})O_3$ systems, respectively, have been synthesized using ball milling prior to solid state sintering. Thermal and mechanical investigations show desirable properties for high-temperature coating applications. On atmospheric plasma spraying, the newly developed thermal barrier coatings reveal promising thermal cycle lifetime up to 1350 °C.

Keywords gas turbine coatings, perovskite ceramics, plasma-sprayed TBCs, thermal cycling lifetime

1. Introduction

In the last two decades, much effort has been devoted to the development of ceramic thermal barrier coatings (TBCs) in order to increase the efficiency of gas turbines (Ref 1). State-of-the-art TBCs are usually based on 7-8 wt.% Y_2O_3-stabilized ZrO_2 (YSZ) prepared by atmospheric plasma spraying (APS) or electron-beam physical vapor deposition (EB-PVD) (Ref 2). The maximum surface temperature of these types of coatings is limited to ~1200 °C for long-term operations due to the phase transformation of zirconia. Such transformation results to change in volume and causes the formation of cracks in the coatings (Ref 3).

In order to remedy such phenomenon, a wide search has been conducted for new TBC materials which allow operation temperatures higher than 1200 °C without phase transformation, aside from thermochemical and thermomechanical compatibility with the substrate layer to be coated. Recently, perovskites (ABO_3) have been considered for this purpose. Such perovskites are usually characterized by melting point higher than 2000 °C. Further, the thermal expansion coefficient (TEC) has values typically greater than 8.5×10^{-6} K^{-1} and thermal conductivity of less than 2.2 W m^{-1} K^{-1}, which are advantageous for use as heat insulating layer (Ref 4). Perovskites additionally offer the possibility for extensive substitution of ions at *A* and/or *B* site, which enables the properties of the materials to be selectively influenced (Ref 5).

Among the perovskites, $SrZrO_3$ of ABO_3 and the complex $Ba(Mg_{1/3}Ta_{2/3})O_3$ of $A(B'_{1/3}B''_{2/3})O_3$ systems have been proven as two of the most refractory oxides known to science with melting temperature as high as 2650 °C (Ref 6) and 3100 °C (Ref 7, 8) respectively.

The $SrZrO_3$ perovskite undergoes three temperature-induced phase transformations, which are as follows (Ref 9):

$$\underset{Pnma}{\text{orthorhombic}} \overset{740^\circ C}{\rightarrow} \underset{Imma}{\text{pseudo-tetragonal}} \overset{840^\circ C}{\rightarrow} \underset{I4/mcm}{\text{tetragonal}} \overset{1130^\circ C}{\rightarrow} \underset{Pm3m}{\text{cubic}},$$

a sequence entirely consistent with both Carlsson's observations (Ref 10). As a candidate TBC material, the phase transformations for $SrZrO_3$ are not desirable. However, there are no distinguishable discontinuities of the unit cell volumes in all of the structural phase transformations of $SrZrO_3$. Only the transformation from orthorhombic *Pnma* to pseudo-tetragonal *Imma* at 700 °C involves a small volume change of ~0.14% (Ref 11). These transformations are thus considered nearly continuous (Ref 12). On the other hand, formation of $Ba(Mg_{1/3}Ta_{2/3})O_3$ (BMT) is usually accompanied by secondary phases which as well are stable even after sintering at 1650 °C for a long time (Ref 13). Hence, such phenomena as phase transformation and occurrence of a secondary phase are not deemed obstacles in the synthesis and design of these potential perovskite ceramics for TBC applications.

This article is an invited paper selected from presentations at the 2008 International Thermal Spray Conference and has been expanded from the original presentation. It is simultaneously published in *Thermal Spray Crossing Borders, Proceedings of the 2008 International Thermal Spray Conference*, Maastricht, The Netherlands, June 2-4, 2008, Basil R. Marple, Margaret M. Hyland, Yuk-Chiu Lau, Chang-Jiu Li, Rogerio S. Lima, and Ghislain Montavon, Ed., ASM International, Materials Park, OH, 2008.

W. Ma, School of Materials Science and Engineering, Inner Mongolia University of Technology, Hohhot, P.R. China; and **M.O. Jarligo, D.E. Mack, D. Pitzer, J. Malzbender, R. Vaßen,** and **D. Stöver**, Institute of Energy Research, Forschungszentrum Jülich GmbH, Juelich, Germany. Contact e-mail: m.o.jarligo@fz-juelich.de.

The thermophysical properties of $SrZrO_3$ modified by doping with 10 mol% Yb_2O_3 and 20 mol% Gd_2O_3 and the complex BMT, together with their thermal cycling lifetime, are presented in this article.

2. Experimental Procedure

Solid-state synthesis of the modified $SrZrO_3$ was conducted through ball milling with ethanol the starting powders of Yb_2O_3 (Treibacher Powdermet, 99.9%), Gd_2O_3 (Treibacher Powdermet, 99.9%), $SrCO_3$ (Aldrich, >98%), and ZrO_2 (Aldrich, 99%). The milled suspension was then dried and calcined at 1400 °C. The synthesis procedure was repeated three times for the purpose of obtaining a single phase material. BMT, on the other hand, followed similar solid state route with starting powders of $BaCO_3$ (Alfa Aesar, 99.8%), MgO (Aldrich, >99%), and Ta_2O_5 (Treibacher Powdermet, 98.95%). After drying, the milled suspension was heat treated at 1500 °C for 48 h. The phases formed after syntheses were determined by XRD (Model D5000, Siemens, Cu K_α radiation, Germany).

Bulk specimens of $SrZrO_3$ and BMT materials were also prepared by cold pressing followed by sintering at 1600 °C for 6 h in air. The densities (ρ) of the sintered samples were measured according to Archimedes principle. Measurements of the thermal expansion coefficients (TECs), thermal diffusivities, and specific heat capacities of the bulk materials were then recorded by a high-temperature dilatometer (Model DIL 402E, Netzsch, Germany), laser flash (Model THETA, Netzsch, Germany), and simultaneous thermal analysis apparatus (Model STA 449C, Netzsch, Germany), respectively. The thermal conductivities of the sintered samples were calculated using the equation:

$$\lambda = D_{\text{th}}(T)C_p(T)\rho(T) \qquad \text{(Eq 1)}$$

where λ, $D_{\text{th}}(T)$, $C_p(T)$, and $\rho(T)$ are the thermal conductivity, thermal diffusivity, specific heat capacity, and measured density, respectively.

For mechanical property measurements, the samples were prepared by cold-pressing and subsequent hot-pressing at 1550 °C with 100 MPa for 2-4 h. The prepared disc-shaped samples were cold mounted using resin and polished to ~1 μm. Young's modulus, hardness, and fracture toughness of the modified $SrZrO_3$ and BMT were determined by a depth-sensing microindentation technique in which the indenter position was precisely recorded at a given applied load. The instrument (Model H-100 Fischerscope, Helmut Fischer GmbH, Germany) displays the indenter displacement with accuracy in the nanometer range and a load resolution of 0.4 mN during the entire loading-unloading period.

For application as TBC topcoats, the synthesized $SrZrO_3$ and BMT powders were further milled with ethanol and 1.8 wt.% dispersing agent, then subsequently spray-dried. Sieved size fractions between 45 and 125 μm were used for plasma spraying. Single layer and double layer systems with 7.8 wt.% yttria-stabilized zirconia powder (Metco 204 NS, Sulzer Metco GmbH, Germany) underlayer were also prepared.

The modified $SrZrO_3$ coating, modified $SrZrO_3$/YSZ double layer coating (DLC), and BMT coating were air plasma sprayed (Triplex I gun, Sulzer Metco, Switzerland) on individual substrates such that a total coating thickness of about 400 μm was deposited. The conditions for the deposition of the coatings were argon and helium plasma gas flow rates of 20 and 13 standard liter per minute (slpm), a plasma current of 300 A at a power of 20 kW, and a spray distance of 100 mm. The DLC consisted of an YSZ coating directly sprayed on the bondcoat and a modified $SrZrO_3$ topcoat. Before deposition of the final topcoat, the substrates were coated with a 150 μm Ni-Co-Cr-Al-Y bondcoat by a vacuum plasma spray (VPS) in a Sulzer Metco facility using an F4 gun. The thickness of each topcoat was about half the total coating thickness.

For thermal cycling, the coatings were sprayed on IN738 substrate with diameter of 30 mm and thickness of 3 mm. The disc-shaped samples had beveled edges to minimize the effect of stresses originating from the free edges of the samples. During the manufacture of the thermal cycling specimens, steel substrates were also coated. These were used to characterize the as-sprayed condition of the coatings. The freestanding coatings, which were produced by removing the steel substrate from the sprayed coats with hydrochloric acid, were used to investigate the pore-size distribution using mercury porosimeters (Models Pascal 140 and 440, CE Instruments, Italy).

Thermal cycling was then carried out in a burner-rig test facility where samples were periodically heated and cooled. On heating, the reverse side of the samples was cooled by compressed air to maintain a temperature gradient of typically 0.4-1.0 K/μm across the samples. After the high-temperature phase, the burner was removed and the sample cooled with compressed air. The cycles usually took 5 min heating and 2 min cooling and were repeated until a clearly visible spallation of about 5×5 mm^2 of the coating occurred. The detailed process description can be found elsewhere (Ref 14).

The microstructure of the coatings was investigated by a scanning electron microscope (SEM) (Model JXA 840, JEOL, Japan).

3. Results and Discussion

3.1 Sintered Samples

The relative densities of the hot-pressed pellets are >95% of the theoretical densities. Moreover, the relative densities of the modified $SrZrO_3$ are slightly higher than that of pure $SrZrO_3$, indicating that doping Yb_2O_3 or Gd_2O_3 is favorable for the densification of $SrZrO_3$. These samples were further used for determination of mechanical properties.

The TECs of the dense $SrZrO_3$, $Sr(Zr_{0.9}Yb_{0.1})O_{2.95}$, $Sr(Zr_{0.8}Gd_{0.2})O_{2.9}$, and BMT are shown subsequently (Fig. 1). The TEC of $SrZrO_3$ is 8.7-10.8 × 10^{-6} K^{-1}

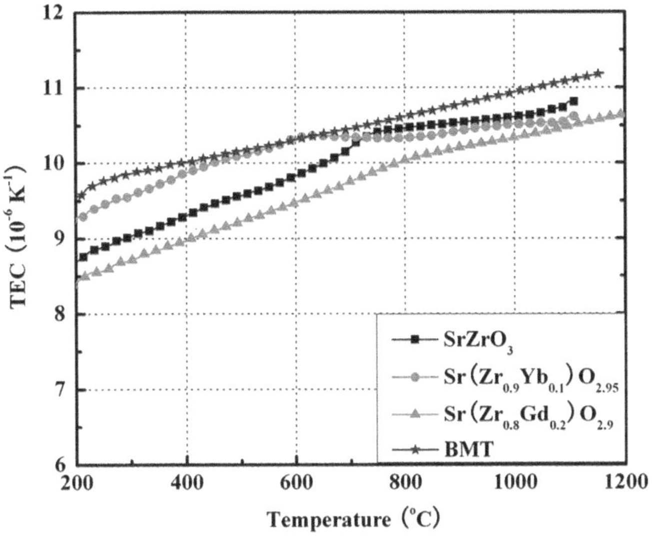

Fig. 1 TEC profiles for $SrZrO_3$, $Sr(Zr_{0.9}Yb_{0.1})O_{2.95}$, $Sr(Zr_{0.8}Gd_{0.2})O_{2.9}$, and BMT bulk materials

(200-1100 °C), which is comparable to that of YSZ (~10.5 × 10^{-6} K^{-1}, 200-1100 °C) (Ref 15). The slope of the TEC profile for $SrZrO_3$ slightly changed at ~750 °C, which is in accordance with a phase transformation from orthorhombic to pseudo-tetragonal. The TEC of $Sr(Zr_{0.9}Yb_{0.1})O_{2.95}$ is higher below ~740 °C and comparable between 740 and 1100 °C to that of $SrZrO_3$. The change in the slope of the TEC profile at ~630 °C is much lower than that of $SrZrO_3$, indicating that doping Yb_2O_3 can shift the phase transformation temperature to a lower temperature range. As for $Sr(Zr_{0.8}Gd_{0.2})O_{2.9}$, TEC is lower than that of $SrZrO_3$ all throughout the measured temperature range. A lower TEC may lead to higher stress generation during thermal cycling, which is a disadvantage for TBC application. However, the TEC of $Sr(Zr_{0.8}Gd_{0.2})O_{2.9}$ has no obvious abnormal change during measurement, indicating that doping with Gd_2O_3 can enhance the phase stability of $SrZrO_3$. This might be due to a larger octahedral factor of r_{Gd}/r_O compared with r_{Yb}/r_O in perovskite structure, r_i being the ionic radius.

BMT, on the other hand, gave the highest and consistent TEC profile from 200 to 1100 °C, indicating no structural transformation despite the presence of minor secondary phases of $BaTa_2O_6$ and $Ba_4Ta_2O_9$ in the test sample. The higher TEC for BMT is beneficial for TBC application, because it can alleviate stress misfit between topcoat and bondcoat, as well as substrate super alloy during thermal cycling. However, at the moment, the data for thermal cycling of BMT are still being generated and will be reported later.

Specific heat capacity (C_p) profiles reveal that BMT has the lowest C_p, followed by $Sr(Zr_{0.8}Gd_{0.2})O_{2.9}$, $Sr(Zr_{0.9}Yb_{0.1})O_{2.95}$, and $SrZrO_3$ consecutively (Fig. 2). Doping with Yb_2O_3 and Gd_2O_3 does not only suppress the phase transformations in $SrZrO_3$, but also lower its C_p value. This doping effect also yields a lower thermal conductivity for a given thermal diffusivity value at the temperature range of 200-1200 °C.

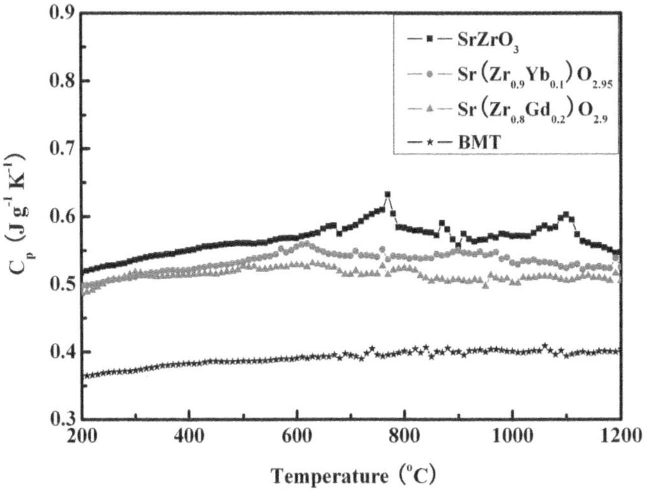

Fig. 2 Specific heat capacity profiles for $SrZrO_3$, $Sr(Zr_{0.9}Yb_{0.1})O_{2.95}$, $Sr(Zr_{0.8}Gd_{0.2})O_{2.9}$, and BMT

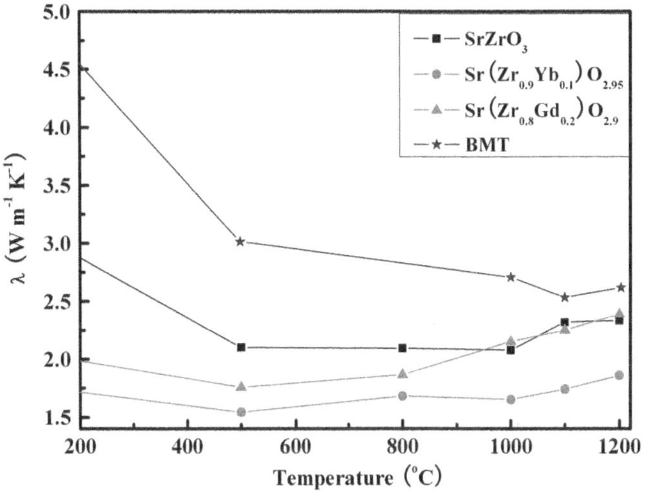

Fig. 3 Thermal conductivities of $SrZrO_3$, $Sr(Zr_{0.9}Yb_{0.1})O_{2.95}$, $Sr(Zr_{0.8}Gd_{0.2})O_{2.9}$, and BMT as a function of temperature

The thermal conductivity of these materials decreases with increase in temperature up to about 500 °C where the profiles suggest a rather constant value up to 1000 °C. After which, the thermal conductivity profiles continue to increase with further temperature rise, except for BMT which further decreases in value (Fig. 3). The thermal conductivity for $Sr(Zr_{0.9}Yb_{0.1})O_{2.95}$ reveals a very promising profile in that it is not only at least ~20% lower than that of $SrZrO_3$, but also ~20% lower than that of YSZ at 1000 °C (2.1-2.2 W m^{-1} K^{-1}, 1000 °C) (Ref 16). The thermal conductivity of $(Zr_{0.8}Gd_{0.2})O_{2.9}$ is lower below ~900 °C and comparable between 900 and 1100 °C to that of $SrZrO_3$. These results reveal that doping with Yb_2O_3 is more efficient in reducing the thermal conductivity of $SrZrO_3$ than doping with Gd_2O_3. BMT has the highest thermal conductivity among the four materials, which could be attributed to its higher thermal diffusivity and bulk density compared to the other three materials.

Peer Reviewed

Table 1 Mechanical properties of bulk materials

Material	Young's modulus, GPa	Hardness, GPa	Fracture toughness, $MPa{\cdot}m^{1/2}$
$SrZrO_3$	170 ± 4	13 ± 1	1.5 ± 0.1
$Sr(Zr_{0.8}Gd_{0.2})O_{2.9}$	160 ± 5	9.2 ± 0.3	1-2
$Sr(Zr_{0.9}Yb_{0.1})O_{2.95}$	150 ± 5	9.2 ± 0.9	1-2
BMT	186 ± 2	12 ± 2	~0.7
YSZ (Ref 17)	210 ± 10	13 ± 1	~3

It is uncertain if the increasing values of the thermal conductivity at high temperatures above 1000 °C in Fig. 3 are realistic. This increase could probably be due to the method of thermal diffusivity measurement, which is the laser flash technique. Such technique requires a supplementary topcoat to make the test sample opaque at measurements above 1000 °C, because ceramic materials are transparent on thermal radiation to a certain extent in this temperature region. The opaque layer, which was graphite in this case, tends to disintegrate at elevated temperatures. This can lead to erroneous values of the thermal diffusivity, and usually the value at high temperatures could therefore be too high.

The mechanical properties of $SrZrO_3$ and BMT were determined by an indentation technique (Table 1). Values for YSZ are also listed for reference. The Young's modulus of the modified $SrZrO_3$ was found to be lower than that of $SrZrO_3$. A low elastic modulus value is advantageous with respect to thermal stresses. The fracture toughness of the modified $SrZrO_3$ is between 1.0 and 2.0 $MPa{\cdot}m^{1/2}$, which is comparable to that of YSZ (1-2 $MPa{\cdot}m^{1/2}$) (Ref 17) and $SrZrO_3$ (1.5 ± 0.1 $MPa{\cdot}m^{1/2}$). BMT, on the other hand, has the lowest value of fracture toughness, which might be associated with an inhomogeneity of the material due to the secondary phases present.

3.2 Plasma-Sprayed Coatings

Plasma-sprayed coatings of $Sr(Zr_{0.9}Yb_{0.1})O_{2.95}$, $Sr(Zr_{0.8}Gd_{0.2})O_{2.9}$, and BMT were first optimized mainly with respect to homogeneity of the microstructure. The porosity distribution of modified $SrZrO_3$ and BMT coatings used for thermal cycling is shown in Fig. 4. The porosity distribution of $SrZrO_3$ coating is also presented for comparison. The results reveal a typical bimodal pore size distribution. The larger defects corresponding to radii above 1 μm are believed to have resulted from macrocracks and voids. The fine pores smaller than 1 μm are mainly attributed to microcracks such as intersplat gaps and intrasplat cracks. The cumulative porosities of the freestanding coatings are ~20.0, ~16.5, and ~15.0% for $Sr(Zr_{0.9}Yb_{0.1})O_{2.95}$, $Sr(Zr_{0.8}Gd_{0.2})O_{2.9}$, and BMT, respectively. The modified $SrZrO_3$ coatings have slightly larger values than the optimized porosity level (~15%) normally used in conventional YSZ-based TBCs (Ref 18). $SrZrO_3$ coating has a lower cumulative porosity than the other three coatings with almost no macrocracks and voids. This could be attributed to the smaller particle size used for plasma spraying (56-90 μm). The smaller particles of $SrZrO_3$ must have been well melted and ultimately formed a denser coating.

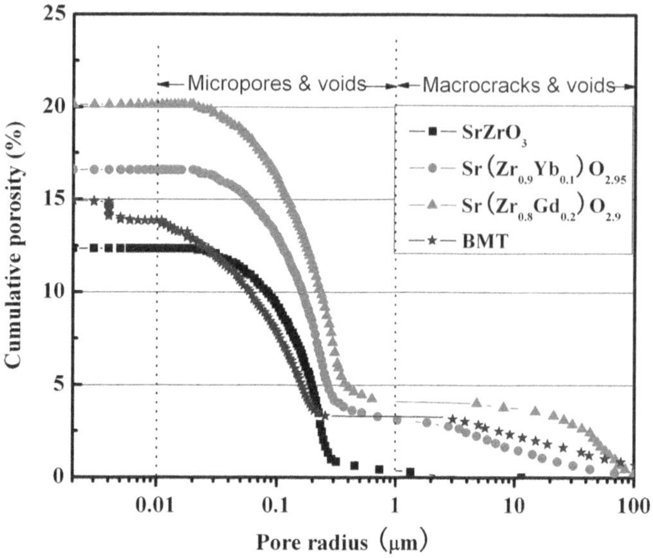

Fig. 4 Porosity distribution of $Sr(Zr_{0.9}Yb_{0.1})O_{2.95}$, $Sr(Zr_{0.8}Gd_{0.2})O_{2.9}$, and BMT freestanding coatings in the as-sprayed state

Table 2 Thermal cycling test results

TBC system	$T_{surf.}$, °C	$T_{sub.}$, °C	Cycles to failure
$SrZrO_3$	1251	965	1514
$Sr(Zr_{0.8}Gd_{0.2})O_{2.9}$	1130	984	13
$Sr(Zr_{0.8}Gd_{0.2})O_{2.9}$/YSZ	1249	965	3853
	1348	1015	208
$Sr(Zr_{0.9}Yb_{0.1})O_{2.95}$/YSZ	1246	974	3443
	1329	1011	1285
$Sr(Zr_{0.9}Yb_{0.1})O_{2.95}$	1239	969	806
$Ba(Mg_{0.33}Ta_{0.67})O_3$/YSZ	1242	1038	560
Typical YSZ (Ref 20)	1320-1350	1000-1035	250-1000

$T_{surf.}$ is the surface temperature, while $T_{sub.}$ is the substrate temperature of a sample

The thermal cycling results for $Sr(Zr_{0.9}Yb_{0.1})O_{2.95}$/YSZ and $Sr(Zr_{0.8}Gd_{0.2})O_{2.9}$/YSZ DLC are listed in Table 2. Typical values for YSZ coatings are also presented for comparison. It is obvious that the single layer $Sr(Zr_{0.9}Yb_{0.1})O_{2.95}$ and $Sr(Zr_{0.8}Gd_{0.2})O_{2.9}$ coatings have shorter thermal cycling lifetime compared to that of the lone YSZ coating. However, the thermal cycling lifetimes of the DLCs $Sr(Zr_{0.9}Yb_{0.1})O_{2.95}$/YSZ and $Sr(Zr_{0.8}Gd_{0.2})O_{2.9}$/YSZ at a surface temperature of ~1250 °C are 3443 cycles and 3853 cycles, respectively, which are comparable to that of YSZ and much higher than that of $SrZrO_3$. This indicates that the DLC system can indeed prolong the thermal cycling lifetime of new TBC materials.

At a surface temperature >1300 °C, the thermal cycling lifetime of $Sr(Zr_{0.9}Yb_{0.1})O_{2.95}$/YSZ coating is 1285 cycles, which appears rather long, whereas the lifetime of $Sr(Zr_{0.8}Gd_{0.2})O_{2.9}$/YSZ coating is much shorter. The shorter coating lifetime of the newly developed materials might be mainly due to the lower critical energy release rate as compared to that of YSZ. Lower critical energy release rate has a higher potential for bondcoat-ceramic delamination (Ref 19).

The surface morphology of the $Sr(Zr_{0.9}Yb_{0.1})O_{2.95}$/YSZ coating after cycling at a surface temperature of 1329 °C reveals some spalling of the topcoat from the bondcoat (Fig. 5a). Its corresponding cross-section micrograph also reveals some vertical cracks that might have developed due to the sintering of the coating surface, which is also observed in YSZ coating after cycling at surface temperature range between 1320 and 1350 °C (Ref 20).

The surface morphology of $Sr(Zr_{0.8}Gd_{0.2})O_{2.9}$/YSZ coating after thermal cycling at a surface temperature of 1348 °C reveals that sintering of the coating surface occurred (Fig. 5b). The corresponding cross-section micrograph shows the formation of parallel cracks at the YSZ-$Sr(Zr_{0.8}Gd_{0.2})O_{2.9}$ interface and vertical cracks in the $Sr(Zr_{0.8}Gd_{0.2})O_{2.9}$ layer. The development of parallel cracks is mainly attributed to the lower TEC of $Sr(Zr_{0.8}Gd_{0.2})O_{2.9}$ as compared to YSZ. The surface coating spalled from the $Sr(Zr_{0.8}Gd_{0.2})O_{2.9}$ layer as confirmed by the XRD profile in Fig. 6(A), which reveals that YSZ diffraction peak after thermal cycling is absent.

The surface XRD patterns of $Sr(Zr_{0.9}Yb_{0.1})O_{2.95}$/YSZ and $Sr(Zr_{0.8}Gd_{0.2})O_{2.9}$/YSZ DLC as-sprayed and after cycling at different surface temperatures are shown in Fig. 6(a). The DLC is single phase in the as-sprayed condition. However, after thermal cycling at high surface temperatures (~1250 or ~1350 °C), secondary phases appeared, which are Yb_2O_3 for $Sr(Zr_{0.9}Yb_{0.1})O_{2.95}$/YSZ and Gd_2O_3 for the $Sr(Zr_{0.8}Gd_{0.2})O_{2.9}$/YSZ. The precipitation of the secondary phase in the coatings is detrimental for TBC applications.

During thermal cycling, different vapor pressures for ZrO_2, SrO, Yb_2O_3, and Gd_2O_3, which are 1×10^{-6} atm (2500 °C), 2×10^{-5} atm (2500 °C), 6×10^{-8} atm (2500 °C), and 9×10^{-6} atm (2500 °C), respectively, are believed to be the cause of the precipitation phenomenon (Ref 21, 22). According to the ZrO_2-SrO phase diagram (Ref 6), during thermal spraying, SrO volatilizes more than ZrO_2 due to its higher vapor pressure, resulting in the deviation of the coating composition from stoichiometric $SrZrO_3$. Zr is partially substituted in $SrZrO_3$ by Yb or Gd, and the vapor pressure of SrO is higher than that of both Yb_2O_3 and Gd_2O_3. Therefore the mixture of $SrZrO_3$ and the small amount of non-equilibrium Yb_2O_3 or Gd_2O_3 is believed to be present in the as-sprayed coatings. The non-equilibrium Yb_2O_3 or Gd_2O_3 transforms into equilibrium phase upon cycling, as well as the amount of Yb_2O_3 or Gd_2O_3 in the coatings increases simultaneously, which might be due to their limited

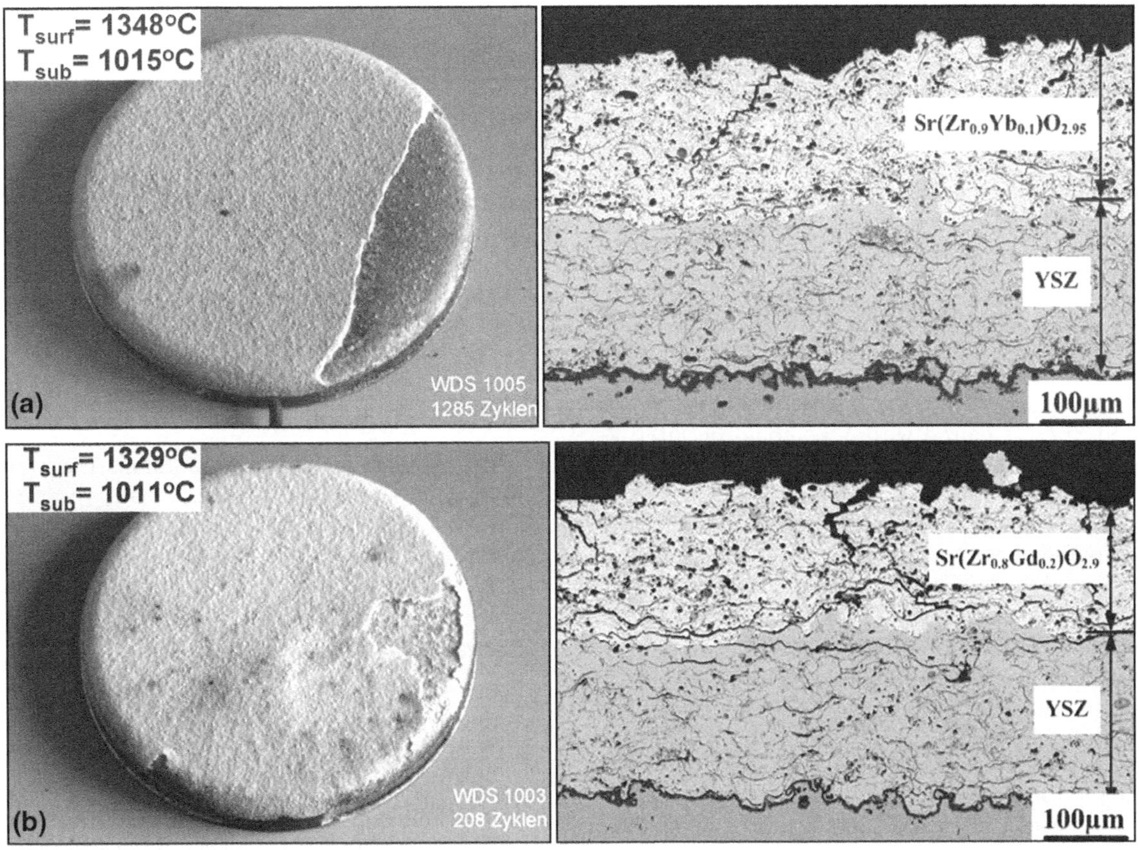

Fig. 5 Surface morphology and corresponding cross-section micrographs after thermal cycling of the double layer coatings of (a) $Sr(Zr_{0.9}Yb_{0.1})O_{2.95}$/YSZ and (b) $Sr(Zr_{0.8}Gd_{0.2})O_{2.9}$/YSZ

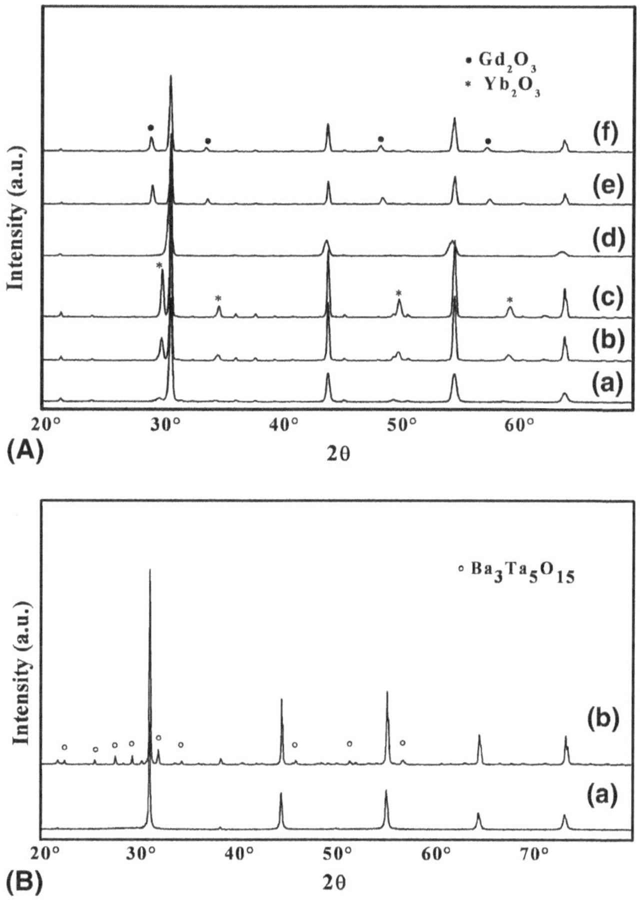

Fig. 6 (A) XRD profiles of the surface of $Sr(Zr_{0.9}Yb_{0.1})O_{2.95}$/YSZ coating in (a) as-sprayed, after cycling at surface temperature of (b) 1246 °C and (c) 1329 °C, and for $Sr(Zr_{0.8}Gd_{0.2})O_{2.9}$/YSZ coating (d) as-sprayed, after cycling at surface temperature of (e) 1249 °C and (f) 1348 °C. (B) XRD profiles of the surface of BMT/YSZ coating in the (a) as-sprayed and (b) after cycling at surface temperature of 1242 °C

solubility in $SrZrO_3$ at lower temperatures as compared to the synthesis temperature of $Sr(Zr_{0.9}Yb_{0.1})O_{2.95}$ and $Sr(Zr_{0.8}Gd_{0.2})O_{2.9}$. The lifetime values of the designed TBCs, however, suggest that the double layer systems of the modified $SrZrO_3$ paired with YSZ could be alternatives for high-temperature applications than possible with lone YSZ coating.

For the double layer BMT/YSZ system, the thermal cycling lifetime is lower compared to the other DLC systems of the modified $SrZrO_3$ owing to its much lower fracture toughness, higher elastic modulus, and higher thermal diffusivity. Such factors greatly influence the material response to stresses resulting from thermal mismatch between the different layer materials (Ref 23). Moreover, some thermal instabilities were observed during the thermal cycling test of the BMT TBC, which could be due to the formation of $Ba_3Ta_5O_{15}$ as revealed in the XRD profile of the cycled sample in Fig. 6(B). The thermal cycling lifetime of 560 cycles at ~1250 °C is nevertheless promising at this early stage of investigation for BMT as a thermal barrier coating material.

4. Conclusions

Perovskite $Sr(Zr_{0.9}Yb_{0.1})O_{2.95}$, $Sr(Zr_{0.8}Gd_{0.2})O_{2.9}$, and BMT were investigated for their potential as new generation TBC materials. The TECs of $Sr(Zr_{0.9}Yb_{0.1})O_{2.95}$ and BMT are 9.2-10.6 × 10^{-6} K^{-1} (200-1100 °C) and 9.5-11.1 × 10^{-6} K^{-1} (200-1100 °C), respectively, which are comparable to that of YSZ, while $Sr(Zr_{0.8}Gd_{0.2})O_{2.9}$ on the other hand has a lower TEC value. The results of specific heat capacity and thermal conductivity measurements reveal that doping Yb_2O_3 is more effective in reducing the specific heat capacity and thermal conductivity of $SrZrO_3$ than doping Gd_2O_3. The thermal conductivity of $Sr(Zr_{0.9}Yb_{0.1})O_{2.95}$ is ~20% lower than that of $SrZrO_3$ within the temperature range of 200-1200 °C, which is beneficial for thermal insulation. Whereas BMT has the highest thermal conductivity among the four materials which could be attributed to its higher thermal diffusivity and bulk density, even with the lowest specific heat capacity. Nevertheless, the potential for its applicability in TBC systems is very promising due to (comparable or better) properties than YSZ.

The Young's moduli of the modified $SrZrO_3$ were found to be lower than that of YSZ, which might yield higher strain tolerance for coatings. Furthermore, compared to thermal cycling lifetime of the single layer $Sr(Zr_{0.9}Yb_{0.1})O_{2.95}$ and $Sr(Zr_{0.8}Gd_{0.2})O_{2.9}$ coatings, $Sr(Zr_{0.9}Yb_{0.1})O_{2.95}$/YSZ and $Sr(Zr_{0.8}Gd_{0.2})O_{2.9}$/YSZ DLCs have much longer thermal cycling lifetimes at a surface temperature of ~1250 °C, which are comparable to that of the YSZ coating at similar surface temperature. A major advantage of the $Sr(Zr_{0.9}Yb_{0.1})O_{2.95}$/YSZ DLC is that its thermal cycling lifetime is better compared to that of the optimized single layer YSZ topcoat at a surface temperature of ~1350 °C. This suggests that double layer coatings with Yb-modified $SrZrO_3$ could be alternatives for lone YSZ coating at higher temperature applications.

Compared to the $SrZrO_3$ coatings, low fracture toughness, higher elastic modulus, and decomposition of BMT lead to its earlier failure at ~1250 °C.

Acknowledgments

The authors would like to thank the staff of Forschungszentrum Juelich most especially those from IEF-1, IEF-2, and B-NM who in one way or another contributed to the fulfillment of this study.

References

1. R.A. Miller, Thermal Barrier Coatings for Aircraft Engines: History and Directions, *J. Therm. Spray Technol.*, 1997, **6**(13), p 35-42
2. W.A. Nelson and R.M. Orenstein, TBC Experience in Land/Based Gas Turbines, *J. Therm. Spray Technol.*, 1997, **6**(2), p 176-180
3. R.A. Miller, J.L. Smialek, and R.G. Garlick, Phase Stability in Plasma-Sprayed Partially Stabilized Zirconia-Yttria. Advances in Ceramics, Vol. 3, Science and Technology of Zirconia, A.H. Heuer and L.W. Hobbs, Eds., Columbus, OH: American Ceramic Society, 1981, p 241-251

4. R. Vassen, S. Schwartz-Luckge, W. Jungen, and D. Stoever, Heat-Insulating Layer Made of Complex Perovskite, US Patent 20050260435 A1, 2005
5. B. Heimberg, W. Beele, K. Kempter, U. Bast, T. Haubold, M. Hoffmann, A. Endriss, P. Greil, C. Hong, F. Aldinger, and H. Seifert, Process for Production a Ceramic Thermal Barrier Layer for Gas Turbine Engine Component, US Patent 6602553 B2, 2003
6. T. Noguchi, T. Okubo, and O. Yonemochi, Reactions in the System ZrO_2-SrO, *J. Am. Ceram. Soc.*, 1969, **52**(4), p 178-181
7. R. Guo, A.S. Bhalla, and L.E. Cross, $Ba(Mg_{1/3}Ta_{2/3})O_3$ Single Crystal Fiber Grown by the Laser Heated Pedestal Growth Technique, *J. Appl. Phys.*, 1994, **75**(9), p 4704-4708
8. A.S. Bhalla and R. Guo, Design of Dielectric Substrates for High TC Superconductor Films, *Acta Phys. Pol. A*, 1997, **92**, p 7-21
9. C.J. Howard, K.S. Knight, B.J. Kennedy, and E.H. Kisi, The Structural Phase Transitions in Strontium Zirconate Revisited, *J. Phys.: Condens. Matter.*, 2000, **12**, p L677-L683
10. L. Carlsson, High-Temperature Phase Transitions in $SrZrO_3$, *Acta Crystallogr.*, 1967, **23**, p 901-905
11. D. Ligny and P. Richet, High-Temperature Heat Capacity and Thermal Expansion of $SrTiO_3$ and $SrZrO_3$ Perovskites, *Phys. Rev. B*, 1996, **53**(6), p 3013-3022
12. Y. Zhao and D.J. Weidner, Thermal Expansion of $SrZrO_3$ and $BaZrO_3$ Perovskites, *Phys. Chem. Miner.*, 1991, **18**, p 294-301
13. Y. Fang, A. Hu, S. Ouyang, and J. Jei Oh, The Effect of Calcination on the Microwave Dielectric Properties of $Ba(Mg_{1/3}Ta_{2/3})O_3$, *J. Eur. Ceram. Soc.*, 2001, **21**, p 2745-2750
14. F. Traeger, R. Vassen, K.-H. Rauwald, and D. Stöver, Thermal Cycling Setup for Testing Thermal Barrier Coatings, *Adv. Eng. Mater.*, 2003, **5**(6), p 429-433
15. R. Morrell, Handbook of Properties of Technical and Engineering Ceramics, Part 1. Her Majesty's Stationery Office, London, 1989
16. R. Vassen, F. Tietz, G. Kerkhoff, R. Wilkenhoener, and D. Stoever, New Materials for Advanced Thermal Barrier Coatings, *Proceedings of the 6th Liège Conference, Part III, Materials for Advanced Power Engineering*, J. Lecomte-Beckers, F. Schubert, and P.J. Ennis, Eds., Forschungszentrum Jülich GmbH, Jülich, Germany, 1998, p 162-165
17. C. Mercer, J.R. Williams, D.R. Clarke, and A.G. Evans, On a Ferroelastic Mechanism Governing the Toughness of Metastable Tetragonal Prime (t′) Yttria-Stabilized Zirconia, *Proc. R. Soc. A*, 2007, **463**, p 1393-1408
18. S. Stecura, Optimization of the Ni-Cr-Al-Y/ZrO_2-Y_2O_3 Thermal Barrier System, *Adv. Ceram. Mater.*, 1986, **1**(1), p 68-76
19. U. Bast and E. Schumann, Development of Novel Oxide Materials for TBCs, *Ceram. Eng. Sci. Proc.*, 2002, **23**, p 525-532
20. F. Traeger, M. Ahrens, R. Vassen, and D. Stoever, A Life Time Model for Ceramic Thermal Barrier Coatings, *Mater. Sci. Eng. A*, 2003, **358**, p 255-265
21. N. Jacobson, Thermodynamic Properties of Some Metal Oxide-Zirconia Systems, NASA TM 102351, 1989
22. U. Schulz, B. Saruhan, K. Fritscher, and C. Leyens, Review on Advanced EB-PVD Ceramic Topcoats for TBC Applications, *Int. J. Appl. Ceram. Technol.*, 2004, **1**(4), p 302-315
23. D. Schwingel, R. Taylor, T. Haubold, J. Wigren, and C. Gualco, Mechanical and Thermophysical Properties of Thick PYSZ Thermal Barrier Coatings: Correlation with Microstructure and Spraying Parameters, *Surf. Coat. Technol.*, 1998, **108-109**(1-3), p 99-106

Peer Reviewed

JTTEE5 17:838–845
DOI: 10.1007/s11666-008-9223-z
1059-9630/$19.00 © ASM International

Peer Reviewed

Thermal Failure of Nanostructured Thermal Barrier Coatings with Cold-Sprayed Nanostructured NiCrAlY Bond Coat

Qiang Zhang, Chang-Jiu Li, Yong Li, Shao-Ling Zhang, Xiu-Ru Wang, Guan-Jun Yang, and Cheng-Xin Li

(Submitted May 13, 2008; in revised form August 9, 2008)

Nanostructured thermal barrier coatings (TBCs) were deposited by plasma spraying using agglomerated nanostructured YSZ powder on Inconel 738 substrate with cold-sprayed nanostructured NiCrAlY powder as bond coat. The isothermal oxidation and thermal cycling tests were applied to examine failure modes of plasma-sprayed nanostructured TBCs. For comparison, the TBC consisting of conventional microstructure YSZ and conventional NiCrAlY bond coat was also deposited and subjected to the thermal shock test. The results showed that nanostructured YSZ coating contained two kinds of microstructures; nanosized zirconia particles embedded in the matrix and microcolumnar grain structures of zirconia similar to those of conventional YSZ. Although, after thermal cyclic test, a continuous, uniform thermally grown oxide (TGO) was formed, cracks were observed at the interface between TGO/BC or TGO/YSZ after thermal cyclic test. However, the failure of nanostructured and conventional TBCs mainly occurred through spalling of YSZ. Compared with conventional TBCs, nanostructured TBCs exhibited better thermal shock resistance.

Keywords bond coat, cold spraying, nanostructure, plasma spraying, thermal barrier coating, YSZ

1. Introduction

Thermal barrier coatings (TBCs) are favorably used as protective coatings on hot section components in advanced gas turbine engines to withstand increased inlet temperatures and thus improving engine performance (Ref 1, 2). Failure of a TBC system is often associated with the formation of a thermally grown oxide (TGO) layer resulting in Al depletion in the bond coat. It has been shown that failure of plasma-sprayed TBCs often occurs through spalling of the ceramic layer near the ceramic/bond coat interface but mostly within the ceramic layer (Ref 3-5).

This article is an invited paper selected from presentations at the 2008 International Thermal Spray Conference and has been expanded from the original presentation. It is simultaneously published in *Thermal Spray Crossing Borders, Proceedings of the 2008 International Thermal Spray Conference*, Maastricht, The Netherlands, June 2-4, 2008, Basil R. Marple, Margaret M. Hyland, Yuk-Chiu Lau, Chang-Jiu Li, Rogerio S. Lima, and Ghislain Montavon, Ed., ASM International, Materials Park, OH, 2008.

Qiang Zhang, Chang-Jiu Li, Yong Li, Shao-Ling Zhang, Xiu-Ru Wang, Guan-Jun Yang, and **Cheng-Xin Li**, State Key Laboratory for Mechanical Behavior of Materials, School of Materials Science and Engineering, Xi'an Jiaotong University, Xi'an, Shaanxi 710049, P.R. China. Contact e-mail: licj@mail.xjtu.edu.cn.

Recently, nanostructured coatings have received wide interest because of their extraordinary properties including hardness, strength, ductility, and toughness. Accordingly, nanostructured materials are expected to be applied to advanced thermal barrier coatings. In the last decade, several papers (Ref 6-12) have dedicated to systematic study of nanostructured materials in the area of surfaces and coatings, including thermal spray coatings.

One of the biggest challenges in thermal spraying nanomaterials is to retain the nanostructure of the feedstock to deposit. If nanostructured powder particles are completely melted during spraying, the traditional behavior of thermal spray particles, such as solidification, nucleation, and growth, will take place. Such processes will destroy the original nanostructured features of the feedstock. During plasma spraying of nanostructured ceramics it is necessary to partially melt the powder particles to achieve the necessary physical conditions for cohesion and adhesion. Therefore, a bimodal microstructure will be present in the deposited coating. The low elastic properties of nanostructured YSZ fraction may benefit from the tolerance of TBCs to the thermal strain (Ref 8). However, the performance of a superior TBC is greatly associated with the features of the bond coat underlying YSZ topcoat. In a previous study (Ref 10), it was found that a uniform oxide can be formed on a cold-sprayed nanostructured NiCrAlY coating surface.

In this study, the nanostructured YSZ and conventional YSZ were deposited on a cold-sprayed nanostructured NiCrAlY bond coat and a conventional NiCrAlY bond coat, respectively. The failure behavior of the plasma-sprayed nanostructured TBCs was investigated in comparison with that of the conventional TBCs. The features of TGO and cracking in the vicinity of TGO during thermal cyclic test were examined.

2. Experimental Materials and Procedures

The conventional and nanostructured Ni-20Cr-10Al-1Y and ZrO_2-8Y_2O_3 powder were used to deposit the bond coat and the ceramic coat, respectively. The nanostructured NiCrAlY powder was produced by the mechanical milling process in a planetary ball miller (ND6-2L, Nanjing University Tianzun Electron Co. Ltd., Nanjing, China) using conventional Ni-20Cr-10Al-1Y powder (KF-343, BGRIMM, Beijing, China). The mean particle size of the nanostructured and conventional NiCrAlY powder were 37 and 23 μm determined by laser size measurement (MASTERSIZER 2000, Malvern Instruments Ltd., UK), respectively. The oxygen contents of the conventional NiCrAlY powders, nanostructured NiCrAlY powders, and as-sprayed NiCrAlY bond coat were determined using an oxygen determinator (RO-316, LECO, USA). The oxygen contents of the conventional and nanostructured NiCrAlY powders were 0.028 and 0.75 wt.%, respectively. The particle size of the agglomerated nanostructured YSZ (DiDa Nano-Materials Co. Ltd., Hubei, China) is nominally from 40 to 90 μm, and the particle size of the conventional YSZ (XianDao Plasma Materials Co. Ltd., YiYang, China) is from 30 to 45 μm. An IN738 superalloy was used as a substrate.

The nanostructured and conventional bond coating in a thickness of 100-150 μm was deposited on the IN738 substrate by cold spraying (CS) using a homemade system (Ref 13). Helium was used as accelerating gas, operating at a pressure of 2.5 MPa. The spray nozzle has a circular shape with a throat diameter of 2 mm and exit diameter of 6 mm. The nanostructured and conventional YSZ of 200-250 μm thick was deposited on the nanostructured and conventional NiCrAlY bond coat, respectively, by air plasma spraying (APS) using a commercial plasma spray system (GDP-K, Jiujiang, 80 kW class). Argon was used as the primary plasma operating gas and hydrogen was used as an auxiliary gas. The pressures of argon and hydrogen were operated at 0.7 and 0.5 MPa, respectively. The flow rate of argon and hydrogen was fixed to 33 and 2 L/min, respectively. Nitrogen was used as powder feed gas. Plasma jet was operated to deposit the coating at the power of 40 kW.

Crystalline structure of YSZ powder and coating was characterized by X-ray diffraction diffractometer (XRD-6000, Shimadzu, Kyoto, Japan) using Cu-Kα radiation. The mean grain size of YSZ powders was calculated using the Scherrer equation (Ref 9). The average grain size of the YSZ powder was approximately 36 nm.

Figure 1 shows morphology of the nanostructured YSZ powder. Clearly, powder particles are spherical (Fig. 1a). The examination of the fractured particle cross-section at a high magnification suggests that the particle is porous (Fig. 1b). The grain size of the nanostructured NiCrAlY coating was reported elsewhere (Ref 13). The estimation yielded a grain size ranging from 15 to 50 nm.

Fig. 1 General surface morphology (a) and the morphology (b) of the fractured nanostructured YSZ powder

The surface layer of the deposited nanostructured bond coat was densified through shot-peening treatment. Subsequently, the bare bond coatings were heat treated at a temperature of 1000 °C for 4 h in an argon atmosphere prior to YSZ deposition. The thermal cyclic test of TBCs samples consisted of a 2 min oven-preheated heating at 1000 °C, 1000 °C × 27 min holding and cooling down to room temperature in 3 min by a forced air cooling in each cycle. To speed up the thermal cyclic test, after 300 cycles at a furnace temperature of 1000 °C, the same thermal cyclic test was continued except that the holding temperature was raised to 1150 °C. Isothermal oxidation test was carried out under a temperature of 1000 °C. The coating microstructure and compositions were characterized using a scanning electron microscope (SEM, VEGA II-XMU, TESCAN, Czech) with an energy dispersive spectrometer (EDS) through the cross-section. The TGO thickness was measured from the SEM micrographs. TGO thickness measurements were made at different positions for 30 times using three SEM micrographs along the interface between YSZ and the bond coat.

Peer Reviewed

3. Results and Discussion

3.1 Microstructure of the As-Sprayed Nanostructured TBC

Figure 2(a) shows the microstructure of the TBC consisting of the as-sprayed nanostructured YSZ. It can be noted in Fig. 2(b) that two types of microstructure are present in the nanostructured YSZ coating: columnar grains solidified from the melted fraction of YSZ powder and the loose microstructure retained from the unmelted YSZ powder. Therefore, YSZ coating exhibits a bimodal microstructure consisting of nanosized particles retained from the powder and microcolumnar grains formed through the solidification of the melted fraction in spray particles. The porous structures will enhance the thermal insulation effect of TBCs (Ref 6-12, 14-16). The coating porosity depends on the melting degree of powder particles during deposition. The as-sprayed bond coat was compacted and oxides were not present in the bond coat, as shown inn Fig. 2(c) and (e), although some pores were present in the nanostructured bond coating. The oxygen contents of the nanostructured and conventional bond coats were 0.72 and 0.024 wt.%, respectively, which were similar to those of the nanostructured and conventional powders. This fact indicates that the in-flight particles are subjected to little oxidation. It can be observed from Fig. 2(a), (d), and (f) that a uniform and exclusive oxide layer was formed at the interface between the YSZ and bond coating. The average thickness of oxides on the nanostructured and conventional bond coats was 1.1 ± 0.3 and 0.7 ± 0.1 μm, respectively. EDS analysis shows that the coating surface layer is rich in Al and O. This fact indicates that a dense Al_2O_3 was preferably nucleated during the heat treatment in argon atmosphere of both nanostructured NiCrAlY and conventional NiCrAlY coatings prior to YSZ deposition. Moreover, the oxide scale formed on the nanostructured NiCrAlY bond coat was denser than that on the conventional NiCrAlY bond coat. This fact implies that the uniform oxide film on the nanostructured NiCrAlY coating surface grew very rapidly during the initial stage of oxidation. This may be attributed to the nanostructure features of the nanostructured NiCrAlY coating. The small grains promote fast diffusion of elemental Al toward the surface, to provide a sufficient Al source for Al_2O_3 growth (Ref 17-19).

3.2 Microstructure of TGO Formed by Isothermal Oxidation of TBC

The typical microstructure of the nanostructured TBC sample and conventional sample isothermally oxidized for 100 h at 1000 °C are shown in Fig. 3(a) and (b). The average thickness of the oxide layer on the nanostructured and conventional bond coat, measured on polished cross-sections, was approximately 2.2 ± 0.3 and 2.0 ± 0.4 μm, respectively. It can be seen in Fig. 3(a) and (b) that the oxide layer on the bond coatings was uniform and continuous. EDS analysis shows that the oxide film mainly is composed of Al and O elements, indicating the formation of Al_2O_3. In addition, the adhesion of the bond coating to YSZ coating is enhanced by forming an exclusive α-Al_2O_3 scale at the interface between the bond coat and YSZ. No cracks were observed in the vicinity of the TGO.

3.3 Cracking of the TBCs During Thermal Cycling Test at 1000 °C

Figure 4 shows a typical cross-sectional view of a nanostructured TBC specimen after 300 times thermal cycles at the holding temperature of 1000 °C. The YSZ coating was intact after the test. A highly undulated bond coat/TGO interface was observed. These undulations are due to the deliberate roughening of the bond coat surface prior to topcoat deposition for improving the topcoat adhesion by mechanical keying. However, these undulations may give rise to an out-of-plane tension, which is primarily responsible for the ultimate failure of TBCs (Ref 1, 2).

It is worth noting in Fig. 5 that the TGO formed at the interface between the top YSZ coat and the bond coat exhibits a uniform morphology along the interface, which depends evidently on the surface morphology of the as-deposited bond coat. Microstructural observations showed cracks in the vicinity of the TGO. Cracking occurred within the topcoat at the interface between TGO and YSZ or near the crest of the bond coat undulations. The linkage of such cracks through propagation likely results in the ultimate failure of the TBC. Crack was also present at the interface between TGO and bond coat, where the TGO is thick, which is possibly due to thermal expansion mismatching, as shown in Fig. 6(a) and (b). It can be observed from the higher magnification view of the bond coat undulation crest in Fig. 6 that the separation occurred between the bond coat and the TGO at the crest or between the TGO and the topcoat. Crack also runs across the TGO. These features were commonly observed around the undulations in these TBCs. However, in local domains cracking also occurred within the topcoat and no crack was presented at the interface of the TGO/YSZ. This fact suggests that the nanostructured YSZ exhibits a high strain tolerability.

3.4 Failure Mode After Thermal Cycling at 1150 °C

Table 1 shows thermal cyclic life of two different TBCs under different thermal cyclings. It was found that the number of thermal shock cycles to failure at 1150 °C of the nanostructured TBCs was about two to four times larger than the conventional TBCs after the thermal cyclic test for 300 cycles at 1000 °C, and exhibited a little increase after the isothermal oxidation for 100 h at 1000 °C. These facts indicate that the influence of the low-temperature thermal cycling on the thermal cyclic lifetime is larger for the conventional TBCs than the nanostructured TBCs.

Figure 7 shows the cross-sectional micrograph of the failed nanostructured TBCs subjected to thermal cyclic test at 1150 °C. The sample shown in Fig. 7(a) has experienced thermal cyclic test for 300 cycles at 1000 °C before the thermal cyclic test at 1150 °C, while the sample in Fig. 7(b) experienced the isothermal oxidation test at

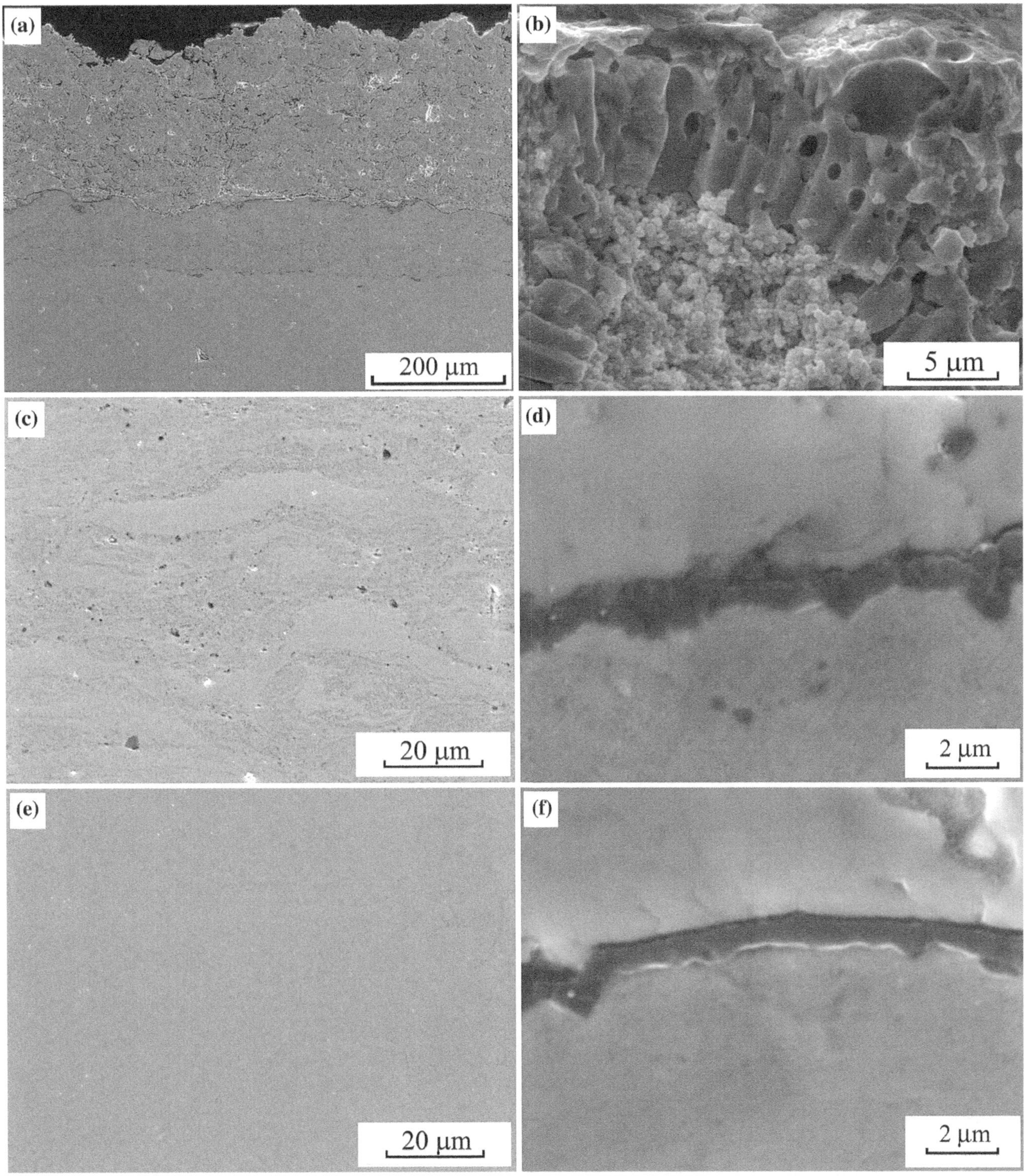

Fig. 2 Cross-sectional microstructure of intact TBCs with nanostructured NiCrAlY/nanostructured YSZ (a); fractured surface of nanostructured YSZ in the TBCs (b); microstructure of nanostructured bond coat (c); typical morphology of TGO formed on the nanostructured bond coat prior to YSZ deposition (d); microstructure of conventional bond coat (e); and typical morphology of TGO formed on the conventional bond coat (f)

1000 °C for 100 h. Two nanostructured TBCs samples withstood further 38 and 36 thermal cycles at 1150 °C, respectively. The failure of the coatings in two nanostructured TBC samples occurred within the topcoat through spalling of YSZ within YSZ coating close to the YSZ/thermal growth oxide interface. The failed cross-sectional microstructure of the conventional TBCs was shown in Fig. 8. It was noted that the failure also occurred

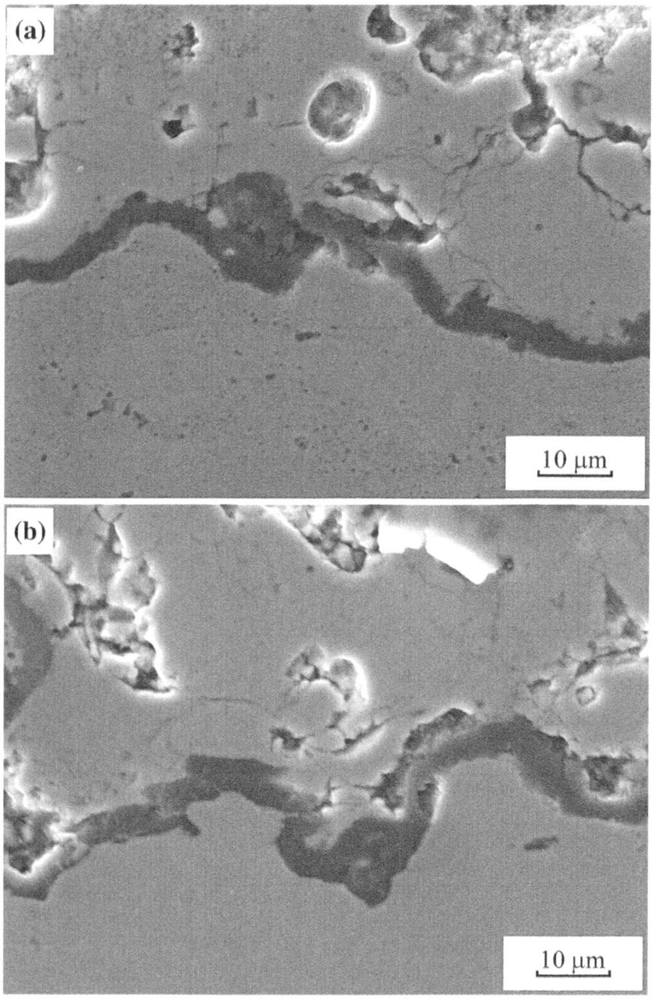

Fig. 3 Cross-sectional microstructure of nanostructured TBCs (a) and conventional TBCs (b) after the isothermal oxidation for 100 h at 1000 °C

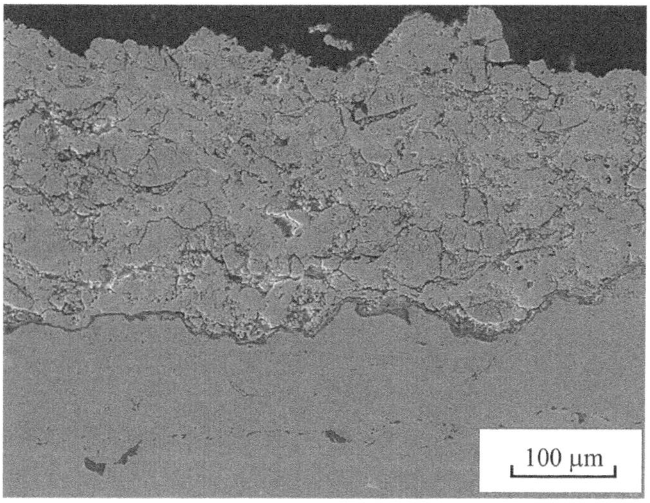

Fig. 4 Cross-sectional micrograph of nanostructured TBCs after thermal cyclic test for 300 cycles at 1000 °C

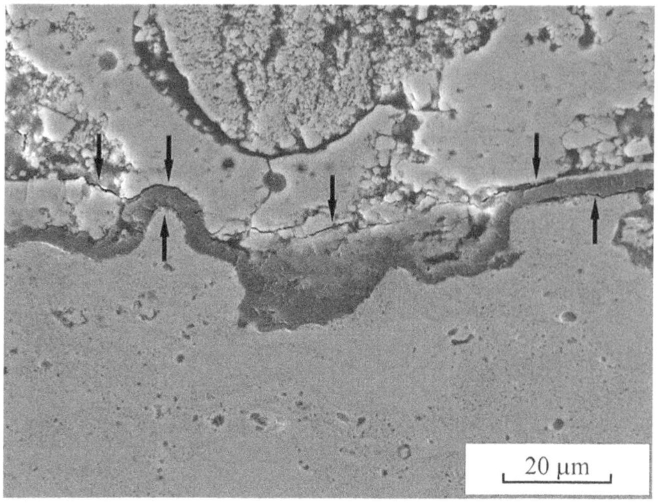

Fig. 5 Cross-sectional micrograph of nanostructured TBCs showing crack formed in the vicinity of the TGO after thermal cyclic test for 300 cycles at 1000 °C

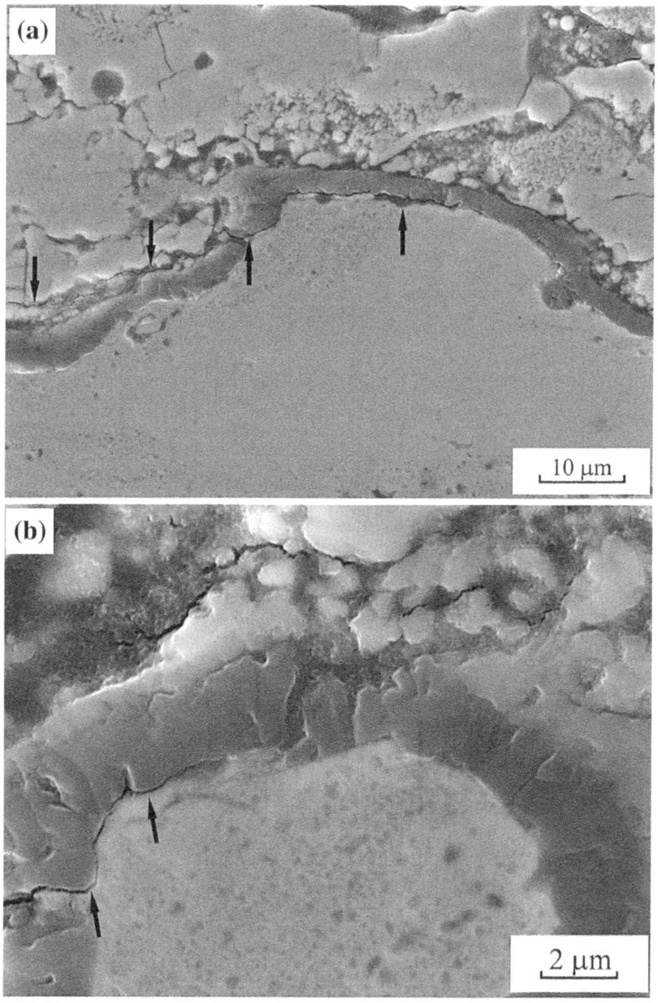

Fig. 6 Cross-sectional micrograph of nanostructured TBCs showing crack formed near the crest after thermal cyclic test for 300 cycles at 1000 °C. (a) and (b) taken by different zone

Table 1 Thermal cycles of the nanostructured and conventional TBCs

Type:	Nanostructured TBCs						Conventional TBCs					
Pretreatment:	1000 °C × 300 cycles			1000 °C × 100 h			1000 °C × 300 cycles			1000 °C × 100 h		
Life	19	38	38	23	36	45	7	10	10	24	29	35
Mean life	32			34			9			29		

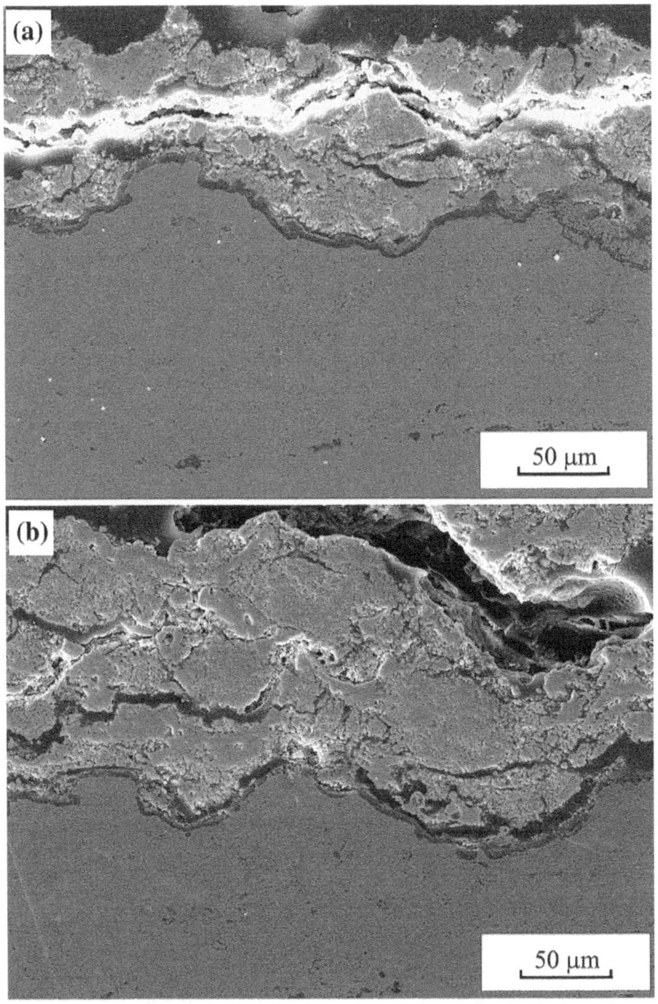

Fig. 7 Cross-sectional micrograph of the failed nanostructured TBCs. (a) The sample experienced 38 cycles at 1150 °C after 300 cycles at 1000 °C; (b) the sample experienced 36 cycles at 1150 °C after isothermal oxidation for 100 h at 1000 °C

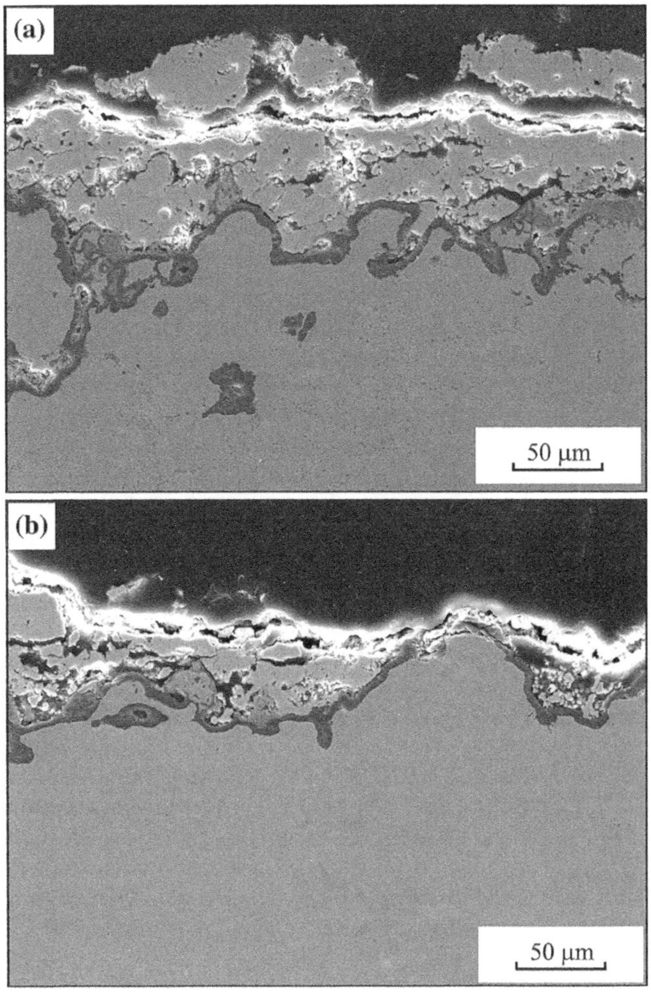

Fig. 8 Cross-sectional micrographs of the failed conventional TBCs. The sample experienced 29 cycles at 1150 °C after isothermal oxidation for 100 h at 1000 °C. (a) and (b) taken by different zone in the failed interface

within the topcoat above TGO, which indicates that the failure of the nanostructured TBCs is similar to the failure scenario of the conventional TBCs (Ref 3-5). However, it was found that the TGO adheres well to the underlying bond coat. Possibly, such well-adhered TGO is attributed to the nanostructure of the bond coat, which promotes the rapid formation of a uniform Al_2O_3 scale on the bond coat surface during the pre-heat treatment (Ref 20-24). Moreover, it was observed in Fig. 9 that the buckling of the nanostructured YSZ topcoat occurred before it spalled out. The buckling may suggest the possibly sealing of YSZ

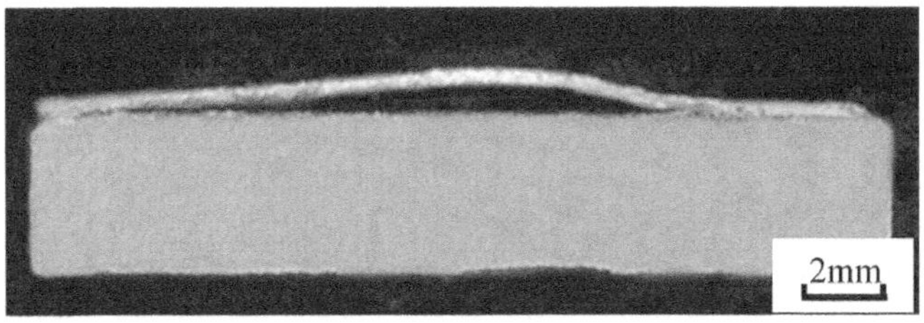

Fig. 9 Cross-sectional micrograph of the failed nanostructured TBCs showing the overview of a buckle. The samples experienced 36 cycles at 1150 °C after isothermal oxidation for 100 h at 1000 °C

topcoat surface through sintering. Although the failure mechanism of the TBC deposited by nanostructured YSZ in the present study is not yet clear, therefore, possibly the sintering of the nanosized YSZ fraction in the coating during high-temperature cyclic test is responsible for the failure of top YSZ coat (Ref 8, 25). This effect is possibly associated with the influence of the thermal cycling at 1000 °C on the thermal shock life of two different TBCs owing to much less sintering and possibly sustainable higher strain tolerability of the nanostructured TBCs. Further detailed study is necessary to reveal the influence of sintering effect and strain tolerability on the thermal failure of the nanostructured TBCs and different failure behavior of two types of the TBCs.

4. Conclusions

Nanostructured YSZ coating was deposited by partially melted YSZ particles. The nonmelted fraction of spray particles retained the porous nanostructure of the starting powder into the deposit. As a result, the nanostructured YSZ coating exhibits a bimodal microstructure consisting of nanosized particles retained from the powder and microcolumnar grains formed through the solidification of the melted fraction in spray particles. The oxidation of the cold-sprayed nanostructured and conventional NiCrAlY bond coat occurred during the heat treatment in Ar atmosphere, and the denser oxides were formed on the nanostructured NiCrAlY bond coat compared with the conventional bond coat. The uniform oxide can be formed during isothermal test at the interface between the bond coat and YSZ in two different TBCs. On the other hand, the cracks were observed at the interface between TGO/BC or TGO/YSZ after thermal cyclic test. However, the failure of both TBCs mainly occurred within the ceramic TBC near the ceramic/TGO coating interface through spalling of YSZ within YSZ coating. Nanostructured thermal barrier coatings exhibits relatively longer lifetime than the conventional coating at cyclic testing at 1150 °C.

Acknowledgments

The present work is supported by National Basic Research Program (Grant No. 2007CB707702), National Science Fund for Distinguished Young Scholars (Grant No. 50725101), and Research Foundation of National Key Laboratory for High Energy Density Beam Processing Technology (Grant No. 51461070705JW0805).

References

1. A.G. Evans, D.R. Mumm, J.W. Hutchinson, G.H. Meier, and F.S. Pettit, Mechanisms Controlling Durability of Thermal Barrier Coating, *Prog. Mater. Sci.*, 2001, **46**(5), p 505-553
2. N.P. Padture, M. Gell, and E.H. Jordan, Thermal Barrier Coatings for Gas-Turbine Engine Applications, *Science*, 2002, **296**, p 280-284
3. K.W. Schlichting, N.P. Padture, E.H. Jordan, and M. Gell, Failure Modes in Plasma-Sprayed Thermal Barrier Coatings, *Mater. Sci. Eng. A*, 2003, **342**(1-2), p 120-130
4. A. Rabiei and A.G. Evans, Failure Mechanisms Associated with the Thermally Grown Oxide in Plasma-Sprayed Thermal Barrier Coatings, *Acta Mater.*, 2000, **48**(15), p 3963-3976
5. E.P. Busso, J. Lin, and S. Sakurai, A Mechanistic Study of Oxidation-Induced Degradation in a Plasma-Sprayed Thermal Barrier Coating System: Part II: Life Prediction Model, *Acta Mater.*, 2001, **49**(9), p 1529-1536
6. C.G. Zhou, N. Wang, Z.B. Wang, S.K. Gong, and H.B. Xu, Thermal Cycling Life and Thermal Diffusivity of a Plasma-Sprayed Nanostructured Thermal Barrier Coating, *Scr. Mater.*, 2004, **51**(10), p 945-948
7. C.G. Zhou, N. Wang, and H.B. Xu, Comparison of Thermal Cycling Behavior Plasma-Sprayed Nanostructured and Traditional Thermal Barrier Coatings, *Mater. Sci. Eng. A*, 2007, **452-453**, p 569-574
8. R.S. Lima and B.R. Marple, Nanostructured YSZ Thermal Barrier Coatings Engineered to Counteract Sintering Effects, *Mater. Sci. Eng. A*, 2008, **485**(1-2), p 182-193
9. R.S. Lima, A. Kucuk, and C.C. Berndt, Integrity of Nanostructured Partially Stabilized Zirconia after Plasma Spray Processing, *Mater. Sci. Eng. A*, 2001, **313**(1-2), p 75-82
10. B. Liang and C.X. Ding, Thermal Shock Resistances of Nanostructured and Conventional Zirconia Coatings Deposited by Atmospheric Plasma Spraying, *Surf. Coat. Technol.*, 2005, **197**(1), p 185-192
11. C.X. Zhang, C.G. Zhou, H. Peng, S.K. Gong, and H.B. Xu, Influence of Thermal Shock on Insulation Effect of Nano-multilayer Thermal Barrier Coatings, *Surf. Coat. Technol.*, 2007, **201**(14), p 6340-6344
12. W.Q. Wang, C.K. Sha, D.Q. Sun, and X.Y. Gu, Microstructural Feature, Thermal Shock Resistance and Isothermal Oxidation Resistance of Nanostructured Zirconia Coating, *Mater. Sci. Eng. A*, 2006, **424**(1-2), p 1-5
13. Q. Zhang, C.J. Li, C.X. Li, G.J. Yang, and S.C. Lui, Study of Oxidation Behavior of Nanostructured NiCrAlY Bond Coatings Deposited by Cold Spraying, *Surf. Coat. Technol.*, 2008, **202**(14), p 3378-3384
14. R.S. Lima, A. Kucuk, and C.C. Berndt, Evaluation of Microhardness and Elastic Modulus of Thermally Sprayed Nanostructured Zirconia Coatings, *Surf. Coat. Technol.*, 2001, **135**(2-3), p 166-172
15. J. Karthikeyan, C.C. Berndt, J. Tikkanen, S. Reddy, and H. Herman, Plasma Spray Synthesis of Nanomaterial Powders and Deposits, *Mater. Sci. Eng. A*, 1997, **238**(2), p 275-286
16. R.S. Lima, A. Kucuk, and C.C. Berndt, Bimodal Distribution of Mechanical Properties on Plasma Sprayed Nanostructured Partially Stabilized Zirconia, *Mater. Sci. Eng. A*, 2002, **327**(2), p 224-232
17. Z.Y. Liu, W. Gao, K.L. Dahm, and F.H. Wang, Improved Oxide Spallation Resistance of Microcrystalline Ni-Cr-Al Coating, *Oxid. Met.*, 1998, **50**(1-2), p 51-69
18. L. Ajdelsztajn, F. Tang, G.E. Kim, V. Provenzano, and J.M. Schoenung, Synthesis and Oxidation Behavior of Nanocrystalline MCrAlY Bond Coatings, *J. Therm. Spray Technol.*, 2005, **14**(1), p 23-30
19. D. Toma, W. Brandl, and U. Köster, The Characteristics of Alumina Scales Formed on HVOF-Sprayed MCrAlY Coatings, *Oxid. Met.*, 2000, **53**(1), p 125-137
20. H. Lau, C. Leyens, U. Schulz, and C. Friedrich, Influence of Bond Coat Pre-Treatment and Surface Topology on the Lifetime of EB-PVD TBCs, *Surf. Coat. Technol.*, 2003, **165**(3), p 217-223
21. M. Matsumoto, T. Kato, K. Hayakawa, N. Yamaguchi, S. Kitaoka, and H. Matsubara, The Effect of Pre-Oxidation Atmosphere on the Durability of EB-PVD Thermal Barrier Coatings with CoNiCrAlY Bond Coats, *Surf. Coat. Technol.*, 2008, **202**(12), p 2743-2748
22. R. Darolia, Post-Deposition Oxidation of a Nickel-Base Superalloy Protected by a Thermal Barrier Coating, US Patent 6,607,611, 19 August 2003

23. M.J. Weimer, B.A. Nagaraj, A. Bangalore, J.C. Schaeffer, and J.A. Heaney, Pre-Service Oxidation of Gas Turbine Disk and Seal, US Patent 6,532,657, 18 March 2003
24. T.J. Nijdam, L.P.H. Jeurgens, J.H. Chen, and W.G. Sloof, On the Microstructure of the Initial Oxide Grown by Controlled Annealing and Oxidation on a NiCoCrAlY Bond Coating, *Oxid. Met.*, 2005, **64**(5-6), p 355-377
25. O. Racek, C.C. Berndt, D.N. Guru, and J. Heberlein, Nanostructured and Conventional YSZ Coatings Deposited Using APS and TTPR Techniques, *Surf. Coat. Technol.*, 2006, **201**(1-2), p 338-346

JTTEE5 17:846–852
DOI: 10.1007/s11666-008-9217-x
1059-9630/$19.00 © ASM International

Peer Reviewed

Toward Highly Sintering-Resistant Nanostructured ZrO_2-7wt.%Y_2O_3 Coatings for TBC Applications by Employing Differential Sintering

R.S. Lima and B.R. Marple

(*Submitted May 1, 2008; in revised form July 16, 2008*)

There are still concerns in the scientific community about the stability of nanostructured YSZ coatings at high temperatures. Questions have been raised about the possibility of accelerated sintering of these ultrafine materials and the associated changes in properties that could accompany this sintering. In this work, nanostructured YSZ coatings were engineered to counteract sintering effects by tailoring the coatings to exhibit a bimodal microstructure formed by (i) a matrix of dense YSZ zones (produced from molten YSZ particles) and (ii) large porous nanostructured YSZ zones (produced from semimolten nanostructured YSZ particles) that were embedded in the coating microstructure during thermal spraying. These coatings were subjected to heat treatment in air at 1400 °C for 1, 5, and 20 h. The superior driving force for sintering exhibited by the porous nanozones, when compared to that of the dense zones, caused the nanozones to shrink at much faster rates than those exhibited by the denser matrix zones (i.e., differential sintering), thereby creating a significant network of voids in the coating microstructure. Due to these effects, after 20 h exposure at 1400 °C, the thermal conductivity and elastic modulus values of the conventional coatings were approximately two times higher than those of the nanostructured ones.

Keywords differential sintering, elastic modulus, nanostructured ZrO_2-7wt.%Y_2O_3 (YSZ), thermal barrier coatings (TBCs), thermal conductivity

1. Introduction

1.1 Nanostructured YSZ Coatings

There is an ongoing need to develop materials and strategies to enhance the performance of thermal barrier coatings (TBCs). Different approaches have been employed to achieve this goal. One of the possible approaches is based on the thermal spraying of porous nanostructured agglomerated ZrO_2-7-8wt.%Y_2O_3 (YSZ) powders, in which individual nanostructured YSZ particles are agglomerated via spray-drying into microscopic clusters.

By using this type of feedstock, it is possible to engineer coatings that exhibit a bimodal microstructure, which is formed by particles that were (i) fully molten and (ii) semimolten in the spray jet (Ref 1, 2). The particles that were fully or almost fully molten in the spray jet will form the matrix that will hold the semimolten particles in the coating microstructure (Ref 1, 2). The semimolten agglomerates, if embedded in porous form in the coating microstructure (i.e., without the complete infiltration of molten material into the capillary network of each agglomerate), will provide the coating a distinct bimodal behavior concerning its elasto-plastic response (Ref 1).

The preliminary high temperature evaluations of this type of nanostructured YSZ coating (produced from agglomerated powders) have been promising. Liang and Ding (Ref 3) evaluated the thermal shock resistance of nano and conventional YSZ coatings deposited via air plasma spray (APS). The coatings were heated at different temperatures up to 1300 °C for 30 min, followed by subsequent quenching in cool water for 10 min. In the temperature range of 1000 to 1300 °C, the number of cycles to failure of the nanostructured coatings was approximately 2-3 times higher than those of the conventional ones. Wang et al. (Ref 4) heated nanostructured and conventional YSZ APS coatings at 1200 °C for 5 min in a furnace. The coatings were quenched in water (room temperature). The number of cycles to failure of the

This article is an invited paper selected from presentations at the 2008 International Thermal Spray Conference and has been expanded from the original presentation. It is simultaneously published in *Thermal Spray Crossing Borders, Proceedings of the 2008 International Thermal Spray Conference*, Maastricht, The Netherlands, June 2-4, 2008, Basil R. Marple, Margaret M. Hyland, Yuk-Chiu Lau, Chang-Jiu Li, Rogerio S. Lima, and Ghislain Montavon, Ed., ASM International, Materials Park, OH, 2008.

R.S. Lima and **B.R. Marple**, National Research Council of Canada, 75 de Mortagne Blvd., Boucherville J4B 6Y4, QC, Canada. Contact e-mail: rogerio.lima@cnrc-nrc.gc.ca.

nanostructured coatings was 2-4 times higher than those of the conventional ones.

1.2 Differential Sintering and Nanomaterials

Despite these interesting preliminary results, there are still open questions concerning the stability of these coatings at high temperatures. Based on sintering principles, it is frequently hypothesized that these nanostructured YSZ coatings will exhibit excessive sintering rates at temperatures above 1200 °C, thereby impeding the application of these coatings as TBCs due to significant increases of thermal conductivity and elastic modulus (stiffness) values, which will lead to premature coating failure.

However, previous work has shown that the application of the concept of differential sintering may be an important strategy to counteract sintering effects in TBCs (Ref 5). This concept is based on the differential sintering rates exhibited by coatings that possess bimodal microstructures engineered from porous nanostructured agglomerated powders, and will be explained in more detail in the following sections. The present work aims at further investigating these effects by determining with a higher degree of precision the thermal conductivity and elastic modulus (E) values of nanostructured and conventional YSZ APS coatings subjected to heat treatment in air at 1400 °C for 1, 5, and 20 h.

2. Experimental Procedure

2.1 Thermal Spraying and Feedstock Powders

Two powders, one nanostructured and one conventional YSZ (ZrO_2-7-8wt.%Y_2O_3), were employed in this study. The nanostructured YSZ (Nanox S4007, Inframat Corp., Farmington, CT) was sprayed using an Ar/H_2 APS torch (F4-MB, Sulzer-Metco, Westbury, NY). The conventional YSZ (Metco 204B-NS, Sulzer-Metco, Westbury, NY) was also sprayed using an Ar/H_2 APS torch (9MB (GH nozzle), Sulzer-Metco, Westbury, NY). The particle size distribution of both powders was evaluated by a laser diffraction particle analyzer (Beckman Coulter LS 13320, Beckman Coulter, Miami, FL). The as-received nanostructured powder exhibited a powder size distribution varying from approximately 10 to 150 μm. Sieving was employed to remove smaller particles from this initial size distribution to produce a distribution containing coarser particles. The as-received nanostructured powder was sieved using a 53 μm (Mesh 270) USA Standard Testing Sieve in order to try to obtain a particle size distribution range of ~50 to 150 μm.

The conventional powder was sprayed based on the standard spray conditions recommended by the manufacturer of the powder and torch (i.e., Sulzer-Metco). The spray parameters employed to deposit the nanostructured coating were developed internally. Before coating deposition, the temperature and velocity of the sprayed particles were measured via a diagnostic tool (Accuraspray, Tecnar Automation, Saint-Bruno, QC, Canada). The particle detector was placed at the same spray distance as used to deposit the coatings, i.e., 10 and 11 cm from the torch nozzle for the nanostructured and conventional powders, respectively. Both powders were sprayed on low-carbon steel substrates. The maximum coating surface temperature for all depositions was approximately 160 °C (measured via a pyrometer). Coating thickness was approximately 450-500 μm. The powder feed rate (for both powders) used for coating deposition was 30 g/min.

2.2 Heat Treatment

Before heat treatment the low-carbon steel substrates were dissolved in an acidic liquid to produce freestanding YSZ coatings. The heat treatment was carried out in a furnace (in air) at 1400 °C for periods of 1, 5, and 20 h. The furnace reached the plateau of 1400 °C after a dwell time of 60-90 min. After the heat treatment, the samples were removed from the furnace and allowed to cool at room temperature, which was reached after approximately 30 min.

2.3 Microstructural Evaluation

The structural characteristics of powders and coatings were evaluated using a field-emission scanning electron microscope (FE-SEM) (S-4700, Hitachi Ltd., Tokyo, Japan). During the metallographic preparation, to better preserve and reveal the true structural features of the coatings, the following procedure was carried out. The coatings were first vacuum impregnated and mounted in epoxy resin, then cut using a diamond saw. Subsequently, the pre-impregnated coatings were remounted in epoxy resin using again vacuum impregnation and polished using standard metallographic procedures.

2.4 Mercury Intrusion Porosimetry

The mercury intrusion porosimetry (MIP) technique was used to determine the density and porosity distribution of the as-sprayed and heat-treated freestanding coatings. The MIP evaluation was carried out by a private laboratory (Poremaster 60, Quantachrome Instruments, Boynton Beach, FL). Just one sample per coating was evaluated.

2.5 Thermal Conductivity

The thermal conductivity value (k) (W/mK) can be calculated by the following equation: $k = \alpha \times C_p \times \rho$; where α, C_p, and ρ are, respectively, the thermal diffusivity (m^2/s), heat capacity (J/kg · K), and density (kg/m^3) values of the coating. The thermal diffusivity values were obtained via the laser-flash method. The experimental conditions in which these values were obtained are described in more detail in a previous reference (Ref 5). One sample per coating was evaluated. The density values of the coatings were obtained via MIP. The value of heat capacity for the YSZ coatings was taken from the

Peer Reviewed

literature. Antou et al. (Ref 6) measured the heat capacity of a YSZ APS coating continuously from room temperature up to ~550 °C. It was observed that the specific heat of the coating was approximately constant within this temperature range: average value (three samples) of 454 ± 17 J/kg · K.

2.6 Elastic Modulus

The *E* values of as-sprayed and heat-treated free-standing coatings were measured via laser-ultrasonics on the in-plane direction, i.e., parallel to the (original) substrate surface. This measurement is done in the following way. A laser pulse on the coating surface generates an ultrasound wave that propagates and is detected by a second laser located on the same surface but positioned away from the generating source of pulses. The velocity of the acoustic wave on the coating is measured based on the known distance between the source laser and the detection laser, and the time taken from the pulse generation to its detection. The *E* value (Pa) of a material (coating) can then be calculated by using the following equation: $E = \rho \times v^2$; where ρ and *v* are, respectively, coating density (kg/m^3) (measured via MIP in this work) and laser ultrasound velocity values in the coating (m/s). Just one sample per coating was evaluated; however, each velocity value represents an average of 5 to 10 measurements repeated twice. More details on these measurements can be found in previous work (Ref 5).

3. Results and Discussion

3.1 Particle Size Distribution

The particle size distributions of the sieved nanostructured and as-received conventional YSZ powders are shown in Fig. 1. It is possible to observe that the nanostructured powder exhibits a particle size distribution significantly larger than those of typical ceramic thermal spray powders, i.e., almost 50% of the particles exhibit diameters larger than 100 μm. As shown in the next sections, this characteristic is paramount for engineering the differential sintering phenomenon.

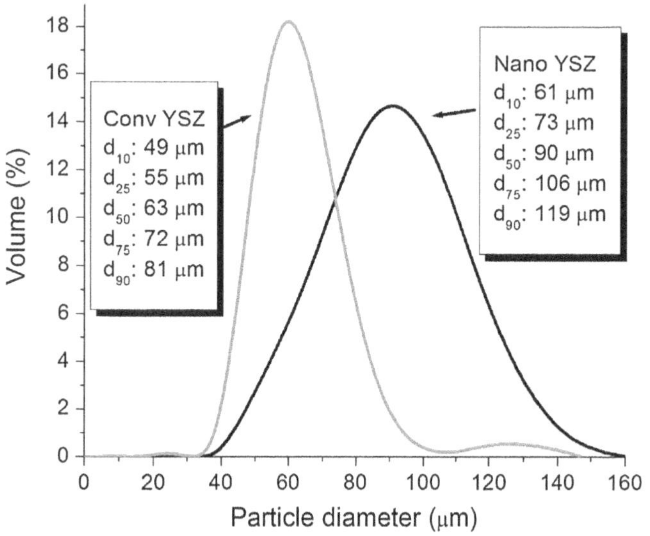

Fig. 1 Particle size distribution of the conventional and sieved nanostructured YSZ powders

3.2 Powder Morphology

The surface characteristics of the nanostructured and conventional YSZ powders can be observed in Fig. 2 and 3, respectively. The nanostructured powder was produced (agglomerated) via spray-drying and sintering of individual nanostructured YSZ particles. It is possible to distinguish the porous and ultrafine characters of the agglomerates.

The conventional YSZ powder was produced via spray-drying and subsequently plasma spheroidization to form hollow spherical particles. Due to this process, the

Fig. 2 Surface of the nanostructured YSZ powder

Fig. 3 Surface of the conventional YSZ powder

individual spray-dried particles were almost completely molten, and without any significant ultrafine character (Fig. 3).

3.3 In-flight Particle Characteristics

The average particle temperatures for the nanostructured and conventional YSZ powders were 2670 °C and 2700 °C, respectively. It is important to point out that the melting point of ZrO_2-7-8wt.%Y_2O_3 is ~2700 °C (Ref 7). Therefore, it is expected that the YSZ powders arriving at the substrate were composed of fully molten, semimolten, and even nonmolten particles. This is particularly true for the nanostructured feedstock due to its larger overall diameter (Fig. 1). This is an important characteristic to preserve and embed part of the original porous ultrafine structure of the feedstock (Fig. 2) in the coating microstructure. The average particle velocities for the nanostructured and conventional YSZ powders were 210 m/s and 148 m/s, respectively.

3.4 As-sprayed Coatings: Microstructure

The microstructure of the conventional YSZ coating is shown in Fig. 4. It exhibits the typical lamellar characteristics and pore structure of YSZ coatings deposited by APS for TBC purposes. The microstructure of the nanostructured YSZ coating contains distinct features (Fig. 5). It is possible to observe a bimodal microstructure, consisting of lighter-colored (dense) and darker-colored regions (Fig. 5a). When the darker zones of the microstructure are observed at higher magnifications, it is possible to distinguish a porous ultrafine structure (Fig. 5b) similar to that of the nanostructured feedstock (Fig. 2). Therefore, these regions represent the semimolten porous nanostructured agglomerated particles that were embedded in the coating microstructure (i.e., the porous nanozones) during thermal spraying. The lighter (dense) zones of Fig. 5(a) represent the previously fully molten nanostructured YSZ particles that form the matrix of the coating microstructure. The percentage of porous nanozones (in cross-sectional area) embedded in the coating microstructure was estimated to be ~35%. A total of 10 SEM pictures were taken (at 500×) and the nanozones were manually selected via image analysis to perform this evaluation. As shown in the next sections, these porous nanozones have a fundamental role during the process of differential sintering.

3.5 Heat-treated Coatings: Microstructure

The microstructures of the heat-treated nanostructured YSZ coatings are shown in Fig. 6. It is possible to observe that the size and percentage in the cross-sectional area of coarse pores/voids (dark zones) tended to increase with increasing heat treatment time. By comparing the microstructural evolution of these coatings with the microstructural characteristics of the as-sprayed one (Fig. 5), it is somewhat evident that the opening of the coarse pores and voids occurred in the porous nanozones at the interface with the denser matrix regions.

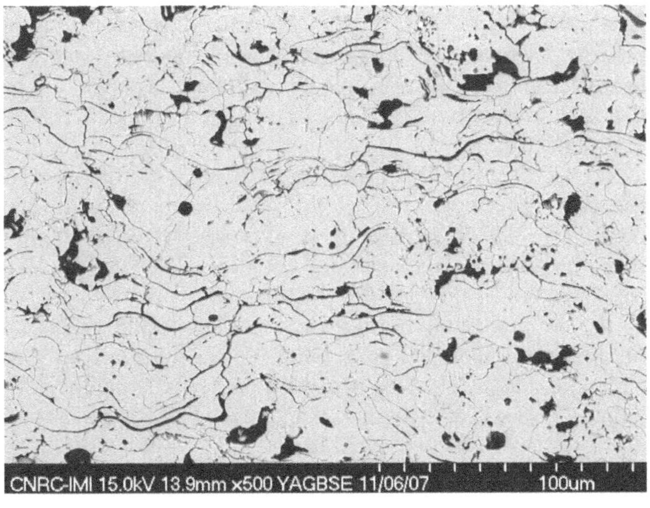

Fig. 4 As-sprayed conventional YSZ coating

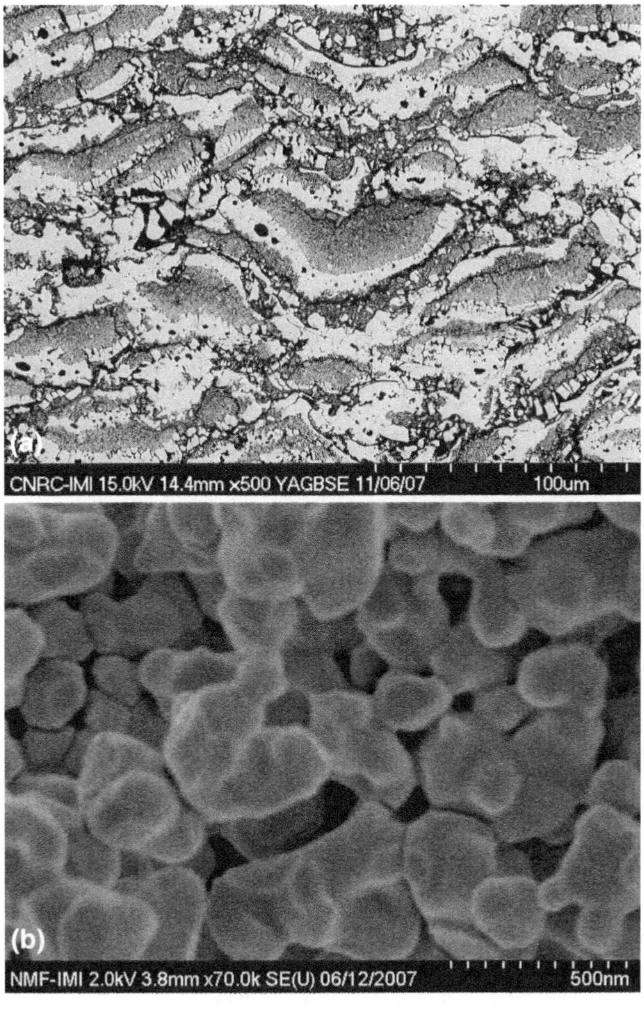

Fig. 5 As-sprayed nanostructured YSZ coating observed at low (a) and high magnifications (b)

Peer Reviewed

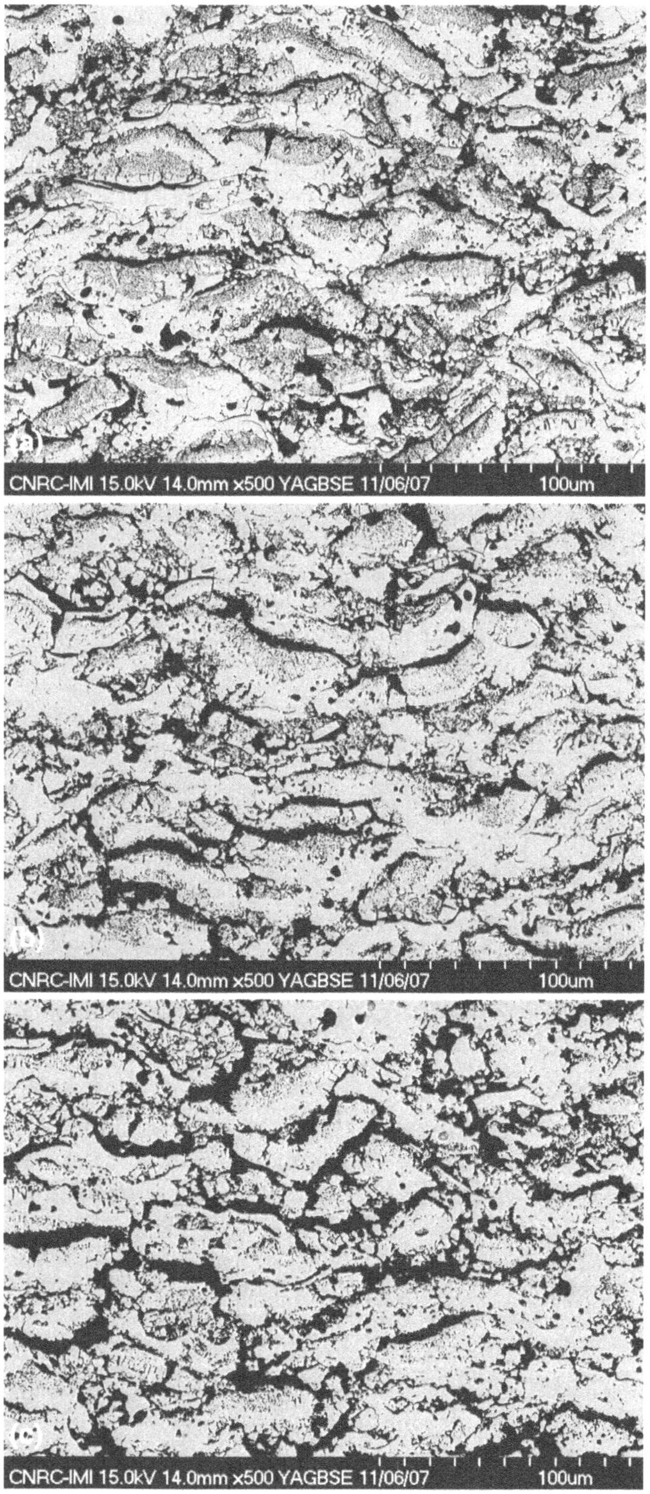

Fig. 6 Heat-treated (1400 °C) nanostructured YSZ coatings: (a) 1 h, (b) 5 h, and (c) 20 h

The porous nanozones exhibit a much higher driving force for sintering (higher surface area) than that of the dense matrix, which was formed by the particles that were fully or almost fully molten in the spray jet. The difference in the driving force for sintering between the two zones gives rise to a differential sintering effect.

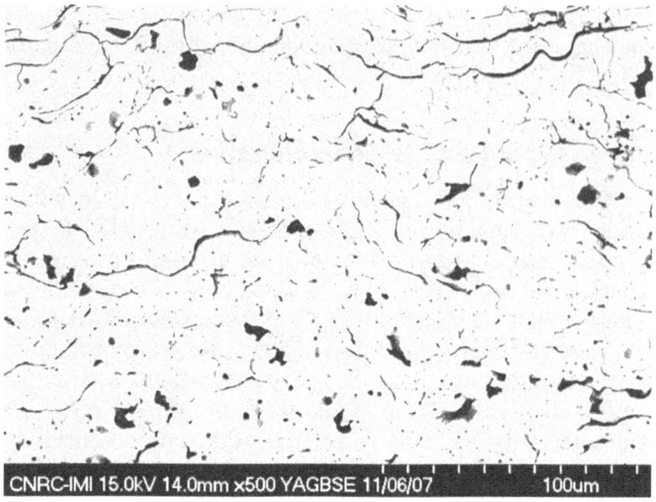

Fig. 7 Heat-treated (1400 °C/20 h) conventional YSZ coating

The conventional YSZ coating exhibited the typical behavior of a ceramic material exposed at high temperatures for long periods, i.e., the size of coarse pores and voids decreased with heat treatment time. This characteristic can be observed when comparing the microstructures of the as-sprayed (Fig. 4) and heat-treated (1400 °C/20 h) (Fig. 7) conventional coatings.

The porosity values for the as-sprayed and heat-treated YSZ coatings exhibited a range of 11 to 16% for the conventional coatings and 28 to 32% for the nanostructured ones, respectively. This difference in the sintering behavior between nanostructured and conventional YSZ coatings will have a significant impact on the evolution of the elastic and thermal conductivity properties at high temperatures, as depicted in the next sections.

3.6 Evolution of Thermal Conductivity

The evolution of thermal conductivity values caused by high temperature exposure for nanostructured and conventional YSZ coatings is shown in Fig. 8. The thermal conductivity value of the as-sprayed conventional coating is almost twice that of the as-sprayed nanostructured coating. Moreover, this difference remains after high temperature exposure. After 1 h exposure at 1400 °C, the thermal conductivity values of both coatings increased substantially.

However, after this point up to 20 h of heat exposure, the thermal conductivity growth rate of the conventional coating [~0.024 (W/mK)/h] is almost five times higher than that of the nanostructured coating [~0.005 (W/mK)/h]. After the 20 h exposure, the thermal conductivity value of the conventional coating is two times higher than that of the nanostructured one. It has to be pointed out that Zhu and Miller (Ref 8) reported a thermal conductivity value of ~1.45 W/mK for a conventional APS YSZ coating after an exposure of 20 h at 1320 °C, i.e., within the range found in this study for the conventional coating after a 20 h exposure at 1400 °C (Fig. 8). In fact, the thermal conductivity value of the nanostructured coating after 20 h of heat treatment (~0.9 W/mK) is within the range of values generally reported for as-sprayed conventional YSZ coatings, i.e., ~1 W/mK (Ref 8).

3.7 Evolution of Elastic Modulus

The evolution of *E* values caused by high temperature exposure for nanostructured and conventional YSZ coatings is shown in Fig. 9. The *E* value of the as-sprayed conventional coating is about eight times higher than that of the as-sprayed nanostructured one. After 1 h of heat treatment, the *E* values of both coatings increase significantly; however, for longer heat treatment times the *E* values of the conventional coating continue to increase almost linearly at a rate of ~2.4 GPa/h, whereas the *E* values of the nanostructured coating exhibit a much smaller rate of increase of ~0.5 GPa/h. After a 20 h exposure at 1400 °C, the *E* value of the conventional coating is about twice that of the nanostructured one.

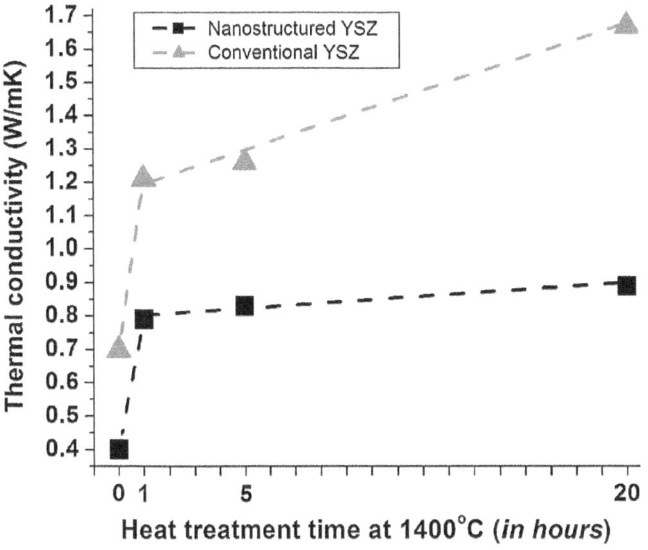

Fig. 8 Evolution of thermal conductivity values from as-sprayed to heat-treated coatings

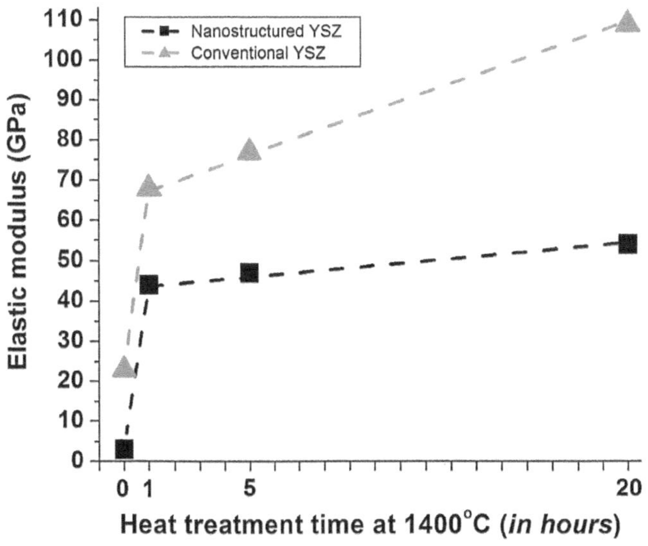

Fig. 9 Evolution of elastic modulus values from as-sprayed to heat-treated coatings

3.8 Role of Differential Sintering

The process of differential sintering has a major role in the evolution of thermal conductivity and elastic modulus values of the YSZ coatings observed in this study. Concerning the nanostructured coating, porous nanostructured agglomerates (nanozones) embedded in the coating microstructure (Fig. 5) during thermal spraying exhibited much higher sintering rates than those of the fully molten ones, which formed the dense matrix of the system. The driving force for this higher sintering rate was strongly associated with the higher surface area provided by the nonmolten and noninfiltrated YSZ nanoparticles. At high temperature exposure, these porous nanozones shrank much faster than the matrix, which led to the opening of large micron-sized voids in the coating structure (Fig. 6). It is important to point out that both coatings exhibit an overall densification of their microstructures. The density values of the conventional coating increased from 4.35 g/cm^3 (as-sprayed) to 4.79 g/cm^3 (1400 °C/20 h), whereas those of the nanostructured coating increased from 3.92 g/cm^3 (as-sprayed) to 4.05 g/cm^3 (1400 °C/20 h). However, the opening of the coarse micron-sized voids in the nanostructured coating as shown in Fig. 5 and 6, counteracted sintering effects on thermal conductivity and stiffness, as observed in Fig. 8 and 9. MIP analysis confirms this trend (Fig. 10). The overall size of the pores of the nanostructured coating increased after the heat treatment, whereas the overall size of the pores of the conventional coating decreased after heat treatment (Fig. 10), as also observed in Fig. 4 and 7, i.e., the typical behavior of conventional materials.

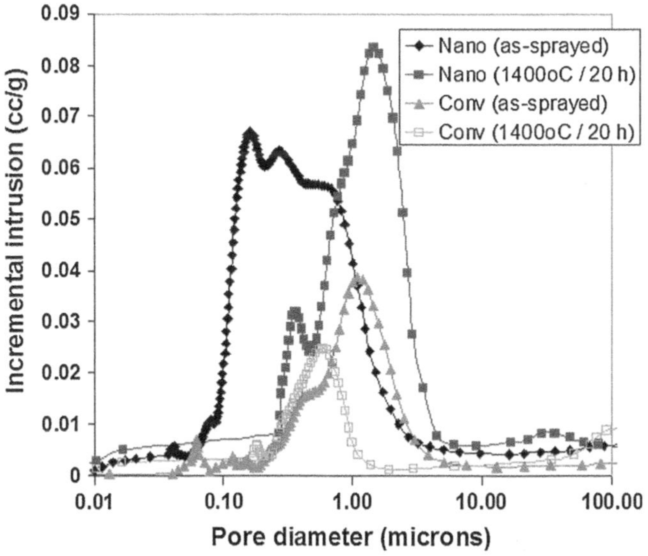

Fig. 10 Evolution of pore diameter values measured via MIP from as-sprayed to heat-treated coatings at 1400 °C/20 h

It is important to point out that Racek et al. (Ref 9) also observed void opening caused by heat treatment due to the shrinkage of porous nanozones in nanostructured YSZ thermal spray coatings. However, in that study the evolution of thermal conductivity and elastic properties with temperature were not investigated.

3.9 Final Comments

Kulkarni et al. (Ref 10) carried out a detailed comparison about the effect of the morphology of YSZ powders on the microstructure, thermal conductivity, and elastic modulus values of APS YSZ coatings. Four powder morphologies were employed: fused and crushed (F&C), sol gel manufactured (SG), spray-dried and sintered (SD), and hollow sphere (HS).

The YSZ coatings produced from the HS powder exhibited the lowest values of thermal conductivity and elastic modulus of all coatings produced, even though they were not the most porous coatings. In fact, the SD coating was the most porous of the four types of coatings, but due to a series of microstructural factors, this higher porosity did not translate into lower thermal conductivity and elastic modulus values.

In another paper of Kulkarni et al. (Ref 11), conventional APS YSZ coatings produced from SD and HS powders were heated-up to 1400 °C (the same temperature used in this study) and their thermal conductivity values were measured. The same trend was observed, i.e., the thermal conductivity values of the HS coatings were lower than those of the SD coatings after high temperature exposure.

Therefore, based on thermal conductivity and elastic modulus values, it seems to be logical to choose HS YSZ coatings (the best performing material) to be compared to these nanostructured YSZ coatings. In addition, the HS YSZ powders are among the most employed powders today for the production of YSZ coatings for TBC applications. Therefore, it is the current benchmark of the turbine industry. It is for these reasons that the conventional YSZ powder employed in this work is also a HS type.

It has to be stressed that the microstructures of the conventional YSZ coating (Fig. 4) of this study exhibited the traditional lamellar structure regularly observed for ceramic thermal spray coatings, whereas that of the nanostructured YSZ coating is a novel microstructure (Fig. 5). To carry out this comparison of microstructures was one of the main objective of the authors.

Finally, it is recognized that it would have been interesting to include a micro-SD YSZ powder with similar particle size distribution to that of the nanostructured YSZ powder (i.e., ~50-150 μm) (Fig. 1) in this study. It could help in improving the understanding of the nanoeffect in the coating properties and high temperature behavior. This is an ongoing work and this type of experiment should be done in the future.

4. Conclusions

It is possible to engineer nanostructured YSZ coatings for TBC applications that counteract sintering effects via differential sintering. This effect is achieved by embedding semimolten porous nanostructured YSZ agglomerates in the coating microstructure during thermal spraying under a careful control of particle size distribution and in-flight particle characteristics. These porous nanozones exhibit a higher driving force for sintering when compared to that of the coating matrix due to their larger surface area, thereby shrinking at faster rates and opening voids within the coating microstructure at elevated temperatures. This localized void creation counteracts overall sintering (densification). After a temperature exposure of 1400 °C for 20 h, the thermal conductivity and *E* values of the conventional YSZ coating are about two times higher than those of the nanostructured material. This result is considered as very promising.

References

1. R.S. Lima, A. Kucuk, and C.C. Berndt, Bimodal Distribution of Mechanical Properties on Plasma Sprayed Nanostructured Partially Stabilized Zirconia, *Mater. Sci. Eng. A*, 2002, **327**, p 224-232
2. R.S. Lima and B.R. Marple, Thermal Spray Coatings Engineered from Nanostructured Ceramic Agglomerated Powders for Structural, Thermal Barrier and Biomedical Applications: A Review, *J. Therm. Spray Technol.*, 2007, **16**(1), p 40-63
3. B. Liang and C. Ding, Thermal Shock Resistances of Nanostructured and Conventional Zirconia Coatings Deposited by Atmospheric Plasma Spraying, *Surf. Coat. Technol.*, 2005, **197**, p 185-192
4. W.Q. Wang, C.K. Sha, D.Q. Sun, and X.Y. Gu, Microstructural Feature, Thermal Shock Resistance and Isothermal Oxidation Resistance of Nanostructured Zirconia Coating, *Mater. Sci. Eng. A*, 2006, **424**, p 1-5
5. R.S. Lima and B.R. Marple, Nanostructured YSZ Thermal Barrier Coatings Engineered to Counteract Sintering Effects, *Mater. Sci. Eng. A*, 2008, **485**, p 182-193
6. G. Antou, F. Hlawka, A. Cornet, C. Becker, D. Ruch, and A. Richie, In Situ Laser Remelted Thermal Barrier Coatings: Thermophysical Properties, *Surf. Coat. Technol.*, 2006, **200**, p 6062-6072
7. K. Muraleedharan, J. Subrahmanyam, and S.B. Bhaduri, Identification of t' Phase in ZrO_2-7.5wt%Y_2O_3 Thermal-Barrier Coatings, *J. Am. Ceram. Soc.*, 1988, **71**(5), p C-226-227
8. D. Zhu and R.A. Miller, Thermal Conductivity and Elastic Modulus Evolution of Thermal Barrier Coatings under High Heat Flux Conditions, *J. Therm. Spray Technol.*, 2000, **9**(2), p 175-180
9. O. Racek, C.C. Berndt, D.N. Guru, and J. Heberlein, Nanostructured and Conventional YSZ Coatings Deposited Using APS and TTPR Techniques, *Surf. Coat. Technol.*, 2006, **201**, p 338-346
10. A. Kulkarni, Z. Wang, T. Nakamura, S. Sampath, A. Goland, H. Herman, J. Allen, J. Ilavsky, G. Long, J. Frahm, and R.W. Steinbrech, Comprehensive Microstructural Characterization and Predictive Property Modeling of Plasma-Sprayed Zirconia Coatings, *Acta Mater.*, 2003, **51**, p 2457-2475
11. A. Kulkarni, A. Vaidya, A. Goland, S. Sampath, and H. Herman, Processing Effects on Porosity-Property Correlations in Plasma Sprayed Yttria-Stabilized Zirconia Coatings, *Mater. Sci. Eng. A*, 2003, **359**, p 100-111

JTTEE5 17:853–857
DOI: 10.1007/s11666-008-9244-7
1059-9630/$19.00 © ASM International

Peer Reviewed

Development of Oxide Dispersion Strengthened MCrAlY Coatings

K. Bobzin, T. Schläfer, K. Richardt, and M. Brühl

(Submitted June 16, 2008; in revised form August 18, 2008)

MCrAlY materials are widely used as bond coats for thermal barrier coatings on turbine blades. The aim of this work is to improve mechanical properties and wear resistance of thermal sprayed NiCoCrAlY-coatings by strengthening the coating with hard phase particles. In order to retain the effect of the dispersion reinforcement at high temperatures, the use of temperature-stable oxide hard phases such as ZrO_2-Y_2O_3 is necessary. To realize this new material structure, the high-energy ball-milling process is applied and analyzed. The mixture ratio between NiCoCrAlY and ZrO_2-Y_2O_3 was varied between 5 and 10 wt.% ZrO_2-Y_2O_3. The influences of the milling time of the high-energy ball-milling process on the distribution of the hard phases in the metal matrix were analyzed. After spraying with a HVOF system the mechanical properties of the coatings are measured and compared with conventional NiCoCrAlY coatings.

Keywords coatings for gas turbine components, composite materials, feedstock, HVOF spray parameters, novel materials, production/preparation technology, wear, wear and corrosion

1. Introduction

MCrAlYs are among the most important protective coating materials applied to counteract hot corrosion and high-temperature oxidation in hot parts of turbine like blades or duct segments. Due to the particulate nature of the aspirated air and the high centrifugal force at the blades, a high mechanical strength and hardness are important features of the protective MCrAlY coatings. To increase the mechanical strength of coatings, metal matrix composites (MMC) were developed during the recent years (Ref 1). In most cases, carbides such as Cr_3C_2 or WC were used as strengthening hard phases. But these materials pyrolyze at temperatures of about 500-800 °C and therefore applications in turbines are very limited.

Because of that MMC with oxide hard phases were developed for high-temperature applications. The high-energy ball-milling technique is an effective method for producing powders for MMCs (Ref 2, 3). In the milling process, for ductile/brittle material combinations like NiCoCrAlY and ZrO_2, the particles of the brittle powder are broken into fine particles, cut into the metal matrix, and surrounded by the metal matrix.

In order to achieve a good hot gas corrosion and oxidation resistance for the MCrAlY coatings, a very dense coating is essential. Due to the high kinetic energy of the particles, the high velocity oxy-fuel (HVOF) process leads to dense coatings. Furthermore, it has been confirmed that MCrAlY coatings can be produced by HVOF spraying with equivalent mechanical properties to vacuum plasma sprayed MCrAlY coatings (Ref 4, 5). Therefore the developed powder will be sprayed by the HVOF system.

This article is an invited paper selected from presentations at the 2008 International Thermal Spray Conference and has been expanded from the original presentation. It is simultaneously published in *Thermal Spray Crossing Borders, Proceedings of the 2008 International Thermal Spray Conference*, Maastricht, The Netherlands, June 2-4, 2008, Basil R. Marple, Margaret M. Hyland, Yuk-Chiu Lau, Chang-Jiu Li, Rogerio S. Lima, and Ghislain Montavon, Ed., ASM International, Materials Park, OH, 2008.

K. Bobzin, T. Schläfer, K. Richardt, and **M. Brühl**, Surface Engineering Institute, RWTH Aachen University, Aachen, Germany. Contact e-mail: bruehl@iot.rwth-aachen.de.

2. Experimental Procedure

2.1 Materials

For the production of the dispersion strengthened powder, a commercial spray powder (NiCoCrAlY −45 + 22.5 μm, H.C. Starck, Goslar, Germany) was used as starting matrix material. The basic material for the hard phases was a commercial yttria partially stabilized ZrO_2 (YPSZ) spray powder (ZrO_2-Y_2O_3 93-7 wt.%/−45 + 10 μm, H.C. Starck, Goslar, Germany).

2.2 Dispersion Powder Preparation

The dispersion strengthened powder was prepared by a high-energy ball mill of type Simoloyer CM08 (Zoz, Wenden, Germany). The mixture ratio between NiCoCrAlY and YPSZ was varied between 5 and 10 wt.% YPSZ. Both powders were mixed and milled in air with various milling times. In order to achieve a high relative velocity between the stainless steel balls (100Cr6) within the mill, the revolution speed was varied during milling.

The ground revolution speed was 900 rev/min and every 5 min the speed was reduced for 1 min to 500 rev/min. Based on the changing of the rotation rate, a balance between the balls can be avoided. The ball-to-powder ratio in mass relation of 10:1 was selected according to previous experiences (Ref 6). The stainless steel chamber, which was filled with 8 kg balls, was water cooled during the milling process in order to keep the powder temperature at approximately 37 °C. The progress of the milling process was measured by taking powder samples every hour. The maximum milling time was 10 h. The selection of the optimal milling time for the two different powders was based on particle morphology, particle size distribution, and homogenous distribution of the hard phase.

2.3 Particle Size Measurement

The particle size distribution was measured with an optical Particle Image Analyser Morphologi G2 (Malvern Instruments Ltd., Herrenberg, Germany). The Morphologi G2 particle analyzer provides repeatable and routine characterization of particle size. The system is equipped with a microscope coupled with a high-resolution digital camera for high definition aberration-free images. The diameter of single particles is measured in the following steps. The 3D image of the particle is captured as a 2D image and converted to a circle of equivalent area to the 2D image. The diameter of this circle is then reported as the diameter of that particle.

Table 1 HVOF spray parameters

Coating parameters	Mechanical alloyed NiCoCrAlY - YPSZ	Commercial NiCoCrAlY
Total gas flow, slpm	880	880
Stoichiometry (O_2/H_2)	0.22	0.26
Shroud gas, slpm	300	300
Spray distance, mm	270	270
Powder feed rate, g/min	25	25

2.4 Microhardness Measurement

The microhardness of the powder and of the coating was measured by the Bühler Micromet 1 microhardness tester (Bühler GmbH, Düsseldorf, Germany). This system calculates the Vickers hardness (HV) from the quotient of the test load and surface of the quadrilateral diamond pyramid indentation left on the material. Within this work a test load of 25 g for the powder and 100 g for the coating was used.

2.5 Thermal Spraying

The YPSZ dispersion strengthened NiCoCrAlY powder, produced by using the deduced optimal milling parameter, was sprayed by the HVOF system Sulzer Metco Diamond Jet Hybrid™ DJ2600 (Sulzer Metco, Wohlen, Switzerland). Before spraying, the powder was sieved with a 45 and 20 μm sieve to achieve a suitable particle size distribution of −45 + 20 μm. The same commercial NiCoCrAlY powder, which was used as matrix material, was also sprayed as reference. The developed spray parameters are listed in Table 1. Due to the high-temperature analysis, the coatings were sprayed onto alloy 600 substrates, which were grit blasted and cleaned. The coatings were heat-treated for 6 h at 1050 °C in a vacuum furnace.

2.6 Wear Behavior

The wear resistance respectively the abrasive wear behavior of the coatings was measured by the pin-on disc test CSEM HT Tribometer (CSEM, Neuchâtel, Switzerland). The pin-on disc tests were performed at a temperature of 700 °C. As counter part an Al_2O_3 pin with 6 mm diameter was used. Prior to the testing, the coatings were ground to an 800-grit finish. The load force was 5 N and the total sliding distance 200 m. The sliding speed was adjusted to 5 cm/s. The surface line profiles of the wear track were analyzed by an UBM laser profilometer (UBM Messtechnik GmbH, Ettlingen, Germany). Based on the average of six profilometer measurements, the wear volume was calculated.

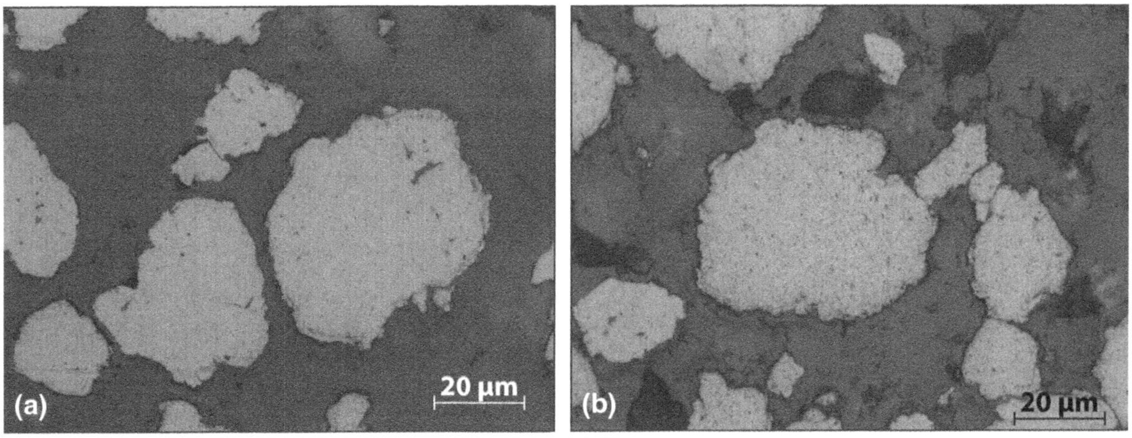

Fig. 1 SEM image of NiCoCrAlY + 5 wt.% YPSZ after 5 h milling (a) and NiCoCrAlY + 10 wt.% YPSZ after 4 h milling (b). The dark small points within the bright NiCoCrAlY matrix are YPSZ

3. Results and Discussion

The powder was analyzed every hour concerning the particle size, powder morphology, and the homogeneous distribution of the oxide hard phases in the metal matrix. Every powder batch was embedded and analyzed by cross sections. Figure 1 shows the cross section of the powder NiCoCrAlY + 5 wt.% YPSZ after 5 h milling. It can be seen that the hard phases (dark phase) are fine and homogeneously dispersed in the metal matrix (bright phase). In the milled powder with 10 wt.% YPSZ, a homogeneous distribution of YPSZ in the metal matrix could be achieved after 4 h.

The particle size distribution of every powder batch was measured by using the optical analyzing system Morphologi G2 produced by Malvern Instruments Ltd. Figure 2 shows the distribution of the two powder batches at different states of milling. In this figure, two different lines for each powder are displayed. After 4 h milling, 90 vol.% of the powder NiCoCrAlY + 10 wt.% YPSZ is finer than 71 μm and 90 vol.% of the powder including 5 wt.% YPSZ is finer than 80 μm after 5 h milling. For the same milling times, half of the volume of the powder NiCoCrAlY + 5 wt.% YPSZ exhibits a particle size under 45 μm and 50 vol.% of the powder NiCoCrAlY + 10 wt.% YPSZ is below 38 μm. The powder was sieved to receive a fraction between 45 and 20 μm and the yield was approximately 40%. The leftover can be filled back into the mill for milling again.

The morphology of the powder is another important property for a suitable thermal spraying powder, as it influences the flowability of the powder. Figure 3 shows particles of the milled NiCoCrAlY + 10 wt.% YPSZ after 2 and 4 h milling. The powder after 4 h milling exhibited a more spherical shape than the powder after 2 h. In this way, the 4 h milled powder was more convenient for feeding. A suitable spherical shape for NiCoCrAlY + 5 wt.% YPSZ was achieved after 5 h milling.

Figure 4 shows Vickers hardness at room temperature measured on the three different powders. It can be seen that the dispersed strengthened powder owned drastically higher hardness than the powder without hard phases. The hardness of the non-dispersed commercial NiCoCrAlY powder was approximately 420 HV0.025. The powder strengthened with 5 wt.% YPSZ exhibited with 660 HV0.025 a higher hardness than the commercial powder. An increase of the dispersion to 10 wt.% YPSZ did not

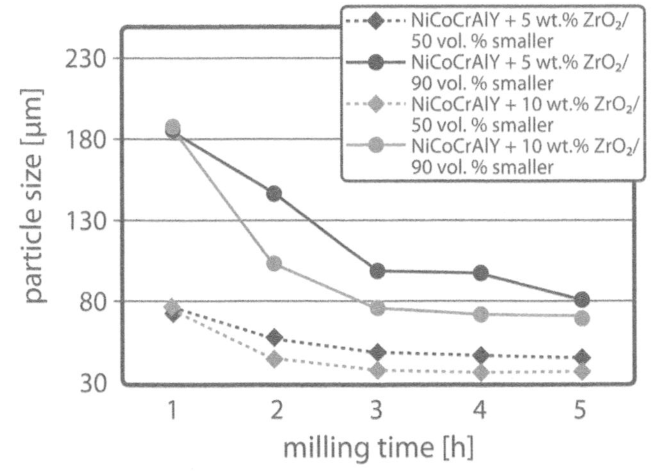

Fig. 2 Particle size of the powder at different milling times

Fig. 3 Milled NiCoCrAlY + 10 wt.% YPSZ after 2 h (a) and 4 h (b) milling time

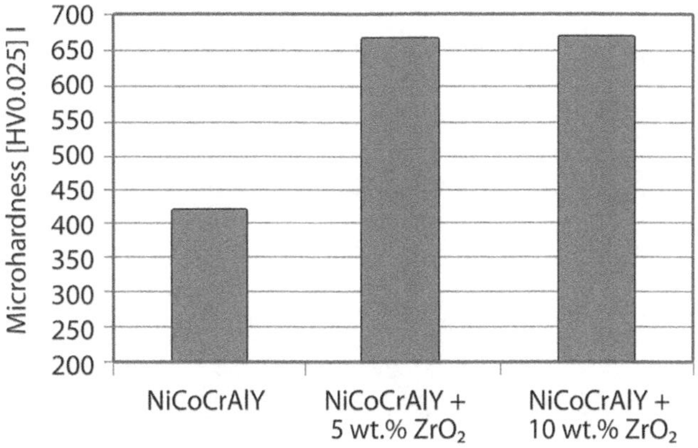

Fig. 4 Vickers hardness at room temperature for the three different powders

lead to a difference in the microhardness of the powder in comparison to the powder strengthened with 5 wt.% YPSZ.

Cross sections of the HVOF sprayed coatings after heat treatment are displayed in Fig. 5. Apart from the coating with 10 wt.% YPSZ, the coatings show a dense structure after diffusion process. The unstrengthened coating possessed a low porosity with 1.15% and the coating strengthened with 5 wt.% YPSZ had a porosity of 2.59%. In contrast to this, the coating with 10 wt.% YPSZ exhibited a porosity of 7.9%.

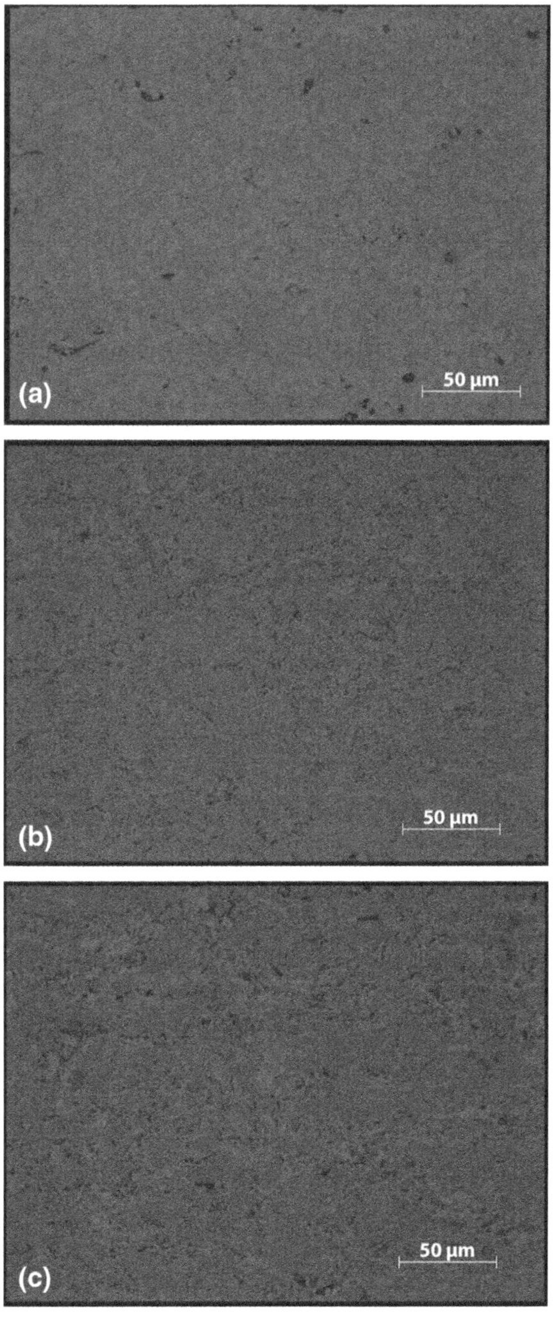

Fig. 5 Cross section of the conventional NiCoCrAlY coating (a), the strengthened NiCoCrAlY coatings with 5 wt.% YPSZ (b), and with 10 wt.% YPSZ (c)

Figure 6 shows the Vickers microhardness of the three different coatings. It can be observed that the microhardness of the unstrengthened conventional NiCoCrAlY coating is 100 HV0.1 lower than dispersed strengthened coatings. The microhardness of the 5 wt.% YPSZ coating and the 10 wt.% YPSZ coating slightly differs.

For the coating strengthened with 10 wt.% YPSZ, the microhardness did not correspond to the values for the abrasion rating of the coating. The abrasion rating of this coating was nearly the same range as the unstrengthened coating. Both coatings exhibited a volume loss of 3.7×10^{-4} mm^3/Nm (Fig. 7). In contrast, the coating strengthened with 5 wt.% YPSZ had just the half of the abrasion rating than the other two coatings (1.75×10^{-4} mm^3/Nm). A picture of the sliding track is shown in Fig. 8. Due to the high temperature, the pin ploughed through the coating without larger disruptions. In Fig. 8 it can also be observed that only less of the coating material was adhered on the pin after the sliding test. Another indication for this wear behavior can be seen in Fig. 9, which shows the track profile of the same coating. At the edge of the sliding track the material was plastically deformed and extruded outside of the sliding track.

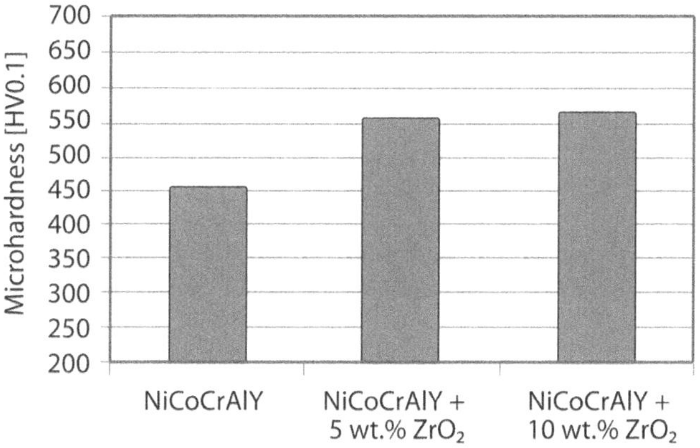

Fig. 6 Vickers hardness at room temperature for the three different coatings

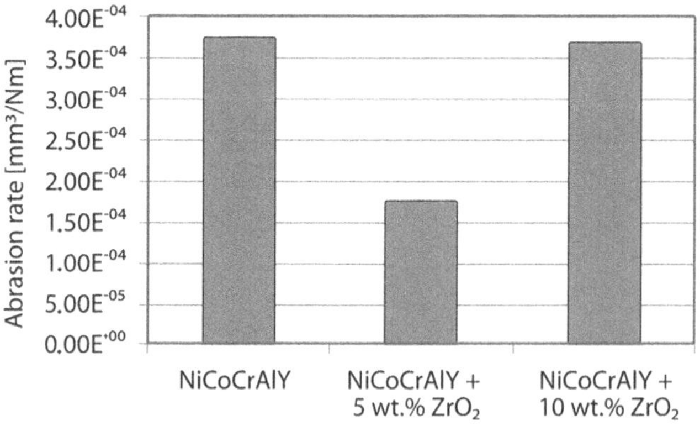

Fig. 7 Abrasion rating of the coating during the pin-on disc test at 700 °C. The testing parameters were sliding speed 5 cm/s, sliding distance 200 m, and load force 5 N. The counter part was an Al_2O_3 pin

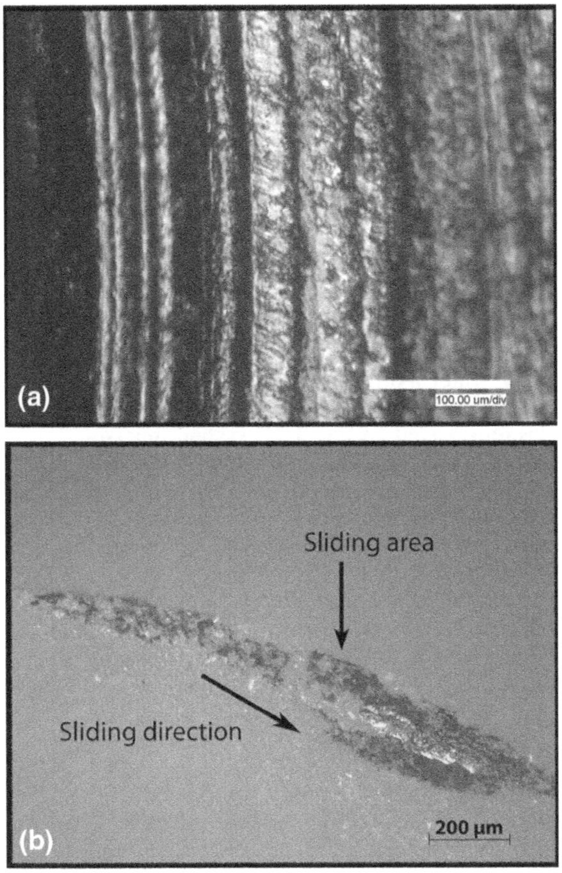

Fig. 8 Sliding track of the coating strengthened with 5 wt.% YPSZ (a) and the sliding area on the Al_2O_3 pin (b)

The higher abrasion rating of the strengthened coating with 10 wt.% YPSZ can be explained by the higher coating porosity after heat treatment. The higher porosity led to a lower cohesion between the particles within the coating in comparison to the other strengthened coating.

4. Conclusion

The presented study has demonstrated that oxide dispersion strengthened NiCoCrAlY powder can be prepared by using a high-energy ball-milling process. A homogeneous dispersion of the oxides can be achieved after 5 h milling for 5 wt.% YPSZ and after 4 h milling for 10 wt.% YPSZ. After this milling time, the powder has a suitable morphology for thermal spraying. By using oxide hard phases, the hardness of the powder can be increased from 420 HV0.025 for the commercial NiCoCrAlY powder to approximately 660 HV0.025 for the strengthened powder.

The powder can be thermally sprayed with a HVOF system. After heat treatment, a dense structure can be achieved. The microhardness of the strengthened coatings

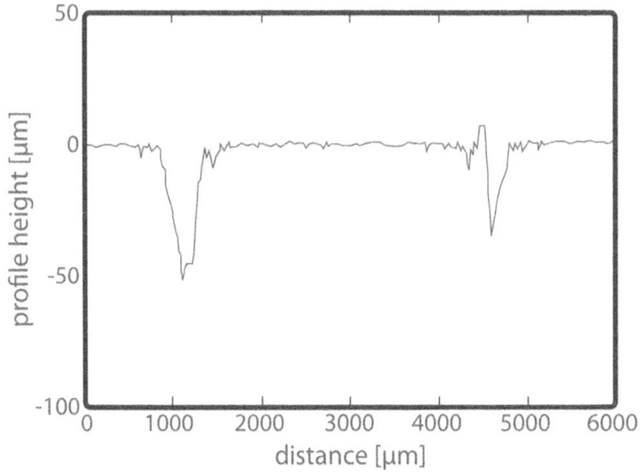

Fig. 9 Track profile on the coating NiCoCrAlY + 5 wt.% YPSZ

is 100 HV0.1 higher than that of the coating produced with the commercial NiCoCrAlY powder.

By strengthening the coating with oxide hard phases, the wear resistance can be improved. The abrasion rating of the coating produced with the NiCoCrAlY + 5 wt.% YPSZ powder is half of the rating of the commercial NiCoCrAlY coating. These results lead to the supposition that the strengthened coatings exhibit better erosion resistance than the commercial coatings. But this has to be investigated by further measurements.

Acknowledgments

The investigations were carried out as part of a research project founded by the German Research Foundation (LU 232/95-2). Our thanks are due for this support.

References

1. E. Lugscheider and C. Herbst, Thermisch gespritzte Verschleißschutzschichten heute und morgen, Pulvermetallurgie in Wissenschaft und Praxis Band, 1997, **13**, p 163-188 (in German)
2. H. Zoz and H. Ren, Processing of Ceramic Powders Using High-Energy Milling, *Interceram.*, 2000, **49**(1), p 24-34
3. I.S. Polkin, A.B. Borzov, F.H. Froes, and C. Suryanarayana, New Materials Produced by Mechanical Alloying, *Proceedings of the 2nd International Conference on Structural Applications of Mechanical Alloying*, Vancouver, British Columbia, Canada, 1993, p 157-164
4. M. Dunlap and G. Meng, Oxygen in Atomized MCrAlY Powder and Its Effect on HVOF Coatings, *International Thermal Spray Conference*, E. Lugscheider and C.C. Berndt, Eds., March 4-6 (Essen, Germany), DVS Deutscher Verband für Scheißen, 2002, p 465-469
5. Y. Itoh, M. Saitoh, and M. Tamura, Characteristics of MCrAlY Coatings Sprayed by High Velocity Oxygen-Fuel Spraying System, *J. Eng. Gas Turb. Power*, 2000, **122**(1), p 43-49
6. L. Zhao, J. Zwick, and E. Lugscheider, The Influence of Milling Parameters on the Properties of the Milled Powders and the Resultant Coatings, *Surf. Coat. Technol.*, 2003, **168**(2-3), p 179-185

JTTEE5 17:858–864
DOI: 10.1007/s11666-008-9251-8
1059-9630/$19.00 © ASM International

Peer Reviewed

TGO Growth and Crack Propagation in a Thermal Barrier Coating

W.R. Chen, R. Archer, X. Huang, and B.R. Marple

(Submitted May 8, 2008; in revised form September 22, 2008)

In thermal barrier coating (TBC) systems, a continuous alumina layer developed at the ceramic topcoat/bond coat interface helps to protect the metallic bond coat from further oxidation and improve the durability of the TBC system under service conditions. However, other oxides such as spinel and nickel oxide, formed in the oxidizing environment, are believed to be detrimental to TBC durability during service at high temperatures. It was shown that in an air-plasma-sprayed (APS) TBC system, postspraying heat treatments in low-pressure oxygen environments could suppress the formation of the detrimental oxides by promoting the formation of an alumina layer at the ceramic topcoat/bond coat interface, leading to an improved TBC durability. This work presents the influence of postspraying heat treatments in low-pressure oxygen environments on the oxidation behavior and durability of a thermally sprayed TBC system with high-velocity oxy-fuel (HVOF)-produced Co-32Ni-21Cr-8Al-0.5Y (wt.%) bond coat. Oxidation behavior of the TBCs is evaluated by examining their microstructural evolution, growth kinetics of the thermally grown oxide (TGO) layers, and crack propagation during low-frequency thermal cycling at 1050 °C. The relationship between the TGO growth and crack propagation will also be discussed.

Keywords crack propagation, HVOF, TBC, TGO growth

1. Introduction

The metallic bond coat (BC) is an important constituent of thermally sprayed TBC systems. It enhances the adhesion of the ceramic thermal barrier layer (the topcoat or TC) to the substrate and also provides oxidation and corrosion protection to the substrate metal. At elevated temperatures, however, oxidation of the bond coat results in the formation of a TGO layer at the ceramic/bond coat interface. If the TGO was composed of a continuous scale of Al_2O_3, it would act as a diffusion barrier during an extended thermal exposure in service, thus helping to protect the substrate from further oxidation and improving the durability of the system under service conditions. However, it has been reported that some other oxides, such as chromia $(Cr,Al)_2O_3$, spinel $Ni(Cr,Al)_2O_4$, and nickel oxide NiO (Ref 1-6), may form along with this TGO layer in thermally sprayed TBC systems. Since oxidation of the bond coat has been recognized as the cause for separation of the ceramic layer from the substrate (Ref 2, 3, 7-10), the formation of the detrimental oxides may accelerate crack nucleation during thermal exposure, leading to premature TBC failure. Studies have shown that, when heat treated in low-pressure oxygen conditions, a continuous Al_2O_3 layer may develop at the ceramic/bond coat interface in both thermally sprayed (Ref 6, 11-13) and electron beam physical vapor deposited (EB-PVD) TBC systems (Ref 14-18). Therefore, appropriate postspraying heat treatment(s) may suppress the formation of detrimental oxides by developing a TGO composed of a predominantly Al_2O_3 layer, leading to improved TBC durability.

In the present investigation, heat treatments in environments with low-pressure oxygen were performed to produce an Al_2O_3 layer at the interface between the ceramic topcoat and the bond coat prior to prolonged thermal exposure in air. The influence of the preoxidation treatments on oxidation behavior was then studied in terms of TGO growth, crack nucleation, and propagation in the TBC system having an HVOF-produced bond coat.

This article is an invited paper selected from presentations at the 2008 International Thermal Spray Conference and has been expanded from the original presentation. It is simultaneously published in *Thermal Spray Crossing Borders, Proceedings of the 2008 International Thermal Spray Conference*, Maastricht, The Netherlands, June 2-4, 2008, Basil R. Marple, Margaret M. Hyland, Yuk-Chiu Lau, Chang-Jiu Li, Rogerio S. Lima, and Ghislain Montavon, Ed., ASM International, Materials Park, OH, 2008.

W.R. Chen, Institute for Aerospace Research, SMPL, National Research Council Canada, Ottawa, ON, Canada K1A 0R6; **R. Archer** and **X. Huang**, Department of Mechanical and Aerospace Engineering, Carleton University, Ottawa, ON, Canada K1S 5B6; and **B.R. Marple**, Industrial Materials Institute, National Research Council Canada, Boucherville QC, Canada J4B 6Y4. Contact e-mail: Weijie.Chen@nrc-cnrc.gc.ca.

2. Experimental

The TBC samples consisted of a CoNiCrAlY bond coat and a ZrO_2-8wt.%Y_2O_3 topcoat. The bond coat was deposited to a thickness of 140 to 180 μm by the HVOF technique (Tafa JP5000, Praxair, CT, USA), with powders of Co-32Ni-21Cr-8Al-0.5Y (wt.%) (CO-210-24, Praxair, CT, USA), onto Ø16 × 10 mm Inconel 625 disks. On top of the CoNiCrAlY, the topcoat was deposited to a thickness of 250 to 280 μm by the APS technique (F4, Sulzer Metco, Westbury, NY, USA), with powders of ZrO_2-8wt.%Y_2O_3 (Metco 204NS, Sulzer Metco, Westbury, NY, USA). Grit blasting was not applied prior to the deposition of the topcoat. The nominal particle size distributions for the BC and TC powders were 20-45 μm and 11-125 μm, respectively. The BC and TC were produced based on the spray parameters recommended by the manufacturers of the torch and feedstock.

Some of the as-sprayed samples were heat-treated in a vacuum furnace at $<10^{-5}$ torr ($P_{O_2} < 0.00028$ Pa) and 1080 °C for 4 h (vacuum heat treatment or VHT), and some others were heat-treated in a low-pressure oxygen environment of about 2×10^{-3} torr ($P_{O_2} \approx 0.056$ Pa) at 1080 °C for 24 h (low-pressure heat treatment or LPOT). All samples were subsequently subjected to low-frequency thermal cycling in air. The thermal cycle consisted of 10 ~ 15-min ramping, 100-h holding at 1050 °C, and >40-min cooling to the ambient temperature (25 °C). The oxidized samples were sectioned after completion of a predetermined time of thermal exposure between 100 and 2500 h, mounted using epoxy, and mechanically polished. The specimens were then examined using a Philips XL30S FEG scanning electron microscope (SEM) with an energy-dispersive spectrometer (EDS). The area and length of TGO, as well as the crack length in the TBC, were measured using ImageTool software, based on the micrographs taken from the cross sections of the tested samples. TGO area was measured by drawing a polygon along the ceramic/TGO and TGO/bond coat interfaces, while TGO length by drawing a line as short as possible along the TGO. About 20-70 micrographs were taken from each sample for the TGO measurement. The various oxides were identified by SEM-EDS, as $NiAl_2O_4$ contains approximately 13-16 at.% Ni and 27-30 at.% Al, while Al_2O_3 and NiO are very easy to be determined.

3. Results and Discussion

3.1 As-Sprayed and Preoxidized Coating Microstructures

The ceramic topcoat contained porosity and some crack-like discontinuities (Fig. 1a). The maximum dimension of the crack-like discontinuities in the ceramic was in the range of 70 to 100 μm. The CoNiCrAlY bond coat, on the other hand, was rather uniform and contained less porosity compared to the ceramic topcoat. Some porosity also appeared at the topcoat/bond coat interface. SEM/EDS semiquantitative analysis showed that the aluminum concentration in the bond coat at the ceramic/bond coat interface was 13 to 19 at.%. After the VHT and LPOT, a nearly continuous thin Al_2O_3 layer was developed at the original ceramic/bond coat interface (Fig. 1b and c), while the aluminum level in the bond coat remained nearly unchanged within the interface region. Meanwhile, some mixed oxide clusters of chromia, spinel, and nickel oxide (CSN) can also be observed at the ceramic/bond coat interface.

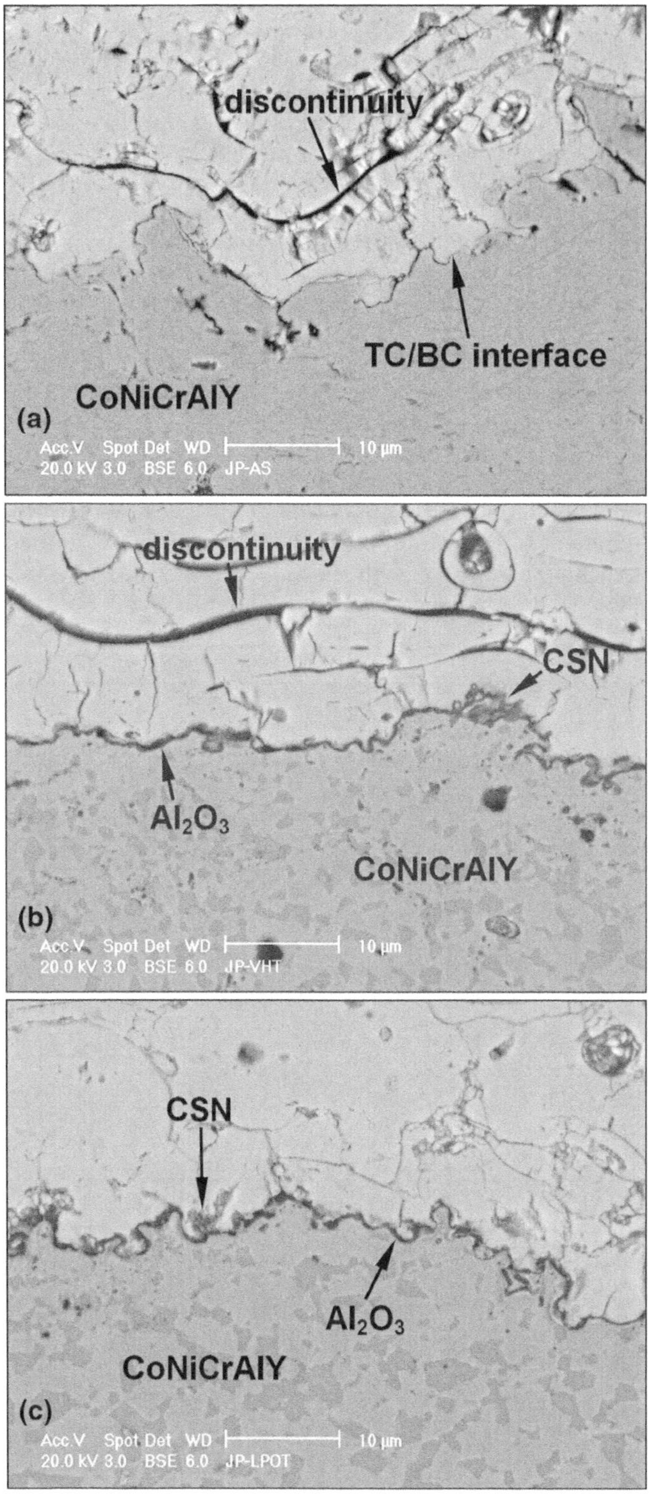

Fig. 1 Microstructures of (a) as-sprayed (i.e., no preoxidation), (b) VHT and (c) LPOT TBCs with HVOF-CoNiCrAlY bond coat

3.2 Oxidation of TBCs

Upon thermal exposure in air, an oxide layer of predominantly Al_2O_3 formed along the interface between the ceramic topcoat and the bond coat (Fig. 2a). In addition, some mixed oxides, CSN clusters, also formed at the ceramic/bond coat interface. The formation of the CSNs was attributed to the localized low aluminum concentration at the ceramic/bond coat interface (Ref 13). As such, the TGO formed in the as-sprayed TBC was composed of an Al_2O_3 layer and a number of CSN clusters. As thermal exposure continued, the TGO layer grew thicker (Fig. 2b). The amount of CSN clusters appeared to be barely changed as exposure time increased, and cracking in the TBC appeared to be very limited.

With a further increase of thermal exposure, preexisting crack-like discontinuities in the ceramic started to open and develop into cracks (Fig. 3a); however, the number of cracks in the TBC was still very limited. Cracking associated with the CSN clusters was also limited, while cracking associated with the TGO occurred mostly at the TGO/bond coat interface (Fig. 3b). Yttrium-rich oxides were observed within the Al_2O_3 as the TGO grew, and the formation of these Y-rich oxides appeared to have no influence on crack nucleation and/or void formation within the TGO. After prolonged thermal exposure, the number of cracks remained limited; however, within 150 μm of the ceramic/bond coat interface, some individual cracks became rather long (Fig. 4), extending up to ~200 μm in length.

In the VHT TBC, however, much less CSNs were formed at the ceramic/bond coat interface (Fig. 5a), compared to those in the as-sprayed TBC (Fig. 2a). This suggests that the VHT could, to a certain level, suppress the formation of mixed oxides during thermal exposure via developing a thin Al_2O_3 layer in vacuum at high temperature. This is consistent with previous observations reported by many other researchers (Ref 6, 11-18). Similar to the as-sprayed TBC, cracking in the VHT TBC was very limited after up to twenty-five 100-h cycles (Fig. 6). The formation of some Cr-rich oxide was observed during

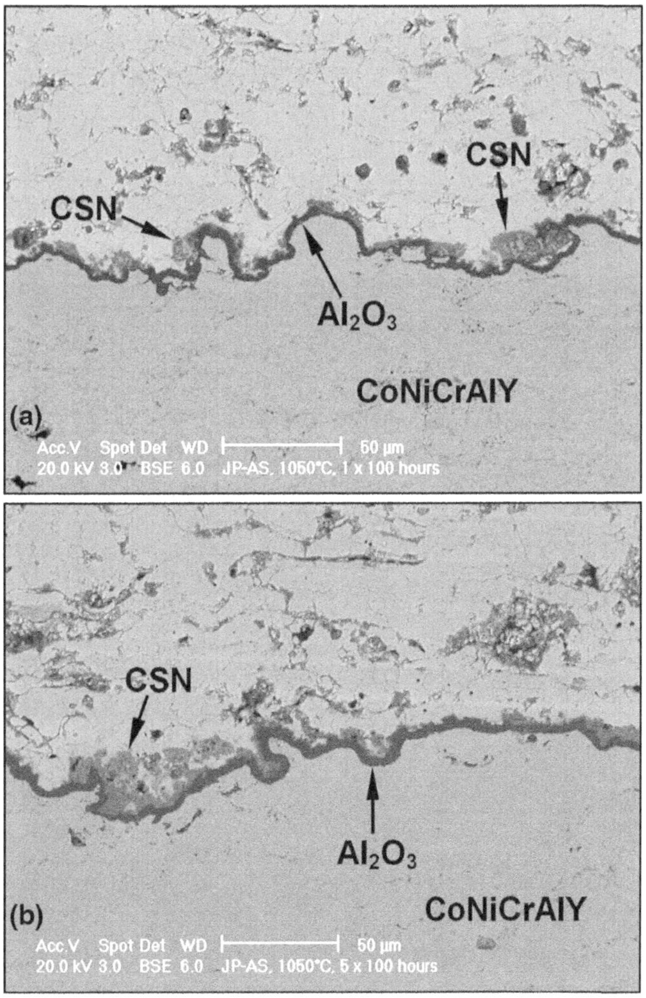

Fig. 2 As-sprayed HVOF-CoNiCrAlY shows (a) a TGO layer of predominantly alumina with CSN clusters between the ceramic topcoat and alumina layer after one 100-h cycle and (b) the increased thickness of the TGO after five 100-h cycles

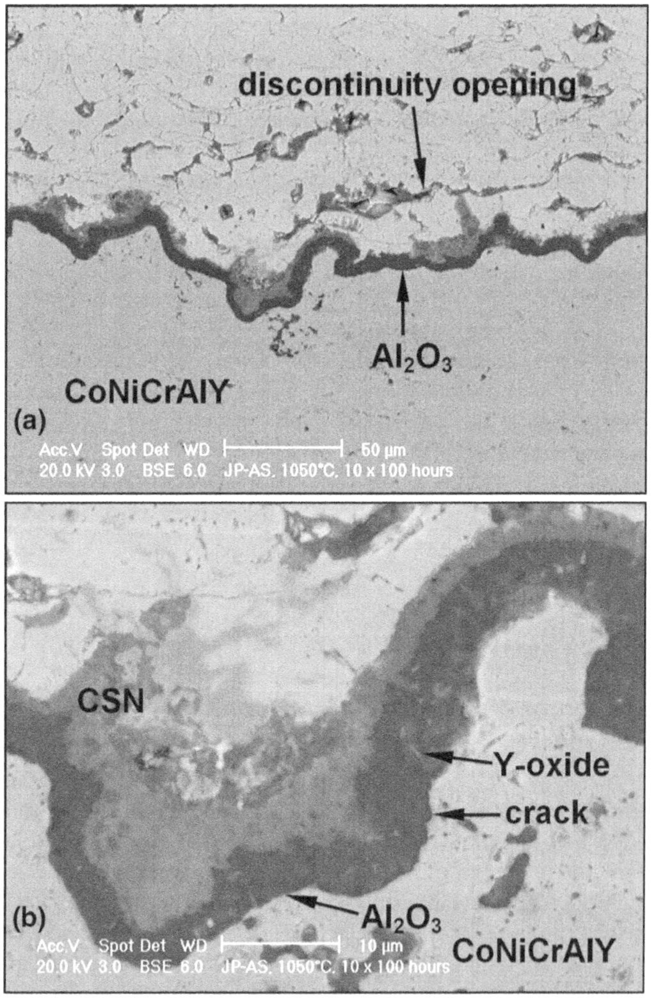

Fig. 3 In the as-sprayed HVOF-CoNiCrAlY, after ten 100-h cycles, a discontinuity in the ceramic/bond coat interface region developed into a crack (a), while cracking associated with the TGO occurred mostly at the TGO/bond coat interface (b)

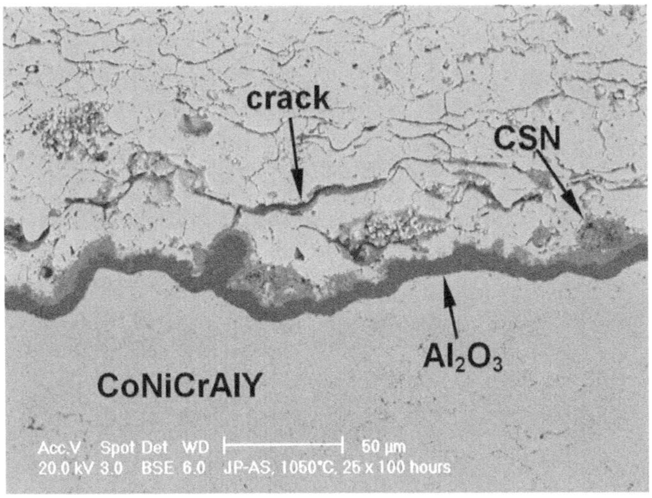

Fig. 4 A discontinuity opened and propagated into a long crack within 100 µm of the ceramic/bond coat interface in the as-sprayed HVOF-CoNiCrAlY after twenty-five 100-h cycles. There are still relatively few cracks in the TBC

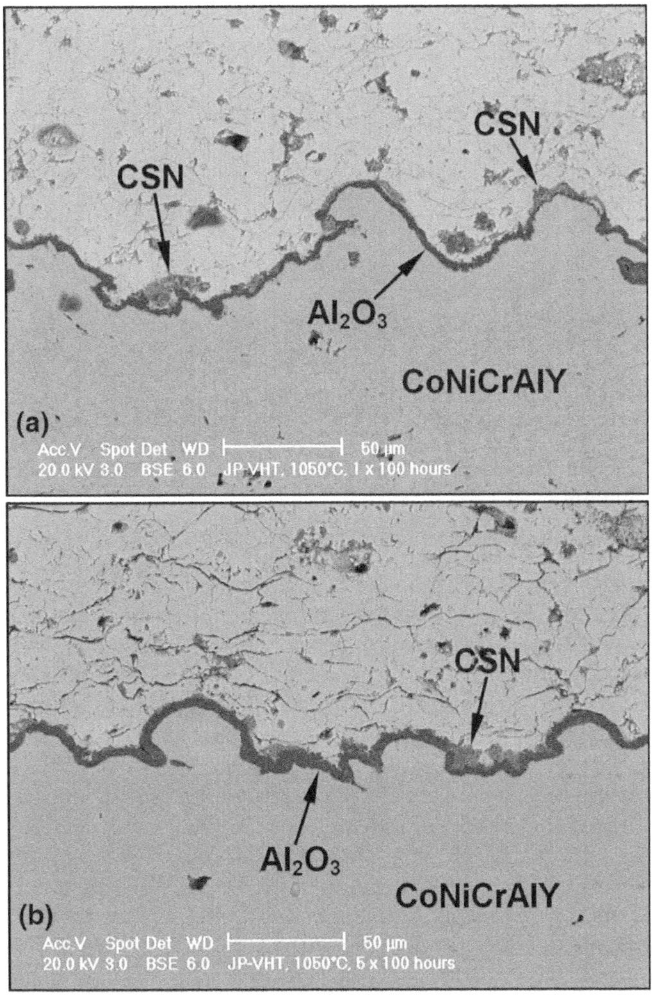

Fig. 5 VHT HVOF-CoNiCrAlY shows (a) a TGO layer of predominantly alumina with limited CSN clusters between the ceramic topcoat and alumina layer after one 100-h cycle and (b) thickening of the TGO after five 100-h cycles

the thickening of the TGO (Fig. 7), where crack nucleation/propagation could be seen. The formation of Cr-rich oxide can be attributed to the aluminum depletion in the bond coat such that a composition was reached at which other oxide(s) could form (Ref 19). The microstructure of the LPOT TBC during thermal exposure was essentially the same as that of the VHT TBC, which had fewer mixed oxides than the as-sprayed TBC, along with a slower TGO thickening process. Cracking associated with mixed oxide clusters at the ceramic/bond coat interface and Cr-rich oxide within the alumina layer could be seen after extended thermal exposure (Fig. 8).

No spallation was observed with any of the three samples (i.e., as-sprayed, VHT, and LPOT) after thirty-five 100-h thermal cycles, and the testing is still in progress.

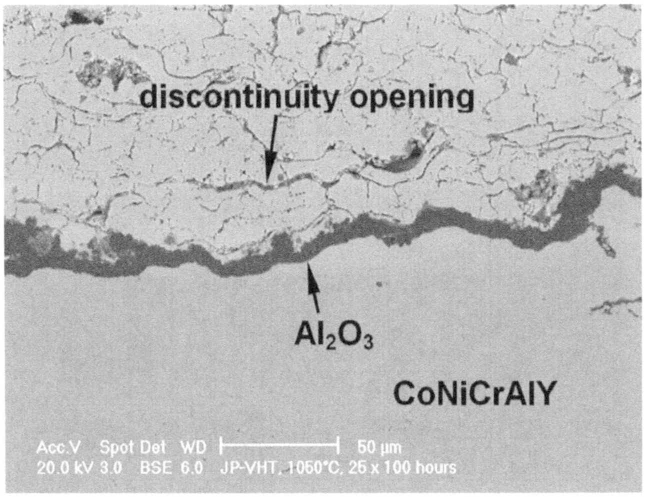

Fig. 6 A discontinuity opened and propagated to become a long crack within 50 µm of the ceramic/bond coat interface in the VHT HVOF-CoNiCrAlY after twenty-five 100-h cycles. Cracking in the TBC remains quite limited

Fig. 7 After twenty-five 100-h cycles, Cr-rich oxide formed during TGO growth, leading to cracking in the VHT HVOF-CoNiCrAlY

Peer Reviewed

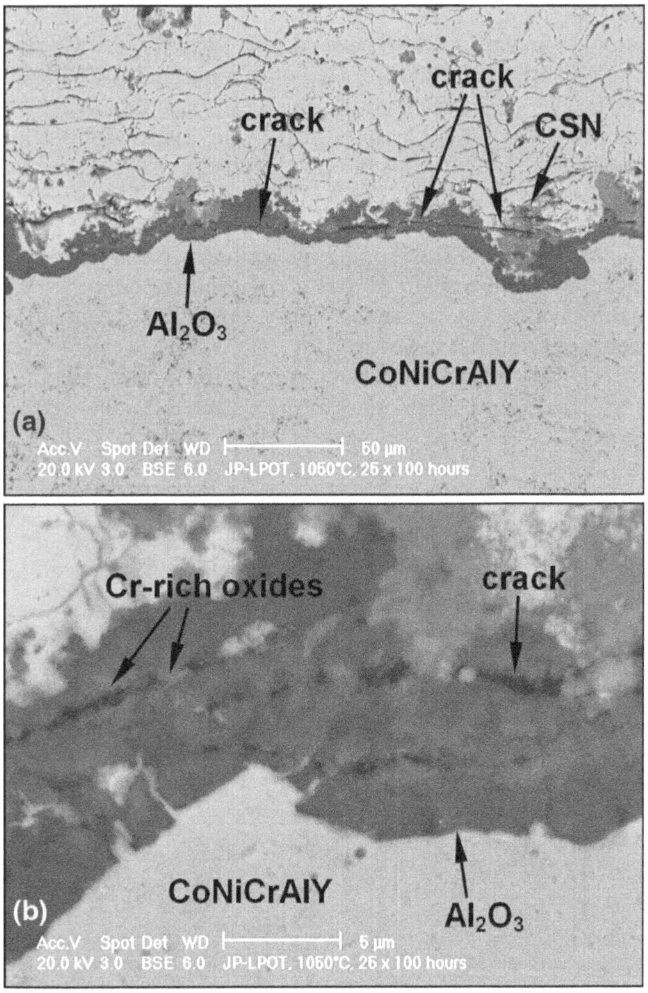

Fig. 8 The LPOT HVOF-CoNiCrAlY TBC coating system after twenty-five 100-h thermal cycles showing (a) cracking associated with CSNs and the alumina layer and (b) cracks nucleated as a result of Cr-rich oxide formation within the alumina layer

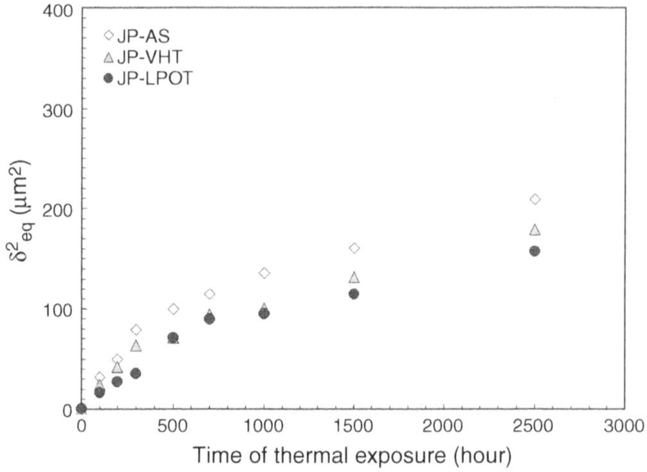

Fig. 9 TGO growth in TBCs with a HVOF-CoNiCrAlY bond coat during low-frequency thermal exposure

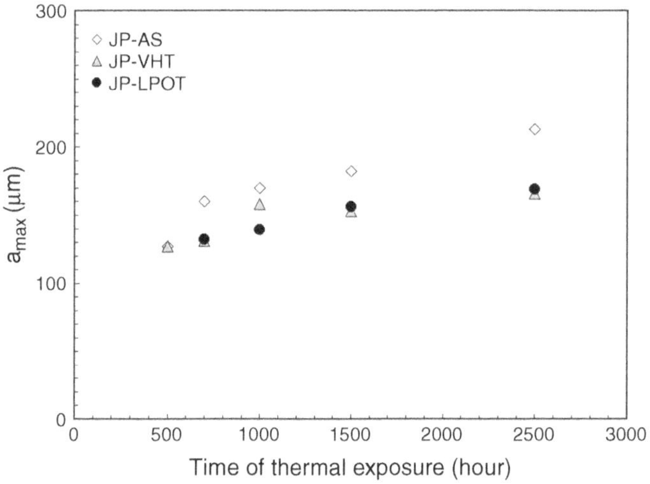

Fig. 10 The maximum crack length in the TBC as a function of the time of thermal exposure

3.3 TGO Growth and Crack Propagation in the TBCs

To take into account the entire TGO, an equivalent TGO thickness, δ_{eq}, is defined as (Ref 20):

$$\delta_{eq} = \frac{\Sigma(\text{cross sectional TGO area})}{\Sigma(\text{cross sectional length TC/BC interface})} \quad \text{(Eq 1)}$$

Use of this approach was due to the rough ceramic/bond coat interface, as well as heterogeneous TGO growth especially when CSN clusters were formed between the ceramic topcoat and the bond coat (Fig. 2 and 5). The relationship between δ_{eq}^2 and exposure time (Fig. 9) shows a parabolic growth of TGO, which is consistent with previous observations (Ref 1, 21-23). It can also be seen that both VHT and LPOT can reduce the growth rate of the TGO layer; this can be attributed to the reduced formation of mixed oxides during thermal exposure. However, the difference between the as-sprayed and VHT/LPOT HVOF-CoNiCrAlY is not nearly as pronounced as that observed for APS-CoNiCrAlY studied in earlier work (Ref 24). This is due to the smaller amount of the mixed oxides formed in the coating as a result of relatively higher Al content in HVOF-CoNiCrAlY. In the case of APS, oxidation during spraying results in lower Al contents in the deposited CoNiCrAlY coating.

In general, the maximum crack length in the TBCs increased as exposure time increased (Fig. 10). It may be noticed that cracks propagate faster in the as-sprayed TBC than in the VHT and LPOT TBCs. When plotting the maximum crack length as a function of TGO thickness (Fig. 11), it appears that crack length increases with TGO thickness, even though the microstructure of the TGOs varies. Regression analysis shows that the relationship between the maximum crack length in the TBC and TGO thickness can be expressed by a power law, i.e., $a_{max} = k\delta_{eq}{}^n$, where k and n are constants. The value of n is determined to be 0.891, which is quite different from that

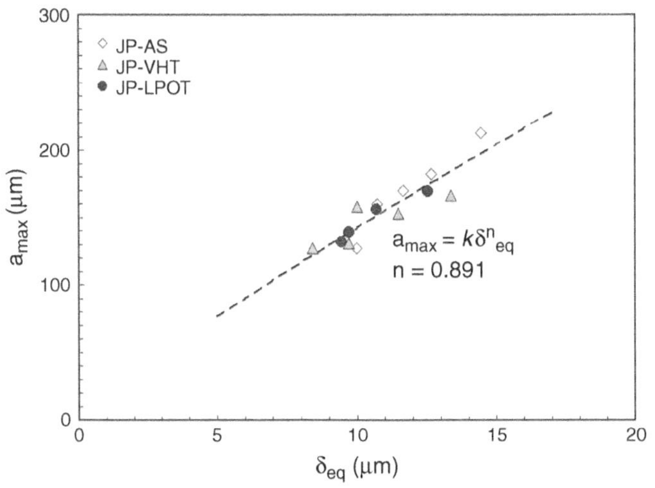

Fig. 11 The maximum crack length in the TBC as a function of TGO thickness

determined in a TBC with an APS-CoNiCrAlY bond coat under a 45-min thermal cycling condition at the same temperature (Ref 24), where n was 0.597. This could be due to the narrow range of TGO thicknesses in the present study, which was from about 8 to 15 μm. Also, the increased holding time in the present study would have an influence on the maximum crack length, since the number of thermal cycles will definitely have an impact on crack propagation. It was reported that longer holding times resulted in a slower mass gain rate during thermal cycling (Ref 25); however, the relationship between holding time and crack length has not been reported in the open literature. Ongoing work by the authors is investigating this relationship.

It has been shown that the critical crack size for initiating TBC spallation is somehow related to the topcoat thickness (Ref 26). It is possible that TBCs with different topcoat thickness values would fail at different TGO thickness levels. As such, determination of TGO growth kinetics, as well as the relationship between crack length and TGO thickness, may be helpful in TBC life prediction. This suggests that the nature of such a relationship merits further investigation in both numerical and experimental studies.

4. Conclusions

The present work investigated the influence of preoxidation treatments in low-pressure oxygen environments on the low-frequency cyclic oxidation behavior of a high-velocity oxy-fuel Co-32Ni-21Cr-8Al-0.5Y (wt.%) bond coat. It was found that both VHT and LPOT could suppress the formation of detrimental mixed oxides to a certain level, and develop a TGO composed of a predominantly Al_2O_3 layer. The growth of the TGO showed a parabolic growth phenomenon up to 2500 h. Crack propagation in the TBC during thermal exposure proceeded via discontinuity opening and growth in the ceramic, as well as limited crack nucleation and propagation associated with the TGO, whereas VHT and LPOT decreased the crack growth rate. A power law relationship between the maximum crack length in the TBC and TGO thickness likely exists, which may be useful to the life prediction of the thermally sprayed TBCs.

Acknowledgments

The authors thank Sylvain Bélanger of the Industrial Materials Institute of NRC Canada for thermal spraying the TBC samples and Peter L'Abbe of the Institute for Chemical Process and Environmental Technology of NRC Canada for preparing the samples for LPOT. They are also grateful to Robert McKellar and Ryan MacNeil of NRC-IAR-SMPL for their help in VHT and oxidation testing. The support from the SURFTEC consortium is also appreciated.

References

1. J.A. Haynes, M.K. Ferber, W.D. Porter, and E.D. Rigney, Characterization of Alumina Scales Formed During Isothermal and Cyclic Oxidation of Plasma-Sprayed TBC Systems at 1150 °C, *Oxid. Met.*, 1999, **52**, p 31-76
2. E.A.G. Shillington and D.R. Clarke, Spalling Failure of a Thermal Barrier Coating Associated with Aluminum Depletion in the Bond-Coat, *Acta Mater.*, 1999, **47**, p 1297-1305
3. A. Rabiei and A.G. Evans, Failure Mechanisms Associated with the Thermally Grown Oxide in Plasma-Sprayed Thermal Barrier Coatings, *Acta Mater.*, 2000, **48**, p 3963-3976
4. C.H. Lee, H.K. Kim, H.S. Choi, and H.S. Ahn, Phase Transformation and Bond Coat Oxidation Behavior of Plasma-Sprayed Zirconia Thermal Barrier Coating, *Surf. Coat. Technol.*, 2000, **124**, p 1-12
5. L. Ajdelsztajn, J.A. Picas, G.E. Kim, F.L. Bastian, J. Schoenung, and V. Provenzano, Oxidation Behavior of HVOF Sprayed Nanocrystalline NiCrAlY Powder, *Mater. Sci. Eng.*, 2002, **A338**, p 33-43
6. W.R. Chen, X. Wu, P.C. Patnaik, and J.-P. Immarigeon, Oxidation Behaviour of a Plasma-Sprayed Thermal Barrier Coating System, *Proceedings from the 1st International Surface Engineering Congress and the 13th IFHTSE Congress* (Columbus, Ohio), 2002, p 535-543
7. R.A. Miller and C.E. Lowell, Failure Mechanisms of Thermal Barrier Coatings Exposed to Elevated Temperatures, *Thin Solid Films*, 1982, **95**, p 265-273
8. C.H. Hsueh, P.F. Becher, E.R. Fuller, S.A. Langer, and W.C. Carter, Surface-Roughness Induced Residual Stresses in Thermal Barrier Coatings: Computer Simulations, *Mater. Sci. Forum*, 1999, **308-311**, p 442-449
9. A.G. Evans, D.R. Mumm, J.W. Hutchinson, G.H. Meier, and F.S. Pettit, Mechanisms Controlling the Durability of Thermal Barrier Coatings, *Prog. Mater. Sci.*, 2001, **46**, p 505-553
10. K.W. Schlichting, N.P. Padture, E.H. Jordan, and M. Gell, Failure Modes in Plasma-Sprayed Thermal Barrier Coatings, *Mater. Sci. Eng.*, 2003, **A342**, p 120-130
11. D. Monceau, F. Crabos, A. Malié, and B. Pieraggi, Effects of Bond-Coat Preoxidation and Surface Finish on Isothermal and Cyclic Oxidation, High Temperature Corrosion and Thermal Shock Resistance of TBC Systems, *Mater. Sci. Forum*, 2001, **369-372**, p 607-614
12. S. Takahashi, M. Yoshiba, and Y. Harada, Failure Analysis of High-Temperature Oxidation for Plasma Sprayed Thermal Barrier Coating Systems with Different Coating Characteristics, *Mater. Sci. Forum*, 2004, **461-464**, p 367-374

Peer Reviewed

13. W.R. Chen, X. Wu, D. Dudzinski, and P.C. Patnaik, Modification of Oxide Layer in Plasma-Sprayed Thermal Barrier Coatings, *Surf. Coat. Technol.*, 2006, **200**, p 5863-5868
14. C. Leyens, K. Fritscher, R. Gehrling, M. Peters, and W.A. Kaysser, Oxide Scale Formation on an MCrAlY Coating in Various H_2-H_2O Atmospheres, *Surf. Coat. Technol.*, 1996, **82**, p 133-144
15. H. Lau, C. Leyens, U. Schulz, and C. Friedrich, Influence of Bondcoat Pre-treatment and Surface Topology on the Lifetime of EB-PVD TBCs, *Surf. Coat. Technol.*, 2003, **165**, p 217-223
16. V.K. Tolpygo and D.R. Clarke, The Effect of Oxidation Pre-Treatment on the Cyclic Life of EB-PVD Thermal Barrier Coatings with Platinum-Aluminide Bond Coats, *Surf. Coat. Technol.*, 2005, **200**, p 1276-1281
17. T.J. Nijdam, L.P.H. Jeurgens, J.H. Chen, and W.G. Sloof, On the Microstructure of the Initial Oxide Grown by Controlled Annealing and Oxidation on a NiCoCrAlY Bond Coating, *Oxid. Met.*, 2005, **64**, p 355-377
18. T.J. Nijdam, G.H. Marijnissen, E. Vergeldt, A.B. Kloosterman, and W.G. Sloof, Development of a Pre-Oxidation Treatment to Improve the Adhesion between Thermal Barrier Coatings and NiCoCrAlY Bond Coatings, *Oxid. Met.*, 2006, **66**, p 269-294
19. F.S. Pettit, Oxidation Mechanisms for Nickel-Aluminum Alloys at Temperatures Between 900 and 1300 °C, *Trans. Metall. Soc. AIME*, 1967, **239**, p 1296-1305
20. W.R. Chen, X. Wu, B.R. Marple, and P.C. Patnaik, The Growth and Influence of Thermally Grown Oxide in a Thermal Barrier Coating, *Surf. Coat. Technol.*, 2006, **201**, p 1074-1079
21. S.M. Meier, D.M. Nissley, and K.D. Sheffler, Thermal Barrier Coating Life Prediction Model Development, ASME 91-GT-40, *International Gas Turbine and Aeroengine Congress and Exposition* (Orlando, FL), 1991
22. W. Brandl, H.J. Grabke, D. Toma, and J. Krüger, The Oxidation Behaviour of Sprayed MCrAlY Coatings, *Surf. Coat. Technol.*, 1996, **86-87**, p 41-47
23. K.S. Chan and N·S. Cheruvu, Degradation Mechanism Characterization and Remaining Life Prediction for NiCoCrAlY Coatings, GT2004-53383, *Proceedings of ASME Turbo Expo 2004, Power for Land, Sea, and Air* (Vienna, Austria), 2004
24. W.R. Chen, X. Wu, B.R. Marple, R.S. Lima, and P.C. Patnaik, Pre-Oxidation and TGO Growth Behaviour of an Air-Plasma-Sprayed Thermal Barrier Coating, *Surf. Coat. Technol.*, 2008, **202**, p 3787-3796
25. Y. Zhang, B.A. Pint, G.W. Garner, K.M. Cooley, and J.A. Haynes, Effect of Cycle Length on the Oxidation Performance of Iron Aluminide Coatings, *Surf. Coat. Technol.*, 2004, **188-189**, p 35-40
26. D. Zhu, S.R. Choi, and R.A. Miller, Development and Thermal Fatigue Testing of Ceramic Thermal Barrier Coatings, *Surf. Coat. Technol.*, 2004, **188-189**, p 146-152

JTTEE5 17:865–871
DOI: 10.1007/s11666-008-9224-y
1059-9630/$19.00 © ASM International

Peer Reviewed

Thermally Sprayed $CaZrO_3$ Coatings

E. Garcia, C. Cano, T.W. Coyle, M.I. Osendi, and P. Miranzo

(Submitted May 14, 2008; in revised form July 31, 2008)

A comparative study of $CaZrO_3$ coatings prepared by air plasma and flame spray processes is presented. The microstructural characteristics and phases were analyzed and discussed as a function of the spraying temperature achieved with each technique. Vickers hardness tests performed on polished cross sections of the different coatings were used to estimate their porosity. Thermal diffusivity values measured on freestanding thick coatings (500 μm) using the laser flash technique were very low and depended strongly on the pore shape. Crystallization was promoted by treating the coatings at 1200 °C, and the results are explained with aid of the ZrO_2-CaO equilibrium phase diagram.

Keywords calcium zirconate, flame spray, hardness, microstructure, plasma spray, thermal barrier coatings, thermal conductivity

1. Introduction

Thermal spraying includes several powerful processing techniques for producing coatings of diverse types such as wear and erosion resistant coatings, protective coatings against oxidation and corrosion, and thermal barrier coatings for high temperature environments. In most of these processes, particles are injected into a high temperature gas where they are heated, melted, accelerated, and sprayed onto a prepared substrate. The temperature distribution in the torch and the particle velocities depend on the particular technique used, both parameters affecting the final microstructure and the properties of the coatings (Ref 1).

Flame spraying (FS) is a widely and commercially used method for low-energy consuming processes that can be adapted to several combinations of gases, such as acetylene, propane, hydrogen, together with oxygen. Flame spray uses the thermal energy released during the combustion of fuel gases to generate heat. The oxy-acetylene torches are the most common, using acetylene as the main fuel in combination with oxygen to generate the highest combustion temperatures.

Atmospheric plasma spray (APS) uses the energy of the ionized plasma forming gas produced by an electric arc as the heating source. In this case, the temperature is higher than that produced by the oxy-acetylene flame spraying. Therefore, higher melting point materials can be deposited by atmospheric plasma spraying. This technique also provides higher particle velocities (800 m s^{-1}) than those achieved by flame spraying (below 100 m s^{-1}) (Ref 2, 3).

The production of thermal barrier coatings (TBCs) by these techniques is of considerable importance in applications such as aerospace and aircraft engines, advanced gas turbines, and diesel engines components. By using TBCs on high temperature components, the efficiency of the industrial processes can be increased due to the higher operating temperatures and reduced cooling air requirements. In the same way, the lifetime of the coated components is prolonged due to the lower metal surface temperatures.

The material commonly used in TBCs has been yttria-stabilized zirconia (7 wt.% Y_2O_3), mainly because of its low thermal conductivity, high thermal expansion coefficient, and good erosion resistance (Ref 4). In comparison, calcium zirconate ($CaZrO_3$) has high thermal and chemical stabilities, and good thermal shock resistance, rendering this to be a convenient candidate for some thermal barrier coating applications (Ref 5). In addition, due to the perovskite-type structure of $CaZrO_3$, a low thermal conductivity can be expected (Ref 6-8). Compared to YSZ, $CaZrO_3$ has a lower melting point (2550 °C), which allows its processing not only by atmospheric plasma spraying but also by flame spraying deposition technique.

In the present work, the properties and microstructures of coatings produced by the two different spraying processes, flame spraying (FS) and atmospheric plasma spray (APS), were compared. These coatings were analyzed in terms of microstructure, phases, composition, hardness, porosity, and thermal conduction properties. Post-heat treatments were carried out to evaluate the phase evolution and the properties of the coatings.

This article is an invited paper selected from presentations at the 2008 International Thermal Spray Conference and has been expanded from the original presentation. It is simultaneously published in *Thermal Spray Crossing Borders, Proceedings of the 2008 International Thermal Spray Conference*, Maastricht, The Netherlands, June 2-4, 2008, Basil R. Marple, Margaret M. Hyland, Yuk-Chiu Lau, Chang-Jiu Li, Rogerio S. Lima, and Ghislain Montavon, Ed., ASM International, Materials Park, OH, 2008.

E. Garcia, C. Cano, M.I. Osendi, and **P. Miranzo**, Institute for Ceramics and Glass (C.S.I.C.), C/Kelsen 5, Campus de Cantoblanco, Madrid 28049, Spain; and **T.W. Coyle**, Centre for Advanced Coating Technologies, University of Toronto, Toronto, ON, Canada M5S 3E4. Contact e-mail: garcia@icv.csic.es.

2. Experimental Procedure

The coatings were deposited over AISI 304 stainless steel substrates. The surface of the substrates was grit-blasted with corundum particles having a mean particle size of 530 μm to get a maximum roughness of 27.6 ± 2.4 μm. The ceramic feedstock was commercial calcium zirconate ($CaZrO_3$) (Metaceram 28085, Eutectic Castolin, Spain) with an average particle size of 47.9 μm.

The FS coatings of $CaZrO_3$ were achieved using an oxygen-acetylene gun (model CastoDyn DS 8000, Eutectic Castolin, Spain). The torch input power was 28 kW and the powder feed rate was ~20 g min^{-1}. The standoff distance was fixed at 15 cm according to previous experiments (Ref 9).

The plasma gun is a homemade gun developed at the Centre for Advanced Coating Technologies of the University of Toronto (CACT) (Ref 10). The plasma forming gases were CO_2/CH_4 mixtures, giving an output power of 40 kW. The coatings were deposited at standoff distances of 10, 17, and 22 cm with a feed rate of 17 g min^{-1}. The substrates were air-cooled during deposition.

In-flight particle properties were measured for the APS processes with a DPV 2000 monitoring system (Tecnar, Canada) to establish interrelations between spraying parameters and particle in-flight properties.

The microstructures of the coatings were analyzed on polished cross sections and on top surfaces by scanning electron microscopy (SEM) (DMS-950 Carl-Zeiss, Germany). Quantitative compositions were calculated by energy-dispersive x-ray spectroscopy (EDS) from standardless point analyses, using Proza $\varphi(\rho z)$ corrections. The crystalline phases in the specimens were identified by x-ray diffraction analysis (XRD) (Siemens D5000, Germany) performed on the coating surface. Chemical analyses were done using an x-ray fluorescence spectrometer (PW-2424, Philips, NL) equipped with an ultrathin window x-ray tube and high power (2.4 kW) rhodium anode. The hardness of the ceramic coatings was measured on the polished cross section of the specimens by microhardness Vickers indentation using loads of 1.9 and 2.9 N for 15 s (ZHU 2,5; Zwick GmbH & Co. KG, Germany). At least 5 measurements were taken for each coating.

Porosity, θ, was calculated from hardness using the following expression (Ref 11):

$$\frac{H}{H_0} = (1-\theta)^2 e^{-\beta\theta} \qquad \text{(Eq 1)}$$

where H is the measured hardness, H_0 is the hardness for the dense material (10.7 GPa (Ref 12)), and β is a constant parameter that depends on the material. The same β parameter calculated for a ZrO_2/Ni coating was used, which was 2.6, as it has been proved to be valid by comparing to imaging analysis data for the present case (Ref 12).

Thermal diffusivity (α) of freestanding coatings was measured using the laser-flash technique (Thermaflash 2000, NESTCH group, MA, USA). The thermal conductivity (k) was obtained through the expression $k = \alpha \cdot \rho \cdot C_p$, where ρ is the coating density and C_p is the heat capacity of the ceramic material taken from thermodynamic tabulated sources (Ref 13).

3. Results and Discussion

Adherent porous coatings were obtained by both methods (FS, APS) with thicknesses between 200 and 400 μm. The average temperatures and velocities of the APS particles at the different standoff distances are shown in Table 1. The highest average temperature (3028 ± 279 °C) and velocity (238 ± 114 m s^{-1}) were reached for the shorter standoff distance (10 cm, APS_10).

3.1 As-Sprayed Coatings

An example of the cross sections of the coatings can be seen in Fig. 1. Both types of coatings are porous but

Table 1 Average temperatures and velocities measured for the APS particles at the different standoff distances

	Standoff distance, cm	Temperature, °C	Velocity, m s^{-1}
APS_10	10	3028 ± 279	238 ± 114
APS_17	17	2598 ± 175	155 ± 51
APS_22	22	2504 ± 75	97 ± 21

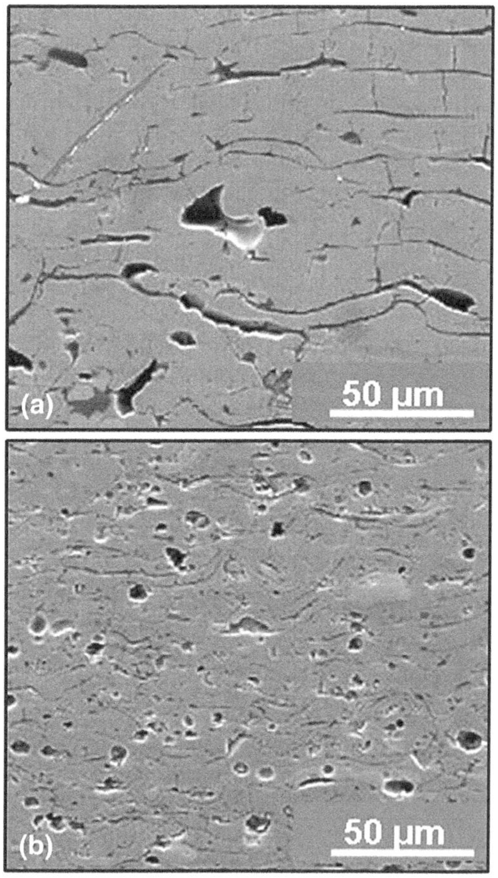

Fig. 1 SEM micrographs of polished cross sections of microstructures of the coatings for both spray processes: (a) FS coating with neutral flame, (b) APS coating sprayed at 17 cm

present different morphologies. The coating obtained by the FS process (Fig. 1a) exhibits a structure with thicker lamellas compared to the APS_17 coating (Fig. 1b). The microstructures of the APS coatings were similar for the three standoff distances and all contain more spherical pores of larger size than those observed in the FS coating.

A comparison of the top surfaces of both $CaZrO_3$ coatings is shown in Fig. 2. The FS coating (Fig. 2a) had a very smooth top surface with a few unmolten particles. The APS $CaZrO_3$ coatings (Fig. 2b) evidenced a rougher top surface and, apparently, higher level of porosity. No appreciable differences among the APS coatings were observed, although probably more overheated particles could be expected for the coating sprayed at 10 cm.

The morphology of the APS coatings indicates formation of bubbles in the splats (Ref 14, 15). During the particle impinging and flattening steps, entrapped hot gases can expand and escape through the flattened particles. Depending on the flattening particle viscosity at impact and the escaping gas energy, the gases could completely pass through the flattening particle generating volcano-sort-bubbles (Ref 16).

The x-ray diffraction patterns recorded for the original $CaZrO_3$ powder and for the coatings sprayed with both processes are shown in Fig. 3. The diffraction patterns showed that all analyzed samples contained orthorhombic $CaZrO_3$. Peaks that can be associated to cubic ZrO_2 phase were also clearly found in the APS coatings, although small ZrO_2 peaks were also found in the FS coating. Some preferential orientation was observed on the XRD pattern of the FS coating as there was an increase in the intensity of the peaks corresponding to the (101) and (202) $CaZrO_3$ planes, compared to that of the (121) plane (Ref 9). The APS coatings did not show a preferential orientation since the $I_{(101)}/I_{(121)}$ ratio was somewhat similar to that of the original powder. This less important texture effect can also be observed at the microstructure level (Fig. 4). While a columnar growth was observed inside the splats for the FS coatings (Fig. 4a), the APS coatings presented only a piling up of thin splats (Fig. 4b). This fact could be explained by the faster cooling rate of the splats in the APS, due to the air-cooling of the substrates, as compared to the FS, where no forced cooling was employed.

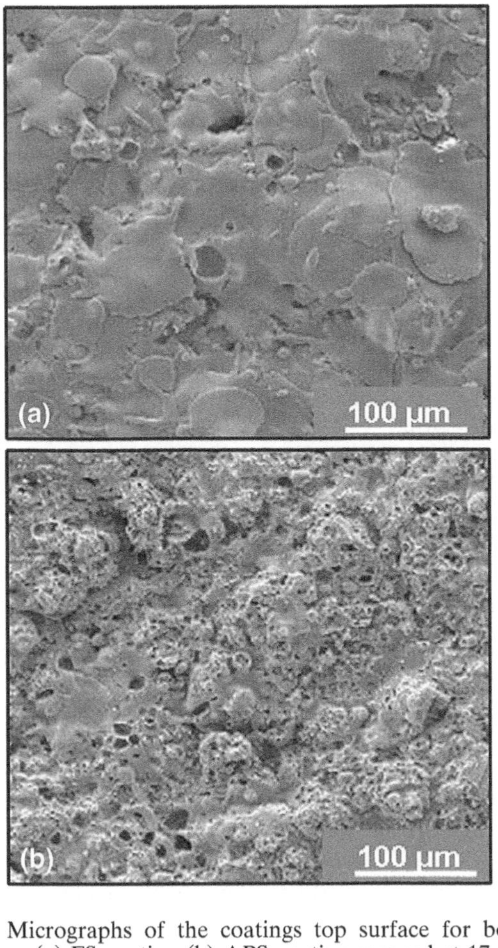

Fig. 2 Micrographs of the coatings top surface for both the processes: (a) FS coating (b) APS coating sprayed at 17 cm

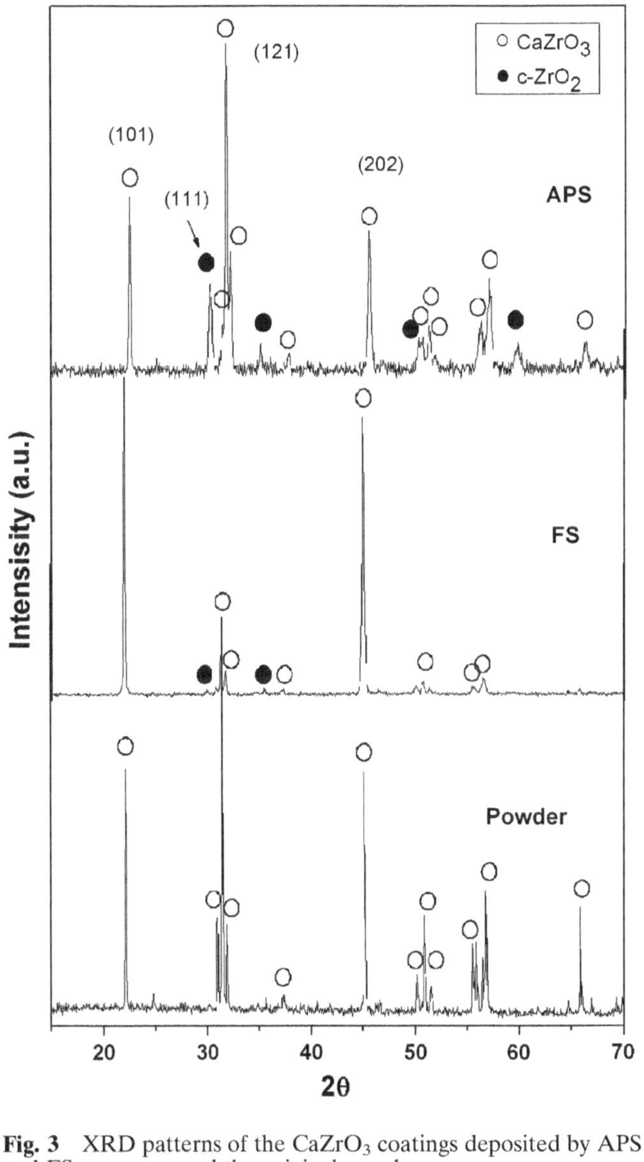

Fig. 3 XRD patterns of the $CaZrO_3$ coatings deposited by APS and FS processes and the original powder

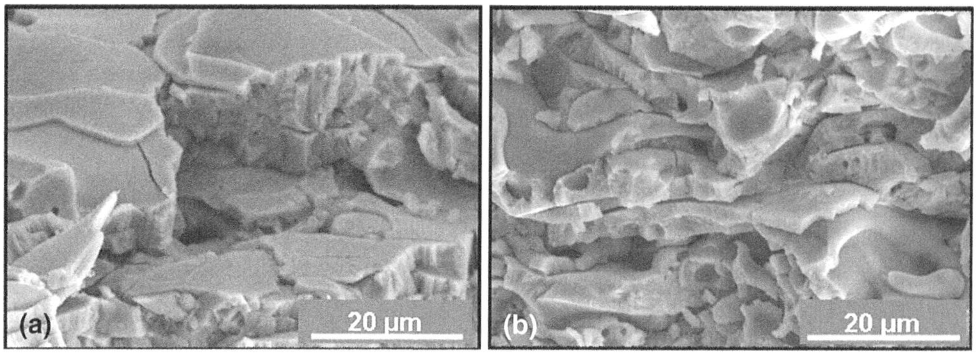

Fig. 4 Fresh fracture micrograph of FS (a) and APS (b) coatings

Table 2 Chemical composition of FS and APS coatings measured by x-ray fluorescence (XRF)

Compound	Feedstock powder, wt.%	FS coating, wt.%	APS coating, wt.%
CaO	30.0	29.7	27.6
MgO	0.29	0.26	0.19
SiO_2	0.52	0.48	0.20
ZrO_2	67.7	67.8	70.1

The relatively higher amount of c-ZrO_2 observed in the APS coatings, compared to the FS coating, can be associated to the CaO volatilization caused by the higher temperatures and reducing atmosphere generated by the CO_2/CH_4 plasma during spraying. This CaO volatilization was confirmed by the x-ray fluorescence (XRF) chemical analyses given in Table 2. About 3 wt.% reduction in CaO concentration was observed in the APS coating, whereas the FS coating showed similar composition to the starting calcium zirconate powders.

As seen in Fig. 5, the microstructure of the coatings showed areas of different gray levels. The average EDS analyses corresponding to those phases showed significant differences on the ZrO_2 and CaO mol.% content. The different compositions related to these phases have been localized in the ZrO_2-CaO phase equilibrium diagram (Ref 17), shown in Fig. 6, to aid the discussion. The matrix main phase (CZ in Fig. 5) has a composition 50:50 mol.% ZrO_2:CaO (Fig. 6, area "CZ"), and therefore, it would correspond to the calcium zirconate identified by XRD. Dark gray phases were found in FS coating ("D" in Fig. 5a), but not in APS coatings. These phases presented compositions with up to 10% of other elements such as aluminum and silicon, although maintaining the same ZrO_2:CaO ratio of ~1. Therefore, these could be identified as amorphous metastable phases that on equilibrium will crystallize on $CaZrO_3$. These phases are not detected for APS coatings because SiO_2 volatilization occurs in these coatings (Table 2). Finally, the main differences among coatings refer to the amount and composition of the clearest phases (C). Those had enriched ZrO_2 compositions with CaO contents ranging from 41 to 20 mol.%, extending from the eutectic point of the system (41 mol.%) down to the limit of CaO solid solution in c-ZrO_2 (Fig. 6 area "C-a"). The largest standoff distance of APS coatings showed composition with even lesser CaO content, in the range 0-6 mol.% (Fig. 6 area "C-b"). According to the XRD analyses and the information supplied by the corresponding phase equilibrium diagram, clear phases with CaO content less than 20 mol.% may be associated to crystalline c-ZrO_2, whereas those with CaO contents >20 mol.% would be amorphous metastable phases.

The presence of ZrO_2-enriched phases was explained because the volatilization of CaO in the coatings was a local phenomenon, which depended on the location of the particles inside the flame or plasma torch. The APS 22 cm coating showed the lowest CaO content because $CaZrO_3$ particles are inside the high temperature plasma plume for a longer time before reaching the substrate, which promoted higher volatilization of the CaO.

The average microhardness and porosity values measured in all $CaZrO_3$ coatings for both the FS and APS processes are provided in Table 3. The FS coating showed a slightly higher hardness value than the APS coatings. This result is consistent with the higher porosity measured for the APS coatings. No appreciable differences in hardness or porosity were found among APS coatings deposited at different standoff distances.

The thermal conductivity data obtained for all coatings were lower than those typically reported in the literature for ZrO_2-based TBCs (Ref 18, 19). This could be explained not just by the porosity amount but also by the pore shape (Ref 20). The FS coating is remarkable since, having a lower porosity, its thermal conductivity is much less than in the APS coatings. One possible cause is the different microstructure of both coatings (Fig. 1). Longer intersplat pores were found in the case of the FS coatings, which means that the heat propagation through this coating is hindered as compared to the APS coatings.

3.2 Heat-Treated Coatings

FS, APS_10, and APS_22 coatings were subjected to thermal treatments at 1200 °C during 12 h, to study phase and properties evolution. As there are not significant differences in the microstructures of the APS coatings, only the longest and shortest standoff distances have been selected for the heat treatments. The microstructure of the heat-treated coatings was characterized by the presence of

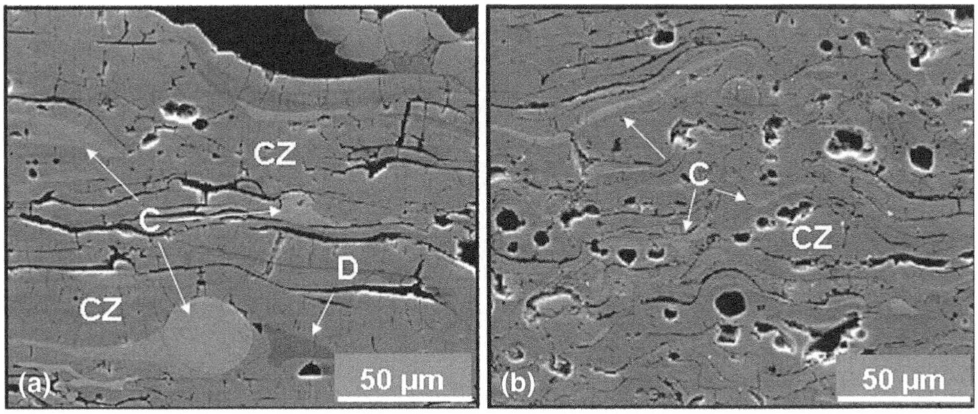

Fig. 5 SEM micrographs of details of the microstructure of FS (a) and APS (b) coatings. "CZ" shows matrix regions of $CaZrO_3$ composition, "C", clear phases, and "D" dark phases

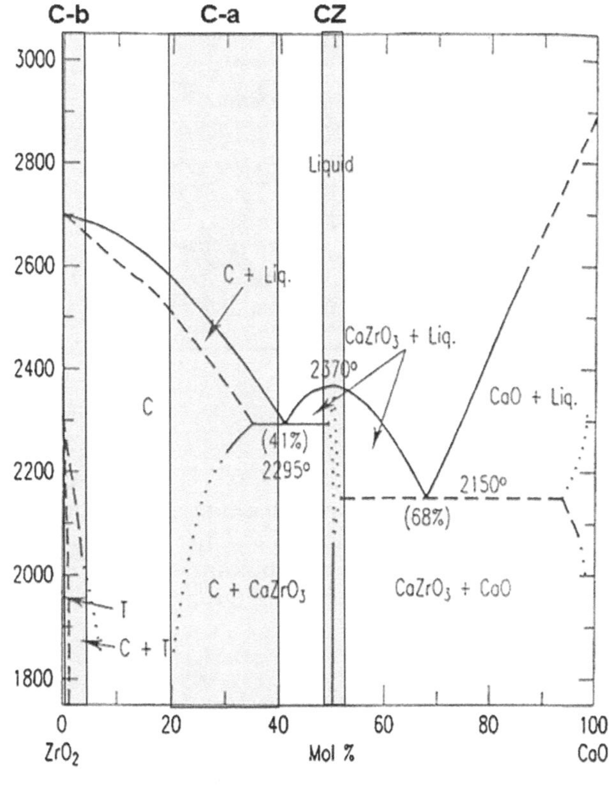

Fig. 6 ZrO_2-CaO phase equilibrium diagram (Ref 17). The shadowed areas are related with the different phase compositions found throughout the coatings

longer intersplat cracks (Fig. 7) compared to the untreated coating (Fig. 1), especially for the APS coatings. A crystallization process with micropores formation is clearly observed at higher magnification (Fig. 7c, d) This crystallization probably involves volume changes and then is likely responsible for the increase in the intersplat crack length.

The XRD analysis identified a small relative increase of the cubic ZrO_2 peaks compared to the $CaZrO_3$ peaks in all the treated coatings (Table 4). When melted particles reach the substrate at very high temperature, they remain partially amorphous; and then, crystallize later on c-ZrO_2 and $CaZrO_3$ when the coating is annealed at 1200 °C. Although no differences in crystalline phases were observed among the different coatings by XRD, the as-sprayed FS coating seemed to be more crystalline as it showed less degradation after heat-treating (Fig 7). This is a consequence of the lower temperatures and slower cooling rates reached by the particles.

In Table 3, the hardness values obtained for the heat-treated samples are also shown. The reduction in more than 25% of the corresponding original value is noticeable. The increase of the intersplat cracks length and the increase in apparent total porosity of the coatings could be the main reason for this reduction.

4. Conclusions

$CaZrO_3$ coatings produced by flame and air plasma spray techniques present different microstructures and

Table 3 Vickers microhardness, porosity, and thermal conductivity measured at room temperature of as-sprayed coatings and Vickers microhardness and porosity of heat-treated coatings

	As-sprayed coatings			Heat-treated coatings, 1200 °C	
	Microhardness, GPa	Porosity, vol.%	Thermal conductivity, Wm^{-1} K^{-1}	Microhardness, GPa	Porosity, vol.%
FS	4.8 ± 1.5	16.5	0.29	2.9 ± 0.6	26.5
APS (10 cm)	4.2 ± 1.0	19.2	0.72	3.1 ± 0.9	25.2
APS (17 cm)	4.2 ± 1.1	19.2	0.91	-	-
APS (22 cm)	4.6 ± 1.0	17.4	0.80	2.4 ± 1.2	30.0

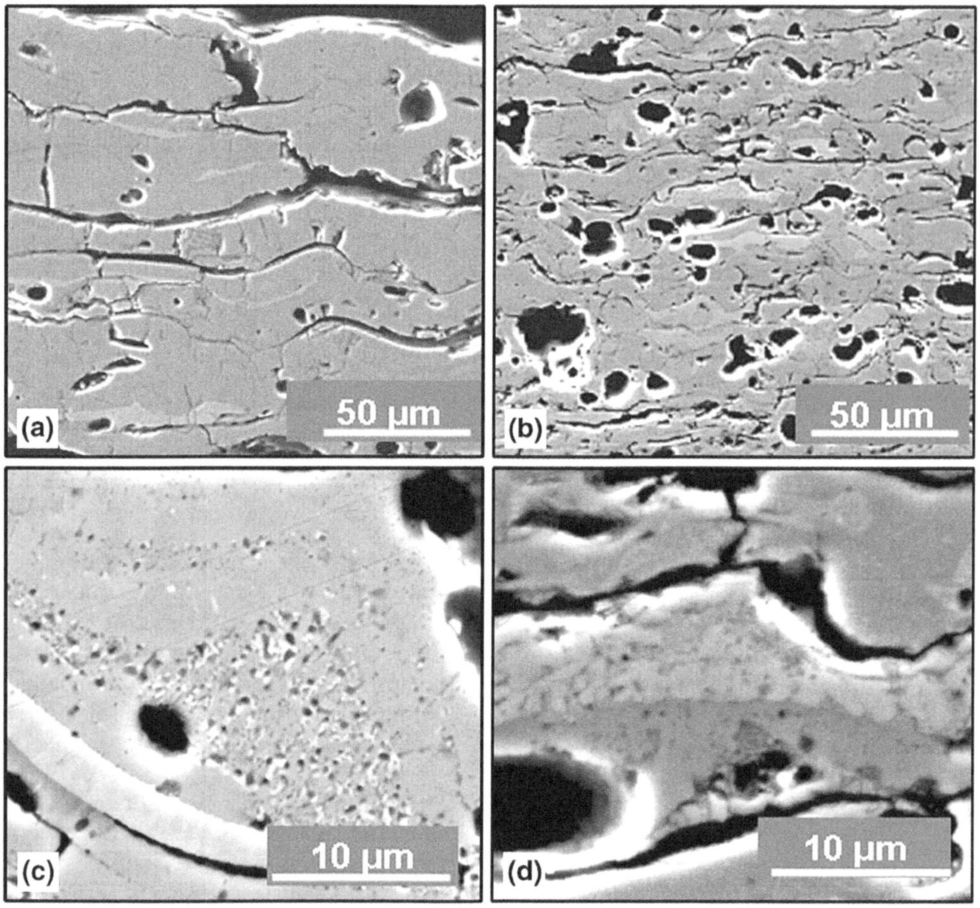

Fig. 7 Micrographs of the heat-treated FS (a, c) and APS 22 cm sprayed coatings (b, d)

Table 4 XRD peak intensity ratios of the c-ZrO_2 (111) and $CaZrO_3$ (121) diffraction planes obtained for the as-sprayed and heat-treated coatings

	I_{ZrO_2}/I_{CaZrO_3}	
	As-sprayed coatings	**Heat-treated coatings, 1200 °C**
FS	0.25	0.37
APS_10	0.33	0.38
APS_22	0.25	0.33

properties. All of them are porous, but flame sprayed coating shows longer intersplat cracks and APS coatings larger round pores. They are formed by a major matrix of $CaO:ZrO_2$ with a composition close to that of o-$CaZrO_3$ (50:50 mol.%), and other ZrO_2-enriched minor phases, which is the consequence of a preferential volatilization of CaO. Some of the latter have compositions with CaO contents within the solid solution limit of the Ca-ZrO_2. Phases with compositions in the range 20-40 mol.% CaO are identified mainly as amorphous metastable phases.

All the $CaZrO_3$ coatings show very low thermal conductivity, which is very attractive for their use as TBCs, much lower in FS than in APS coatings due to the particular microstructure of intersplat cracks.

The thermal treatments at 1200 °C induce some crystallization of the amorphous phases present in the coating, modifying the microstructure and the mechanical properties. This process is more significant in the APS coatings because they were more amorphous as sprayed.

Acknowledgments

The authors would like to acknowledge the financial support from the MEC-MAT2006-7118. E. Garcia acknowledges financial support from the I3P-PC2005L program. Thanks are given to F.J. Valle for ED-XRF analyses.

References

1. R.B. Heimann, *Plasma Spray Coatings, Principles and Applications*, VCH, Weinheim, New York, USA, 1996
2. J.R. Davis, *Handbook of Thermal Spray Technology*, ASM International, Materials Park, OH, EEUU, 2004
3. L. Pawlowski, *The Science and Engineering of Thermal Spray Coatings*, Ed. John Wiley & Sons Ltd., Chichester, UK, 1995
4. P.G. Klemens and M. Gell, Thermal Conductivity of Thermal Barrier Coatings, *Mater. Sci. Eng. A Struct.*, 1998, **245**, p 143-149
5. A. De Petris, F. Ricciardiello, and O. Sbaizero, Mechanical Properties of Polycrystalline $CaZrO_3$, *Powder Metall. Int.*, 1986, **18**, p 427-443

6. D.R. Clarke and S.R. Phillpot, Thermal Barrier Coating Materials, *Mater. Today*, 2005, **8**, p 22-29
7. S. Yamanaka, K. Kurosaki, T. Maekawa, T. Matsuda, S. Kobayashi, and M. Uno, Thermochemical and Thermophysical Properties of Alkaline-Earth Perovskites, *J. Nucl. Mater.*, 2005, **344**, p 61-66
8. P.S. Murti and M.V. Krishnaiah, Investigation of the Thermal Conductivity of Calcium Cerate and Calcium Zirconate, *Mater. Chem. Phys.*, 1992, **31**, p 347-350
9. C. Cano, M. Belmonte, M.I. Osendi, and P. Miranzo, Effect of the Type of Flame on the Microstructure of $CaZrO_3$ Combustion Flame Sprayed Coatings, *J. Surf. Coat. Technol.*, 2007, **201**, p 3307-3313
10. L. Pershin, L. Cheng, and J. Mostaghimi, Comparison of Molecular and Argon Gases for Plasma Spraying, *Thermal Spray 2007: Global Coating Solutions*, B.R. Marple, M.M. Hyland, Y.C. Lau, C.J. Li, R.S. Lima, and G. Montavon, Ed., ASM International Materials Park, Ohio, USA, e-version
11. R. Polanco, P. Miranzo, and M.I. Osendi, Fabrication and Microstructure of a ZrO_2-Ni Functionally Graded Bonding Interlayer using the Airbrush Spraying Method, *Acta Mater.*, 2006, **54**, p 2215-2222
12. C. Cano, "Ceramic Coatings for Thermal and Environmental Barrier Applications," Ph. D. Thesis, Autonomous University of Madrid, 2008 (Spanish)
13. *Landolt-Börnstein, Thermodynamic Properties of Inorganic Materials*, Scientific Group Thermodata Europe (SGTE), Part I, Springer-Verlag, Berlin-Heidelberg, 1999
14. Y. Tanaka and M. Fukumoto, Investigation of Dominating Factors on Flattening Behaviour of Plasma Sprayed Ceramic Particles, *Surf. Coat. Technol.*, 1999, **120-121**, p 124-130
15. A. Kulkarni, A. Vaidya, A. Goland, S. Sampath, and H. Herman, Processing Effects on Porosity-Property Correlations in Plasma Sprayed Ytria-Stabilised Zirconia Coatings, *Mater. Sci. Eng. A Struct.*, 2003, **359**, p 100-111
16. A.A. Syed, A. Denoirjean, B. Hannoyer, P. Fauchais, P. Denoirjean, A.A. Khan, and J.C. Labbe, Influence of Substrate Surface Conditions on the Plasma Sprayed Ceramic and Metallic Particles Flattening, *Surf. Coat. Technol.*, 2005, **200**, p 2317-2331
17. J.P. Traverse and M. Foex, Studies of the Systems Formed by Zirconium with Lime and Strontium Oxide, *High Temp. High Pressures*, 1969, **1**, p 409-427
18. J. Wu, X. Wei, N.P. Padture, P.G. Klemens, M. Gell, E. Garcia, P. Miranzo, and M.I. Osendi, Low Thermal Conductivity Rare Earth Zirconates for Potential Thermal Barrier Coating Applications, *J. Am. Ceram. Soc.*, 2002, **85**, p 3031-3035
19. S. Ahmaniemi, P. Vuoristo, T. Mantyla, F. Cernuschi, and L. Lorenzoni, Modified Thick Barrier Coatings: Thermophysical Characterization, *J. Eur. Ceram. Soc.*, 2004, **24**, p 2669-2679
20. A. D. Jadhav, N.P. Padture, E.H. Jordan, M. Gell, P. Miranzo, and E.R. Fuller, Low-Thermal-Conductivity Plasma-Sprayed Thermal Barrier Coatings with Engineered Microstructures, *Acta Mater.*, 2006, **54**, p 3343-3349

Peer Reviewed

JTTEE5 17:872–877
DOI: 10.1007/s11666-008-9241-x
1059-9630/$19.00 © ASM International

Peer Reviewed

Fabrication of Bulk Nanocrystalline Ceramic Materials

T. Chraska, K. Neufuss, J. Dubsky, P. Ctibor, and M. Klementova

(Submitted June 13, 2008; in revised form August 28, 2008)

An alternative method to produce bulk nanocrystalline materials is to produce amorphous material by rapid solidification followed by controlled solid state crystallization. Four different materials studied in this work are based on near-eutectic mixtures of alumina, zirconia, and silica. Powder feedstock materials have been plasma sprayed using a water stabilized plasma torch (WSP®) and subsequently heat-treated to prepare materials with nanocrystallites of 12 nm diameter on average. The as-sprayed materials have very low porosity and are mostly amorphous. The as-sprayed amorphous materials crystallize at temperatures around 950 °C. The resulting structure is best described as nanocomposite with the very small crystallites embedded in an amorphous matrix. The role of silica on phase composition, microstructure, and mechanical properties of the as-sprayed and annealed materials is discussed. The as-sprayed amorphous materials exhibit high hardness and high abrasion resistance. Both properties are significantly improved in the heat-treated nanocrystalline samples.

Keywords abrasion resistance, amorphous phase, heat treatment, nanocrystalline ceramics

1. Introduction

There is a wide potential for applications of structural nanocrystalline ceramics due to their significantly improved properties such as hardness, strength, and abrasion resistance. However, limited ability to fabricate large three-dimensional parts of dense nanocrystalline materials is the major obstacle to their greater use. Ceramic nanocrystalline powder can be conventionally synthesized in large quantities but the consolidation of nanoparticles into mesoscopic structures and large bulk parts remains a challenge (Ref 1, 2). In an alternative approach, amorphous material is produced by rapid solidification followed by heat treatment to introduce nanocrystalline structure in the bulk by controlled crystallization. One technique capable of achieving rapid solidification in materials is plasma spraying (PS) (Ref 3, 4).

This article is an invited paper selected from presentations at the 2008 International Thermal Spray Conference and has been expanded from the original presentation. It is simultaneously published in *Thermal Spray Crossing Borders, Proceedings of the 2008 International Thermal Spray Conference*, Maastricht, The Netherlands, June 2-4, 2008, Basil R. Marple, Margaret M. Hyland, Yuk-Chiu Lau, Chang-Jiu Li, Rogerio S. Lima, and Ghislain Montavon, Ed., ASM International, Materials Park, OH, 2008.

T. Chraska, K. Neufuss, J. Dubsky, and **P. Ctibor**, Institute of Plasma Physics AS CR, Za Slovankou 3, 18200, Prague, Czech Republic; and **M. Klementova**, Institute of Inorganic Chemistry AS CR, Husinec-Řež 1001, 25068, Řež, Czech Republic. Contact e-mail: tchraska@ipp.cas.cz.

In our study, the ternary Al_2O_3-ZrO_2-SiO_2 system was chosen as it contains binary and ternary eutectic points. It has been previously reported that materials with near eutectic composition can solidify as amorphous or partially nanostructured solids due to rapid solidification (e.g. Ref 5-10). Four different compositions of feedstock material for plasma spraying were selected in order to study the influence of varying silica content on amorphization during rapid solidification, subsequent crystallization, and the final properties of nanocomposite ceramics.

2. Experimental

The four selected feedstock powder compositions (labeled as NZ, E-I, E-II, and E-III) are shown in a partial ternary phase diagram (Fig. 1, Table 1). The NZ+ powder with particle size 65-120 μm is commercially available as fused and crushed material. The other three feedstock powders are based on the bulk cast ceramic material called Eucor™ (Ref 11) and were produced by pouring compositionally modified molten material into water. After drying, the solidified material was crushed to powder and sieved to a particle size of 63-100 μm. Feedstock powders were plasma-sprayed in air at atmospheric pressure using the water stabilized plasma torch WSP® 500 operating at 160 kW power. The powders were injected into the plasma jet by compressed air at a constant feed rate of about 250 g/min and deposited onto mild steel substrates positioned 350 mm away from the torch nozzle. The mild steel surfaces were either sand blasted to assist in the production of well adhered thick coatings or treated for easy deposit removal from substrates to produce freestanding parts (e.g. 25 × 100 × 3 mm test coupons or pipes with 40-80 mm diameters).

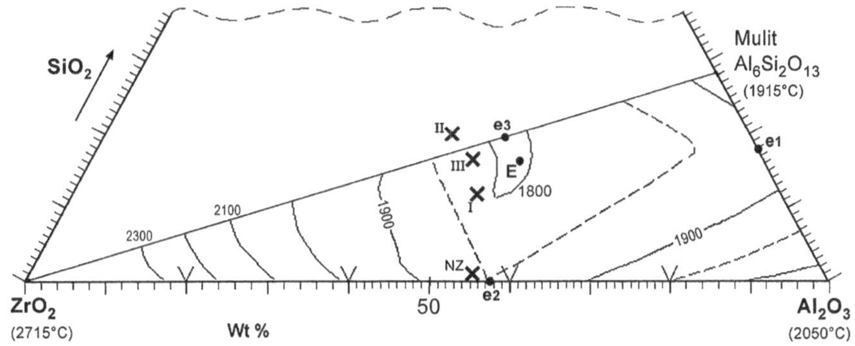

Fig. 1 Part of ternary equilibrium phase diagram showing different compositions of feedstock powder used in this study (marked by crosses). Compositional values shown from Table 1 were normalized to 100% with respect to the three main compounds to facilitate plotting in the ternary diagram. Circles denote binary (e1, e2, e3) and ternary (E) eutectic points. Diagram from ACerS-NIST Phase Equilibria Diagram, CD-ROM Database, Version 3.0

Fig. 2 SEM image (backscattered electrons) of un-etched polished cross-section of E-II feedstock powder embedded in epoxy (black regions). Zirconia is represented by white regions and alumina by gray regions

Table 1 Composition of feedstock powders

Material	Al_2O_3, wt.%	ZrO_2, wt.%	SiO_2, wt.%	Others, wt.%
NZ+	53	42.5	1	3.5
Eucor I	48	37	11.5	3.5
Eucor II	41	35	18.5	5.5
Eucor III	44.5	35	15.5	5

Microstructure was characterized by a variety of experimental techniques, namely light microscopy, scanning electron microscopy (SEM) and energy dispersive X-ray spectrometry (EDS), transmission electron microscopy (TEM), and high-resolution TEM (HRTEM) with EDS. Thin foil cross-sectional samples for TEM were prepared by the wedge polishing technique (Ref 12) and by crushing into very fine particles. X-ray diffraction (XRD) was used to determine crystallinity and phase content of the bulk samples and line-profile analysis was employed to estimate crystallite size. Porosity was measured by Archimedean (water immersion) technique and helium pycnometry. Thermal properties were measured on finely crushed samples by differential thermal analysis (DTA) and on oblong parallelepiped samples by

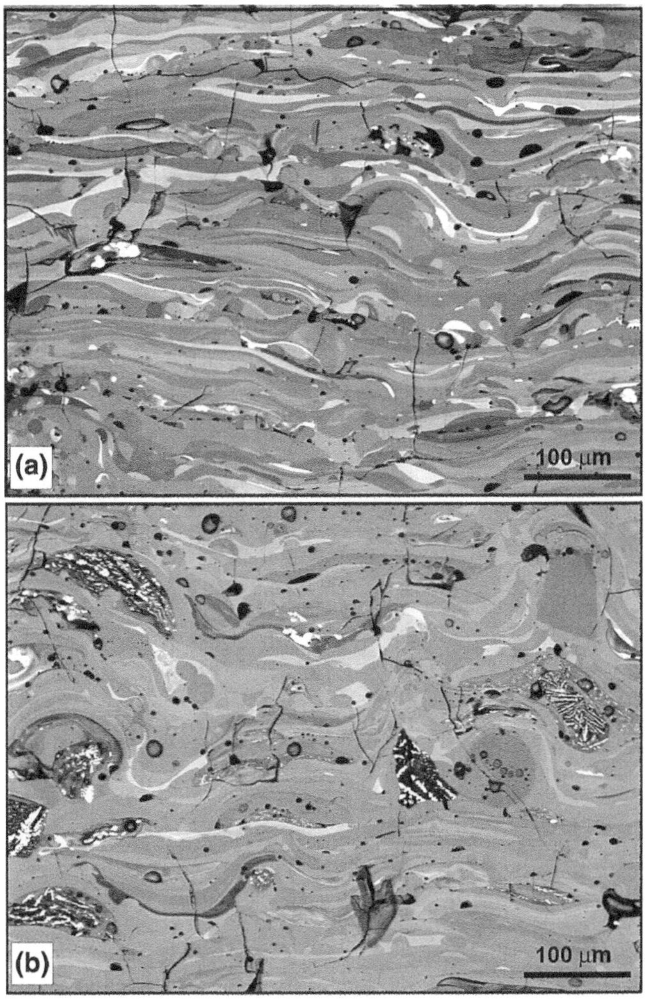

Fig. 3 SEM image (backscattered electrons) of un-etched polished cross-section of as-sprayed samples: (a) NZ+ and (b) Eucor II

dilatometry (i.e. thermomechanical analysis (TMA)). Microhardness of the coatings was measured by Vickers indenter using 1 N load applied over 15 s. The mean value of microhardness was calculated as an average from 20 indentations. Finally, slurry abrasion response (SAR) tests were performed according to a modification of the ASTM G75 standard using alumina slurry.

3. Results and Discussion

All four feedstock powders are fully crystalline as determined by XRD. The following phases were identified in the powders: monoclinic ZrO_2 (baddeleyite), α-Al_2O_3 (corundum), and tetragonal ZrO_2 (t-ZrO_2). Eucor powders also exhibit a small peak that belongs to crystalline SiO_2 (quartz). SEM examination of materialographic cross-sections of the powder particles determined that each particle in all four feedstock powders always contains both of the main constituents (i.e. alumina and zirconia) embedded in a eutectic microstructure (Fig. 2).

The typical microstructure of as-sprayed samples is presented in Fig. 3. Most of the sample consists of thin splats with just a few visible cracks and a few unmelted particles. Total porosity of the as-sprayed materials was found to be between 3.5% for material E-II and 6.8% for material NZ+. Each splat in the as-sprayed materials is a homogeneous blend of Al, Zr, and Si oxides. Each individual splat is chemically homogeneous but there is a significant chemical variation among splats (Fig. 3) due to different proportions of Al and Zr oxides in the individual feedstock particles. XRD analysis established the as-sprayed samples of Eucor materials to be fully amorphous, whereas the as-sprayed NZ+ material does contain wide peaks of tetragonal zirconia and delta alumina (Fig. 4). It was confirmed by TEM that the as-sprayed NZ+ material contains nanocrystalline regions of tetragonal zirconia and delta alumina (Fig. 5).

DTA measurement of two consecutive identical heating runs at 5 K/min indicates exothermic transformation during the first run. XRD patterns taken from annealed samples verified that the exothermic transformation corresponds to crystallization. Further DTA measurements were carried in the differential scanning calorimetry mode at several heating rates. The heat evolved per mass unit of sample was moderately decreasing with the increasing content of SiO_2 (Table 2). Even though the temperatures of the crystallization onset were very similar for the four materials, the rate of crystallization was larger for the NZ+ sample than for Eucor samples (containing a significant amount of SiO_2). This can be inferred from the widths of crystallization interval, i.e. the temperature interval of heat evolution during the linear temperature scan (Table 2). TMA measurements confirmed the onset of crystallization at the above mentioned temperatures

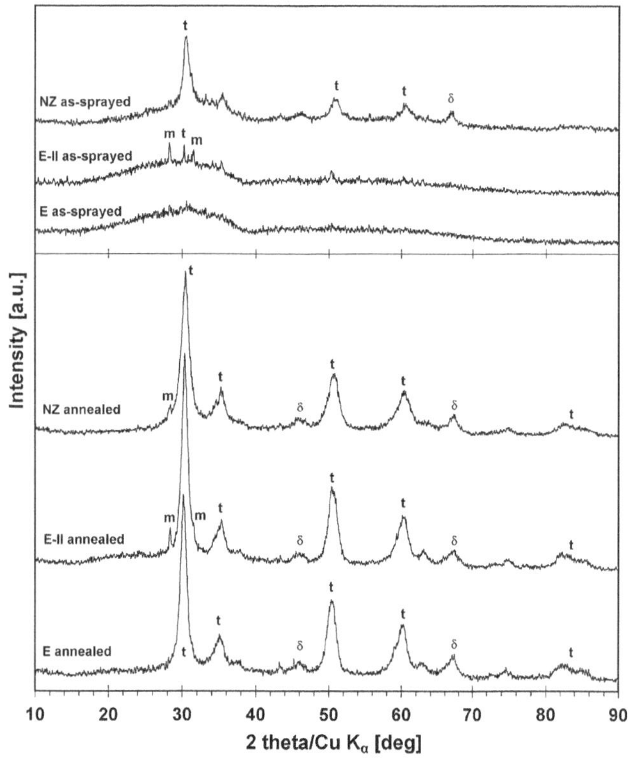

Fig. 4 XRD patterns of as-sprayed and annealed samples: t = tetragonal ZrO_2; m = monoclinic ZrO_2; δ = δ-Al_2O_3

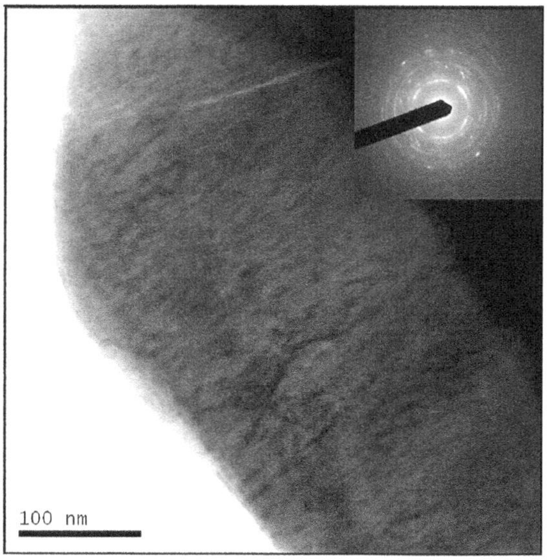

Fig. 5 Bright field TEM micrograph of the as-sprayed NZ+ material. Corresponding selected diffraction pattern is shown in the inset

Table 2 Results of differential thermal analysis of the as-sprayed samples at the heating rate of 5 K/min

Material	Temperature of crystallization onset, °C	Crystallization interval, K	Heat evolution per sample mass (sample NZ+ = 100%), %
NZ+	942 ± 3	32 ± 5	100
Eucor I	947 ± 3	53 ± 5	90
Eucor II	940 ± 3	56 ± 5	81
Eucor III	941 ± 3	52 ± 5	93

and detected material shrinkage during crystallization. Silica containing materials shrink by 1.5-2% during crystallization and that does not cause material breakup. NZ+ material, however, shrinks by almost 5%. As a result, the annealed NZ+ material becomes extremely fragile and cannot be used for any further mechanical tests. SEM observations confirmed that all of the annealed samples possess the same microstructure composed of tightly packed splats of varying compositions as in the case of as-sprayed samples.

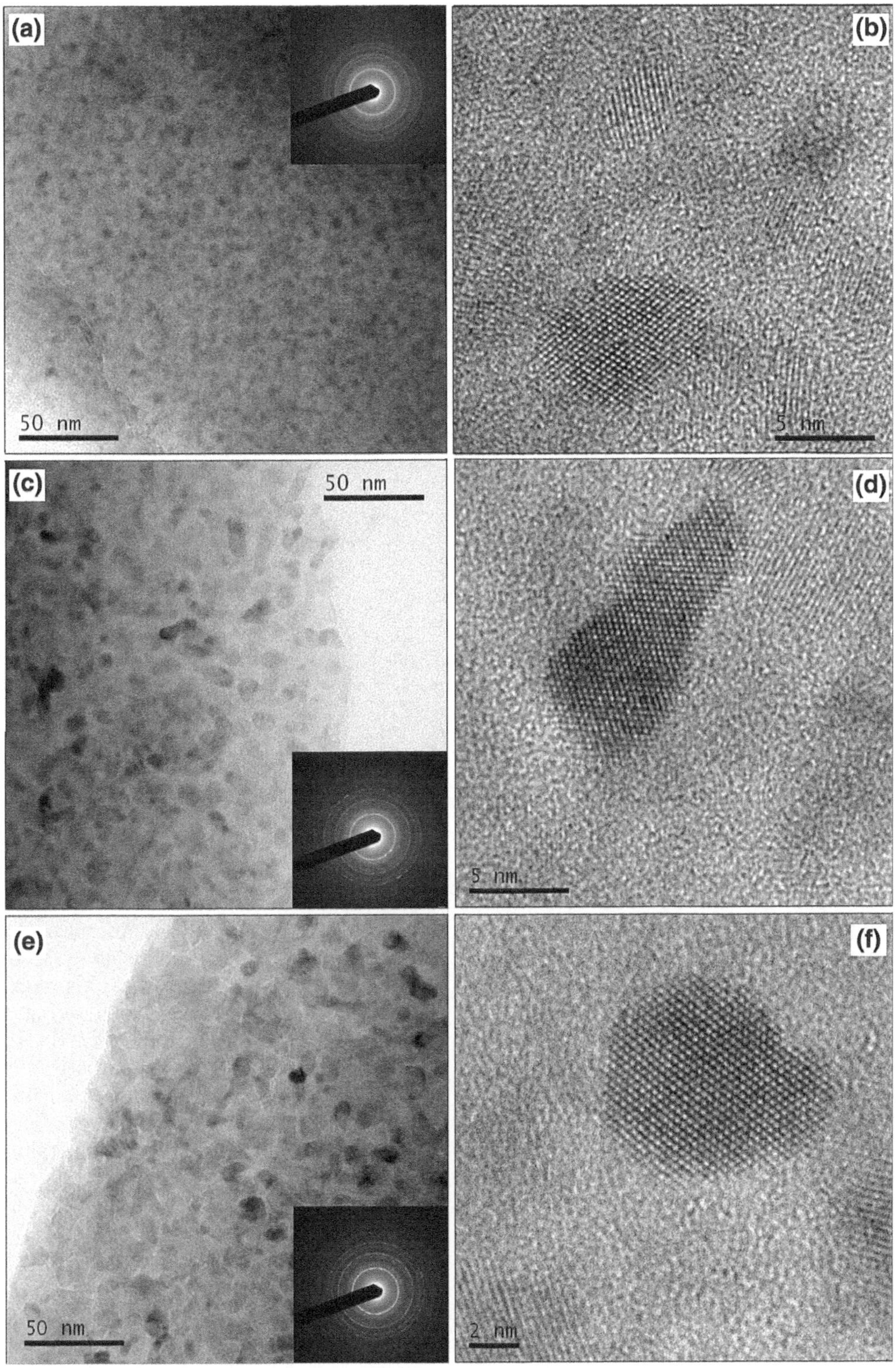

Fig. 6 Bright field TEM and HRTEM micrographs of annealed samples with corresponding diffraction patterns shown as insets. (a, b) NZ+, (c, d) Eucor I, (e, f) Eucor II. High-resolution images (b, d, f) show tetragonal ZrO_2 nanocrystals embedded in a weak contrast amorphous phase

Table 3 Comparison of different properties of as-sprayed and annealed materials

Material	state	Crystallite size, nm	Hardness, GPa	SAR coefficient
NZ+	As-sprayed	9	11.5	2.3
	Annealed	10.5	n/a	n/a
Eucor I	As-sprayed	Fully amorphous	12.8	5.4
	Annealed	12.5	16.5	2.7
Eucor II	As-sprayed	Fully amorphous	10.0	4.5
	Annealed	16	13.7	3.1
Eucor III	As-sprayed	Fully amorphous	10.7	5.9
	Annealed	14.5	15.0	3.9

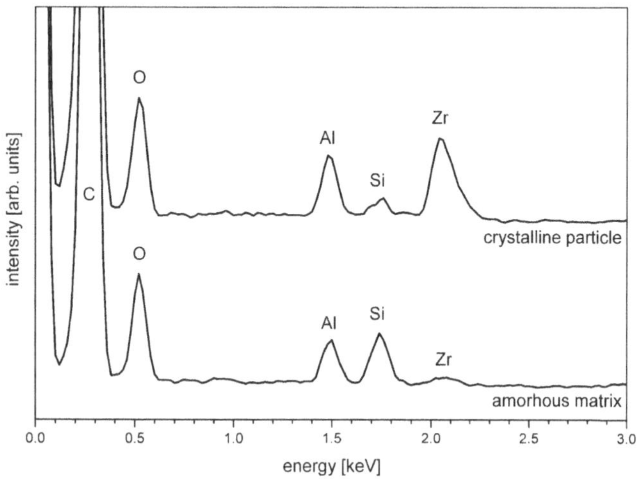

Fig. 7 EDX analysis of the nanocomposite structure (oxide ceramics) in HRTEM: nanocrystallites are rich in Zr whereas the amorphous matrix has Zr concentration depleted and is enriched in Si

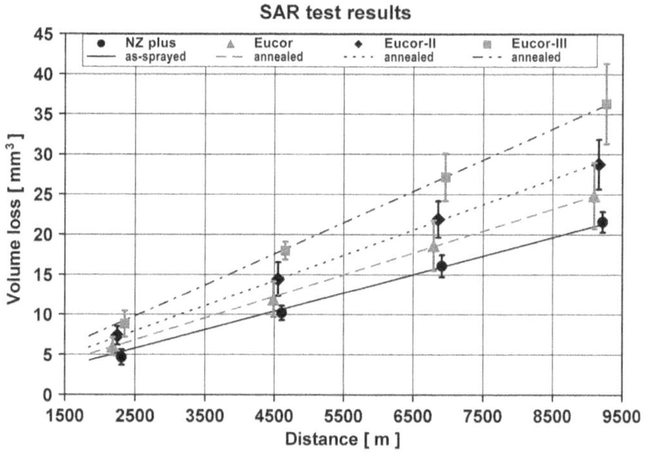

Fig. 8 Graphical presentation of SAR test results for annealed Eucor samples and as-sprayed NZ+ sample. The error bars represent the standard deviations of the measured values after each sliding distance. The data x-positions are slightly offset in the graph to improve clarity

Additional heat treatments were carried out with heating and cooling rates fixed at 10 K/min and short dwell times just above the crystallization temperature T_C. XRD patterns of the annealed samples (Fig. 4) contain wide peaks of t-ZrO_2 as the predominant phase and also wide, but small, peaks of δ-Al_2O_3 which crystallizes simultaneously within the amorphous splats. Verification of the annealed microstructure was carried out in TEM and HRTEM (Fig. 6). Several hundreds of nanocrystallites were analyzed on several TEM micrographs (e.g. Fig. 6). The TEM results are in good agreement with the XRD line width analysis (Table 3). Detailed observations of bright field TEM images reveal a fine network of a weak contrast phase in between the individual t-ZrO_2 crystallites (Fig. 6). This fine network surrounding individual crystallites is shown to be an amorphous phase on HRTEM images (Fig. 6). The nanocrystallites thus do not form standard grain boundaries and appear to be rounded rather than angular. The predominant t-ZrO_2 crystallites within each splat are embedded in remains of the original amorphous matrix. EDS analysis in HRTEM showed that each nanocrystallite is rich in Zr oxide whereas the amorphous matrix is enriched by Si and Al oxides and contains only a small amount of Zr oxide (Fig. 7). In some splats, the amorphous matrix has partially transformed to δ-Al_2O_3 phase.

The predominance of t-ZrO_2 phase in the heat-treated samples indicates that the material is still far from equilibrium and both alumina and silica are in solid solution with tetragonal zirconia. The ternary phase diagram (Fig. 1) predicts mullite as the third equilibrium phase but this phase was not detected. SiO_2 is dissolved both in the inter-crystallite phase and in the t-ZrO_2 solid solution nanocrystallites. TEM image analysis combined with EDS analysis of the all Eucor materials indicated that a higher fraction of silica within a splat tends to produce larger nanocrystallites and a thicker matrix. In other words, the average size of nanocrystallites within each splat depends on silica content in such a way that for higher silica concentration the average size is larger. This is also true when comparing the overall average crystallite size for the four tested materials (Table 3). It can be concluded that increasing SiO_2 concentration in the amorphous samples results in the decrease of crystal nucleation and growth rates. Since the average crystallite size is larger in samples with higher silica fraction in the material, it is mainly the nucleation rate that is responsible. Higher silica content thus reduces nucleation rate, leading to fewer nucleated crystallites that can grow larger. This is in accordance with the above mentioned tendency of the NZ+ samples, which contain almost no silica, to partially crystallize during the process of spraying and deposition.

Microhardness measurements of annealed samples, as measured by Vickers indenter, are summarized in Table 3. The microhardness measured in the as-sprayed samples is lower by 30-40%. The significant increase of hardness values in samples annealed for 1 min at the transformation temperature T_C can be attributed to the fine nanocomposite structure. When comparing the silica containing samples (Eucor), a higher content of silica tends to reduce the microhardness values for both as-sprayed and heat-treated samples.

Slurry abrasion response (SAR) tests were performed for as-sprayed and annealed (to T_C) samples according to a modified ASTM G75 standard. In each case, four samples (12 × 25 mm) were slid back and forth on rubber strips immersed in a slurry of fine alumina powder and water. Weight loss had been measured four times after 2 h intervals which correspond to ~2000 m sliding distance. Average volume loss after each interval was calculated using the sample density. The slope of a linear fit to the four measured points (Fig. 8) determines the overall abrasion loss and is presented as the SAR coefficient in Table 3. Samples annealed to the crystallization temperature exhibit improvement of the abrasion resistance by 60-80%. The improvement of abrasion resistance is consistent with the hardness increase and can both be attributed to the fine nanocomposite structure of the heat-treated samples. However, the best abrasion resistance is exhibited by the partially nanocrystalline as-sprayed sample of NZ+, which contains almost no silica. Abrasion resistance of the nanocomposite samples is extremely high and comparable or higher to that of pure chromia (Ref 13) sprayed by WSP®.

4. Conclusions

Four different crystalline feedstock powders of near eutectic composition containing alumina, zirconia, and silica have been successfully transformed into bulk nanocomposite ceramic parts in a two step process involving plasma spraying by WSP® and controlled heat treatment. The as-sprayed materials in the form of coatings and freestanding parts exhibited low porosity and were either fully amorphous or partially nanocrystalline depending on their silica content. Higher content of silica in feedstock material completely inhibits crystallization during rapid solidification, resulting in amorphous as-sprayed samples.

On annealing, these as-sprayed materials crystallize around 950 °C with shrinkage of 1.5-2% in the case of silica containing samples and 5% in no silica material. The presence of silica tends to lower the nucleation rate in solid state crystallization, leading to larger nanocrystallites for higher silica content. Short annealing to the crystallization temperature produces a nanocomposite structure with average crystallite size as low as 10 nm. Nanocrystallites of the dominant phase, t-ZrO_2, are embedded in fine amorphous matrix that is rich in Al_2O_3 and especially SiO_2. Both alumina and silica are also in solid solution with tetragonal zirconia. Microhardness and abrasion resistance values of the as-sprayed materials are significantly increased in the annealed nanocomposite samples. Nanocrystalline or nanocomposite ceramic parts with such a high abrasion resistance will be commercially exploited.

Acknowledgment

This work was carried out under the "Nanotechnologies for Society" program and supported by project no. KAN300430651 administered by the Academy of Sciences of the Czech Republic.

References

1. C. Suryanarayana, The Structure and Properties of Nanocrystalline Materials: Issues and Concerns, *J. Miner. Met. Mater. Soc.*, 2002, **54**(9), p 24-27
2. M. Winterer, *Nanocrystalline Ceramics: Synthesis and Structure*, Springer Series in Materials Science, vol. 53, Springer, Berlin, 2002
3. T. Chraska and A.H. King, Transmission Electron Microscopy Study of Rapid Solidification of Plasma Sprayed Zirconia – Part II. Interfaces and Subsequent Splat Solidification, *Thin Solid Films*, 2001, **397**(1-2), p 40-48
4. L. Li, B. Kharas, H. Zhang, and S. Sampath, Suppression of Crystallization During High Velocity Impact Quenching of Alumina Droplets: Observations and Characterization, *Mater. Sci. Eng. A*, 2007, **456**(1-2), p 35-42
5. A. Pawlowski, J. Morgiel, and T. Czeppe, Amorphisation and Crystallisation of Phases in Plasma Sprayed Al_2O_3 and ZrO_2 Based Ceramics, *Arch. Metall. Mater.*, 2007, **52**(4), p 635-639
6. H.J. Kim and Y.J. Kim, Amorphous Phase Formation of the Pseudo-Binary Al_2O_3-ZrO_2 Alloy During Plasma Spray Processing, *J. Mater. Sci.*, 1999, **34**(1), p 29-33
7. X.Z. Zhou, V. Shukla, W.R. Cannon, and B.H. Kear, Metastable Phase Formation in Plasma-Sprayed ZrO_2 (Y_2O_3)-Al_2O_3, *J. Am. Ceram. Soc.*, 2003, **86**(8), p 1415-1420
8. A. Petersson, H. Keshavan, and W.R. Cannon, Nanostructured Ceramics Processed from Coarse Powder, *Mater. Sci. Eng. A*, 2008, **475**(1-2), p 62-67
9. A.L. Vasiliev, N.P. Padture, and X.Q. Ma, Coatings of Metastable Ceramics Deposited by Solution-Precursor Plasma Spray: I. Binary ZrO_2-Al_2O_3 System, *Acta Mater.*, 2006, **54**(18), p 4913-4920
10. G. Bayrak and S. Yilmaz, Crystallization Kinetics of Plasma Sprayed Basalt Coatings, *Ceram. Int.*, 2006, **32**(4), p 441-446
11. Eutit Ltd., www.eutit.cz
12. T. Chraska, Transmission Electron Microscopy of Plasma Sprayed Ceramic Deposits, *Acta Tech. CSAV*, 2006, **51**(4), p 403-413
13. J. Nohava and P. Chráska, Wear Resistance of Al_2O_3 and Cr_2O_3 Coatings Deposited by Water-Stabilized Plasma Spraying, *Proceedings of the International Symposium on Plasma Chemistry*, R. d'Agostino, Ed., Int. Plasma Chemistry Society, 2003

JTTEE5 17:878–886
DOI: 10.1007/s11666-008-9285-y
1059-9630/$19.00 © ASM International

Peer Reviewed

Plasma Sprayed Coatings of High-Purity Ceramics for Semiconductor and Flat-Panel-Display Production Equipment

Junya Kitamura, Hiroyuki Ibe, Fumi Yuasa, and Hiroaki Mizuno

(Submitted May 15, 2008; in revised form August 12, 2008)

High-purity oxide ceramic powders of alumina (Al_2O_3) and yttria (Y_2O_3) have been developed to apply to semiconductor and flat-panel-display (FPD) production equipment. The ceramic coatings on the inside chamber wall of the equipment are required to have high erosion resistance against CFx plasma in dry etching process for microfabrications of the devices. It is found that the yttria coating formed from agglomerated-and-sintered powder consisting of large primary particles has smoother eroded surface with high erosion resistance. Considering the particle deposition on the devices, this coating will be effective in decreasing generation of large-sized particles, which easily deposit on the devices. Electric insulating properties of the coatings are also investigated to apply to electrostatic chuck. Electric breakdown voltage of yttria coatings is comparable to that of alumina coatings. Smaller powder is effective for improving the electric properties, and the influence of coating purity is lower than that of the powder size.

Keywords alumina, APS coatings, erosion and abrasion resistance, high-purity ceramics, semiconductor applications, yttria

1. Introduction

The size of semiconductor and flat-panel-display (FPD) production equipment for dry etching, sputtering, and ashing has been increasing due to the increasing size of Si wafer (from 200 to 300 mm in diameter) and of the FPD, where plasma treatment is effectively used for micro fabrication especially in dry etching. Applied power for generating plasma is also increasing to fabricate the Si and LCD devices uniformly onto the large-scaled substrates. This trend strongly promotes application of plasma sprayed coatings by high-purity ceramics for anti-plasma erosion at inside chamber wall and for high electric strength (high breakdown voltage) of electrostatic chuck (ESC) to replace from conventional techniques, such as anodized aluminum (alumite film) and sintered bulk ceramics (Ref 1, 2). For example, inner diameter of the chamber wall is increasing from 400 to 500-600 mm in the Si device production equipment due to the enlargement of the wafer size.

Using the alumite film as a shield or an ESC to protect the chamber parts has been difficult because the halogen-containing plasma with high power erodes the film at a high rate. This intense erosion generates a large amount of particles and results in frequent maintenance of the production equipment and decrease of yield ratio of the devices. Sintered bulk ceramics are also difficult to be used as the shielding parts by the enlargement of the equipment because production of large-scaled ceramics is technically difficult and its cost tends to become high.

Plasma sprayed ceramics coatings have technical and commercial advantages to overcome these problems, such as no limitation of the equipment size, relatively higher anti-plasma erosion resistance, higher breakdown voltage, and relatively low cost to make thick coating of about a few hundred micrometers. As ceramic materials for plasma spraying, aluminum oxide (alumina, Al_2O_3) and yttrium oxide (yttria, Y_2O_3) have been utilized presently due to their high durability against the halogen-containing plasma (Ref 2, 3). In particular, spraying technique seems to be better for yttria because of its high brittleness and high cost compared with alumina.

Our recent systematic investigations have revealed that plasma sprayed Y_2O_3 coatings have higher plasma erosion resistance than Al_2O_3 coatings as well as sintered-bulk Al_2O_3 against CF_4/O_2 at wide-ranged plasma power conditions including the actual conditions in semiconductor fabrication processes (Ref 4, 5). However, the resistance is still inferior to sintered-bulk Y_2O_3 showing that further

This article is an invited paper selected from presentations at the 2008 International Thermal Spray Conference and has been expanded from the original presentation. It is simultaneously published in *Thermal Spray Crossing Borders, Proceedings of the 2008 International Thermal Spray Conference*, Maastricht, The Netherlands, June 2-4, 2008, Basil R. Marple, Margaret M. Hyland, Yuk-Chiu Lau, Chang-Jiu Li, Rogerio S. Lima, and Ghislain Montavon, Ed., ASM International, Materials Park, OH, 2008.

Junya Kitamura, Hiroyuki Ibe, Fumi Yuasa, and **Hiroaki Mizuno**, Thermal Spray Materials Department, Fujimi Incorporated, Kakamigahara, Gifu Pref. 509-0103, Japan. Contact e-mail: kitamuraj@fujimiinc.co.jp.

improvement is required for plasma spray coating. Recently, $Ar/CF_4/O_2$ plasma is increasingly being used in the actual device fabrication at plasma etching process (Ref 3). Ar addition enhances physical etching effect as well as reduces chemical etching effect, which results in increasing etching rate and high capability of making deep and narrow structures that are required to fabricate highly integrated circuits for next generation.

In this paper, the structural properties of Al_2O_3 and Y_2O_3 coatings have been investigated to understand the mechanism of the plasma erosion resistance under $Ar/CF_4/O_2$ and CF_4/O_2 plasma conditions using reactive ion etching (RIE) system, which is generally used for dry etching process. The influence of the plasma gas composition and mechanism of the erosion are also discussed with the help of the microstructural analysis using scanning electron microscopy (SEM). Dielectric breakdown properties, key properties for ESC, have also been measured to investigate the effect of coating structure.

2. Experimental Procedure

2.1 Spray Powders

Al_2O_3 powders used in this study are summarized in Table 1. Coarse (A1) and fine (A2) powders with purity of more than 99.9% were prepared by fused-and-crushed process. A3 powder produced by Sumitomo Chemical Co., Ltd. (Japan), whose size was comparable to A2, was also selected. This powder was prepared by vaporization method and its purity was more than 99.99%. Fused-and-crushed A4 (99.6%) was a much finer powder with an average size of 3.1 μm, which was used for fine powder high velocity oxy-fuel (HVOF) spraying system (Ref 6).

Table 2 summarizes agglomerated-and-sintered Y_2O_3 powders used in this study, in which the primary-particle diameter was changed ranging from 0.6 to 5.3 μm (Y1-Y3). The powder size distribution and purity were

Table 1 Spray powders and bulk of Al_2O_3

Name	Purity, %	Manufacturing methods	Average diameter, μm	Powder size distribution, μm
A1	99.9	Fused & crushed	29.6	−45 + 15
A2	99.9		25.8	−38 + 8
A3	99.99	Vaporization	24.9	−38 + 10
A4	99.6	Fused & crushed	3.1	−5 + 1
AB	99.9	Sintered bulk (as a reference)		

Table 2 Spray powders and bulk of Y_2O_3

Name	Purity, %	Manufacturing methods	Average diameter, μm	Primary-particle diameter, μm
Y1	99.9	Agglomerated & sintered	39	0.6
Y2			36.3	2.9
Y3			42.8	5.3
YB	99	Sintered bulk (as a reference)		

−63 + 10 μm and 99.9%, respectively. All Al_2O_3 and Y_2O_3 powders except for A3 were prepared by Fujimi Incorporated (Japan). SEM images of the spray powders are shown in Fig. 1. Both sintered-bulk Al_2O_3 (AB) and Y_2O_3 (YB), whose sizes were 15 × 15 × 2.0^t mm, were used for the plasma erosion test, as a reference. Sintered-bulk ceramics were prepared by Tsukuba Ceramic Works (Japan).

2.2 Preparation of the Specimens

Plasma spray and HVOF conditions are summarized in Table 3. Atmospheric plasma spraying by SG-100 (Praxair, USA) was conducted to prepare the coatings except for A4 powder. HVOF spraying by theta-gun (WHITCO, Japan) was used to prepare Al_2O_3 coating by A4 powder, where acetylene injection ports were added at the downstream of powder ports to generate higher temperature flame (Ref 6). Substrates of aluminum alloy (A6061) were sand-blasted by alumina grit of F40 prior to plasma spraying. The size of the substrates was 15 × 15 × 2.0^t mm for plasma erosion test and 50 × 70 × 5.0^t mm for dielectric breakdown test. The surfaces of sprayed coatings and sintered-bulk ceramics were mirror-polished using colloidal silica with an average diameter of 0.1 μm to study both the durability against the processing plasma and the erosion mechanism in detail for plasma erosion tests.

2.3 Plasma Erosion Test

Conditions of the plasma erosion tests under low and high power are summarized in Table 4 and 5, respectively. Durability of the coatings against CF_4/O_2 plasma was compared under the low and medium conditions (80 and 175 W). Effect of argon (Ar) addition in the plasma was studied at high power (400 W).

Before introducing into the reactive ion etching (RIE) chambers, the surface of the mirror-polished specimens was partially masked by polyimide tape at both sides. The eroded area was about 5 × 15 mm. The erosion rate was estimated by measuring the step height between masked area and eroded area using stylus method (Surfcom2000SD, Tokyo Seimitsu Co. Ltd., Japan). The Possibility of generating large-sized particle was studied through microstructural analysis of the eroded surface by SEM and the stylus method.

2.4 Dielectric Breakdown Test

A schematic of the dielectric breakdown test is illustrated in Fig. 2. A cylindrical electrode with 25 mm in diameter was placed on the specimen, which was followed by applying AC voltage with a ramp rate of 500 V/s at ambient air. Electric strength was estimated by dividing breakdown voltage by coating thickness. It is expected that electrical weak spot limits the measured breakdown voltage by such a large area contact than point contact. Considering the lower uniformity of plasma spray coatings and large ESC size in actual applications, this contact configuration will be better to compare the electric strength of the specimen quantitatively.

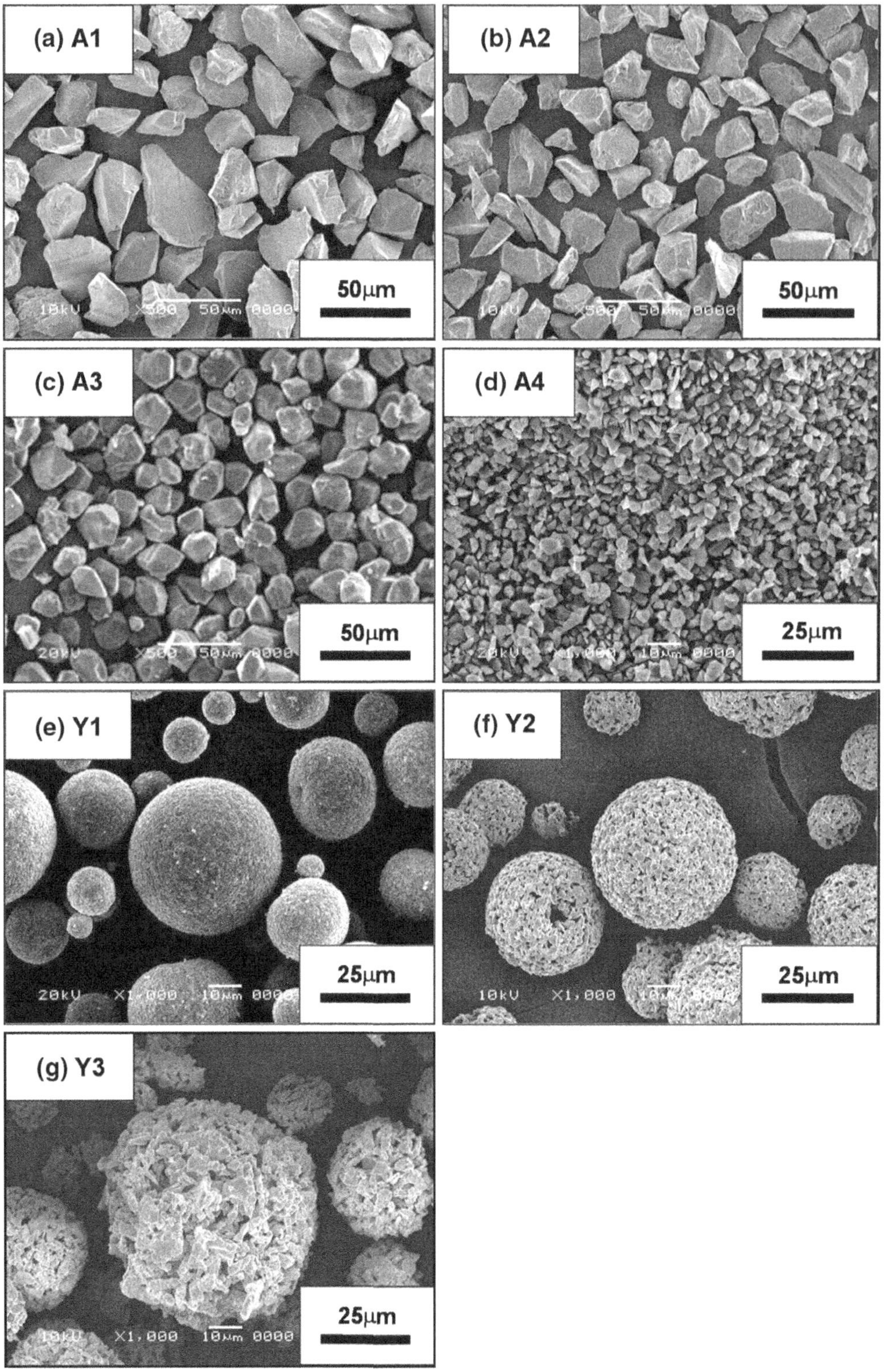

Fig. 1 SEM images of the (a-d) Al_2O_3 and (e-g) Y_2O_3 powders

3. Results and Discussion

3.1 Coating Structure of Al_2O_3 and Y_2O_3 Coatings

Figure 3 shows the cross-sectional SEM images of as-sprayed coatings. Lamellar structure is clearly seen in plasma sprayed Al_2O_3 coating of A1, which is consistent with previous report (Ref 7-12). A4 coating by fine powder HVOF system has quite a dense structure with high wear resistance (Suga abrasion test) and is mainly composed of α-phase for crystal structure (Ref 6). Density of all Y_2O_3 coatings seems to be higher in the SEM images and the porosity is observed by image analysis from

Table 3 Plasma and HVOF spray conditions

Equipments	Conditions	
Plasma SG-100 (Praxair)	Ar/He pressure, MPa	0.34/0.34
	Ar/He flow rate, L/min	39/7.9
	Arc current, A/voltage, V	900/36
	Spray distance, mm	120
HVOF θ-gun (Whitco)	Oxygen flow rate, L/min	893
	Kerosene flow rate, L/min	0.32
	Acetylene flow rate, L/min	43
	Spray distance, mm	150

Table 4 Plasma erosion test at low power

RIE equipment		RIE-200L (Samco Inc.)
CF_4 flow rate, L/min		0.054
O_2 flow rate, L/min		0.005
Chamber pressure, Pa		5
Plasma power, W		80
Exposure area, mm		200φ
Exposure time, min		480
Exposure cycle, min	Exposure	120
	Interval	60

Table 5 Plasma erosion test at medium and high power

Conditions	Medium	High	High
Gas compositions	CF_4/O_2	CF_4/O_2	$Ar/CF_4/O_2$
RIE equipment	RIE-200L	NLD-800 (ULVAC)	NLD-800
Ar flow rate, L/min	...	...	0.095
CF_4 flow rate, L/min	0.053	0.053	0.0095
O_2 flow rate, L/min	0.005	0.005	0.001
Chamber pressure, Pa	5	1	1
Plasma power, W	175	400	400
Exposure area, mm	200φ	100φ	100φ
Exposure time, min	480	60	60
Exposure cycle, min			
Exposure	120	0.5	0.5
Interval	60	3	3

optical micrographs. The A1 coating has higher porosity of 5% compared to the about 3% porosity of the Y1-Y3 coatings. It is suggested that Y_2O_3 melts easily than Al_2O_3 because of the porous shape of agglomerated-and-sintered powder during spraying process although melting point of Y_2O_3 (2410 °C) is higher than that of Al_2O_3 (2050 °C). The influence of primary Y_2O_3 particle size on the coating and crystal structure is unclear within this SEM observation because of the almost same porosity.

3.2 Plasma Erosion Test at Low Power (80 W)

Erosion rate of the spray coatings against CF_4/O_2 plasma at the low power (80 W) condition is shown in Fig. 4. Y_2O_3 coatings show lower erosion rate of about 1/5 times or less compared to the Al_2O_3 coatings suggesting that this large difference is mainly caused by material properties. Within the Al_2O_3 coatings, A2 and A3 coatings show high anti-erosion resistance suggesting that the effect of powder purity is small, ranging between 99.9 and 99.99%. Lower resistance of A1 coating might be derived from higher porosity because of using coarser powder.

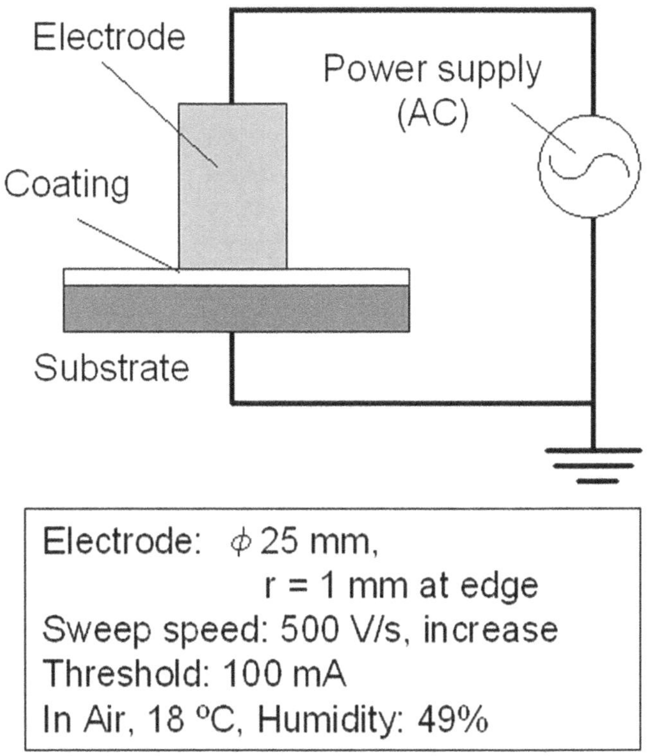

Fig. 2 A schematic of the dielectric breakdown test

On the other hand, other mechanisms should be considered for dense A4 coating. The effect of primary particle for Y_2O_3 coatings is low at the test condition.

3.3 Influence of Plasma Power (CF_4/O_2)

Figure 5 shows the erosion rates of the plasma sprayed coatings and the sintered-bulk by CF_4/O_2 plasma at both medium (175 W) and high (400 W) power conditions. At medium power, the order of durability is Y_2O_3 bulk (YB) > Y_2O_3 coatings (Y1-Y3) > Al_2O_3 bulk (AB) > Al_2O_3 coating (A2), showing the superiority of Y_2O_3 and sintered-bulk. The same tendency is seen at high power, showing that the order of the durability is independent of the plasma power within the conditions studied.

The Y_2O_3 coatings (Y1-Y3) have slightly better durability than the Al_2O_3 bulk (AB) at low plasma power. Y3, best within the Y_2O_3 coating, has higher durability of 2 times that of Al_2O_3 bulk. When increasing the plasma power, the durability of the Y_2O_3 coatings compared to the Al_2O_3 bulk becomes much better—about 3-4 times. The same tendency is obtained between Y_2O_3 coatings and Y_2O_3 bulk (YB). Although Y_2O_3 bulk has higher durability at both the powers, the durability difference becomes smaller. Considering that there is a trend toward increasing the plasma power in the actual manufacturing of Si devices, it is likely that the above tendencies support the superiority of use of the Y_2O_3 coatings. Within the Y_2O_3 coatings, use of larger primary particle (Y3) is effective in forming a highly durable coating at low plasma power. The influence of the primary particle size on the durability at high power, however, is found to be small.

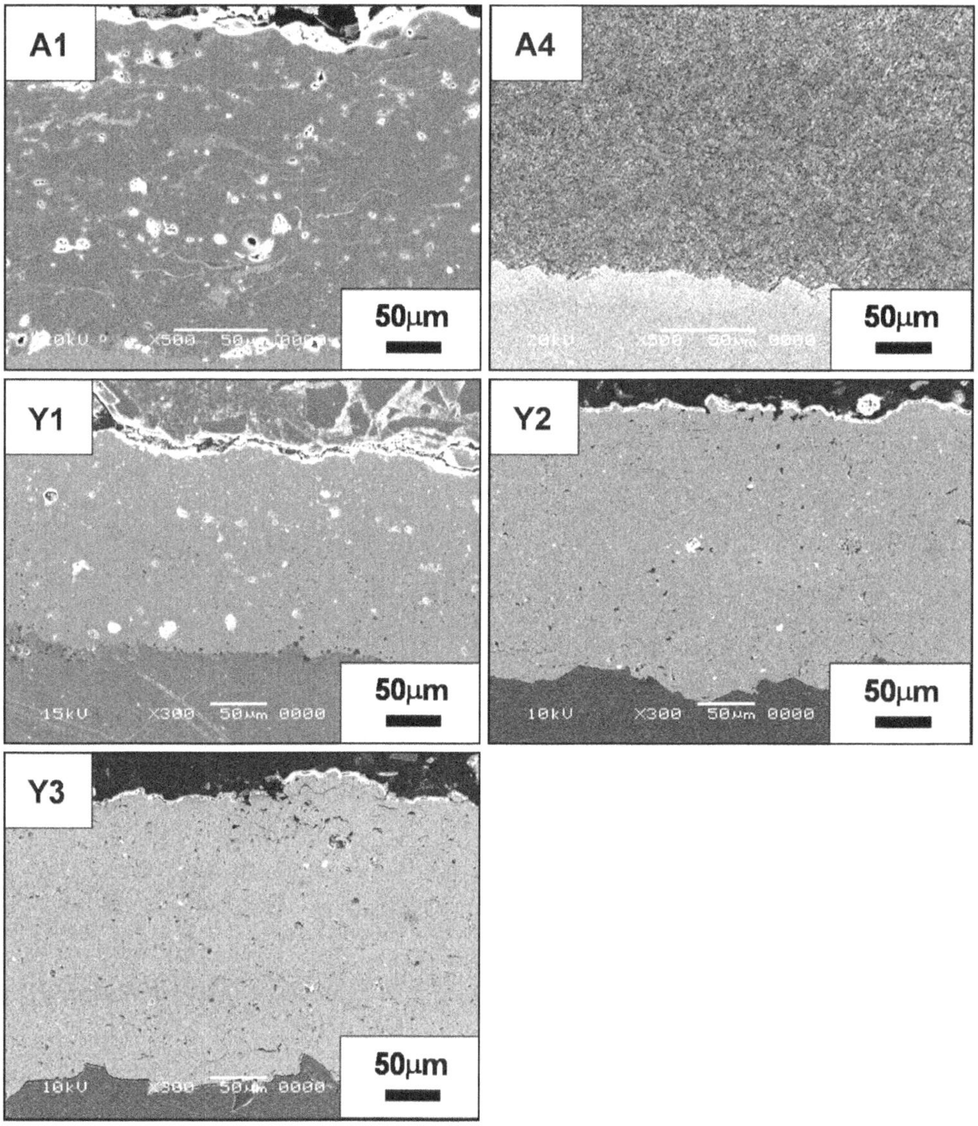

Fig. 3 Cross-sectional SEM images of as-sprayed coatings

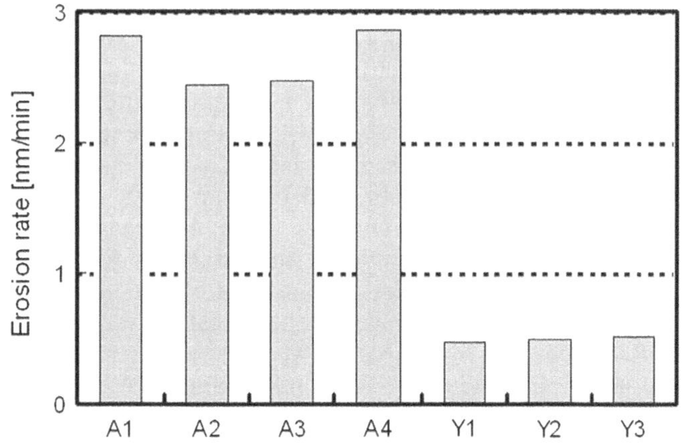

Fig. 4 Erosion rate of the spray coatings against CF_4/O_2 plasma at the low power of 80 W

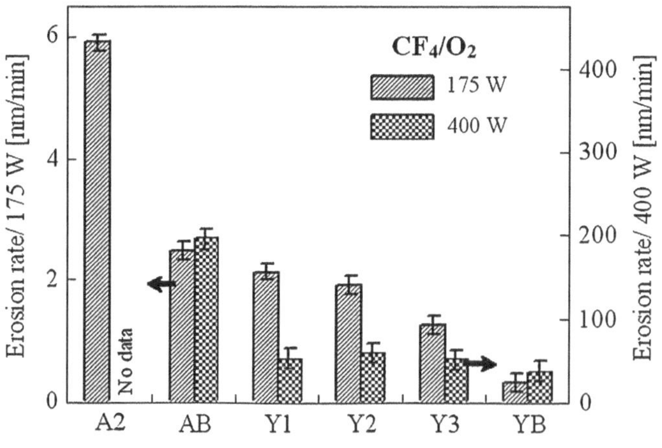

Fig. 5 Erosion rates of the plasma sprayed coatings and the sintered-bulk ceramics by CF_4/O_2 plasma at both medium (175 W) and high (400 W) power conditions

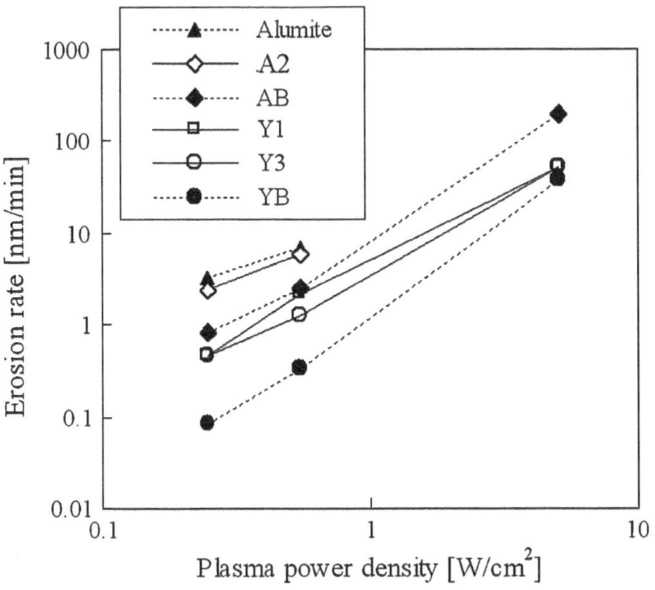

Fig. 6 Erosion rates of the spray coatings, sintered-bulk ceramics, and alumite film as a function of plasma power density

Figure 6 shows the logarithmic scaled erosion rates of the spray coatings, the sintered bulk, and conventional alumite film as a function of plasma power from 80 to 400 W, where horizontal axis is normalized as the plasma power density because effective exposure area is different from low, medium, and high power conditions. As for sintered-bulk, Y_2O_3 and Al_2O_3 have a similar tendency (similar slope), with the erosion rates being almost proportional to the square of plasma power density. The similar tendency is considered to be derived from the intrinsic properties of the ceramic materials. In the Y_2O_3 coatings, the increasing rate of the erosion rates as a function of plasma power density is lower than that of bulk ceramics. As a result, although the durability of Y_2O_3 coatings is still inferior to Y_2O_3 bulk at high power density, its difference becomes smaller with increasing power density.

3.4 Influence of Plasma Gases (CF_4/O_2 and $Ar/CF_4/O_2$)

Influence of plasma gas compositions between CF_4/O_2 and $Ar/CF_4/O_2$ is shown in Fig. 7. Little effect of primary particle size is seen at both gas conditions in the Y_2O_3 coatings (Y1-Y3). On the other hand, the influence of plasma gas composition on the erosion rates of Al_2O_3 and Y_2O_3 is clearly observed. In all Y_2O_3 specimens of the coatings and bulk, the erosion rate becomes higher at $Ar/CF_4/O_2$ plasma condition. In contrast, higher erosion rate is observed at CF_4/O_2 plasma condition in Al_2O_3 case. The difference in the influence of the gases is considered to be caused by both erosion processes by the plasma and properties of Al_2O_3 and Y_2O_3.

Coburn et al. (Ref 13) reported that the RIE etching process is an ion-assisted plasma etching process by combination of chemical etching by radicals and physical etching by ion sputtering. It is known that the etching rate

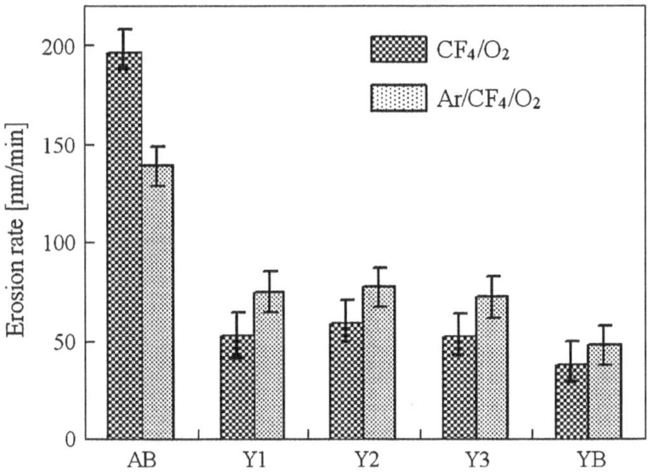

Fig. 7 Erosion rates of the spray coatings and sintered-bulk ceramics against CF_4/O_2 and $Ar/CF_4/O_2$ plasma gases at higher plasma power (400 W)

Table 6 Properties of alumina and yttria

	Standard free energy of formation ΔG (kJ/mol @373 K)	Vapor pressure (atm @1273 K)	Sputtering yield (Ar^+, 500 eV)
Al_2O_3	−1038	2×10^{-17} (AlF_3: 7×10^{-3})	1.05
Y_2O_3	−1791	5×10^{-19} (YF_3: 4×10^{-8})	0.68

increases 10 times or more with the combination compared to only physical or chemical etching in the case of Si wafer. As for gases, it is well known that CF_4 affects both chemical and physical etching. O_2 is mainly used to remove the carbon from the Si wafer by forming CO or CO_2 gases, where carbon is generated by decomposition of CF_4. Ar only affects physical etching because of its chemical stability. Properties of the Al_2O_3 and Y_2O_3, which will be related to the chemical and physical etching phenomenon, are summarized in Table 6. From the properties, it is observed that Y_2O_3 is chemically stable with a higher standard free energy of formation (ΔG), lower vapor pressure, and physically stable, low ion sputtering yield, compared to Al_2O_3, which is consistent with the results obtained in this study.

Therefore, it is suggested that difference in both ΔG and vapor pressure strongly affects in the CF_4/O_2 plasma. Yttrium has much lower vapor pressure than aluminum, about $1/10^2$ times for oxides and $1/10^5$ times for fluorides. This large difference may contribute to the large difference of the erosion rate between yttria and alumina. In the $Ar/CF_4/O_2$ plasma, on the other hand, ion sputtering effect is suggested to be stronger. Since durability of Y_2O_3 against ion sputtering is less than 2 times that of Al_2O_3, the difference of the erosion rate may be smaller.

3.5 Analysis of Eroded Surfaces by CF_4/O_2 and $Ar/CF_4/O_2$ Plasma

SEM images of the surface of the specimens after the plasma erosion test (400 W) are shown in Fig. 8.

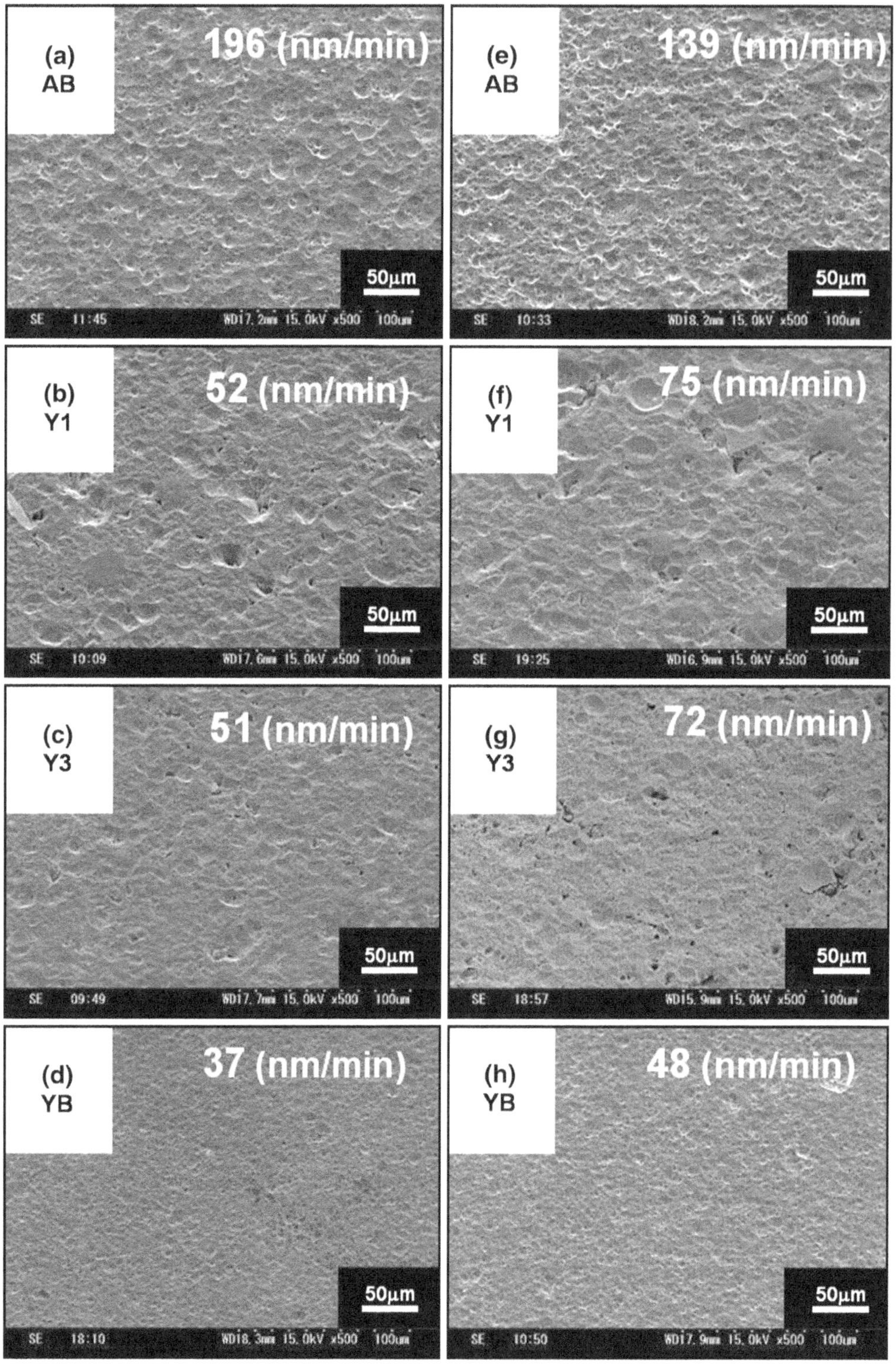

Fig. 8 SEM images of the eroded surface of the specimens: (a-d) CF_4/O_2 and (e-h) $Ar/CF_4/O_2$ showing the erosion rates

Crater-shaped pits are observed on the sintered-bulk specimens and pit size is smaller for Y_2O_3 than Al_2O_3. Within spray coatings, although the erosion rates are almost same at both conditions, surface morphologies are different, where eroded surface seems to become smoother from Y1 to Y3. Relation between roughness of the eroded surface measured by stylus method and plasma erosion rate is investigated as shown in Fig. 9. It is clear that average surface roughness (R_a) of bulk materials is lower than that of the coatings. In spite of high erosion rate for bulk Al_2O_3, the R_a is almost comparable to Y3 coating, which is smoother than Y1 coating.

It can be assumed that maintaining a smooth surface during plasma erosion prevents generating large-sized

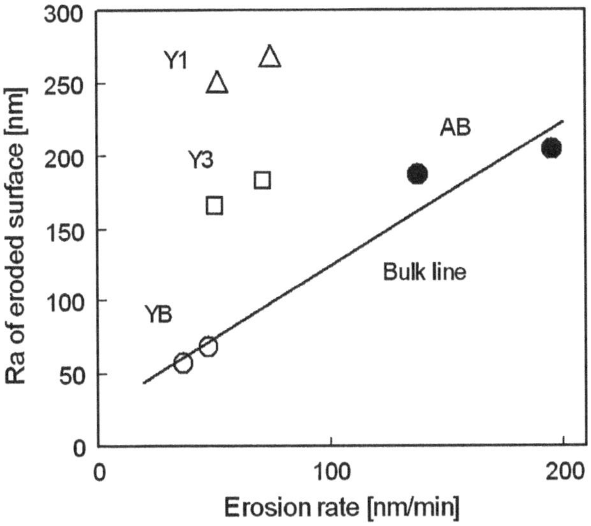

Fig. 9 Relation between average roughness (R_a) of the eroded surface measured by stylus method and plasma erosion rate

particle. As the RIE keeps evacuating using vacuum pump at etching process, small particles are easily exhausted from the chamber, which avoids deposition onto the device. Therefore, it is considered that Y3 is the most suitable coating in terms of both lower erosion rate and retaining smoother surface within the spray coatings in this study. However, Y3 coating has still higher roughness of about 3 times and higher erosion rate of about 1.5 times compared to bulk Y_2O_3. Increase of bonding strength between lamellae as well as decreasing pores should still be necessary by optimization of powder and spray parameters for improvement.

3.6 Electric Strength of the Coatings

Electric strength of the spray coatings is shown in Fig. 10, measured by typical withstand voltage tester as illustrated in Fig. 2, where the coating thickness is adjusted ranging from 200 to 300 μm. A4 coating shows the highest electric strength of about 20 kV/mm. Within plasma sprayed Al_2O_3 coatings, use of smaller powders (A2 and A3) is effective to increase the strength and a possible reason for this is the increasing density (decreasing porosity). On the other hand, purity is less affected than porosity, suggested from the result of A2 and A3 coatings. Furthermore, in spite of low purity due to both low purity powder (99.6%) and inclusion of carbon from HVOF flame, A4 coating has the highest electric strength. This could be attained from high coating density as shown in Fig. 2.

It should be noted that the use of Y_2O_3 coating for ESC is difficult without solving the problem of mechanical strength, for example, low Vicker's hardness of about 3-400HV. However, electric strength of the Y_2O_3 coatings (Y2 and Y3) is slightly higher (about 1 kV/mm) than that of A2 and A3 coatings. Considering the energy gap, which is one of the factors to define insulating properties, Y_2O_3 and Al_2O_3 have almost same value of about 6 eV (Ref 14, 15), which suggests comparable electric strength. Thus, the slight difference in electric strength may be due to the

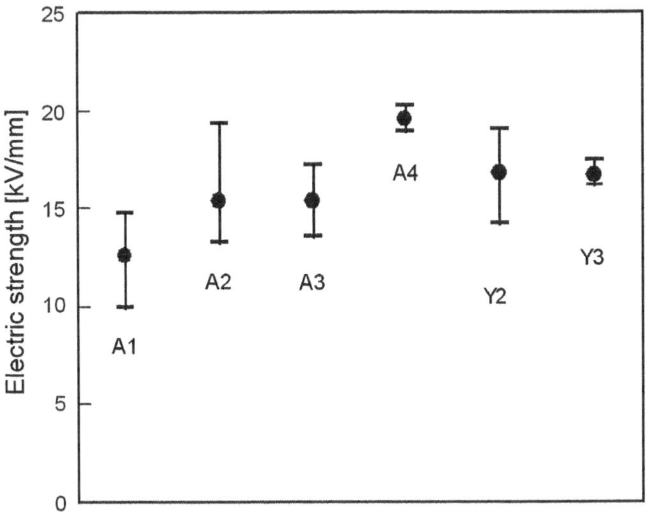

Fig. 10 Electric strength of the spray coatings

difference in coating density, which is higher for Y_2O_3 coatings as shown in Fig. 3.

Low fluctuation of the electric strength for Y3 coating suggests higher uniformity compared to Y2 coating in spite of no observation of clear difference between SEM images. This high uniformity may also result in preservation of smoother surface after the plasma erosion. It can be considered that Y3 powder produces denser coating because of low density of the agglomerated-and-sintered powder, in other word, porous powder structure as shown in Fig. 1. The porous structured powder of Y3 will easily melt during plasma flame and followed decrease of inclusion of no melting particle that degrades coating uniformity.

4. Conclusions

Structural properties of Al_2O_3 and Y_2O_3 coatings have been investigated in this study to understand the mechanism that degrades the plasma erosion resistance and electric strength. The study has arrived at the following conclusions for improving plasma erosion resistance and electric strength, which are major required properties for chamber shield and electrostatic chuck, respectively, in semiconductor and flat-panel-display production equipment.

1. Use of agglomerated-and-sintered Y_2O_3 powder consisting of larger primary particles (Y3) is effective for producing high anti-plasma erosion resistance retaining smooth surface, which may prevent generation of large-sized particles.
2. Y3 coating also shows higher electric resistance with its lower fluctuation, which suggests high uniformity in terms of low porosity as well as crystallinity.
3. Coating density strongly affects electric strength for both Al_2O_3 and Y_2O_3 coatings and the influence of purity is found to be much lower in the study. As a result, densest Al_2O_3 coating of A4 by HVOF shows best electric strength in spite of high impurity.

References

1. *Status and Perspective of Thermal Spray Markets 2004*, Digital Research Institute Inc., Nagoya, Japan, 2004, p 20-33 (in Japanese)
2. Y. Kobayashi, Current Status and Needs in the Future of Ceramics Used for Semiconductor Production Equipment, *Proc. 37th Seminar on High-Temperature Ceramics* (Osaka, Japan), Japan Ceramics Society, 2005, p 1-7 (in Japanese)
3. R. Ohtsuki, Publication of Japanese Patent Application, 2001-226773
4. J. Kitamura, H. Mizuno, N. Kato, and I. Aoki, Ceramic Coatings Prepared by Plasma Spraying for Semiconductor Production Equipments, *Mater. Trans.*, 2006, **47**(7), p 1677-1683
5. J. Kitamura, H. Mizuno, H. Ibe, and I. Aoki, Erosion Properties of Plasma Sprayed Ceramic Coatings Against Process Plasma in Semiconductor Production Equipment, *Proc. International Thermal Spray Conference 2007* (Beijing, China), ASM International, 2007, p 943-947
6. T. Morishita, S. Osawa, and T. Itsukaichi, HVOF Ceramic Coatings, *Proc. International Thermal Spray Conference 2004* (Osaka, Japan), DVS, 2004 (CD-ROM)
7. R. Westergård, L.C. Erickson, N. Axén, H.M. Hawthorne, and S. Hogmak, The Erosion and Abrasion Characteristics of Alumina Coatings Plasma Sprayed Under Different Spraying Conditions, *Tribol. Int.*, 1998, **31**(5), p 271-279
8. Y. Xie and H.M. Hawthorne, Wear Mechanism of Plasma-Sprayed Alumina Coating in Sliding Contacts with Harder Asperities, *Wear*, 1999, **225-229**, p 90-103
9. D.I. Pantelis, P. Psyllaki, and N. Alexopoulos, Tribological Behaviour of Plasma-Sprayed Al_2O_3 Coatings Under Severe Wear Conditions, *Wear*, 2000, **237**, p 197-204
10. R. Westergård, N. Axén, U. Wiklund, and S. Hogmak, An Evaluation of Plasma Sprayed Ceramic Coatings by Erosion, Abrasion and Bend Testing, *Wear*, 2000, **246**, p 12-19
11. J.R. Mawdsley, Y.J. Su, K.T. Faber, and T.F. Bernecki, Optimization of Small-Particle Plasma-Sprayed Alumina Coatings Using Designed Experiments, *Mater. Sci. Eng. A*, 2001, **308**, p 189-199
12. P.P. Psyllaki, M. Jeandin, and D.I. Pantelis, Microstructure and Wear Mechanisms of Thermal-Sprayed Alumina Coatings, *Mater. Lett.*, 2001, **47**(1-2), p 77-82
13. J.W. Coburn and H.F. Winters, Plasma Etching—A Discussion of Mechanisms, *J. Vac. Sci. Technol.*, 1979, **16**(2), p 391-403
14. I.P. Batra, Electronic Structure of α-Al_2O_3, *J. Phys. C: Solid State Phys.*, 1982, **15**, p 5399-5410
15. A. Ohta, M. Yamaoka, and S. Miyazaki, Photoelectron Spectroscopy of Ultrathin Yttrium Oxide Films on Si(100), *Microelectron. Eng.*, 2004, **72**(1-4), p 154-159

JTTEE5 17:887–894
DOI: 10.1007/s11666-008-9268-z

A Comparative Microstructural Investigation of Nanostructured and Conventional Al_2O_3 Coatings Deposited by Plasma Spraying

D. Zois, A. Lekatou, M. Vardavoulias, I. Panagiotopoulos, and A. Vazdirvanidis

(Submitted June 12, 2008; in revised form September 19, 2008)

Nanostructured and conventional Al_2O_3 powders have been plasma sprayed to produce coatings. The parameters for retaining a fraction of the nanostructure were investigated. Dissimilarities were observed between the two types of coating, regarding properties and phase proportions, which are related with the different percentages of semimolten particles in the coatings. The nanocoatings retained a higher percentage of semimolten particles than the conventional coatings owing to the higher porosity of the nanoparticle agglomerates. The molten part of both conventional and nanostructured coatings consisted of γ-Al_2O_3 of columnar morphology. In order to investigate the mechanism of the melting front advance into the particle interior, the particles were sprayed directly into deionized water. The nanoparticles mainly formed hollow spheres, whereas the conventional particles mainly formed compact spheres. The internal porosity of the solidified nanoparticle agglomerates, which affected the overall coating porosity and, consequently, coating properties such as hardness, adhesion, and surface roughness, was linked to the hollow sphere phenomenon.

Keywords APS coatings, influence of spray parameters, nanopowders, nanostructured coatings, porosity of coatings

1. Introduction

Nanocrystalline bulk materials (having a grain size smaller than 100 nm) are experiencing a rapid development in recent years due to their existing and/or potential applications to a wide variety of technological areas, such as electronics, catalysis, ceramics, magnetic data storage, structural components, etc. As the grain size is reduced to the nanometer range, materials exhibit peculiar and interesting mechanical and physical properties, e.g. increased mechanical strength, enhanced diffusivity, higher specific heat and electrical resistivity, in comparison with conventional coarse-grained counterparts. Nanostructured bulk materials and coatings are reported to demonstrate superior resistance to wear, erosion, cracking, and spallation (Ref 1-3).

Thermal spraying has recently been employed to prepare nanostructured coatings (Ref 4-8). Plasma sprayed ceramic coatings using nanostructured powder feed have been deposited and studied. Nanostructured titania and alumina-titania coatings have been obtained by vacuum plasma and atmospheric plasma spraying (APS) (Ref 9-12). Nanostructured zirconia has also been APS deposited (Ref 13). Certain conditions have to prevail for retaining a portion of the original nanostructure in the coating during the spraying process, thus maintaining the beneficial properties of the nanostructure. The retained nanostructure in the coatings is considered to be the decisive factor for the enhancement of the mechanical properties. Works have shown that nanostructured thermal spray ceramic oxide coatings like Al_2O_3-TiO_2 can have superior wear performance when compared to similar coatings produced from conventional ceramic oxide powder (Ref 1, 14). However, Liu et al. (Ref 15) have not observed this same improvement in the wear behavior when using nanostructured feedstocks, but they observed dependence on the spray method applied. Dependence of the mechanical properties on the spraying technique and the nanostructure feedstock has been suggested by Lima and Marple (Ref 16). Besides the retained nanostructure (depending on proper spraying conditions), the coating must have integrity, stability, and adhesion on the substrate, which can be attained by a sufficient degree of particle melting.

This article is an invited paper selected from presentations at the 2008 International Thermal Spray Conference and has been expanded from the original presentation. It is simultaneously published in *Thermal Spray Crossing Borders, Proceedings of the 2008 International Thermal Spray Conference*, Maastricht, The Netherlands, June 2-4, 2008, Basil R. Marple, Margaret M. Hyland, Yuk-Chiu Lau, Chang-Jiu Li, Rogerio S. Lima, and Ghislain Montavon, Ed., ASM International, Materials Park, OH, 2008.

D. Zois and **M. Vardavoulias**, Pyrogenesis SA, Technological Park of Lavrion, 19500 Lavrion, Greece; **D. Zois, A. Lekatou,** and **I. Panagiotopoulos**, Department of Materials Science and Engineering, University of Ioannina, 45111 Ioannina, Greece; and **A. Vazdirvanidis**, Hellenic Centre for Metal Research, Pireos 252, 17778 Athens, Greece. Contact e-mail: dzois@pyrogenesis-sa.gr.

Peer Reviewed

In the present study, the factors for achieving the above-mentioned requirements are examined for a High Energy Ball Milled nanostructured alumina powder. Moreover, an effort is made to associate the melting degree of the powder with the portions of the phases present in the deposited coating and its metallographic properties. A conventional counterpart is used for comparison reasons.

Here, it should be noted that this study was induced by the Pyrogenesis SA need to produce coatings for fiber and thread wear resistant applications at high temperatures. These coatings should possess excellent wear behavior and excellent erosion resistance at high temperatures. Up to now, the raw material for these coatings is conventional Al_2O_3 powder.

2. Experimental

The Al_2O_3 nanopowder, hereafter named NP, employed in this work was composed of agglomerated nanoparticles, in order to enable the flow, through the powder feeder, into the torch (Ref 17). The powder was produced by a high-energy ball mill manufactured by MBN and functioning according to the kinematic characteristics of a Spex® 8000 model (Ref 18). After milling, the powder was centrifugally classified to exclude particles finer than 10 μm. Rounder particle shapes were obtained by low energy milling (Jet Mill). Finally, the appropriate particle distribution for spraying was attained by mechanical sieving. No heat treatment process was used. The received powder's nominal size distribution was $(-53+10)$ μm. The conventional powder, hereafter named CP, was Sulzer Metco's Al_2O_3-105NS of $(-45+15)$ μm particle distribution. The powders were sprayed by a MiniGun torch with an external powder feeder developed by Pyrogenesis Inc. The substrates used were 50 × 50 × 5 mm 304 stainless steel coupons.

The apparent porosity and thickness of the coatings were evaluated by optical microscopy (Leica DM4000M) and scanning electron microscopy (Philips XL 40 SFEG). Microhardness was measured by a Shimadzu tester (average of 14 measurements on each coating) using a load of 0.1 kg for 10 s. The fracture toughness of the coatings was qualitatively examined by indentation with a 1.0 kg load. The adhesion strength of the coatings (average of three measurements on each coating) was measured by a portable elcometer (110 P.A.T.T.I.®). Surface roughness was determined by a TR200 Roughness tester. The phase composition of the feedstocks and coatings was established by X-ray diffraction (XRD) (Bruker D8 Advance, Standard Slit). XRD was also employed to estimate the average crystal size of the feedstocks and coatings (Ref 19, 20). The crystal size and lattice strain in the coatings can be determined by the X-ray broadening techniques, since XRD peaks are broadened due to three factors: (i) instrumental effects, (ii) small crystal size, and (iii) lattice strain in the material (Ref 21). The instrumental effects were measured and subtracted by diffracting a SiO_2 standard. The crystal size was determined by the use of Williamson-Hall equation (Ref 22):

$$D = \frac{K\lambda}{\beta_{size} \cos\theta} + 2\varepsilon \sin\theta \qquad \text{(Eq 1)}$$

where D is the crystal size in Å, $K=0.9$, β_{size} the peak width in rad, λ the wavelength of the X-ray radiation in Å, θ the Bragg angle, and ε is the internal strain ($\varepsilon = \Delta d/d$, where d is the lattice spacing and Δd the lattice strain).

The X-ray Rietveld Method was applied to determine the phase percentages in the original powders and their coatings by refining the crystalline parameters and scale factors (Ref 23). The software used was PowderCell v2.4.

3. Results

3.1 Feedstock Characterization

The morphologies of the conventional (CP) and nanostructured (NP) feedstocks are shown in Fig. 1(a) and (b), respectively. CP is in the form of solid granules due to the fusing and crushing manufacturing process, whereas NP is in the form of agglomerates of irregular spherical shapes. Individual nanoparticles on the surface of the agglomerated granules can be discerned in Fig. 1(c). Figure 2 shows that the main phase in both powders is α-Al_2O_3. NP also contains some iron. Probably, the powder was contaminated by the tool steel balls and vials of the high-energy ball mill. The percentage of iron was estimated by EDS analysis to be ~0.74 wt.%, while by Rietveld analysis it was estimated to be 1.14 wt.%.

In Fig. 2, NP presents broader peaks than the conventional powder due to nanocrystalline dimensions and presence of non-uniform strain. The strain originates from the consecutive high-energy impacts during the milling process introducing high stress and distortion into the grains. The calculated strain was 1.15%. The estimated crystal size of the nanostructured powder was 105 nm.

3.2 Thermal Spraying as a Function of the Torch Power

The effect of the torch power on characteristic coating properties (microporosity, surface roughness, microhardness, adhesion, and crack propagation resistance) was investigated by conducting three series of spraying experiments. The series are characterized as low (25 kW, coatings obtained designated as C_1 and N_1, conventional and nanostructured, respectively), medium (32 kW, C_2 and N_2), and high (39 kW, C_3 and N_3) torch power sprayings.

The coating porosity (estimated by image analysis and SEM at 500× on cross sections) as a function of the torch power is shown in Fig. 3. The nanocoatings exhibit higher porosity than the conventional ones when sprayed at the same power. A decrease in porosity with an increase in power is noticed for both types of coatings. For the nano-feedstock, the reduction rate of the porosity versus power is higher. It should be emphasized that the image analysis techniques can only identify the coarser pores of thermal spray coatings and give an overall view of the coating

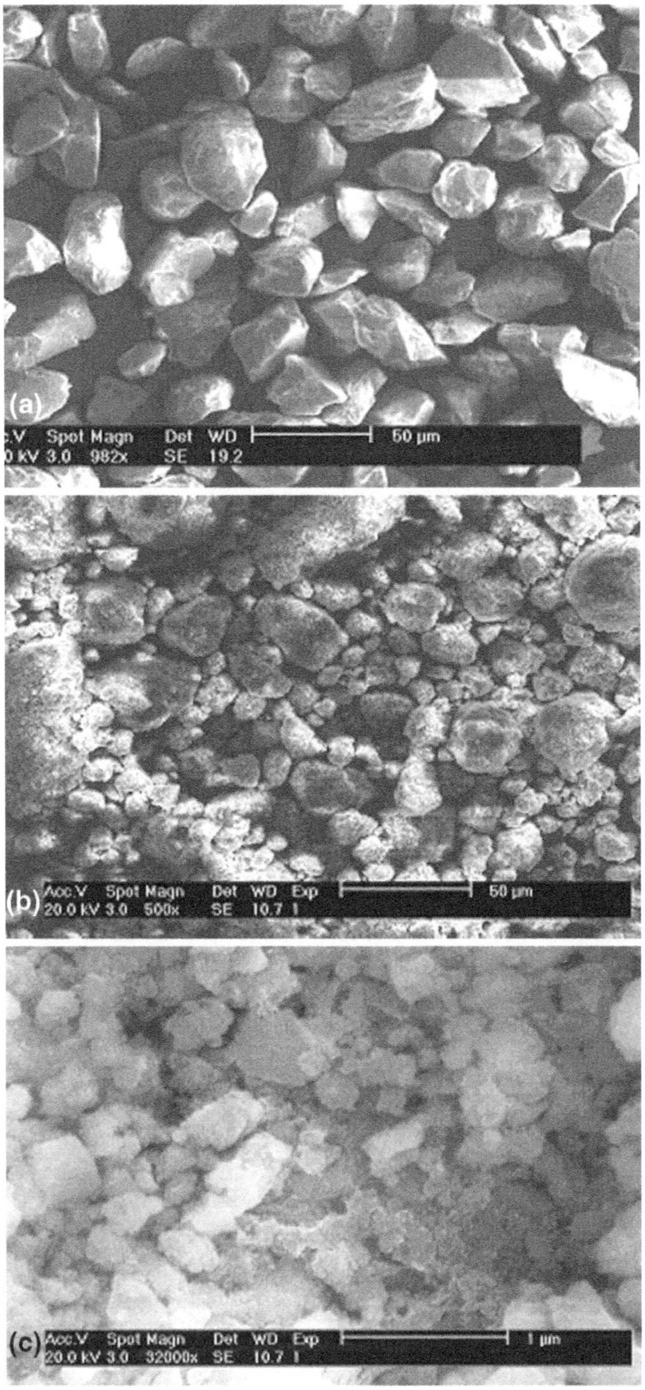

Fig. 1 SEM images of (a) CP, (b) NP, and (c) surface of the agglomerated NP particles

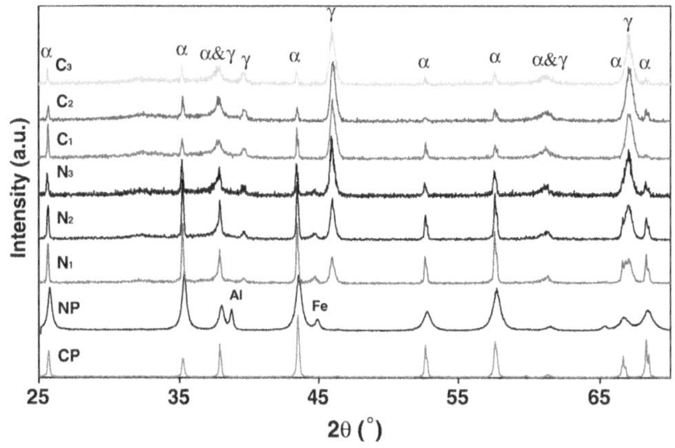

Fig. 2 XRD patterns of the initial powders and obtained coatings: α, α-Al_2O_3; γ, γ-Al_2O_3

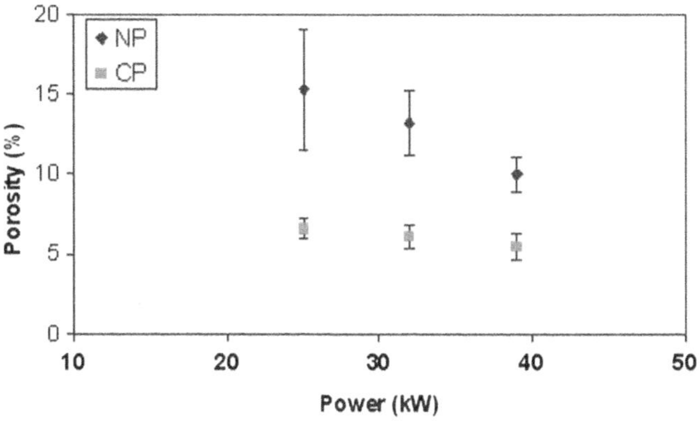

Fig. 3 The porosity for nano and conventional coatings as a function of the spray power

porosity. Mercury intrusion porosimetry (MIP) is investigated as an additional porosity calculating method for future work.

The surface state of the coatings deposited at 32 kW is illustrated in Fig. 4. The CP coatings present smoother surface than the NP coatings at all power levels, which is also verified by the surface roughness values in Table 1. The irregular/rough morphology of the nanocoatings suggests that the agglomerates retain an unmelted core. Therefore, the pile-up of the splats is not dense due to a large fraction of deposited semimolten particles, possibly accounting for the higher porosity of the nanocoatings. Similarly, the decreasing trend of roughness with torch power increasing suggests a decrease in the fraction of unmelted deposited particles and consequently an increase in the fraction of well-melted low viscosity material capable of filling the asperities of previously deposited particles. In order to verify that the agglomerated nanoparticles remain semimolten during spraying, the feedstocks were sprayed directly into deionized water.

CP and NP coatings deposited at the same torch power present similar microhardness values. Since NP coatings exhibit approximately the double porosity of the respective conventional ones, it is expected that they would have lower hardness, as the carrying load solid area is decreased (Ref 24). This contradictory result might be elaborated by two facts: first, the relatively low indentation load leads to microhardness values mostly depending on the intrinsic material properties and less depending on the material extrinsic properties, such as coarse pores, splat boundaries, etc. (Ref 25). Second, as mentioned above, the NP coatings contain a higher percentage of unmelted material, which has retained its nano or submicron grain size. High hardness values in nanostructured materials are closely related to their low grain size, as suggested by the Hall-Petch relationship (Ref 26).

The CP coatings exhibit better adhesion than the NP coatings, as presented in Table 1. However, the adhesion value of the nanostructured coatings is considerably increased at the highest torch power (39 kW).

Figure 5(a) and (b) shows the indentation impressions on the conventional (C_3) and nanocoating (N_3) cross sections, respectively. For conventional coatings, cracks tend to propagate along splat borders parallel to the coating-substrate interface irrespectively of the torch power. No cracks or very short cracks were generated on directions vertical to the coating surface. Such an anisotropic behavior is related to accumulation of micropores between lamellae that decreases the real area of contact (Ref 27). That was not the case for the nanostructured coatings. In the NP coatings, more often crack branching was encountered. Cracks did not propagated exclusively along splat borders but were arrested or deflected along microstructure features. Such features were unmelted nanostructured areas, interfaces between fully melted and partially melted particles and pores. In Fig. 5(b), unmelted areas of retained nanostructure (marked by an arrow) are

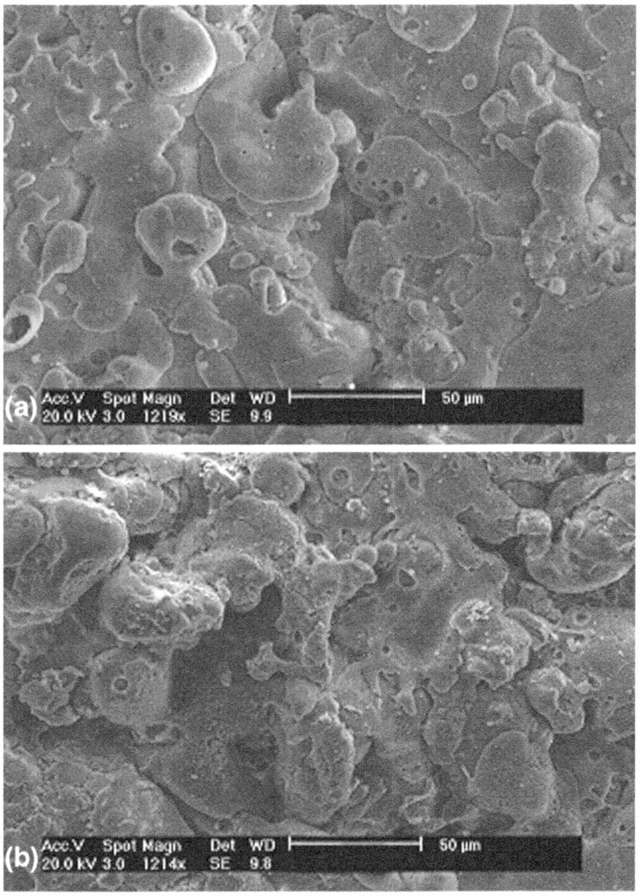

Fig. 4 The surface state of (a) CP and (b) NP coatings obtained at 32 kW spray power

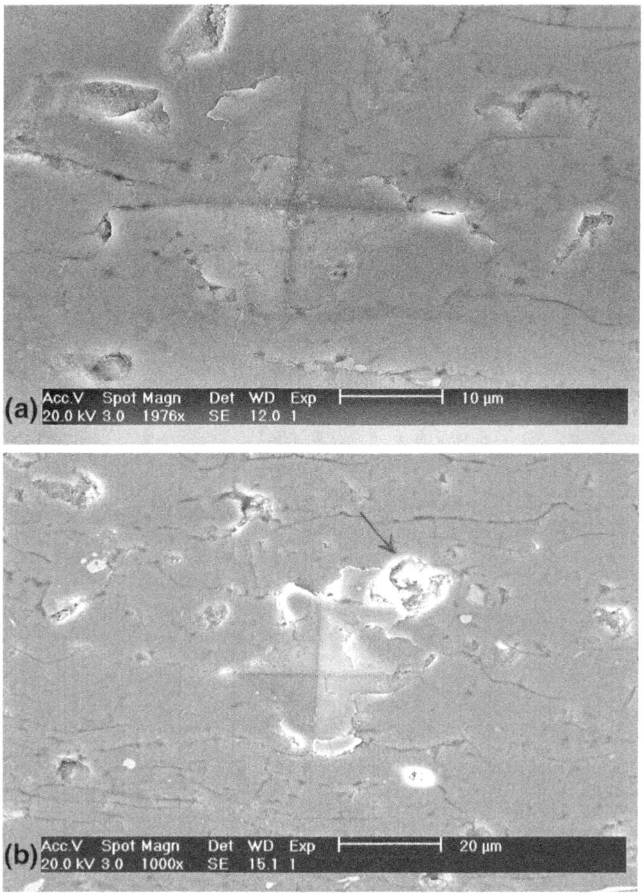

Fig. 5 The indentation impressions on coating (39 kW) cross sections. (a) CP and (b) NP; the arrow points at a crack arresting nanozone

Table 1 Coating properties

Designation	Torch power, kW	Porosity, %	Surface roughness (R_a), µm	Hardness, $HV_{100g/10s}$	Adhesion, MPa	Indentation crack length (a), µm
C_1	25	6.6 ± 0.6	7.4 ± 1.0	1063 ± 145	17.6 ± 0.9	114 ± 32
C_2	32	6.1 ± 0.7	5.9 ± 0.8	1078 ± 259	18.1 ± 1.5	78 ± 28
C_3	39	5.5 ± 0.8	5.8 ± 0.7	1147 ± 94	19.8 ± 1.6	59 ± 16
N_1	25	15.3 ± 3.8	10.2 ± 2.4	975 ± 126	10.5 ± 0.8	(b)
N_2	32	13.2 ± 2.0	10.8 ± 0.7	1070 ± 166	11.4 ± 1.3	94 ± 31
N_3	39	10.0 ± 1.1	8.4 ± 1.4	1141 ± 111	17.7 ± 2.4	55 ± 5

(a) Length of cracks at the edges of the indentation diagonal that is parallel to the coating surface
(b) Discarded measurements due to material crushing under indentation

discerned to act as crack arresters, indicating a toughening mechanism. That complicated the measurements of the cracks lengths; however, for comparison reasons, the two longest cracks propagating parallel to the coating-substrate interface were measured. For coating N_1, material crushing occurred under the indentation load due to the high level of porosity; therefore, measurements were regarded as invalid and discarded. The measured cracks for each coating are shown in Table 1. Despite the higher percentage of retained nanostructure at the lower torch powers, the higher porosity reduces fracture toughness by decreasing the solid area that carries load. However, as porosity is decreased, due to increasing torch power, and nanostructure is retained at a major fraction, a considerable increase in fracture propagation resistance is noticed.

3.3 Water-Quenched Particles

Both powders were sprayed directly into deionized water (at 39 kW) in order to investigate the mode of the melting front advance into the particle interior during their presence in the plasma stream. SEM examination shows that both water-quenched powders display a spherical microstructure; however, CP particles are compact, whereas most of the NP particles are hollow, as seen in Fig. 6 and 7, respectively. The hollow spheres consist of a dense molten wall enveloping a large void core. Since, at moderate power levels, a significant fraction of the nanoparticle agglomerates is deposited semimolten, their central voids contribute to the higher porosity (Fig. 8). As the spray power is increased, the fraction of semimolten particles is reduced, which is consistent with the rapid decrease in the porosity of the nanocoatings (Fig. 3). The higher porosity of the nanocoatings is also amplified by the low deformation ability of the semimolten particles, restricting material flow into surface asperities and irregularities of the previously deposited layer (Ref 28). Figure 9 shows that the semimolten agglomerates retain part of the initial nanostructure in the core (marked with "1" in Fig. 9).

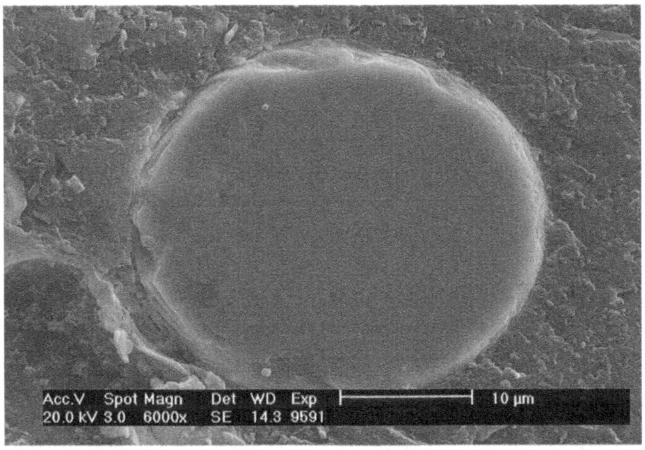

Fig. 6 CP feedstock particle, water-quenched (39 kW)

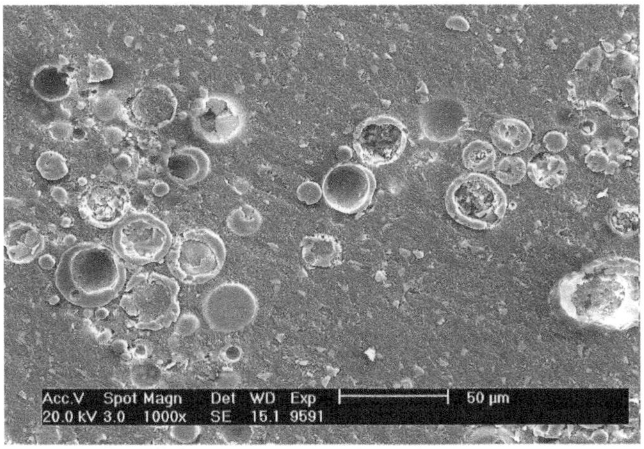

Fig. 7 NP feedstock particles, water-quenched (39 kW)

3.4 Quantitative Phase Analysis

XRD analysis (Fig. 2) shows that the deposited coatings consist of α-Al_2O_3 and γ-Al_2O_3, in compatibility with

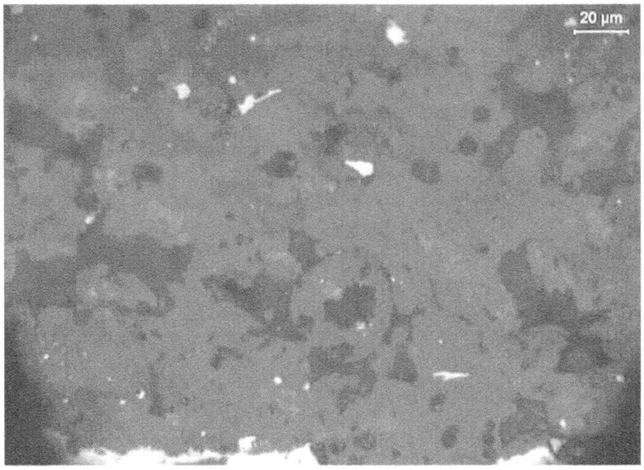

Fig. 8 Semimolten hollow spheres in the nanocoating (25 kW)

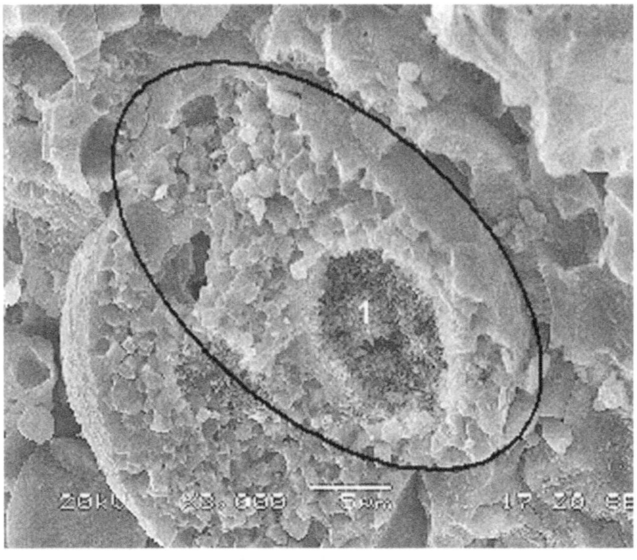

Fig. 9 Fracture surface of a semimolten particle in the nanocoating (39 kW)

previous works (Ref 8, 29). The fast cooling rates of thermal sprayed particles (10^6-10^7 K/s) can cause the stabilization of originally metastable phases. Thus, the formation of γ-Al_2O_3 in coatings is the result of rapid quench from molten Al_2O_3. α-Al_2O_3 in the coating stems from the unmelted α-Al_2O_3 fraction in the pristine powders (Ref 30).

The phase percentages for both types of coatings, as estimated by Rietveld analysis, are presented in Fig. 10. As the power increases, more γ-phase is produced, which implies that an increasing proportion of the powder particles has been melted. For the same power levels, more γ-phase is produced by the conventional powder than the nano, suggesting a higher degree of melting for the CP feedstock.

4. Discussion

During spraying, as the particles penetrate the flame, a melting front is formed that moves toward the core of the particle. As the agglomerated nanoparticles are characterized by a porous structure, air is captivated in the voids. The melting front impels the air bubbles from the voids toward the core due to the high surface tension of the molten alumina (Ref 31). This way, a central large void is developed. The formation of hollow Al_2O_3 microspheres has been attributed to the gas expansion inside the molten spheres, on temperature increasing (Ref 31).

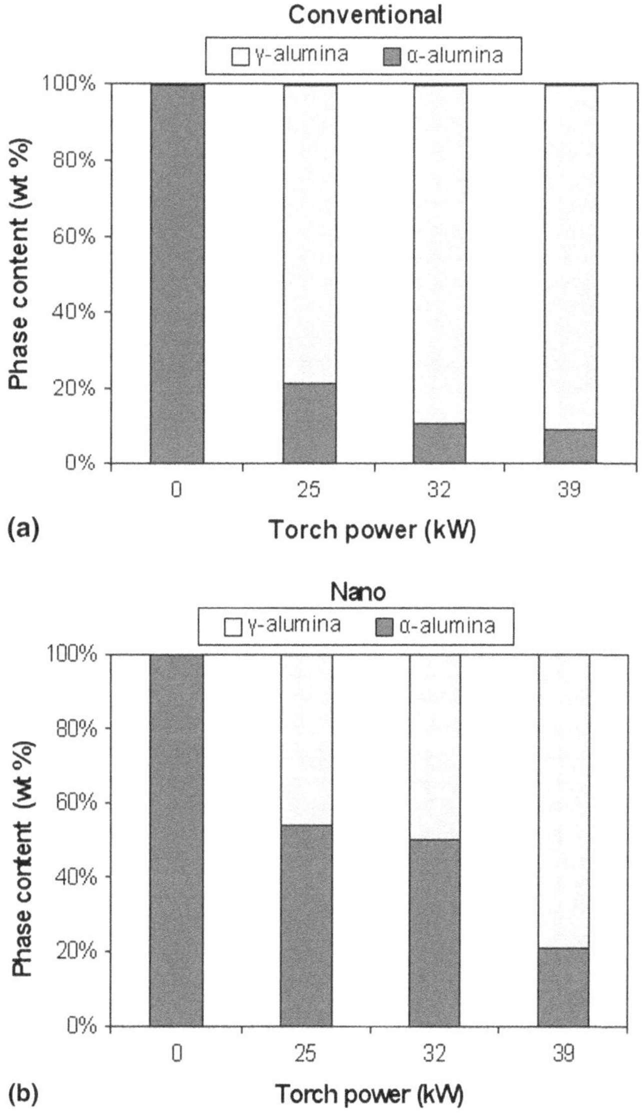

Fig. 10 Phase percentages in the coatings versus torch power

On the other hand, conventional fused and crushed powder particles are not capable of air entrapping in their interior. Furthermore, being compact bodies, conventional particles exhibit higher thermal conductivity than their porous counterparts (Eq 2), a fact that explains their higher melting capability.

$$\lambda_p = \frac{\lambda_m}{1 + P^* \frac{\lambda_m}{\lambda_f - 1}} \quad \text{(Eq 2)}$$

where λ_p, λ_m, λ_f are the thermal conductivities of the porous coating, solid matrix, and gas (air) in the pores, respectively; P is the porosity of the coating (Ref 32).

Figures 3 and 10 indicate that a higher spraying energy is required for the nanopowder to receive porosities and phase proportions that are comparable to those obtained by spraying of the conventional powder. However, retaining part of the nanostructure is necessary for property improvement, as reported in Sect 1; too high energy levels would jeopardise this.

The decrease in the adhesion strength of the coatings (NP and CP) with spraying power decreasing is related to the respective increase in the fraction of material deposited unmelted as semimolten/unmelted particles. The large increase in the γ-fraction of the NP coatings at 39 kW may account for the large increase in the adhesion strength of the coatings deposited at this power. This energy seems sufficient to melt most of the sprayed material and increase the coating adhesion. Figure 10 shows that the N_3 coating (sprayed at 39 kW) and C_1 coating (sprayed at 25 kW) exhibit the same phase fractions, suggesting equal melting degrees. Also, Table 1 shows that the N_3 coating presents slightly lower adhesion strength than the C_1 coating. This could be related to (i) its higher porosity, as NP semimolten particles tend to form hollow spheres, and (ii) the lower cohesion of the unmelted particles compared to conventional unmelted ones, as no heat treatment of the powder was conducted prior to spraying. Figure 9 exhibits splitted semimolten particles on a fractured surface. Despite that, the N_3 coating hardness is higher than that of C_1, and taking into consideration its higher porosity, it can be claimed that its superior hardness is due to its embedded nanostructure in the coating.

Here, it should be noted that an agglomerated structure is generally expected to be more susceptible to thermal processes than a dense bulk structure, since it is a more open structure. In the present case, this did not occur (regarding the melting degree) due to the inherent porosity of the agglomerated nanoparticles and the lower thermal conductivity of captivated air in the pores as mentioned above. Related studies, confirming this, have been mentioned also elsewhere (Ref 24). NP powder manufacturing was not followed by any post-treatments, such as spray drying or sintering, which could lead to tighter structures and higher nanoparticle bond strengths (Ref 29). Sintering prior to thermal spraying was not conducted in order to avoid any grain size increase. Previous work by the authors on similar powders demonstrated that the introduction of a sintering step into the manufacturing process significantly improved the adhesion properties at all power levels (Ref 33). This improvement was attributed to sintering consequences, such as intraparticle cohesion, particle coalescence, and strain relief. Increasing particle bond strength can prevent particle dissociation when entering the plasma flame and increase the agglomerated particles density. This way, the particles may retain the momentum and access the plasma hot zone and melt more sufficiently (Ref 34).

Zones of different melting degree can be observed in the fractured surface of a semimolten particle (Fig. 9). These distinguishable zones in the particle can give information about its thermal history during its trajectory inside the plasma flame. So, from the exterior to the core of a semimolten particle, (i) complete melting of the surface has occurred that has led to γ-Al_2O_3 solidification. (ii) A liquid phase sintering zone follows formed by the flow of the molten material into the voids and the consolidation of the nanoparticles to submicron particles. This reminds of the dense nanozones described by Lima and Marple (Ref 26); though, in the present case, a major degree of consolidation has taken place, converting the nanoparticles into submicron ones. (iii) Then, a solid state sintering zone is formed, evidence that there was no sufficient time for the melt to penetrate the voids at such a distance. (iv) Finally, the central zone is free of sintering; the time and temperature were not sufficient to induce any sintering in the core, as Al_2O_3 is a moderate to bad conductor (5.9 W/mK at 1000 °C) (Ref 34).

In this work, it is possible that the thermal processes occurring during thermal spraying of the nanopowder cannot be classified into strict categories of melting or sintering, as claimed by previous researchers (Ref 28, 34). However, a combination of these may occur during thermal spraying, depending on the trajectory of the particle in a heterogeneous plasma flame, with temperature rapidly decreasing from higher than 15,000 °C (center of plasma) to ~1000 °C (end of plasma) (Ref 12).

It should be noted that the employment of this certain nano-feedstock did not yield coatings with retained nanoparticles scattered in the molten material; instead, the particles were dispersed as submicron agglomerates. However, part of the original nanograins has been retained in the central voids of the semimolten particles.

5. Conclusions

Nanostructured and conventional Al_2O_3 powders have been plasma sprayed on 304 stainless steel. The following conclusions are extracted:

1. The nanopowder consists of agglomerated nanoparticles of average size of 105 nm and lattice strain of 1.15%.
2. Less degree of melting for the nanostructured feedstock is attributed to the inherent porosity of the agglomerates.
3. The high porosity of the nanostructured coatings is partly related to the large voids formed in the interior of the semimolten particles and partly to their low deformation ability.
4. The nano and conventional coatings sprayed at the same torch power levels exhibited similar hardness values despite the higher porosity of the nanocoatings.
5. Nanostructure coatings of similar phase fractions to the conventional but with higher porosity exhibited higher microhardness due to their retained nanostructure.
6. The nano and conventional coatings displayed different crack propagation modes under indentation. The excessively increased porosity of the NP coatings plays a major role in reducing fracture toughness despite the presence of unmelted areas of retained nanostructure in the nanocoatings that can act as crack arresters. Porosity in lower levels may act beneficiary as crack arrester or crack deflector.
7. The nanocoatings demonstrated lower adhesion than conventional coatings which was related to insufficient melting and low intraparticle cohesion.
8. Increasing the torch power led to an increase in the powder melting degree and consequent decrease in the coating porosity; as a result, the coating surface roughness decreased, and the coating microhardness, adhesion strength, and fracture toughness increased with power increasing.
9. The original nanograins are mainly retained in the voids of the semimolten particles. In other parts of the structure, nanograins are consolidated to submicron particles.
10. α- and γ-Al_2O_3 proportions in the coatings are associated with the melting degree of the particles. γ-Al_2O_3 results from rapid quenching of molten particles. α-Al_2O_3 corresponds to the unmelted α-Al_2O_3 fraction in the feed powders.
11. From the periphery to the core of a semimolten particle, different structure zones can be discerned, in analogy with the transition from complete melting and rapid solidification, to liquid-state sintering, to solid-state sintering, and finally to intact areas.

Acknowledgment

D. Zois is grateful for the financial support from the Greek State Scholarships Foundation.

References

1. M. Gell, E.H. Jordan, Y.H. Sohn, D. Goberman, L. Shaw, and T.D. Xiao, Development and Implementation of Plasma Sprayed Nanostructured Ceramic Coatings, *Surf. Coat. Technol.*, 2001, **146-147**, p 48-54

Peer Reviewed

Peer Reviewed

2. K. Jia and T.E. Fischer, Abrasion Resistance of Nanostructured and Conventional Cemented Carbides, *Wear*, 1996, **200**, p 206-214
3. H.H. Nersisyan, H.I. Won, C.W. Won, and J.H. Lee, Study of the Combustion Synthesis Process of Nanostructure WC and WC-Co, *Mater. Chem. Phys.*, 2005, **94**, p 153-158
4. C. Berndt and E.J. Larernia, Thermal Spray Processing of Nanoscale Materials—A Conference Report with Extended Abstract, *J. Therm. Spray Technol.*, 1998, **7**, p 411-440
5. B.H. Kear and G. Skandan, Thermal Spray Processing of Nanoscale Materials, *Nanostruct. Mater.*, 1997, **8**, p 765-769
6. J. Karthikeyan, C.C. Berndt, J. Tikkanen, J.Y. Wang, A.H. King, and H. Herman, Preparation of Nanophase Materials by Thermal Spray Processing of Liquid Precursors, *Nanostruct. Mater.*, 1997, **9**, p 137-140
7. J.P. Celis, A. Basak, S. Achanta, M. Vardavoulias, and P. Mateazzi, Tribology of Nanostructured Plasma Sprayed Coatings, *2nd International Conference on Nanomaterials*, Sept 2005, Aver, Portugal, published in the proceedings
8. A. Tsetsekou, E. Tsioutsioulikli, N. Petsas, and M. Vardavoulias, Plasma Spray Coatings of Sol-Gel Ceramic Nanopowders, *20th International Conference on Surface Modification Technologies—SMT 20*, Sept 25-29, 2006, Vienna, Austria, published in the proceedings
9. Y. Zhu, M. Huang, J. Huang, and C. Ding, Vacuum-Plasma Sprayed Nanostructured Titanium Oxide Films, *J. Therm. Spray Technol.*, 1999, **8**, p 219-222
10. B.H. Kear, Z. Kalman, R.K. Sadangi, G. Skandan, J. Colaizzi, and W.E. Mayo, Plasma-Sprayed Nanostructured Al_2O_3/TiO_2 Powders and Coatings, *J. Therm. Spray Technol.*, 2000, **9**, p 483-487
11. L.L. Shaw, D. Goberman, R. Ren, M. Gell, S. Jiang, Y. Wang, T.D. Xiao, and P.R. Strutt, The Dependency of Microstructure and Properties of Nanostructured Coatings on Plasma Spray Conditions, *Surf. Coat. Technol.*, 2000, **130**, p 1-8
12. X. Lin, Y. Zeng, X. Zheng, and C. Ding, Thermal Diffusivity of Plasma-Sprayed Monolithic Coating of Alumina-3wt.% Titania Produced with Nanostructured Powder, *Surf. Coat. Technol.*, 2005, **195**, p 85-90
13. H. Chen and C.X. Ding, Nanostructure Zirconia Coating Prepared by Atmospheric Plasma Spraying, *Surf. Coat. Technol.*, 2002, **150**, p 31-36
14. E.H. Jordan, M. Gell, Y.H. Sohn, D. Goberman, L. Shaw, S. Jiang, M. Wang, T.D. Xiao, Y. Wang, and P. Strutt, Fabrication and Evaluation of Plasma Sprayed Nanostructured Alumina–Titania Coatings with Superior Properties, *Mater. Sci. Eng. A*, 2001, **301**, p 80-89
15. Y. Liu, T.E. Fischer, and A. Dent, Comparison of HVOF and Plasma-Sprayed Alumina/Titania Coatings—Microstructure, Mechanical Properties and Abrasion Behavior, *Surf. Coat. Technol.*, 2003, **167**, p 68-76
16. R.S. Lima and B.R. Marple, From APS to HVOF Spraying of Conventional and Nanostructured Titania Feedstock Powders: A Study on the Enhancement of the Mechanical Properties, *Surf. Coat. Technol.*, 2006, **200**, p 3428-3437
17. S. Eidelman and X. Yang, Three Dimensional Simulation of HVOF Spray Deposition of Nanoscale Materials, *Nanostuct. Mater.*, 1997, **9**, p 79-84
18. D. Basset, P. Mateazzi, and F. Miani, Measuring the Impact Velocities of Balls in High Energy Ball Mills, *Mater. Sci. Eng. A*, 1994, **174**, p 71-74
19. H.P. Klug and L.E. Alexander, X-Ray Diffraction Procedures, 2nd ed., Wiley, New York, 1974, p 618
20. M.E. Rabanal, A. Varez, B. Levenfeld, and J.M. Torralba, Magnetic Properties of Mg-Ferrite After Milling Process, *J. Mater. Process. Technol.*, 2003, **143-144**, p 470-474
21. C. Suryanarayana, Mechanical Alloying and Milling, *Prog. Mater. Sci.*, 2001, **46**, p 1-184
22. G.K. Williamson and W.H. Hall, X-Ray Line Broadening from Filed Aluminium and Wolfram, *Acta Metall.*, 1953, **1**, p 22-31
23. D.L. Bish and S.A. Howard, Quantitative Analysis Using the Rietveld Method, *J. Appl. Crystallogr.*, 1988, **21**, p 86-91
24. H. Luo, D. Goberman, L. Shaw, and M. Gell, Indentation Fracture Behaviour of Plasma-Sprayed Nanostructured Al_2O_3-13wt.% TiO_2 Coatings, *Mater. Sci. Eng. A*, 2003, **345**, p 237-245
25. R.S. Lima, A. Kucuk, and C.C. Berndt, Bimodal Distribution of Mechanical Properties on Plasma Sprayed Nanostructure Partially Stabilized Zirconia, *Mater. Sci. Eng. A*, 2002, **327**, p 224-232
26. R.S. Lima and B.R. Marple, Thermal Spray Coatings Engineered from Nanostructured Ceramic Agglomerated Powders for Structural, Thermal Barrier and Biomedical Applications: A Review, *J. Therm. Spray Technol.*, 2007, **16**(1), p 40-63
27. R. McPherson, A Review of Microstructure and Properties of Plasma Sprayed Ceramic Coatings, *Surf. Coat. Technol.*, 1989, **39/40**, p 173-181
28. B.H. Kear and P.R. Strutt, Chemical Processing and Application for Nanostructured Materials, *Nanostruct. Mater.*, 1995, **6**, p 227-236
29. J. Zhang, J. He, Y. Dong, X. Li, and D. Yan, Microstructure Characteristics of Al_2O_3-13wt.% TiO_2 Coating Plasma Spray Deposited with Nanocrystalline Powders, *J. Mater. Process. Technol.*, 2008, **197**(1-3), p 31-35
30. M. Wang and L.L. Shaw, Effects of the Powder Manufacturing Method on Microstructure and Wear Performance of Plasma Sprayed Alumina-Titania Coatings, *Surf. Coat. Technol.*, 2007, **202**, p 34-44
31. Z. Karoly and J. Szepvolgyi, Hollow Alumina Microspheres Prepared by RF Thermal Plasma, *Powder Technol.*, 2003, **132**, p 211-215
32. V. Hurevich, L. Smurov, and L. Pawlowski, Theoretical Study of the Powder Behaviour of Porous Particles in a Flame During Plasma Spraying, *Surf. Coat. Technol.*, 2002, **151-152**, p 370-376
33. D. Zois, A. Lekatou, M. Vardavoulias, and A. Vazdirvanidis, Nanostructured Alumina Coatings Manufactured by Air Plasma Spraying: Correlation of Properties with the Raw Powder Microstructure, ISMANAM, *15th International Symposium on Metastable, Amorphous and Nanostructured Materials*, July 6-10 2008, Buenos Aires, Argentina
34. A. Agarwal, T. McKechnie, and S. Seal, Net Shape Nanostructure Aluminium Oxide Structures Fabricated by Plasma Spray Forming, *J. Therm. Spray Technol.*, 2003, **12**, p 350-359

JTTEE5 17:895–901
DOI: 10.1007/s11666-008-9270-5
1059-9630/$19.00 © ASM International

Peer Reviewed

Particle Temperature Fluctuations in Plasma Spraying

S. Goutier, E. Noguès-Delbos, M. Vardelle, and P. Fauchais

(*Submitted June 25, 2008; in revised form September 16, 2008*)

In direct current (d.c.) plasma spray torches, the dynamic behavior of the arc attachment at the anode nozzle results in arc voltage fluctuations and correspondingly power fluctuations. The resulting plasma jet instabilities affect the treatment (heat and momentum) of particles injected in the plasma flow and, thus, the coating quality. However, it is not clear if the experimentally observed fluctuations of particle temperatures are a major phenomena and if their frequencies are always in unison with those of voltage. In this study, two online techniques are used to investigate, respectively, the time variation of particle temperatures and its correlation with voltage variations; the first technique makes it possible to analyze plasma voltage instabilities and the second one to investigate those of particle temperatures. Both allow determining the frequencies and amplitude variations of voltage and particle temperature. Experiments were carried out with three plasma torches (F4-type and two 3MB-type) using, respectively, argon-hydrogen (F4-type and 3MB) and nitrogen-hydrogen (3MB) mixtures (all with restrike mode for the voltage fluctuations) as plasma-forming gases. A good correlation between arc voltage and particle temperature fluctuations is observed when the plasma torch is operated with argon-hydrogen mixtures and high mass flow rate. However, it is not the case when the torch is operated with nitrogen-hydrogen mixtures even if the amplitudes of voltage fluctuations are two to three times higher than those obtained with Ar-H_2.

Keywords arc voltage, diagnostic, particle temperature, plasma fluctuations, zirconia

1. Introduction

It is well known that the quality of plasma sprayed coatings depends on the thermal treatment of particles in the plasma jet. However, the effects of arc root fluctuations on particles are not yet obvious.

The simulation of the plasma spray process integrating the arc root fluctuations phenomenon and the comparison with experimental results (particle temperatures and velocities) have shown a significant scattering of these parameters ($\Delta v \approx 25$ m/s and $\Delta T \approx 1000$ °C) (Ref 1-3).

Actually, the temperature of particles can be measured using DPV-2000, Spray Watch, or Accuraspray systems. These techniques are based on the measurement of the thermal radiation emitted by hot particles. Such measurements give mean values, except DPV-2000, which can give parameters related to a single particle, and the imaging acquisition frequency is too low (about 30 Hz) for time-dependent measurements.

The aim of this study is to establish the relationship, if any, between the arc root fluctuations and the particles heat treatment. So, an experimental setup was developed to detect, for each single particle, its in-flight temperature within the plasma jet. This system is coupled with the device allowing the measurement of arc fluctuations. For the analysis of single particle radiation, the powder mass flow rate must be such that it allows at the maximum the presence of only one particle in the measuring volume each ten or twenty microseconds. The flow fluctuates at frequencies between 1000 and 10000 Hz, and therefore measurements must be carried out in a few μs range.

Very few studies were undertaken to analyze the influence of arc voltage fluctuations on those of particle temperatures. It was shown (Ref 4, 5) by using a modified DPV-2000 and spraying with Sulzer Metco PTF4 torch, working with Ar-H_2 plasma-forming gas mixtures, that alumina particle temperatures and velocities depended strongly on the transient voltage level at which they penetrate into the plasma jet. Correspondingly, their recorded fluctuations had the same frequency (4500 Hz) as that of the arc root. These results were obtained by synchronizing the acquisition of particle signals (using DPV-2000) with the arc voltage fluctuations. The data processing technique required particle signals to be very well defined and separated from the others. So Bisson et al. (Ref 5) have used a low powder mass flow rate (1.5 g/mn).

This article is an invited paper selected from presentations at the 2008 International Thermal Spray Conference and has been expanded from the original presentation. It is simultaneously published in *Thermal Spray Crossing Borders, Proceedings of the 2008 International Thermal Spray Conference*, Maastricht, The Netherlands, June 2-4, 2008, Basil R. Marple, Margaret M. Hyland, Yuk-Chiu Lau, Chang-Jiu Li, Rogerio S. Lima, and Ghislain Montavon, Ed., ASM International, Materials Park, OH, 2008.

S. Goutier, E. Noguès-Delbos, M. Vardelle, and **P. Fauchais**, SPCTS Laboratory, UMR6638, University of Limoges, 123 av A. Thomas, 87060 Limoges Cedex, France. Contact e-mail: simon.goutier@unilim.fr.

2. Experimental Setup

2.1 Plasma Torches Working Conditions

The investigations presented here were carried out with three Sulzer-Metco plasma spray torches: first a PTF4 torch working with Ar-H_2 and then two 3MB torches, one working with Ar-H_2 and the other with N_2-H_2. The principal torch configurations are summarized in Table 1.

The sprayed powder was 8 wt.% yttria partially stabilized zirconia, with 5-25 μm size distribution. The powder feed rate was 25 g/mn, i.e., close to spray conditions.

Different parameters were studied to determine their effects on particle temperature fluctuations. For each plasma torch, these parameters were the total mass flow rate, the hydrogen flow rate, and the arc current intensity.

2.2 Particle Temperature and Voltage Fluctuations Measurement Systems

The experimental setup for measuring the particle temperature fluctuations and voltage fluctuations is presented in Fig. 1.

It consists of a fast pyrometer (50 ns response time) and a voltmeter. The bichromatic pyrometer (Ref 6) collects the thermal radiation emitted by in-flight particles. Signals are filtered by a monochromator, H10IR-Jobin Yvon with holographic gratings, 600 grooves/mm, with a bandwidth of 16 nm, at two different wavelength bands adapted to the sprayed materials, for example, respectively, 528 and 684 nm for zirconia powders. The two signals obtained are then transmitted by optical fiber to two photomultipliers (Hamamatsu R928) (PM).

An interface (National Instrument) allows the fast acquisition of signals and the data processing under Labview®.

The program running under Labview® includes:

- fast acquisition of the signals (2 μs response time);
- detection of signal peaks that cross a threshold defined by the electric and photonic noise of plasma;
- calculation of the ratio of the intensities of the peaks over the two wavelengths;
- calculation of the evolution of the particle temperature and its Fast Fourier Transformation (FFT).

The fast voltmeter (one megahertz) (Ref 7) includes a digitalization of the arc voltage and data processing (FFT and the root mean square, RMS, of the arc voltage).

Table 1 Principal plasma torch configurations

	Plasma torches	
	PTF4	3MB
Anode shape	Cylindrical	
Nozzle internal diameter, mm	6	5.5
Injector location	External	
Internal diameter, mm	1.8	2
Injector location		
Versus torch axis, mm	9	8.5
Versus torch exit, mm	6	5

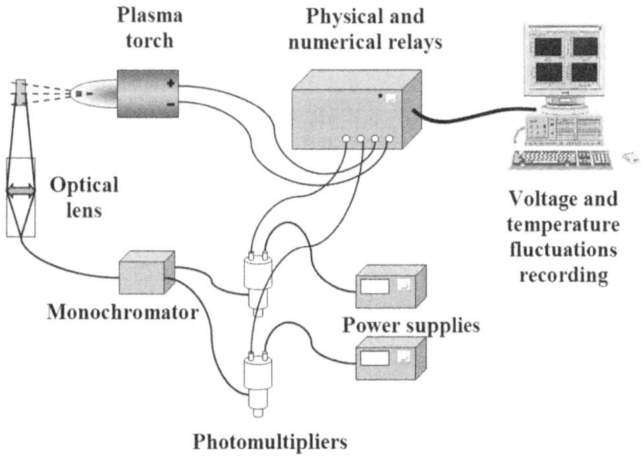

Fig. 1 Experimental setup for measuring the particle temperature and voltage fluctuations

3. Results

3.1 Comparison of Temperature and Voltage Signals

The spray parameters with the PTF4 torch are summarized in Table 2.

Particle temperatures obtained at 110 mm from the nozzle exit along the mean particle trajectory (i.e., 4 degrees relatively to the torch axis (Ref 8)) are shown for case 2 of Table 2 in Fig. 2(a) and (b); the arc voltage signal is represented for the same parameters.

Both signals are periodic and to compare their fluctuation frequencies the corresponding FFT signals are drawn as shown in Fig. 3. The first peak observed at 300 Hz is probably related to the powder feeder working frequency.

These two spectra are identical for the low frequency peak (4300 Hz). But there is only one peak on the particle temperature spectra; thus it can probably be deduced that

Table 2 Particle temperature, voltage fineness, and spray parameters for PTF4 torch

Case	Ar, L/mn	H_2, L/mn	Gas mass flow rate, kg/h	Current, A	$F_{temperature}$, s	$F_{voltage}$, s
1	33	10	3.22	600	0.20	0.03
2	45		4.37		0.55	0.12
3	60		5.81		1.24	0.16
4	70		6.77		2.32	0.17
5	33	10	3.22	300	0.71	0.11
6				350	0.58	0.06
7				500	0.25	0.03
8	33	3.7	3.19	500	0.50	0.08
9		8.5	3.21		0.28	0.02

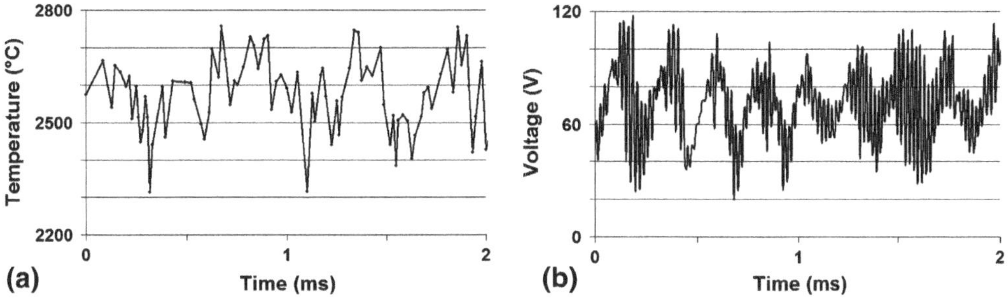

Fig. 2 Comparison between (a) particle temperature and (b) arc voltage signals [case 2/Table 2: Ar = 45 slm-H_2 = 10 slm-I = 600 A]

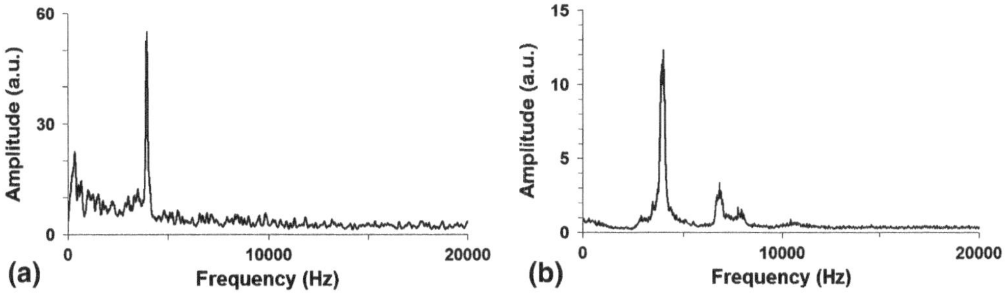

Fig. 3 FFT signals of (a) particle temperatures and (b) arc voltage signals [case 2/Table 2: Ar = 45 slm-H_2 = 10 slm-I = 600 A]

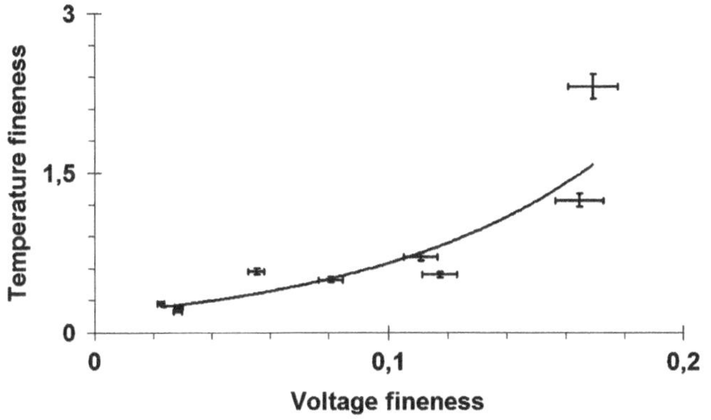

Fig. 4 Particle temperature FFT peak fineness versus arc voltage FFT peak fineness

particles follow mainly the arc fluctuations at the lower frequency.

To compare the periodicity quality of both signals, the fineness parameter F is introduced. It is defined as the ratio of the peak intensity to its width at mid-height. This very sensitive factor is associated to the signal periodicity quality.

In Fig. 4 the particle temperature fineness has been represented versus the voltage fineness for different plasma spray parameters.

It can be noted that when the voltage fineness is large a high particle temperature fineness is obtained. This means that when the arc voltage fluctuations frequency is strongly marked, the particle temperature follows better this fluctuation. It also means that particles are not influenced by the arc voltage frequency below a certain value of the F parameter (about 0.02). In viewing Table 2, the following question is raised: why does increase in argon flow rate induce an increase of particle temperature fineness?

3.2 Influence of Argon Flow Rate

For different argon flow rates (keeping the hydrogen flow rate constant at 10 L/mn), the time variations of particle temperatures is measured at 110 mm downstream of the nozzle exit. The detector is focused along the mean trajectory (Ref 8). Results are shown in Fig. 5 as a function of time.

First, it can be noted that the periodicity of particle temperatures time evolution is better defined with higher fluctuation amplitudes for high argon flow rate (70 L/mn) than for low argon flow rate (33 L/mn). It also seems that particle temperatures dispersion is weaker for low argon flow rate and so the particle heat treatment is more stable.

To determine whether this effect is due to the arc root fluctuations, variations of the voltage RMS and particle temperature RMS have been represented versus the argon flow rate in Fig. 6.

It can be seen from this figure that the increase in the argon mass flow rate increases the RMS of particle temperatures as well as that of arc voltage. This increase in the RMS of the arc voltage is due to the fact that arc voltage fluctuations depend very strongly upon plasma flow properties. Indeed, when the argon mass flow rate increases, the arc column diameter decreases slightly and the arc attachment at the anode nozzle is pushed farther downstream (Ref 9). The mass enthalpy depends strongly on the arc column length. Its fluctuations are

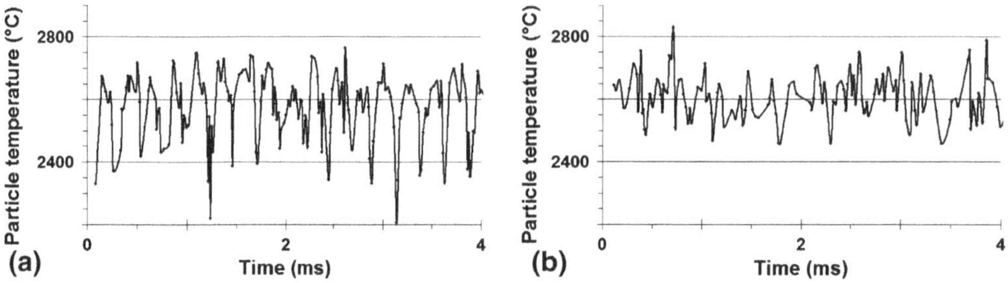

Fig. 5 Particle temperatures: (a) 70 L/mn Argon flow rate [case 4/Table 2: Ar = 70 slm-H_2 = 10 slm-I = 600 A] and (b) 33 L/mn Argon flow rate [case 1/Table 2: Ar = 33 slm-H_2 = 10 slm-I = 600 A]

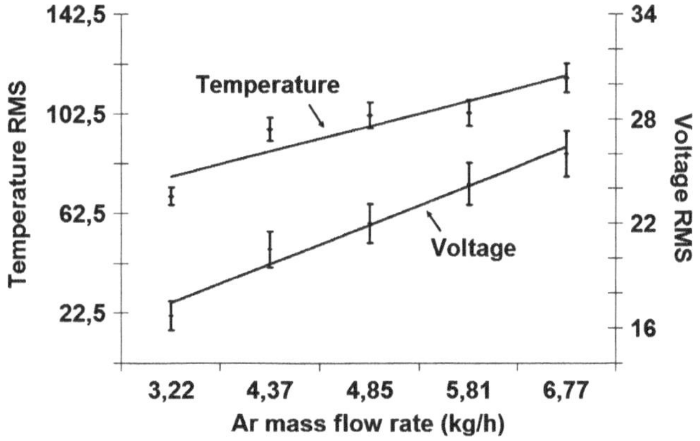

Fig. 6 Particle temperature and arc voltage RMS versus argon mass flow rate

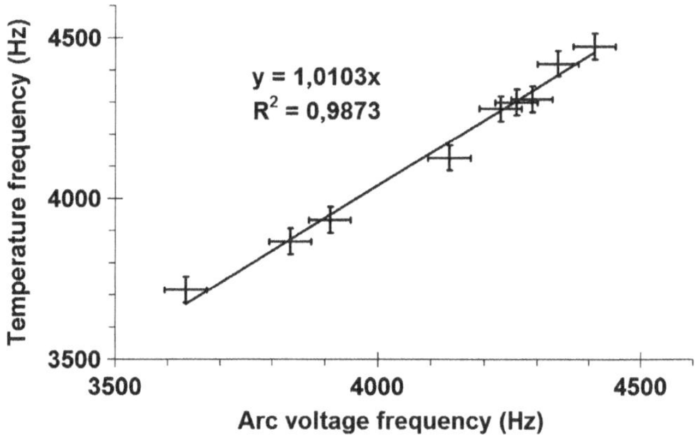

Fig. 8 Particle temperature frequency versus arc voltage frequency [case PTF4]

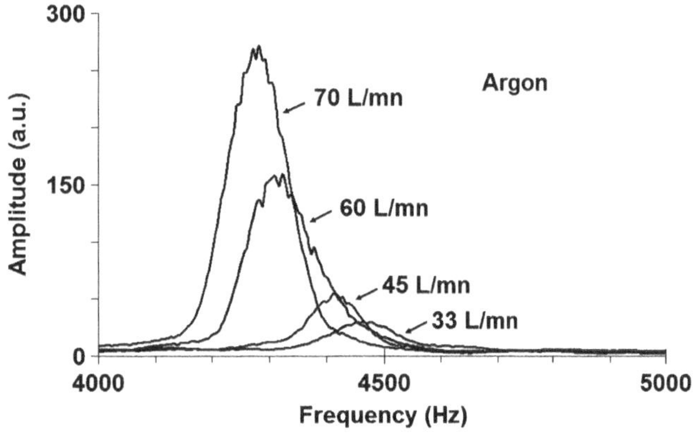

Fig. 7 Particle temperature FFT peak evolution versus argon flow rate

instantaneously followed by the gas velocities and plasma jet length almost as well by particle temperatures due to their rather low inertia (Ref 5).

The evolution of the FFT peak frequency with the argon flow rate is represented in Fig. 7.

A progressive frequency shift from 4500 to 4250 Hz can be noted when the argon flow rate is increased. This phenomenon was previously noted for the arc voltage FFT (Ref 10). Indeed, with the increase of the mass flow rate the cold boundary layer becomes thicker, the time between two successive breakdowns grows, and so the arc voltage frequency diminishes.

In addition, it can be noted that with high mass flow rates, the peak frequency is much more marked than that for low mass flow rates, which confirms the observation carried out on the study of the RMS. This phenomenon indicates that the particle temperature fluctuations seem to be more linked to the arc voltage fluctuations with high argon gas flow rates. In fact, when the argon mass flow rate increases, the intensity of the arc voltage FFT is higher (Ref 9).

3.3 Relationship Between Particle Temperature and Arc Voltage Fluctuations

To check if the fluctuation frequency of particle temperatures is identical to the one of the arc voltage, the evolution of the particle temperature frequency versus that of arc voltage is shown in Fig. 8. The resulting curve of these two parameters is linear with a slope of one. It can be concluded that the particle temperature frequency is identical to the one of the arc voltage.

To determine the parameters that influence the particle temperature periodicity quality, the particle temperature and arc voltage finenesses for different torch working parameters are summarized in Table 2.

Correlatively, the evolution of particle temperature fineness versus the particle temperature frequency for

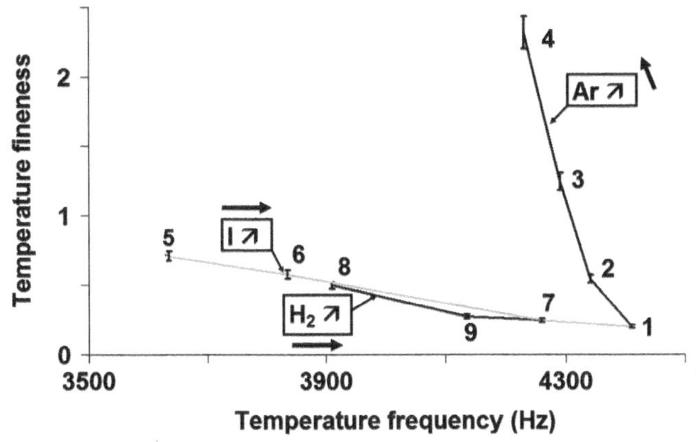

Fig. 9 Particle temperature fineness versus fluctuation frequency [case PTF4, Table 2]

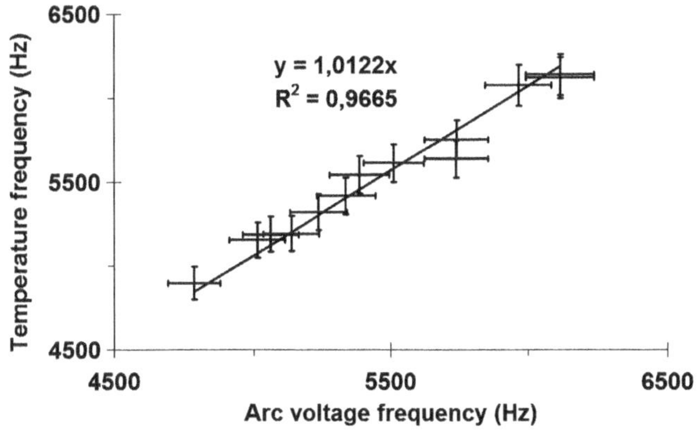

Fig. 10 Particle temperature frequency versus arc voltage frequency [case 3MB, running on argon]

different torch working parameters are represented in Fig. 9.

Figure 9 shows that, for the different torch working parameters (argon flow rate, hydrogen percentage, and arc current) considered, particle temperature fluctuations FFT peak fineness decreases when the arc voltage FFT frequency directly linked to that of particle temperature increases. It is probably due to the better particle heat transfer for a lower frequency of plasma flow pulsations.

It must also be noted that for almost the same arc voltage frequency ($\Delta F = 30$ Hz, between cases 7 and 4), different particle temperature finenesses are obtained. The high particle temperature fineness corresponds to a high argon gas flow rate (70 L/mn). This phenomenon is probably induced by the higher plasma jet velocities when the plasma mass flow rate is increased. As the particle time of flight is lower, particles undergo less plasma puffs, and hence their temperatures depend strongly on the voltage level at which they are injected in the plasma jet core. Thus, their heating directly depends on the arc voltage fluctuations.

On the other hand, when the arc current increases, the plasma velocity also goes up, although the particle temperature fineness goes down. This phenomenon is due to the fact that the increase of the arc current, increasing the arc column diameter $d_a \approx \sqrt{I}$, causes more random voltage breakdowns spoiling the signal periodicity. Thus it induces a decrease of the voltage fineness, and as a consequence, the particle temperature fineness decreases.

However, the following question is now raised: is such a behavior similar when using another plasma spray torch or different plasma-forming gases?

3.4 Comparison Between PTF4 and 3MB

The torch 3MB running on argon and hydrogen and having only one hole vortex injector of plasma-forming gases (against eight for the PTF4 torch) has been used to check whether the particle heating follows the same behavior as that obtained with the PTF4 torch.

As can been seen from Fig. 10, there is an excellent correlation between both fluctuation frequencies. The particle temperature fineness and the arc voltage fineness for different torch working parameters have been summarized in Table 3. For both torches the same evolution of particle temperature fineness is obtained (compare Fig. 9 and 11). However, the fineness variation with the 3MB torch is lower. It can also be seen that for an argon flow rate of 25 L/mn there is no periodic signal for particle temperature with the 3MB torch.

Table 3 Particle temperature, voltage fineness, and spray parameters for 3MB torch (running on argon)

Case	Ar, L/mn	H_2, L/mn	Gas mass flow rate, kg/h	Current, A	$F_{temperature}$, s	$F_{voltage}$, s
1	25	4.5	2.42	500	...	0.003
2	35		3.38		0.016	0.006
3	40		3.86		0.049	0.01
4	45		4.34		0.315	0.017
5	50		4.82		0.583	0.032
6	55		5.3		0.857	0.04
7	35	4.5	3.38	350	0.165	0.032
8				400	0.021	0.024
9				450	0.03	0.012

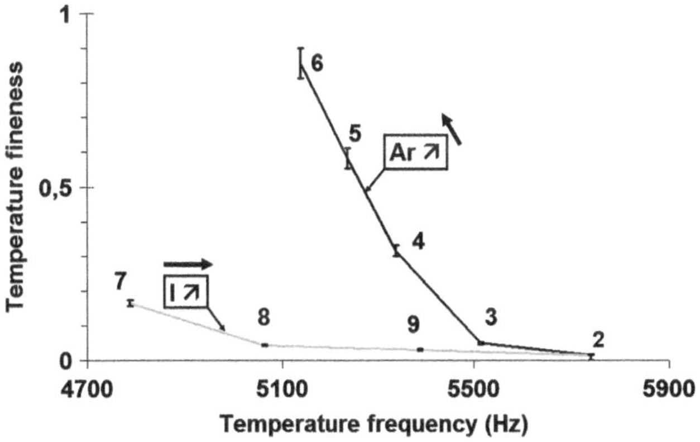

Fig. 11 Particle temperature fineness versus fluctuation frequency [case 3MB, Table 3]

In general, for this torch the periodicity quality of particle temperature is rather poor. This phenomenon is probably due to the fact that in this torch (3MB), the plasma gas injection is made only with one vortex hole. Indeed for PTF4 torches, the plasma arc column shrouding, thanks to the eight-hole vortex gas injection, stabilizes well the fluctuations of the plasma flow, which is not the case with the one hole of 3MB torch and a bad periodicity of arc voltage induces a bad periodicity for particle temperature.

3.5 Comparison Between Nitrogen and Argon as Primary Gases

The arc voltage fineness for the 3MB torch working with nitrogen (to be compared with those obtained when it is working with Ar-H_2: Table 3) have been summarized in Table 4.

Table 4 Voltage fineness and spray parameters for 3MB torch (running on nitrogen)

Case	N_2, L/mn	H_2, L/mn	Gas mass flow rate, kg/h	Current, A	$F_{voltage}$, s
1	25	4.5	2.42	500	0.002
2	35		3.38		0.001
3	40		3.86		0.001
4	45		4.34		0.001
5	50		4.82		0.002
6	55		5.3		0.003
7	35	4.5	3.38	350	0.001
8				400	0.001
9				450	0.001

It can be seen from Fig. 12(a) that the particle temperature FFT peak, corresponding to the arc voltage frequency (see Fig. 12b), does not exist anymore in spite of the high plasma primary forming gas flow rate (see Table 4). The only peak that can be observed is at very low frequency (300 Hz), which could correspond to the fluctuation of powder mass flow rate.

The calculation of the parameter F in this case gives a value of 0.003, which is very weak.

Torches working with nitrogen as primary gas have quite a different behavior from those working with argon. With nitrogen at atmospheric pressure, the minimum temperature to sustain the arc column is 7500 K, and it corresponds to a strong increase of the thermal conductivity. Thus the nitrogen arc column is self-constricted, and its diameter is almost independent of the gas mass flow rate and hydrogen percentage (Ref 9). Moreover, the radial plasma-forming gas injector used results in no plasma shrouding at all. Thus, the mass flow rate of the nitrogen primary forming gas has no effect on the particle temperature fluctuations for the 3MB torch running with nitrogen.

As can be seen from Fig. 13, there is no correlation between the arc voltage fluctuation and the particle temperature time evolution when nitrogen is used. In fact, for the 3MB torch running on argon, the voltage RMS is much lower (8.32 V) than that obtained with the 3MB torch running on nitrogen (25 V). But the particle temperature RMS values are similar for both torches.

For the 3MB torch running on nitrogen, the arc voltage frequency is high (near 7000 Hz). Thus, particles undergo more plasma puffs, and as the voltage fineness is weak (below 0.01), the plasma jet fluctuations are more random.

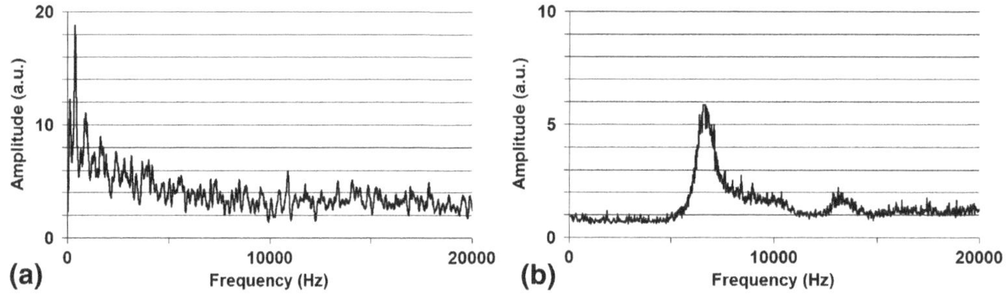

Fig. 12 (a) Particle temperature FFT and (b) arc voltage FFT [case 6/Table 4: $N_2=55$ slm-$H_2=4.5$ slm-I = 500 A]

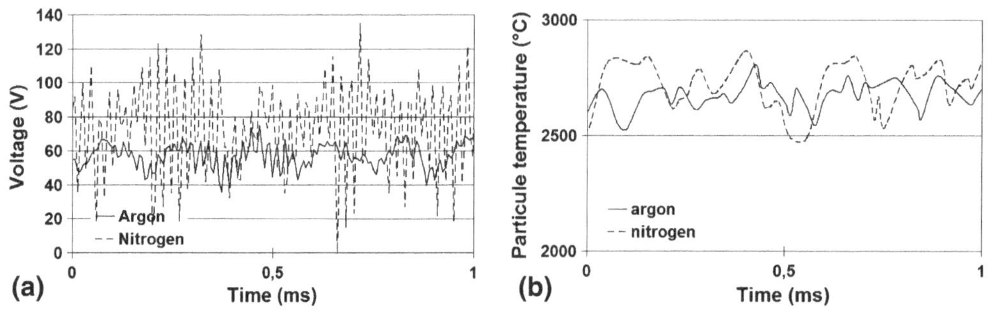

Fig. 13 (a) Arc voltage signals and (b) particle temperature evolution for 3MB torches [Ar/$N_2=35$ slm-$H_2=4.5$ slm-I = 500 A]

As a consequence, the large voltage RMS does not induce a high particle temperature RMS.

4. Conclusion

For the PTF4 plasma torch, the arc voltage fluctuations strongly influence the heat treatment of the particles. However, experimental results have shown that this influence strongly depends on the arc voltage signal quality. Thus, this influence is linked to the torch working parameters and design. For example, the 3MB torch, running on nitrogen with its arc column self constricted, does not impose its arc voltage fluctuations frequency on the particle heating. In spite of its high voltage RMS (about 25 V), the particle temperature RMS is close to that of the 3MB torch running on argon. So there is no link between the arc voltage fluctuation and the heat treatment for the 3MB torch running on nitrogen.

For the PTF4 and 3MB torch running on argon, the parameter that has the main influence on the particle temperature fluctuations is the plasma-forming gas mass flow rate. This is because on one hand the plasma shrouding is more important and stabilizes the plasma jet fluctuations, and on the other hand the velocity of the plasma jet being more important, particles also undergo less plasma puffs.

With these torches, the influence of the other plasma spray parameters (current, hydrogen percentage) depends first on the quality of arc voltage fluctuation and then on the value of arc voltage fluctuation frequency. For example, with the PTF4, the more stable heat treatment is obtained for the following plasma spray parameters: Ar = 33 slm, H_2 = 10 slm, I = 600 A.

References

1. J. Fincke, R. Williamson, and C. Chang, Plasma Spraying of Functionally Graded Materials: Measured and Simulated Results, *Proceedings of the International Thermal Spray Conference*, 2000, p 141-148
2. J. Park, J. Heberlein, E. Pfender, Y. Lau, J. Ruud, and H. Wang, Particle Behavior in a Fluctuating Plasma Jet, *Ann. N. Y. Acad. Sci.*, 1999, **891**, p 417-424
3. A. Vardelle, P. Fauchais, M. Vardelle, and G. Mariaux, Direct Current Plasma Spraying: Diagnostics and Process Simulation, *Adv. Eng. Mater.*, 2006, **8**, p 599-610
4. J. Bisson, B. Gauthier, and C. Moreau, Effect of Plasma Fluctuations on In-Flight Particle Parameters, *J. Therm. Spray Technol.*, 2003, **12**, p 38-43
5. J. Bisson, B. Gauthier, and C. Moreau, Effect of Plasma Fluctuations on In-Flight Particle Parameters, *Thermal Spray 2001: New Surfaces for a New Millennium*, 2001, p 715-721
6. J. Mishin, M. Vardelle, J. Lesinski, and P. Fauchais, Two-Colour Pyrometer for the Statistical Measurement of the Surface Temperature of Particles Under Thermal Plasma Conditions, *J. Phys. E: Sci. Instrum.*, 1987, **20**, p 620-625
7. E. Nogues, M. Vardelle, P. Fauchais, and P. Granger, Influence of Voltage Fluctuations Related to Plasma Torch Working Conditions on Zirconia Particle Thermal Treatment, *High Temp. Mater. Process.*, 2007, **11**, p 161-174
8. P. Fauchais, A. Vardelle, and B. Dussoubs, Quo Vadis Thermal Spraying?, *J. Therm. Spray Technol.*, 2001, **10**, p 44-66
9. E. Noguès, P. Fauchais, M. Vardelle, and P. Granger, Relation Between the Arc-Root Fluctuations, the Cold Boundary Layer Thickness and the Particle Thermal Treatment, *J. Therm. Spray Technol.*, 2007, **16**(5-6), p 919-926
10. Z. Duan and J. Heberlein, Anode Boundary Layer Effects in Plasma Spray Torches, *Thermal Spray: Surface Engineering Via Applied Research*, 2000, p 1-7

Peer Reviewed

JTTEE5 17:902–907
DOI: 10.1007/s11666-008-9262-5
1059-9630/$19.00 © ASM International

Peer Reviewed

Gas Permeability of Porous Plasma-Sprayed Coatings

K. Wittmann-Ténèze, N. Caron, and S. Alexandre

(*Submitted May 28, 2008; in revised form September 23, 2008*)

For different applications, such as solid oxide fuel cells, there is an interest in understanding the relationship between the microstructure and the gas permeability of plasma-sprayed coatings. Nevertheless, plasma spraying processes allow to elaborate coatings with singular microstructures, depending strongly on the initial material and plasma operating conditions. And so, the evolution of permeability is not directly linked to the porosity. In this work, coatings were manufactured using different initial feedstock and spray parameters to obtain various microporous structures. Measurements of their permeation with the pressure drop method and their open porosity just as the observation of the morphology and the structure by optical microscopy were achieved. The different data show that the evolution of the gas permeability with the open porosity follows the Kozeny-Carman equation. This result correlated with the microstructural observation highlights the relationship between the permeability and the physical properties of porous plasma-sprayed layers.

Keywords Kozeny-Carman equation, permeability, plasma spraying, porous coating, spraying parameters

1. Introduction

Porous materials attracted much attention due to their wide use in many modern applications such as fuel cell or biological materials. They are very often characterized in terms of porosity but not in terms of gas permeability. Though, the permeability of electrode layers in the solid oxide fuel cell (SOFC) field, for example, is an essential quality for their performance at the operating conditions of the SOFC system: flow resistance in the material determines the power losses, particularly at high power outputs and high gas exchange rate (Ref 1). Nevertheless, little information about gas permeability of porous ceramics exists in the literature. Most of studies deal with water permeability through ceramic membranes for environmental, petrochemical, and food industries (Ref 2-4) or with gas permeability of dense ceramic layers for SOFC or thermal barrier coatings applications (Ref 5-10). Those works underline that permeability is dependent of the microstructural properties of a porous medium. It can be estimated by the Kozeny-Carman (KC) equation, a classical semiempirical relation applied in various fields such as chemical engineering, medicine, biochemical, and electrochemical engineering (Ref 11).

The microstructure of thermally sprayed coatings is very peculiar, lamellar, and anisotropic, depending strongly on the initial material and plasma conditions. During the manufacturing step, a natural porosity is created. The resulting pores can be arranged in three categories (Ref 12-14): globular pores, interlamellar pores, and interlamellar cracks (in the case of ceramics). To this natural porosity, it is possible to add an artificial one increasing the porosity level by using an agent favoring the generation of globular pores. The characterization of such microstructure is more pertinent using a permeability factor rather than a porosity level. Very few studies were done on this topic and most of them were achieved in the SOFC field (Ref 1, 15).

The present work aims at establishing the relationship between gas permeability and microstructure of porous plasma-sprayed coatings, which may be helpful to underline the optimized coating microstructure for better permeation performance.

This article is an invited paper selected from presentations at the 2008 International Thermal Spray Conference and has been expanded from the original presentation. It is simultaneously published in *Thermal Spray Crossing Borders, Proceedings of the 2008 International Thermal Spray Conference*, Maastricht, The Netherlands, June 2-4, 2008, Basil R. Marple, Margaret M. Hyland, Yuk-Chiu Lau, Chang-Jiu Li, Rogerio S. Lima, and Ghislain Montavon, Ed., ASM International, Materials Park, OH, 2008.

K. Wittmann-Ténèze, N. Caron, and **S. Alexandre**, Plasma Spraying Laboratory, CEA Le Ripault, BP 16, 37260 Monts, France. Contact e-mails: karine.wittmann-teneze@cea.fr; nadege.caron@cea.fr and samuel.alexandre@cea.fr.

2. Experiments

2.1 Starting Feedstock Material

Commercial ceramic and metallic powders were used as feedstock material. Ceram 1 and Ceram 2 are both a mixture of two ceramic oxides having the same composition. Ceram 2 and Metal have the same particle size and

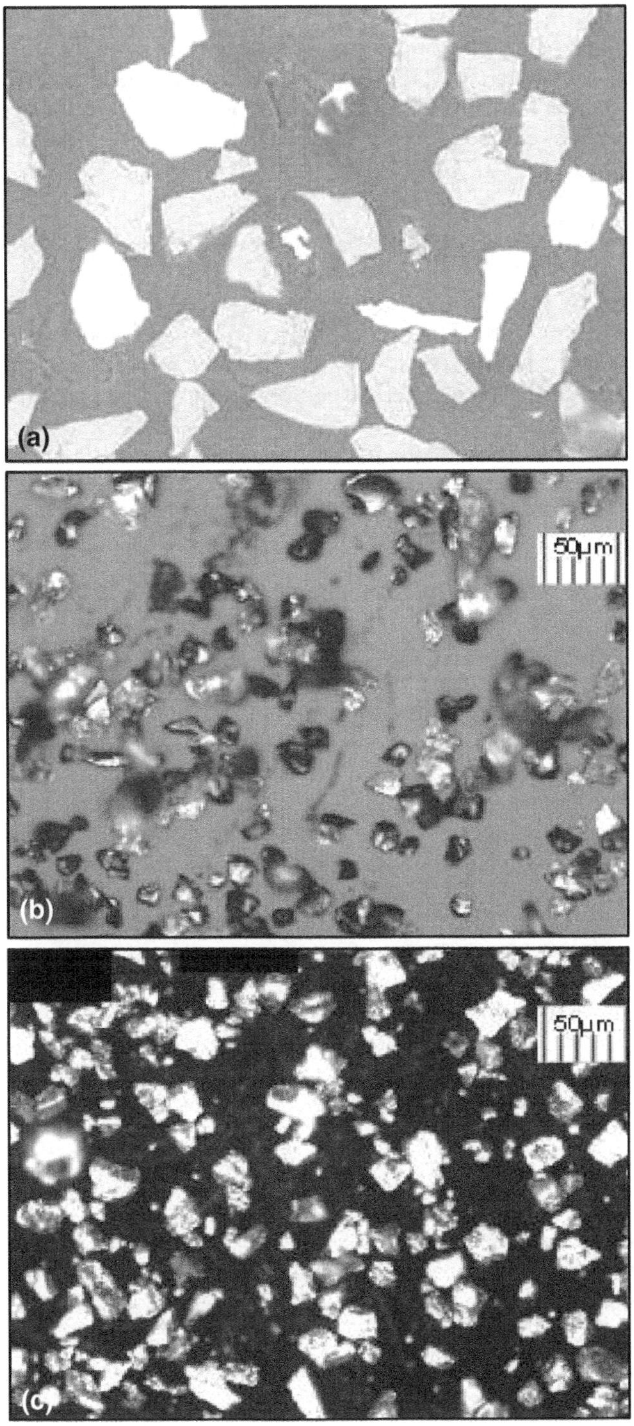

Fig. 1 Optical images of the fused and crushed powders: (a) Ceram 1, (b) Ceram 2, and (c) Metal

morphology (Fig. 1a, c) whereas Ceram 1 consists of higher diameter particles (Fig. 1b). The different particle size distributions given by the supplier are presented in Table 1.

A solid agent favoring the generation of pores was co-sprayed to generate the open porosity of coatings. Thus, by increasing its proportion, open porosity up to 55% was achieved in this work.

Table 1 Feedstock particle sizes

Powder	Ceram 1	Ceram 2	Metal
Particle size, µm			
d_{10}	40	15	15
d_{50}	49	24	25
d_{90}	90	40	45

2.2 Spray Parameters

Plasma spraying was performed with a F100-Connex torch (Sulzer-Metco, Wohlen, Switzerland) under inert atmosphere (argon at atmospheric pressure) to avoid metallic particles oxidation. Table 2 summarizes the five plasma conditions used; standoff distance, injector diameter, powder feed rate, gun traverse speed, and substrate temperature were maintained constant for all the spraying conditions.

The condition C1 was used for Metal powder, C2 for Ceram 2, and the three others for Ceram 1. Conditions C1 to C3 were adapted to the different powder characteristics; each of them resulting from a parametric study that aimed at identifying operating parameters permitting to form a cohesive coating with a high deposition rate.

For the powder Ceram 1, two other conditions were used: C4 having a higher plasma jet velocity than C3 and C5 generating a higher enthalpy plasma by adjusting appropriately the operating parameters.

The centerline velocity of the plasma jet at the nozzle exit, V_{plasma}, was predicted by the formula (Ref 16):

$$V_{plasma} \approx R_{th} \cdot \frac{U \cdot I}{p \cdot S_{anode}} \cdot \frac{\gamma - 1}{\gamma} \qquad \text{(Eq 1)}$$

where p is the atmospheric pressure, S_{anode} is the cross section area of the channel, R_{th} is the thermal efficiency of the torch, V is the arc voltage, I is the arc current, and γ is the isentropic coefficient.

The calculated values for conditions C1 to C5 are given in Table 3.

Aluminum substrates, 30 mm in diameter and 2 mm in thickness, were prepared by grit blasting and degreasing before spraying. To obtain, in a second time, self-supported coatings for permeation measurements, the substrates were covered with a soluble layer that was solubilized in a second time. During coating manufacturing, the substrate temperature was maintained at a temperature lower than 100 °C by means of liquid argon cooling.

2.3 Porosity Measurements

The open porosity of the self-supported coatings was measured using the hydrostatic pressure method based on the Archimedes principle. The sample is carefully dried in a furnace and its weight, w_d, is measured. It is then impregnated with water under soft vacuum. It is placed on a support in water. The immerged weight, w_i, is measured. At last, the sample is withdrawn from the water and is delicately wiped to measure the humid weight, w_h. The open porosity, ε, of the sample is given by Eq. 2:

Table 2 Plasma spraying conditions

Conditions: Powder:	C1 Metal	C2 Ceram 2	C3 Ceram 1	C4 Ceram 1	C5 Ceram 1
Injector diameter, mm	1.8	1.8	1.8	1.8	1.8
Anode internal diameter, mm	6	6	6	6	6
Ar, NL/min	20	24	25	34	14
He, NL/min	30	16	0	0	24
H_2, NL/min	0	3	5	6	4
Arc current intensity, A	450	450	450	550	500
Plasma flow mass enthalpy, kJ/g	10	18	19	18	33
Spray distance, mm	100	100	100	100	100

Table 3 Centerline velocities of the different plasma jets at the nozzle exit

	C1	C2	C3	C4	C5
V_{plasma}, m/s	500	790	800	1000	940

$$\varepsilon\,(\%) = \frac{w_h - w_d}{w_h - w_i} \times 100 \qquad \text{(Eq 2)}$$

The determination of the open porosity is made three times on the same coating and the given ε in this work is the average of those three values.

This method to determine the porosity, even if the quantitative value remains hard to measure with accuracy, avoids the polishing step, necessary in the case of porosity measurement by image analysis, which is very often a source of mistakes. Moreover, for this work, it is the open porosity which is the more pertinent value (instead of the total or closed porosity) because it participates in the fluid permeability.

2.4 Gas Permeability Measurements

As-sprayed coating permeability was assessed by the pressure drop method. Several forced air flow rates are applied on one face of the measured porous coating, the other face being at atmospheric pressure. The resulting pressure drop is measured for each air flow rate. After checking the linear evolution of the pressure drop with the gas flow rate, the permeability can be measured with the Darcy's law (Ref 17):

$$Q = k \cdot \frac{A \cdot \Delta P}{\eta \cdot e} \qquad \text{(Eq 3)}$$

where Q is the air mass flow rate (m^3/s) through a cross-sectional area A (m^2), η is the air cinematic viscosity (Pa.s), ΔP is the pressure drop (Pa) over a coating thickness e (m), and k is the permeability of the material (m^2).

3. Permeability Evolution of Coatings

In this work, it is on interest to correlate the permeability of a plasma-sprayed coating to its microstructure (specific area, particle and pore geometry, particle nature, etc.). A relation proposed by Kozeny and later modified by Carman (Ref 18, 19) has been used to describe a laminar flow through a porous medium. This semiempirical KC equation is the most used permeability-porosity relation:

$$k = \frac{1}{k_c \cdot S_v^2} \cdot \frac{\varepsilon^3}{(1-\varepsilon)^2} \qquad \text{(Eq 4)}$$

where ε is the porosity level (between 0 and 1) of the medium, S_v is the specific surface area per volume (m^{-1}) of this medium, k_c is the Kozeny factor which depends itself on porosity, microstructures of pores, and capillaries (Ref 11).

The Kozeny factor is defined by:

$$k_c = 2\gamma\left(\frac{L_p}{e}\right)^2 \qquad \text{(Eq 5)}$$

where γ is a constant defined by the shape of the cross section of the flow channel, L_p is the real length of the flow channel (m), and e is the thickness of the coating (m). The ratio (L_p/e) is also called the tortuosity, a nonindependent factor, varying with many parameters, such as porosity or geometry of porous media (Ref 20). Its value has a theoretical minimum of 1 and a reasonable maximum of 1.4.

Increasing the coating porosity level is the first reflex that favors permeability. Nevertheless, a too high porosity (particularly for ceramics) impedes the mechanical performances of the porous material. So, it is on interest to understand other parameters which can influence the coating permeability (pores and particles geometry, particles size and spreading, etc.). The KC equation can be a pertinent tool to correlate plasma coating microstructures (and so, spray parameters) to its gas permeability.

To be easily interpreted, the KC equation for the same initial powder can be expressed as follows:

$$k = A_i \cdot \frac{\varepsilon^3}{(1-\varepsilon)^2} \qquad \text{(Eq 6)}$$

where $A_i = e^2 \big/ (2\gamma \cdot L_p^2 \cdot S_v^2)$ is a proportional coefficient for the powder i, depending on the sprayed material microstructure.

3.1 Correlation Between Porosity and Permeability

Coatings with different porosity levels adjusted by varying the porous agent proportion were realized with the three initial powders to characterize their permeability.

Figure 2 shows the evolution of the air permeability, measured and calculated with the KC equation, of coatings with their porosity. Here, Ceram 1 was sprayed with condition C3. The value of A_i is estimated by correlating the KC equation (following Eq. 6) to the experimental data.

From Fig. 2, several observations were made:

- The KC relation can be applied on thermal-sprayed porous coatings (with a porosity rate between 20%

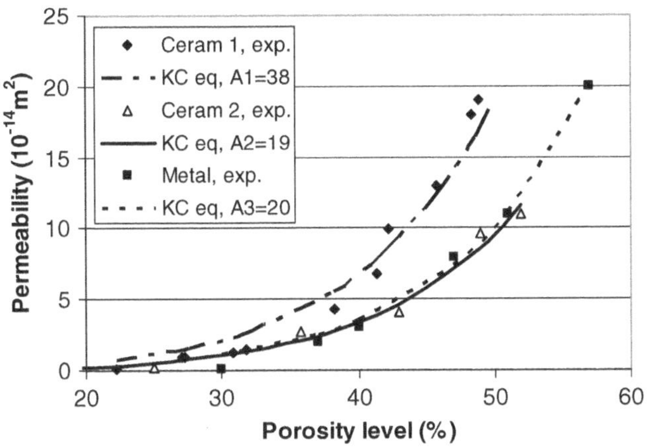

Fig. 2 Influence of the open porosity on the permeability of plasma-sprayed porous coatings elaborated with three different initial powders

and 55%) for the three initial powders, regardless of their nature and size, because of the well-fitting of the experimental results with the theoretical curb.

- The same evolution of the permeability with the porosity is observed for coatings elaborated with Ceram 2 and Metal, having the same initial particle size distribution.
- The higher the powder particle average size, the higher is the value of A_i: Ceram 1 having a d_{50} of 49 μm, A_i is equal to 38 and Ceram 2 and Metal having a d_{50} of 24 and 25 μm, A_i values come close to 20 and 19, respectively.

The relation between porosity and permeability seems to be more influenced by the initial particle size than their nature in the evaluated range of porosities (between 20% and 55%).

The higher A_i, the more advantageous is the coating microstructure to favor high permeability values. Microstructure of coatings elaborated with the three different powders but having a same porosity rate (about 30%) were observed in cross section by optical microscopy (Fig. 3).

In Fig. 3, the black zones correspond to the porosity and the gray zones to ceramic or metallic-sprayed particles.

The most significant result is that particles are more well spread in the case of coatings elaborated with the finer initial particles (Fig. 3b, c). It leads to a higher value of tortuosity (and so a lower value of A_i) and the gas path being longer.

On the contrary, the Ceram 1 coating (Fig. 3a) is constituted of not very well spread splats and almost circular-shaped particles. In that case, the tortuosity is lower (and so the value of A_i is higher) and the gas path through the layer is shorter.

To confirm these first partial conclusions, coatings were sprayed with different plasma jet enthalpies and velocities which directly influence the particle spreading by affecting their viscosity and momentum upon impact.

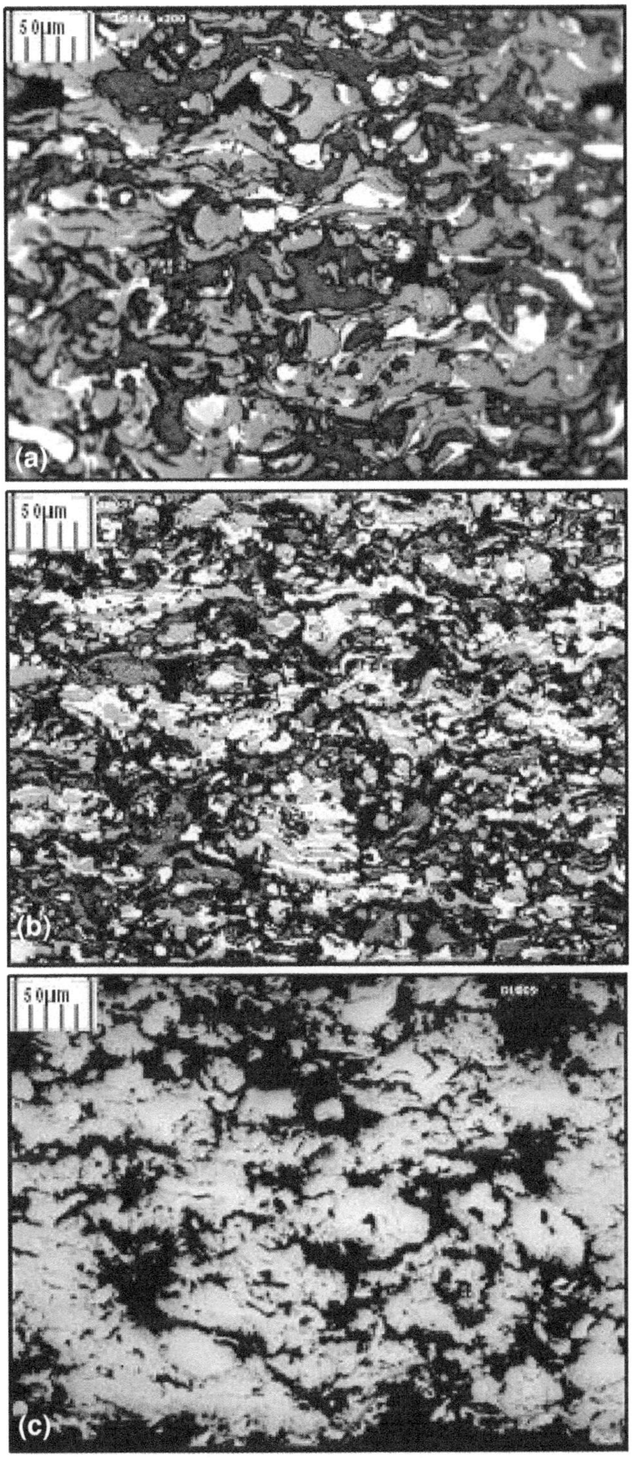

Fig. 3 Optical microscopy of polished cross-section microstructure of coatings elaborated with the three different powders: (a) Ceram 1, (b) Ceram 2, and (c) Metal following conditions C3, C2, and C1, respectively

3.2 Influence of Plasma Jet Properties on Permeability

The influence of the average plasma mass enthalpy, h, on the permeability was studied with the higher average particle size powder (Ceram 1). This powder was sprayed

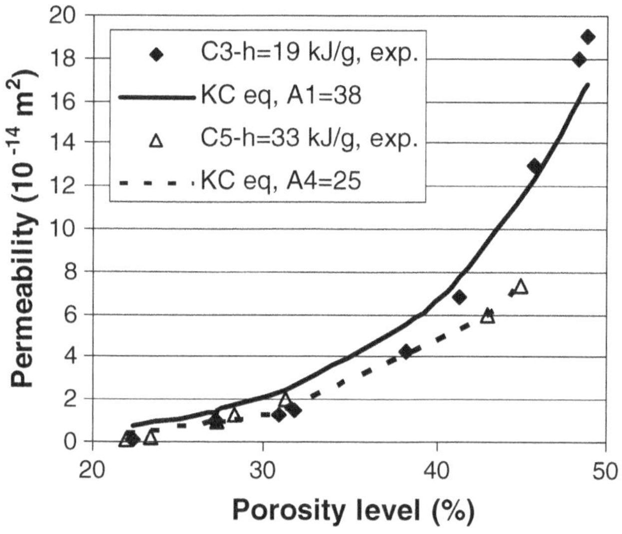

Fig. 4 Influence of the open porosity on the permeability of plasma-sprayed porous coatings elaborated with the same Ceram 1 powder and two plasma jets having different enthalpy values

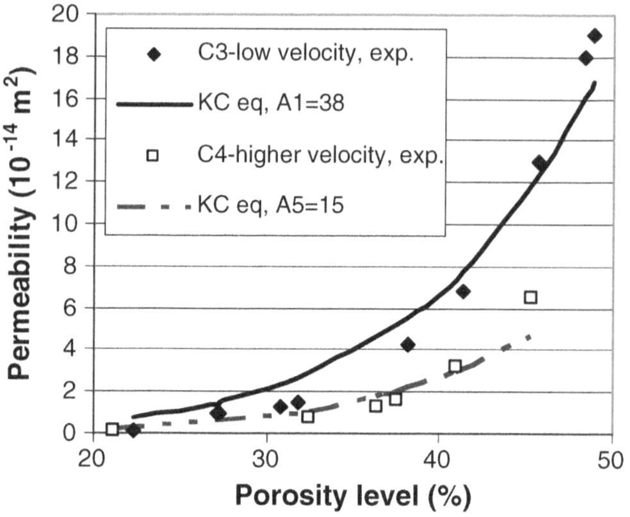

Fig. 5 Influence of the open porosity on the permeability of plasma-sprayed porous coatings elaborated with the same Ceram 1 powder and two plasma jets having different velocities (C3—lowest velocity and C4—highest velocity)

with C3 ($h = 19$ kJ/g) and C5 ($h = 33$ kJ/g) conditions (Table 2).

Figure 4 shows that A_i is lowered when the mass enthalpy is increased. In that example, a variation of the mass enthalpy of 40% induces a variation of the permeability of 35%. This can be explained quite easily: a higher plasma enthalpy is favorable to a good particle treatment in flight in the plasma jet conducting to well-spread particles. It results in a quite high tortuosity increasing the gas path through the coating and so a lower value of A_i and so a smaller permeability.

The influence of plasma jet velocity on the permeability was also studied with Ceram 1. The plasma conditions that were used are C3 and C4. As mentioned in Table 3, the centerline velocity of the plasma jet at the nozzle exit is 800 m/s for C3 and 1000 m/s for C4. Figure 5 shows that A_i is higher when plasma velocity is lowered. A possible explanation is that particles accelerated more given a more tortuous microstructure when they impact on the substrate.

Though, this result can be explained by comparing Fig. 6 with Fig. 3(b): the microstructure of the coating elaborated with condition C4 is composed of shredded particles. It results in a higher tortuosity and so a lower value of A_i (and so a lower permeability).

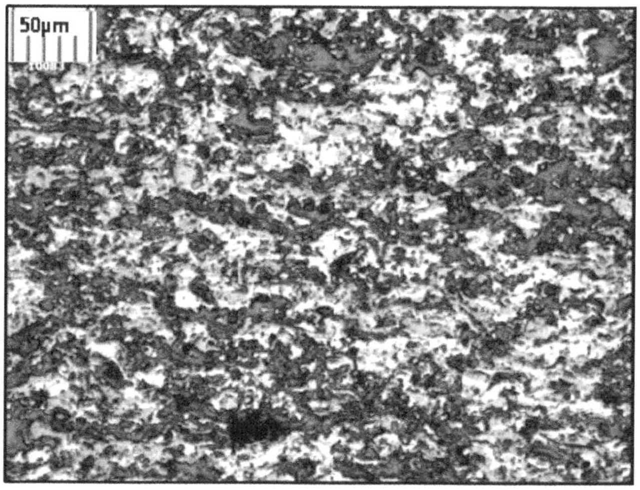

Fig. 6 Optical microscopy of polished cross-section microstructure of a coating elaborated with powder Ceram 1 and condition C4

4. Conclusion

To increase the gas permeability of a plasma spray coating, the first possibility is to increase its porosity level, for example, by adding a porous agent. Nevertheless, a too high porosity, particularly for ceramics, can affect its mechanical resistance.

The present work, supported by the KC relation, showed that microstructure can also be modified by varying the thermal spray parameters to control the permeability. A well-adapted coating seems to be composed of particles that are weakly spread to a low tortuous structure. Following this study, the main spraying parameters that can increase the permeability of a porous coating are a high initial particle size, a low plasma mass enthalpy, and a low plasma jet velocity.

Acknowledgments

The authors gratefully acknowledge J.M. Pereira for the spraying operations, P. Brelivet for the optical microscopy observations, and P. Tourenne for the help in the understanding of the correlation between porosity and permeability.

References

1. H. Weckmann, A. Syed, Z. Ilhan, and J. Arnold, Development of Porous Anode Layers for Solid Oxide Fuel Cell by Plasma Spraying, *Proceedings of the 2006 International Thermal Spray Conference*, R. Marple, M. Hyland, Y. Lau, R. Lima, and J. Voyer, Eds., May 15-18, 2006 (Seattle, WA), ASM International, 2006
2. W. Li, W. Xing, and N. Xu, Modeling of Relationship Between Water Permeability and Microstructure Parameters of Ceramic Membranes, *Destination*, 2006, **192**, p 340-345
3. I.H. Huisman, B. Dutre, K.M. Persson, and G. Tragardh, Water Permeability in Ultrafiltration and Microfiltration: Viscous and Electroviscous Effects, *Destination*, 1997, **113**, p 95-103
4. K.M. Persson, V. Gekas, and G. Tragardh, Study of Membrane Compaction and Its Influence on Ultrafiltration Water Permeability, *J. Membr. Sci.*, 1995, **100**, p 155-162
5. C.J. Li, C.X. Li, and X.J. Ning, Performance of YSZ Electrolyte Layer Deposited by Atmospheric Plasma Spraying for Cermet-Supported Tubular SOFC, *Vacuum*, 2006, **73**, p 699-703
6. C. Zhang, C.J. Li, H. Liao, M.P. Planche, C.X. Li, and C. Coddet, Effect of In-Flight Particle Velocity on the Performance of Plasma-Sprayed YSZ Electrolyte Coating for Solid Oxide Fuel Cells, *Surf. Coat. Technol.*, 2008, **202**, p 2654-2660
7. A.C. Fox and T.W. Clyne, Oxygen Transport by Gas Permeation Through the Zirconia Layer in Plasma Sprayed Thermal Barrier Coatings, *Surf. Coat. Technol.*, 2004, **184**, p 311-321
8. K. Marumoto, S. Aya, and Y. Matsui, Characterization of Solid Oxide Fuel Cell Components by Gas Permeability Measurement, *JSME Int. J., Series 2*, 1992, **35**, p 293-299
9. J.R. Mawdsley, Y.J. Su, K.T. Faber, and T.F. Bernecki, Optimization of Small-Particle Plasma-Sprayed Alumina Coatings Using Design Experiments, *Mater. Sci. Eng.*, 2001, **A308**, p 189-199
10. S.J. Oh, J. Joseph, T. Otagawa, and M. Madou, Multilayer Ionic Devices Fabricated by the Plasma Spray Method, *Solid State Ionics*, 1992, **53**(6), p 90-94
11. P. Xu and B. Yu, Developing a New Form of Permeability and Kozeny-Carman Constant for Homogeneous Porous Media by Means of Fractal Geometry, *Adv. Water Resour.*, 2008, **31**, p 74-81
12. C.J. Li, Y. He, and A. Ohmori, Characterization of Structure of Thermally Sprayed Coating, *Proceedings of the 15th International Thermal Spray Conference*, C. Coddet, Ed., May 25-29, 1998 (Nice, France), ASM International, 1998, p 717-722
13. A.J. Allen, J. Ilavsky, G.G. Long, J.S. Wallace, C.C. Berndt, and H. Herman, Microstructural Characterization of Yttria Stabilized Zirconia Plasma-Sprayed Deposits Using Multiple Small-Angle Neutron Scattering, *Acta Mater.*, 2001, **49**, p 1661-1675
14. G. Antou, G. Montavon, F. Hlawska, A. Cornet, and C. Coddet, Exploring Thermal Spray Grey Alumina Coating Pore Network Architecture by Combining Stereological Protocols and Impedance Electrochemical Spectroscopy, *Proceedings of the 2006 International Thermal Spray Conference*, B.R. Marple, M.M. Hyland, et al., Ed. (Seattle, WA), ASM International, 2006, CD ROM
15. Y. Kaga, Y. Ohno, K. Tsukamoto, and F. Uchiyama, Relationships Between the Gas Permeabilities and the Microstructures of Plasma Sprayed Oxide Layers, *Solid State Ionics*, 1990, **40-41**, p 1000-1002
16. J. Fazilleau, "Contribution to the Understanding of the Phenomena Implied in the Achievement of Finely Structured Oxides Coatings by Suspension Plasma Spraying." Ph.D. Thesis, University of Limoges, France, 2003 (in French)
17. S. Zhu, R.H. Pelton, and K. Collver, Mechanistic Modelling of Fluid Permeation Through Compressible Fiber Beds, *Chem. Eng. Sci.*, 1995, **50**(22), p 3557-3572
18. P.C. Carman, Fluid Flow Through Granular Beds, *Trans. Inst. Chem. Eng.*, 1937, **15**, p 150-167
19. P.C. Carman, Permeability of Saturated Sands, Soils and Clays, *J. Agric. Sci.*, 1938, **29**, p 263-273
20. J.H. Azar, A. Javaherian, M.R. Pishvaie, and M. Nabi-Bidhendi, An Approach to Defining Tortuosity and Cementation Factor in Carbonate Reservoir Rocks, *J. Petrol. Sci. Eng.*, 2008, **60**, p 125-131

JTTEE5 17:908–914
DOI: 10.1007/s11666-008-9274-1
1059-9630/$19.00 © ASM International

Peer Reviewed

Predicting Quenching and Cooling Stresses within HVOF Deposits

J. Stokes and L. Looney

(Submitted June 13, 2008; in revised form September 22, 2008)

Due to the nature of the HVOF and other thermal spray processes, residual stress build up in thick deposits is a significant and limiting problem since it impedes the coating behavior in service. The residual stress-state that evolves in a deposit is largely dependent on the thermal conditions to which the substrate/coating system has been subjected, and is a combination of quenching stresses, peening stresses that develop in some cases in HVOF, both of which arise during deposition, and cooling stresses, postdeposition. It follows that precise control of these phenomena is essential, if a thick deposit or one with low levels of residual stress are to be thermally sprayed. This paper applies looks at analytical and finite element techniques used to predict quenching and cooling stresses within tungsten carbide-cobalt thermally sprayed deposits. The analysis investigates and predicts the quenching and cooling stresses using improved analytical and finite element analysis techniques by validating the models with experimental results such as X-ray diffraction and the hole drilling method. The result of this paper is a thermo-mechanical equation for quenching stress which includes the effects of misfit strain, the Poisson's effect, variation of coating and substrate thicknesses, thermal expansion, and process temperature effects.

Keywords cooling stresses, HVOF thermal spraying, quenching stresses, residual stress

1. Introduction

The HVOF thermal spray system is most often used for deposition wear resistant coatings, WC-Co type coatings are typical of such an application. Challenges do exist in using HVOF thermal spraying for the purpose of depositing thicknesses greater than 1 mm without causing cracks to form or delamination of the deposit. It is important to be able to predict the level of residual stress within a thermal spray deposit in order to control and reduce its build-up by adjusting the operating spray parameters. It has been shown in previous research (Ref 1, 2) that increasing the thickness of the deposit increases the residual stress promoting fracture or failure of such coatings. Thermal spraying techniques not only produce thin deposits, but have also been shown to achieve thicknesses of several millimeters with the use of forced cooling (Ref 1-5). Due to the recent advances in thermal spraying technology, considerable research emphasis has been placed on the development of models capable of predicting what happens at various stages during the process. In order to gain a deeper knowledge of the residual stress, one needs to rely on simple models or equations to provide us with the necessary residual stress predictions required to reduce residual stresses.

Nomenclature	
σ_{RS}	residual stress (MPa)
σ_q	quenching stress (MPa)
σ_c	cooling stress (MPa)
$\sigma_{coating}$	coating stress (MPa)
$\sigma_{Substrate}$	substrate stress (MPa)
α_c	coefficient of thermal expansion of the coating (/K)
α_s	coefficient of thermal expansion of the substrate (/K)
T_m	lamella melting temperature (K)
T_s	substrate temperature (K)
T_R	room temperature (K)
E_c	coating Young's modulus (GPa)
E_s	substrate Young's modulus (GPa)
ν_c	coatings Poisson's ratio
ν_s	substrate Poisson's ratio
t_c	coating thickness (m)
t_s	substrate thickness (m)
κ	coatings curvature (m)
δ	shift of neutral axis (m)
$\Delta\varepsilon$	misfit strain

This article is an invited paper selected from presentations at the 2008 International Thermal Spray Conference and has been expanded from the original presentation. It is simultaneously published in *Thermal Spray Crossing Borders, Proceedings of the 2008 International Thermal Spray Conference*, Maastricht, The Netherlands, June 2-4, 2008, Basil R. Marple, Margaret M. Hyland, Yuk-Chiu Lau, Chang-Jiu Li, Rogerio S. Lima, and Ghislain Montavon, Ed., ASM International, Materials Park, OH, 2008.

J. Stokes and **L. Looney**, Materials Processing Research Centre & National Centre for Plasma Science and Technology, Dublin City University, Glasnevin, Dublin 9, Ireland. Contact e-mail: joseph.t.stokes@dcu.ie.

Residual stress build up within the coating structure has been modeled by several authors (Ref 6-14). All of these models treat the residual stresses as differential thermal contraction only that is due to the mismatch between the deposit and the substrate. However, residual stresses arise from two main sources, initially due to 'intrinsic' or quenching stresses developed during deposition and then due to cooling stresses (due to the mismatch of the materials). As quenching stresses were ignored, therefore differences between the experimental and finite element analysis (FEA) results were expected. The difference between the deposit properties (such as Young's apparent modulus and Poisson's ratio) used in the simulation, often differ from the actual properties of the material, thus adding to the differences found between results (Ref 8). Stokes (Ref 4) produced both a generated moment and thermal models, which were combined and compare this with both analytical and experiment results. Finite element analysis modeling is a useful tool, used to predict an outcome of a particular engineering situation, however it can be time consuming to implement, hence the ideal situation is to have analytical equations to predict the residual stresses and individual quenching and cooling stresses which can predict the stresses quickly and directly. However, it can be difficult to find equations to predict quenching stresses for both thick and thin deposits, as certain derivations are dependant on the ratio of the coating to substrate thickness. Hence one equation which could be used to determine the residual stress of a thin deposit often is unsuitable for that of thick deposits. This paper describes the equations used in the literature to predict residual stress and both quenching and cooling stresses. These analytical equations are applied to experimental data and compared to the FEA models (Ref 4) and experimental residual stress results (Ref 2).

2. Residual Stress Determination

In the current study, WC-Co was deposited using the Sulzer Metco DJ HVOF process onto 30 strips (generally 20 mm wide × 80 mm long) of AISI 316L stainless steel, according to Table 1 (Ref 1, 2), where coating thickness (0.2-3 mm) and substrate thickness (0.075-2 mm thick) were varied. The Young's modulus values used in the residual stress equations, for the coating and the substrate were 185 GPa (found as part of the present study using a cantilever test, Ref 15) and 200 GPa (Ref 16), respectively.

Table 1 Spraying parameters for WC-Co material (Ref 1, 2)

Parameters Gases:	Diamalloy 2003 WC-Co Oxygen	Propylene	Air
Pressure, bar, psi	10.4 (150)	6.9 (100)	5.2 (75)
Flow, SLPM	265.4	73.1	325
Spray rate, g/min (lb/h)	38 (5)		
Spraying distance, mm (inches)	200 (8)		

SLPM, standard liters per minute

2.1 Analytical Methods for Predicting Residual Stresses

Residual stress may be calculated analytically using the curvature (by measuring the resulting deflection of the sample) and the physical properties of the sample. Various equations exist and they can be classified according to the prediction of quenching and cooling and the final residual stress of a deposit. The research paper looks at the most popular equations used.

2.1.1 Quenching Stresses. The quenching stress (deposition stress) caused on impact and quenching of individual lamella, has been shown to be (Ref 17, 18):

$$\sigma_q \approx \alpha_c (T_m - T_s) E_c \quad \text{(Eq 1)}$$

where α_c, T_m, T_s, E_c, are the deposit coefficient of thermal expansion, lamella melting and substrate temperatures, and deposit apparent Young's modulus values, respectively. Another equation often used is (Ref 19):

$$\sigma_q = \frac{[E_c (T_m - T_s)(\alpha_c)]}{[1 - \nu_c]} \quad \text{(Eq 2)}$$

This equation is similar to Eq 1, except the effective coating Young's modulus E'_c is used:

$$E'_c = \frac{E_c}{(1 - \nu_c)} \quad \text{(Eq 3)}$$

where ν_c represents the coatings Poisson's ratio.

2.1.2 Cooling Stresses. The cooling stress (postdeposition stress) caused due to thermal mismatch between the substrate and deposit post spraying can be estimated by the following (Ref 20, 21):

$$\sigma_{\text{cooling}} = \frac{[E_c (T_s - T_R)(\alpha_c - \alpha_s)]}{\left[1 + 2\left(\frac{E_c t_c}{E_s t_s}\right)\right]} \quad \text{(Eq 4)}$$

where E_s is the substrate stiffness, α_s, T_R, t_s, and t_c are the substrate coefficient of thermal expansion, room temperature, and the thicknesses of the substrate/deposit, respectively.

2.1.3 Residual Stress. The final overall stress at the coating surface can be obtained by adding the quenching stress result to the cooling stress result. Whether the cooling stress in the coating is tensile or compressive depends upon the relative values of these expansion coefficients. If, as the temperature decreases, the coating contracts to a greater extent than the substrate ($\alpha_c > \alpha_s$), a tensile stress is generated in the coating. This may lead to adhesion loss and cracking of the coating or formed material (Ref 22). If the coefficients are equal, no cooling stress will develop. However, if the coating contracts by a smaller amount than the substrate ($\alpha_c < \alpha_s$), the resulting cooling stress will be compressive (Ref 18).

A number of equations have been derived to determine both overall coating residual stress and stress distribution through the sample. The Stoney equation (Ref 23) is a natural start, where:

$$\sigma_{RS} = \frac{[E_s \cdot t_s^2]\kappa}{[6t_c]} \quad \text{(Eq 5)}$$

This Stoney equation was modified to take into account the effective substrate Young's modulus E'_s:

$$E'_s = \frac{E_s}{(1 - \nu_s)} \quad \text{(Eq 6)}$$

to form the equation:

$$\sigma_{RS} = \frac{[E_s \cdot t_s^2]\kappa}{[6(1 - \nu_s)t_c]} \quad \text{(Eq 7)}$$

where κ and ν_c represents the coatings curvature and Poisson's ratio, respectively. A general expression was obtained by Clyne (1996) (Ref 22) for the curvature arising from the imposition of a uniform misfit strain, $\Delta\varepsilon$, such as would arise during a change in temperature ($\Delta\varepsilon = \Delta\alpha \cdot \Delta T$). Therefore the analytical curvature of a thermally sprayed sample can be related by the following equation:

$$\kappa = \frac{6E_cE_s(t_c + t_s)t_ct_s\Delta\varepsilon}{E_c^2t_c^4 + 4E_cE_st_c^3t_s + 6E_cE_st_c^2t_s^2 + 4E_cE_st_ct_s^3 + E_s^2t_s^4} \quad \text{(Eq 8)}$$

Shaw (Ref 17) derived an equation to predict residual stress as follows:

$$\sigma_{RS} = (\alpha_s - \alpha_c)(T_m - T_s)\frac{E_c(1 - \nu_c)}{(1 - \nu_c)^2} \quad \text{(Eq 9)}$$

While Senderoff and Brenner's equation for residual stress has a more complicated form (Ref 20, 24):

$$\sigma_{RS} = \frac{E_s}{(1 - \nu_s)} \cdot \left[\frac{\left(t_s + \frac{E_c(1-\nu_s)}{E_s(1-\nu_c)}t_c\right)^3}{6t_st_c\left(\frac{1}{\kappa}\right)}\right] \quad \text{(Eq 10)}$$

Clyne et al. (Ref 22, 25-27) used an analytical method which considers a pair of plates bonded together with a misfit strain $\Delta\varepsilon$ in the x-direction. Distributed residual stress results are derived for the top of the deposit, each side of the deposit/substrate interface and the bottom surface of the substrate, as follows:

$$\sigma_{\text{coating}}\big|_{y=t_c} = -\Delta\varepsilon\left(\frac{E'_ct_sE'_s}{t_cE'_c + t_sE'_s}\right) + E'_c\kappa(t_c - \delta) \quad \text{(Eq 11)}$$

$$\sigma_{\text{coating}}\big|_{y=0} = -\Delta\varepsilon\left(\frac{E'_ct_sE'_s}{t_cE'_c + t_sE'_s}\right) - E'_c\kappa\delta \quad \text{(Eq 12)}$$

$$\sigma_{\text{substrate}}\big|_{y=0} = -\Delta\varepsilon\left(\frac{E'_ct_sE'_s}{t_cE'_c + t_sE'_s}\right) - E'_s\kappa\delta \quad \text{(Eq 13)}$$

$$\sigma_{\text{substrate}}\big|_{y=-t_s} = \Delta\varepsilon\left(\frac{E'_ct_sE'_s}{t_cE'_c + t_sE'_s}\right) - E'_s\kappa(t_s + \delta) \quad \text{(Eq 14)}$$

where δ is the shift of neutral axis and $\Delta\varepsilon$ is the misfit strain given by:

$$\Delta\varepsilon = (\alpha_s - \alpha_c)(T_m - T_R) \quad \text{(Eq 15)}$$

2.2 Comparative FEA Modeling

In this research, an ANSYS finite element program was used to predict the residual stress in the coated sample. A substrate material (stainless steel) is coated with WC-Co to a certain thickness. During deposition, stresses (quenching of lamella and cooling of coating) generate, generating a moment at the ends of the sample, causing the sample to deflect. The simulation of both the quenching and cooling stresses in one system is quite difficult hence the method of simulation used in the present study relied on the deformation of the final sample postspraying. To demonstrate the numerical formulation for this system, two simple approximation methods were analysed.

Hence two-dimensional modeling of the residual stress was carried out in two ways; one to model the quenching stresses as droplets and in turn layers were applied to predict the stresses during deposition. This was based on the modeling technique where various temperatures were applied to a coated sample of known thickness and then measure the stress through the tungsten carbide-cobalt deposit when the resulting deflection measured in the FEA, equaled that found experimentally in the Clyne's method. In this case, the applied temperature in the finite element system generated a thermal load in the center of the beam, and caused the beam to deflect (Fig. 1), similar to that used by Steffens et al. (Ref 8).

Similarly this system was also simulated (second method) by applying a known deflection (measured experimentally) to the sample and compared to the latter. The system was simulated by locking both ends of the sample while subjecting the center of the sample to a 'thermal load' (to generate the deflection) by applying various loads horizontally to the deposit in a tensile manner and similar loads to the substrate in a compressive manner, to generate a moment until a deflection the same as that found experimentally was attained. The choice of tensile or compressive applied to either the deposit/substrate depended on the stress that existed in the experimental sample. The basis of this technique may be found in Ref 4. The load and in turn deflection creates an

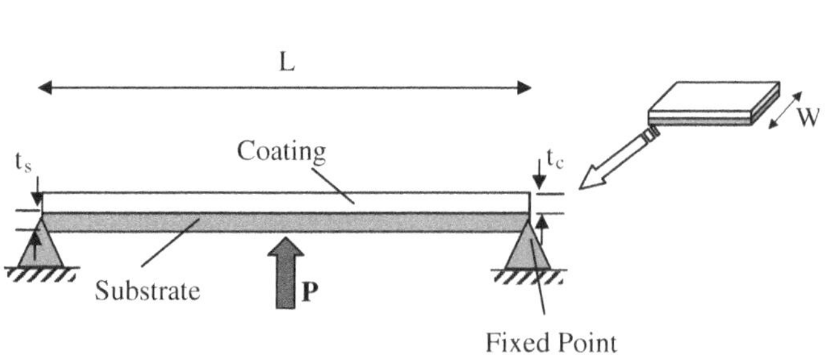

Fig. 1 Coated sample subjected to a central displacement

internal stress in the deposit causing permanent bending in the sample. In each case, the coating was considered as a homogeneous field; however, the dimensions used in the models matched that of the samples produced experimentally. For each model, a coupled field PLANE13 element was used which has 2D thermal and structural field capability. It is defined by four nodes with up to four degrees of freedom per node and has large deflection and stress stiffening capabilities and in some cases large strain capabilities.

2.2.1 Comparative Method. To validate the results found from the analytical method, the experimental results of ten of these samples ranging from a deposit surface compressive stress of 26 MPa to a tensile stress of 82 MPa were selected. Five of these coated samples were polished (plane ground using a P60 SiC abradable grit, working down to a 0.02 μm Nap Cloth) and the residual stress was measured using the Siemens 500 X-ray diffraction system, supported by the Bruker AXS Diffrac software with a diffraction scanning range from −110° to +165° for 2θ. The residual stress was measured for the final five-coated samples using the Vishay Measurements Group RS-200 Milling Guide (hole-drilling technique) (Ref 2, 4) were also measured. The strain gauge rosettes used in the study were the Vishay CEA-06-062UM-120 precision strain gauges, which are constructed of self-temperature-compensated foil on a flexible polymide carrier and incorporate a centering target for use with a precision milling guide as described in ASTM E1561. The test is carried out by connecting the strain gauge to a strain-gauge instrument (strain-recording instrument) P-3500. The hole is drilled through the coating/deposit via the central region of the strain gauge to relax the residual stress in the material being measured. Residual stress of the samples using these two techniques were then compared to Clyne's analytical method and in turn used to validate the models.

3. Results and Discussion

A series of coating conditions were used with the equations described. This mainly involved the scenario from $t_c \gg t_s$ to $t_c \ll t_s$. The objective of this research is to identify the equations best used for thick deposits. Table 2 describes the mechanical properties used and Table 3 describes the results observed.

3.1 Comparison of Quenching Stress Results

Equation 1 estimates a quenching stress of 3.40 GPa for all scenarios as this equation does not take into account the effect of varying coating or substrate thickness. Equation 2 although includes the effective coating

Table 2 Mechanical properties used in the study

Mechanical property	Unit
α_c, Coefficient of thermal expansion of the coating	8.0×10^{-6}/K
α_s, Coefficient of thermal expansion of the substrate	16.0×10^{-6}/K
T_m, Lamella melting temperature	3070 K
T_s, Substrate temperature	775 K
T_R, Room temperature	293 K
E_c, Coating Young's modulus	183.6 GPa
E_s, Substrate Young's modulus	200 GPa
ν_c, Coatings Poisson's ratio (−)	0.26
ν_s, Substrate Poisson's ratio (−)	0.3

Table 3 The effect of stress on various coating/substrate thickness analytically and using the finite element technique

Stress equation used	Sample thickness					
	$t_c = 0.2$ mm $t_s = 0.075$ mm	Thickness ratio $t_c/t_s = 2.67$	$t_c = 1$ mm $t_s = 0.075$ mm	Thickness ratio $t_c/t_s = 13.33$	$t_c = 0.2$ mm $t_s = 1$ mm	Thickness ratio $t_c/t_s = 0.2$
1. Quenching stress (Ref 17, 18)	3.40 GPa		3.40 GPa		3.40 GPa	
2. Kuroda et al., Quenching stress (Ref 19)	4.73 GPa		4.73 GPa		4.73 GPa	
3. Senderoff et al., Cooling stress (Ref 20, 21)	−0.118 GPa		−0.027 GPa		−0.513 GPa	
4. Pawlowski, Residual stress (1. + 3.) (Ref 18)	3.28 GPa		3.37 GPa		2.89 GPa	
5. Pawlowski, Residual stress (2. + 3.) (Ref 18)	4.61 GPa		4.70 GPa		4.22 GPa	
6. Stoney equation (Ref 23)	92.0 MPa		1.60 MPa		26.7 GPa	
7. Modified Stoney equation (Ref 23)	0.13 GPa		2.28 MPa		3.81 GPa	
8. Shaw, Residual stress equation (Ref 17)	8.40 GPa		8.40 GPa		8.40 GPa	
9. Senderoff and Brenner, Residual stress equation (Ref 20, 24)	5.16 GPa		5.00 GPa		6.26 GPa	
10. Clyne et al., distributed residual stress (Ref 22, 25-27)						
Top of deposit	−0.17 GPa		1.75 GPa		−4.02 GPa	
Interface	−5.21 GPa		−0.44 GPa		−4.84 GPa	
Average residual stress	−2.69 GPa		−0.66 GPa		−4.43 GPa	
11. Derived quenching stress, Eq 22	2.86 GPa		0.69 GPa		5.15 GPa	
12. Model coating quenching stress	3.27 GPa		1.22 GPa		4.57 GPa	
13. XRD method surface stress	0.84 GPa		0.76 GPa		0.88 GPa	
14. Hole drilling method surface stress	0.79 GPa		0.68 GPa		0.82 GPa	

Peer Reviewed

Young's modulus E'_c, it too assumes the quenching stresses to be constant for all thicknesses at 4.73 GPa. Therefore, these equations do not adequately determine these stress situations and Kuroda et al. (Ref 19) states that this equation is not suitable unless the substrate is much greater than the coating thickness ($t_c \ll t_s$). However, these equations do not take account the effect of varying coating or substrate thickness.

3.1.1 Comparison of Cooling Stress Results. Equation 4 estimates the cooling stress for all scenarios as it does take into account the effect of varying coating or substrate thickness. It provides a compressive stress result based on the thickness, Young's modulus, coefficients of thermal expansion and change in temperature during cooling. However, this equation does not take into account the effective substrate and coating Young's modulus E'_s and E'_c which is applicable to substrate/coating systems having an equal biaxial stress state (Ref 26).

3.1.2 Comparison of Residual Stress Results. First, combining the results of Eq 1 and 4 (and Eq 2 and 4, respectively) provides an overall calculation for residual stress (Ref 18). However as the quenching stresses provide poor approximations for varying thicknesses, these results are of little importance here.

The Stoney equation (Ref 23) (Eq 5) provides a varying residual stress result for each scenario taking into account all parameters such as; thickness, Young's modulus, coefficients of thermal expansion, and change in temperature during cooling and combining together with Eq 8 misfit strain effects. Equation 7, a modified version of the Stoney equation, although includes the effective substrate Young's modulus E'_s, it does not take into account the effective coating Young's modulus E'_c. Second this equation is accurate if t_c tends toward zero and κ must also tend toward zero. Thus this is unsuitable for $t_c = t_s$ and $t_c \gg t_s$ situations.

Shaw's (Ref 17) equation (9) does not take into account the effect of varying coating or substrate thickness, hence is too general for thermal spray analysis. Senderoff and Brenner's equation (10, Senderoff's formula) seems to provide a satisfactory solution for residual stress; however, Eq 8 does help this solution as it accounts for misfit strain effects. This formula can be used in the case of large deflections (indirectly proportional to κ) and are based on the elastic beam theory. However, this theory is only applicable if the width-to-length ratio of the substrate is small, hence the validity of Senderoff s formula is highly dependent on the substrate/sample geometry used.

Previous research has already described the confidence of using Clyne's equations as a measure of distributed residual stress across a deposit, when this technique was compared to X-ray diffraction and hole drilling experimental techniques (Ref 2, 4). While there were differences in the three measurement methods, correlation between the methods was reasonable, particularly for higher deposit thicknesses. These equations work for both thick and thin deposits, hence provide researchers with solutions for sample residual stress determination.

However if one was to determine the quenching and cooling stresses based on the above group of equations, it would seem that Clyne's equations (11 and 12) combined residual stress equations could be used with a form of Eq 4 (cooling stress) to determine the quenching stress result:

$$\text{Quenching stress} = \text{Residual stress} - \text{Cooling stress} \quad \text{(Eq 16)}$$

For Clyne's equations to be used a mean coating result would have to be used:

$$\left\{-\Delta\varepsilon\left(\frac{E'_c t_s E'_s}{t_c E'_c + t_s E'_s}\right) + E'_c \kappa (t_c - \delta) - \Delta\varepsilon\left(\frac{E'_c t_s E'_s}{t_c E'_c + t_s E'_s}\right) - E'_c \kappa \delta\right\} \Big/ 2 \quad \text{(Eq 17)}$$

Thus Eq 17 reduces to:

$$-\Delta\varepsilon\left(\frac{E'_c t_s E'_s}{t_c E'_c + t_s E'_s}\right) + \frac{E'_c \kappa \cdot t_c}{2} \quad \text{(Eq 18)}$$

If Eq 4 was modified to take into account the effective substrate and coating Young's modulus E'_s and E'_c for an isotropic in-plane stress state, there is effectively another stress equal to the x directional stress state in the z direction, thus inducing a Poisson strain in the x direction, therefore Eq 3 and 6 should be included (Ref 26), hence:

$$\sigma_{\text{cooling}} = \frac{[E'_c \Delta T \Delta\alpha]}{(1 - \nu_c)\left[1 + 2\left(\frac{(1-\nu_s)E_c t_c}{(1-\nu_c)E_s t_s}\right)\right]} \quad \text{(Eq 19)}$$

which reduces to:

$$\sigma_{\text{cooling}} = \frac{[E'_c (T_s - T_R)(\alpha_c - \alpha_s)]}{\left[(1 - \nu_c) + 2\left(\frac{(1-\nu_s)E_c t_c}{E_s t_s}\right)\right]} \quad \text{(Eq 20)}$$

This provides a quenching stress formula of the form:

$$\sigma_{\text{quenching}} = -\Delta\varepsilon\left(\frac{E'_c t_s E'_s}{t_c E'_c + t_s E'_s}\right) + \frac{E'_c \kappa \cdot t_c}{2} - \frac{[E'_c (T_s - T_R)(\alpha_c - \alpha_s)]}{\left[(1 - \nu_c) + 2\left(\frac{(1-\nu_s)E_c t_c}{E_s t_s}\right)\right]} \quad \text{(Eq 21)}$$

which includes misfit strain, effective stiffnesses (Poisson's effect), and thickness for both the coating and substrate, coefficients of expansion and process temperatures.

However, this equation can result in either a compressive or tensile stress value, therefore if one is to achieve the tensile result for quenching stress, then the equation needs to be modified to:

$$\sigma_{\text{quenching}} = \pm\left\{-\Delta\varepsilon\left(\frac{E'_c t_s E'_s}{t_c E'_c + t_s E'_s}\right) + \frac{E'_c \kappa \cdot t_c}{2}\right\} - \frac{[E'_c (T_s - T_R)(\alpha_c - \alpha_s)]}{\left[(1 - \nu_c) + 2\left(\frac{(1-\nu_s)E_c t_c}{E_s t_s}\right)\right]} \quad \text{(Eq 22)}$$

where + is used where the residual stress mean is tensile and – is used where the residual stress mean is compressive.

3.1.3 Finite Element Model. A model was designed to mimic quenching stresses and these results were compared to the quenching result found using Eq 22, Fig. 2. The model predicted a quenching stress of 6.35 GPa compared to 6.63 GPa obtained by the new analytical equation. Similar layered models (Fig. 3) were produced and compared with Eq 22, Table 3. Hence the profile of the results does appear to support each other, where the model slightly over estimates the quenching stress found. Residual stress of samples used in this study was also measured using two other techniques; XRD and the hole drilling method. Table 3 shows the results for each method. While there are differences in the four measurement methods, correlation between the methods is reasonable, as the XRD and hole drilling values are those associated to that of the coating stress state rather than intrinsic stresses. The XRD method indicates, in most cases, marginally higher stress results than the hole drilling method (this finding supports results observed by Tan, Ref 28). In the case of deposits 1 mm or greater, the XRD

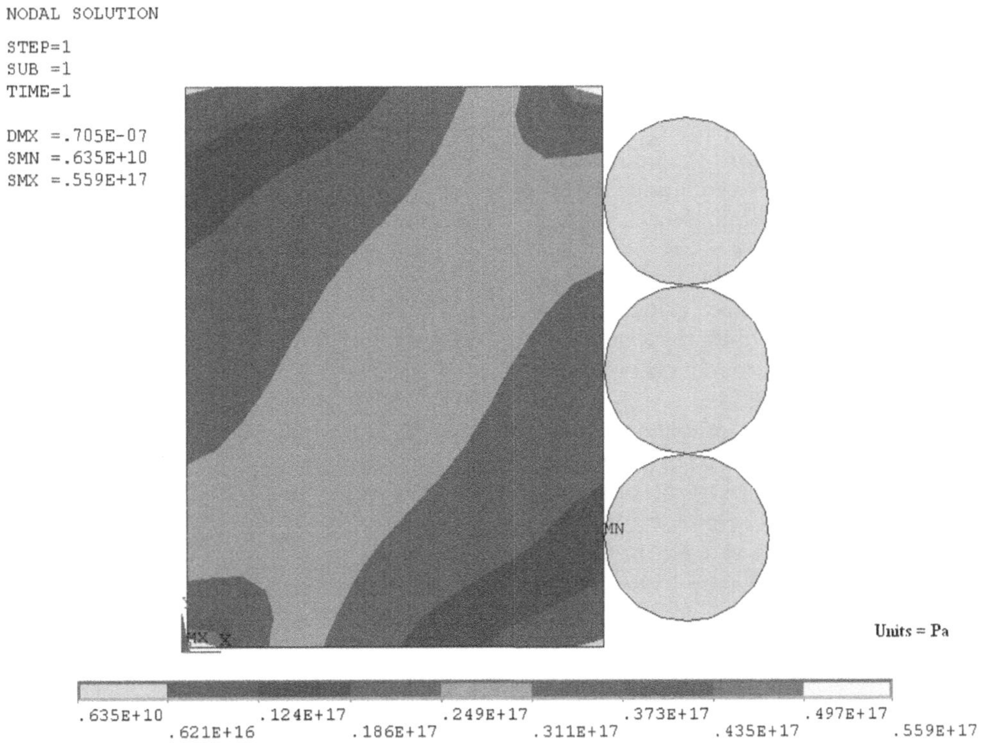

Fig. 2 Resultant *x*-directional stress results for single pass of WC-Co deposit on a 0.075-mm substrate

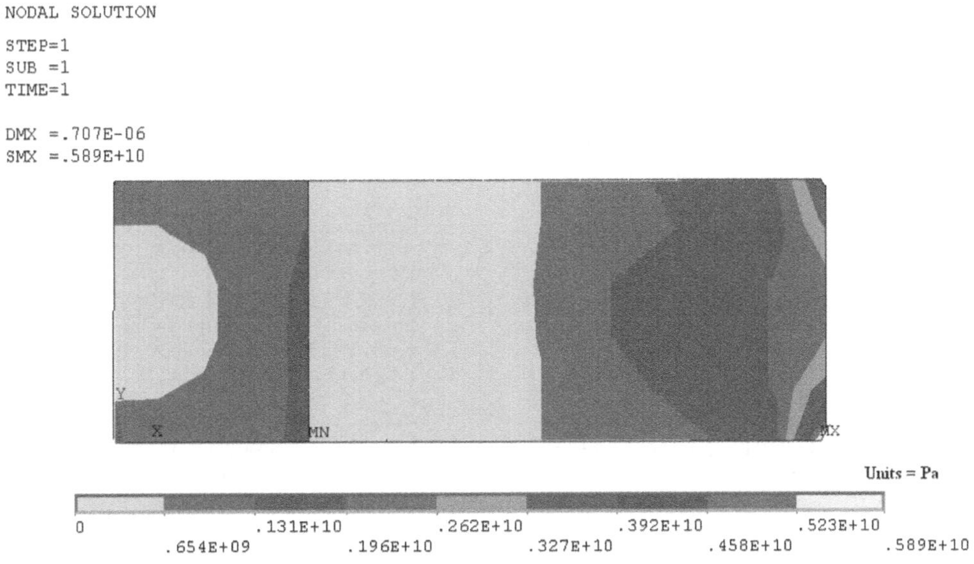

Fig. 3 Resultant x-directional stress results for 0.2 mm WC-Co deposit on a 0.075-mm substrate

Peer Reviewed

results also correlate well with those found using Clyne's method. Previous research carried out by the author showed that Clyne's analytical method produced stress values approximately 30% of that found using the hole drilling and XRD methods, respectively (Ref 4).

4. Summary and Conclusion

In this study, the analysis of residual stress generated in high velocity oxy-fuel thermal spray tungsten carbide-cobalt deposits was analysed, both analytically and with the use of numerical modeling. Hence this technique was used as a benchmark in validating the results of the FE model and in turn the analytical results. The paper reviews equations previously used by various authors to predict quenching, cooling, and residual stress. The research identifies the issues with some of these equations, such as not accounting for various combinations of coating/substrate thicknesses and for the Poisson's effect, and selects two sets of equations which provide reasonable predictions of stress. An equation for the calculation of quenching stresses is deduced for both thick and thin deposits and its results are compared to models previously benchmarked against experimental data.

Although this equation may not provide an exact solution for quenching stresses, it does allow the researcher to interchange various thicknesses of coating and substrate types thus to predict the stresses produced during deposition. This enables the researcher to monitor and control the final thickness of the deposit in situations where coating thickness is desired.

References

1. J. Stokes and L. Looney, HVOF System Definition to Maximise the Thickness of Formed Components, *Surf. Coat. Tech.*, 2001, **148**(1), p 18-24
2. J. Stokes and L. Looney, Residual Stress in HVOF Thermally Sprayed Thick Deposits, *Surf. Coat. Tech.*, 2004, **177-178**, p 18-23
3. J. Stokes and L. Looney, Properties of WC-Co Components Produced Using the HVOF Thermal Spray Process, *Thermal Spray: Surface Engineering Via Applied Research*, C.C. Berndt, Ed., 8-11 May 2000 (Montréal, Québec, Canada), ASM International, 2000, p 263-271
4. J. Stokes and L. Looney, Finite Element Analysis of Residual Stress Generated by the HVOF Process, *Building on 100 Years of Success, Proceedings of the 2006 International Thermal Spray Conference*, B.R. Marple, M.H. Hyland, Y.C. Lau, R.S. Lima, and J. Voyer, Eds., May 15-18, 2006 (Seattle, WA, USA), ASM International, 2006, ISBN: 978-0-87170-836-6
5. J. Stokes and L. Looney, HVOF System Definition to Maximise the Thickness of Formed Components, *International Conference on Advances in Materials and Processing Technology*, M.S.J. Hashmi, Ed., 3-6 Aug 1999 (Dublin, Ireland), 1999, p 775-784
6. K.A. Khor and Y.W. Gu, Effects of Residual Stress on the Performance of Plasma Sprayed Functionally Graded ZrO_2/NiCoCrAlY Coatings, *Mater Sci Eng*, 2000, **A277**, p 64-76
7. X.C. Zhang, B.S. Xu, H.D. Wang, and Y.X. Xu, Modelling of the Residual Stresses in Plasma-Spraying Functionally Graded ZrO_2/NiCoCrAlY Coatings using Finite Element Method, *Mater. Design*, 2005, **27**(4), p 308-315
8. H.D. Steffens and M. Gramlich, FEM-Analysis of Plasma Sprayed Thermal Barrier Coatings, *Thermal Spray: International Advances in Coatings Technology*, C.C. Berndt, Ed., May 25-June 5, 1992 (Orlando, FL), ASM International, 1992, p 531-536
9. J.K. Wright, J.R. Fincke, R.N. Wright, W.D. Swank, D.C. Haggard, Experimental and Finite Element Investigation of Residual Stress Resulting from the Thermal Spray Process, *Advances in Thermal Spray Science & Technology*, C.C. Berndt and S. Sampath, Ed., Sept 11-15, 1995 (Houston, TX), ASM International, 1995, p 187-192
10. H. Gruhn, W. Fischer, C. Funke, W. Mallener, and D. Stover, Residual Stress Calculation by Finite Element Methods, *Thermal Spray: Practical Solutions for Engineering Problems*, C.C. Berndt, Ed., Oct 7-11, 1996 (Cincinnati, OH), ASM International, 1996, p 869-874
11. H. Gruhn, W. Malliner, and D. Stover, Modeling of Residual Stresses in Plasma Sprayed Multilayer Systems, *Advances in Thermal Spray Science and Technology*, C.C. Berndt and S. Sampath, Ed., Sept 11-15, 1995 (Houston, TX), ASM International, 1995, p 231-236
12. C.H. Hsueh and A.G. Evans, Residuals Stresses in Metal/Ceramic Bonded Strips, *J. Am. Ceram. Soc.*, 1985, **68**, p 241-248
13. S. Suresh, A.E. Giannkopoulos, and M. Olsson, Elastoplastic Analysis of Thermal Cycling: Layered Materials with Sharp Interfaces, *J. Mech. Phys. Solids*, 1994, **42**, p 979-1018
14. B.S. Yilbas and A.F.M. Arif, Residual Stress Analysis for HVOF Diamalloy 1005 Coating on Ti-6Al-4V alloy, *Surf. Coat. Tech.*, 2007, **202**(3), p 559-568
15. E.F. Rybicki, J.R. Shadley, Y. Xiong, and D.J. Greving, A Cantilever Beam Method for Evaluation of Young's Modulus and Poisson's Ratio of Thermal Spray Coatings, *J. Therm. Spray Tech.*, 1995, **4**(4), p 377-383
16. R.E. Whan, *ASM Handbook Material Characterization*, 9th Ed., Vol. 10, G.M. Crankovic, Ed., (Metals Park, OH), American Society for Metals, 1992. ISBN: 978-0-87170-016-2
17. L.L. Shaw, Thermal Residual Stresses in Plates and Coatings Composed of Multi-layered and Functionally Graded Materials, *Compos. B Eng.*, 1998, **29**(3), p 199-210
18. L. Pawlowski, *The Science and Engineering of Thermal Spray Coatings*, John Wiley and Sons, Inc., London UK, 1995
19. S. Kuroda and T.W. Clyne, The Quenching Stress in Thermally Sprayed Coatings, *Thin Solid Films*, 1991, **200**, p 273-284
20. S. Senderoff and A. Brenner, Calculation of Stress in Electrodeposits from the Curvature of a Plated Strip, *J. Res. Nat. Bur. Stand.*, 1949, **42**(2), p 105-123
21. A.A. Tipton, The Effect of HVOF Sprayed Coatings on the Elevated Temperature High Cycle Fatigue Behaviour of a Martensitic Stainless Steel, *Proceedings of the 8th National Thermal Spray Conference*, C.C. Berndt and S. Sampath, Eds., 1995 (Houston, Texas), 1995, p 463-468
22. A. Itoh and T.W. Clyne, Initiation and Propagation of Interfacial Cracks During Spontaneous Debonding of Thermally sprayed coatings, *Proceedings of the 8th National Thermal Spray Conference*, C.C. Berndt and S. Sampath, Eds., 1995 (Texas, USA), 1995, p 425-432
23. G.G. Stoney, The Tension of Metallic Films Deposited by Electrolysis, *Proc. Roy. Soc.*, 1909, **A82**, p 172-175
24. R.O.E. Vijgen and J.H. Dautzenberg, Mechanical Measurement of the Residual Stress in thin PVD Films, *Thin Solid Films*, 1995, **270**, p 264-269
25. T.W. Clyne and S.C. Gill, Residual Stresses in Surface Coatings and Their Effects on Interfacial Debonding: A Review of Recent Work, *J. Therm. Spray Tech.*, 1996, **5**(4), p 401-418
26. Y.C. Tsui and T.W. Clyne, An Analytical Model for Predicting Residual Stresses in Progressively Deposited Coatings, Part 1: Planar Geometry, *J. Thin Solid Films*, 1997, **306**, p 23-33
27. T.W. Clyne, Residual Stresses in Surface Coatings and Their Effects on Interfacial Debonding, *Key Eng. Mats.*, 1996, **116-117**, p 307-330
28. J.C. Tan, L. Looney and M.S.J. Hashmi, Residual Stress Analysis of WC-Co Solid Components Formed by the HVOF Thermal Spraying Process, *Adv. Powder Metall. Partic. Mater.*, 1996, **3**, p 9.41-9.52

JTTEE5 17:915–923
DOI: 10.1007/s11666-008-9242-9
1059-9630/$19.00 © ASM International

Peer Reviewed

Residual Stresses Distribution through Thick HVOF Sprayed Inconel 718 Coatings

C. Lyphout, P. Nylén, A. Manescu, and T. Pirling

(Submitted June 13, 2008; in revised form August 20, 2008)

Residual stress buildup in thick thermal spray coatings is a property of concern. The adhesion of these coatings to the substrate is influenced by residual stresses that are generated during the coating deposition process. In the HVOF spray process, significant peening stresses are generated during the impact of semimolten particles on the substrate. The combination of these peening stresses together with quenching and thermal mismatch stresses that arise after deposition can be of significant importance. Both numerical method, i.e., Finite Element Method (FEM), and experimental methods, i.e., the Modified Layer Removal Method (MLRM) and Neutron Diffraction, to calculate peening and quenching stresses have been utilized in this work. The investigation was performed on thick Inconel 718 coatings on Inconel 718 substrates. Combined, these numerical and experimental techniques yield a deeper understanding of residual stress formation in the HVOF process and thus a tool for process optimization. The relationship between the stress state and deposit/substrate thickness ratio is given particular interest.

Keywords FEM simulation analysis, HVOF, Inconel 718, MLRM, neutron diffraction, residual stress, thick coating

1. Introduction

Thermally sprayed thick coatings have mainly been investigated in the perspective to form freestanding components. Less attention has been given to develop thick coatings for repair applications, which is of significant interest for the aerospace industry. One challenge when a coating several millimeters thick is to be sprayed is to control the residual stresses through the deposit thickness, and to understand the relationship between these stresses and coating adhesion. It is commonly known that coating adhesion and residual stress distribution are dependent on several factors such as pretreatment and process conditions during spraying and on post-treatment. In the present study, residual stress distribution of HVOF sprayed Inconel 718 coatings on Inconel 718 substrates has been investigated in order to understand the relationship between coating thickness and residual stresses.

1.1 Development of Residual Stresses

Residual stress is commonly defined as the stress that remains in a body that is not being subjected to external forces. It can be detrimental or beneficial to the performance of a material. Residual stresses in a component arise from mainly three different origins. Mechanically induced stresses are generated during manufacturing processes (such as grinding, blasting, and machining) and can produce nonuniform plastic deformation. Chemically induced stresses arise due to volume changes when chemical reactions, precipitations, or phase transformations occur during manufacturing. Thermally induced stresses are a consequence of nonuniform heating and cooling during manufacturing and processing. In a thermally sprayed coating, significant residual stresses can occur as a consequence of the high thermal and kinetic energies involved in the process and due to the difference in thermophysical and mechanical properties of substrate and coating materials. The stress distribution, its intensity and sign in the coating, is strongly dependent on the specific spray process (Ref 1) and processing conditions (Ref 2). Independent of the used process, coatings are subjected to a stress gradient at the coating-substrate interface. Compressive stresses are usually considered as beneficial increases with the in-flight particle velocity. The final residual stress state through the whole coating/substrate system is determined by superposition of stresses of different nature induced during the spray process: quenching, thermal mismatch, and peening stresses, together with the compressive stress state of the substrate induced during the gritblasting prior to spraying (Ref 3) (Fig. 1). Quenching stresses occur due to the rapid quenching of molten droplets when they solidify upon impact on the substrate or on the previous deposited layer. Thermal mismatch stresses occur during cooling but after solidification due to difference in thermal expansion coefficients between each deposited layer and the substrate material. In HVOF spraying significant peening stresses can occur due to high semimolten particle velocities (600 m/s), which are to be compared with plasma spraying

C. Lyphout and **P. Nylén**, University West, Trollhättan, Sweden; **A. Manescu**, Università Politecnica delle Marche, Ancona, Italy; and **T. Pirling**, Laue-Langevin Institute (ILL), Grenoble, France. Contact e-mail: christophe.lyphout@hv.se.

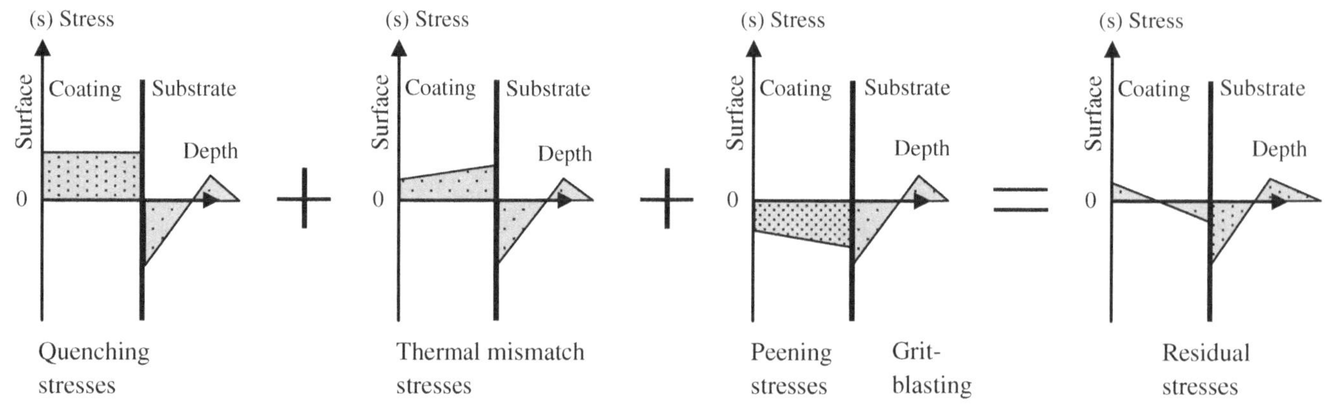

Fig. 1 Schematic representation of radial residual stress distribution in HVOF sprayed coating/substrate system

with corresponding velocities of 200 m/s and particles that are fully molten. The high particle velocity in HVOF spraying may cause a local plastic deformation resulting in local compressive stresses in the substrate as well as in the previous deposited layer. In a repair application the same material with almost identical composition as the substrate is sprayed. The small difference in thermal expansion coefficient between coating and substrate significantly reduces the contribution of the thermal mismatch stresses compared to quenching and peening stresses. The objective of this ongoing study is firstly to evaluate the final stress distribution using both numerical and experimental techniques and secondly to understand how coating thickness and process conditions affect the final stress state.

1.2 Effect of Residual Stresses on Coating Adhesion

The adhesion strength of a coating is dependent on the bonding between the coating and substrate as well as on the coating microstructure (Ref 4). Both the bonding and the microstructure are strongly influenced by residual stress distribution. It is commonly known that the level of residual stresses can significantly change at the coating substrate interface creating delaminations, which in worst cases can cause spallation. Compressive residual stresses at the interface are known to inhibit the formation of through-thickness cracks and to improve adhesion bonding (Ref 5). Stresses determination in thermal sprayed coatings has therefore been extensively studied. In situ monitoring of curvature (Ref 6) has been used to study the stress generation during coating formation. However, it is difficult to relate the in situ coating stress to the final residual stress state since both coating and substrate accommodate stresses during cooling. Godoy et al. (Ref 7) studied the change in curvature using the calculation methodology proposed by Clyne and Gill (Ref 5, 8). Even though the model was presented in the context of plasma sprayed coatings, adhesion strength was related to the shift between $\left(\Delta\sigma_{\text{int}}^{\text{surf}}\right)$ and $\left(\Delta\sigma_{\text{int}}^{\text{S,C}}\right)$, here adapted to HVOF coating according to Lesage et al. (Ref 9) (Fig. 2). $\left(\Delta\sigma_{\text{int}}^{\text{surf}}\right)$ Is the difference between the stress in the coating near the interface $\left(\sigma_{\text{int}}^{\text{C}}\right)$

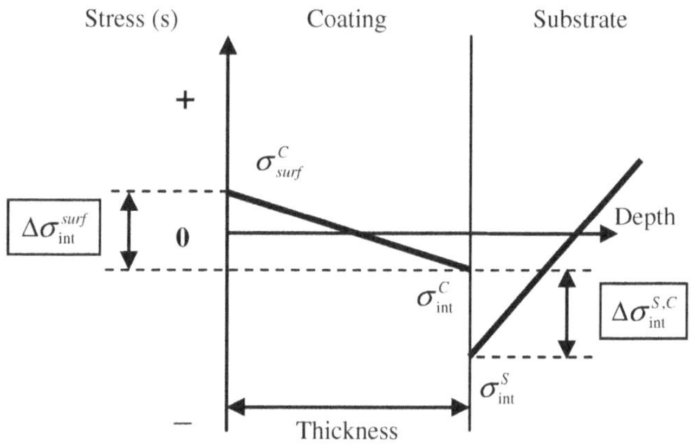

Fig. 2 Representation of radial residual stress distribution in HVOF system adapted from Lesage et al. (Ref 9)

and the stress at the surface $\left(\Delta\sigma_{\text{surf}}^{\text{C}}\right)$, and $\left(\Delta\sigma_{\text{int}}^{\text{S,C}}\right)$ is the difference between the stress in the coating near the interface $\left(\sigma_{\text{int}}^{\text{C}}\right)$ and the stress in the substrate near the interface $\left(\sigma_{\text{int}}^{\text{S}}\right)$. In the current study, such a model may provide a good framework to understand the relationship between the coating adhesion bond strength and the residual stress gap $\left(\Delta\sigma_{\text{int}}^{\text{S,C}}\right)$ at the coating-substrate interface.

2. Measurement of Residual Stresses

A number of techniques have been used in the past decades to measure residual stresses in thermal spray coatings (Ref 10). Limitations with these techniques occur when stress distributions in coatings several millimeters thick are of interest. Limitations also occur when identical materials are used for coating and substrate. A brief description of each method and its limitations for the current application is discussed below (Table 1).

2.1 Destructive Methods

Curvature measurement methods rely on the monitoring of changes in component distortion, either during

Peer Reviewed

Table 1 Comparison of residual stress methods (Ref 10, 11)

Measurements techniques	X-ray diffraction	Neutron diffraction	Synchrotron diffraction	Layer removal
Strain state	Actual strain	Actual strain	Actual strain	Change in strain
Stress state	surface	volume	volume	Surface—volume
Depth profiling	Nondestructive	Nondestructive	Nondestructive	Destructive
Penetration	50 μm (Al); 5 μm (Ti)	200 mm(Al); 4 mm (Ti)	50 mm (Al)	Polishing step
Spatial resolution	1 mm lateral 20 μm in depth	0.5 mm	20 μm lateral 1 mm in depth	50 μm/removed layer
Accuracy	±20 MPa	±50 με	±10 με	±40 MPa

Table 2 HVOF process parameters, composition of the Amdry IN718 powder and IN718 mechanical properties

HVOF—Process parameters		Amdry—Powder								
Oxygen/Hydrogen ratio	0.5	Element	Ni	Cr	Mo	Fe	Ti	Al	Nb
Carrier gas flow rate, slpm	25	%, weight	53	19	3,0	18	0,9	0,5	5,1
Powder feeding rate, g/min	80	Properties	Substrate				Coating		
Spray distance, mm	230	E-modulus	200, GPa				120, GPa		
External cooling, slpm	360	Poisson's ratio	0.30				0.25		

deposition (in situ) or after (postmortem) by successively removing material to allow relaxation of the stresses (Ref 11). In the latter, layers are removed by polishing one side of the specimen; then the stresses become unbalanced and the specimen bends. For coatings, the method is based on the same basic principles as the layer removal technique, usually called Modified Layer Removal Method (Ref 12, 13), which is further described in section 2.2.1 beneath, where strains of successively deposited layers are recorded. The major limitation with this method is that stresses might be induced during polishing.

2.2 Nondestructive Methods

Diffraction methods are based on the elastic deformations within a polycrystalline material to measure internal stresses in a material (Ref 14, 15). The stresses cause deformation i.e., changes in the distance between the lattices, which are used as internal strain gages. Shifts in diffraction peaks are recorded from which the strain distribution is calculated.

2.2.1 X-Ray Stress Evaluation (XSE). The penetration depth of low-energy X-Ray Diffraction using a traditional Kα-Cu radiation wavelength is limited to a few tens of micrometers, depending on the investigated material. Although X-ray scattering offers a high spatial resolution, analyzed thickness is not high enough to cover the depth of thick thermally sprayed coatings. Such a surface diffraction method is more dedicated to surfaces and interfaces structural characterization.

2.2.2 Neutron Diffraction. Compared to low-energy X-rays the main advantage of working with neutrons is the possibility to analyze greater depths, i.e., higher coating thicknesses. Neutrons have the following advantage over X-ray photons: for wavelengths comparable to the atomic spacing, their penetration into engineering materials is in the range of several millimeters, due to their interaction only with nuclei instead of electrons.

2.2.3 High-Energy X-ray Diffraction. High-energy photons using synchrotrons radiation offer the possibility for combining a high penetration power similar to neutron diffraction with the high resolution reached with X-ray scattering. The resulting scattered intensities require smaller correction factors since extremely small diffracted angles are measured, compared to neutron diffraction.

All of these diffraction techniques use the same methods to determine the shift in diffracted peaks to compute strain distribution. Residual stress calculation relies on mechanical properties of both bulk and coating material, and the determination of the unstressed lattice parameter d_0, which can be difficult to determine from the coating material since the particulate feedstock is sprayed.

3. Experimental Procedure

3.1 Spraying Process

Inconel 718 powder was deposited on Inconel 718 substrate using a Sulzer-Metco HVOF hybrid DJ-2600 gun. Oxygen/hydrogen gases were chosen as the oxy-fuel mixture. HVOF parameter settings are listed in Table 2. The substrates used were Inconel 718 coupons with dimensions of 25.4 mm in diameter 3 mm thickness. Substrates were degreased and gritblasted according to usual standard procedures prior to coating deposition. The gritblasted substrates had a mean roughness (Ra) of 2.5 μm. A series of four sets of three specimens each were coated to 1.0, 1.5, 2.0, and 2.5 mm (±40 μm) thickness. Both the Neutron Diffraction method and the Modified Layer Removal Method were utilized.

3.2 Coating Characterization

3.2.1 Destructive methods. The Modified Layer Removal Method was used to determine the through-thickness

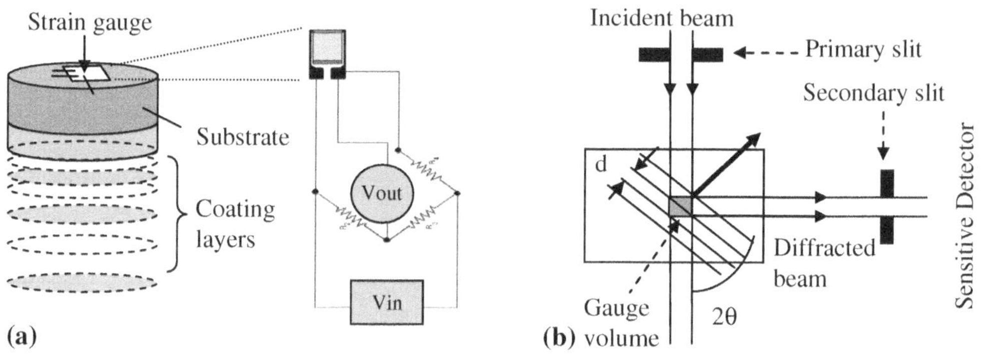

Fig. 3 (a) Modified layer removal method; (b) Neutron diffraction method

residual stress distribution in the coating. The procedure involved gluing a strain gage to the uncoated side of the specimen and then measuring the strain change after removing successive layers by mechanical polishing (Fig. 3a). Diamond particles (9 μm) were used under a constant force of 135 N to remove the material thickness from 20 to 100 μm in each polishing step. Strain and thickness changes were recorded simultaneously as layers were removed, and used for the residual stress determination using the back-computation procedure (Ref 13, 16). Material properties are given in Table 2. Previous studies revealed that the longitudinal and transverse strains were similar, measured via a 90° rosette gage. Therefore, three-wire strain gages were used.

3.2.2 Nondestructive Methods. Neutron Diffraction measurements were carried out at the Hahn-Meitner Institute (HMI) facility located in Berlin, and at the ILL SALSA reactor facility in Grenoble. A monochromatic neutron beam of a cross section of 0.6×0.6 mm^2 was used for residual strain measurements at room temperature, in a spatially resolved mode (Fig. 3b). The Bragg peaks (220) and (111) for Ni-alloy were measured to $2\theta = 86°$ and 47°, using a wavelength (λ) of 0.147 and 0.165 nm, respectively. From the obtained diffracted peaks, the lattice spacing d_{hkl} (where h, k, and l stand for Millers indices of the investigated lattice plane) was evaluated using Bragg's law (Eq. 1) with the corresponding lattice strain (Eq. 2) defined as a function of the "stress free" lattice parameter d_0.

$$\lambda = 2\, d_{hkl} \sin\theta \qquad \text{(Eq 1)}$$

$$\varepsilon_{hkl} = (d_{hkl} - d_0)/d_0 \qquad \text{(Eq 2)}$$

The d_0 values for both deposit and substrate were obtained by balancing the stress, respectively, through a freestanding coating and an annealed gritblasted substrate. Considering the deposition method, measurements were performed only in radial and axial directions, by scanning the sample from the top to the bottom (Fig. 4), using, respectively, the reflection and transmission modes. Corresponding residual stresses were calculated using Eq. 3 and 4 for isotropic Hookean elasticity:

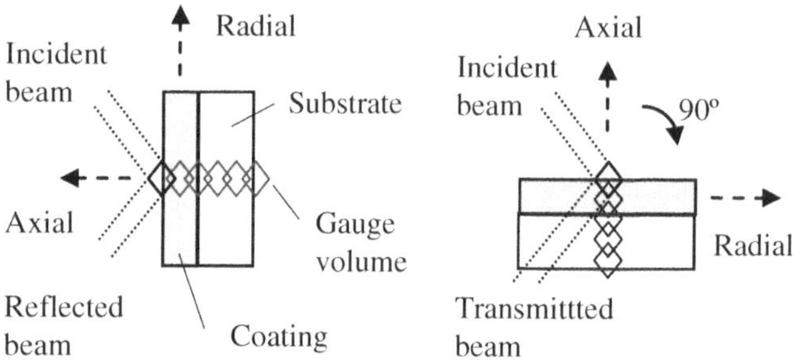

Fig. 4 Reflection and transmission measurement modes

$$\sigma_{11} = \frac{E}{(1+\nu)(1-2\nu)}[\varepsilon_{11} + \varepsilon_{33}] \qquad \text{(Eq 3)}$$

$$\sigma_{33} = \frac{E}{(1+\nu)(1-2\nu)}[2\nu\varepsilon_{11} + (1-\nu)\varepsilon_{33}] \qquad \text{(Eq 4)}$$

3.2.3 Finite Element Analysis. The current ongoing work concerns the development of a finite element (FE) methodology to simulate residual stresses generated in a HVOF coating and substrate system. Preliminary studies on a layer deposition model have been investigated to predict the quenching and thermal mismatch stresses. However, to implement peening stresses into the layer deposition model, the thermomechanical process associated with particle impingement has to be first investigated. The impact of a single particle has been extensively studied using FE modeling for the shot-peening process (Ref 17), and recently extended to simulate the peening contribution of particles impingement during HVOF spraying, using a finite element code ABAQUS (Ref 18, 19). In the present study, the capability of the solid-mechanic Msc.Marc® software was evaluated. The impact of an IN718 solid single particle just below its melting temperature was modeled to predict the peening stresses induced in the radial and axial direction of an IN718 substrate. The impact was defined as a nonlinear dynamic contact event, and studied via a thermomechanical coupled dynamic transient analysis.

A two-dimensional axisymmetric model of a 50 μm diameter particle impacting on a substrate disc of 0.2 mm in radius and 0.2 mm in depth was developed (Fig. 5a). Temperature dependence of the plasticity model obtained from experimental data (Ref 20) was considered (Fig. 5b). In a first approach, a constant strain rate of 10 s^{-1} was considered, keeping in mind that the real rapid impact may lead to a higher and time dependent strain rates. Temperature dependence of specific heat, thermal conductivity, thermal expansion coefficient, and Young's Modulus were also considered (Ref 20). Large deformations at the interface were accommodated using quadratic elements and an adaptative local mesh refinement, based on maximum curvature control. Contact between the impacting particle and the target substrate was controlled by a Coulomb friction model, assuming a friction coefficient of 0.5, which was also compared with maximal friction using a glue contact. The heat transfer at the interface was controlled by introducing a thermal contact resistance of 10^{-7} m^2 K/W. Particle velocities of 500 m/s and 600 m/s were evaluated with a particle temperature of 1270 °C, just below its melting point.

4. Results and Discussion

4.1 Modified Layer Removal Method (MLRM)

Increasing the coating thickness from 1 to 2.5 mm did not significantly change the residual stress level through the coating thickness (Fig. 6a). The profile exhibited a succession of negative and positive low stress values along the coating thickness; resulting from local accommodation between each deposited layer of the different nature of stresses issued while spraying. The top of each coating displayed tensile stresses whereas stresses of compressive nature remained at the interface. High compressive stresses were found at the interface in both coating and substrate, and these stresses rapidly changed into tensile stresses within the first hundred microns of the substrate. The compressive stress amplitude of the substrate at the interface decreased from 420 to 215 MPa for the coating thicknesses 1 and 2.5 mm (Fig. 6a). Continuity of compressive stresses at the coating/substrate interface transfers the stress accommodation from the interface to the first hundreds microns of the substrate itself. A decrease

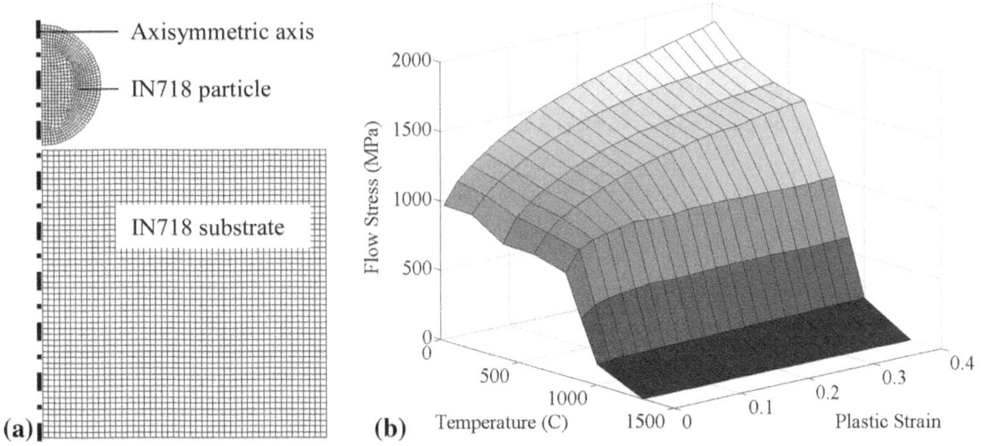

Fig. 5 (a) Axisymmetric model of an IN718 particle impacting on an IN718 substrate; (b) Temperature dependent stress-strain relationship for a 10 s^{-1} strain rate (Ref 20)

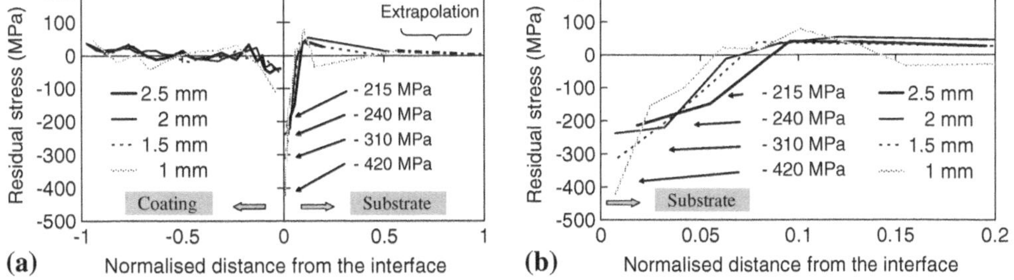

Fig. 6 MLRM radial residual stresses profiles through thick coatings deposited on 3 mm thick substrates

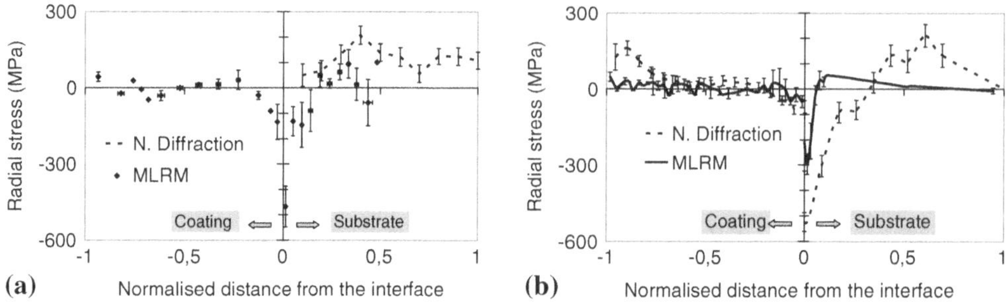

(a) (b)

Fig. 7 Radial residual stress profiles through: (a) 1 mm thick coating on 3 mm substrate—HMI,Berlin; (b) 2 mm thick coating on 3 mm substrate—ILL, Grenoble

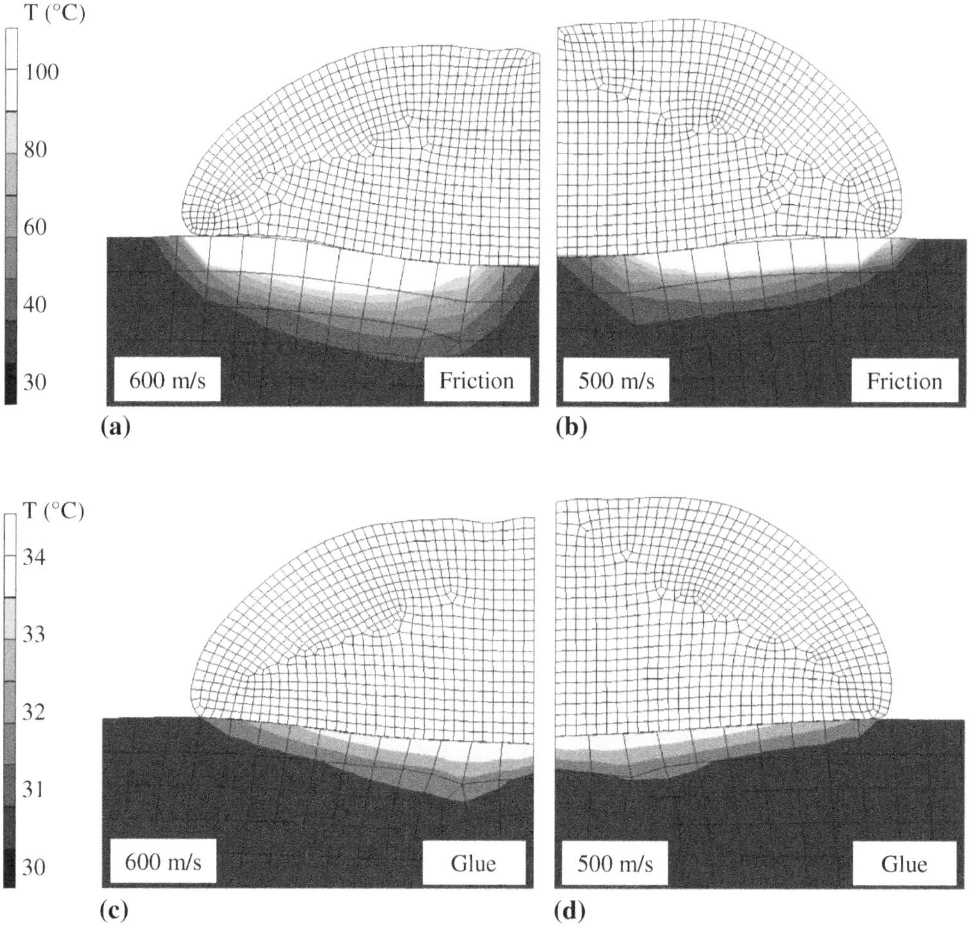

(a) (b) (c) (d)

Fig. 8 Temperature distribution through the particle/substrate system after reaching a flattening time of 0.1 μs and before cooling starts. Contact friction model after impact at (a) 600 m/s and (b) 500 m/s; contact glue model after impact at (c) 600 m/s and (d) 500 m/s

of the stress gap at the coating-substrate interface $\left(\Delta\sigma_{\text{int}}^{S,C}\right)$ for increased coating thickness can be seen in Fig. 6(b). The hypothesis of heat effect as stress relief mechanism at the interface will be investigated in future work. It should be noted that the tensile strength has been previously measured above 80 MPa (Ref 21) for all coating thicknesses.

4.2 Neutron Diffraction Method (ND)

The experiments carried out at the HMI institute in Berlin allowed analyses of only stress profiles in the substrate (Fig. 7a). From this experimental setup, the Ni-(220) diffraction peak was analyzed. The corresponding diffraction angle close to 90° gave high spatial resolution, but not enough signal intensity to measure the broader peaks of the coating. To gain in response intensity, the Ni-(111) most intense peak was considered in the experimental work carried out at the ILL Institute in Grenoble. The full coating/substrate system was scanned (Fig. 7b). Good agreement was found between these results and the MLRM ones with regard to the "stress-free" profile through the coating thickness. However, interface measurements, which superimpose both coating and substrate contribution, resulted in overestimating the local compressive stresses in the vicinity of the interface (±0.2 mm). It has to be noted that the resolution near the coating-substrate interface was highly affected by the relatively large sampling volume. Therefore, the stress level in this region has to be considered as uncertain. The development of a new deconvolution procedure, based on modeling the Gaussian-like distribution of the intensity through the gage volume, may increase the resolution in this region. This will be considered in future work.

4.3 Finite Element Analysis

The previous experimental results with low stress levels in the coating compared to corresponding levels at the coating-substrate interface initiated modeling work to simulate the stresses in this region. Initial simulations with a layer by layer model indicated that the major stress components in the current applications are quenching and peening stresses, the reason why it was decided to concentrate the work to modeling of single particle impacts. The deformed shape of flattened particles, impacting over the target substrate with two different velocities, is presented in Fig. 8. The flattening degree, commonly defined as the radius ratio of the final splat to the initial particle, increased from 1.48 to 1.60 for the corresponding particle velocities of 500 and 600 m/s, independent of the contact model used. The flattening time of the particle ($t_f < 0.1$ μs), defined as the time needed for the impinging particle to reach its final diameter after impact, was found to be significantly shorter than the solidification time, as it has been already reported by Fan et al. (Ref 22). When considering the glue contact model, the heat transfer from particle to substrate during the flattening time was controlled by conduction (Fig. 8c, d). In comparison the introduction of a friction model generated heat into the substrate subsurface during the flattening period at each location point where the contact was initiated (Fig. 8a, b). The heat flux generated by friction increased with the particle kinetic energy. The model predicted axial compressive stress levels in the range of 500 MPa, up to 20 microns in the substrate depth, which implies a dominant contribution of the particle momentum to the axial stresses (Fig. 9a). However, the predicted radial residual stress amplitude exceeds the yield strength of the material (1040 MPa) with unrealistically 80% in the radial direction, when the particle impacts at 600 m/s (Fig. 9b). Such radial distribution is not representative of any physical bonding since the contact is numerically controlled by imposing a constant Coulomb friction coefficient of 0.5. However, these results are comparable with those obtained from shot peening of strain-hardening materials (Ref 23). In the case of a fully molten droplet, assuming a negligible vaporization of the in-flight particle, its momentum may introduce the same order of axial compressive stresses in the substrate. However, the kinetic

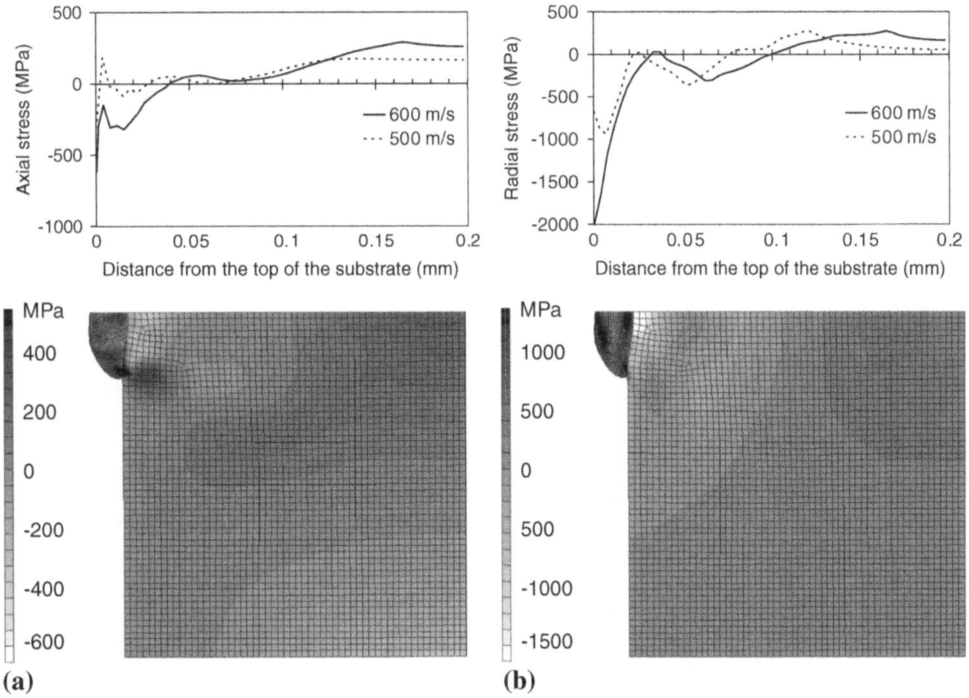

Fig. 9 (a) Axial and (b) radial residual stress distribution along the centerline and through the particle-substrate system, after impact of a single solid particle at 600 m/s

energy may be radially redistributed, since the contact may be essentially influenced by the liquid properties, such as surface tension and wetting angle, lowering both interfacial friction and radial compressive stresses.

5. Summary and Conclusion

Residual stress distributions in HVOF sprayed Inconel 718 coatings on Inconel 718 substrates have been investigated. The purpose with the ongoing study is to gain a deeper understanding of residual stress formation in thick coatings through experimental and modeling techniques. The main results can be briefly summarized:

- Increasing the coating thickness seems not to modify the residual stress distribution through the coating, and corresponding stress values have to be considered as low for all coating thicknesses.
- Compressive stress is achieved both in coating and in substrate at the interface, but the difference in stress amplitude at the interface $\left(\Delta\sigma_{\mathrm{int}}^{\mathrm{S,C}}\right)$ seems to significantly decrease when coating thickness is increased. These compressive stresses may be the reason for the good bonding of the coatings, which was measured to be above 80 MPa for all thicknesses in previous work.
- The Neutron Diffraction method needs to be further developed to be capable of determining the stress levels in the specific coating/substrate system and thicknesses used in this study. Examples of necessary developments are:
  - development of a reliable deconvolution procedure at the interface to distinguish between the two superimposed diffraction peaks, since both coating and substrate materials are Ni-based alloys.
  - a precise method to determine the unstressed lattice plane parameter d_0 for this specific material.
- The capability to predict peening and quenching stresses by the software MSC. Marc that utilizes thermo-mechanical coupled analysis with remeshing was demonstrated by simulating single particle impacts. The model is to be extended to predict thermal mismatch stresses through a multilayer deposition model.

Acknowledgments

The authors acknowledge the financial support of the VINNOVA (Swedish Governmental Agency for Innovation Systems) funded NFFP project and the technical support of the Thermal Spray Department of Volvo Aero Corporation (Trollhättan, Sweden). The research project has been supported by the EC under the 6th Framework Programme through the Key Action: Strengthening the European Research Area, Research Infrastructures. Contract no.: RII3-CT-2003-505925 (NMI3). The authors would like to thank Dr. R. Wympory and Dr. D.J. Hughes for their respective collaboration to carry out Neutron Diffraction measurement at the Hahn-Meitner Institute (Berlin) and at the Laue-Langevin Institute (Grenoble). Particular acknowledgments are addressed to Kjell Niklasson of University West (Trollhättan) for supervision in the FEA simulations.

References

1. S. Sampath, Y.Y. Jiang, J. Matejicek, L. Prchlik, A. Kulkarni, and A. Vaidya, Role of Thermal Spray Processing Method on the Microstructure, Residual Stress and Properties of Coatings: An Integrated Study for NiAl Bond Coats, *Mater Sci. Eng.*, 2004, **364**(1-2), p 216-231
2. J. Matejicek, S. Sampath, D. Gilmore, and R. Neiser, In Situ Measurement of Residual Stresses and Elastic Moduli in thermal sprayed Coatings—Part 2: Processing Effects on Properties of Mo Coatings, *Acta. Mater.*, 2003, **51**, p 873-885
3. C.R.C. Limaa, J. Ninb, and J.M. Guilemany, Evaluation of residual stresses of thermal barrier coatings, *Surf. Coat. Technol.*, 2006, **200**, p 5963-5972
4. M. Ohring, *Materials Science of Thin Films, Deposition and Structures*, Academic Press, 2002
5. T.W. Clyne and S.C. Gill, Residual Stresses in Thermal Spray Coatings and Their Effect on Interfacial Adhesion: a Review of Recent Work, *J. Therm. Spray Tech.*, 1996, **4**, p 401-409
6. R. Moltz, A. Valarezo, and S. Sampath, Comparison of Coating Stresses Produced by High Velocity Liquid/Gas Fuel and Triplex 200 Plasma Processes Using In-Situ Coating Stress Measurement, *Proceedings of the ITSC 2008*, Maastricht, Netherlands, 2008, p 473-478
7. C. Godoy, E.A. Souza, M.M. Lima, and J.C.A. Batista, Correlation Between Residual Stresses and Adhesion of Plasma Sprayed Coatings: Effects of a Post-Annealing Treatment, *Thin Solid Films*, 2002, **420-421**, p 438-445
8. S.C. Gill and T.W. Clyne, Investigation of Residual Stress Generation During Thermal Spraying by Continuous Curvature Measurements, *Thin Solid Films*, 1994, **250**, p 172-180
9. P. Araujo, D. Chicot, M. Staia, and J. Lesage, Residual Stresses and Adhesion of Thermal Spray Coatings, *Surf. Eng.*, 2005, **21**, p 35-40
10. P.J. Withers and H.K.D. Bhadeshia, Residual Stress—Measurement Techniques, *Mater. Sci. Technol.*, 2001, **17**, p 355-365
11. F.A. Kandil, J.D. Lord, A.T Fry, and P.V. Grant, *A Review of Residual Stress Measurement Methods*, NPL Materials Centre, report MATC(A)04, 2000
12. ASM International-TSS. Accepted Practice Mechanical Properties # 1-AP MP001-02 Modified layer Removal Method for Evaluating Residual Stresses in Thermal Spray Coatings, ASM-TSS, Materials Park, OH, USA, 2002, p 25
13. D.J. Greving, E.F. Rybicki, and J.R. Shadley, Through-Thickness Residual Stress Evaluation for Several Industrial Thermal Spray Coatings Using a Modified Layer-Removal Method, *J. Therm. Spray Tech.*, 1994, **3**(4), p 379-388
14. J. Pina, A. Dias, and J.I. Lebrun, Study by X-Ray Diffraction and Mechanical Analysis of the Residual Stress Generation During Thermal Spraying, *Mater. Sci. Eng.*, 2003, **A347**, p 21-31
15. T.C. Totemeier, R.N. Wright, and W.D. Swank, Residual Stresses in High-Velocity Oxy-Fuel Metallic Coatings, *Metall. Mater. Trans.*, 2004, **35A**, p 1807-1814
16. R.T.R. McGrann, E.F. Rybicki, and J.R Shadley, Applications and Theory of the Modified Layer Removal Method, *Proceedings of the Fifth ICRS*, Linköping, Sweden, June 16-18, 1997
17. S.A. Meguid, G. Shagal, and J.C. Stranart, Finite Element Modelling of Shot-Peening Residual Stresses, *J. Mater. Process. Tech.*, 1999, **92-93**, p 401-404
18. K. Yokoyama, M. Watanabe, S. Kuroda, Y. Gotoh, T. Schmidt, and F. Gärtner, Simulation of Solid Particle Impact Behavior for Spray Processes, *Mater. Trans.*, 2006, **47**(7), p 1697-1702

19. P. Bansal, P.H. Shipway, S.B. Leen, *Acta Materialia*, 2007, **55**, 5089-5101
20. Internal Report, Volvo Aero Corporation
21. C. Lyphout, P. Nylén, and J.Wigren, Characterization of Adhesion Strength and Residual Stress of HVOF sprayed Inconel 718, *Proceedings of the ITSC 2007*, Beijing, China, 2007, p 588-593
22. Q. Fan and L. Wang, Modelling of Temperature and Residual Stress Fields, *Proceedings of the ITSC 2005*, Basel, Switzerland, 2005, p 275-279
23. T. Hong et al., A Numerical Simulation to Relate the Shot Peening Parameters to the Induced Residual Stresses, *Eng. Fail. Anal.*, 2008, **15**, p 1097-1110

JTTEE5 17:924–932
DOI: 10.1007/s11666-008-9234-9
1059-9630/$19.00 © ASM International

Peer Reviewed

Influence of the HVOF Gas Composition on the Thermal Spraying of WC-Co Submicron Powders (−8 + 1 μm) to Produce Superfine Structured Cermet Coatings

W. Tillmann, E. Vogli, I. Baumann, G. Matthaeus, and T. Ostrowski

(Submitted May 29, 2008; in revised form July 25, 2008)

Thermal spraying technology represents a novel and promising approach to protect forming tools with complex surfaces and highest shape accuracy against abrasive wear and galling. However, due to high or nonuniform layer thicknesses or inappropriate surface roughness conventional coarse-structured coatings are not suitable to achieve this aim. The application of novel submicron or nanoscaled feedstock materials in the thermal spray process can provide the deposition of cermet coatings with significantly improved characteristics and is recently of great interest in science and industry. In this collaborative study, the feeding and HVOF spraying of WC-Co submicron powders (−8 + 1 μm) have been investigated to manufacture superfine structured, wear resistant, near-net-shape coatings with improved macroscopic properties and smooth surfaces. The influences of varying HVOF gas compositions on the spray process and the coating properties have been analyzed.

Keywords deep-drawing tools, hard material coatings, HVOF, near-net-shape coatings, submicron powders, superfine structured and nanostructured coatings, surface engineering, thermal spraying, fine powder feeding

1. Introduction

Deep-drawing is a widely used sheet metal forming process in the aircraft and automotive industries. The manufacturing of modern parts with complicated shapes and curvatures requires forming tools with highest shape accuracy even at complex surface geometries. However, the application of novel, high-strength sheet metals combined with a continuous increase in productivity impose high tribological demands on forming tools and finally lead to increasing wear. To reduce the high costs for the repair and the maintenance of such tools, the surface has to be modified by an appropriate modification process, which is able to preserve the high dimensional accuracy. Thermal spraying has been regarded as a promising and cost-efficient surface engineering technology, which has emerged as a viable solution for a wide range of wear and corrosion resistance applications to improve the service life of tools and other machine components (Ref 1-3). It is well known that HVOF spraying provides far superior results than any other thermal spraying technique for manufacturing cermet coatings, because of the much higher gas jet velocity and lower flame temperature. These result in coatings with extremely low porosity, low splat oxidization, and low carbide decomposition or dissolution (Ref 4). However, for the protection of deep-drawing tools with complex surface geometries and highest shape accuracy, conventional cermet coatings obtained by HVOF spraying of coarse-grained feedstock powders (>15 μm) are not suitable, as they impose high or nonuniform layer thicknesses or inappropriate surface roughness. A subsequent cost-efficient surface finishing, for example, by grinding, is often necessary to obtain appropriate surface characteristics with high dimensional accuracy. Thermal spraying of novel submicron or nanoscaled feedstock materials provides significant improvements in macromaterial properties for many applications and is recently of great interest in science and industry (Ref 5). Fine powders not only show a distinct potential for the deposition of thin near-net-shape coatings with high surface quality, but are also very promising for the production of dense and homogeneous coatings with improved mechanical properties leading to an enhanced wear behavior (Ref 6, 7).

This article is an invited paper selected from presentations at the 2008 International Thermal Spray Conference and has been expanded from the original presentation. It is simultaneously published in *Thermal Spray Crossing Borders, Proceedings of the 2008 International Thermal Spray Conference*, Maastricht, The Netherlands, June 2-4, 2008, Basil R. Marple, Margaret M. Hyland, Yuk-Chiu Lau, Chang-Jiu Li, Rogerio S. Lima, and Ghislain Montavon, Ed., ASM International, Materials Park, OH, 2008.

W. Tillmann, E. Vogli and **I. Baumann**, Institute of Materials Engineering, Technische Universität Dortmund, Leonhard-Euler-Str. 2, 44227 Dortmund, Germany; and **G. Matthaeus** and **T. Ostrowski**, Thermico GmbH and Co. KG, Franz-Schlueter-Str. 10-16, 44147 Dortmund, Germany. Contact e-mail: wolfgang.tillmann@udo.edu.

The mechanical and functional properties of engineering materials are to a large extent determined by their microstructure, particularly the distribution of phases and their crystallite sizes. Studies by Jia and Fisher (Ref 8, 9) on sintered nanostructured WC-Co materials have been reported to exhibit enhanced performance in both sliding and abrasive wear and substantially higher hardness than their conventional coarse structured counterpart. A similar approach of using sintered and agglomerated micron-sized powders with submicron or nanosized WC particles in a Co matrix to improve the performance of HVOF-sprayed cermet coatings has been pursued by other researchers (Ref 5). These studies concluded that in comparison to coatings with conventional coarse-grained microstructure, the deposition of superfine and nanostructured coatings leads to a reduction of the friction coefficient and the roughness by a factor of two and to a wear reduction by a factor of five (Ref 10). However, these advantages can only be achieved as long as appropriate process technologies are employed, which preserve the submicron or nanoscaled feedstock material structure in the final as-sprayed coating morphology. Due to the large specific surface of fine powders, they show high thermal susceptibility. A sensitive temperature control of the HVOF process is fundamentally required to avoid particle overheating. At the same time, the low specific weight of fine powder particles has to be compensated by higher process gas velocities to provide the particles with adequate kinetic energy. Conventional HVOF techniques cannot completely solve this problem. Therefore, the use of modern technologies to process fine particles from submicron to the nanometric scale for novel macro-applications is becoming more and more important in the spraying technology (Ref 11). In the present study, the feeding and HVOF spraying of WC-Co submicron powders have been investigated to manufacture superfine structured, wear resistant, near-net-shape coatings with improved macroscopic properties. During the spray experiments, the influence of different HVOF gas compositions with varying kerosene, oxygen, and hydrogen levels on the process dynamics and the corresponding coating characteristics have been analyzed. The focus was placed on achieving dense coatings with high hardness and a smooth surface. The velocity and the temperature of the thermally sprayed particles were measured in-flight. A special powder feeder system and a novel HVOF device, designed by Thermico and optimized for processing finest powder fractions, were used (Ref 12).

2. Experimental

2.1 Powder Feedstock

88WC-12Co irregular and broken submicron powder (Thermico SJA 610/48, $-8+1$ μm), separated from a standard powder distribution by cyclone sieving, was employed as feedstock material. A SEM picture of the powder morphology is illustrated in Fig. 1. To minimize agglomeration effects, the fine powder was heated up to 120 °C in a furnace for 1 h before using it in the feeder system.

Fig. 1 88WC-12Co submicron powder used in this study

2.2 Powder Feeding and HVOF Equipment

Thermico CPF2-Twin powder feeder system equipped with special features and optimized for the reliable processing of fine powder fractions was used to feed the fine particles into the HVOF gun. A further description of this powder feeder system is given by Mathaeus et al. (Ref 12-14). To minimize the risk of particle agglomerations, both powder containers were heated up to 60 °C with heating sleeves and were agitated with a vibrator pressure of 6.2 bar. In addition, a cyclical evacuation of the powder containers and a subsequent hot powder gas flushing at 80 °C were performed prior to the feeding operation. Thermico CJS-HVOF gun with novel K5.2 combustion chamber design and dual-radial powder injection was used to conduct the coating experiments. This HVOF system is optimized for the fusibility and the acceleration of fine-scaled powder fractions <15 μm. It works with a two-stage hydrogen-oxygen and a subsequent liquid fuel (kerosene) oxygen combustion as described by Mathaeus et al. (Ref 12), which allows to control the velocity and the temperature of the process gas flow almost independently (Ref 6). The parameters for the powder feeding and the HVOF gun velocity as well as the standoff distance between the gun and substrate are summarized in Table 1.

2.3 Substrate Material and Preparation

Rectangular C45 steel specimens (1.0503) were employed as substrate material for the coating experiments. In preparation for thermal spray experiments, the surfaces were grit-blasted with fine-grained F100 alumina (106-150 μm) using a blasting air pressure of 3 bar. Afterward, the samples were cleaned for 15 min in an ethanol ultrasonic bath and heated to 110 °C in a furnace for about 30 min.

2.4 Coating Process

During the spraying experiments, the kerosene, oxygen, and hydrogen levels were varied in accordance with Fig. 2.

Whereas one gas/fuel level was varied in a set of experiments, the other ones were kept constant to analyze the influence on the process dynamics and the corresponding coating properties. The temperature and the velocity of the particles were measured in-flight at a standoff distance of 150 mm with Accuraspray-g3 device (Tecnar, Canada). Simultaneously, the chamber pressure and the combustion energy, transferred to the coolant, were detected by means of sensors, which are integrated in the HVOF system. During the coating experiments, the backsides of the specimens were constantly cooled with compressed air (2.7 bar). Characterization of the coatings in terms of hardness, porosity, and roughness were performed. In addition, the deposition efficiency was determined. The surface roughness was measured according to the DIN 4760 standard with a T-1000 tracing stylus instrument (Hommel, Germany). An M-400 microhardness tester (LECO, Germany) was employed to ascertain the coating hardness. Cross-section images of the coating and the interface between coating and substrate material were taken by AXIOPHOT light microscope (Zeiss, Germany). In addition, the layer thickness was determined. The porosity was analyzed with integrated image analysis software Axiovision 4.63 Outmess at a magnification level of 200, degressive contrast amplification (using gamma function), and medium illumination level. Cross-section and fracture section images of the coating with higher magnification were made with JXA 840 scanning electron microscope with secondary-electron and backscattered electron detectors (Jeol, Germany). The sliding wear behavior and the friction coefficient were evaluated using a pin-on-disc tribometer (CSEM, Switzerland) with alumina ball (Ø 6 mm, 1985 ± 193 HV 0.1), a load of 10 N, and a distance of 10 km at ambient temperature.

Table 1 Parameters

Parameter	Values	Comment
Number of powder containers	2	
Feeder disc velocity, rpm	3	Per powder container
Carrier gas flow, L/min	13	Per powder container
Vibrator pressure, bar	6.2	Per powder container
Heating sleeve temp., °C	60	Per powder container
Powders gas heating temp., °C	80	Per powder container
Stand-off distance, mm	150	
Gun velocity, mm/min	30,000	In all directions
Step in vertical (Y-)direction, mm	4	
Number of passes	25	

3. Results and Discussion

3.1 In-flight Particle Diagnostics

The in-flight particle measurements in this study showed that the chamber pressure and the thermal energy provided by the HVOF gas combustion allowed control of the temperature and velocity of the thermally sprayed submicron powders. However, the kerosene and oxygen levels showed a primary influence on the process dynamics. With rising kerosene level an almost linear increase in the thermal energy in the combustion chamber (32 kW at 6 L/h and 75 kW at 18 L/h), transferred to the coolant, is obtained (Fig. 3). However, kerosene flows between 6 and 14 L/h exhibited no significant change in the particle temperature around 1600 °C. It seems that the whole thermal energy in the HVOF flame is used for the particle melting. Higher kerosene levels (16-18 L/h) add more thermal energy to the process. A larger portion of the available fuel is burned within the combustion chamber

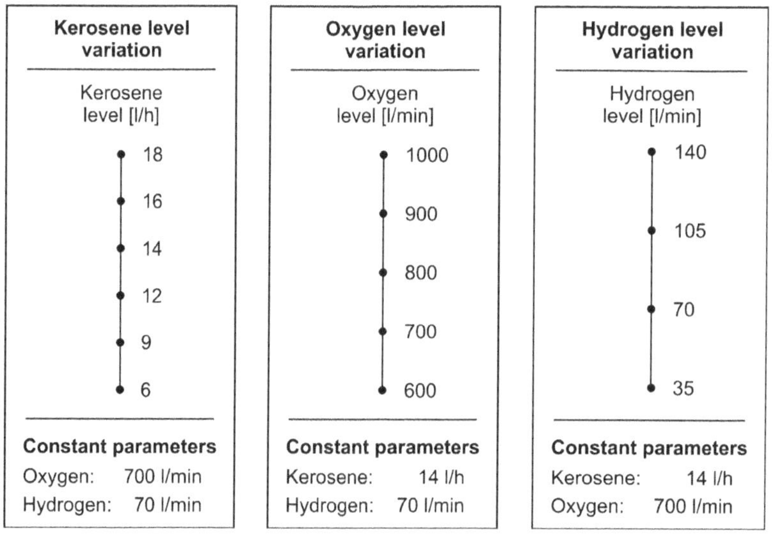

Fig. 2 Varied HVOF gas compositions

and the expansion nozzle. At the same time, a stronger postcombustion in the free jet is obtained, leading to higher particle jet temperatures at a maximum of 1900 °C. As shown in Fig. 3, the intermediate kerosene levels (12-14 L/h) showed highest values (747 ± 6 m/s) in terms of the particle velocity and should be favored to achieve a coating structure with high density. Whereas the oxygen level showed a significant influence on the particle velocity in Fig. 4, only a minor correlation to the particle temperature could be found. Rising oxygen levels from 600 L/min to 1000 L/min correlate with higher combustion chamber pressures from 8.8 bar (600 L/min oxygen) to 13.9 bar (1000 L/min oxygen). Subsequently, higher process gas velocities are achieved, leading to an increase in the particle velocity from 639 ± 3 m/s (600 L/min) to 931 ± 5 m/s (1000 L/min). This tendency has already been observed by Li et al. (Ref 15, 16). Due to higher process gas velocity the particles are supplied with more kinetic energy, which is transformed into higher impact forces on the substrate material. The high combustion chamber pressure also enables a particle adhesion of partially solidified particles, which impinge on the substrate material, as described by Kreye et al. (Ref 17). At the same time, a slight but continuous increase in temperature from 1580 ± 6 °C (at 600 L/min oxygen) to 1665 ± 2 °C (at 800 L/min oxygen), followed by a lower temperature increase to 1675 ± 3 °C (at 1000 L/min oxygen) have been observed with rising oxygen levels, as shown in Fig. 4. Due to corresponding higher chamber pressure and higher process gas velocities, the combustion is further transferred from the combustion chamber into the acceleration nozzle and the free jet.

Particles are exposed to a higher energy generated by the oxygen-kerosene combustion resulting in an increase in the particle temperatures. However, at highest oxygen levels (from 900 to 1000 L/min) the combustion reaction is supersaturated. Hence, the excess of oxygen leads to a slight cooling of the HVOF jet. In contrast to these results, the variation of the hydrogen level showed no measurable influence on the thermokinetic behavior of the particles but an observable effect on the HVOF flame stability. Within the two-stage hydrogen-oxygen and kerosene-oxygen combustion, low kerosene levels (<9 L/h) lead to an unstable and strongly pulsating flame. However, in some cases, a low kerosene level is required to supply the HVOF flame with lower thermal energy in order not to overheat sensitive feedstock powder materials. Consequently, the use of higher hydrogen levels is indispensable to stabilize the kerosene-oxygen combustion. In this case, the hydrogen level has a strong influence on the thermokinetic flame behavior to achieve a homogenous particle melting and a uniform coating deposition by avoiding flame pulsations. However, at higher kerosene levels the two-stage combustion is dominated by kerosene and oxygen. A nonpulsating flame is obtained, which does not have to be stabilized by the addition of higher hydrogen levels. From a scientific point of view, the particle velocity and particle temperature are the main parameters affecting the thermal spraying process. They determine the deposition efficiency of the HVOF process and strongly influence the coating microstructure development (Ref 18). Studies by Li et al. on in-flight phenomena in HVOF spraying of WC-Co particles to deposit cermet coatings (Ref 15, 16) have shown that the particle-melting degree and particle velocity also depend on the morphology and the grain size of the feedstock powder. Furthermore, it has been reported by Bobzin et al. that in

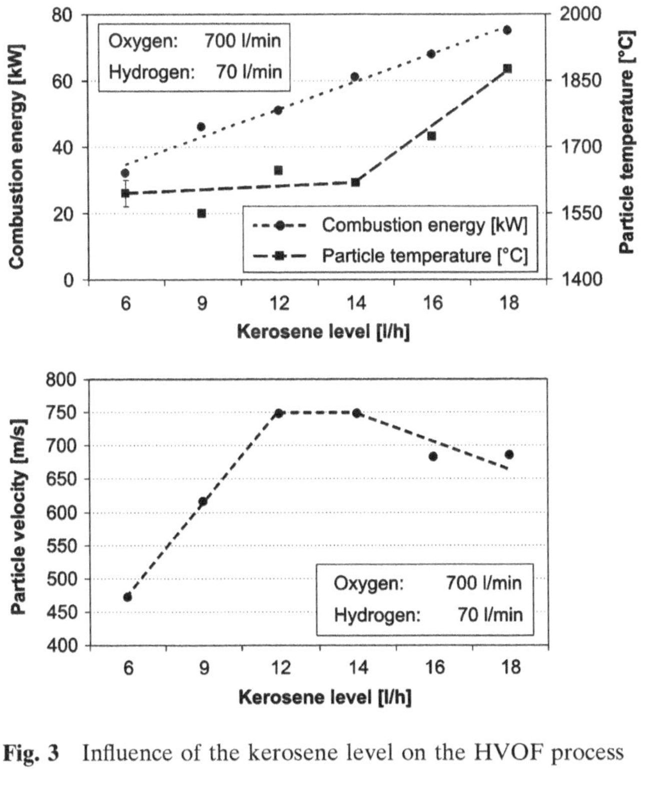

Fig. 3 Influence of the kerosene level on the HVOF process

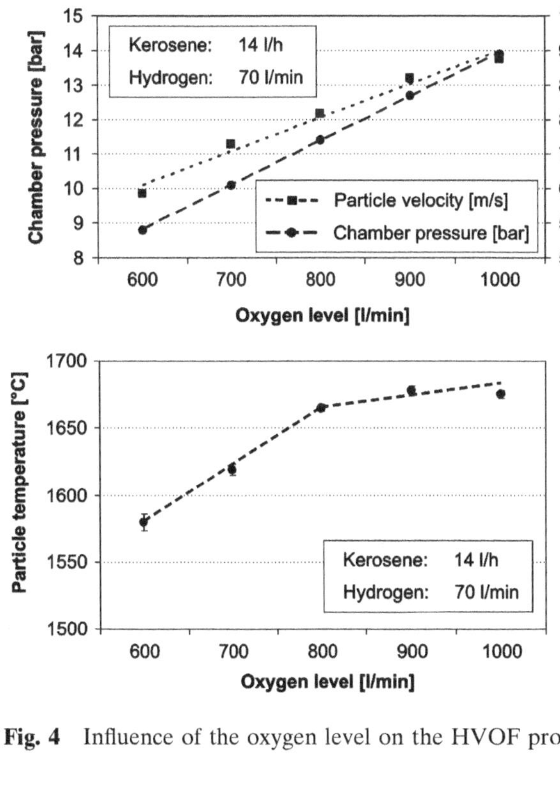

Fig. 4 Influence of the oxygen level on the HVOF process

contrast to conventional coarse-grained WC 10Co 4Cr powders (−30 + 16 μm) finer particles of the same composition (−15 + 1 μm) show continuously higher particle temperatures even at "colder" spray parameters in the HVOF process. They explained this effect by the higher surface energy of finer powders (Ref 6).

3.2 Deposition Efficiency

The deposition efficiency of WC-Co submicron powders sprayed at varying HVOF gas compositions has been determined. Figure 5 shows the results as a function of the kerosene and oxygen levels. The experiments indicate that both gas levels have a detectable influence on the deposition efficiency. Whereas highest values of deposition efficiency (68%) and a layer thickness of 432 ± 8.38 μm could be achieved with 25 passes of the gun at lowest oxygen level (600 L/min), a continuous decrease in deposition efficiency has been observed with increasing oxygen level. At 1000 L/min, the lowest value of deposition efficiency (34%) and a layer thickness of 244 ± 14.66 μm have been determined. Higher oxygen levels lead to faster particle acceleration and consequently to a shorter dwell time of the powder particles in the HVOF jet. This has an effect on the fusing of the particles. A lower number of particles are melted leading to a decrease in deposition efficiency. The kerosene level also showed an influence on the deposition efficiency. With the lowest kerosene level of 6 L/h, a coating deposition could not be achieved using submicron WC-Co powders. The particles are not supplied with enough thermal energy in the HVOF process and cannot be sufficiently transferred in a molten state. Consequently, the predominant solid particles bounce off as they hit the substrates surface, as reported by Li et al. (Ref 15, 16). As demonstrated in Fig. 5, the values of deposition efficiency rise from 48% (at 9 L/h kerosene flow) and a layer thickness of 308 ± 6.03 μm with increasing kerosene level to a maximum deposition efficiency of 57% (14 L/h) and a layer thickness of 385 ± 17.84 μm. However, at higher kerosene levels (14-18 L/h), the particles are exposed to higher thermal energy, which leads to a slight decline in deposition efficiency (54%) and to a layer thickness of 348 ± 10.72 μm. Investigations on varying hydrogen levels showed no significant influence on the deposition efficiency.

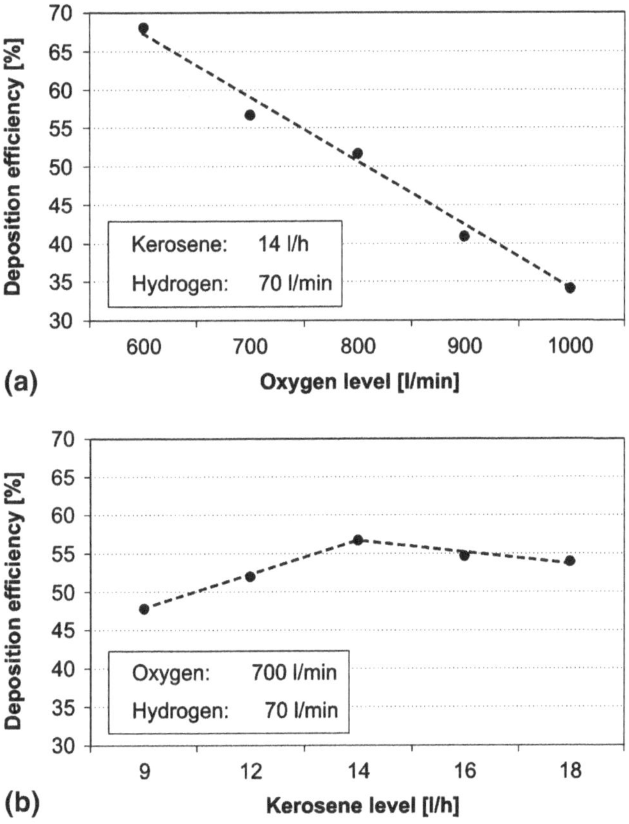

Fig. 5 Deposition efficiency depending on the oxygen (a) and the kerosene (b) levels

3.3 Hardness and Wear Resistance

It is generally known that hardness is an important parameter, which significantly affects the sliding and abrasive wear resistance of coatings. The deposited superfine structured coatings in this study showed Vickers hardnesses between 1027 ± 47 HV 0.1 and 1312 ± 45 HV 0.1. It has been found that variations of the HVOF gas composition have an influence on the coating hardness. Whereas lower hardness values but no clear trend could be determined between a level of 9 to 12 L/h kerosene, highest hardness values of 1244 ± 89 HV 0.1 could be measured on coatings deposited with medium kerosene (14 L/h) and medium hydrogen (70 L/min) flows (Fig. 6a). This effect may be attributed to the larger proportion of available fuel, which is burned in the combustion chamber, the acceleration nozzle, and the free jet using increasing levels of kerosene from 9 to 14 L/h. Consequently, higher thermal energy is applied in the HVOF process leading to better and more homogenous particle fusing. As reported by Li et al., a higher melting degree is considered to be desirable to obtain dense coatings. When the particle-melting ratio is low, many large particles may bounce off as they hit the substrate and the deposited coatings tend to have a high porosity (Ref 15, 16). However, higher levels of kerosene (>14 L/h) resulted in a slight reduction in hardness again to a value of 1163 ± 32 HV 0.1. This may be explained by the fact that further addition of thermal energy to the particles leads to decarburization and thermally activated carbide dissolution, as has been described in many studies. It has also been reported that fine powders show a higher thermal sensitivity in the HVOF process than their conventional coarse-grained counterparts. A reduction of the powder grain size corresponds with higher tendency toward decarburization effects. The higher surface area-to-volume ratio of fine particles leads to enhanced kinetics of the decarburization process (Ref 19).

As shown in Fig. 6(b), only a slight increase in the coating hardness could be observed with increasing oxygen level in the thermal spraying process. Whereas lowest hardness of 1182 ± 84 HV 0.1 has been found at 600 L/min oxygen flow, the highest value of 1312 ± 45 HV 0.1 has

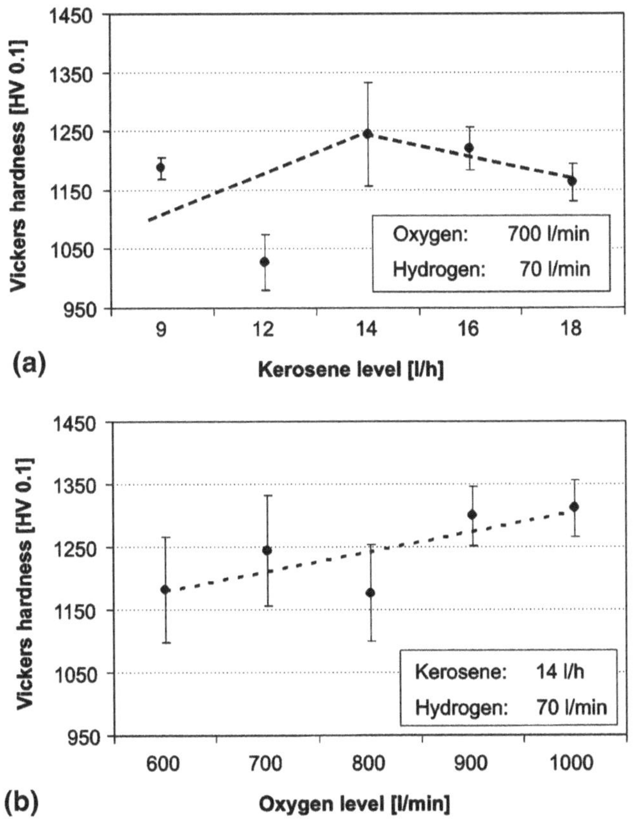

Fig. 6 Coating hardness depending on the kerosene (a) and the oxygen (b) levels

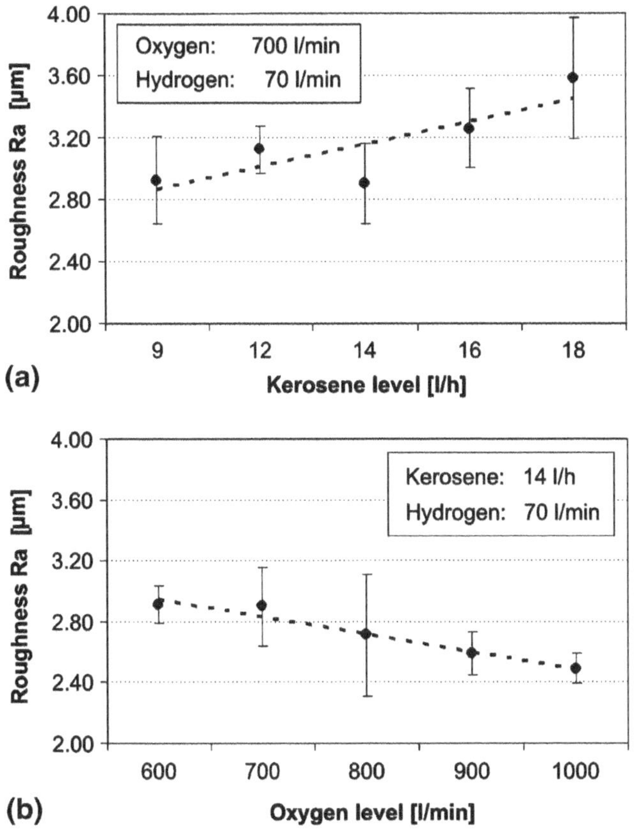

Fig. 7 Influence of the kerosene (a) and oxygen (b) levels on the surface roughness Ra of the as-sprayed coatings

been detected at the highest oxygen level of 1000 L/min. Higher oxygen levels provide the HVOF particle jet with more kinetic energy, which is transformed into higher impact force. This causes strong splat deformation and a subsequent densification. Therefore, a fine and dense coating structure could be achieved. Due to their outstanding combination of high resistance against abrasive, sliding, and erosive wear, high adhesive strength, and high impact stress, WC-Co cermet coatings are mainly used for industrial wear protective applications (Ref 20, 21). Studies on the wear behavior of conventional and nanostructured HVOF-sprayed coatings by Shipway et al. have reported achieving higher hardness with reduced powder grain size leading to enhanced sliding and abrasive wear characteristics of the coatings. They ascribed this effect to greater material strengths and reduced tendency for binder phase extrusions with decreasing grain size in the coatings (Ref 19). In this study, the superfine structured WC-Co coatings show comparable high resistance against sliding wear. The pin-on-disc test did not lead to a measurable weight or volume loss after a wear distance of 10 km and a load of 10 N. The friction coefficient was determined to be 0.74 ± 0.06.

3.4 Surface Roughness

The deposited superfine structured WC-Co coatings show a smooth surface as-sprayed with low roughness Ra values ranging from 2.49 ± 0.10 µm to 3.58 ± 0.39 µm. These results clearly indicate an improvement in surface roughness, which can be provided by thermal spraying of submicron powders. Comparative experiments on HVOF spraying of conventional coarse-grained agglomerated and sintered 88WC-12Co powders (Sulzer Metco WOKA 3102, $-45+15$ µm) with spherical form have been conducted. The cermet coatings, which have been deposited on C45 steel substrates, feature a higher surface roughness Ra of 4.21 ± 0.74 µm as-sprayed. In studies on HVOF spraying of conventional agglomerated and sintered 88WC-12Co powders with spherical morphology performed by Magnani et al. (Ref 22) and Fedrizzi et al. (Ref 23), similar results in terms of the surface roughness were achieved. Magnani et al. used Amperit 518.074 ($-45+15$ µm) powders and Sulzer Metco DJH 2700 gun in their research work on the tribocorrosion behavior of HVOF cermet coatings. The deposited coatings showed an average roughness Ra of 5.50 µm (Ref 22). Fedrizzi et al. investigated the influence of HVOF parameters on the corrosion and wear resistance of cermet coatings deposited onto C45 steel using a Tafa/Praxair JP-5000 HVOF gun and powder grain sizes in the range of 20 to 40 µm. Thereby, an as-sprayed coating roughness Ra between 3.7 and 4.9 was determined (Ref 23).

During thermal spraying of WC-Co submicron powders, it has been found that the combustion and gas dynamics, mechanical and thermal in-flight behavior of the powder particles, and droplet flattening have

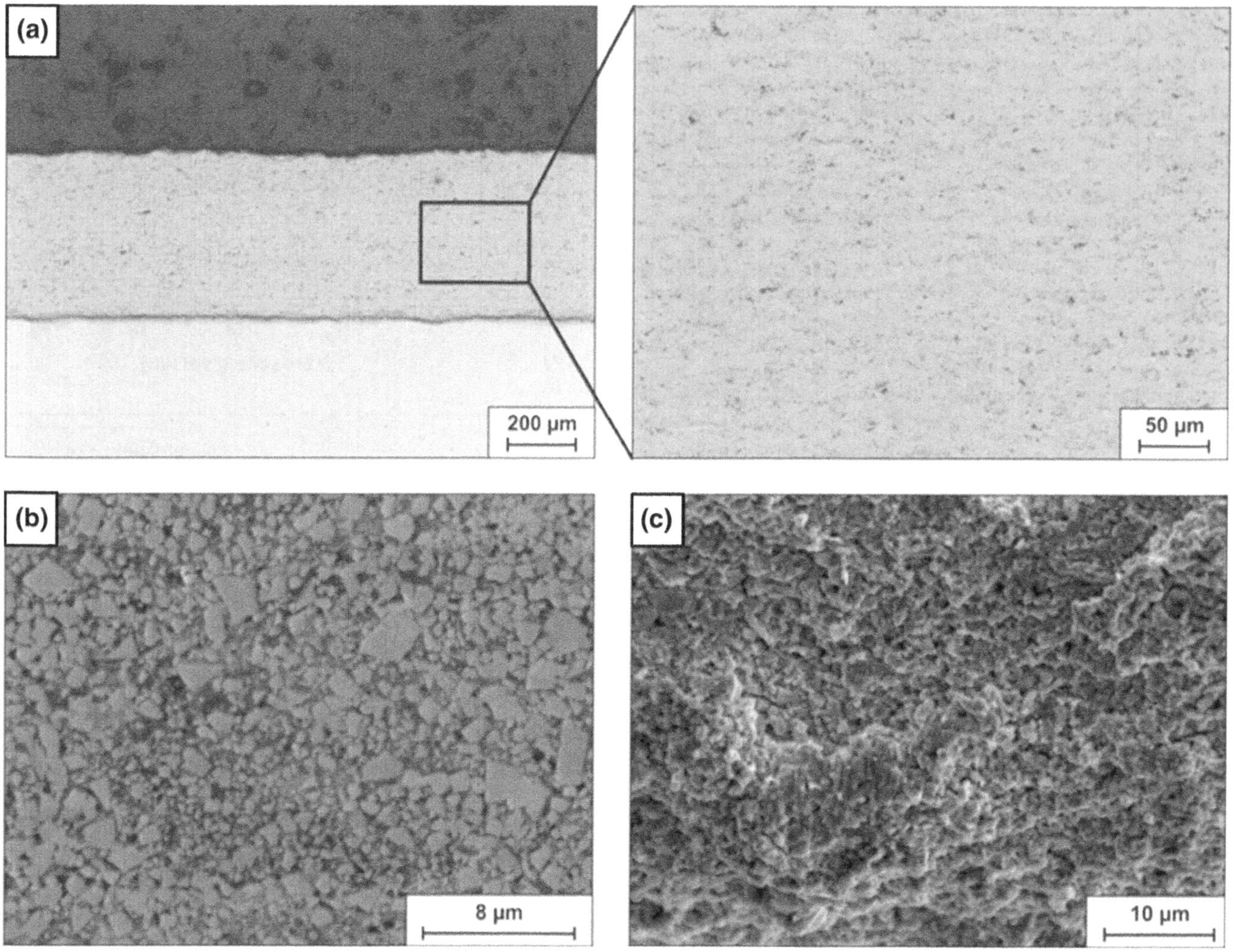

Fig. 8 Polished cross-section images of the coating morphology taken by light microscope (a) and SEM (b) and SEM image of the fracture cross section (c)

considerable influence on the coating quality, as already reported by Sobolev et al. (Ref 24). Whereas rising kerosene levels lead to an increase in surface roughness Ra from 2.93 ± 0.28 μm (at 9 L/h kerosene) to 3.58 ± 0.39 μm (at 18 L/h kerosene), shown in Fig. 7(a), an improvement in roughness can be obtained with higher oxygen levels. The roughness Ra decreases from 2.91 ± 0.12 μm at 600 L/min oxygen to 2.49 ± 0.10 μm at 1000 L/min (Fig. 7b). Correlated to the in-flight particle measurements, the reduction of the surface roughness can be explained by increasing particle velocity at higher oxygen levels. Consequently, it can be concluded that higher oxygen but lower kerosene levels have to be favored for the HVOF spraying of submicron WC-Co powders to obtain coatings with smooth surfaces.

As reported in literature, the application of fine powders leads to an increase in layer homogeneity attributed to a better fusibility of finescaled materials and a finer dispersed phase distribution in the coating structure (Ref 25). This is typically correlated with a decrease in surface roughness (Ref 7, 8). The splat densification level and the corresponding roughness of the as-sprayed coatings are mainly affected by the particle velocity and the molten state of the sprayed particles. The high velocity of the particles provided by the HVOF process, in combination with a better and more homogenous molten state when using finescaled WC-Co powders, helps to deform or flatten the particle upon impact. Consequently, a very fine-grained and dense coating structure composed of many densified splats is formed. In comparison to HVOF flame spraying of conventional coarse-grained WC-Co powders with a particle velocity of 450 m/s (Ref 17), the novel HVOF system, used in this study, provides fine particles with velocities to a maximum of 931 ± 5 m/s at 1000 L/min oxygen by special K5.2 combustion chamber and nozzle design. The higher velocity may be converted into additional heat aiding the flattening of the sprayed particles (Ref 26, 27). The surface roughness can be regarded as an important indicator of the quality of thermally sprayed near-net-shape coatings. Zimmermann et al. have already reported on lower surface roughness and enhanced wear resistance of the as-sprayed coatings using agglomerated and sintered WC-Co 83/17 powders with finer WC hard materials particles (0.8-1 μm) in the HVOF spraying process (Ref 7). In addition, Landa et al. found that smooth coatings could provide a significant reduction in friction (Ref 10). This is very important for applications, where the surface quality of the coatings plays a significant role, such as calendar rolls and corrugator rolls for the paper industry (Ref 7). It can also be expected that smooth surfaces may reduce material adhesion and galling on tools in the deep-drawing process (Ref 10).

3.5 Coating Morphology

Polished cross-section images of the as-sprayed coatings taken by light microscope and a fracture cross-section image analyzed by SEM are shown in Fig. 8(a) and (c), respectively. The coatings feature an extremely dense, superfine structure and homogenous morphology with porosities <1%. Only very small-sized micropores on a submicron and nanosized level are detectable. The figures also show excellent interface bonding between the coating and the roughened C45 steel surface. It seems that substrate preparation using fine-grained alumina as blasting medium (106-150 μm) leads to an appropriate roughness profile for the subsequent coating operation in terms of the coating adhesion. Figure 8(b) shows a SEM image of the coating morphology with high magnification. It can be seen that the finescaled, spattered WC particle structure is preserved in the coating morphology, well embedded in a homogenous molten Co binder phase. The morphology consists in a primarily submicron structure with partially nanosized WC phases. It has also been found that medium kerosene levels (14 L/h), medium hydrogen levels (70-105 L/min), and oxygen flows ≥700 L/min have a positive influence on the coating morphology in terms of lower porosity and finer and more homogenous phase development. This is attributed to a more homogenously molten state of the particles and an improved coating adhesion on the C45 steel substrate material. The results in these experiments are consistent with studies on HVOF spraying of nanocrystalline WC-18wt.%Co powders, which have been carried out by Ban and Shaw (Ref 28). They have also reported that low process temperatures and short dwell time of the powder particles passing the HVOF flame jet mainly lead to the preservation of the powder structure within the coating morphology.

4. Conclusion and Outlook

Feeding of submicron 88WC-12Co powders (−8 + 1 μm) has been performed with Thermico CPF2 powder feeder system. A novel CJS-HVOF gun, designed by Thermico, has been used to deposit superfine structured cermet coatings. In-flight diagnostics on varying HVOF gas compositions showed a significant influence on the process dynamics. Different HVOF combustion gas compositions enable to adjust the particle velocity and temperature. In this way the deposition efficiency, the coating structure, and their macroscopic properties can be affected. Coatings with smooth surfaces (Ra = 2.49 ± 0.10 μm), extremely dense structures (porosities <1%), and high hardness of 1312 ± 45 HV 0.1 were achieved. A high process efficiency of 68% could also be realized. The deposited superfine structured coatings in this study show a high potential to be used as wear resistant, near-net-shape coatings for deep-drawing tools without any posttreatment or surface finish. It is expected that further improvements will be achieved by reducing the feedstock powder to nanoscale grain size.

Peer Reviewed

Acknowledgments

The authors gratefully acknowledge the financial support of the DFG (German Science Foundation) within the Collaborative Research Centre SFB 708.

References

1. T.N. Rhys-Jones, Thermally Sprayed Coating Systems for Surface Protection, Clearance Control Applications in Aero Engines, *Surf. Coat. Technol.*, 1990, **43-44**, p 402-415
2. R.V. Hillery, Coatings for Performance Retention (in Military Aircraft Engines), *J. Vacuum Sci. Technol.*, 1986, **4A**(6), p 2624-2628
3. J.K.N. Murthy and B. Venkataraman, Abrasive Wear Behaviour of WC-CoCr, Cr(ind 3)C(ind 2)-20(NiCr) Deposited by HVOF and Detonation Spray Processes, *Surf. Coat. Technol.*, 2006, **8**(200), p 2642-2652
4. R. Nieminen, P. Vuoristo, K. Niemi, T. Mäntylä, and G. Barbezat, Rolling Contact Fatigue Characteristics of Thermally Sprayed WC + Co Coatings, *TS96 Thermal Spraying Conference*, Mar 6-8 1996 (Germany), DVS, 1996, p 354-359
5. P. Chivavibul, M. Watanabe, S. Kuroda, and K. Shinoda, Effects of Carbide Size, Co Content on the Microstructure and Mechanical Properties of HVOF-Sprayed WC-Co Coatings, *Surf. Coat. Technol.*, 2007, **3**(202), p 509-521
6. K. Bobzin, F. Ernst, J. Zwick, and G. Matthäus, Analysis of In-Flight Particle Properties of Thermal Sprayed Ultrafine Powders, *Mat. Sci. Eng. Technol.*, 2007, **2**(38), p 149-154 (in German)
7. S. Zimmermann, H. Keller, and G. Schwier, Improved Coating Properties by Optimized Carbide Powders for Modern HVOF Systems, *6th HVOF Spraying Colloquium*, Nov 27-28 2003 (Germany), 2003, p 31-38
8. K. Jia and T.E. Fisher, Abrasion Resistance of Nanostructured, Conventional Cemented Carbides, *Wear*, 1996, **1-2**(200), p 206-214
9. K. Jia and T.E. Fisher, Sliding Wear of Conventional, Nanostructured Cemented Carbides, 11th International Conference on Wear of Mat, *Wear*, 1997, **203-204**, p 310-318
10. J. Landa, I. Illarramendi, J.M. Montalban, A. Igartua, and G. Mendoza, Application of Nano-HVOF to Piston Rings, *11th Nordic Symposium on Tribology*, Jun 1-5 2004 (Norway), p 115-127
11. M. Ignatiev and E. Kovalev, Nanoparticles Processing for Fabrication of Multi-functional Nanostructured Coatings, *Plastic Metal Forming*, 2006, **4**(17), p 37-41
12. G. Matthaeus, M. Kostecki, and O. Dau, The Fully Automatic, Computer Controlled C-CJS (Computerised Carbide Jet System) HVOF System with 25 bar Combustion-Chamber pressure by Thermico, *5th Colloquium on HVOF Flame Spraying*, Nov 16-17 2000 (Germany), GTS e.V., 2000, p 147-158
13. G. Matthaeus and A. Sturgeon, Application of the HVOF Process to Internal Coatings of Cylinders, and Hard-and Software for the Internal Coating of Complex Components, *7th HVOF Spraying Colloquium*, Nov 9-10 2006 (Germany), GTS e.V., 2006, p 199-205

14. G. Matthaeus and G. Stevens, New HVOF Equipment and Technology for the Use of Superfine Powders—10 μm and Internal Coating Applications, *6th HVOF Spraying Colloquium*, Nov 27-28 2003 (Germany), GTS e.V., 2003, p 167-176
15. M.H. Li, D. Shi, and P.D. Christofides, Modeling, Control of HVOF Thermal Spray Processing of WC-Co Coatings, *Powder Technol.*, 2005, **2-3**(156), p 177-194
16. M.H. Li and P.D. Christofides, Multi-scale Modeling, Analysis of an Industrial HVOF Thermal Spray Process, *Chem. Eng. Sci.*, 2005, **13**(60), p 3649-3669
17. H. Kreye, F. Gärtner, and H.J. Richter, High Velocity Oxy-fuel Flame Spraying-state of the Art, New Developments and Alternatives, *6th HVOF Spraying Colloquium*, Nov 27-28, 2003 (Germany), GTS e.V., 2003, p 5-17
18. E. Turunen, "Diagnostic tools for HVOF process optimization," Ph.D. Thesis, Helsinki University of Technology, 2005
19. P.H. Shipway, D.G. McCartney, and T. Sudaprasert, Sliding Wear Behavior of Conventional, Nanostructured HVOF Sprayed WC-Co Coatings, *Wear*, 2005, **7-12**(259), p 820-827
20. A. Wank, A. Schwenk, B. Wielage, T. Grund, E. Friesen, and H. Pokhmurska, Untersuchungen zur Beständigkeit thermisch gespritzter Schichten bei durch Abrasion dominierter tribologischer Beanspruchung (Investigation on the Resistance of Thermally Sprayed Coatings Under Abrasion Dominated Tribologic Stress), Sep 07-08, 2006 (Germany), Werkstoffe und werkstofftechnische Anwendungen **24** (2006), p 301-306 (in German)
21. M.L. Lau, H.G. Jiang, W. Nuchter, and E.J. Lavernia, Thermal Spraying of Nanocrystalline Ni Coatings, *Phys. Stat. Sol.*, 1998, **166**, p 257-268
22. M. Magnani, P.H. Suegama, N. Espallargas, S. Dosta, C.S. Fugivara, J.M. Guilemany, and A.V. Benedetti, Influence of HVOF Parameters on the Corrosion, Wear Resistance of WC-Co Coatings Sprayed on AA7050 T7, *Surf. Coat. Technol.*, 2008, **202**, p 4746-4757
23. L. Fedrizzi, L. Valentinelli, S. Rossi, and S. Segna, Tribocorrosion Behaviour of HVOF Cermet Coatings, *Corros. Sci.*, 2007, **49**, p 2781-2799
24. V.V. Sobolev and J.M. Guilemany, Dynamic Processes During High Velocity Oxyfuel Spraying, *Int. Mater. Rev.*, 1996, **1**(41), p 13-32
25. O.P. Solonenko, *Advanced Thermophysical Fundamentals of Melt-Droplet-Substrate Interaction and its Application in Thermal Spraying*, Novosibirrsk, Russia, 2003
26. H. Du, W. Hua, and J. Liu, Influence of Process Variables on the Qualities of Detonation Gun Sprayed WC-Co Coatings, *Mater. Sci. Eng.*, 2005, **1-2**(408A), p 202-210
27. R.C. Eucker, *Technologies for Films and Coatings, Developments and Applications*, Noyes Publications, 1982
28. Z.G. Ban and L.L. Shaw, Characterisation of Thermal Sprayed Nanostructured WC-Co Coatings Derived from Nanocrystalline WC-18wt.%Co Powders, *J. Therm. Spray Technol.*, 2003, **1**(12), p 112-119

JTTEE5 17:933–941
DOI: 10.1007/s11666-008-9263-4
1059-9630/$19.00 © ASM International

Peer Reviewed

Innovation of Ultrafine Structured Alloy Coatings Having Superior Mechanical Properties and High Temperature Corrosion Resistance

X.Q. Ma, D.W. Gandy, and G.J. Frederick

(Submitted May 30, 2008; in revised form September 19, 2008)

High temperature protection requires full coating density, high adhesion, minor oxide inclusions, and preferably fine grains, which is not achievable in most thermal spray processes. High velocity oxygen fuel (HVOF) thermal spray process has been applied extensively for making such coatings with the highest density and adhesion strength, but the existence of not melted or partially melted particles are usually observed in the HVOF coatings because of relatively low flame temperature and short particle resident time in the process. This work has investigated the development of an innovative HVOF process using a liquid state suspension/slurry containing small alloy powders. The advantages of using small particles in a HVOF process include uniform coating, less defective microstructure, higher cohesion and adhesion, full density, lower internal stress, and higher deposition efficiency. Process investigations have proven the benefits of making alloy coatings with full density and high bond strength attributing to increased melting of the small particles and the very high kinetic energy of particles striking on the substrate. High temperature oxidation and hot corrosion tests at 800 °C have demonstrated that the alloy coatings made by novel LS-HVOF process have superior properties to conventional counterpart coatings in terms of oxidation rates and corrosion penetration depths.

Keywords boiler parts, corrosion of HVOF coatings, high temperature oxidation, HVOF spray parameters, HVOF microstructures, nanostructured materials, scratch or impact loading testing

1. Introduction

Of late, research and development of innovative thermal spray processes using liquid state feedstock, i.e., liquid precursors or suspensions, have attracted more attention and interest. It has been demonstrated that the new processes are very promising for making coatings with unique microstructures and superior properties (Ref 1-3).

So far, superfine-grained ceramic coatings have been manufactured mostly using the liquid-based plasma spray processes. However, it is challenging to form a metallic coating by utilizing the same techniques. The challenges include at least: (1) to avoid severe oxidation of superfine metallic powders during spray; (2) to keep stable suspension of high density metallic powders; and (3) to determine spray window parameters for metallic powders typically having a wide range of size distribution. In a previous study, it was demonstrated that the technical obstacles could be overcome by hybridizing of HVOF spray technique and liquid suspension method, so-called LS-HVOF or SPS-HVOF for the first time (Ref 4).

Of the various thermal spray techniques available, high velocity oxygen fuel (HVOF) process gives the highest coating bond strength to substrate and the highest coating density, and thus has been proven to be more preferable for making the metallic coatings, as well as nanostructured ceramic coatings (Ref 5, 6). With innovative LS-HVOF process, further improvement in these HVOF merits can be achieved. This work will provide extensive investigation on the microstructures, mechanical properties, and especially corrosion performance of LS-HVOF-formed alloy coatings.

2. Experimental

2.1 Materials

An AISI 316L stainless steel was used as substrate in this study. The specimens have the dimensions of

This article is an invited paper selected from presentations at the 2008 International Thermal Spray Conference and has been expanded from the original presentation. It is simultaneously published in *Thermal Spray Crossing Borders, Proceedings of the 2008 International Thermal Spray Conference*, Maastricht, The Netherlands, June 2-4, 2008, Basil R. Marple, Margaret M. Hyland, Yuk-Chiu Lau, Chang-Jiu Li, Rogerio S. Lima, and Ghislain Montavon, Ed., ASM International, Materials Park, OH, 2008.

X.Q. Ma, Inframat Corporation, Farmington, CT; and **D.W. Gandy** and **G.J. Frederick**, Electric Power Research Institute, Charlotte, NC. Contact e-mail: xma@inframat.com.

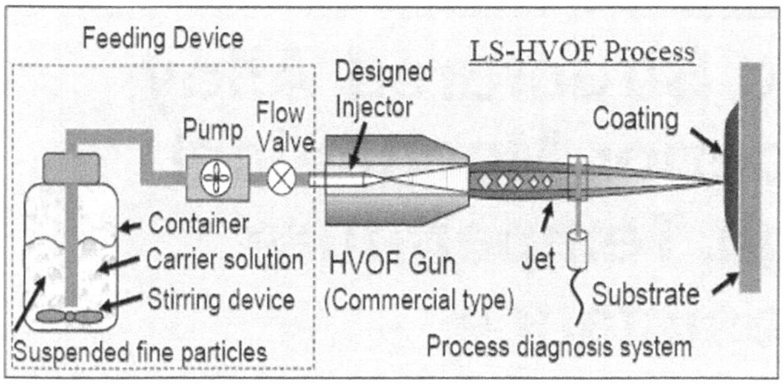

Fig. 1 Schematic of LS-HVOF process setup

100 × 100 × 2.5 mm^3 for process trials, 60 × 20 × 2.5 mm^3 for mechanical property tests, and 10 × 10 × 2.5 mm^3 for oxidation and hot corrosion tests.

Two powders were purchased from a commercial vendor (Praxair, US). The powders have the identical nominal composition (wt.%) of Ni-14.55Cr-3.22B-4.7Si-4.78Fe, but different size distribution of +16-42 μm for conventional HVOF use and of −11 μm for innovative LS-HVOF use, respectively. Reagent grade chemicals were used to form liquid state suspensions (Ref 4).

2.2 LS-HVOF Setup

A Diamond Jet 2700 hybrid high velocity oxygen fuel (HVOF) spray system (Sulzer-Metco, NY, USA) equipped with a HVOF gun, a turntable, and a six-axis robot was used to spray the powder and suspension feedstock. Propylene was used as fuel, and routine surface preparation procedures were applied prior to the HVOF spraying. The spray parameters were optimized based on microstructural analyses and process diagnostic methods (Ref 4). In LP-HVOF process as shown in Fig. 1, the fine alloy powder was mixed well with the liquid solvent and was kept well suspended in a container. Then, the precursor was pumped to a liquid port and coaxially fed into the HVOF gun via a modified liquid injector. The flow of liquid precursor was controlled at a typical flow rate of 60-150 mL/min.

2.3 Test Methods

2.3.1 Microstructure Evaluation. Coating microstructures were evaluated according to a metallographic procedure, mainly on the cross sections using optical microscopy and electronic scanning microscopy (SEM). Coating grain structures were observed using transmission electronic microscopy (TEM). Coating porosity, unmelted particles, and coating/substrate interfaces were examined in these coatings. In addition, the oxygen contents in the coatings were analyzed using an oxygen-nitrogen analyzer (Leco TC-436) according to ASTM E1019-00.

2.3.2 Physical Property Tests. Microhardness (HVN) was measured on the cross sections of the coatings using a microhardness tester. Nanohardness and modulus was tested using a nano-indentation tester.

2.3.3 Mechanical Tests. Bending tests were carried out to determine coating cohesive strength/cracking resistance and flexibility/ductility in accordance with the similar methods to ASTM B489-85 and ASTM D522. 90°-bending tests were conducted under both tensile and compressive loading modes. In 180° U-shape bending tests, a mandrel of 1″ in diameter was used. The thickness of the coatings used in the bending tests is 150-180 μm.

Progressive load scratch test (PLST) was used to determine the critical loads corresponding to major coating damage failure and coating adhesion properties. This test corresponded to a load ramp applied to the indenter during defined displacement of the sample beneath it. In the test, 0.2 mm tip radius Rockwell diamond indenter tip was drawn across the coating with a loading rate of 300 N/min, start load of 0.9 N, and end load of 150 N. Using optical microscopy examination at 200× magnification, the critical loads (LC1, LC2, LC3, and LC4) were assessed.

2.3.4 Oxidation Tests. Short-term oxidation testing was used to determine the oxidation kinetics of the coatings in the early stages of oxide formation via TGA/DSC analyses. The rapid formation of protective oxides and low parabolic constant K_p represents better oxidation resistance. Long-term oxidation testing for 500 hours was conducted in a box furnace at 800 °C. Samples were put in a ceramic crucible, and weight changes were recorded every 50 h. The weight gain was plotted with time for each sample, and a parabolic constant was obtained by curve fitting method.

2.3.5 Hot Corrosion. Hot corrosion test was conducted in Na_2SO_4 + 20wt.%NaCl mixture salt at 800 °C. The salt was precoated evenly over the entire coating surfaces at a weight of 2-3 mg/cm^2. The corrosion resistance of the coatings was measured by weight change and corrosion depth.

3. Results and Discussion

3.1 Microstructures

The typical microstructures and fractured morphologies of the coatings made by HVOF and LS-HVOF processes are shown in Fig. 2. In general, both coatings are well adherent to the substrates, quite dense, and crack-free. The LS-HVOF coating shows better microstructures with higher density and uniformity relative to conventional HVOF coating. The HVOF coating contains a high percentage of unmelted phase (as the arrows indicate in Fig. 2a and c) and at least 30 vol.%, and the average splat size is about 40 μm. In contrast, the LS-HVOF coating contains much less unmelts, and it is difficult to identify individual splats (Fig. 2b). As a result, the LS-HVOF coating is nearly full dense and is more uniform.

Oxygen content results are given in Table 1. The starting powders (with both fine and normal sizes) of NiCrBSi alloy contain about 0.06 wt.% oxygen. After HVOF processing, oxygen contents in the coatings were increased due to the "hot" nature of HVOF process and spraying in air. However, the oxygen content of 0.3% in the LS-HVOF coating is lower than that of 1% in the HVOF coating. It is believed that the use of organic

Fig. 2 Cross-sectional views of HVOF coating (a, c) and LS-HVOF (b, d). *Top row*: optical micrographs. *Bottom row*: SEM fractured morphologies

Table 1 Results of oxygen content analysis

Coatings	Oxygen contain, wt.%
Powder 1 (normal cut)	0.06
HVOF coating	1.02
LS-HVOF coating	0.287

solvent and additive has provided a protective environment to reduce oxygen partial pressure in the process.

The grain structure and oxide phase formation have been investigated using TEM analyses. From Fig. 3, it is verified that HVOF coating consists of dominantly large grains (>1 μm). It is assumed that the large grains correspond to those unmelted or partially melted particles. The higher magnification TEM indicates the presence of nanosized grains within the melted splats (Fig. 3b). Some of the dark phases are identified as oxides by diffraction analyses. The potential oxides include Cr_2O_3, SiO_2, and/or (Ni,Cr,Si)xOy, with average grain sizes of 50-100 nm. As to the LS-HVOF coating shown in Fig. 3(c) and (d), the coating consists of dominant nanosized grains (<100 nm) and some ultrafine grains (<0.2 μm), but no large grains (>1 μm). It is believed that the formation of the nanograins and ultrafine grains is associated with the melting of the particles and the rapid quenching of the melted droplets on the substrate. The oxides were surely formed during HVOF process, but oxide contents were not measured by TEM analyses. By comparing the grain difference in the two coatings, it is further thought that a higher degree of melting and a higher quenching rate can be easily achieved for small size droplets than for large size droplets. Consequently, a high percentage of nanograins are obtained in the LS-HVOF coating. The results indicate that nanograined coatings can be cost-effectively produced from small particles without having to use expensive nanoparticles.

3.2 Physical Property

Vickers microhardness HVN (load: 300 g) was measured. The hardness value is about 530 ± 41 for HVOF-formed coating and about 620 ± 32 for LS-HVOF-formed coating. The increase in hardness in LS-HVOF coating should be related to the increased cohesive strength among the well-melted and bonded splats as well as the increase in coating density.

To confirm the above result of increased hardness in the nanograined LS-HVOF coating compared to the coarse-grained HVOF coating, nanohardness was measured on both the coatings using a nano-indentation technique, and exemplary results for the LS-HVOF coating are shown in Fig. 4. The modulus and hardness were estimated from the load-displacement curves in Fig. 4(a), and their changes as a function of displacement are given in Fig. 4(b) and (c), respectively. For the results of modulus, LS-HVOF coating is about 131 GPa, much higher than HVOF coating of about 85 GPa. For the results of

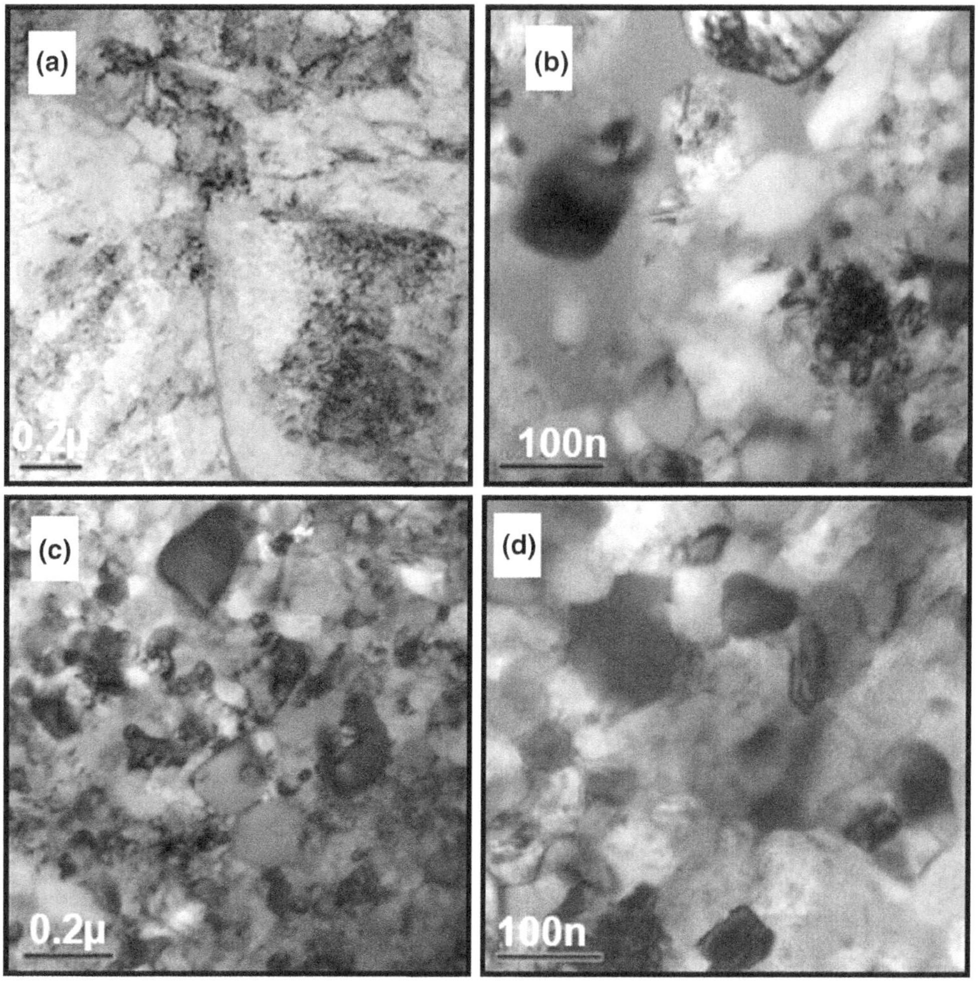

Fig. 3 TEM analyses on the grain sizes and oxide phases in the coatings of (a, b) HVOF coating and (c, d) LS-HVOF coating

nanohardness, LS-HVOF coating is about 7.2 GPa, much higher than HVOF coating of about 5.6 GPa. Therefore, the LS-HVOF coating has higher modulus and hardness than its HVOF counterpart. The results of nanohardness are consistent with those of microhardness. As to the comparison between nano- and microhardness, nanohardness value is about 31% higher than microhardness for the LS-HVOF coating and 8% higher for the HVOF coating. Other studies have shown similar results: nanoindentation hardness is about 10-30% larger than the microhardness for copper, stainless steel, and nickel titanium (Ref 7). It is assumed that the increases in modulus and hardness are attributed to the formation of high percentage of nanograins in the LS-HVOF coating.

3.3 Mechanical Behavior

For coatings in engineering applications, their mechanical behavior under loading and/or strain is of vital importance. As indicated in Fig. 5, the coatings made by HVOF and LS-HVOF processes have been tested under tensile and compressive strain in 90°- and 180°-bend tests, respectively. The results of coating inspection after the tests show obvious difference in coating damages among the coating samples.

3.3.1 Type 1: 90°-tensile Bend Test. HVOF NiCrBSi coating (A1) has one big crack starting on the bending tip (smallest radius), and propagating laterally along the coating/substrate interface to cause a delamination. LS-HVOF NiCrBSi coating (B1) has no major damage, except a few cracks in the edge area.

3.3.2 Type 2: 90°-compressive Bend Test. Under a compressive strain, no visible damage is observed in either HVOF (A2) or LS-HVOF NiCrBSi coating (B2), except a few cracks.

3.3.3 Type 3: 180°-tensile Bend Test. In general, the coating samples suffered less damage in 180°-bend tests than in 90°-bend tests, and again the damages induced in LS-HVOF coating is less severe than those in HVOF coating. HVOF coating (A3) cracked evenly, but was still attached to the substrate. LS-HVOF coating (B3) indicated good integrity without visible cracking or spalling. In summary, the results indicate that the LS-HVOF sprayed coating is more flexible/ductile and strongly adherent under a tensile or compressive strain. The improved strain compliance in the LS-HVOF coating should be attributed

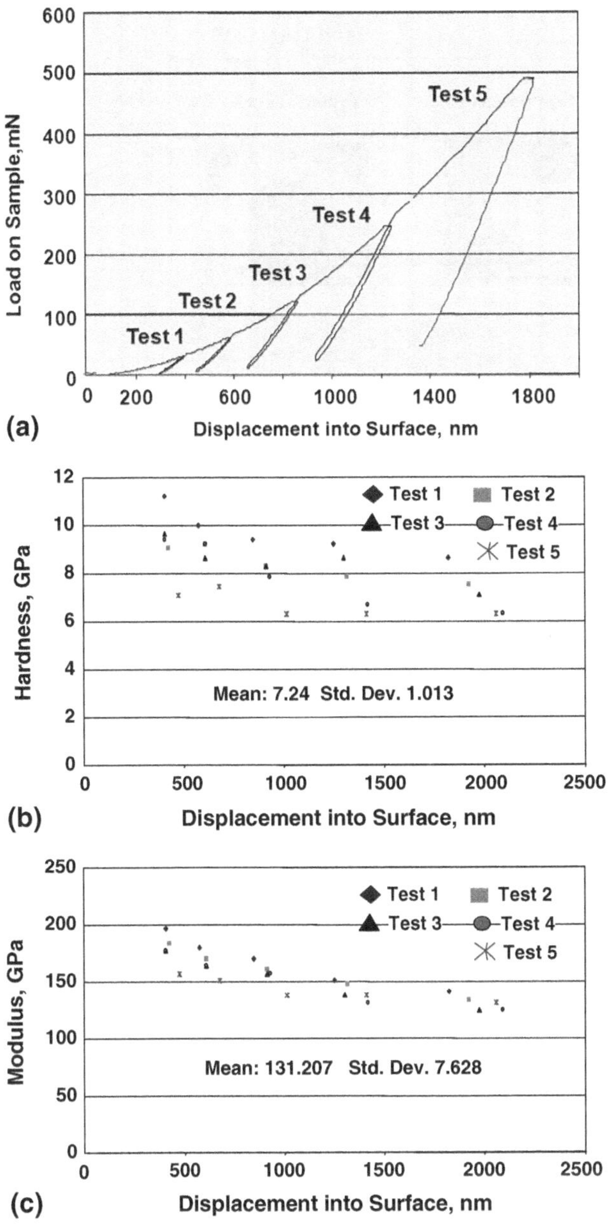

Fig. 4 Nanoindentation test in (a) LS-HVOF NiCrBSi coating and (b, c) the results of estimated hardness and modulus in the coating, respectively

to an increase in coating ductility, adhesion, cohesion, and a high degree of melting of the small particle feedstock in the LS-HVOF process.

Progressive load scratch test was conducted to test coating ductility and adhesion. A typical load-depth curve is illustrated for a HVOF coating in Fig. 6. It is noticed that there are two major coating cracking events observed under optical microscope during the progress of loading. The two critical failure points correspond to the first cracking within the coating starting at about 84 N and the second cracking at the coating/substrate interface at about 100 N. The scratched areas formed on the surface of the tested coatings are indicated as white streaks in Fig. 6(a) and (b). The Fig. 6(a) shows the semicircular cracks induced at the second critical load and the indenter has nearly reached the substrate. The Fig. 6(b) shows the case of no cracking in LS-HVOF coating, indicating a smooth transition from the coating to the substrate at 90 N. No cracking suggests a high ductility and a strong adhesion of the coating with its substrate. Although more data need to be collected for statistic analysis, the tests have indicated that this PLST method provides a powerful tool to determine coating mechanical properties in a quantitative manner.

3.4 Oxidation Behavior

The early-stage oxidation kinetics of the coatings was measured by TGA analyses at 800 °C in air. In Fig. 7(a), the curves of oxidation kinetics indicate that both coatings obey nearly the parabolic law for weight gaining, but LS-HVOF NiCrBSi coating gained weight earlier and faster than its HVOF counterpart. After the early stage of rapid growth (150-200 min), the oxidation of LS-HVOF coating reaches a stable stage at lower rates for thermally grown oxide (TGO) growth. Alloy materials can form a protective oxide scale based on selective oxidation of elements such as Al, Cr, and Si. The rapidly formed oxide scale will provide better protection of the base alloy. Many studies have confirmed that ultrafine/nanograined alloys or coatings can promote selective oxidation of active elements and thus enhance oxidation resistance (Ref 8, 9). Therefore, the TGA test verified that LS-HVOF coating with nanometer grains is advantageous to provide superior oxidation resistance by promoting early and fast formation of protective oxide scale. At this moment, however, there is no data to support the formation of pure oxide scale such as α-alumina.

Long-term oxidation tests were conducted at 800 °C for 500 h in air. The weight gains seem to obey the parabolic law, and the parabolic constants K_p were estimated for the coatings tested by curve fitting the weight gain-time curves shown in Fig. 7(b). The K_p values are 16.9×10^{-7} mg^2/mm^4 hr for the HVOF coating and 0.2×10^{-7} mg^2/mm^4 hr for the LS-HVOF coating, respectively. Obviously, the LS-HVOF coating exhibits much lower long-term TGO rate than its HVOF counterpart. Based on the microstructures observed on the two coatings, it can be concluded that the microstructural features such as density grain size and defect inclusions (unmelts, etc.) play a determinable role on the oxidation behavior of the coatings having identical chemical composition. The lower K_p value with LS-HVOF coating should be greatly attributed to the formation of well-melted finer grained and bonded splat structure in the coating as well as improved coating density and uniformity.

3.5 Hot Corrosion Test

Accelerated hot corrosion tests were performed by prepainting of the coating samples with a mixture of $Na_2SO_4 + 20\%NaCl$ salts, and then tested at 800 °C in air. Figure 8 shows the kinetic behavior of the coatings tested in the salt mixture. Firstly, all the coatings keep losing weight during the test period, indicating the coatings were chemically attacked without forming complete protective

Peer Reviewed

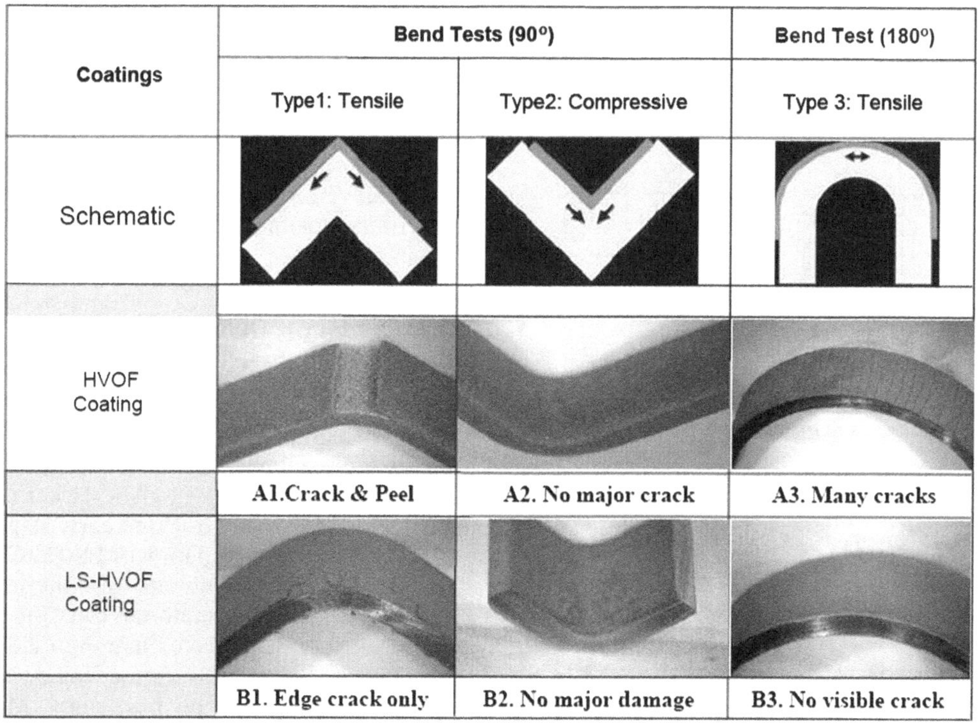

Fig. 5 Macrographs of the coatings after tested in 90°- and 180°-bending tests

oxide scales in the salt mixture. Secondly, LS-HVOF coating seems to be more resistant to this corrosion environment with much less weight loss relative to its HVOF coating counterpart.

The tested coatings were observed on the cross-sectional microstructures, which are shown in Fig. 9. In the HVOF coating (9a), corrosion took place severely in near surface zone and penetrated deeply down to the coating/substrate interface. In the LS-HVOF NiCrBSi coating (9b), the coating was corroded only in the near surface zone, without corrosion penetration deeply into the interface of coating and substrate. Therefore, if the corrosion resistance is measured by the depth of corrosion, undoubtedly, the LS-HVOF coating is more resistant to the hot corrosion than the HVOF coating counterpart. XRD analyses were used to determine the composition of the corroded surface layers. The main phases identified on the corroded surface include SiO_2, Cr_2O_3, Fe_2O_3, and minor FeS or Fe_7S_8.

Many investigations have revealed that the presence of sulfur in the form of sulfates can cause internal sulfidation of alloys beneath the external oxide layer. The addition of chlorides can cause the formation of volatile species, which form voids and pits at grain boundaries, thus forming an easy path for flow of corrodents (Ref 10). The studies on HVOF-sprayed Ni-Cr-W-Mo-B coatings (NiCrWMoB, NiCr, NiCrMo, etc.) found that the corrosion resistance of the HVOF alloy coatings can be improved by heat treatment due to increased microstructural and chemical homogeneity, such as the reduction of porosity and densification (Ref 11). The studies on laser remelting of HVOF coatings reported that the treated coatings suffered much less corrosion damage, whereas as-sprayed coating was penetrated by corrosive species. Laser remelting efficiently removed the interconnected network of voids and oxides at splat boundaries of the HVOF coatings (Ref 12). Therefore, it can be concluded that the corrosion resistance of the LS-HVOF coating has been improved by the formation of well-bonded splats at their boundaries and less porosity structure due to well-melting of the small particles used in LS-HVOF spraying. As a result, the less defective and homogenous coating structure provides fewer paths for the penetration of corrodents. With the presence of the nanograins in the LS-HVOF coating, additionally, enhanced ability for selective oxidation should also contribute to the improvement, at least prolonging the transition time before hot corrosion process moves from an initiation stage to a propagation stage.

It is worthy to point out that this work has been originally performed based on the US DOE specially funded program titled as "Surface Modification of Alloys for Ultra-supercritical Coal-Fired Boilers (USC)." This program aims to seek innovative surface modification methods for a reliable and durable USC boiler system. The extensive tests above have been conducted to evaluate coating performance in simulated USC operating conditions, typically with steam temperatures up to 760 °C and pressures up to 37.9 MPa. Besides the boiler application, the process is applicable to make superior metallic coatings required in other advanced industry applications, such as a bondcoat in thermal barrier coating. The process is capable of producing a fine-structured or nanostructured coating using small-sized (a few microns and submicrons) or nanosized particles in a liquid suspension. As to coating

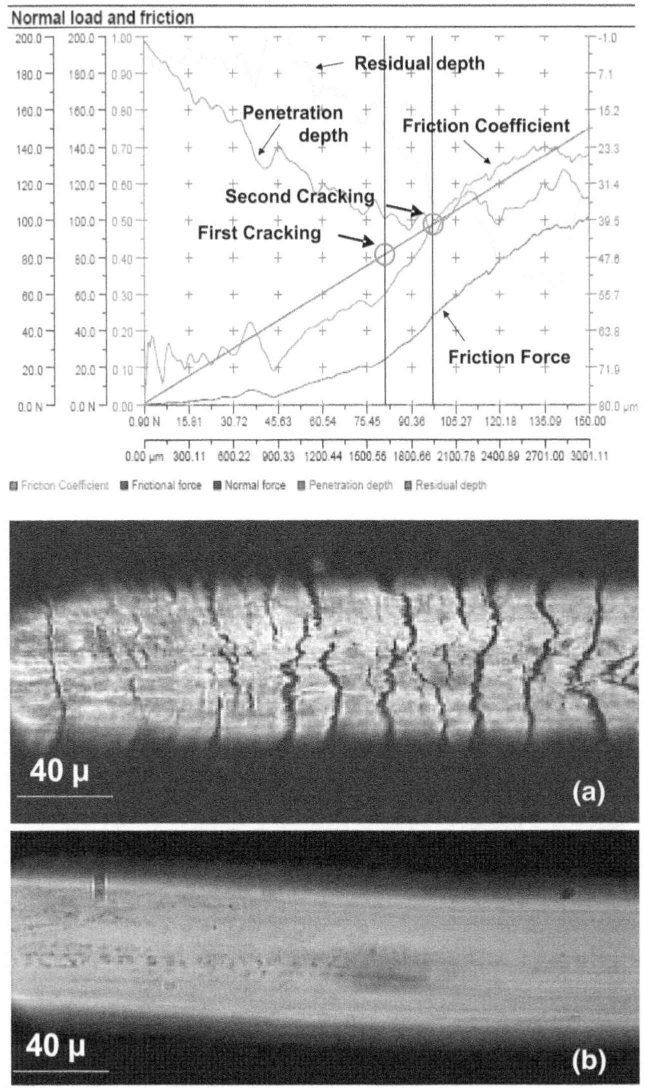

Fig. 6 A typical load-depth curve for HVOF coating tested in progressive load scratch test. Micrographs of (a) tested HVOF coating and (b) tested LS-HVOF coating

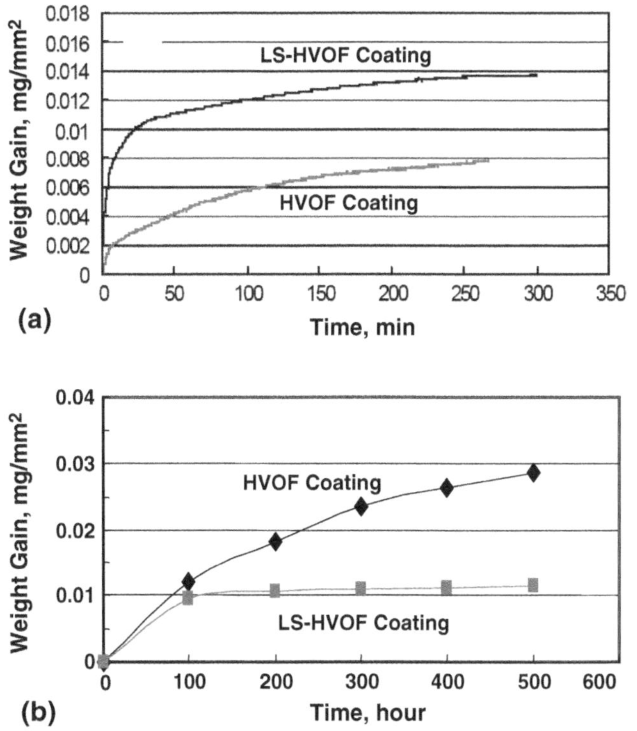

Fig. 7 Curves of isothermal oxidation kinetics. (a) Short-term TGA tests at 800 °C in air and (b) Long-term furnace tests at 800 °C in air

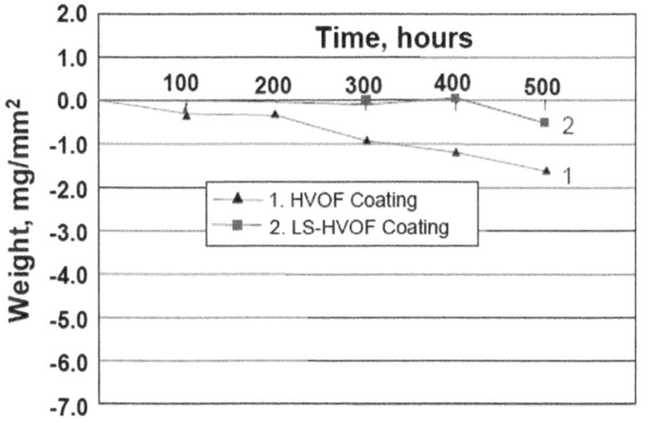

Fig. 8 Kinetic behaviors of hot corrosion for the coatings tested in the mixture of Na_2SO_4 + 20wt.%NaCl salts at 800 °C in air

materials, so far more than five types of alloy coatings including Inconel 625 have been produced by LS-HVOF spray.

4. Conclusions

An innovative LS-HVOF process has been developed to realize ultrafine-grained alloy coating deposition by utilizing liquid suspension containing small metallic particles. Experimental investigations have been conducted to compare physical and mechanical properties, oxidation resistance, and hot corrosion behaviors of the coatings produced by conventional HVOF and innovative LS-HVOF processes. Results have shown that LS-HVOF NiCrBSi coating has superior properties and performance in comparison with its HVOF counterpart: (i) Microstructure evaluation verified that a nanograined microstructure is produced in the LS-HVOF coating with high percentage of melted splats, low defects, and good uniformity; (ii) physical property results indicated that the nanostructured coating has a relative higher hardness and modulus; (iii) mechanical property tests clarified that LS-HVOF coatings have higher ductility and cohesive and adhesive strength and maintain good coating integrity with higher strain resistance when tested under a tensile or compressive load; (iv) oxidation tests indicated that nanostructured LS-HVOF coating can form a protective oxide scale rapidly in the early stage of oxidation and has a lower TGO growth rate for long-term exposure at 800 °C;

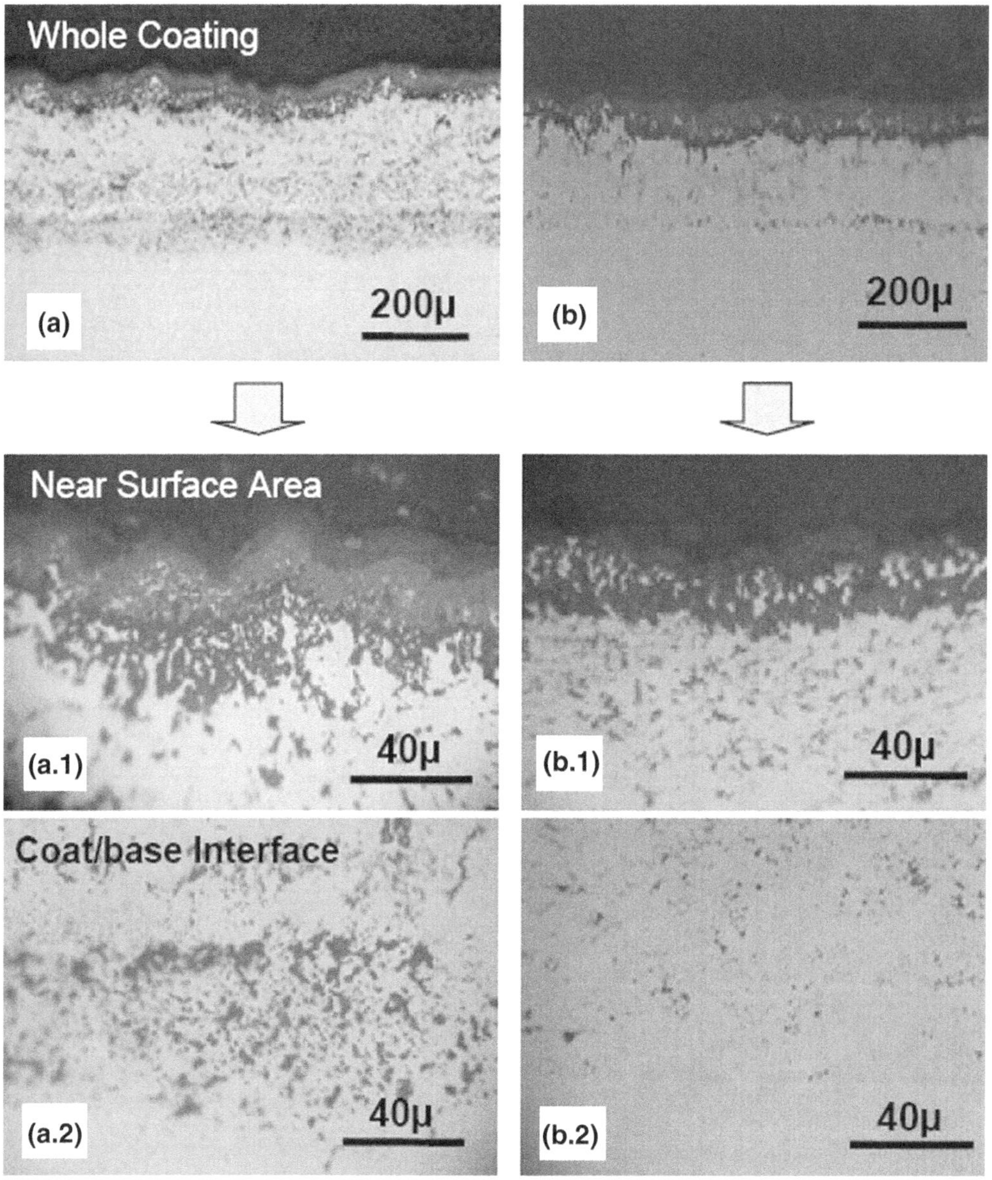

Fig. 9 Microstructures of the coatings tested in the mixture of Na_2SO_4 + 20wt.%NaCl salts at 800 °C for 500 h. (a) HVOF coating and (b) LS-HVOF coating

and (v) hot corrosion tests revealed that the LS-HVOF coating is more resistant to hot corrosion in Na_2SO_4 + 20 wt.%NaCl at 800 °C than its HVOF counterpart. The main contributors to these improvements include higher degree of melting and higher velocity of the small particles. Therefore, LS-HVOF technique is very promising for making superior coatings in many advanced industry applications such as ultrasupercritical boiler components.

Acknowledgments

The authors thank Mr. Jeff Roth at Inframat Corporation for his assistance in conducting HVOF spray trials. This work was supported by US Department of Energy (Grant No. DE-FG02-06ER-84604) managed by Mr. U. Rao and Ms. P. A. Rawls.

References

1. X. Ma, F. Wu, J. Roth, M. Gell, and E.H. Jordan, Low Thermal Conductivity Thermal Barrier Coating Deposited by the Solution Plasma Spray Process, *Surf. Coat. Technol.*, 2006, **201**(7), p 4447-4452
2. M. Gell, E.H. Jordan, M. Teicholz, B.M. Cetegen, N. Padture, L. Xie, D. Chen, X. Ma, and J. Roth, Thermal Barrier Coatings Made by the Solution Precursor Plasma Spray Process, *J. Therm. Spray Technol.*, 2008, **17**(1), p 1-12
3. P. Fauchais, Understanding Plasma Spraying, *J. Phys. D: Appl. Phys.*, 2004, **27**, p R86-R108
4. X. Ma, J. Roth, D.W. Gandy, and G.J. Frederick, A New High-velocity Oxygen Fuel Process for Making Finely Structured and

Highly Bonded Inconel Alloy Layers from Liquid Feedstock, *J. Therm. Spray Technol.*, 2006, **15**, p 670-675
5. M. Gaona, R.S. Lima, and B.R. Marple, Nanostructured Titania/Hydroxyapatite Composite Coatings Deposited by High Velocity Oxy-fuel (HVOF) Spraying, *Mater. Sci. Eng. A*, 2007, **458**, p 141-149
6. A. Killinger, M. Kuhn, and R. Gadow, High-velocity Suspension Flame Spraying (HVSFS), a New Approach for Spraying Nanoparticles with Hypersonic Speed, *Surf. Coat. Technol.*, 2006, **201**, p 1922-1929
7. L.M. Qian, et al., Comparison of Nano-indentation Hardness to Microhardness, *Surf. Coat. Technol.*, 2005, **195**, p 264-271
8. F. Liu, W. Gao, K. Dahm, and F. Wang, The Effect of Coating Grain Size on the Selective Oxidation Behavior of Ni-Cr-Al Alloy, *Scripta Mater.*, 1997, **37**(10), p 1551-1558
9. G. Chen and H. Lou, Effect of Nanocrystallization on the Oxidation Behavior of a Ni-8Cr-3.5Al Alloy, *Oxid. Met.*, 2000, **54**, p 155-162
10. D. Deb, S.R. Lyer, and V.M. Radhakrishnan, A Comparative Study of Oxidation and Hot Corrosion of a Cast Nickel Base Superalloy in Different Corrosive Environments, *Mater. Lett.*, 1996, **29**, p 19-23
11. C.H. Lee and K.O. Min, Effects of Heat Treatment on the Microstructure and Properties of HVOFsprayed Ni-Cr-W-Mo-B Alloy Coatings, *Surf. Coat. Technol.*, 2000, **132**, p 49-57
12. M.A. Uusitalo, P.M.J. Vuoristo, and T.A. Mantula, High Temperature Corrosion of Coatings and Boiler Steels in Reducing Chlorine-containing Atmosphere, *Surf. Coat. Technol.*, 2002, **161**, p 275-285

JTTEE5 17:942–947
DOI: 10.1007/s11666-008-9260-7

Peer Reviewed

Microstructure and Oxidation Behavior of Cr39Ni7C Cermet Coatings Deposited by Diamond Jet Spray Process

F.-X. Ye, S.-H. Wu, and A. Ohmori

(*Submitted May 15, 2008; in revised form September 14, 2008*)

For the promising erosion and oxidation resistance of carbide cermet coating, Cr39Ni7C cermet coatings were deposited by Diamond Jet spray process using a commercial Sulzer Metco 5241 powder in this study. The microstructure, phase compositions, and high-temperature oxidation behavior of the deposited coatings were investigated. The speed and temperature of in-flight particles were measured by DPV-2000. The results revealed that the speed of in-flight particles decreased from 229 to 150 m/s with the increasing of the spraying distance from 100 to 300 mm, whereas the average temperature of in-flight particles increased from 1926 to 2245 K. The decarburization of Cr_3C_2 increased with the increasing of the fuel gas flow from 58 SLPM to 77 SLPM for higher heat enthalpy of the flame. Due to the formation of Cr_2O_3 on/in the coatings at high temperature, the sprayed coatings had good oxidation resistance at 1073 K in an air atmosphere. The oxidation behavior was found to be governed by the parabolic rate law. For the lower porosity of the coating deposited under the spraying distance of 150 mm, its oxidation resistance was better than that of the coating sprayed at the spraying distance of 200 mm. A slight reduction of sprayed Cr39Ni7C cermet coating in micro hardness occurred after high temperature at 723 K and 1073 K high-temperature oxidation.

Keywords cermet coatings, HVOF microstructures, influence of properties

1. Introduction

Thermal spraying is an expanding area within the technology of surface engineering. It is a process that involves the deposition of molten or semi-molten droplets of powder on to a substrate to form a coating (Ref 1). Tungsten carbide and chromium carbide-based coatings are frequently used in various industrial fields such as the power generation industry and aerospace industry to improve the resistance to sliding, abrasive and erosive wear, and high-temperature oxidation (Ref 2-12). Although the wear resistance is lower compared with WC-Co cermet's, the Cr_3C_2-NiCr system exhibits a high oxidation resistance up to over 850 °C, while the WC-Co system will be severely oxidized at the temperature over 500 °C (Ref 13-16). Therefore, the Cr_3C_2-NiCr alloy cermet coatings have been widely used for high-temperature wear resistance and corrosion resistance applications. WC and Cr_3C_2 with different metallic binders like Co, Ni, and Fe have been studied using different amounts of binder contents with Co and Ni most commonly used. Addition of Cr to the matrix has been found to improve the wear and oxidation resistance of these cermets (Ref 17).

Oxygen and a fuel gas in high velocity oxy-fuel (HVOF) spraying flow at high pressures and flow-rates with internal combustion to produce very high particle velocities with relatively low temperatures compared with other thermal spray processes such as air or vacuum plasma spraying. As a result, HVOF is capable of producing dense coatings with low degrees of decomposition, which are well bonded to the substrate (Ref 1). It has been reported that carbide containing coatings deposited by high velocity processes have good wear resistance compared to plasma-sprayed coatings due to the better coating properties achievable using high velocity processes (Ref 17).

In this study, Cr39Ni7C powder that has much more NiCr compared with commonly applied Cr_3C_2-25NiCr powder was used as the feedstock powder for high-temperature applications up to 900 °C. The temperature and velocity of the in-flight particles were measured by DPV-2000, and the microstructure and high-temperature oxidation behavior of the coatings deposited by the Diamond Jet method were characterized and discussed.

This article is an invited paper selected from presentations at the 2008 International Thermal Spray Conference and has been expanded from the original presentation. It is simultaneously published in *Thermal Spray Crossing Borders, Proceedings of the 2008 International Thermal Spray Conference*, Maastricht, The Netherlands, June 2-4, 2008, Basil R. Marple, Margaret M. Hyland, Yuk-Chiu Lau, Chang-Jiu Li, Rogerio S. Lima, and Ghislain Montavon, Ed., ASM International, Materials Park, OH, 2008.

F.-X. Ye, School of Materials Science and Engineering, Tianjin University, Weijin Road No.92, Tianjin 300072, People's Republic of China; **S.-H. Wu**, Suzhou Nuclear Power Research Institute, Suzhou, People's Republic of China; and **A. Ohmori**, TOCALO Co. Ltd., Minamifutami, Futami-Cho, Akashi, Japan. Contact e-mail: yefx@tju.edu.cn.

2. Materials and Experimental Procedure

2.1 Feedstock Powders and Substrate

The feedstock powder was Cr39Ni7C (available from Sulzer Metco 5241 powder). This consisted of fine particles that are obtained by melting a material having a composition of Cr:Ni:C = 54:39:7 (% by mass) and rapidly solidifying the melt, with Cr and C forming chromium carbide and Ni and Cr forming a Ni-Cr alloy by melting and rapid solidification. Sulzer Metco 5241 has a structure in which crystallized chromium carbide particles are dispersed in a Ni-Cr alloy (Ref 18, 19). The X-ray diffraction pattern of Cr39Ni7C powder is shown in Fig. 1. The size distribution of the Cr39Ni7C powder was −53 + 15 μm and its morphology is given in Fig. 2, The substrate for coating preparation was mild steel (JIS SS400) and its thickness was 5.7 mm. Prior to the coating, the mild steel substrates were cleaned with acetone and grit blasted using Al_2O_3 sand with average size of 830 μm.

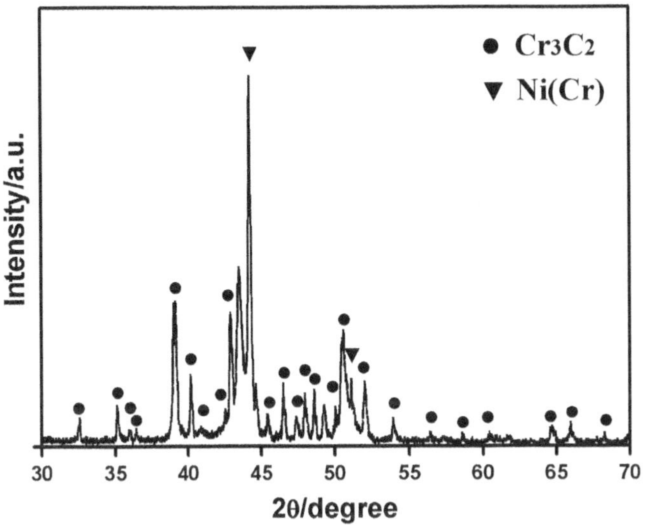

Fig. 1 X-ray diffraction pattern of the Cr39Ni7C feedstock powder

Fig. 2 Typical morphology of Cr39Ni7C feedstock powder

2.2 Thermal Spraying Equipment

The thermal spraying equipment was a commercial HVOF system (Diamond Jet 2600 manufactured by Sulzer Metco). The thermal spraying parameters are given in Table 1.

2.3 Temperature and Velocity of In-flight Particle

The temperature and velocity of the in-flight particles were measured by DPV-2000 equipment (Tecnar Automation Ltd., Canada). DPV-2000 applies infrared pyrometry along with a two-slit photo-mask in order to perform in-flight diagnostics on individual particles. A particle passing in front of the photo-mask will generate a two-peaks signal. The peaks distance and intensity are used to calculate the temperature and velocity of particle.

2.4 Analysis of Microstructure and Compositions

Scanning electron microscope (SEM) and energy dispersive X-ray analysis (EDAX) were used to examine the structure characteristics of the feedstock powder and the sprayed coatings. The phase compositions of the feedstock powder and the sprayed coatings were investigated by X-ray diffraction using Cu-Kα radiation ($\lambda = 1.5406$ Å) and a graphite crystal monochromator (JDX3530, JEOL, Japan). Vickers hardness of the sprayed coatings was measured using an automatic hardness testing machine (AAV-500, Akashi Group, Japan) with a 100 gf load for 15 s. An average hardness was calculated from five indents per specimen. The porosity of the coating was calculated using image software according to the coating's cross section of SEM image.

2.5 High-Temperature Oxidation of the Sprayed Coatings

Isothermal oxidation behavior of the sprayed coating was conducted at 1073 K up to 100 h in a programmable furnace. The coated sample size was 15 × 15 × 6.0 mm for oxidation experiment. The mass gain was measured by an analytical balance with a sensitivity of 0.1 mg.

Table 1 Diamond Jet 2600 spraying parameters

Propylene fuel gas pressure, MPa	0.69
Propylene fuel gas flow, SLPM	58, 67, 77
Oxygen gas pressure, MPa	1.03
Oxygen gas flow, SLPM	316
Air gas pressure, MPa	0.52
Air gas flow, SLPM	207
Argon carrier gas pressure, MPa	0.86
Argon carrier gas flow, SLPM	14
Spraying distance, mm	100-300

The "Peer Reviewed" label runs vertically in the left margin.

3. Results and Discussion

3.1 Temperature and Velocity of the In-flight Particles

Although HVOF processes minimize decomposition of the carbide phase due to lower heat enthalpy and shorter duration involved in the coating processes, the highest temperature of the in-flight particle is also a main factor to evaluate the decarburization of Cr_3C_2 in the thermal spraying of Cr_3C_2-NiCr materials. In addition, higher in-flight particle velocity during deposition provides several advantages such as lower porosity, higher bond strength and hardness. The temperature of the in-flight particle increased with the increasing of spraying distance from 150 to 300 mm, while the velocity decreased from 230 to 150 m/s (Fig. 3). It is reported that the decomposition temperature of Cr_3C_2 is 2168 K (Ref 20, 21). The average temperature of the in-flight particle did not exceed 2168 K with spraying distance shorter than 250 mm.

3.2 Microstructure and Compositions of As-Sprayed Coatings

Figure 4 shows the cross sectional microstructure of Cr39Ni7C coating sprayed with the fuel gas flow of 67 SLPM and spraying distance of 225 mm. According to Fig. 4a, the Cr_3C_2 particles were much finer than that in the general thermal-sprayed Cr_3C_2-25%NiCr coating with average particle size of 2-4 micrometers (Ref 13, 17). The as-sprayed coating had a dense structure with a homogenous distribution of fine-grained carbide. Using SEM-EDAX and XRD analysis, the dark area was composed of a finely dispersed chromium carbide phase (dark) surrounded by layers of metal matrix (lighter). Generally, the spraying distance has been observed to have a great influence on the coating porosity. This conclusion was also observed in this study. With the increasing of spraying distance from 150 to 200 mm, the porosity of the coating increased from 0.4 to 4.6% at the fuel gas flow of 77 SLPM.

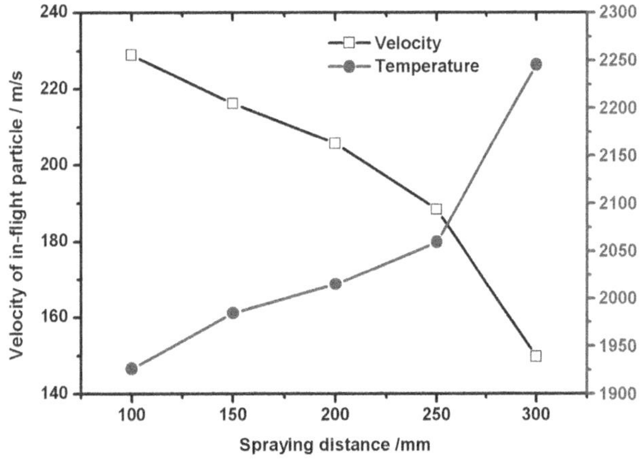

Fig. 3 The dependence of temperature and velocity of in-flight Cr39Ni7C particles with the spraying distance in Diamond Jet spraying

The X-ray diffraction patterns of the Diamond Jet-sprayed Cr39Ni7C coatings under the different fuel gas flow are illustrated in Fig 5. Dissolution of carbide into the matrix and the consequent super saturation of the matrix with Cr and C gave rise to a large amorphous background on the low angle side of the NiCr peak in the XRD scan. The only detectable crystalline phases were Cr_3C_2, NiCr, and Cr_2O_3. While small amounts of Cr_7C_3 may have formed through decarburization of Cr_3C_2, its concentration was too low to be distinguished from other overlapping peaks. A small amount of Cr_2O_3 was formed during thermal spraying. With the increasing of fuel gas flow from 58 SLPM to 77 SLPM, the intensity of Cr_3C_2 peak decreased for the higher heat enthalpy of flame. The Cr_3C_2 decomposed more with higher fuel gas flow in the thermal spraying processes.

3.3 Isothermal Oxidation Behavior of Sprayed Coatings

In all cases, a black tarnish layer had formed on the coatings surface after high-temperature oxidation exposure. The X-ray diffraction pattern of the sprayed coating

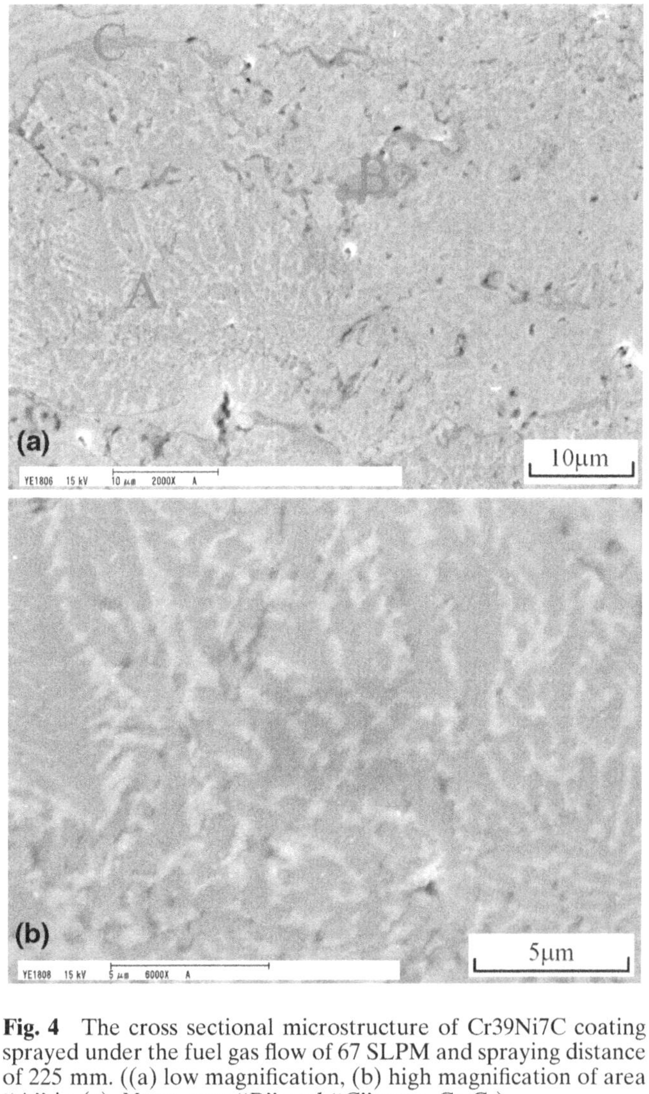

Fig. 4 The cross sectional microstructure of Cr39Ni7C coating sprayed under the fuel gas flow of 67 SLPM and spraying distance of 225 mm. ((a) low magnification, (b) high magnification of area "A" in (a), Note: gray "B" and "C" were Cr_3C_2)

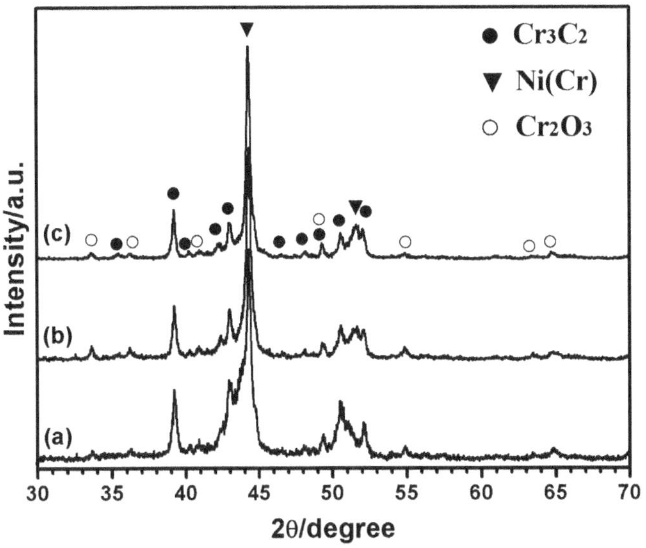

Fig. 5 The X-ray diffraction patterns of Cr39Ni7C coatings sprayed under different fuel gas flow and spraying distance of 225 mm ((a) Fuel gas flow 58 SLPM, (b) Fuel gas flow 67 SLPM, (c) Fuel gas flow 77 SLPM)

after the high-temperature experiment is shown in Fig. 6. It indicated that the patterns of the coatings after oxidation were dominated by narrow crystalline peaks of Cr_3C_2, NiCr, and Cr_2O_3. A large amount of Cr_2O_3 was formed on the coating after 100 h of high-temperature oxidation at 1073 K due to the reaction of Cr with oxygen in air. As a result, the components of the coating structure were changed and the coating was covered by formed Cr_2O_3. Therefore, the corrosion progress of the coating was retarded.

The squares of the mass gain as a function of oxidation time for the coated and bare SS400 at 1073 K are shown in Fig. 7. The weight gains of the coated SS400 are much less than that of the bare SS400 during the oxidation experiments. The roughly straight-line relationship indicates that the oxidation kinetic of the bare and coated SS400 samples was governed by parabolic rate law $y^2 = kt$. The parabolic rate constants (k) for bare SS400, the coated SS400 by Cr39Ni7C coatings prepared with spraying distance of 200 mm and 150 mm were 1.3×10^{-6} mg^{2}mm^{-4}s^{-1}, 3.4×10^{-7} mg^{2}mm^{-4}s^{-1}, 2.5×10^{-7} mg^{2}mm^{-4}s^{-1}, respectively. It is apparent that for the Cr39Ni7C coated SS400, the parabolic rate constants for oxidation decreased by about one order of magnitude at temperature of 1073 K. The oxidation resistance of Cr39Ni7C coating sprayed at the spraying distance of 150 mm was better than that of the sprayed at 200 mm. This was attributed partially to its lower porosity as discussed in section 3.2.

3.4 Microhardness Variation of Cr39Ni7C Coatings

The complex metastable state of the as-sprayed coating means that there is a large driving force for microstructural and compositional transformations when exposed to elevated temperatures. The most notable change is the precipitation of fine carbides or possibly oxides accompanied by recrystalization of the matrix. The mechanical

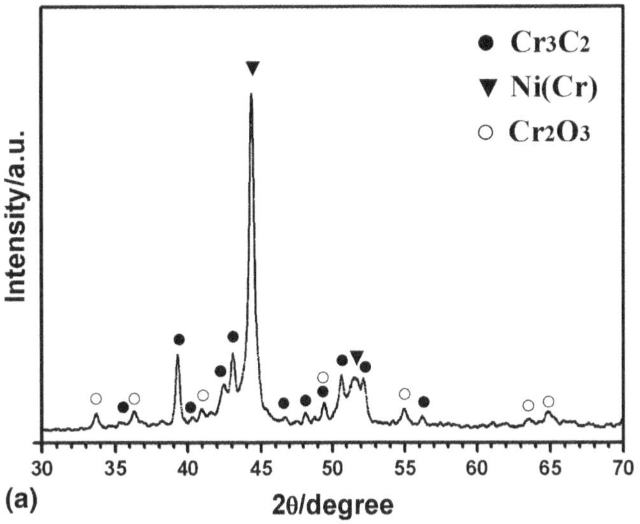

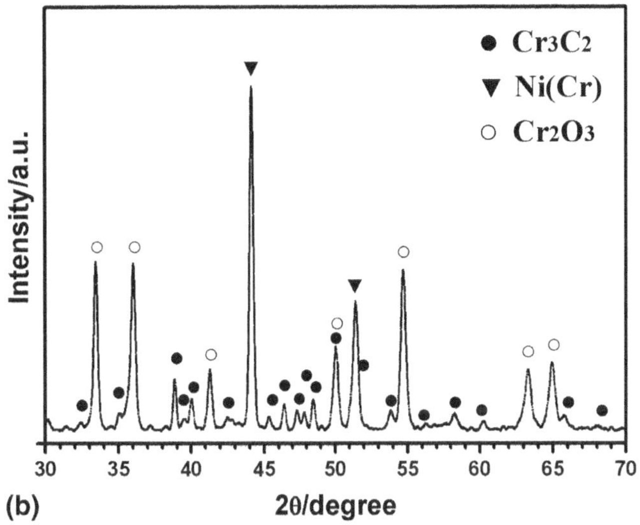

Fig. 6 The X-ray diffraction patterns of as sprayed (a) and 1073 K during 100 h heat treated (b) Cr39Ni7C coatings (Fuel gas flow: 77 SLPM, Spraying distance: 150 mm)

properties also change, in particular the microhardness, which is commonly related to the performance of carbide coatings under wear conditions (Ref 22). The Vickers hardness of the sprayed coatings after high-temperature oxidation is given in Fig. 8. The average Vickers hardness of the coatings sprayed under a fuel gas flow of 77 SLPM and spraying distance of 150 mm was approximated to 1126 $HV_{0.1}$. A slight reduction in micro hardness occurred after 115 h at 723 K high-temperature oxidation and after 100 h at 1073 K high-temperature oxidation. This may result from the residual stress relief, carbide dissolution, and matrix phase refinement, which were also reported by S. Matthews (Ref 22).

4. Conclusions

The microstructure, phase compositions, and high-temperature oxidation behavior of the deposited cermet

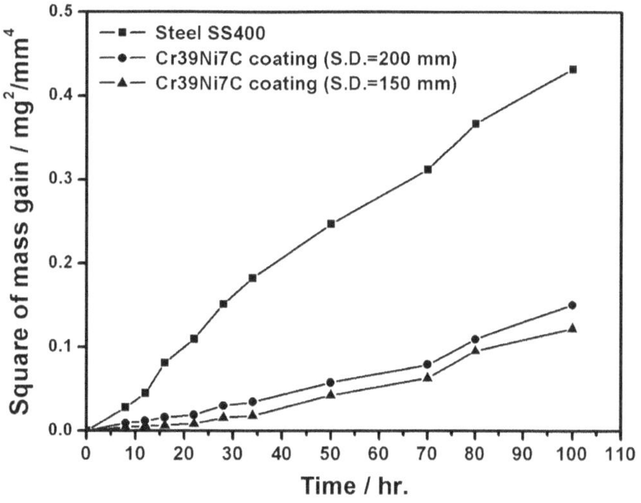

Fig. 7 Squares of the mass gain as a function of oxidation time at 1073 K for bare SS400 and Cr39Ni7C coating sprayed at the spraying distance (S.D.) of 150 mm and 200 mm

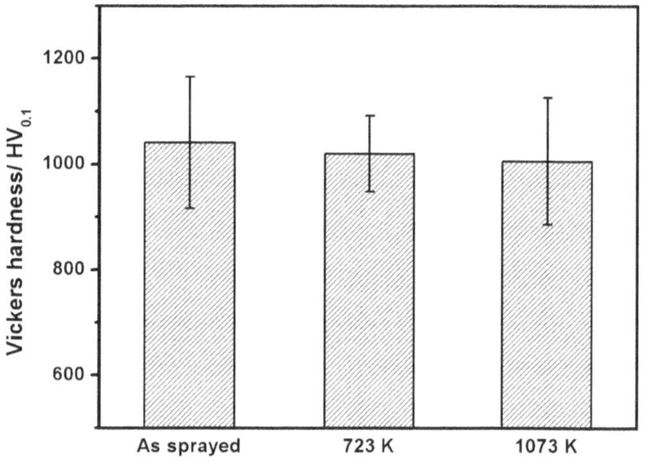

Fig. 8 Vickers hardness of the Cr39Ni7C-sprayed coatings after 115 h at 723 K and 100 h at 1073 K high-temperature oxidation

Cr39Ni7C coatings were investigated. The speed of in-flight particles decreased from 229 to 150 m/s with the increasing of spraying distance from 100 to 300 mm, whereas the average temperature of in-flight particles increased from 1926 to 2245 K. The decarburization of Cr_3C_2 increased with the increasing of fuel gas flow from 58 SLPM to 77 SLPM for higher heat enthalpy of flame. Due to the formation of Cr_2O_3 on/in the coatings at high temperature, the sprayed coatings had good oxidation resistance at 1073 K in air atmosphere. The oxidation behavior was found to be governed by the parabolic rate law. For the lower porosity of the coating deposited under the spraying distance of 150 mm, its oxidation resistance was better than that sprayed under the spraying distance of 200 mm. The average Vickers hardness of the coatings sprayed under fuel gas flow of 77 SLPM and spraying distance of 150 mm was approximated to 1126 $HV_{0.1}$. A slight reduction of sprayed Cr39Ni7C cermet coating in micro hardness occurred after 115 h in 723 K high-temperature oxidation and after 100 h in 1073 K high-temperature oxidation.

Acknowledgments

This work was partly supported by the Tianjin Natural Scientific Research Foundation (No. 07JCYBJC18200), and by the Research Fund for the Doctoral Program of Higher Education of China (No. 20070056097) and by the National Natural Science Foundation of China (No. 50805104).

References

1. S. Wirojanupatump, P.H. Shipway, and D.G. McCartney, The Influence of HVOF Powder Feedstock Characteristics on the Abrasive Wear Behaviour of Cr_xC_y-NiCr Coatings, *Wear*, 2001, **249**, p 829-837
2. M.H. Staia, T. Valente, C. Bartuli, D.B. Lewis, and C.P. Constable, Part I: Characterization of Cr_3C_2-25%NiCr Reactive Plasma Sprayed Coatings Produced at Different Pressures, *Surf. Coat. Technol.*, 2001, **146-147**, p 553-562
3. M.H. Staia, T. Valente, C. Bartuli, D.B. Lewis, C.P. Constable, A. Romand, J. Lesage, D. Chicot, and G. Mesmacque, Part II: Tribological Performance of Cr_3C_2-25%NiCr Reactive Plasma Sprayed Coatings Deposited at Different Pressures, *Surf. Coat. Technol.*, 2001, **146-147**, p 563-570
4. M. Mohanty, R.W. Smith, M. De Bonte, L.P. Celis, and E. Lugscheider, Sliding Wear Behaviour of Thermally Sprayed 75/25 Cr_3C_2/NiCr Wear Resistant Coatings, *Wear*, 1996, **198**, p 251-266
5. G.Z. Xie, Y.J. Lu, Z.Y. He, B. Hu, K.Y Wang, X.Y Mo, Y.P. Wu, and P.H Lin, Microstructure and Corrosion Properties of Plasma-Sprayed NiCr-Cr_3C_2 Coatings Comparison with Different Post Treatment, *Surf. Coat. Technol.*, 2008, **202**, p 2885-2890
6. K.J. Stein, B.S. Schorr, and A.R. Marder, Erosion of Thermal Spray MCr-CrC Cermet Coatings, *Wear*, 1999, **224**, p 153-159
7. B.Q. Wang and A. Verstak, Elevated Temperature Erosion of HVOF Cr_3C_2/TiC-NiCrMo Cermet Coating, *Wear*, 1999, **233-235**, p 342-351
8. J. Wang, L. Zhang, B.D. Sun, and Y.H Zhou, Study of the Cr_3C_2-NiCr Detonation Spray Coating, *Surf. Coat. Technol.*, 2000, **130**, p 69-73
9. Y. Ishikawa, S. Kuroda, J. Kawakita, Y. Sakamoto, and M. Takaya, Sliding Wear Properties of HVOF Sprayed WC-20%Cr_3C_2-7%Ni Cermet Coatings, *Surf. Coat. Technol.*, 2007, **201**, p 4718-4727
10. D. Toma, W. Brandl, and G. Marginean, Wear and Corrosion Behaviour of Thermally Sprayed Cermet Coatings, *Surf. Coat. Technol.*, 2001, **138**, p 149-158
11. X.M. Li, Y.Y. Yang, T.M. Shao, Y.S. Jin, and G. Barbezat, Impact Wear Performance of Cr_3C_2-NiCr Coatings by Plasma and HVOF Spraying, *Wear*, 1997, **202**, p 208-214
12. N. Espallargas, J. Berget, J.M. Guilemany, A.V. Benedetti, and P.H. Suegama, Cr_3C_2-NiCr and WC-Ni Thermal Spray Coatings as Alternatives to Hard Chromium for Erosion-Corrosion Resistance, *Surf. Coat. Technol.*, 2008, **202**, p 1405-1417
13. G.C. Ji, C.J. Li, Y.Y. Wang, and W.Y. Li, Microstructural Characterization and Abrasive Wear Performance of HVOF Sprayed Cr_3C_2-NiCr Coating, *Surf. Coat. Technol.*, 2006, **200**, p 6749-6757
14. C.J. Li, Y.Y. Wang, T. Wu, G.C. Ji, and A. Ohmori, Effect of Types of Ceramic Materials in Aggregated Powder on the Adhesive Strength of High Velocity Oxy-Fuel Sprayed Cermet Coatings, *Surf. Coat. Technol.*, 2001, **145**, p 113-120
15. C.J. Li, G.C. Ji, Y.Y. Wang, and K. Sonoya, Dominant Effect of Carbide Rebounding on the Carbon Loss During High Velocity Oxy-Fuel Spraying of Cr_3C_2-NiCr, *Thin Solid Films*, 2002, **419**, p 137-143

16. T.S. Sidhu, S. Prakash, and R.D. Agrawal, Hot Corrosion Studies of HVOF Sprayed Cr_3C_2-NiCr and Ni-20Cr Coatings on Nickel-Based Superalloy at 900°C, *Surf. Coat. Technol.*, 2006, **201**, p 792-800
17. J.K.N. Murthy and B. Venkataraman, Abrasive Wear Behaviour of WC-CoCr and Cr_3C_2-20(NiCr) Deposited by HVOF and Detonation Spray Processes, *Surf. Coat. Technol.*, 2006, **200**, p 2642-2652
18. http://www.freepatentsonline.com/20060040125.html, "Piston Ring and Thermal Spray Coating Used Therein, and Method for Manufacturing Thereof," United States Patent 20060040125. Accessed 10 May 2007
19. http://www.sulzermetco.com/en/Portaldata/13/Resources/2_Products_Services/Thermal_Spray_Materials_Guide_022006.pdf, *Sulzer Metco Thermal Spray Materials Guide*, Issued April 2005, p HVOF-13. Accessed 10 May 2007
20. E. Yun and S. Lee, Correlation of Microstructure with Hardness and Wear Resistance in Cr_3C_2/Stainless Steel Surface Composites Fabricated by High-Energy Electron Beam Irradiation, *Mater. Sci. Eng., A*, 2005, **405**, p 163-172
21. K. Hirota, K. Mitani, M. Yoshinaka, and O. Yamaguchi, Simultaneous Synthesis and Consolidation of Chromium Carbides (Cr_3C_2, Cr_7C_3 and $Cr_{23}C_6$) by Pulsed Electric-Current Pressure Sintering, *Mater. Sci. Eng., A*, 2005, **399**, p 154-160
22. S. Matthews, M. Hyland, and B. James, Microhardness Variation in Relation to Carbide Development in Heat Treated Cr_3C_2-NiCr Thermal Spray Coatings, *Acta Mater.*, 2003, **51**, p 4267-4277

Peer Reviewed

JTTEE5 17:948–955
DOI: 10.1007/s11666-008-9261-6
1059-9630/$19.00 © ASM International

Peer Reviewed

Influence of the Spray Angle on the Characteristics of Atmospheric Plasma Sprayed Hard Material Based Coatings

Wolfgang Tillmann, Evelina Vogli, and Benjamin Krebs

(Submitted May 15, 2008; in revised form September 16, 2008)

This paper presents an investigation of the influence of the spray angle on thermally sprayed coatings. Spray beads were manufactured with different spray angles between 90 and 20° by means of atmospheric plasma spraying (APS) on heat-treated mild steel (1.0503). WC-12Co and Cr_3C_2-10(Ni20Cr) powders were employed as feedstock materials. Every spray bead was characterized by a Gaussian fit. This opens the opportunity to analyze the influence of the spray angle on coating properties. Furthermore, metallographic studies of the surface roughness, porosity, hardness, and morphology were carried out and the deposition efficiency as well as the tensile strength was measured. The thermally sprayed coatings show a clear dependence on the spray angle. A decrease in spray angle changes the thickness, width, and form of the spray beads. The coatings become rougher and their quality decreases.

Keywords Cr_3C_2-NiCr, hard material, morphology, off-axis, spray angle, thermal spraying, WC-12Co

1. Introduction

Sculptured sheets in the automotive industry require forming tools that are able to meet highest demands regarding shape accuracy, even at high complexities, in particular for high-strength steels. These demands are influenced by many uncontrollable variables and partly insufficiently investigated boundary conditions like metallurgical phenomena and tribological effects (Ref 1).

A commonly used process for forming of metal sheets is deep drawing. In recent years, the requirements for the productivity of deep drawing tools have particularly increased in the field of the automotive industry. At the same time, the application of high strength steels has increased due to their potential in the fields of light weight design and advancement in strength (Ref 2). During the forming process of high strength steels enormous press capacities operate, which significantly lead to tool wear. Therefore, it is becoming more and more important to enhance the wear resistance of deep drawing tools by an appropriate surface modification (Ref 3).

A novel approach to surface modification of deep drawing tools is the thermal spraying technology. However, thermally sprayed coatings normally feature undesirable properties like high porosity and roughness especially on complex geometries. In recent years, the complexity of deep drawing tools has highly increased. Consequently, it is often not possible to realize a perpendicular spray angle on the whole component surface during the spray process (Fig. 1). This can lead to significant changes of the coating properties (Ref 4-7). Therefore, it is of high importance to study the influence of the spray angle on the coating properties. This can provide fundamentals to determine the tolerance to which coatings exhibit reproducible properties.

In the present study, experiments have been conducted to investigate the angular dependence of the spray profile, roughness, porosity, hardness, morphology, tensile strength and deposition efficiency of two hard materials in an atmospheric plasma spray process.

2. Experimental

2.1 Substrate Preparation

Rectangular heat-treated mild steel substrates (1.0503) with a dimension of 200 × 100 × 5 mm^3 were employed as base material. The substrates were cut from sheet material. In preparation for thermal spray experiments, the substrates were grit-blasted with coarse-grained alumina (ø = 1180-1700 μm), at a pressure of 4 bar, an

This article is an invited paper selected from presentations at the 2008 International Thermal Spray Conference and has been expanded from the original presentation. It is simultaneously published in *Thermal Spray Crossing Borders, Proceedings of the 2008 International Thermal Spray Conference*, Maastricht, The Netherlands, June 2-4, 2008, Basil R. Marple, Margaret M. Hyland, Yuk-Chiu Lau, Chang-Jiu Li, Rogerio S. Lima, and Ghislain Montavon, Ed., ASM International, Materials Park, OH, 2008.

Wolfgang Tillmann, Evelina Vogli, and **Benjamin Krebs**, Institute of Materials Engineering, Technical University of Dortmund, Leonhard-Euler-Str. 2, 44227, Dortmund, Germany. Contact e-mail: wolfgang.tillmann@udo.edu.

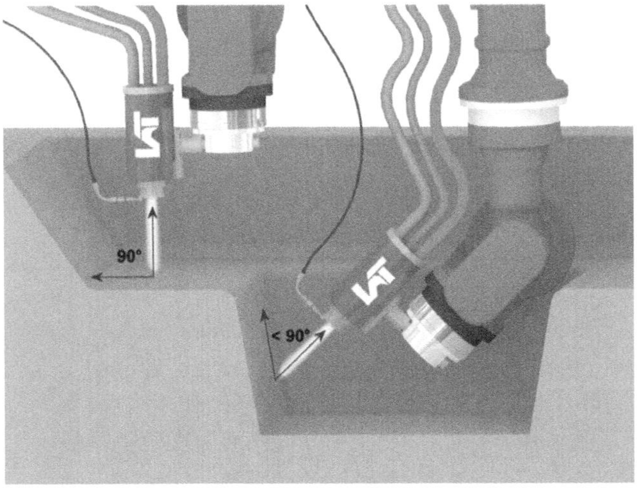

Fig. 1 Scheme of a thermal spray process of a deep drawing tool. Some regions with a high geometrical complexity cannot be coated with a perpendicular spray angle

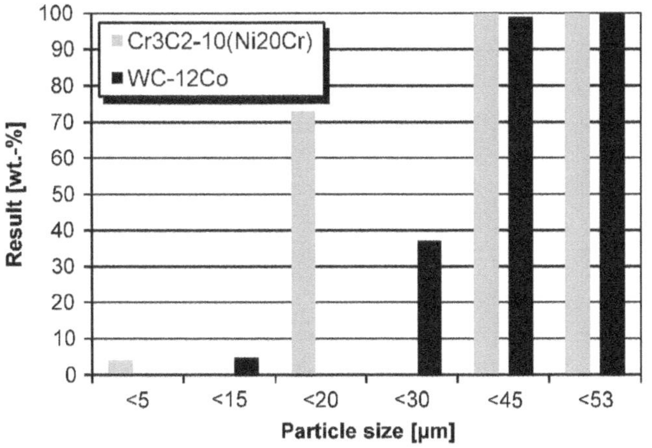

Fig. 2 Cumulative particle size distribution of WC-12Co and Cr_3C_2-10(Ni20Cr) powders

inclination angle of 45°, and a distance of 100 mm. The substrates showed an average roughness R_Z of 50 µm after grit blasting. Afterward, the samples were cleaned in an ethanol ultrasonic bath (15 min) and heated in a furnace (30 min) to approx. 100 °C.

2.2 Feedstock Powder

Two commercially available powders, a WC-12Co (Sulzer Metco, Metco 72F-NS) powder and a Cr_3C_2-10(Ni20Cr) (Sulzer Metco, AMDRY 367) powder, were employed as plasma spray feedstocks. The particle size distribution of the powders was measured by means of laser scattering. The WC-12Co powder possesses particle sizes of 15 to 45 µm with a mode of 30 µm, the Cr_3C_2-10(Ni20Cr) powder below 45 µm with a mode of 20 µm. Their cumulative particle size distribution is shown in Fig. 2. The WC-12Co powder was manufactured by an agglomeration and sintering process and shows a spheroidal shape (Fig. 3a). The Cr_3C_2-10(Ni20Cr) powder was spattered and broken and exhibits a blended shape (Fig. 3b). The chemical composition was measured by inductive coupled plasma—mass spectroscopy (ICP-MS) (Co, Fe, Ni, Cr), the difference method (W), and gravimetry (C). The results are shown in Table 1.

2.3 Experimental Setup

The deposits were manufactured using atmospheric plasma spray equipment Sulzer Metco, model Multi Coat equipped with a F4-MB torch. The main spray parameters are listed in Table 2. The plasma gun moved at a velocity of 0.5 ms^{-1} for 60 overruns accomplished in front of each substrate. The distance between the spray gun and substrate was 130 mm. To prevent overheating of the substrates, the specimens were cooled from the backside with compressed air with a pressure of 2.76 bar.

The experimental setup is shown in Fig. 4. A spray procedure was performed by moving the spray gun horizontally over the substrate. This results in a spray bead in the middle of the substrate. The substrates were coated with spray angles between 90 and 20°. The spray angle is defined as 90° when the spray jet is parallel to the substrate normal and 0° when the jet is perpendicular to the normal. Directly after each coating procedure the temperature on the coating surface was measured by pyrometry. The average temperature of the coatings amounted to 70 °C irrespective of the spray angle.

2.4 Analytics

The profile and the roughness of the spray bead were measured with a 3D surface profilometer (Alicona, model Infinite Focus). The topography of the profile was analyzed by scanning across the spray bead (in Fig. 4 into *z*-direction). The deposition was characterized by the relative deposition efficiency, which is defined as the ratio of weight gain of specimen coated with a spray angle of 90° to the weight gain of specimen coated with shallower spray angles. Cross sections of the coated samples were investigated by light microscopy (Zeiss, Axiophot) and porosity measurements were done by image processing software (Zeiss, Axiovision 4.6). Microhardness measurements were performed by a Vickers indentation with 0.1 kp (0.981 N) (Leco, model M-400). Each indentation was done for 15 s. Tensile stress tests were accomplished according to DIN EN 582 with a tensile testing devise (Mohr & Federhoff). The roughness, porosity, deposition efficiency, and tensile strength measurements were repeated three times, and the mean of the three values including the standard deviation were determined. The microhardness test was repeated five times.

2.5 Theoretical Background

The central limit theorem states that a sum of n independent, identically distributed random variables at the limit ($n \rightarrow \infty$) can be described by a continuous probability density function, the normal or Gaussian distribution (Ref 8). According to this, a random variable is normally distributed when it is developed from an interaction of a

(a) (b)

Fig. 3 SEM images of morphology and polished cross section of employed spray powders (a) WC-12Co and (b) Cr_3C_2-10(Ni20Cr)

Table 1 Chemical composition of investigated powders

		Cr_3C_2-10(Ni20Cr)	
[wt.%]	**WC-12Co**	**Cr_3C_2**	**Ni-20Cr**
W	81.52	...	...
Co	12.07	...	...
C	6.15	13.57	...
Ni	...	...	78.40
Cr	...	86.22	20.02

Table 2 Spray parameters

Powder	WC-12Co	Cr_3C_2-10(Ni20Cr)
Plasma gun	F4-MB	F4-MB
Argon flow rate	65 L min^{-1}	65 L min^{-1}
Hydrogen flow rate	7 L min^{-1}	15 L min^{-1}
Current intensity	550 A	550 A
Powder carrier gas flow rate	3.2 L min^{-1}	3.2 L min^{-1}
Powder feed rate	37 g min^{-1}	28 g min^{-1}

large number of influences and the separate influencing value only delivers a minor contribution (e.g., particle impact during thermal spraying). A summation of the separate influencing values is the normal or Gaussian distribution. Many terms of science and engineering such as the spray jet during thermal spraying and the resulting coating profile can be described exactly or in good approximation by this distribution (Ref 9). The mathematical description of the Gaussian distribution is:

$$G(x) := y_0 + \frac{A}{\sigma\sqrt{2\pi}} \cdot \exp\left[-\frac{1}{2}\left(\frac{x - x_0}{\sigma}\right)^2\right] \qquad \text{(Eq 1)}$$

where A is an amplitude scale factor, σ is the standard deviation, x_0 the mean, and y_0 an additive constant. The characteristic bell shape of a normal distribution is shown in Fig. 5. The deviation between an experimental measured distribution and the ideal norm distribution is mathematically determined by the skewness s, the kurtosis k, and the full width at half maximum FWHM. The

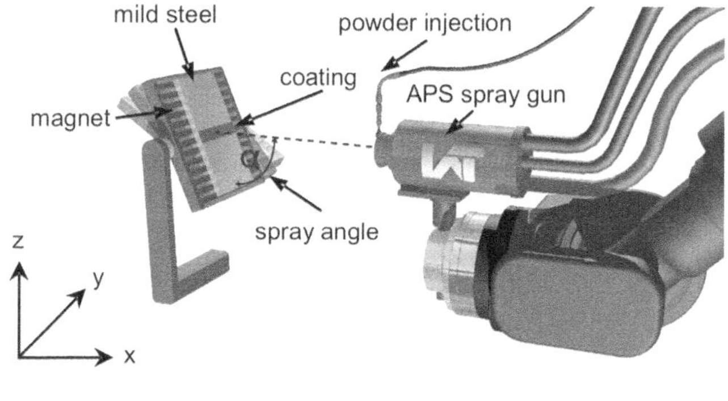

Fig. 4 Scheme of a spray procedure

skewness s is a measure of the symmetry of a distribution in reference to the mean. The skewness can be used to declare if a distribution is concentrated more on the left ($s < 0$) or on the right ($s > 0$) side. The kurtosis k indicates whether a distribution features a higher ($k > 0$) or

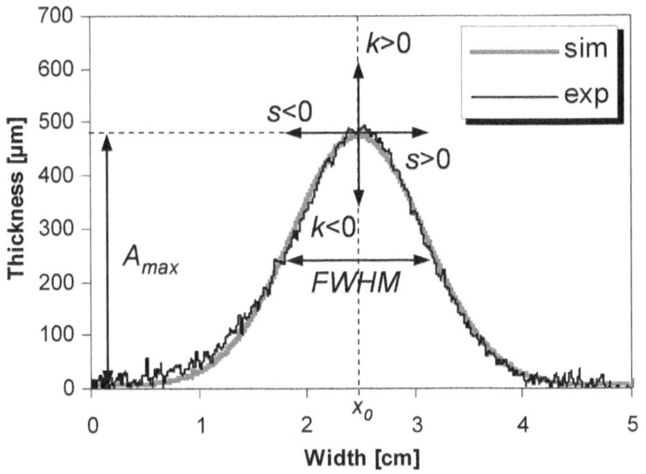

Fig. 5 Experimentally measured Gaussian profile and fit (WC-12Co, $\alpha = 90°$). The profile can be described by the skewness s, the kurtosis k, the full width at half scale FWHM, the mean x_0, and the maximal coating thickness A_{max}

lower ($k < 0$) peakness regarding the normal distribution. The full width at half maximum FWHM is defined as the distance between two abscissa values, where the ordinate values decreased to half of the distribution's maximum. The full width at half maximum is proportional to the standard deviation σ.

3. Results and Discussion

3.1 Topography of the Spray Bead

The spray beads generated at spray angles between 90 and 20° exhibit Gaussian profiles (Fig. 6). For investigation of the profiles Gaussian fits as stated in paragraph 2.5 were used and the relevant parameters, maximal thickness, full width at half maximum, skewness, and kurtosis, were calculated. The maximal thickness A_{max} of the profile increases between 90 and 80° (Fig. 7). Below 80° the thickness decreases with increasing spray angle for both

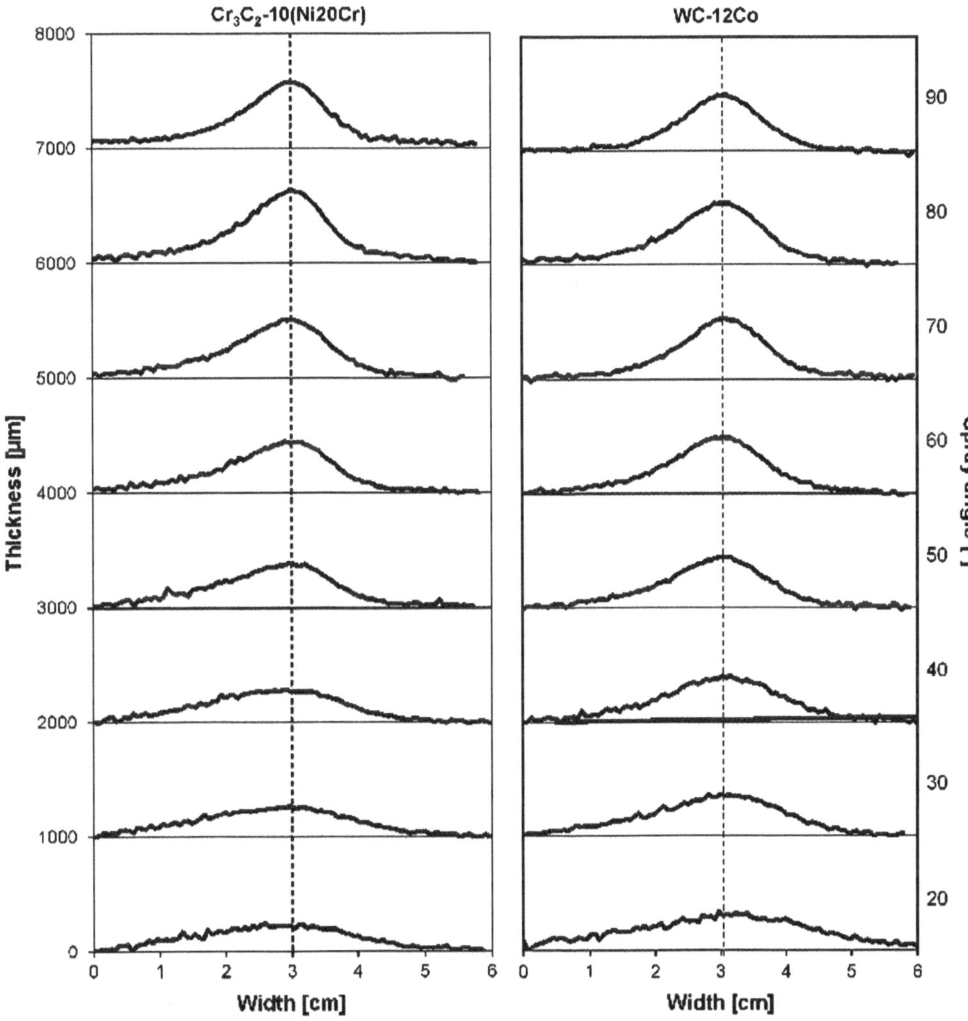

Fig. 6 Topography profiles of WC-12Co and Cr_3C_2-10(Ni20Cr) spray beads made with spray angles between 90 and 20°

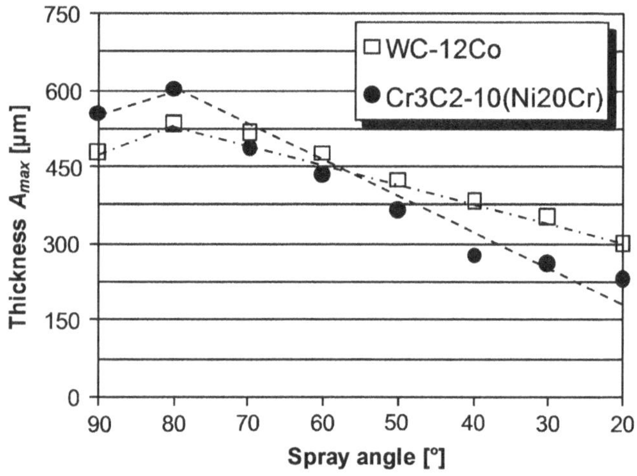

Fig. 7 Thickness A_{max} of spray bead

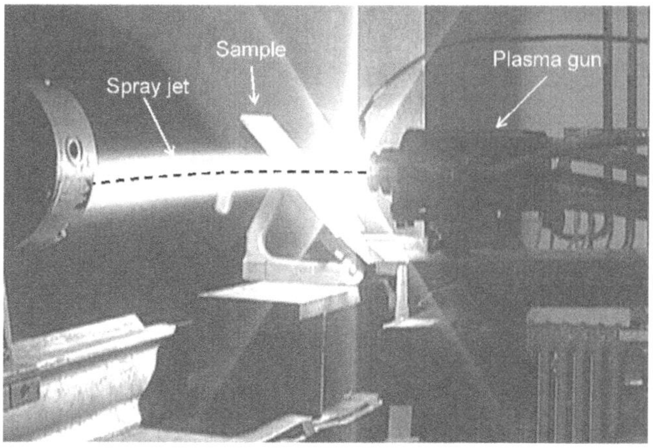

Fig. 8 Experimental setup. The skew spray jet influences the maximal thickness A_{max} of the spray beads and the deposition efficiency DE

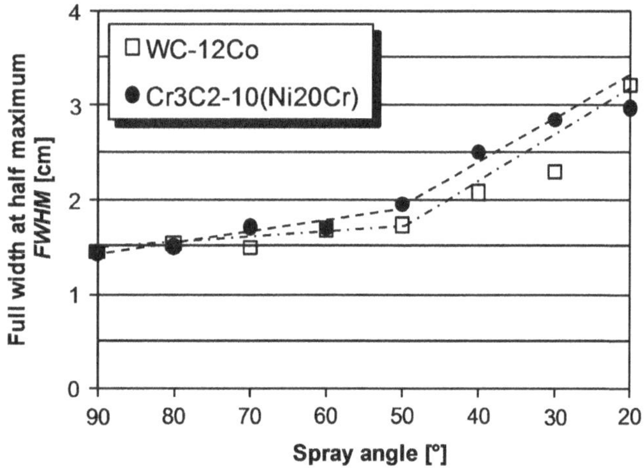

Fig. 9 Full width at half maximum FWHM of spray bead

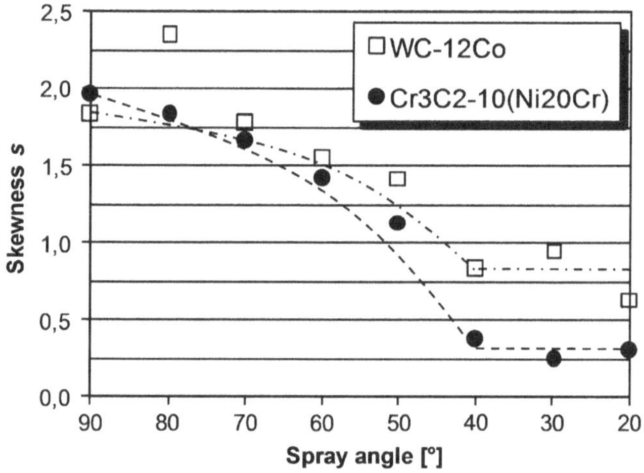

Fig. 10 Skewness s of spray bead

materials. Kang et al. describe the decline in coating thickness at smaller spray angles with a decrease of the normal component of the particle velocity regarding the substrates surface. At the same time the tangential component increases and the probability rises to bounce off a sprayed particle from the specimen's surface. This results in a lower layer thickness (Ref 10). The increase of A_{max} between 90 and 80° is due to a skew spray jet affected by a perpendicular powder injection (Fig. 8). At 90° the sprayed particles exhibit a tangential component with respect to the substrates surface. This leads to a real spray angle smaller than 90° and an increased probability to bounce off sprayed particles. However, at 80° the skewness of the spray jet regarding the substrates surface is compensated and a smaller amount of particles rebound. Consequently, the maximal thickness A_{max} increases.

The full width at half maximum FWHM increases by decreasing the spray angle for both investigated materials (Fig. 9). A slight increase is established between 90 and 50°, whereas the width of the profile increases steeply below 50°. This behavior is due to a growth of the tangential component regarding the substrates surface when decreasing the spray angle and results in a broadening of the spray bead.

The skewness s of the spray beads (Fig. 10) is positive at a spray angle of 90° for both materials. Between 90 and 40° the skewness decreases relatively steeply. Therefore, the symmetry of the spray bead increases. Below 40° only a slight decrease was observed. A positive skewness states a concentration of the spray bead at its right side (in Fig. 4 bottom). The positive skewness at 90° is caused by the perpendicular powder injection. A rotation of the substrate to smaller spray angles leads to a decrease in skewness as long as the substrate rotates in the direction of the profile's concentration. A rotation in the other direction should result in an increase in skewness. This simple behavior is caused by the geometric parameters at off normal angles (Ref 11). However, there are further influencing values such as interaction between sprayed (hard) material and substrates surface. An evidence is the relative constant skewness at spraying angles less than 40°. If the skewness was only influenced by the geometrical

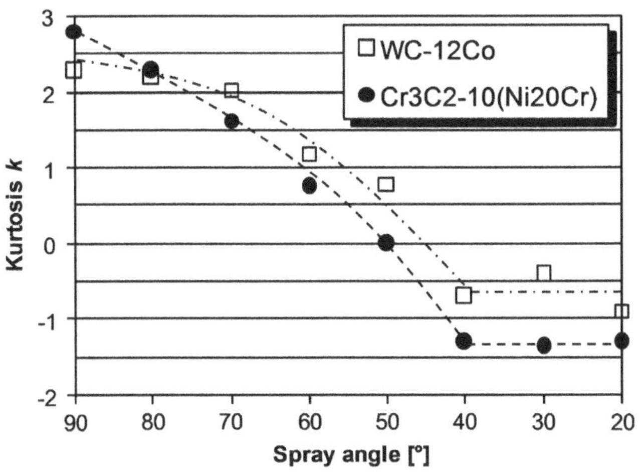

Fig. 11 Kurtosis k of spray bead

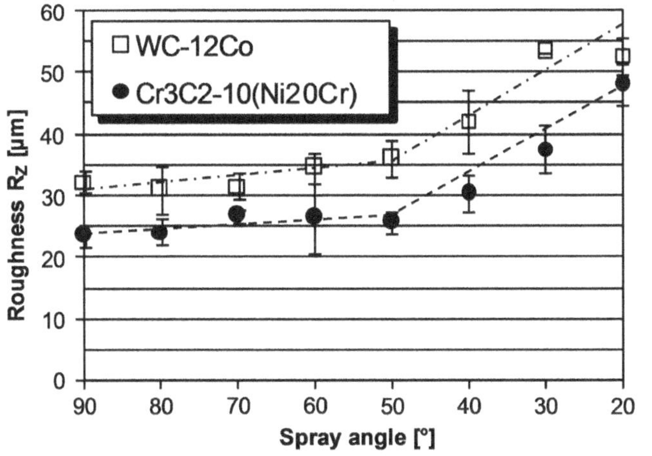

Fig. 12 Surface roughness R_Z versus spray angle

parameters, it should also decrease at small spray angles. The kurtosis k possesses a comparable run as the measured skewness (Fig. 11). At 90° the kurtosis is positive; hence the spray beads show sharp shapes. On reducing the spray angle the kurtosis decreases to negative, and below 40° the kurtosis remains relatively constant. Consequently, the spray bead is broadened and flattened on decreasing the spray angle.

3.2 Roughness

Both spray beads show an increase in the average surface roughness R_Z at decreasing spray angle (Fig. 12). Especially, below 50° the roughness increases steeply. A similar behavior was observed by Smith et al. (Ref 6). This phenomenon is attributed to an increased probability of sprayed particles at low spray angles to interlock with surface inhomogeneities. Consequently, the surface roughness increases.

3.3 Porosity

Within the margin of error the coatings exhibit constant porosities between 90 and 50°. Below 50° a significant

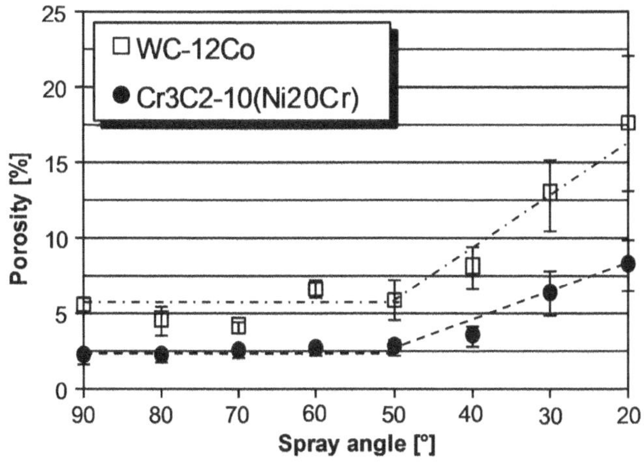

Fig. 13 Porosity versus spray angle

increase of porosity was measured (Fig. 13). Furthermore, the WC-12Co coatings show a more significant tendency to increase in porosity at lower spray angles. An explanation was provided by Smith et al. (Ref 6). Spraying with off normal angles leads to a surface profile with the shape of a sawtooth. Consequently, regions in the sawtooth are shadowed at lower spray angles and therefore striking particles cannot cover the complete surface. As a result pores arise.

The higher porosity and dependency on the spray angle of the WC-12Co coatings compared to the Cr_3C_2-10(Ni20Cr) ones can be explained by the shape, size, and manufacturing method of the employed powder. The WC-12Co powder exhibits a spheroidal shape with a mode of 30 μm, whereas the Cr_3C_2-10(Ni20Cr) is spattered and broken with a mode of 20 μm. Due to the spherical shape of the WC-12Co powder, off normal sprayed particles are likely to slip when they hit the surface. Thus, growing pores are not filled up during spray process. In contrast, it is more likely that a spattered Cr_3C_2-10(Ni20Cr) particle interlocks with the surface of the coating. This can prevent a growth of pores. The lower particle size mode of the Cr_3C_2-10(Ni20Cr) effects smaller spikes at the coating surface and thereby denser coatings. Additionally, the sintered WC-12Co powder possesses a higher porosity than the spattered and broken Cr_3C_2-10(Ni20Cr) powder (cp. Fig. 3). Consequently, the WC-12Co coatings are more porous.

3.4 Relative Deposition Efficiency

The relative deposition efficiency DE increases app. 5% from 90° to 80° (Fig. 14), whereas below 80° the deposition efficiency decreases rather linearly. A decrease in spray angle results in a reduction of the normal component and provides a benefit to the tangential component of the spray particles regarding the substrates surface. An increase in the tangential component results in a higher probability that particles bounce off from the surface of the substrate. Therefore, the deposition efficiency declines. The increase between 90 and 80° is attributed to the perpendicular powder injection into the spray flame. This leads to a skew spray jet with respect to the substrates surface (Fig. 8). The effect is compensated up to 80°.

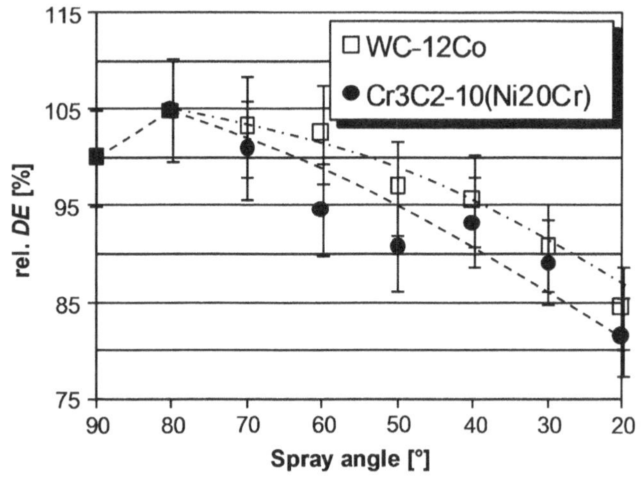

Fig. 14 Variation of relative deposition efficiency with spray angle

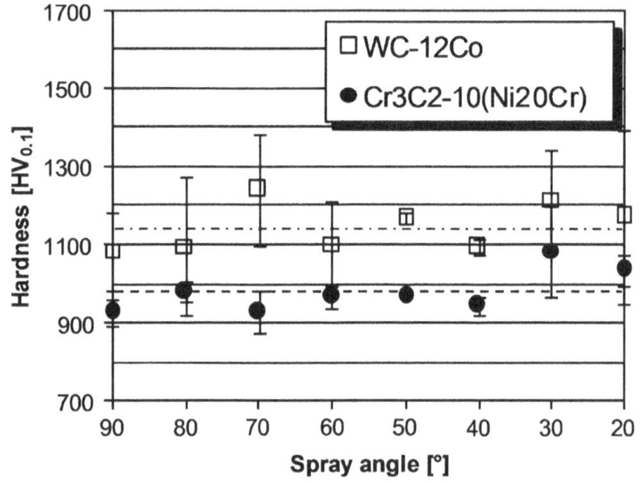

Fig. 15 Relation between microhardness and spray angle

3.5 Hardness

Figure 15 illustrates the results of the microhardness tests. The WC-12Co coatings exhibit hardness between 1080 and 1240$HV_{0.1}$, while Cr_3C_2-10(Ni20Cr) show hardness values between 930 and 1080$HV_{0.1}$. The WC-12Co data points are more widely distributed. This hardness variation is attributed to a more inhomogeneous structure of the coating, which results from the larger powder particle size of the WC-12Co powder. Within the error of margin no dependency of microhardness and spray angle can be stated. The microhardness test affects only small indentations into the coating. Therefore, the influence of pores and changes in microstructure cannot be measured. A macrohardness test should deliver more reliable results.

3.6 Tensile Strength

Figure 16 illustrates the results of the tensile tests. The coatings show a high tensile strength of 67 ± 8 N mm^{-2} for WC-12Co and 65 ± 1 N mm^{-2} for Cr_3C_2-10(Ni20Cr)

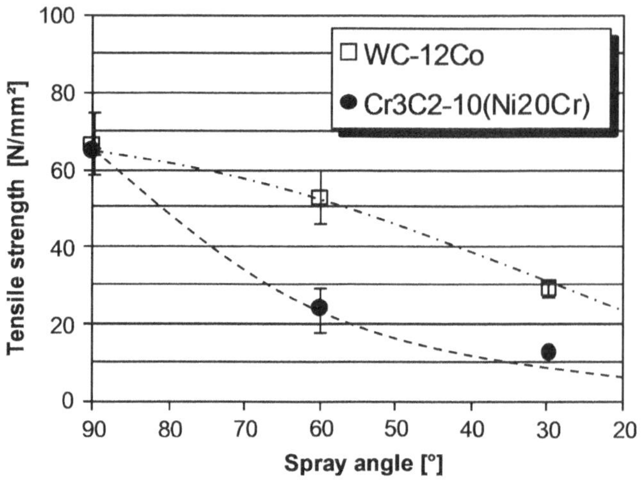

Fig. 16 Effect of spray angle on the tensile strength

at a spray angle of 90°. Both coatings exhibit a significant decrease of the tensile strength on decreasing the spray angle. Particularly, the tensile strength of the WC-12Co coating is strongly influenced by the spray angle.

The coatings sprayed with an angle of 90° delaminated during the tensile tests at the interface between substrate and coating. A decrease in spray angle resulted in an increased probability of a mixed fracture composed of interfacial and cohesive failure. At 30° all coatings failed cohesively. This behavior is due to the rough coating microstructure at smaller spray angles and especially at 30° caused by the high porosity of the coatings.

3.7 Morphology

Polished cross sections of the Cr_3C_2-10(Ni20Cr) and the WC-12Co coatings are illustrated in Fig. 17. The Cr_3C_2-10(Ni20Cr) coatings show a dense structure between 90 and 50°. Below 50° large pores arise. In contrast, the WC-12Co coatings exhibit relative large pores even at 90°. The larger diameter of these pores is due to the particle size of the applied WC-12Co powder. Furthermore, the WC-12Co coating is more sensitive to a change in spray angle. Already at 50° a decline in coating quality is observed.

4. Summary and Conclusion

The results of this study demonstrate that both investigated spray materials show a significant dependence on the spray angle. Independent of the type of powder a change of thickness, width, and profile of the plasma sprayed bead could be established. Especially, from 50° onward the geometry of the spray bead changes considerably. This is connected to an increase in surface roughness and porosity as well as a decrease of tensile strength. The cross sections show a decline in microstructure from 30° for Cr_2C_2-10(Ni20Cr) and from 50° for WC-12Co, whereas a relative linear decrease in deposition efficiency is observed below 80°.

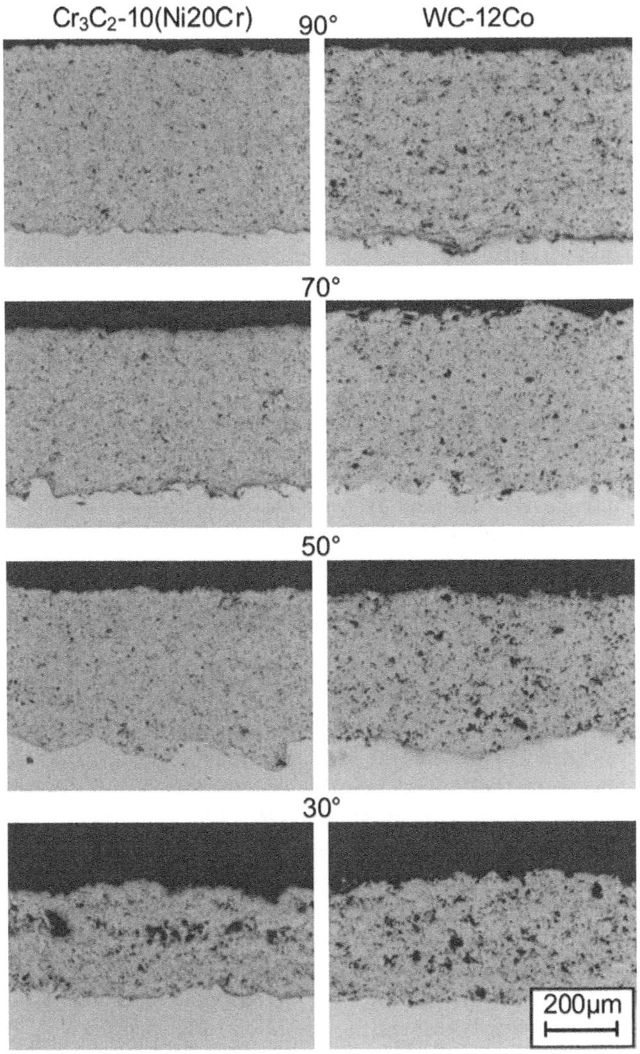

Fig. 17 Cross sections of Cr_3C_2-10(Ni20Cr) and WC-12Co coatings deposited with spray angles between 90 and 30°

Our results suggest that within the traditional 90-45° spray window coatings with only a marginal loss of coating quality can be manufactured. In this window mainly the thickness of the coating and the deposition efficiency decrease. However, for highest shape accuracy the influence of the spray angle must be taken into consideration. Below 45° the coatings show a clear decrease in quality. If possible, these shallow spray angles shall be avoided. If it is not procedurally possible, it must be accepted that the coatings are thinner, with a higher porosity and a higher surface roughness. Only the thickness of the coating can partially be influenced by transferring the spray gun more slowly or numerously over the specimen.

Acknowledgment

This work is supported by the German Research Foundation (DFG) within the Research Centre SFB708.

References

1. A. Hrivnak and L. Sobotova, The Influence of the Deformation Aging and the Conditions of Stress on the Properties of the Deep Drawing Steel Sheet, *J. Mater. Process. Technol.*, 1992, **34**, p 425-130
2. D. Raabe, Werkstücke hoher Qualität und Produktivität durch Verfahrenskombination beim Kaltmassivumformen, *16. Jahrestreffen der Kaltmassivumformer*, Düsseldorf, 2001, p. 1-11
3. H. Schulz-Marner and S. Schulte, Verschleißprüfstand für Blechumformwerkzeuge, *Bänder Bleche Rohre*, 1997, **38**, p 50-53
4. V. Sobolev and J.M. Guilemany, Influence of Droplet Impact Angle on Droplet-substrate Mechanical Interaction in Thermal Spraying, *Mater. Lett.*, 1998, **33**, p 315-319
5. A.J. Allen, G. Long, and H. Herman, Influence of Spray Angle on the Pore and Crack Microstructure of Plasma-Sprayed Deposits, *J. Am. Ceram. Soc.*, 1997, **80**(3), p 733-742
6. M.F. Smith, R.A. Neiser, and R.C. Dykhuizen, An Investigation of the Effects of Droplet Impact Angle in Thermal Spray Deposition, *Proceedings of the 7th National Thermal Spray Conference*, 1994, p. 603-608
7. M.R. Kanouff, R.A. Neiser, and T.J. Roemer, Surface Roughness of Thermal Spray Coating Made with Off-Normal Spray Angles, *J. Therm. Spray Technol.*, 1998, **7**(2), p 219-228
8. L. Sachs, *Angewandte Statistik*, Springer-Verlag, Berlin, 1992, p. 109
9. T. Florin-Iuliu, G. Montavon, and C. Coddet, On the Relationships Between the Geometric Processing Parameters of APS and the Al_2O_3-TiO_2 Deposit Shapes, *Surf. Coat. Technol.*, 2005, **195**, p 54-69
10. C.W. Kang and H.W. Ng, Imaging Diagnostics Study on Obliquely Impacting Plasma-Sprayed Particules Near to the Substrate, *J. Therm. Spray Technol.*, 2006, **15**(1), p 118-130
11. S.H. Leigh and C.C. Berndt, Evaluation of Off-angle Thermal Spray, *Surf. Coat. Technol.*, 1997, **89**, p 213-224

JTTEE5 17:956–965
DOI: 10.1007/s11666-008-9240-y

Peer Reviewed

A Co-Condensation Model for In-Flight Synthesis of Metal-Carbide Nanoparticles in Thermal Plasma Jet

A. Vorobev, O. Zikanov, and P. Mohanty

(Submitted June 11, 2008; in revised form July 28, 2008)

We present a theoretical analysis of the formation, growth, and transport of two-component nanoparticles in thermal plasma jet. The approach of the aerosol science and the idea of multicomponent co-condensation are employed for the analysis. The processes of homogeneous nucleation, heterogeneous growth, and coagulations due to Brownian collisions are considered in combination with the convective and diffusive transport of particles and the reacting gases within an axisymmetric domain. As a particular example, the authors study multicomponent co-condensation of metal-carbide nanoparticles from various precursors in a DC plasma gun operating in an argon atmosphere.

Keywords Brownian collisions, heterogeneous condensation, homogeneous nucleation, nanoparticles, numerical modeling, refractory coating, thermal plasma spraying

1. Introduction

Metal carbides such as TaC, SiC, TiC are used in a wide range of industrial applications due to their excellent high temperature properties as well as chemical/wear resistance. The majority of these carbide powders are produced by carbothermic reduction. Alternatively, thermal plasma spraying is an excellent process for the preparation of metal carbide nanopowders from liquid/gaseous precursors due to its unique characteristics such as high temperature (>10,000 K), high chemical activity, and rapid quenching (~$10^{6\text{-}7}$ K/s). The use of thermal plasma spraying technique for production of nanopowders has been experimentally explored by several investigators (e.g., Ref 1-4). Several theoretical analyses also have focused on the synthesis of nanoparticles in different plasma reactors (Ref 1, 5-11). Most of the theoretical work devoted to the synthesis of nanoparticles is, however, focused on the nucleation and growth of particles consisting of one component.

Recently, the authors developed an approach for modeling the in-flight synthesis of nanoparticles based on the concept of multicomponent co-condensation and descriptive tools of the aerosol science (Ref 12). Synthesis of TaC nanoparticles in atmospheric condition was considered. Practical applicability of the open atmosphere synthesis is diminished by the oxidation problem (Ref 2). Because of that, in the present paper, the authors extend the analysis to a promising configuration of an argon-filled chamber. Furthermore, the model is applied to examine the synthesis of other practically interesting metal-carbide systems.

Saturation vapor pressures of the condensing materials play an important role in the co-condensation process. In this paper, the authors consider metals that possess lower saturation vapor pressure (e.g., Si and Ti) as well as higher saturation vapor pressure (Ta and W) compared to that of carbon. The nucleation and condensation sequence based on their relative saturation vapor pressures and phase diagrams are discussed.

2. Coating System and Operational Parameters

The system under consideration has certain specific features since the parameters of the DC plasma gun installed in the laboratory at the University of Michigan-Dearborn, are used for simulations. However, the general modeling approach to the particle synthesis can be extended to other systems based on the same principal mechanism, namely, fast quenching leading to homogeneous nucleation and subsequent heterogeneous condensation.

The coating system uses a DC plasma gun operating at 38-40 V and 600 A. Pure argon is injected into the plasma gun at the rate of 1.68×10^{-3} kg s^{-1}. The plasma gun operates in a chamber filled with argon. Suction is applied

This article is an invited paper selected from presentations at the 2008 International Thermal Spray Conference and has been expanded from the original presentation. It is simultaneously published in *Thermal Spray Crossing Borders, Proceedings of the 2008 International Thermal Spray Conference*, Maastricht, The Netherlands, June 2-4, 2008, Basil R. Marple, Margaret M. Hyland, Yuk-Chiu Lau, Chang-Jiu Li, Rogerio S. Lima, and Ghislain Montavon, Ed., ASM International, Materials Park, OH, 2008.

A. Vorobev, University of Southampton, Southampton, UK; and **O. Zikanov** and **P. Mohanty**, University of Michigan-Dearborn, Dearborn, MI. Contact e-mail: A.Vorobev@soton.ac.uk.

to maintain constant argon pressure approximately equal to atmospheric pressure in the chamber. Measurements near the chamber walls show temperature of about 300 K.

The formation of nanoparticles is initiated by injection of precursors (such as liquid tantalum ethoxide $Ta(OC_2H_5)_5$ used for production of TaC nanoparticles (Ref 2)) into the nozzle. The actual details of injection are unimportant, since it can be assumed that the plasma jet is sufficiently hot to completely evaporate and chemically decompose the precursor and dissociate its atoms. As a result, it is assumed that monoatomic gases of Ta and C are formed before the jet exists from the nozzle. The only limiting assumption made in the following is that the precursor loading is low, so that evaporation does not significantly affect the argon flow.

3. Numerical Model of Gas Flow

The electromagnetic, thermodynamic, and gas dynamic processes governing the flow within a thermal plasma coating device are extremely complex. A complete model should, in principle, include description of discharge arc, turbulent energy and mass transport in a magnetohydrodynamic flow of real plasma, processes of ionization and deionization, etc. Such models, especially when realistic motions of discharge arc are considered, are unfeasible at today's level of physical understanding and computational capabilities. Even the most advanced models, for example (Ref 13), where dynamic simulations of the flow induced by a torch operating in the takeover mode were conducted, achieved only qualitative agreement with experiments. Since the focus of this investigation is on the formation of nanoparticles rather than on the arc and jet behavior the authors obtain the needed properties of the flow by using a simple hydrodynamic model correlated with experimental data. Similar approach was taken in other studies of nanoparticle synthesis (for example, Ref 1, 6, 8).

The Navier-Stokes equations for a compressible, viscous, and heat-conducting ideal argon gas are solved numerically using the CFD software Fluent. Mean (ensemble averaged) steady-state flow is calculated. The effect of turbulent fluctuations on the mean fields is taken into account via the turbulent eddy viscosity and heat conduction coefficients modeled by the realizable k-ε turbulence model (Ref 14). The thermodynamic and molecular transport coefficients of argon are computed as functions of temperature following the formulas provided in Ref 15 and 16. The numerical scheme used by Fluent is based on a finite-volume discretization and is of the second order of accuracy. To identify the gird providing acceptably small numerical errors, the authors have conducted simulations with different degrees of grid refinement and clustering in the areas of strong flow gradients.

The computational domain is shown in Fig. 1(a). Axial symmetry is assumed. In accordance with the focus on the synthesis of nanoparticles rather than on the actual coating process, the target wall, which is typically located at the distance of 10-15 cm from the nozzle, is removed in the model. The chamber walls are also replaced by the free outlet boundaries located far from the nozzle and maintained at constant pressure $p = p_{atm}$ and temperature $T = 300$ K. This simplification is supported by the measurements of actual pressure and temperature conditions and justified by the fact that the particle formation occurs in a zone close to the nozzle and at significant distance to the walls. Throughout the entire domain outside the nozzle, the gas pressure is found to be close to

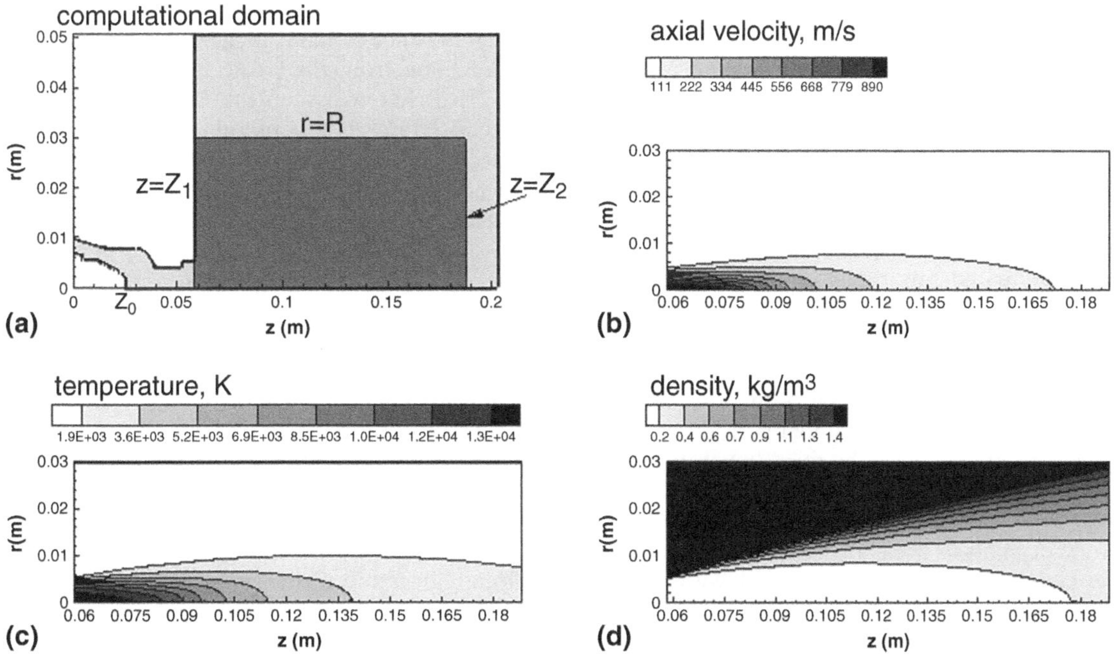

Fig. 1 Results of the simulations of the gas flow. Full computational domain (light gray) and the domain where the evolution of particles is studied (dark gray) (a). Distributions of axial velocity (b), temperature (c), and gas density (d)

atmospheric. The inlet boundary conditions include the mass flux of 1.68×10^{-3} kg s^{-1} of pure argon at temperature of 300 K and pressure of 1.5 p_{atm}.

The effect of arc discharge is imitated by the energy input from a heat source located within the plasma gun. This approach, albeit being a drastic simplification of the real flow physics, can be justified by the fact that the nanoparticle formation occurs entirely outside the nozzle. It can be assumed that the parameters of the jet flow in that area are primarily determined by the rate of energy generation within the nozzle and, as an approximation, neglect the details of the energy generation. The power of the heat source is chosen as 16.2 kW. This is small compared with the typical powers at which the DC plasma reactors operate (22.8-24 kW for the setup in the authors laboratory). The difference accounts for the losses not taken into account directly in the model. The energy generation rate is distributed within the nozzle interior with the density: $E = E_0 \exp\left(-2(s/s_0)^5\right)$, where $E_0 = 1.35 \times 10^{10}$ J m^{-3}, s is the distance from the tip of the cathode ($z = z_0$, $r = 0$) (see Fig. 1a), $z_0 = 2.5$ cm, and $s_0 = z_0/3$.

To account for the energy losses due to cooling of the anode, the convective boundary conditions with the heat transfer coefficient 10^5 W m^{-2} K^{-1} are imposed at its walls. The cathode boundary temperature is constant in time and varies from 300 K at the inlet to 1000 K at the cathode tip, according to the following profile: $T = 300 + 700 \exp\left(-2(s/s_0)^5\right)$. It should be understood that such temperature boundary conditions represent only an attempt of approximation needed in the absence of reliable experimental or computational data. The possible effect of inaccuracy in the thermal boundary conditions on the temperature and velocity of the jet by varying the parameters, such as the temperature of the cathode surface, heat transfer coefficient for the anode wall, value of the inlet pressure, is investigated. No significant effect was found for plausible (not very large) variations.

Yet another simplifying assumption used in this model was needed because of the difficulty of modeling the effect of the precursor droplets and the heat of their evaporation on the gas flow. Although feasible in principle, accurate modeling of this effect is impossible in this case, since essential thermodynamic properties of liquid precursors (e.g., tantalum ethoxide) are largely unknown for considered temperature ranges. The stages of chemical decomposition, which is likely to include hundreds of chemical reactions, are also difficult to predict. For these reasons the authors chose to limit the cases under consideration to those with low loading, i.e., with the mass flow rate of the precursor much smaller than that of the argon gas.

The results of the numerical calculations for the case chosen for further analysis are shown in Fig. 1. The jet of hot argon is emitted into a chamber filled with cooled argon gas. The turbulent mixing provides a rapid fall in the gas temperature. The rate of the temperature fall is, however, smaller compared with the results obtained for jet emitted into the air atmosphere (Ref 12).

4. Theoretical and Numerical Model of Formation and Growth of Nanoparticles

The key elements of the analysis of the particle synthesis are the model of particle formation via the multicomponent co-condensation description using the particle distribution function, and the approximate solution based on the assumption of nearly monodisperse distribution. Only a brief discussion is given here. For a detailed discussion, the reader is referred to the authors' earlier paper (Ref 12).

The nanoparticle formation is modeled within the computational domain marked by the dark gray area in Fig. 1(a). As mentioned above, it is assumed that the liquid precursor molecules are completely dissociated within the nozzle so concentrations of monoatomic gases forming the particles (e.g., Ta and C) can be prescribed at the inlet into the marked zone. Other chemical elements resulting from the precursor dissociation are disregarded in the following analysis. Their existence is partially accounted for by taking the initial concentration of the carbon vapor to be less than the amount of carbon produced by precursor dissociation. This reflects the fact that part of carbon is bounded in, e.g., CO gas that passively leaves the system. This simplification is inevitable, since analysis of the entire set of chemical reactions expected to occur (more than 100 as shown in Ref 6) is very difficult to conduct for multidimensional spatially variable distributions of reagent concentrations, pressure, and temperature. Further difficulty is due to the lack of reliable information on the reaction rates in the temperature range relevant to this problem.

4.1 Multicomponent Co-Condensation

The formation of nanoparticle is described in the framework of the multicomponent co-condensation model (Ref 9, 10). The main idea of the model is based on two facts. First, from the phase diagrams of TaC (Ref 17) or SiC (Ref 18), it is seen that neither TaC nor SiC exist in gaseous state. The compounds are formed after the elements are mixed in liquid droplets. Furthermore, gaseous Ta and C (and other combinations of metal and carbon considered below) have strongly different vapor pressures or, in other words, strongly different temperatures of nucleation of liquid droplets at a given partial pressure.

The resulting scenario of particle formation can be summarized as follows. The first step is the homogeneous nucleation of the element with higher condensation temperature, which is Ta in the case of Ta-C system. As soon as stable clusters of this element appear, the remaining atoms of this and another elements start precipitating on the clusters in the process of heterogeneous condensation. The particles collide and coalesce and are transported by the flow. The elements form compounds while being already in liquid form within a particle.

Because of their small size, the droplets solidify at significantly lower temperature than the bulk melting point of the solid (Ref 19). This allows us to assume that the particles stay in the liquid form until their final deposition. Furthermore, one can make conclusions about the chemical substance to be formed on the substrate based on the final composition of the particles. The reduction in the melting point also allows us to assume fast coalescence of two particles after their collision.

4.2 Governing Equations

The composition of a particle is defined by two internal coordinates x_1 and x_2, which denote the numbers of composing atoms (such as Ta and C). The distribution of compositions at a space location $\mathbf{r}$ and time moment t is given by the particle density function $n(x_1, x_2, \mathbf{r}, t)$. The general dynamic equation for n is (Ref 19)

$$\frac{\partial n}{\partial t} + \nabla\cdot(n\mathbf{v}) + \frac{\partial}{\partial x_1}[G_1 n] + \frac{\partial}{\partial x_2}[G_2 n]$$
$$= \nabla\cdot(D_t \nabla n) + J + \frac{1}{2}\int_0^{x_1}\int_0^{x_2} d\tilde{x}_1 d\tilde{x}_2 \beta(x_1-\tilde{x}_1, x_2-\tilde{x}_2, \tilde{x}_1, \tilde{x}_2)$$
$$\times\, n(\tilde{x}_1,\tilde{x}_2)\, n(x_1-\tilde{x}_1, x_2-\tilde{x}_2)$$
$$-\int_0^{\infty}\int_0^{\infty} d\tilde{x}_1 d\tilde{x}_2 \beta(x_1,x_2,\tilde{x}_1,\tilde{x}_2)\, n(x_1,x_2)\, n(\tilde{x}_1,\tilde{x}_2), \quad \text{(Eq 1)}$$

where $\nabla = \partial/\partial\mathbf{r}$. According to the equation, the particle distribution changes due to transport by the flow velocity $\mathbf{v}$ (the second term in the left-hand side), growth by heterogeneous condensation of element 1 and element 2 (the third and fourth terms in the left-hand side), turbulent diffusion (the first term in the right-hand side), birth by homogeneous nucleation J, and, finally, coagulation of particles due to collisions (the integral terms in the right-hand side). Models for the nucleation, condensation, and coagulation terms are presented below. The turbulent diffusivity D_t is defined as $D_t = \mu_t/(\rho Sc_t)$, where μ_t is the turbulent viscosity calculated by the k-ε model and Sc_t is the Schmidt number taken to be 0.7.

The evolution of molecular concentrations of gaseous reagents is described by the conservation equations:

$$\frac{\partial C_1}{\partial t} + \nabla\cdot(C_1\mathbf{v}) = \nabla\cdot((D_1 + D_t)\nabla C_1) - x_p^* J$$
$$-\int_0^{\infty}\int_0^{\infty} dx_1 dx_2 [G_1 n],$$
$$\frac{\partial C_2}{\partial t} + \nabla\cdot(C_2\mathbf{v}) = \nabla\cdot((D_2 + D_t)\nabla C_2) \quad \text{(Eq 2)}$$
$$-\int_0^{\infty}\int_0^{\infty} dx_1 dx_2 [G_2 n].$$

Here, it is assumed that element 1 is causing the homogeneous nucleation. x_p^* is the critical number of atoms in a nucleating particle. D_1 and D_2 are the coefficients of molecular diffusion estimated using the Stokes-Einstein expression (Ref 19): $D_i = kT/f = \frac{3kT}{2d_i^2}[2\pi p\rho]^{-1/2}$, where the friction coefficient f is evaluated using the kinetic energy formula with zero accommodation coefficient, ρ and p are the density and pressure of surrounding argon gas, and d_i is the diameter of the reagent gas atom. It is found that the molecular diffusion is comparable with the turbulent diffusion for the reagent gases. The situation is different in the case of nanoparticles, where the molecular diffusion is much smaller than D_t and, thus, can be neglected in (1).

The authors do not include the thermophoresis effect into their first-order model defined by (1) and (2). For the small particles (much smaller than the mean free path), the ratio between the thermodiffusion and the simple diffusion can be reduced to v/D, where v is the kinematic viscosity coefficient and D is the diffusion coefficient. The estimations show that this ratio is small even if D is the molecular diffusivity. The ratio turns out to be much less than 1 if the turbulent diffusivity is important. Indeed, the kinematic viscosity of argon at 2000 K is about 10^{-3} m^2 s^{-1} (see Fig. 1 and data in Ref 15), while the coefficient of molecular diffusion is of the order of 10^{-2} m^2 s^{-1} and the coefficient of turbulent diffusion reaches 10^{-1} m^2 s^{-1} (see Ref 12).

The system of governing equations needs to be complemented by the appropriate conditions on the physical boundaries of the subdomain. It is assumed that concentrations of reagent species are uniformly distributed along the inlet boundary at $r \le R_{in}$, where $R_{in} = 0.42$ cm is chosen as a boundary of a zone with sufficiently high temperature so that no nucleation of nanoparticles occurs outside the domain of investigation. Correspondingly, zero concentration of particles at the inlet boundary is assumed. Zero diffusive fluxes of particle density function n and concentrations of chemical species C_1 and C_2 are imposed at other physical boundaries. In the internal coordinates, the conditions of zero flux $G_1 n = 0$ and $G_2 n = 0$ at $x_1 = 0$ and $x_2 = 0$, respectively, are applied. The condition $n \to 0$ at $x_1, x_2 \to \infty$ is used for evaluation of integrals in (2) and the derivation of approximate model in Sect. 4.4.

4.3 Models for Homogeneous Nucleation, Heterogeneous Condensation, and Coagulation

There are numerous theoretical and semiempirical models for the process of homogeneous nucleation. Their common goal is to predict the size of stable nuclei and the nucleation rate given by physical properties and partial pressure of the nucleating gas and ambient temperature. All these theories are based on strong simplifying assumptions and can only give qualitative predictions.

The self-consistent model (Ref 20) is used, according to which the rate of homogeneous nucleation of gas 1 is defined as

$$J = 2C_1 \frac{\sigma^{1/2} v_{11}}{(2\pi m_{11})^{1/2}} C_{S,1} \exp\left(\Theta - \frac{4\Theta^3}{27(\ln(S_1))^2}\right), \quad \text{(Eq 3)}$$

where $\Theta = \sigma s_{11}/(kT)$ is the nondimensional coefficient of surface tension. $S_1 = C_1/C_{S,1} = p_1/p_{S,1}$ is the oversaturation level, p_1 is the partial pressure of the gas, and $C_{S,1}$ and $p_{S,1}$ are the values of saturation concentration and pressure for this material at the local temperature. s_{11}, v_{11}, and m_{11}

Peer Reviewed

stand for the surface, volume, and mass of one atom of the gas, respectively. The droplets appearing as a result of nucleation are stable if their size, measured by the number of atoms, is larger than the critical size

$$x_p^* = (2\Theta/(3\ln(S_1)))^3.$$

The formula (3) is derived under the assumptions that the typical time of nucleation is much smaller than the typical time of the evolution of particles, and that the critical nucleus consists of sufficiently many atoms, generally not fewer than 10. The calculations show that the first assumption is justified in the coating system. On the other hand, the size of a critical nucleus x_p^* can become even smaller than the size of one atom. This is the typical result for the nucleation of materials with very low saturation vapor pressure (Ref 19). In the calculations discussed below, all gas molecules of Ta are treated as stable nuclei if x_p^* is smaller than 2.

The rate of the heterogeneous condensation of material i on the stable nucleus is determined by the growth law, for which the kinetic theory expression (Ref 19) is:

$$G_i = \frac{\alpha_i p_{S,i}}{(2\pi m_{1i} kT)^{1/2}} (S_i - 1)\pi d(x_1, x_2)^2, \quad i = 1, 2, \qquad \text{(Eq 4)}$$

where $d(x_1, x_2)$ is the particle diameter and α_i is the accommodation coefficient assumed to be equal to 1.

Frequency of the particle collision and subsequent coagulation can be described by the expression derived from the kinetic theory of gases (Ref 19):

$$\beta(x_1,x_2,\tilde{x}_1,\tilde{x}_2) = (3kT/\rho_p)^{1/2} \times \left(d(x_1,x_2)^{-3} + d(\tilde{x}_1,\tilde{x}_2)^{-3}\right)^{1/2} \times (d(x_1,x_2) + d(\tilde{x}_1,\tilde{x}_2))^2. \qquad \text{(Eq 5)}$$

Here ρ_p is the particle density, which, as an approximation, is assumed to be constant equal to the density of final material (e.g., TaC or SiC). It is also assumed that, because of the small size and liquid state of the particles, all collisions are 100% effective and the coalescence is rapid, so no fractal agglomeration occurs.

4.4 Method of Solution

The system of equations and boundary conditions outlined above is useful as a theoretical framework but cannot be solved directly. The main reason is that the adequate resolution of the five-dimensional computational domain is hardly possible at today's level of computational power. Furthermore, even a perfectly accurate numerical solution of the full problem would only present a qualitative picture of the actual process due to the approximate manner in which the key factors, such as thermodynamic and transport coefficients, rate of homogeneous nucleation, growth laws, and the collision frequency function, are determined.

For these reasons, an approximate solution approach conceptually similar to the method of moments is used (Ref 19). The system is reformulated into a system of evolution equations for the statistical parameters of the particle distribution: the number concentration of particles, $N \equiv \int_0^\infty \int_0^\infty dx_1 dx_2(n)$, their mean size, $(A_1 + A_2)/N$, and composition, A_2/A_1. Here, the first moments of the particle density function, $A_1 = \int_0^\infty \int_0^\infty dx_1 dx_2(x_1 n)$ and $A_2 = \int_0^\infty \int_0^\infty dx_1 dx_2(x_2 n)$, which represent the mean values of the content of elements 1 and 2 in the particles, are introduced. Applying integration in internal coordinates, it is easy to derive

$$\begin{aligned}
&\frac{\partial N}{\partial t} + \nabla\cdot(N\mathbf{v}) = \nabla\cdot(D_t \nabla N) + J\left(x_p^*\right) + J_{col},\\
&\frac{\partial A_1}{\partial t} + \nabla\cdot(A_1\mathbf{v}) - \int_0^\infty\int_0^\infty dx_1 dx_2 [G_1 n]\\
&\quad = \nabla\cdot((D_1 + D_t)\nabla A_1) + x_p^* J\left(x_p^*\right),\\
&\frac{\partial A_2}{\partial t} + \nabla\cdot(A_2\mathbf{v}) - \int_0^\infty\int_0^\infty dx_1 dx_2 [G_2 n]\\
&\quad = \nabla\cdot((D_2 + D_t)\nabla A_2).
\end{aligned} \qquad \text{(Eq 6)}$$

The Eq 6, together with the transport equations for free gas concentrations (2), represent a system for three-dimensional (depending on two physical space coordinates and time) fields, which is tractable by standard methods of numerical analysis. One problem remains, though. The term J_{col} representing the loss of the number of particles due to coagulations and the integrals containing the growth laws, G_1 and G_2, cannot be computed without knowledge of the full distribution $n(x_1, x_2, \mathbf{r}, t)$ and have to be modeled. This is done in the current work on the basis of the assumption of *monodisperse* distribution. As discussed in detail in Ref 12, the nucleation rate peaks at physical space locations characterized by strong oversaturation S_1. This means that on every streamline of the flow, the particles are formed in a localized area and, thus, are characterized by approximately the same initial size x_p^*. As they follow the streamline, the particles grow in the process of heterogeneous condensation, the rates of which G_i are determined by the local flow conditions and the particle size (see (4)). This means that their distribution in internal coordinate space remains nearly monodisperse.

Assuming the monodisperse distribution

$$\begin{aligned}
n(x_1, x_2, \mathbf{r}, t) &\approx N(\mathbf{r}, t)\,\delta(x_1 - x_1^*)\,\delta(x_2 - x_2^*),\\
A_1(\mathbf{r}, t) &= N x_1^*,\\
A_2(\mathbf{r}, t) &= N x_2^*,
\end{aligned} \qquad \text{(Eq 7)}$$

it is straightforward to derive the following approximations

$$J_{col} = \frac{1}{2}\beta N^2, \quad \beta = (3kT/\rho_p)^{1/2}((A_1 + A_2)/N)^{1/6} \qquad \text{(Eq 8)}$$

and

$$\int_0^\infty\int_0^\infty dx_1 dx_2 [G_i n] = N G_i^*,$$

$$G_i^* = \frac{\alpha_i p_{S,i}\pi(S_i - 1)}{(2\pi m_i kT)^{1/2}} \left(\frac{A_1 d_{i1}^3}{N} + \frac{A_2 d_{i2}^3}{N}\right)^{2/3}, \quad i = 1, 2, \qquad \text{(Eq 9)}$$

where the numbers of atoms used to determine the volume fractions d_{i1} and d_{i2} are given by x_1^* and x_2^*.

Two factors neglected in this model tend to widen the peak and destroy the monodisperse character of the distribution and, therefore, should always be taken into careful consideration. The first is the turbulent diffusion (note that the parameters (x_1^*, x_2^*) of the distribution peak vary strongly between the streamlines and, thus, between locations in physical space). This effect can be neglected if the typical residence time of particles within the system is smaller than the typical time of diffusion. As illustrated in Ref 12, this requirement is satisfied by the current system. The second effect is the coagulation due to random collisions of particles. A detailed discussion of its role in the in-flight synthesis and verification analysis of this approach was provided in Ref 12. It was shown using a one-dimensional one-component model system that the collisions affect the solution during a short period of time immediately after the nucleation. After that, collisions led to rapid decrease of the total number of particles and, correspondingly, to reduction of the collision rate J_{col} (see (8)). The cumulative impact on the particle distribution parameters N, A_1, and A_2 was within the error margin of the other physical models used in this analysis.

The system of Eq 6 with assumptions (7)-(9) was solved numerically in the computation domain shown by the dark gray region in Fig. 1(a) by using the finite-difference approach based on the central-difference scheme of the second order implemented on a uniform rectangular grid. The size of the domain is $(Z_1 - Z_2) \times R = 13 \times 3$ cm. The numerical resolution of 450×120 nodes was found sufficient, with the error of numerical approximation being significantly lower than the errors potentially introduced by the adopted simplifications of the physical model. To use the main flow field for the simulations, the Fluent data for the main gas flow were interpolated onto the new rectangular mesh. The flow data were taken from the computations made on the finest finite-volume mesh with cell size comparable to the step of the finite-difference grid used for the particle simulation.

The last comment concerns verification of the simplified model and numerical approach used in the current study for the process of particle formation. Both were verified in an earlier work (Ref 12) using *a posteriori* comparison with the direct solution of the full problem (1)-(2) available in the case of one-dimensional one-component synthesis. The verification is considered sufficient for the presently studied system since the system retains the principal features that warranted the model validity in Ref 12.

5. Results

5.1 Synthesis of TaC Nanoparticles

As mentioned earlier, the analysis of TaC particle synthesis is conducted for the case of a plasma reactor filled with argon. The formation of nanoparticles is illustrated by Fig. 2 for the inlet concentration of Ta and C gases equal to $C_{10} = C_{20} = 5 \times 10^{21}$ m^{-3}. Figure 2 contains the evolution plots of the concentrations of gaseous Ta and C species, C_1 (Ta) and C_2 (C) (Fig. 2a and b); of concentrations of the materials in their condensed form, A_1 (Ta) and A_2 (C) (Fig. 2c and d); of the nucleation rate J (Fig. 2e), of the number concentration of particles, N (Fig. 2f); and of the mean diameter and mean composition of TaC particles (Fig. 2g and h).

The concentration of atoms of Ta and C are imposed at the left inlet boundary of the computational domain; the evolution of the free atoms is illustrated in Fig. 2(a) and (b). As temperature of the jet decreases, the partial pressure of Ta vapor overcomes its saturation limit. This starts the process of homogeneous nucleation, which rapidly consumes the free atoms of Ta gas. In the course of the process, the nucleation conditions change rapidly so that soon the critical size x_p^* becomes smaller than the size of two Ta atoms. After this moment, all yet free atoms of this material are considered as stable nuclei. The appearance of the nanoparticles marks the initial point of two simultaneous processes. First is the heterogeneous condensation, which is primarily essential for the gas of carbon atoms. The second is the Brownian collisions between the already formed particles.

The temperature of the jet decreases slowly as compared with the results obtained for the plasma jet emitted into the open air atmosphere (Ref 12). As a result, the heterogeneous condensation occurs slowly (see Fig. 3, where the results illustrating the formation of TaC nanoparticles in the plasma jet injected into the air atmosphere are provided for reference). Based on Fig. 2(b), it can be stated that the processes of homogeneous nucleation and heterogeneous condensation occur entirely within the considered computational domain. However, the particles continue their growth due to collisions at the outlet boundary of the domain, which also leads to a rapid fall in their number (Fig. 2f).

As in Ref 12, one should note that the zone where the nucleation occurs (Fig. 2e) is shifted to the outer boundaries of the jet, which is explained by the effect of strong turbulent diffusion generated in the turbulent jet flow. The radial component of the velocity is very small and cannot explain the observed shift in the nucleation zone. The results reported in Fig. 2 are based on the full velocity field of the main gas flow as obtained in Sect. 3. The results remain the same if the radial velocity is manually taken to be 0.

The chosen computational domain covers only the initial moment of the particle development; the particles continue to change because of collisions. The effect of collisions is proportional to the square of the particle number concentration and its importance diminishes as the particles move downstream. At the right outlet boundary, the size of the particles is 2.6 nm, which is small, compared with the experimental data (Ref 2). The discrepancy can be explained, firstly, by the fact that particles continue growing, and, secondly, in Ref 2, the particles are collected from the chamber walls or from the gas that leaves the chamber, i.e., after a much longer particle evolution. In addition, the RF plasma gun used for nanoparticle formation in Ref 2 generates less intensive velocity field. The authors of Ref 1, 4, and 11 found that

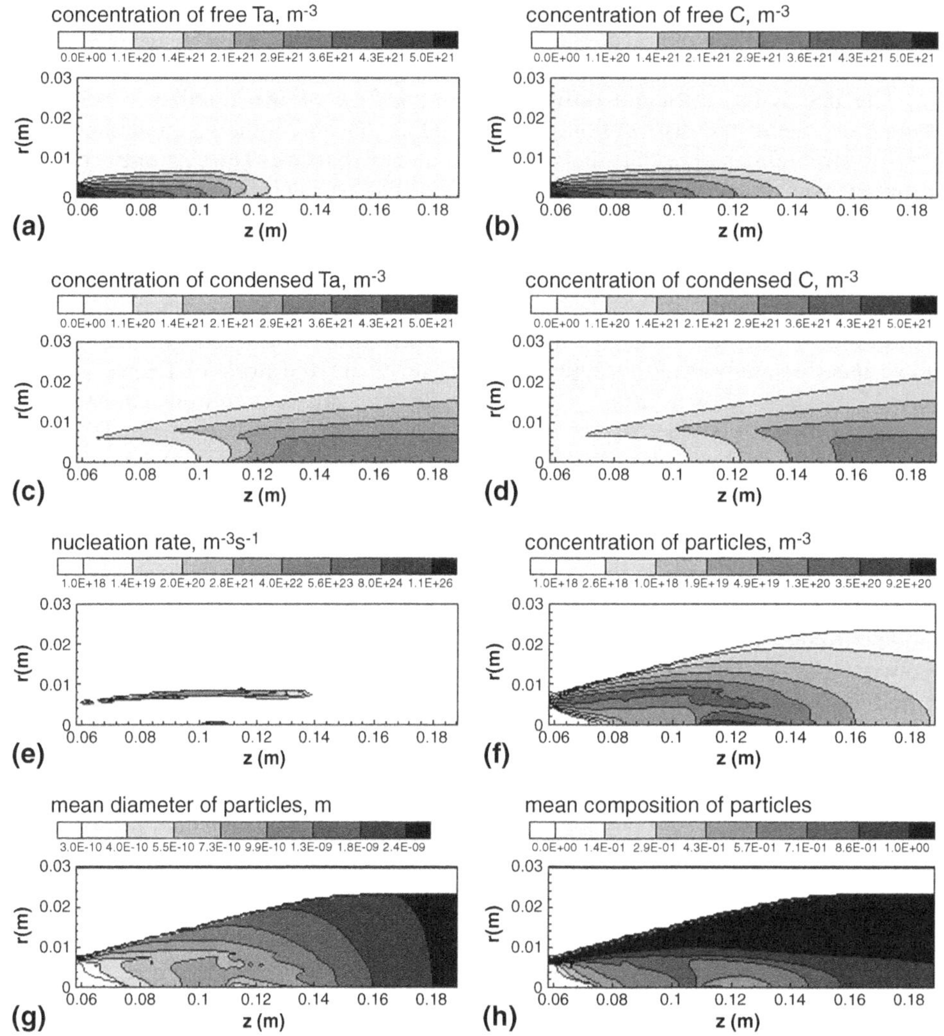

Fig. 2 Formation of TaC nanoparticles in argon-filled plasma reactor. (a) Concentration of free, not-yet-nucleated, Ta atoms, C_1, (b) concentration of free C atoms, C_2, (c) concentration of the condensed atoms of Ta, A_1, (d) concentration of the condensed atoms of C, A_2, (e) rate of homogeneous nucleation, J, (f) concentration of nanoparticles, N, (g) mean diameter and (h) mean composition of nanoparticles. The inlet concentrations of Ta and C vapors are $C_{10} = C_{20} = 5 \times 10^{21}$ m^{-3}. The composition is determined as the ratio of the number of carbon atoms to the number of atoms of metal (Ta) in a particle. The data in plots (g) and (h) are shown for the region with particle concentration $N > 10^{18}$ m^{-3}

the particle size becomes smaller as the cooling rate is increases (by introducing the additional cooling quench to the main jet flow), i.e., the small particle sizes obtained in the present paper could be also explained by the fact that a DC plasma jet has much higher velocity causing higher cooling rate.

The particles at the right outlet boundary have the ratio of Ta to C atoms equal to 1 (Fig. 2h), similar to what was observed in Ref 12, where the final particle composition was determined by the initial inlet concentration of the Ta and C vapors. This is explained by the complete transition of initially injected atoms of Ta and C into a condensed state.

5.2 Other Metal-Carbide Systems

We also analyzed the in-flight formation of other metal-carbide particles of interest, in particular, SiC, TiC, and WC. All these materials can be prepared in similar manner, by using different precursor materials. For instance, SiC particles can be obtained by injection of the mixture of silicon tetrachloride ($SiCl_4$) and methane (CH_4) into the plasma jet (Ref 3).

The analysis of the saturation vapor pressure (see Appendix) reveals that there is a principal difference in the in-flight nucleation of TaC and WC particles from the synthesis of SiC and TiC. Formation of the WC particles follows the scenario discussed in the preceding section, and can be illustrated by the set of plots totally similar to Fig. 2.

In the case of SiC and TiC systems, the nucleation occurs differently, as in this case metal (Si and Ti) possesses lower saturation vapor pressure. For such systems, the nucleation should start with formation of carbon nanoparticles, with the metal precipitating on the surface

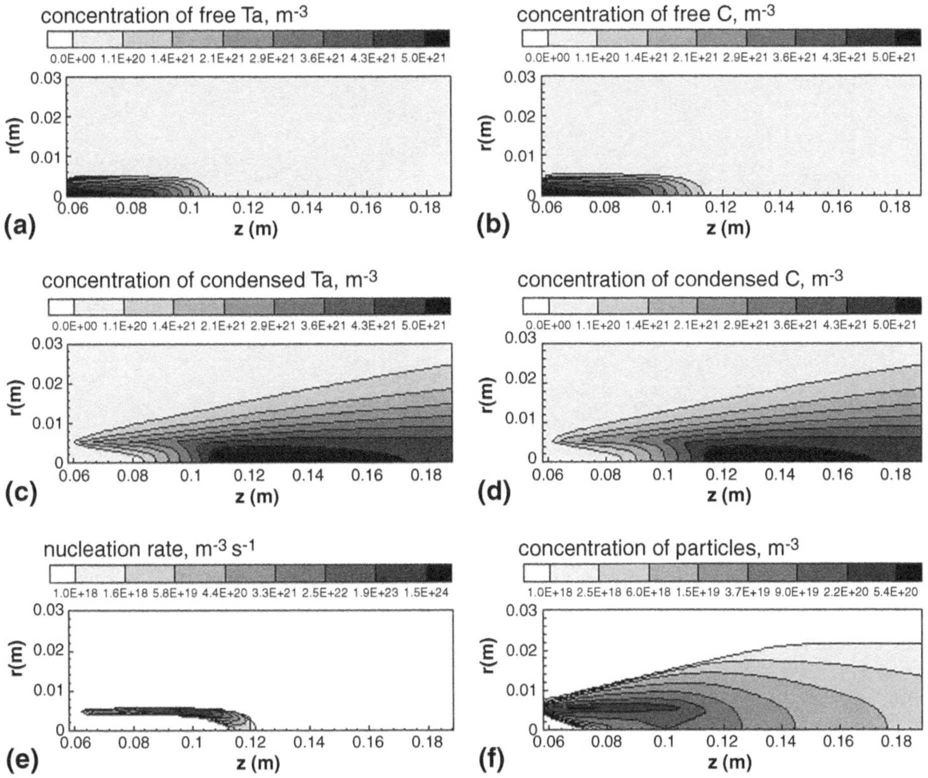

Fig. 3 Formation of TaC nanoparticles in an open air atmosphere. (a) Concentration of free Ta atoms, C_1, (b) concentration of free C atoms, C_2, (c) concentration of the condensed atoms of Ta, A_1, (d) concentration of condensed atoms of C, A_2, (e) rate of homogeneous nucleation, J, (f) concentration of nanoparticles, N. The inlet concentrations of Ta and C vapors are $C_{10} = C_{20} = 5 \times 10^{21}$ m^{-3}

of the nuclei. The analysis becomes more complicated, taking into account the fact that carbon does not exist in the liquid state, at least in bulk volumes, under atmospheric pressure (Ref 21). The simulations of the preceding section can be accomplished by using the free surface energy for a solid material instead of the surface tension.

The results of numerical simulations fulfilled for the Si-C system are summarized in Fig. 4. The synthesis of TiC nanoparticles would occur very similar to this case. Figure 4 is organized in complete correspondence to Fig. 2 and contains the set of plots obtained for the equal inlet concentrations of C and Si, $C_{10} = C_{20} = 5 \times 10^{21}$ m^{-3}. In Fig. 4(h), the particle composition is defined as the ratio of number of carbon atoms to the number of atoms of silicon. At the initial moment the particles are made up of carbon, i.e., the composition ratio is infinite. Later on, the mean particle composition decreases and approaches 1.

The analysis of formation of SiC and TiC nanoparticles shows that lower saturation pressures of nucleating gases delay the processes of homogeneous nucleation and heterogeneous condensation. The condensation of free atoms of metal remains nonnegligible even at the right outlet border of the computational domain. Even at this border, not all atoms of Si gas, injected into the plasma reactor, are condensed (Fig. 4b). This also explains the variation in the mean particle composition (from 1.3 at the symmetry axis to 1 at the outer regions) along the right outlet boundary of the considered domain.

6. Conclusions

The model for the in-flight synthesis of the two-component nanoparticles developed in Ref 12 can be easily adapted for the analysis of the nucleation of various compounds from different precursors that are injected into the thermal plasma jet. The model is based on several simplifying assumptions and can only be used for the qualitative description of the nucleation and growth processes. A more quantitative approach is hindered by the lack of understanding and reliable information concerning the essential physical properties of various processes on the nanoscale. This, in particular, concerns the process of nucleation subject to very high supersaturation levels, when the nucleus size reaches the size of an atom. In the current work, the authors extrapolate the classical approach on this process.

Albeit of approximate nature, this model produces interesting and qualitatively adequate results regarding the in-flight synthesis of metal-carbide nanoparticles. The model predicts the essential role of turbulence on the nucleating dynamics. The turbulent nature of the high-velocity jet shifts the zone where the nucleation takes place and has effect on the size of nuclei. In addition, the co-condensation model allows us to consider how the particle composition varies during the in-flight synthesis. This helps to correlate the amount of the precursor used and the size

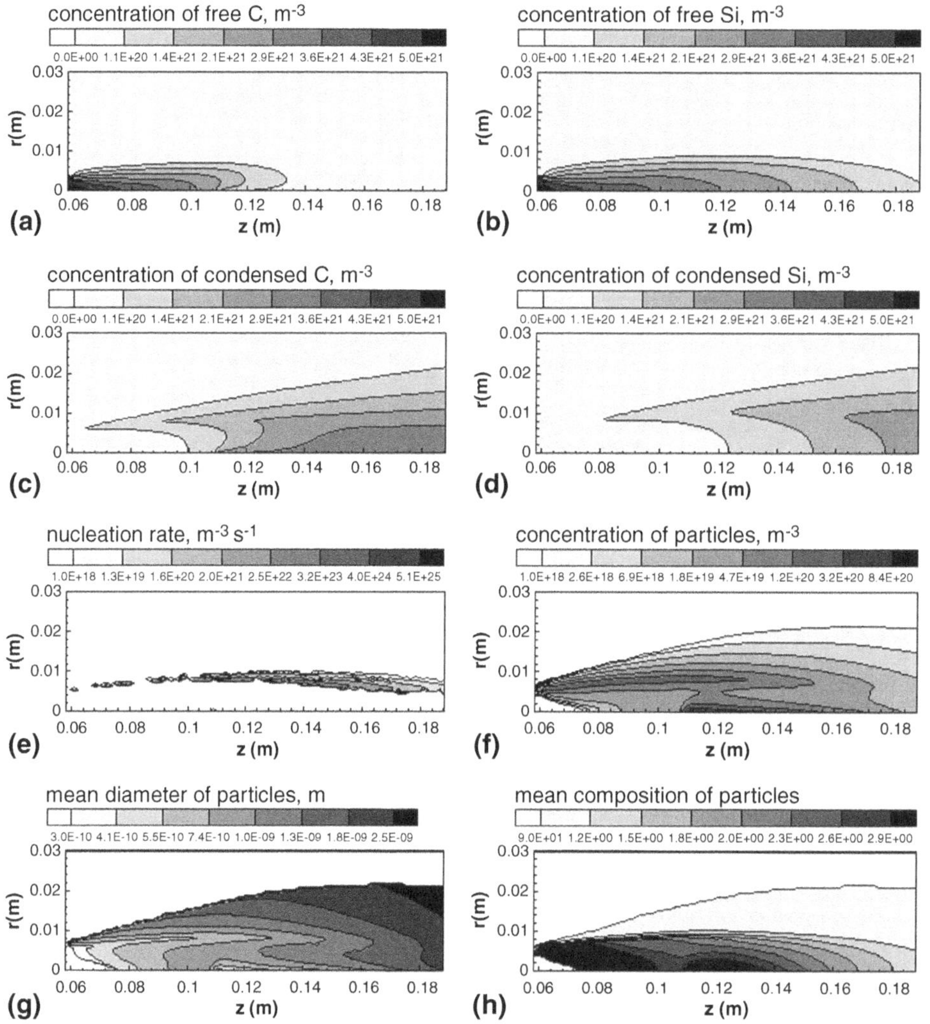

Fig. 4 Formation of SiC nanoparticles in an argon-filled plasma reactor. (a) Concentration of free C atoms, C_1, (b) concentration of free Si atoms, C_2, (c) concentration of the condensed atoms of C, A_1, (d) concentration of the condensed atoms of Si, A_2, (e) rate of homogeneous nucleation, J, (f) concentration of nanoparticles, N, (g) mean diameter, and (h) mean composition of nanoparticles. The inlet concentrations of C and Si vapors are $C_{10}=C_{20}=5\times10^{21}$ m^{-3}. The composition is determined as the ratio of the number of carbon atoms to the number of atoms of metal (Si) in a particle. The data in plots (g) and (h) are shown for the region with particle concentration $N>10^{18}$ m^{-3}

and composition of the nanoparticles, as well as the preferable zone where the nanoparticles should be collected.

One of the most important conclusions is underlining of the principal differences in the nucleation scenarios between TaC and WC systems from SiC and TiC. The co-condensation approach is based on the idea that formation of the nanoparticles begins with nucleation of the element that has lower vapor pressure. For Ta-C and W-C systems, metal initiates the formation of particles. The synthesis of SiC and TiC nanoparticles starts with sublimation of carbon. The analysis becomes even more complex taking into account the fact that the classical theories of the homogeneous nucleation presuppose that the vapor nucleates into liquid state, which does not happen for carbon as carbon does not exist in liquid state under atmospheric pressure.

The theoretical predictions of the co-condensation theory could be experimentally verified by collecting the nanoparticles at different distances from the plasma gun's outlet. For the system of the first kind (TaC and WC), the particles should have smaller carbon content at short distances, and the carbon content should increase downstream; the particles of the second kind (SiC and TiC) should be carbon-rich at shorter distances and the carbon content would decrease downstream.

Acknowledgments

The authors are thankful to the reviewers for insightful and constructive comments.

Appendix

7.1 Physical Properties

The referenced points for Ta, C, and Si gases used for the simulations are summarized in Table 1. Between the

Table 1 Vapor pressure reference points

P, Pa	1	10	10^2	10^3	10^4	10^5	p_{atm}
Carbon: *T*, K	-	2839	3048	3289	3572	3908	-
Tantalum: *T*, K	3297	3597	3957	4395	4939	5634	5731
Tungsten: *T*, K	3477	3773	4137	4579	5127	5823	5828
Silicon: *T*, K	1908	2102	2339	2636	3021	3537	3538
Titanium: *T*, K	1982	2171	2403	2692	3064	3558	3560

The data are taken from Ref 22 and 23

data points, the vapor pressure was approximated by the Clausius-Clapeyron equation:

$$p = p_{\text{ref}} \exp\left(-\frac{Q}{R}\left(\frac{1}{T} - \frac{1}{T_{\text{ref}}}\right)\right).$$

Here p_{ref} and T_{ref} are the values given in Table 1, Q is the latent heat of vaporization per mole, and R is the universal gas constant. For the latent heat of vaporization the following values were used: for Ta, $Q = 3.65 \times 10^5$ J mol^{-1} (Ref 24), for C, $Q = 3.56 \times 10^5$ J mol^{-1} (Ref 25), and for Si, $Q = 3.59 \times 10^5$ J mol^{-1} (Ref 24).

For the homogeneous nucleation of Ta gas, the value of the surface tension coefficient $\sigma = 2.07$ N m^{-1}, reported in Ref 26, was used. In the calculations in the current study, the temperature variations exceed the reported measurement interval. The authors chose to extrapolate the constant value for the surface tension coefficient on the entire temperature range where the particles exist. Following the common assumption of the classical nucleation theories the dependence of the surface tension on the particle size is not considered. To describe the process of nucleation of Si-C, the surface tension of *C* is needed. No measurements of the surface tension for carbon were found in the literature, which is likely because of the fact that carbon is not observed in a liquid form under atmospheric pressure (Ref 21). To estimate the value of the surface tension, the results of measurements of the free surface energy of solid carbon fibers (Ref 27), $\sigma = 0.046$ J m^{-2}, were used.

The density of TaC in solid form is 1.39×10^4 kg m^{-3}, and the density of SiC is 3.22×10^3 kg m^{-3}. To calculate the size of particles, the following diameters of Ta and C atoms were assumed, 2.7×10^{-10} m and 2.6×10^{-10} m; the diameter of Si atoms was taken as 2.2×10^{-10} m.

References

1. B.M. Goortani, P. Proulx, S. Xue, and N.Y. Mendoza-Gonzalez, Controlling Nanostructure in Thermal Plasma Processing: Moving from Highly Aggregated Porous Structure to Spherical Silica Nanoparticles, *Powder Technol.*, 2007, **175**, p 22-32
2. T. Ishigaki, S.-M. Oh, J.-G. Li, and D.-W. Park, Controlling the Synthesis of TaC Nanopowders by Injecting Liquid Precursor into RF Induction Plasma, *Sci. Technol. Adv. Mater.*, 2005, **6**, p 111-118
3. S.-M. Oh, M. Cappelli, and D.-W. Park, Preparation of Nano-Sized Silicon Carbide Powder Using Thermal Plasma, *Korean J. Chem. Eng.*, 2002, **19**(5), p 903-907
4. R. Ye, J.-G. Li, and T. Ishigaki, Controlled Synthesis of Alumina Nanoparticles Using Inductively Coupled Thermal Plasma with Enhanced Quenching, *Thin Solid Films*, 2007, **515**, p 4251-4257
5. M. Desilets, J.-F. Bilodeau, and P. Proulx, Modelling of the Refractive Synthesis of Ultra-Fine Powders in a Thermal Plasma Reactor, *J. Phys. D: Appl. Phys.*, 1997, **30**, p 1951-1960
6. J.S. McFeaters, R.L. Stephens, P. Schwerdtfeger, and M. Liddell, Numerical Modeling of Titanium Carbide Synthesis in Thermal Plasma Reactors, *Plasma Chem. Plasma Proc.*, 1994, **14**(3), p 333-360
7. N.Y. Mendoza-Gonzalez, B.M. Goortani, and P. Proulx, Numerical Simulation of Silica Nanoparticles Production in an RF Plasma Reactor: Effect of Quench, *Mater. Sci. Eng. C*, 2007, **27**(5-8), p 1265-1269
8. A.B. Murphy, Formation of Titanium Nanoparticles from a Titanium Tetrachloride Plasma, *J. Phys. D: Appl. Phys.*, 2004, **37**, p 2841-2847
9. M. Shigeta and T. Watanabe, Multi-Component Co-Condensation Model of Ti-Based Boride/Silicide Nanoparticle Growth in Induction Thermal Plasmas, *Thin Solid Films*, 2007, **515**, p 4217-4250
10. M. Shigeta and T. Watanabe, Growth Mechanism of Silicon-Based Functional Nanoparticles Fabricated by Inductively Coupled Thermal Plasmas, *J. Phys. D: Appl. Phys.*, 2007, **40**, p 2407-2419
11. M. Shigeta and T. Watanabe, Numerical Investigation of Cooling Effect on Platinum Nanoparticle formation in Inductively Coupled Thermal Plasmas, *J. Appl. Phys.*, 2008, **103**, p 074903
12. A. Vorobev, O. Zikanov, and P. Mohanty, Modelling of the In-Flight Synthesis of TaC Nanoparticles from Liquid Precursor in Thermal Plasma Jet, *J. Phys. D: Appl. Phys.*, 2008, **41**, 085302 (15p)
13. J.P. Trelles, E. Pfender, and J. Heberlein, Multiscale Finite Element Modeling of Arc Dynamics in a DC Plasma Torch, *Plasma Chem. Plasma Proc.*, 2006, **26**(6), p 557-575
14. T.-H. Shih, W.W. Liou, A. Shabbir, Z. Yang, and J. Zhu, A New k-ε Eddy Viscosity Model for High Reynolds Number Turbulent Flows, *Comput. Fluids*, 1995, **24**(3), p 227-238
15. W.L.T. Chen, J. Heberlein, and E. Pfender, Critical Analysis of Viscosity Data of Thermal Argon Plasmas at Atmospheric Pressure, *Plasma Chem. Plasma Proc.*, 1996, **16**(4), p 635-650
16. B. Liu, T. Zhang, and D.T. Gawne, Computational Analysis of the Influence of Process Parameters on the Flow Field of a Plasma Jet, *Surf. Coat. Technol.*, 2000, **132**, p 202-216
17. *A Critical Review of Refractories*, Los Alamos Scientific Laboratory of the University of California, Los Alamos, NM, 1964
18. Yu.M. Tairov and V.F. Tsvetkov, Semiconductor Compounds $A^{IV} B^{IV}$, *Handbook on Electrotechnical Materials*, Vol. 3, Yu.V. Koritskii, V.V. Pasynkov, and B.M. Tareev, Ed., Energomashizdat, Leningrad, 1988, p 446-471 (in Russian)
19. S.K. Friedlander, *Smoke, Dust, and Haze. Fundamentals of Aerosol Dynamics*, Oxford University Press, 2000
20. S.L. Girshick, C.-P. Chiu, and P.H. McMurry, Time-Dependent Aerosols Models and Homogeneous Nucleation Rates, *Aerosol Sci. Technol.*, 1990, **13**, p 465-477
21. J. Zazula, *On Graphite Transformations at High Temperature and Pressure Induced by Absorption of the LHC Beam*, CERN-LHC-Project-Note-78, 1997
22. H.R. Leider, O.H. Krikorian, and D.A. Young, Thermodynamic Properties of Carbon up to the Critical Point, *Carbon*, 1973, **11**(5), p 555-563
23. D. Stull, *American Institute of Physics Handbook*, 3rd ed., D.E. Gray, Ed., McGraw-Hill, New York, 1972
24. J.A. Dean, *Lange's Handbook of Chemistry*, 15th ed., McGraw-Hill, New York, 1999
25. R.H. Schumm, D.D. Wagman, S. Bailey, W.H. Evans, and V.B. Parker, National Bureau of Standards (USA), *Technical Notes* 270-1 to 270-8, 1973
26. A.P. Miiller and A. Cezairliyan, Measurement of Surface Tension of Tantalum by a Dynamic Technique in a Microgravity Environment, *Int. J. Thermophys.*, 1993, **14**(5), p 1063-1075
27. S. Nuriel, L. Liu, A.H. Barber, and H.D. Wagner, Direct Measurement of Multiwall Nanotube Surface Tension, *Chem. Phys. Lett.*, 2005, **404**, p 263-266

JTTEE5 17:966–973
DOI: 10.1007/s11666-008-9272-3
1059-9630/$19.00 © ASM International

Peer Reviewed

Particle Behavior During the Arc Spraying Process with Cored Wires

W. Tillmann, E. Vogli, M. Abdulgader, M. Gurris, D. Kuzmin, and S. Turek

(Submitted May 8, 2008; in revised form September 29, 2008)

To use the manifold possibilities that arc spraying offers to deposit wear resistance layers, knowledge of the particle formation and their behavior is necessary. This work is focused on studying the particle properties during arc spraying with cored wires. Different cored wires under various spraying parameters are investigated by means of a high speed camera. Particle properties in-flight, such as velocity and temperature, are determined. Correlation between particle behavior and particle characteristics at different spraying conditions is established. At the same time, the particle-laden gas flow is simulated numerically and the computed solutions are used to illustrate the utility of the proposed CFD model and compared with experimental results. The employed mathematical model represents a system of macroscopic conservation laws for the continuous gas phase and for the gas-solid mixture. This approach formulation makes it possible to circumvent the numerical difficulties associated with the implementation of a (potentially ill-posed) two-fluid model. The discretization in space is performed using a high-resolution finite element scheme based on algebraic flux correction in terms of local characteristic variables. The artificial diffusion operator is constructed on the discrete level and fitted to the local solution behavior using a multidimensional flux limiter of TVD type.

Keywords arc spraying, CFD model, cored wires, multiphase flow, particle properties

1. Introduction

The *t*win *w*ire *a*rc *s*praying (TWAS) technique is regarded as a versatile spraying process, where the materials to be sprayed in wire form are melted (Ref 1). The melted materials are atomized and accelerated toward the substrate by means of pressurized air. The coating properties are considered to be a sum of the process parameters, which directly or indirectly influence the particle formation, size, velocity, and temperature (Ref 2, 3). Lampann (Ref 4) found out by spraying metallic wires through TWAS that particle formation is strongly affected by the atomization gas pressure. By applying a low atomization gas pressure (<2 bars), thread-shaped particles are predominantly generated. With the increase of atomization gas pressure the molten wire materials are directly atomized at the electrode spikes. The generated primary particles will further atomize under the influence of the high atomization gas pressure and the turbulent

W. Tillmann, E. Vogli, and **M. Abdulgader**, Institute of Materials Engineering, Dortmund University of Technology, Leonhard-Euler-Str. 2, 44227 Dortmund, Germany; and **M. Gurris, D. Kuzmin,** and **S. Turek**, Insitute of Applied Mathematics, Dortmund University of Technology, Vogelpothsweg 87, 44227 Dortmund, Germany. Contact e-mail: wolfgang.tillmann@udo.edu.

Nomenclature	
Latin Alphabet	
$\overline{u}_g$	Velocity of the inviscid gas, m/s
P_g	Pressure of the inviscid gas, Mpa
E_g	Total energy of the inviscid gas per unit volume, J/m^3
$\overline{u}_g$	Velocity of the disperse particle phase, m/s
E_p	Total energy of the disperse particle phase per unit volume, J/m^3
E	Energy of the mixture per unit volume, J/m^3
I	Unit tensor
M	Mach number
Greek Symbols	
α_g	Volume fraction of the inviscid gas
ρ_g	Density of the inviscid gas, kg/m^3
γ_D	Rate of interphase momentum transfer, kg/(s·m^3)
α_p	Volume fraction of the disperse particle phase
ρ_p	Density of the disperse particle phase, kg/m^3
ρ	Effective density of the mixture, kg/m^3
∇	Del operator
Abbreviations	
TWAS	Twin wire arc spraying
TVD	Total variation diminishing
CFD	Computational fluid dynamics
PDE	Partial differential equation
CCD	Charge-coupled device
SEM	Scanning electron microscopy

stream in so-called secondary particles, which are smaller in size than the primary particles. Thus, the particle size distribution exhibits a broad spectrum associated with an inhomogeneous coating morphology. Wen et al. (Ref 5), Watanabe et al. (Ref 6), and Hussary et al. (Ref 7) came to the same conclusion.

Recently, the use of cored wires has extended the field of TWAS application. The inhomogeneous composition of the cored wires causes an additional uncertainty in particle formation and characteristic features of particles in flight (Ref 8, 9). To have a good coating quality produced by means of cored wires and to secure the reproducibility of a desired coating, an accurate interpretation of the particle formation and particles in-flight characteristics is needed.

In addition to process and coatings characterization, effort has been made to describe the spraying processes by mathematical models (Ref 10, 11). For example, the flow behavior in a plasma torch and plasma flow after powder feeding in a plasma spraying process are described with multiphase compressible flow models, while the particle broadening and solidification as well as the determination of their size and form are characterized by employing the incompressible Navier-Stokes equations. Kamnis et al. (Ref 12) developed mathematical models to predict particle dynamic behavior in a liquid fuel high velocity oxy-fuel thermal spray gun. They applied a 3-dimensional CFD model, which employed a Lagrangian particle tracking frame coupled with a steady-state gas flow field.

In this work, an online diagnostic of particles was utilized to measure their velocity and temperature along the spray jet. The particle sizes were examined by analyzing sprayed particles, which are quenched in water during flying. In addition to the experimental work, a mathematical model was developed and applied to the TWAS process. Because of the by principal measurement restrictions, the behavior of the particles, especially the velocity, cannot be measured at a standoff distance of less then four centimeters of the nozzle exit (Ref 13). On the other hand this area is very important to understand the spray generation. Due to this fact, a computational model is developed to predict the properties of the gas and the particulate phase. Moreover, the gas properties inside the nozzle and the influence of the injected particles are rather unknown and difficult to measure. Such properties are also predicted by the proposed model. Furthermore, the physical processes controlling the TWAS are complex due to the turbulent fluid flow as well as the unsteady arcs formed between the wire tips. Modeling the process is therefore difficult and requires extensive validation. Finally, this model may help in the future to optimize the geometry and shape of the nozzle, which would require a very huge amount of work to be done by experiments. In contrast to (Ref 12), the mathematical modeling used in the present work was based on a Euler/Euler approach. The theoretical and experimental results were compared and aligned for further optimization of the mathematical model.

2. Experimental Setup

2.1 Mathematical Model

Mathematical models of particle-laden gas flows are usually based on macroscopic conservation laws that can be postulated or derived using a suitable averaging procedure (Ref 14-18). Neglecting molecular diffusion one obtains a set of partial differential equations for the mass, momentum, and energy of the continuous gas phase:

$$\partial_t(\alpha_g\rho_g) + \nabla\cdot(\alpha_g\rho_g\overline{u}_g) = 0 \qquad \text{(Eq 1)}$$

$$\partial_t(\alpha_g\rho_g\overline{u}_g) + \nabla\cdot(\alpha_g\rho_g\overline{u}_g\otimes\overline{u}_g + \alpha_g P_g I) = \gamma_D\alpha_p(\overline{u}_p - \overline{u}_g) \qquad \text{(Eq 2)}$$

$$\partial_t(\alpha_g\rho_g E_g) + \nabla\cdot(\alpha_g\overline{u}_g(\rho_g E_g + P_g)) = \gamma_D\alpha_p\overline{u}_p(\overline{u}_p - \overline{u}_g) \qquad \text{(Eq 3)}$$

where α_g, ρ_g, $\overline{u}_g$, P_g, and E_g are the volume fraction, density, velocity, pressure, and total energy of the inviscid gas. The constant coefficient γ_D denotes the rate of interphase momentum transfer due to the viscous drag force and includes information about the particles diameter (Ref 16). For simplicity, all other forces (virtual mass and lift) as well as mass and heat transfer effects are neglected. A similar system of conservation laws can be formulated for the disperse particle phase (index p):

$$\partial_t(\alpha_p\rho_p) + \nabla\cdot(\alpha_p\rho_p\overline{u}_p) = 0 \qquad \text{(Eq 4)}$$

$$\partial_t(\alpha_p\rho_p\overline{u}_p) + \nabla\cdot(\alpha_p\rho_p\overline{u}_p\otimes\overline{u}_p + \alpha_p P_g I) = \gamma_D\alpha_p(\overline{u}_g - \overline{u}_p) \qquad \text{(Eq 5)}$$

$$\partial_t(\alpha_p\rho_p E_p) + \nabla\cdot(\alpha_p\overline{u}_p(\rho_p E_p + P_g)) = \gamma_D\alpha_p\overline{u}_p(\overline{u}_g - \overline{u}_p) \qquad \text{(Eq 6)}$$

It is commonly assumed that the pressure P_g is the same for both phases and satisfies the equation of state:

$$P_g = (\gamma - 1)\rho_g\left(E_g - \frac{|\overline{u}_g|^2}{2}\right) \qquad \text{(Eq 7)}$$

Furthermore, the volume fractions of the continuous and disperse phase are coupled by the constraint:

$$\alpha_g + \alpha_p \equiv 1 \qquad \text{(Eq 8)}$$

Inviscid two-fluid models such as (1)-(6) are particularly difficult to solve numerically due to the lack of hyperbolicity and uncertainty regarding the modeling of the disperse phase in particle-free regions. If the volume fraction of particles equals zero, the effective density $\alpha_p\rho_p$ and momentum $\alpha_g\rho_g\overline{u}_g$ show the same behavior. As a consequence, the velocity of particles $\overline{u}_p$ is not uniquely defined, which may give rise to nonphysical solutions and waves moving at wrong speeds. Clearly, such a potentially degenerated two-fluid model is of little practical utility. To circumvent the above mathematical and numerical difficulties, the gasparticle mixture was considered as a single

Peer Reviewed

fluid. The effective density, momentum, and energy of the mixture are given by the following equations:

$$\rho = \alpha_p \rho_p + \alpha_g \rho_g \qquad \text{(Eq 9)}$$

$$\rho \bar{u} = \alpha_p \rho_p \bar{u}_p + \alpha_g \rho_g \bar{u}_g \qquad \text{(Eq 10)}$$

$$\rho E = \alpha_p \rho_p E_p + \alpha_g \rho_g E_g \qquad \text{(Eq 11)}$$

The so-defined conservative variables ρ, $\rho\bar{u}$, and ρE are assumed to satisfy the compressible Euler equations:

$$\partial_t(\rho) + \nabla \cdot (\rho \bar{u}) = 0 \qquad \text{(Eq 12)}$$

$$\partial_t(\rho \bar{u}) + \nabla \cdot (\rho \bar{u} \otimes \bar{u} + P_g I) = 0 \qquad \text{(Eq 13)}$$

$$\partial_t(\rho E) + \nabla \cdot (\bar{u}(\rho E + P_g)) = 0 \qquad \text{(Eq 14)}$$

By definition of the mixture density ρ, the continuity Eq 12 is the sum of (1) and (4). The effective pressure of the mixture can also be defined as P_g plus a correction describing the influence of the particles (Ref 19, 20).

The derivation of drift-flux/mixture models can be found in (Ref 21) among others. The model used in the present work does not incorporate turbulence explicitly. The modeling of turbulence phenomena is rather controversial especially in the framework of multiphase flow. Moreover, turbulence modeling results in overdiffusive schemes due to the additional turbulent diffusion and increases the computational complexity. On the other hand, the numerical diffusion, which is added by the nonlinear high-resolution scheme, can be viewed as an implicit large eddy simulation. Therefore, many researchers shift their focus to the implicit modeling of turbulence, provided by the nonlinear discretization scheme (Ref 22). The implicit treatment of turbulence is adopted in this study as well.

In the present work, Eq. 12 to 14 are solved instead of (1)-(6), whereby the density, momentum, and energy of the particulate phase can be readily inferred from algebraic relations (9)-(11). This formulation can be classified as a drift-flux model. The PDE system to be solved consists of subproblems (1)-(3) and (12)-(14) both of which reduce to the Euler equations of gas dynamics in particle-free regions, where the standard two-fluid model (1)-(6) is not applicable.

2.2 Spraying Experiments

To study the particle behaviors by arc spraying, a device with a closed nozzle system Smart Arc 350 PPG (Sulzer Metco, Switzerland) was used. As feedstock a Fe-based cared wise and a Ni-based cored wire with a diameter of 1.6 mm and hard particle grain size of 40-70 μm were employed (Fig. 1). The composition of the used cored wires is shown in Table 1 and the experimental setup is shown in Fig. 2.

In an aim to analyze the particles formation as well as their temperature and velocity along the spray jet, the standoff distance was varied between 45 and 150 mm, while other spraying parameters were kept constant (Table 2). The electric current and voltage levels were

Fig. 1 SE micrograph of hard particles in Fe-based cored wire

Table 1 Composition of the employed wires

Wire	Content, wt.%	Grain size, µm	Velum material	Velum thickness, µm
FeCrBSiMn	Cr-29; Mn-1,6; Si-1,6; B-3,8; Fe-Bal.	40-70	Fe	200
NiCrBSi+PC	Cr-24; Si-5,25; B-1,8; PC-0,4; Ni-Bal.	40-70	Ni	200

taken from prior works, in which the arc ignition stability and arc fluctuations were analyzed (Ref 23).

The particle temperature and velocity at different standoff distances were measured by means of Accuraspray-g3 (Tecnar, Canada). The properties are measured within two measuring volumes with a dimension of 3 mm in diameter and 25 mm in length. Through a dual fiber optical system the flow of particles at two different points along the spray jet were detected, and the velocity is determined as a ratio of the gap between two measuring points and the time delay. Based on the twin wave length pyrometer principle, the mean particle temperature is measured. In this case, it was assumed that the emissivity of the particles is the same for the two wave lengths.

The particle formation and atomization were analyzed by means of a high speed camera (PCO, Germany), which is equipped with a TV-zoom lens with a 0.9× to 6.3× magnification (Opto Sonderbedarf GmbH, Germany). This camera system comprises an image splitter unit, four intensified CCD camera modules with fast switchable MCP image intensifiers, and high-resolution CCD image sensors. Each module with its 12 bit dynamic range and a high-resolution CCD image sensor features a good signal-to-noise-ratio and the ability of single photon detection. Four high speed serial fiber optic data links connect the system to a PC. All images in this study have been captured at a framing rate of 333 megaframes/s with an image resolution of 1280 × 1024 pixels. To investigate the particle size during arc spraying experiments, a custom-made

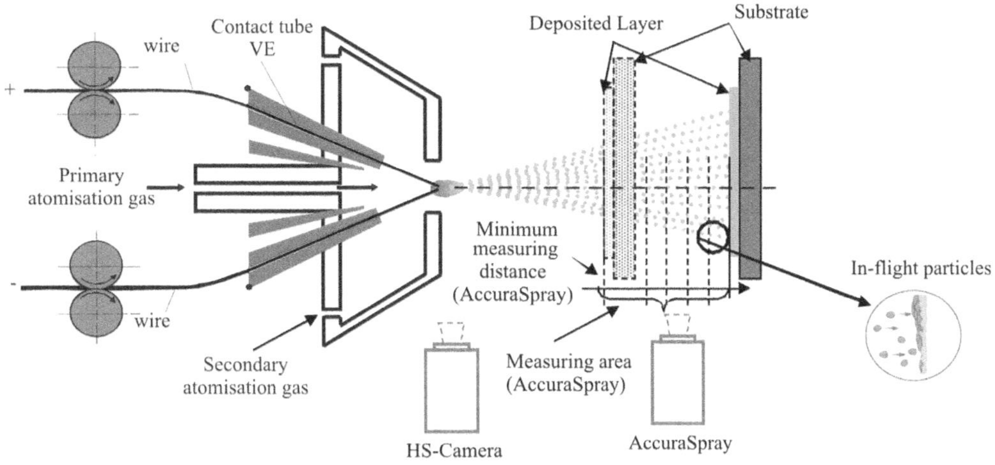

Fig. 2 Experimental setup

Table 2 Arc spraying process parameters

Parameter	Parameter value
Current, A	320
Voltage, V	32
Primary air pressure, bar	6
Secondary air pressure, bar	0
Spraying distance, mm	45-150

experimental setup was designed and applied (for more details, see Ref 23). The generated particles were analyzed by means of SEM (SEM, JSM-840, Co. Jeol, Japan) with respect to their size and shape. Correlation between particle sizes, their temperatures, and velocities were studied. Finally, the results from the mathematical model were compared to the experimental results.

3. Results and Discussion

3.1 Numerical Algorithm and Results

The numerical treatment of coupled PDE systems that constitute an inviscid two-fluid/drift-flux model is a very challenging task. Most of the numerical algorithms published to date are based on a finite difference or finite volume discretization, while the use of finite elements is rather uncommon in this area of CFD research.

Our implementation of the new drift-flux model (1)-(3), (12)-(14) is based on a high-resolution finite element scheme, which is applicable to unstructured meshes and implicit time-stepping schemes. A multidimensional generalization of Roe's approximate Riemann solver and an algebraic flux correction scheme of TVD type are constructed building on the methodology described in (Ref 24-26). The design of the artificial viscosity operator, flux limiting, and artificial implementation of the numerical boundary conditions are performed in terms of local characteristic variables. Particle-laden jets may give rise to low Mach number regions that coexist with sonic points and local supersonic regions. The stiffness associated with the wide range of Mach numbers to be resolved calls for the use of an unconditionally stable implicit time discretization. Explicit schemes turn out rather inefficient, unless characteristic time stepping or local preconditioning is employed. The stationary counterpart of the above drift-flux model is solved using the fully implicit backward Euler method to match the solution to a steady state. This pseudo-time-stepping scheme can be interpreted as an iterative solver, whereby the artificial time step serves as a variable under relaxation parameter.

The validation of the proposed drift-flux model in the context of thermal spraying processes is complicated by the lack of experimental data and reliable benchmarks for this class of multiphase flows. The numerical results for a prototypical 2D configuration with a simplified nozzle are depicted in Fig. 3.

The simplified nozzle consists of two wires, which are modeled by solid walls with a gas-particle inlet in-between. Since no model of particle formation and atomization is available so far, the particles are injected through the boundary in the vicinity of the wires. The volume fraction of particles equals $\alpha_p = 0.001$ at the inlet. The dimensions of the computational domain are 0.1 m × 0.1 m. Numerical solutions are computed on an unstructured triangular mesh with 48,785 nodes and a total of 390,280 unknowns.

The presented simulation results were computed using the supersonic boundary conditions:

$$M_g = 1.1,\ P_g = 0.2\,\text{MPa},\ u_{1P} = u_{1g}$$

$$u_{2g} = u_2 = 0\,\text{m/s},\ \rho_g = 2.3767\,\text{kg/m}^3$$

in the region between the wires and the slip condition on the wires. The slip conditions are implemented by projecting the residual and the solution on the tangent to the boundary. At the remaining boundaries, atmospheric conditions are prescribed for the incoming Riemann invariants. The density of the particles (nickel) is $\rho_p = 8908$ kg/m^3 and the viscous drag is given by $\gamma_D = 1.26 \times 10^6$ kg/(s m^3). At the postprocessing step, it is assumed that the particles move with the gas velocity if the

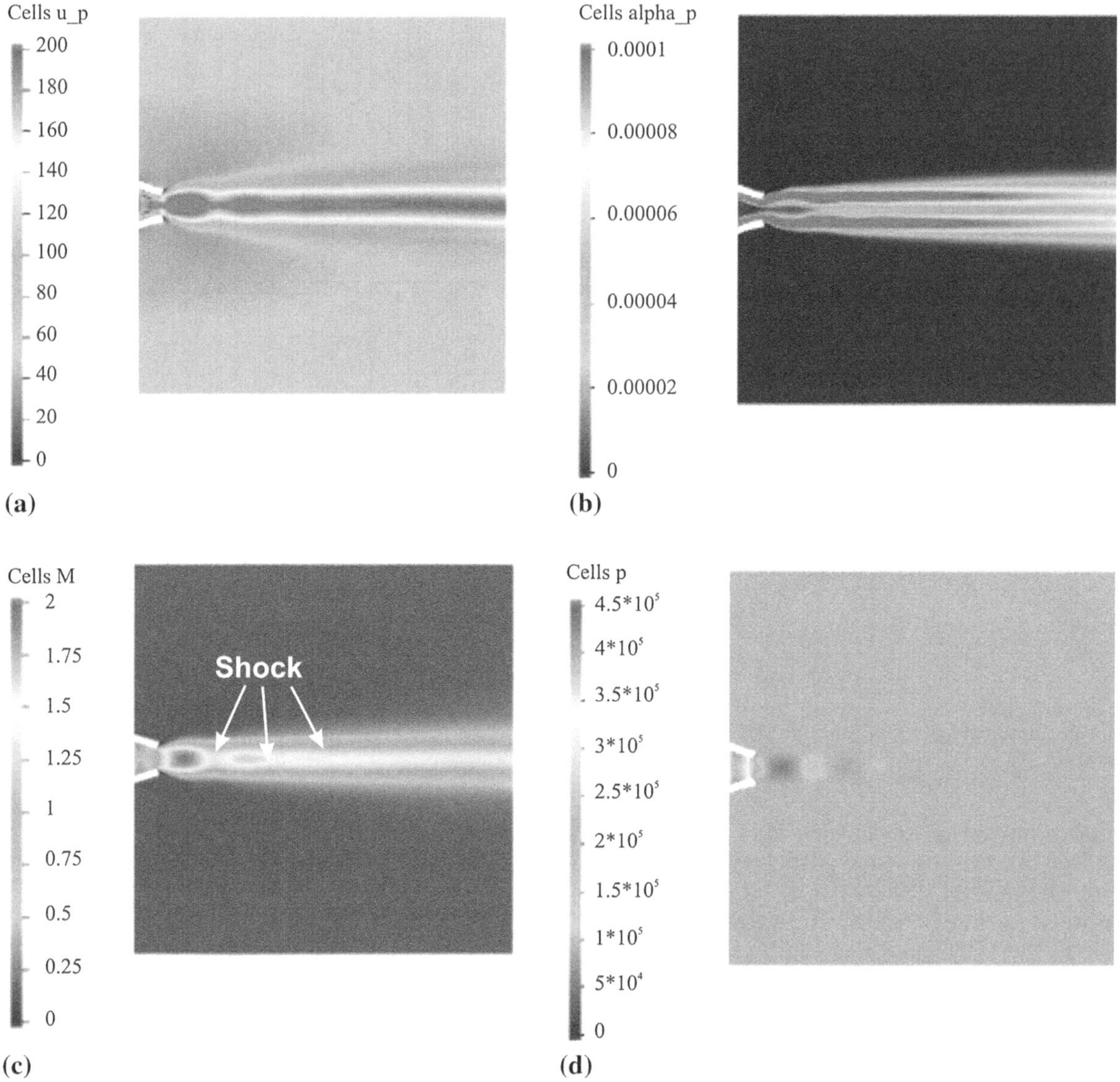

Fig. 3 Numerical results of (a) particle velocity in m/s; (b) volume fraction of particles in %; (c) Mach number of the gas; and (d) gas pressure in Pa

volume fraction is less than 10^{-7}. This prevents division by zero and does not affect the computed solution.

Due to the lack of experimental measurements of the gas inlet data and the particle mass flux at the current state, these quantities are only estimated for the computations. This leads to an overestimation of the velocities and the Mach number. Therefore, more accurate estimates of the inlet boundary conditions will give more accurate velocities. Note that these overestimations arise from the estimation of the boundary values and they are not caused by the computational model.

On the other hand, these results compare qualitatively well with the ones computed by Pourmousa (Ref 27). Especially the pressure profile at a standoff distance of more than five centimetres is approximately the same. This indicates an accurate prediction of the pressure by this model, which is an important issue in multiphase flow modeling. Furthermore the velocities decrease in a short standoff distance in front of the nozzle, remain rather constant near the outlet boundary (see Ref 27), and depend on the pressure (a high gas pressure results in a high particle velocity and vice versa Ref 7). Moreover, three shocks with a very high Mach number at a standoff distance of about 18, 35, and 50 mm in front of the nozzle are observed. These high Mach number regions cause the turbulences, which are observed in the experimental results. Wilden comes to the same conclusion (Ref 28). Taking everything into consideration, this model is able to predict the main properties of arc spraying, which are also observed by other researchers. To improve the results, more accurate estimates of the inlet boundary conditions are necessary.

3.2 Spraying Results

The results of particle velocities along the spray jet for both kinds of wires employed are shown in the Fig. 4. Particles generated from both wires exhibit the same

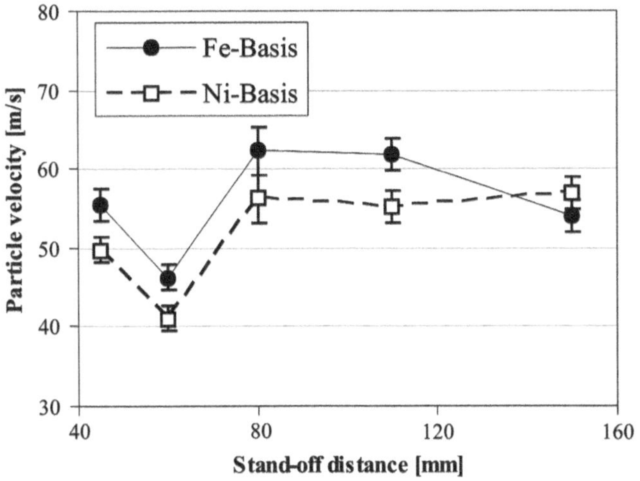

Fig. 4 Variation of particle velocity along the spray jet

behavior. The velocity of particles decreases up to a standoff distance of 60 mm. This is in accordance with the mathematical modeling results. At a standoff distance of 50 mm a shock wave is predicted in the mathematical model, which is associated with turbulences (see Fig. 3d). Due to the high gas pressure of the spray jet and the existing turbulences in this area, the particles are decelerated and therefore have low velocities. No turbulences were predicted in the mathematical model at standoff distances larger than 50 mm, which is verified in experimental measurements. The particle velocities posses approximately a constant value after a standoff distance of 80 mm.

It is worth nothing that the particles generated from Fe-based wires possess higher velocities than the particles generated from Ni-based wires (Fig. 4). To explain this different behavior the particles generated from both wires during spraying were analyzed by means of SEM (Fig. 5). The Fe-based particles show a mean particle size of 6 μm (Fig. 5a), while the mean particle size of Ni-based particles is 8.4 μm (Fig. 5b). The different particle sizes are explained with different melting points of the wires velum. Ni-based wires have a Ni-velum with a melting point at 1455 °C, while the Fe-based wires have a Fe-velum with a melting point at 1538 °C (Ref 29). Therefore by applying the same spraying parameter (voltage and current) the melting rate of Ni-based wires is larger than that of the Fe-based wires. Consequently, more material is molten on Ni-based wire tips than on Fe-based wire tips. This is associated with a higher atomization rate of Fe-based melting bath and smaller particle sizes compared to Ni-based wires during continuous removal of generated melted bath from the wire tips. Hussary and Heberlein (Ref 7) also came to the same conclusion by arc spraying of steel wires.

The turbulences developed in the spray jet favour in turn the atomization process of generated particles as a result of aerodynamic forces. By analyzing the arc jet with the high speed camera, it was found that after the first atomization of the melting bath large particles are

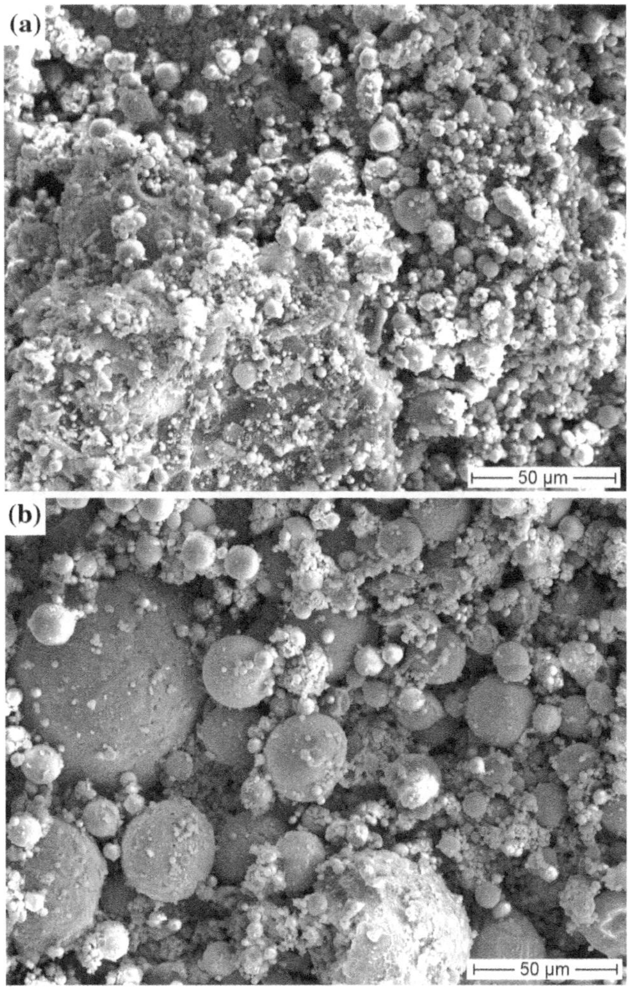

Fig. 5 SE micrographs of (a) Fe- and (b) Ni-based particles

produced (Fig. 6). Along the spray jet they are fragmented in to smaller particles. Hussary and Heberlein (Ref 7) and Newbery et al. (Ref 30) came to the same conclusion by studying arc spraying of metallic wires and spraying of steel wires, respectively. They explain this behavior with the aerodynamic forces and breakup mechanisms of molten bath.

After atomization, particles with smaller sizes are produced, which move with higher velocities than the larger particles at the same gas pressure levels. This explains the higher velocities of small Fe-based particles compared to the large Ni-based particles.

However, the movement of particles in the spray jet depends on the spray jet properties. At high turbulences (high Mach and Reynolds number), the small particles can be more easily stopped during flight due to their low weight, which explains the slight decrease of the velocities of the Fe-based particles at 150 mm standoff distance Mahesh et al. (Ref 31) came to the same conclusion by modeling and analyzing droplet dynamic and thermal behavior of particles during thermal spray deposition. They found out that the smaller particles attain maximum velocity at a shorter time, and subsequently they travel

Peer Reviewed

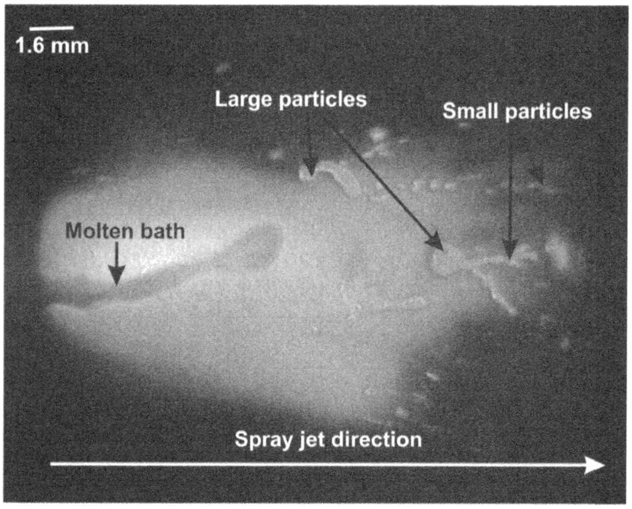

Fig. 6 High speed camera image of Fe-based wire

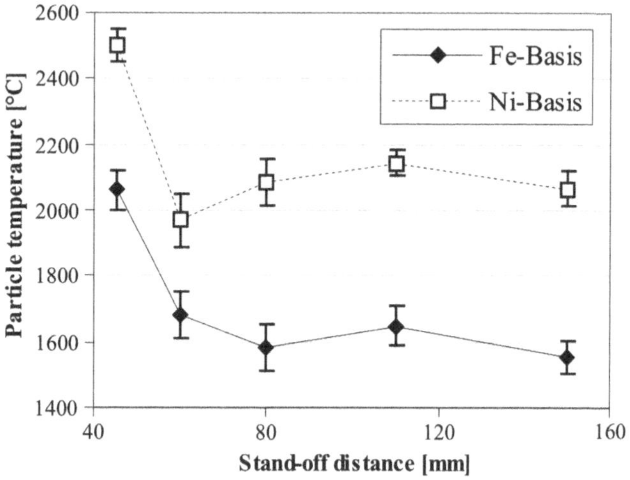

Fig. 7 Variation of particle temperature along the spray jet

faster after atomization than the larger particles. On the other hand the drag forces of small particles are increased during flight within the spray jet, and as a result small particles are decelerated. On the contrary, the drag coefficient for the large particles is low and approximately constant and therefore they do not show a significant deceleration during flight. Watanabe et al. (Ref 32) came to the same conclusion by spraying Ti-Al intermetallic compounds via TWAS process.

The temperature of the particles shows a decrease of approximately 26% with the standoff distance independent of the kind of wires (Fig. 7). The atomization gas and the environmental air impose a cooling effect on the accelerated particles. The smaller the particles, are the larger the contact area with the gas is, which leads to a higher heat transfer. Therefore, the Ni-based particles with larger sizes feature higher temperatures than the Fe-based particles Mahesh et al. (Ref 31) came to the same conclusion by studying Al-Si-Mg-alloy spraying.

4. Conclusions

A new mathematical model of Euler-Euler type was presented for the process of arc spraying. This model represents a generalization of the compressible Euler equations to particle-laden gas flows. Unlike the usual two-fluid model, the equations to be solved remain valid in particle-free regions. Promising results were obtained for a simplified 2D model problem. Further research is required to improve the robustness and efficiency of the numerical algorithm. Work is under way to impose characteristic boundary conditions in a weak sense and accelerate convergence to the steady state using a nonlinear multigrid solver. Furthermore, an Euler-Lagrange model of particulate flow is to be implemented for validation purposes.

Experimental results show a deceleration of particles up to a stand off distance of 60 mm due to the turbulences in this area. This is also associated with a further atomization of particles. It is also found that the higher the wires melting point is, the less the material melts, and smaller particles are generated. Due to the high surface area of smaller particles compared to the larger particles, the smaller particles cool down along the spray jet faster than the larger particles. Further analyses and measurements must be conducted at different pressure levels, voltages, and electric currents to explain the melting behavior and breakup mechanisms of cored wires during arc spraying.

Acknowledgments

The authors gratefully acknowledge the financial support of the DFG (German Science Foundation) within the Collaborative Research Centre SFB 708.

References

1. Fr.-W. Bach, K. Möhwald, A. Laarmann, and Th. Wenz, *Modern Surface Technologies*, Wiley-VCH Verlag, Germany, 2005, p 370
2. J. Wilden, A. Wank, and F. Schreiber, Wires for Rrc- and High Velocity Flame Spraying—Wire Design, Materials and Coatings Properties, *Thermal Spray: Surface Engineering via Applied Research*, C.C. Berndt, Ed., May 8-11, 2000 (Montréal, Québec, Canada), ASM International, Materials Park, OH, USA, 2000, p 609-617
3. M. Nakagawa, K. Shimoda, T. Tomoda, M. Koyama, Y. Ishikawa, and T. Nakajima, Development of Mass Production Technology of Arc Spraying for Automotive Engine Aluminum Alloy Valve Lifters, *Thermal Spray Research and Applications*, T.F. Bernecki, Ed., May 20-25, 1990 (Long Beach, California), ASM International, Materials Park, OH, USA, 1991, p 457-464
4. F. Lampmann, Entwicklung einer Hybridbeschichtungstechnologie zur Herstellung von Hochleistungs-Diesel-Pleuellagern, Dissertation TU Berlin, 2005, 187 p (in German)
5. J. Wen and S. Wen, Aerodynamic Behavior Within the Arc Gap During Arc Spraying, *Thermal Spray Industrial Applications*, C.C. Berndt and S. Sampath, Eds., June 20-24, 1994 (Boston, MA), ASM International, Materials Park, OH, USA, 1994
6. T. Watanabe, X. Wang, E. Pfender, and J. Heberlein, Correlations Between Electrode Phenomena and Coating Properties in Wire Arc Spraying, *Thin Solid Films*, 1998, **316**, p 169-173

7. N. Hussary and J. Heberlein, Metal Droplet Formation Mechanisms in the Wire Arc Spraying Process, *Proc. 48th International Research Colloquium* (Illmenau, Germany), Sept 22-25, 2003
8. M.P. Planche, H. Liao, and C. Coddet, Relations Between In-Flight Particle Characteristics and Coating Microstructure with a Twin Wire Arc Spray Process and Different Working Conditions, *Surf. Coat. Technol*, 2004, **182**, p 215-226
9. F. Schreiber, Karbidische Werkstoffe zum Thermischen Spritzen, http://www.durmat.com/PDF-Files/d_Karbidische_Werkstoffe.pdf, DURUM-VERSCHLEISS-SCHUTZ GMBH, Willich (in German)
10. H.H. Tawfik and F. Zimmerman, Mathematical Modelling of the Gas and Powder Flow in HVOF Systems, *J. Thermal Spray Technol.*, 1997, **6**(3), p 345-252
11. H. Hu and S.A. Argyropoulos, Mathematical Modelling of Solidification and Melting: A Review, *Model. Simul. Mater. Sci. Eng.*, 1996, **4**(4), p 371-396
12. S. Kamnis, S. Gu, and N. Zeoli, Mathematical Modeling of Inconel 718 Particles in HVOF Thermal Spraying, *Surf. Coat. Technol.*, 2008, **202**(12), p 2715-2724
13. http://www.tecnar.com/DATA/DOCUMENT/Accuraspray-g3%20Reference%20Manual.pdf
14. H. Städke, *Gas Dynamic Aspects of Two-Phase Flow*, Wiley-VCH, 2006, 450 p
15. D. Drew, Mathematical Modeling of Two-Phase Flow, *Annu. Rev. Fluid Mech.*, 1983, **15**, p 261-291
16. C.E. Brennen, *Fundamentals of Multiphase Flows*, Cambridge University Press, 2005, 342 p
17. M. Pelanti and R.J. Leveque, High-Resolution Finite Volume Methods for Dusty Gas Jets and Plumes, *SIAM J. Sci. Comp.*, 2006, **28**(4), p 1335-1360
18. T. Flatten and S.T. Munkejord, The Approximate Riemann Solver of Roe Applied to a Drift-Flux Two-Phase Flow Model, *ESAIM*, 2006, **40**(4), p 735-764
19. F.H. Harlow and A.A. Amsden, Numerical Calculation of Multiphase Fluid Flow, *J. Comput. Phys.*, 1975, **17**, p 19-52
20. T. Saito, Numerical Analysis of Dusty Gas Flows, *J. Comput. Phys.*, 2002, **176**, p 129-144
21. M. Manninnen and V. Taivassalo, On the mixture model for multiphase flow, *VTT Publications 288*, Technical Research Centre of Finnland, 1996
22. F.F. Grinstein and C. Fureby, On Monotonically Integrated Large Eddy Simulation of Turbulent Flows Based on FCT Algorithms. *Flux Corrected Transprot: Principles, Algorithms and Applications*, D. Kuzmin, R. Lohner, and S. Turek, Eds., Springer, Germany, 2005, p 79-130
23. W. Tillmann, E. Vogli and M. Abdulgader, Influence of Atomisation Gas on the Particle Formation During Arc Spraying with Cored Wires, *Thermal Spray 2007, Global Coating Solutions*, B.R. Marple, M.M. Hyland, Y.-C. Lau, C.-J. Li, R.S. Lima and G. Montavon, Eds., ASM International, Materials Park, Ohio/USA, 2007, p 837-842
24. D. Kuzmin and M. Möller, Algebraic Flux Correction I. Scalar Conservation Laws. *Flux-Corrected Transport: Principles, Algorithms, and Applications*, D. Kuzmin, R. Lohner, and S. Turek, Eds., Springer, Germany, 2005, p 155-206
25. D. Kuzmin and M. Möller, Algebraic Flux Correction II. Compressible Euler Equations. *Flux-Corrected Transport: Principles, Algorithms, and Applications*, D. Kuzmin, R. Löhner, and S. Turek, Eds., Springer, Germany, 2005, p 207-250
26. D. Kuzmin, Algebraic Flux Correction for Finite Element Discretizations of Coupled Systems. *Computational Methods for Coupled Problems in Science and Engineering II*, E. Önate, M. Papadrakakis, and B. Schrefler, Eds., CIMNE, Barcelona, 2007, p 653-656
27. A. Pormousa, "Wire-Arc Spraying System: Particle Production, Transport and Deposition," Dissertation, University of Toronto, 2007, 148 p
28. J. Wilden, Virtuelles Werkstoff- und Prozessdesign funktioneller Schichten, *Mat.-wiss. u. Werkstofftech.*, 2007, **38**(2), p 131-138 (in German)
29. http://nt-systemloesungen.de/index.php?opt-ion=com_content&task=view&id=209&Itemid=171.N.S
30. A.P. Newbery, P.S. Grant, and T.R.A. Neiser, The Velocity and Temperature of Steel Droplets During Electric Arc Spraying, *Surf. Coat. Technol.*, 2005, **195**, p 91-101
31. M.J. Mendonca, M.K. Muralidhara, B.K. Muralidhara, and C. Ramachandra, Modeling of Droplet Dynamic and Thermal Behaviour During Spray Deposition, *Bull. Mater. Sci.*, 2003, **26**(3), p 355-364
32. T. Watanabe, T. Sato, and A. Nezu, Electrode Phenomena Investigation of Wire Arc Spraying for Preparation of Ti-Al Intermetallic Compounds, *Thin Solid Films*, 2002, **407**, p 98-103

Peer Reviewed

JTTEE5 17:974–982
DOI: 10.1007/s11666-008-9269-y
1059-9630/$19.00 © ASM International

Peer Reviewed

Asymmetric Melting Behavior in Twin Wire Arc Spraying with Cored Wires

W. Tillmann, E. Vogli, and M. Abdulgader

(*Submitted June 9, 2008; in revised form September 26, 2008*)

Asymmetric melting behavior of the electrodes is a process-related feature of the Twin Wire Arc Spraying (TWAS) technique since the heating of the negative connected wire is different from that of the positive connected wire. Due to these differences in melting behavior, a tracking of particle velocity and temperature for both electrodes individually is very important. Particle velocity and temperature have been recorded from anode side and cathode side by positioning the tracking device respectively. To draw the whole picture of the spraying, jet, particles have been tracked also from the top side of the spray gun. The goal of this study is to have an experimental data setup for model building and simulation of depositing process in TWAS. Corresponding measuring devices have been employed to investigate the TWAS process by spraying of massive and cored wires.

Keywords diagnostics and control, influence of process parameters, spray deposition

1. Introduction

Thermal spraying is a generic term for a group of coating processes that deposit finely divided (like in plasma or HVOF) or atomized (like in twin wire arc spraying) metallic or nonmetallic spraying material. These processes differ in their thermal source and in their feedstock. The three major categories of feedstock define the form of the spraying material used: powder, massive wires, and cored wires. The size of in-flight particles is predefined by powder size distribution in case of powder feedstock. The powder particles are finely divided by means of carrier gas and subsequently through the heat source in plasma or HVOF. In twin wire arc spraying (TWAS), a spray is formed by atomization of molten or semimolten droplets by the impingement of fast moving and continuously flowing atomization gas upon melting tips of consumable and electrical conductive wires. The wires are connected individually as anode and the other one as cathode and fed together to ignite an arc at the shortest distance between the electrical conductive part in wire tips.

The asymmetric melting behavior of massive wires was observed using similar or different wires in many studies (Ref 1-5). It is well known that the anode and cathode are heated differently in a TWAS process (Ref 2, 4, 5). The arc attaches to the anode over a larger area than the cathode. The heating is more localized at cathode and appears as cathode spot (Ref 3, 4, 6, 7). At the tip of the anode wire a large area is heated due to diffuse arc-anode attachment, melting a layer of metal that is pushed off the edge of the wire tip by the atomizing gas, creating an "anode sheet." At the cathode, constricted arc attachment causes much more localized heating and melting (Ref 1-3). The asymmetric melting behavior of anode and cathode together with the arc fluctuation, due to periodic removal of molten material, affects consequently the in-flight particle characteristics. In-flight particle characteristics such as velocity, temperature, and size are controlling factors in splat formation, layer evolution, and in determination of the coating quality in thermal spraying technique (Ref 6-12).

Since in TWAS the particles are atomized from the bath of melting material in the arc zone, their size is not predictable and is dependent on the process parameters and the material properties of the wires. The use of cored wires has enhanced the coating quality of TWAS regarding wear resistance. The composition of the cored wires is inhomogeneous. It contains electrical conductive solid velum and a powder filling, which strongly influences the particle formation and in-flight particle characteristic data (Ref 13, 14).

The aim of this study is to develop a more fundamental understanding of the interactions between asymmetric melting behavior in TWAS process, independent process parameter, and in-flight particle catachrestic data by using cored wires. A comparison with massive wire is made to

This article is an invited paper selected from presentations at the 2008 International Thermal Spray Conference and has been expanded from the original presentation. It is simultaneously published in *Thermal Spray Crossing Borders, Proceedings of the 2008 International Thermal Spray Conference*, Maastricht, The Netherlands, June 2-4, 2008, Basil R. Marple, Margaret M. Hyland, Yuk-Chiu Lau, Chang-Jiu Li, Rogerio S. Lima, and Ghislain Montavon, Ed., ASM International, Materials Park, OH, 2008.

W. Tillmann, E. Vogli, and **M. Abdulgader**, Institute of Materials Engineering, Dortmund University of Technology, Dortmund, Germany. Contact e-mail: wolfgang.tillmann@udo.edu.

see the effect of the filling powder in cored wires on these phenomena.

The correlations between the abovementioned subjects of study will help to better predict the coating quality, make the spraying process more reliable, improve the reproducibility of desired coating, and give the conditions of spraying process simulation.

2. Materials and Experiments

2.1 Materials

Smart Arc 350 PG (Sulzer Metco, Switzerland) was used to investigate the particle formation and in-flight particle characteristics. The investigations were carried out on two types of wires with a diameter of 1.6 mm: cored wires (AS850, Co. Durum, Cr 4%, Mn <1%, Si 1,4%, C 2%, WSC 50%, Fe Bal, particle grain sizes 35-75 μm of the filling powder) and massive wires (Co.GTV, Ni. 99.9%).

2.2 Experiments

Design of Experiment (DoE) is useful in reducing the number of experiments required to optimize thermal spraying process (Ref 1, 12, 13). The primary purpose of the DOE screening objective is to select or to screen out the few important main effects from the many less important ones. These screening designs are also termed main effect designs. The approach chosen in this study is described as two-level full factorial designs with center point (Fig. 1).

In this study the Independent Controllable Process Parameters (ICPP) were assigned to atomization, arc voltage, and arc current (higher current higher wire feed rate). Atomization is the arrangement between Primary and Secondary atomization Gas pressure (PG_SG). Three evenly spaced low, center and high preset values denoted

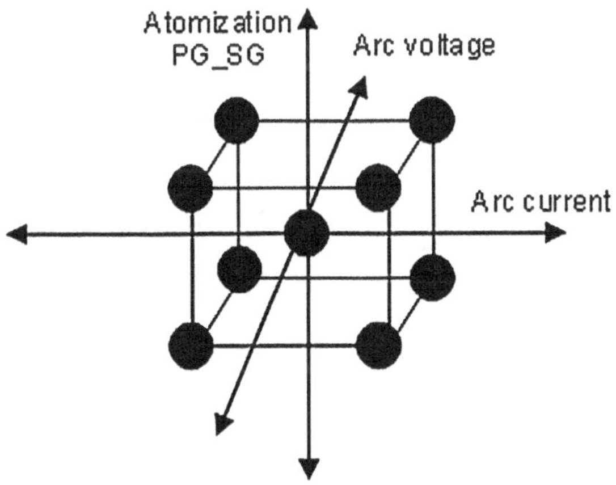

Fig. 1 Schematic diagram showing two-level full factorial design with center point used for this study

−1, 0, and 1, respectively, were chosen for each ICPP. The values for ICPP are as following:

- Atomization gas pressure in Bar (PG_SG) (−1, 0, +1) (2_0, 4_3, 6_6), respectively
- Arc voltage in Volt (−1, 0, +1) (28, 31, 34), respectively
- Arc current in Ampere (−1, 0, +1) (180, 220, 260), respectively

Experiments were performed using the combinations of ICPP settings corresponding to the eight points at the cube cell corners (−1, 1, −1), (1, 1, −1), (−1, −1, −1), (1, −1, −1), (−1, 1, 1), (1, 1, 1), (−1, −1, 1), (1, −1, 1) and the cube center point (0, 0, 0). The center point setting was executed at the middle and at the end of the experimental setup. Three complete experimental setups were implemented at standoff distances of 45, 100, and 150 mm along the axial spray cone on Anode (A) side as well as on Cathode (C) side. The center point setting was performed at axial standoff distances beginning from 30 to 150 mm in a 10 mm step. To verify the asymmetry in spray jet, a radial scanning of in-flight particles along the axial standoff distances was carried out. In doing so, an additional measurement with the center point setting was taken at the intersection between anode and cathode at axial standoff distances in the rage of 30 to 150 mm with a 10 mm step.

The numbering of experiments consists of a capital letter. It gives the measuring device position: "A" for anode side, "C" for cathode side, and "O" for the intersection between anode and cathode. The letter followed by a number combination reveals the ICPP setting for arc voltage-Atomization gas pressure-arc current.

For example A_28_2-0_180 means anode side, voltage 28 V, primary gas 2 bar, secondary gas 0 bar, and current 180 A.

2.3 Diagnostic

To determine the in-flight particle velocity and temperature AccuraSpray-g3 (Tecnar, Canada) was used, which consists of sensor head and controller unit (Fig. 2).

The principle behind AccuraSpray is based on optical means. Particle velocities are obtained from cross-correlation of signals which are recorded at two locations. In addition, the detectors are filtered at two different colors allowing the mean particle temperature to be measured using the very well-known twin wavelength pyrometry principle (assumes that the emissivity of the particles is the same for the two wavelengths). The AccuraSpray-g3 provides ensemble average data representing the particle temperature and velocity in a measurement volume of approx. Ø3 mm * 25 mm (Ref 15). The accuracy of the AccuraSpray measurements reaches 2% for particle temperature and 0.5% for particle velocity according to the product data sheet of the manufacturer.

As shown in Fig. 3 the AccuraSpray sensor head was located alternatively along the positively charged wire (anode) side, negatively charged wire (cathode) side, and along the intersection between electrodes. Different

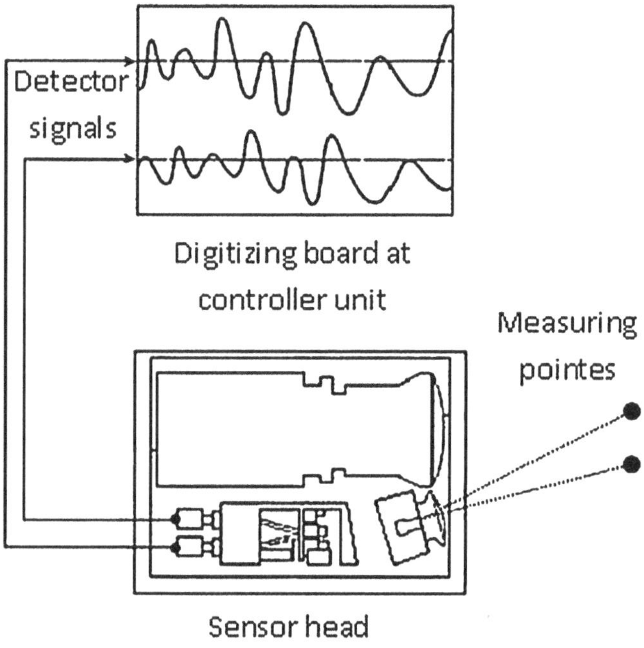

Fig. 2 Principles behind AccuraSpray

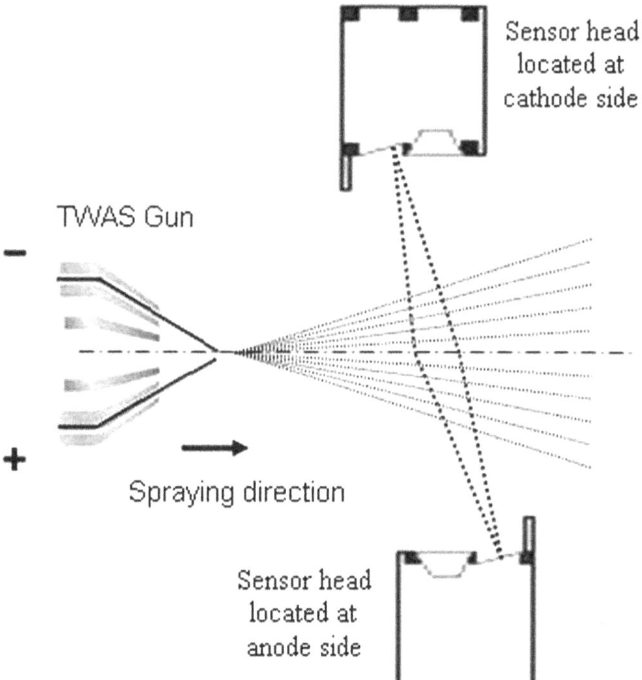

Fig. 3 Positioning of sensor head along the spraying axes on cathode side, anode side, and on intersection of electrodes at different standoff distances

parameter setups were tested at several standoff distances along the main spraying axis. Particle velocity and temperature have been recorded for every single experimental setup. The collected data have been analysed and conclusions of the different evaluated experimental setups have been drawn.

3. Results

Because of asymmetry of the electrical arc itself and the resulting differences in melting behavior of anode and cathode, the spray jet exhibited asymmetric radial variation of both particle velocity and temperature, which make a radial investigation by additionally measuring along the intersection between anode and cathode unavoidable.

3.1 Variations in Particle Temperature and Velocity for all ICPP Settings

Figure 4 shows a decrease in the temperature of cathode side particles with increasing standoff distances, and a simultaneous increase in the temperature of anode side particles. Due to the heat exchange between the particles in-flight, a declination of the temperature discrepancy between anode side particles and cathode side particles was revealed at higher standoff distances. The deposited particles are conveyed in an air jet, which has a conical shape. The air jet moves with high velocity through the surrounding environment and can be regarded as adiabatic conveying cells. Each cell consists of particles with different temperature surrounded by conveying air jet. Inside the adiabatic conveying cells a heat exchange occurred between the particles. The particles from anode side and cathode side reached approximately the same temperature at 150 mm standoff distance. In general the variation in particle temperature from anode side and cathode side was higher in the case of massive wire. The particles from massive wire had a slightly higher temperature at lower standoff distances between 45 and 100 mm. In higher standoff distances, a higher decrease in particle temperature was recorded using massive wire. Particle temperature of cored wire was slightly higher at 150 mm standoff distance. A variation in temperature was observed between anode side particles and cathode side particles. At standoff distances smaller than 100 mm, a higher particle temperature variation (200 to 400 °C) was recorded at the different ICPP settings and for both wire types.

The particle velocity increased at all ICPP settings with increasing axial standoff distance. A maximum velocity was reached at a standoff distance between 100 and 150 mm. The particles generated from massive wires (MW) attained about 20 to 30% higher velocity then the ones from cored wires (CW). At a lower ICPP setting, a variation in particle velocity between anode side and cathode side was recorded for both wire types. At a higher ICPP setting, a deviation in particle velocity appeared clearly. It was stated that the particles from cathode side has a 10% higher velocity compared to the ones from anode side for both wire types. There was a clear velocity gap for the particle velocity of cored wire at different ICPP settings. Particles from massive wire showed a different behavior. A higher velocity gap was recorded by changing the ICPP setting from low to middle, while a slight variation in particle velocity was obtained in lower standoff distance by changing the ICPP from middle to high.

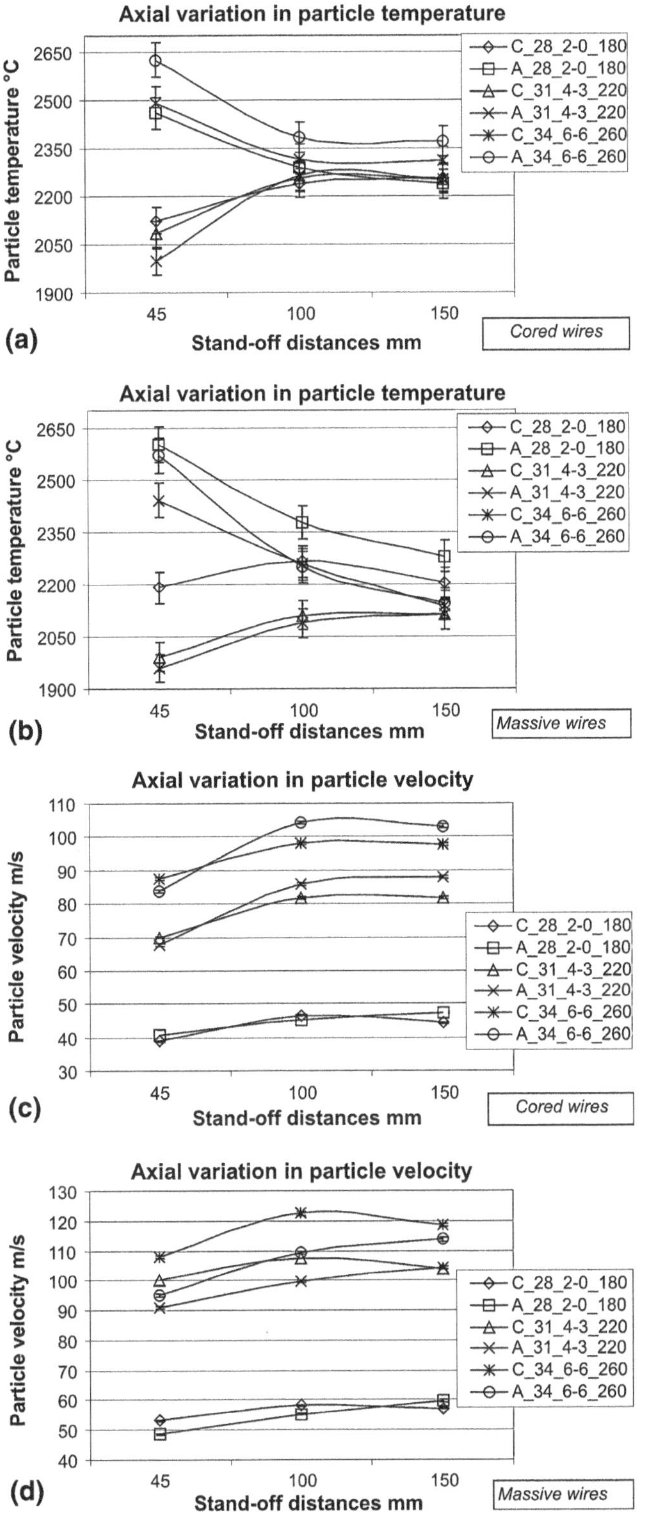

Fig. 4 Behaviour of in-flight particles by the different settings (a) and (b) particle temperature (c) and (d) particle velocity (A anode side/C cathode side/_voltage_primary gas-secondary gas_current)

3.2 Axial Variations Due to Arc Voltage

As expected, an increase of particle temperature (Fig. 5) was recorded as arc voltage was increased for both cathode side and anode side, as well as for both wire types. The temperature divergence between anode side particles and cathode side particles was higher by spraying massive wires compared to cored wires.

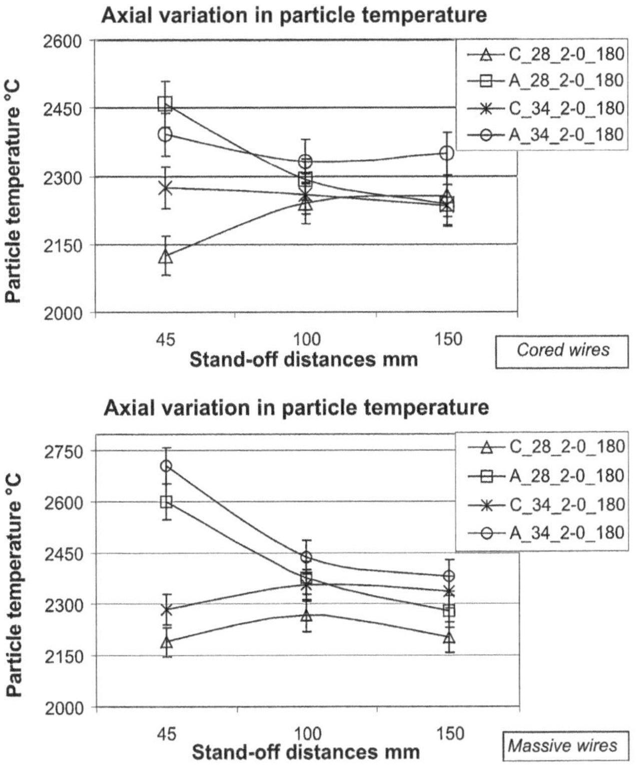

Fig. 5 Variation in particle temperature due to the change in arc voltage (A anode side/C cathode side/_voltage_primary gas-secondary gas_current)

At higher arc voltage and at smaller standoff distances, a lower heat exchange takes place compared to the lower voltage setting. This can be explained to be due to higher starting temperatures at higher voltage settings. The particle temperature increased about 50 to 100 °C at a standoff distance of 150 mm by increasing the voltage from 28 to 34 V. Lower arc voltage results in a higher particle temperature difference between anode side and cathode side for both wires.

Figure 6 shows a reduction of particle velocity at a higher arc voltage setting by using massive wires for particles generated in the anode side and cathode side. The velocity decreases approximately 10 m/s along the axial standoff distances. The velocity reduction can be explained to be due to larger particle size generated by higher voltage setting. In the case of massive wires, the melted (heated) area is larger than in cored wires. In cored wires, the powder inside the velum affects strongly the size of initiated particles.

3.3 Axial Variations due to Arc Current

It is seen from Fig. 7(a) and (b) that by increasing of arc current has a negligible effect on the temperature of anode side particles and cathode side particles for both wire types. The temperature of cathode side particles

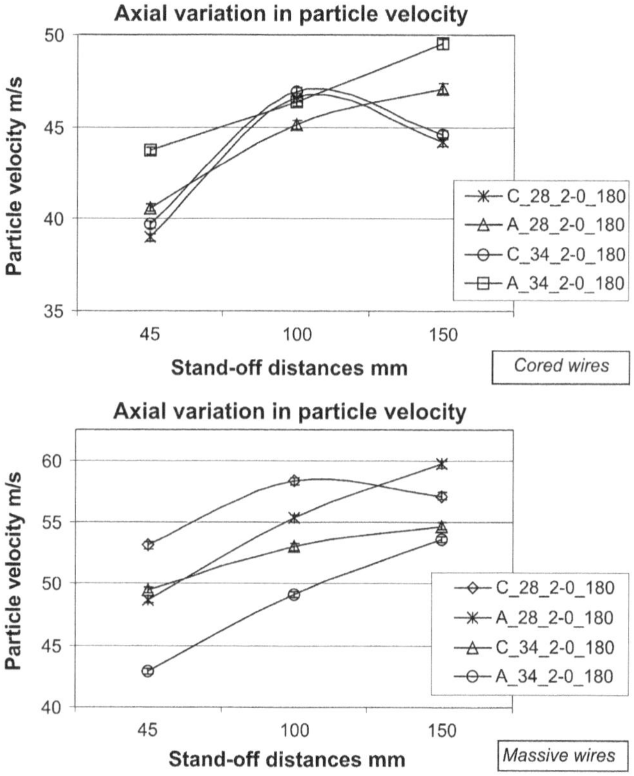

Fig. 6 Variation in particle velocity due to the change in arc voltage (A anode side/C cathode side/_voltage_primary gas-secondary gas_current)

obtained is found to be similar at higher arc current 260 A for both wires at lower standoff distances (45 to 60 mm). The same results were reported by Hale et al. in (Ref 7). The magnitude of current changes is not big enough to record high changes in particle temperature.

The particle velocity of both wires and both electrodes was decreased (Fig. 7c and d) with the increase of arc current from 180 to 260 A. No significant changes were recorded for both wire types and both electrodes with the increase of arc current. A slight decrease in particle velocity was obtained for both wires at 150 mm standoff distance.

It is observable that the particles from cathode side and anode side sprayed by massive wires showed higher temperatures and higher velocities then those sprayed by cored wires.

3.4 Effects of the Atomization Gas Pressure

The atomization was realized by combining primary and secondary atomization gases together (PG_SG). Figure 8 shows that increasing of atomization gas pressure considerably increases particle velocity at anode side and at cathode side. The particle velocity at anode side for both wire types increased linearly along the axial distances between 45 and 150 mm and reached its maximum at 150 mm. A linear increase at standoff distances between 45 and 100 mm was recorded for the cathode side too. The particles reached their highest velocity at 100 mm,

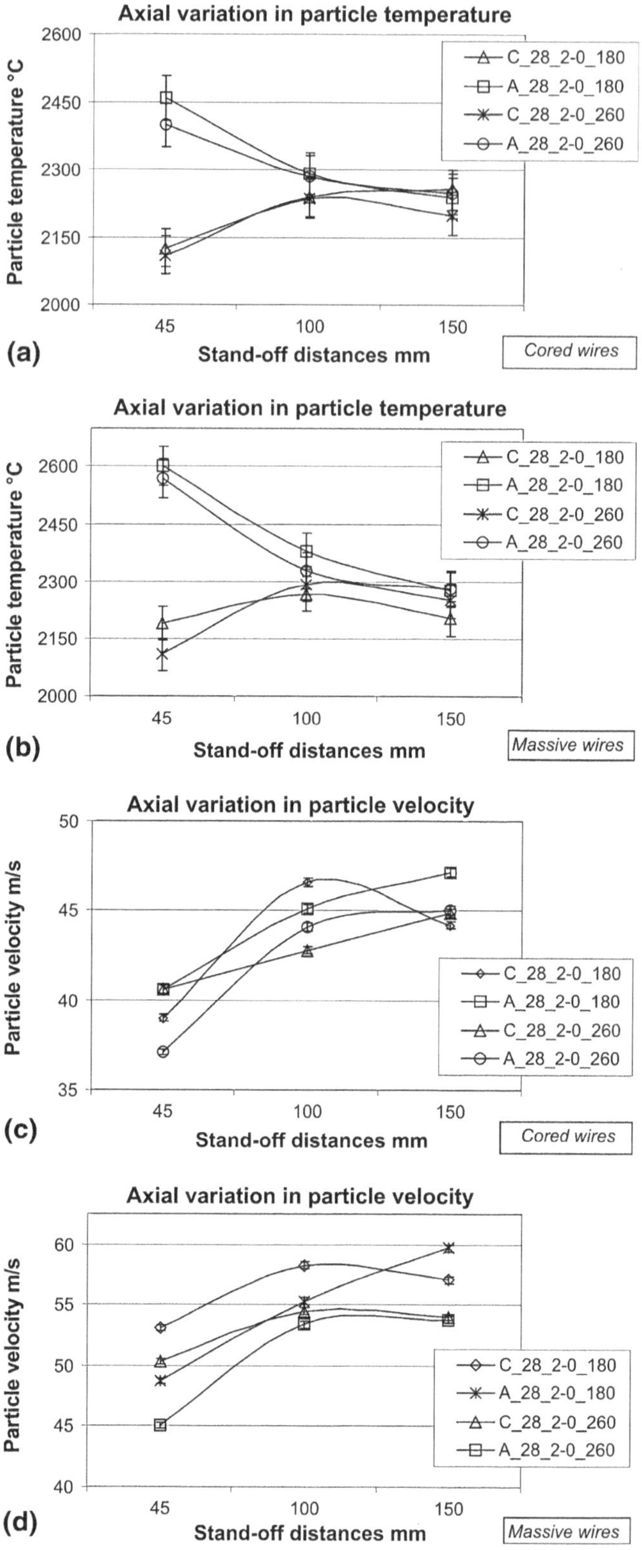

Fig. 7 Behavior of in-flight particles due to the change in arc current (A anode side/C cathode side/_voltage_primary gas-secondary gas_current)

followed by a slight decrease at 150 mm standoff distance in the case of cored wires and remained constant in massive wires. There was a negligible velocity gap between

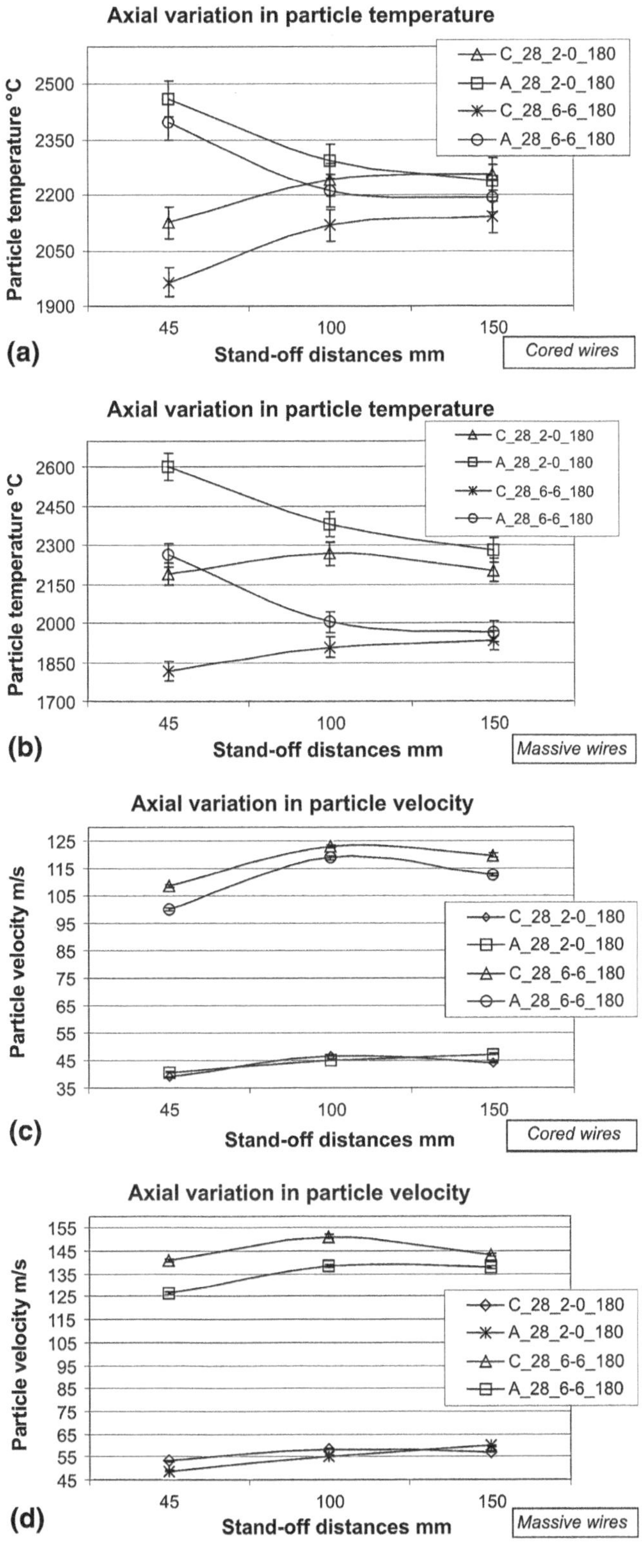

Fig. 8 Behavior of in-flight particles due to the change in atomization pressure PG_SG (A anode side/C cathode side/ _voltage_primary gas-secondary gas_current)

anode particles and cathode particles for both wire types at lower atomization gas pressure.

At a higher atomization setup, a higher velocity gap was obtained between anode side particles and cathode side particles for both wires. The particles from cored wires reached their highest velocity at 100 mm standoff distance; afterward a decrease of about 10% occurred. Approximately 20 to 25 m/s higher particle velocity was obtained by using massive wires for particles at anode side as well as at cathode side.

A lower particle temperature was reported for the higher atomization setup. The reason was the higher cooling rate due to the higher gas flow rate. In general, the cooling rate of the particles sprayed with lower atomization setup along the standoff distances was higher compared to particles sprayed by higher setting. The reason was the lower particle velocity of the lower atomization setup resulted in longer flying time.

Particles from cored wire revealed a higher temperature than the particles from massive wire at the higher atomization setting for both electrodes. In lower it setting was the other way around.

The velocity discrepancy at higher setting was larger for massive wire than for cored wires. The velocity variation at 100 mm standoff distance between particles generated from anode and the ones from cathode was 5 and 14 m/s for cored wire and massive wire, respectively.

3.5 Behavior of In-Flight Particles in Axial and Radial Direction

The setting of the cube center point was used for these experiments with the following combination: arc voltage 31 V, atomization primary gas 4 bar, secondary gas 3 bar and, arc current 220 A (31_4-3_220). Recording of particle velocity and temperature was made at different standoff distances from 30 to 150 mm in 10 mm step. The recording device was located one along the anode, one along the cathode, and one along the intersection point between anode and cathode. The axial setup confirmed the previous results. In addition to the axial variation, there was a radial variation at every standoff distance. The radial variation decreased with increasing axial standoff distances.

Figure 9 shows a temperature gap between anode side particles and cathode side particles. In the case of cored wires a smaller gap was recorded. It decreased at standoff distance of 60 mm and reached its minimum at about 90 mm. The particles from massive wires revealed a larger discrepancy between anode side and cathode side particles. The discrepancy got smaller at higher standoff distances when compared to cored wires (at about 90 mm) and reached its minimum at 130 mm standoff distance.

The particle velocity of cored wires and massive wires showed the same behavior as in the case of temperature. The velocity gap between anode side particles and cathode side particles was smaller by using cored wires. In general, it can be stated that the particles generated from cored wires revealed a lower radial variation in temperature and velocity along the entire axial standoff distances than those generated from the massive wires. The spray jet of cored wires was supposed to be more homogeneous with respect to the smaller discrepancy in particle temperature and velocity.

Peer Reviewed

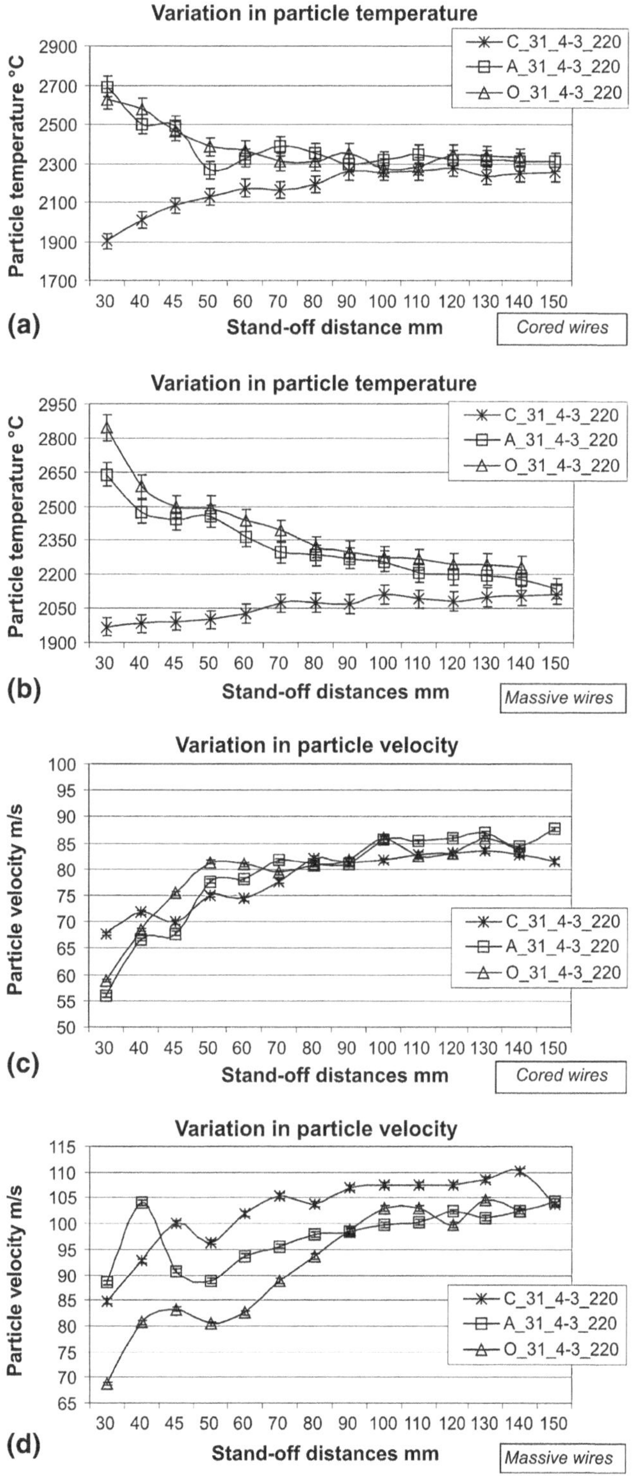

Fig. 9 Axial and radial behavior of in-flight particles (A anode side/C cathode side/O intersection voltage_primary gas-secondary gas_current)

4. Discussion

With the use of design of experiment (DoE), the interactive effects of ICPP on the axial and radial variation of particle velocity and temperature for anode and cathode can be thoroughly explored (Table 1). The variation in particle temperature and particle velocity between cathode side and anode side for both wires was computed as follows:

$$\text{CAi} = (1/4)\,(\text{S responses at high setting}) - (1/4)\,(\text{S responses at low setting})$$

$$\text{CCi} = (1/4)\,(\text{S responses at high setting}) - (1/4)\,(\text{S responses at low setting})$$

where (i) indicates one of the ICPP parameters:

CA1: Responses at anode side for voltage setting

CA2: Responses at anode side for atomization gas pressure setting

CA3: Responses at anode side for atomization gas pressure setting

CC1: Responses at cathode side for voltage setting

CC2: Responses at cathode side for atomization gas pressure setting

CC3: Responses at cathode side for atomization gas pressure setting

Variance factor for each ICPP = $|\text{CAi} - \text{CCi}|$.

The calculations were done for all standoff distances for both wires and at the cathode side as well as at anode side. The values in Table 1 are the results obtained at 45 and 100 mm standoff distances and show the effect of ICPP setting on the velocity gap and temperature gap between particles from cathode and particles from anode. The values in Table 1 show the inhomogeneity factor (the higher the inhomogeneity factor, the bigger the gap between anode and cathode) for particle temperature and particle velocity in the spraying jet at different ICPP settings. The results in Table 1 confirmed that the TWAS spraying jet is an inhomogeneous spray jet with respect to particle velocity and temperature.

The results clearly showed that the atomization pressure was the predominant ICPP affecting particle velocity; it was found that high atomization pressure exerted a cooling effect at lower axial standoff distances. An increase of particle velocity along the axial standoff distance was revealed by all ICPP settings. The increase of particle velocity is related to the secondary atomization at standoff distances close to the arc zone up to 50 mm away from the arc, which is in agreement with the results from earlier studies (Ref 3, 4, 6, 8).

Particles from cored wires revealed a higher temperature than the particles from massive wires with higher atomization setting for both electrodes. In lower setting it was the other way around.

On wire tips in twin wire arc spraying, there are two kinds of energy impact. Firstly, a continuously dynamic force was applied by atomization gas. Its magnitude is directly affected by gas pressure. Secondly, a heating energy caused by a fluctuating electrical arc stroked between the anode and cathode and its magnitude is affected by arc

Peer Reviewed

Table 1 Effect of ICPP setting on particle temperature gap and on particle velocity gap between (A) anode and (C) cathode for CW cored wire and MW massive wire (inhomogeneity value in spraying jet)

	T-gap A-C				V-gap A-C			
	CW		MW		CW		MW	
Variance value	**@45**	**@100**	**@45**	**@100**	**@45**	**@100**	**@45**	**@100**
ICPP setting								
Arc voltage	113.5	91.8	492.2	127.3	3.8	2.7	9.5	8.4
Atomization gas pressure	563.5	112.3	509.5	130.9	6.5	3.4	13.7	11.7
Arc current	442.1	78.8	499.6	113.5	3.9	2.1	10.1	7.1

voltage and arc current. The arc fluctuation is caused by the material removal from wire tips by means of atomization gas and the continuous feed of wire into the arc zone. This leads to ignition followed by extinction and reignition in periodic time of milliseconds. The atomization gas force deforms the wire tips material as soon as its tensile strength decreased by arc heating. The dynamic forces due to continuously flowing atomization gas causes elongation of wire tips material. The material must not be fully molten. The material sheared and formed the so-called anode sheet and cathode sheet (Ref 1, 2). The anode sheet is different from the cathode sheet due to the different attachment of arc to the anode and cathode. The length and form of the abovementioned sheets are strongly affected by the atomization gas pressure, arc voltage, and arc current. The heat impact on the formed sheets continued due to arc reignition.

In the case of cored wires, the material removal occurs at electrical conductive velum on the inside of wires at first. Therefore, the electrical nonconductive filling powder, which is not a part of arc ignition, is exposed to excessive heating by arc ignition and to erosion by continuously flowing atomization gas. Because of arc fluctuation and continuous erosion, a considerable part of filling powder does not melt and will be wetted or impregnated into molten velum material during their flight to the substrate. It is supposed that this phenomena is different in the cathode and the anode.

5. Conclusions

The evaluation of particle velocity and temperature by twin wire arc spraying was carried out on cored and massive wires. The recordings were taken by measuring at different standoff distances at the anode side and at the cathode side. The radial variations of both particle velocity and temperature were revealed by adding an axial measurement setup along the intersection point between the anode and cathode. It was found that the velocity did not increase in the same magnitude as the primary gas and/or secondary gas. The highest particle acceleration took place from arc ignition up to 50 mm standoff distance. At standoff distances between 100 mm and 150 mm, the velocity reached its maximum value. Furthermore the velocity of particles generated by spraying of cored wires is about 30% smaller than that of particles generated by spraying of massive wire.

Cooling effects of PG and SG have been recorded on measurements carried out at both wires. The temperature of particles generated from massive wire is higher than that of particles generated from of cored wire at almost all ICCP settings. An exception is the particles generated at high atomization gas setup, in which the particle temperature of cored wires is higher than that of the massive wires.

Based on the results of the diagnostics of TWAS process an accurate simulation of spraying process can be established. This will lead to an improvement of the coating properties and surface quality, which can be used to improve the coating quality for tool surfacing.

Acknowledgments

The authors gratefully acknowledge the financial support of the DFG (German Science Foundation) within the Collaborative Research Center SFB 708.

References

1. N.A. Hussary and J.V.R. Heberlein, Effect of System Parameters on Metal Break-Up and Particle Formation in the Wire Arc Spray Process, *J. Thermal Spray Technol.*, 2007, **16**(1), p 1-13
2. N.A. Hussary and J.V.R. Heberlein, Atomization and Particle-Jet Interactions in the Wire-Arc Spraying Process, *J. Thermal Spray Technol.*, 2001, **10**(4), p 604-610
3. T. Watanabe, T. Sato, and A. Nazi, Electrode Phenomena Investigation of Wire Arc Spraying for Preparation of Ti-Al Intermetallic Compounds, *Thin Solid Films*, 2002, **40**(7), p 98-103
4. T. Watanabe, X. Wang, E. Pfender, and J. Heberlein, Correlations Between Electrode Phenomena and Coating Properties in Wire Arc Spraying, *Thin Solid Films*, 1998, **316**, p 169-173
5. Y.L. Zhu, H.L. Liao, C. Coddet, and B.S. Xu, Characterization Via Image Analysis of Cross-Over Trajectories and Inhomogeneity in Twin Wire Arc Spraying, *Surf. Coat. Technol.*, 2003, **162**, p 301-108
6. A. Pourmousa, J. Mostaghimi, A. Abedini, and S. Chandra, Particle Size Distribution in a Wire-Arc Spraying System, *J. Thermal Spray Technol.*, 2005, **14**(4), p 502-510
7. D.L. Hale, W.D. Swank, and D.C. Haggard, In-Flight Particle Measurements of Twin Wire Electric Arc Sprayed Aluminium, *J. Thermal Spray Technol.*, 1998, **7**(1), p 58-63
8. A.P. Newbery, P.S. Grant, and R.A. Neiser, The Velocity and Temperature of Steel Droplets During Electric Arc Spraying, *Surf. Coat. Technol.*, 2005, **195**, p 91-101

Peer Reviewed

9. A.P. Newbery, T. Rayment, and P.S. Grant, A particle Image Velocimetry Investigation of In-Flight and Deposition Behaviour of Steel Droplets During Electric Arc Sprayforming, *Mater. Sci. Eng.*, 2004, **A383**, p 137-145
10. I. Gedzevicius and A.V. Valiulis, Influence of the Particles Velocity on the Arc Spraying Coating Adhesion, *Mater. Sci.*, 2003, **9**(4), p 334-337
11. J.J. Fang, Z.X. Li, M. Qian, C.L. Ren, and Y.W. Shi, Effects of Powder Size in Cored Wire on Arc-Sprayed Metalceramic Coatings, ITSC, Beijing, May 14-16, 2007, p. 365-370
12. S. Sampath, X. Jiang, A. Kulkarni, J. Matejicek, D.L. Gilmore, and R.A. Neiser, Development of Process Maps for Plasma Spray: Case Study for Molybdenum, *Mater. Sci. Eng.*, 2003, **A348**, p 54-66
13. I.K. Hui, M. Hua, and H.C.W. Lau, A Parametric Investigation of Arc Spraying Process for rapid mould making, *Int. J. Adv. Manuf. Technol.*, 2003, **22**, p 786-795
14. J.P. Fang, Z.X. Li, J.M. Jiang, and Y.W. Shi, Difference in Particle Characteristics and Coating Properties Between Spraying Metallic and Ceramic Powder Cored Wires, *Trans. Nonferrous Metals Soc. China*, 2007, **17**, p 537-642
15. G. Mauer, R. Vaßen, and D. Stöver, Comparison and Applications of DPV-2000 and Accuraspray-g3 Diagnostic Systems, *J. Thermal Spray Technol.*, 2007, **16**(3), p 414-424